ELECTRIC CIRCUITS
FUNDAMENTALS

ELECTRIC CIRCUITS
FUNDAMENTALS

SERGIO FRANCO

San Francisco State University

New York   Oxford
OXFORD UNIVERSITY PRESS
1999

Oxford University Press

Oxford   New York
Athens   Auckland   Bangkok   Bogotá   Buenos Aires   Calcutta
Cape Town   Chennai   Dar es Salaam   Delhi   Florence   Hong Kong   Istanbul
Karachi   Kuala Lumpur   Madrid   Melbourne   Mexico City   Mumbai
Nairobi   Paris   São Paulo   Singapore   Taipei   Tokyo   Toronto   Warsaw

and associated companies in
Berlin   Ibadan

Copyright © 1995 by Oxford University Press, Inc.

Published by Oxford University Press, Inc.,
198 Madison Avenue, New York, New York 10016
http://www.oup-usa.org

Oxford is a registered trademark of Oxford University Press

Library of Congress Cataloging-in-Publication Data
available upon request

ISBN 978-0-19-513613-5

Printing (last digit): 9 8 7

Printed in the United States of America
on acid-free paper

For My Mother and Father
Vigiuta Braidota and *Gigi Mulinâr*

———————————————

FOREWORD

Perhaps the most important of the electrical engineering curriculum, the circuits course serves many purposes. First and foremost, it presents electrical and computer engineering students with a subject fundamental to their chosen discipline. Circuit theory is a subject of intrinsic beauty, which, if properly taught, can excite and motivate the student. Although based on the laws of physics, circuit theory need not deal with the non-idealities of practical electrical elements, at least not initially. Rather, circuit theory provides the abstraction that future electrical engineers need to deal with practical circuits and systems at a higher, conceptual level, somewhere between physics and mathematics.

Second, the circuits course has an immediate utilitarian value: It is required to prepare the students for the study of electronic circuits. In most electrical and computer engineering curricula, the first course in electronics occurs immediately following the circuits course and relies heavily on the techniques of circuit analysis learned in the circuits course.

This dual role of the circuits course—laying one of the important theoretical foundations of electrical engineering and teaching practical circuit analysis techniques—makes for not just an important subject but also one that is somewhat difficult to teach. Sergio Franco's *Electric Circuits Fundamentals* does an outstanding job of fulfilling this role. It strikes an appropriate balance between the abstractions necessary for the proper understanding of circuit theory and the specifics needed to master the techniques of circuit analysis. The author accomplishes this difficult task by drawing on his extensive experience both as a circuit designer and as an educator of electrical engineering. The result is a presentation that moves from the practical to the theoretical and back to the practical with ease and elegance. It does so while keeping the student motivated and eager to learn the theory, for the application is always in sight and the pay-off appears to be substantial.

A third purpose of the circuits course is introducing the student to the language and symbolism of electrical engineering. Here again Franco does an excellent job. He utilizes the pictorial nature of circuits to the utmost advantage, illustrating difficult concepts with clear and extensive diagrams and instilling in the student the habit of sketching correct diagrams to represent various circuits situations.

From the vantage point of a long-time teacher of electronic circuits and researcher in theory and design of circuits, I found Franco's text to combine the fundamentals of circuit theory with the intuition and insight required for circuit design. For this reason, I believe that *Electric Circuits Fundamentals* should make an excellent text for the electric circuits course and I am pleased to recommend it for this purpose.

Adel S. Sedra
Toronto, Canada
July 1994

PREFACE

This book is designed to serve as a text for the first course or course sequence in circuits in an electrical engineering curriculum. The prerequisites or corequisites are basic *physics* and *calculus*. Special topics such as *differential equations* and *complex algebra* are treated in the text as they arise. It can be used in a two-semester or a three-quarter course sequence, or in a single-semester course.

The goal of the book is twofold: *(a)* to teach the **foundations of electric circuits**, and *(b)* to develop a thinking style and a problem-solving methodology based on **physical insight**. This approach is designed to benefit students not only in the rest of the curriculum, but also in the rapidly changing technology that they will face after graduation.

Since the study of circuits is the foundation for most other courses in the electrical curriculum, it is critical that we guide the reader in developing an **engineering approach** from the very beginning. Armed, on the one hand, with the desire to finally apply the background of prerequisite math courses, and overwhelmed, on the other, by the novelty and wealth of the subject matter, students tend to focus on the mathematics governing a circuit without developing a feel for its physical operation. After all, circuit theory has a mathematical beauty of its own, and many a textbook emphasizes its abstract aspects without sufficient attention to its underlying physical principles, not to mention applications. Quite to the contrary, we believe that the role of a circuits course is to provide an engineering balance between **mathematical conceptualization** and **physical insight**. Whether analyzing existing circuits or designing new ones, engineers start from a physical basis and resort to mathematical tools only to gain a deeper and more systematic understanding. But, in the true spirit

of engineering, results are ultimately checked against physical substance, not mere mathematical abstraction. This is especially true in computer-aided analysis and design, where results must be interpreted in terms of known physical behavior. Students who are not given an early opportunity to develop physical intuition will subsequently have to do so on their own in order to function as engineers. This book was written to provide this early opportunity without necessarily watering down the mathematical rigor.

As an enthusiastic practitioner of circuit theory, the author firmly believes that a circuits course should be accompanied by a **laboratory** to provide experimental reinforcement of the theory as well as to introduce the student to instrumentation and measurement. Whenever the opportunity arises we attempt to relate concepts and principles to laboratory practice, but we do so without disrupting the flow of the material. Lacking a laboratory, the use of SPICE should be promoted as a **software breadboard** alternative for trying out the circuits seen in class. Moreover, the Probe graphics post-processor available with PSpice provides an effective form of **software oscilloscope** for the visualization of waveforms and transfer curves. Ultimately, however, the student must be stimulated to interpret SPICE results in terms of known physical behavior.

Pedagogy and Approach

The material covered is fairly standard in terms of topic selection. To prevent excessive conceptualization from obfuscating the physical picture, we try to be as mathematically rigorous *as needed*. Moreover, to maintain student interest and motivation, we attempt, whenever possible, to relate the theory to real-life situations. In this respect we introduce ideal *transformers* and *amplifiers* early enough to stimulate the reader with a foretaste of engineering practice. We also make extensive use of the operational amplifier to provide a practical illustration of abstract but fundamental concepts such as *impedance transformation* and *root location control*, always with a vigilant eye on the underlying physical basis. To this end, we often draw analogies to nonelectrical systems such as mechanical and hydraulic systems.

To comply with ABET's emphasis on integrating **computer tools** in the curriculum, the book promotes the use of SPICE as a means for checking the results of hand calculations. The features of SPICE are introduced gradually, as the need arises. We have kept SPICE coverage separate from the rest of the material, in end-of-chapter sections that can be omitted without loss of continuity.

Each chapter begins with an **introduction** outlining its objectives and its relation to the rest of the subject matter, and it ends with a **summary** of its key points. Whenever possible, chapters and sections have been designed to proceed from the *elementary* to the more *complex*. Subjects preceded by the symbol • are optional and may be skipped without loss of continuity.

The text is interspersed with 351 worked **examples**, 402 **exercises**, and 1000 end-of-chapter **problems**. The goal of the examples is not only to demonstrate the direct application of the theory, but also to develop an engineering approach to problem solving based on *conceptual understanding* and *physical intuition* rather than on *rote procedures*. We feel that if we help the student identify a core of basic principles and a common thread of logic, the student will in turn be able to apply these principles and logic to the solution of a wide range of new and more complex problems.

In solving the numerous examples, we stress the importance of *labeling*, *inspection*, *physical dimensions*, and *checking*, and we highlight critical points by means of **remarks**. Students are continually challenged to question whether their findings make *physical sense*, and to use alternative approaches to *check* the correctness of their results.

The exercises are designed to provide immediate reinforcement of the student's grasp of principles and methods, and the end-of-chapter problems are designed to offer additional practice opportunities. For the convenience of students and instructors alike, the problems have been keyed to individual sections.

To give students a better appreciation for the human and historical sides of electrical engineering, each chapter contains an **essay** or **interview**. Subjects include biographies of famous inventors, interviews with well-established electrical engineers and recent graduates, and topics of current interest.

CONTENT

Following is a course content description. Additional details and rationale considerations can be found in the introductions to the individual chapters.

Chapters 1–4 introduce basic circuit concepts, theorems, and analysis techniques for circuits consisting of *resistances* and *independent sources*. Dependent sources and energy-storage elements have been deferred to later chapters because we believe that the student should first develop an intuitive feel for simpler circuits. To facilitate this task we have tried to contain circuit complexity, but without sacrificing generality. In fact, many of the circuits and procedures are of the type practicing engineers see every day.

Anticipating a concurrent laboratory, we introduce basic signal concepts such as *average* and *rms* values early in the book. The emphasis on *i-v* curves to characterize not only individual elements but also one-ports is designed to provide a more intuitive framework for Thévenin and Norton theorems. Much emphasis is placed on the concepts of *equivalence*, *modeling*, *loading*, and the use of *test sources* to find equivalent resistances. The analytical advantages of circuit theorems are demonstrated with one-ports driving *linear* as well as *nonlinear* loads such as diodes and diode-connected BJTs and MOSFETs. However, this brief exposure to nonlinear techniques, designed to provide also a foretaste of electronics, can be omitted without loss of continuity.

Chapter 5 discusses *dependent* sources and their application to the modeling of simple two-ports such as *ideal transformers* and *amplifiers*. This material is designed to motivate the reader with practical applications and also to provide further opportunities for practicing with the analytical tools of the previous chapters. Additional applications are presented in **Chapter 6**, but using the op amp as a vehicle. We have chosen basic instrumentation circuits that are of potential interest to EE majors and nonmajors alike, and that can easily be tried out in the lab or simulated via PSpice.

Chapters 7–9 deal with the *time-domain analysis* of circuits containing *energy-storage elements*. The discussion of the *natural response* emphasizes energetic considerations. The *complete response* is investigated via the *integrating* factor method in order to provide a rigorous treatment of its *natural, forced, transient, dc steady-state*, and *ac steady-state* components. Chapter 8 deals with the transient response of first-order circuits, including switched networks and the step, pulse, and pulse-train responses of *RC* and *RL* pairs; Chapter 9 investigates transients in *RLC* and other second-order circuits.

Significant attention is devoted to correlating response characteristics to *root location* in the *s* plane. In this respect, we demonstrate the use of the op amp to effect *root location control* both for first-order and second-order circuits, and we investigate the physical basis of *converging, steady*, and *diverging* responses. We feel that combining simple physical insight with the pictorial immediacy of *s*-plane diagrams should help demystify the material in anticipation of the network functions and transform theory of subsequent chapters. However, depending on curricular emphasis, the *s*-plane material can be omitted without loss of continuity.

Chapters 10–13 cover *ac circuits*. After introducing *phasors* as graphic means to emphasize amplitude and phase angle, we use *time-domain* techniques to develop a basic understanding of ac circuit behavior, especially the *frequency response* and its relationship to the transient response. Then, we develop *phasor techniques* and the concept of *ac impedance* to investigate *ac power, polyphase systems*, and *ac resonance*, and to illustrate practical applications thereof. To provide the reader with a foretaste of electronics, we now use the op amp to demonstrate *impedance transformations* as well as *resonant characteristics control* in inductorless circuits; but, again, this material can be omitted without loss of continuity.

Recognizing the difficulties encountered by the beginner in trying to reconcile *complex* variables and *physical circuits*, we use the correspondences $dx/dt \leftrightarrow j\omega X$ and $\int x\,dt \leftrightarrow X/j\omega$ to introduce phasor algebra from a time-domain perspective, and we develop the subject from a decidedly circuital viewpoint. Moreover, we give equal weight to *single-frequency* phasor analysis for ac power and *variable-frequency* analysis for communication and signal processing.

Chapter 14 generalizes *ac techniques* to *network-function techniques* in order to provide a unified approach to the study of the *natural, forced, transient, steady-state*, and *complete* response, as well as the *frequency* response. The physical significance of zeros and poles is investigated in depth, along with the effect of their *s*-plane location upon the characteristics of the various responses. The chapter emphasizes *dimensional* and *asymptotic* verifications, and provides an exhaustive treatment of *Bode diagrams*, including the use of PSpice to generate frequency plots of network functions.

We have chosen to introduce network functions via *complex exponential signals*, rather than via Laplace transforms, to comply with the many curricula that defer Laplace to courses other than circuits courses. But, if desired, one can rearrange the material by omitting the first part of Chapter 14 and appending the remainder to the end of Chapter 16, which covers Laplace transforms.

Chapter 15 discusses *two-port networks*, and it applies the two-port concept as well as network-function techniques to the study of *coupled coils*. Two-port measurements and modeling are illustrated via a variety of examples, including transistor amplifiers and linear transformers.

Chapters 16 and **17** provide an introduction to *Laplace* and *Fourier* techniques. True to our intent to proceed from the elementary to the more complex, we present the Laplace approach as a generalization of the network-function approach of Chapter 14. Using a variety of examples, we demonstrate its ability to accommodate a wider range of *forcing functions*, to account for the *initial conditions* automatically, and to provide a more systematic link between the *time-domain* and the *frequency-domain* behavior of a circuit via the initial-value and final-value theorems, the impulse response, and convolution. The treatment stresses the physical basis of the *step* and *impulse* responses, and of the *natural* and *forced* response components.

Fourier techniques are presented as a better alternative to Laplace techniques in those situations in which the focus is on the information and energy content of a *signal* as well as the manner in which a circuit *processes* this content. Though some authors introduce Fourier before Laplace, our desire to emphasize Laplace as a *circuit analysis* tool and Fourier as a *signals processing* tool has led us to introduce Laplace first, as a generalization of the network-function method. Moreover, even though the Fourier transform could be introduced on its own terms, we find it physically more enlightening—if mathematically somewhat less rigorous—to introduce it as a limiting case of the Fourier series.

Applications of both the Fourier series and transform are demonstrated via a variety of signal and circuit examples emphasizing *input* and *output spectra*. Moreover, the effects of filtering are illustrated both in the time domain and the frequency domain. We conclude with practical comparisons of the Laplace and Fourier transforms.

COURSE OPTIONS

This book can be used in a two-semester or a three-quarter course sequence, or in a single-semester course.

In a *two-semester* sequence the first semester typically covers the material up to ac power (Section 12.2), and the second semester covers the remainder of the text. In a *three-quarter* sequence the material can be subdivided as follows: Chapters 1–8, Chapters 9–14, and Chapters 15–17.

For a *single-semester* course we identify two options: An option emphasizing *analytical techniques* and covering approximately the same material as the first course of the two-semester sequence; an option emphasizing specialized areas and thus including specific chapters or sections. An emphasis in *power* would include Section 5.3, all of Chapter 12, and the second half of Chapter 15; an emphasis in *electronics* would include Section 5.4, Chapter 6, and parts of Chapter 14; an emphasis in *systems* would include Chapters 8, 9, 14, 16, and 17. To facilitate omissions without loss of continuity, the optional material identified by the symbol • usually appears at the end of sections or chapters.

SUPPLEMENTS

The following aids for instructor and student are available from the publisher.

Student SPICE Manual by James S. Kang (California Polytechnic State University—Ponoma) introduces students to SPICE commands and programming and includes numerous examples, exercises, and problems tied to the text. It is available for purchase by students.

Student Problems Book by Reza Nahvi and Michael Soderstrand (University of California—Davis) is also available for purchase. It contains about 600 problems arranged by chapter to help students practice and develop their problem-solving skills. About 15% of the problems require students to use SPICE. Complete solutions for all problems are included.

Instructor's Manual contains complete solutions for all exercises and end-of-chapter problems in the text.

Transparency Pack contains transparency masters of the most important figures and graphs in the text to help instructors prepare their lectures.

ACKNOWLEDGMENTS

The development of this text has involved a thorough program of reviewing, for both pedagogical and topical content as well as for accuracy. Many colleagues were involved in this process. Their contributions are gratefully acknowledged:

Douglas B. Brumm, *Michigan Technological University*
Arthur R. Butz, *Northwestern University*
Peter Dorato, *University of New Mexico*
John A. Fleming, *Texas A&M University*
Gary E. Ford, *University of California—Davis*
William J. Jameson, Jr., *Montana State University*
Bruce E. Johansen, *Ohio Northern University*
James S. Kang, *California Polytechnic State University—Pomona*
Robert F. Lambert, *University of Minnesota*
Robert J. Mayhan, *The Ohio State University*
Mahmood Nahvi, *California Polytechnic State University, San Luis Obispo*

Many thanks to the focus group participants for their helpful suggestions:

Roger Conant, *University of Illinois—Chicago*
Ibrahim Hajj, *University of Illinois—Urbana*
Sayfe Kiaei, *Oregon State University*
Michael Lightner, *University of Colorado*
Hari Reddy, *California State University—Long Beach*
Michael Soderstrand, *University of California—Davis*

The accuracy of the examples, exercises, and problems was carefully checked during manuscript, galleys, and page proof stages by James Kang (*California Polytechnic State University—Pomona*), Mahmood Nahvi (*California Polytechnic State University—San Luis Obispo*), and James Rowland (*University of Kansas*). Their careful and diligent work is greatly appreciated.

In addition, I wish to thank Adel Sedra of the University of Toronto for his encouragement and guidance, and Irene Nunes for her research and writing of the historical and biographical essays. I am grateful for the help and dedication of the Saunders staff, especially

Emily Barrosse, Executive Editor; Alexa Barnes, Developmental Editor; Anne Gibby, Project Editor; Anne Muldrow, Art Director; Susan Blaker, Art Director; and Monica Wilson, Marketing Manager. My gratitude goes also to my colleagues T. Holton and W. Stadler at San Francisco State University for their stimulating discussions, and to my wife Diana for her encouragement and steadfast support.

Sergio Franco
San Rafael, California
July 1994

BRIEF TABLE OF CONTENTS

CONTENTS

BASIC CONCEPTS

B efore undertaking a systematic study of electrical engineering principles, we review the basic concepts and laws underlying this discipline: *charge, electric field, voltage, current, energy,* and *power,* as well as electrical *signal, circuit* and *circuit element, circuit laws,* and *element laws.*

Though most of these topics are usually covered in prerequisite courses such as sophomore physics, we are devoting an entire chapter to their review because they are critical for developing physical insight into the operation of the circuits we are about to study. Depending on student background and curricular emphasis, parts of this chapter may be covered as reading assignments.

This chapter also introduces a consistent system of units and notation, and familiarizes the student with the style and approach that will be pursued in the remainder of the book. Whenever possible, *analogies* are drawn to nonelectrical physical systems (such as mechanical and hydraulic systems). Concepts are illustrated by means of practical examples, followed by exercises that you are urged to solve before proceeding.

As you study a worked example, be observant of how the problem is approached, how the framework for its solution is laid out, and how the results are interpreted. You are interested in reaching an answer, but also in developing a **method** and a **style** of doing things. This is an important asset that will accompany you throughout your career.

An integral part of this style is **always questioning** whether your results are correct, or at least whether they make *physical sense.* As an applied discipline, engineering bears the enviable responsibility of not only finding answers, but also ensuring that they are correct. Mercifully, there is always a **physical basis** against which engineers can assess the validity of their mathematical manipulations and of their numerical results.

If finding a solution is fundamental to the engineering discipline, striving to minimize the time and effort involved is just as important as the solution itself. Thus, as you check the solutions to the exercises and problems provided by your instructor, be observant of the mistakes you may have made and of the unnecessarily tortuous paths you may have followed. Make note of them; **study your own way of solving problems.** This will help you improve your approach the next time around. None of us likes making mistakes, but we can exploit them to enhance our understanding and improve our performance.

1.1 UNITS AND NOTATION

Electrical engineering is a quantitative discipline that seeks to describe electrical phenomena in terms of mathematical equations. For this task to succeed, the quantities appearing in these equations must be expressed in a consistent and reproducible system of units. Moreover, it is desirable that the notation be clear, consistent, and self-explanatory whenever possible.

SI Units

In this book we use the *Système Internationale* (SI) *d'Unités* (International System of Units), which is based on the *meter* as the unit of length, the *kilogram* as the unit of mass, the *second* as the unit of time, the *kelvin* as the unit of temperature, the *ampere* as the unit of current, and the *candela* as the unit of light intensity. Table 1.1 provides a summary of these units (boldfaced), as well as other important units.

TABLE 1.1 SI Units

Quantity	Symbol	Unit	Abbreviation
Length	ℓ	meter	m
Mass	m	kilogram	kg
Time	t	second	s
Temperature	T	kelvin	K
Force	f	newton	N
Energy	w	joule	J
Power	p	watt	W
Charge	q	coulomb	C
Current	i	ampere	A
Voltage	v	volt	V
Electric field	E	volt/meter	V/m
Magnetic flux density	B	tesla	T
Magnetic flux	ϕ	weber	Wb
Resistance	R	ohm	Ω
Conductance	G	siemens	S
Capacitance	C	farad	F
Inductance	L	henry	H
Luminous intensity	I	candela	cd

TABLE 1.2 Magnitude Prefixes

Prefix	Abbreviation	Magnitude
femto-	f	10^{-15}
pico-	p	10^{-12}
nano-	n	10^{-9}
micro-	μ	10^{-6}
milli-	m	10^{-3}
kilo-	k	10^{3}
mega-	M	10^{6}
giga-	G	10^{9}
tera-	T	10^{12}

Unit Prefixes

Electrical quantities may range in value over many orders of magnitude. To simplify notation, it is convenient to use the standard prefixes of Table 1.2, which you are encouraged to start applying right away. For example, suppose the result of your calculations is 1.5×10^{-4} A. Since this can be written either as 0.15×10^{-3} A or as 150×10^{-6} A, you should learn to express your result either as 0.15 mA or as 150 μA.

Consistent Sets of Units

We say that a system of units is *consistent* if an equation as expressed in the SI system remains unchanged when expressed in the new system. For instance, shortly we shall study Ohm's Law, which relates voltage v, resistance R, and current i as $v = Ri$. In SI units this law is expressed as $[V] = [\Omega][A]$. Suppose we now multiply and divide the right-hand side by 10^3. This yields $[V] = [10^3\,\Omega][10^{-3}\,A]$, or $[V] = [k\Omega][mA]$, indicating that if resistance is expressed in kΩ and current in mA, then Ohm's Law remains unchanged. Thus, $[V]$, $[k\Omega]$, and $[mA]$ form a consistent set of units, just like $[V]$, $[\Omega]$, and $[A]$. On the other hand, $[V]$, $[k\Omega]$, and $[A]$ *do not* form a consistent set because we would have to include the term 10^{-3} to express Ohm's Law correctly, namely $[V] = [\Omega][A] = 10^{-3}[k\Omega][A]$.

The advantage of working with a consistent set is that calculations can be speeded up considerably. For instance, suppose $R = 2$ kΩ and $i = 3$ mA. Then we can write directly $v = 2 \times 3 = 6$ V, which is quicker and less prone to error than writing $v = 2 \times 10^3 \times 3 \times 10^{-3} = 6$ V. Likewise, if $v = 10$ V and $R = 4$ kΩ, then i must be in mA. In fact, $i = 10/4 = 2.5$ mA. You are encouraged to start working with consistent sets right away. The most frequently encountered consistent sets are:

$$[V] = [\Omega][A] \tag{1.1a}$$

$$[W] = [V][A] \tag{1.1b}$$

$$[V] = [k\Omega][mA] \tag{1.2a}$$

$$[mW] = [V][mA] \tag{1.2b}$$

$$[V] = [M\Omega][\mu A] \qquad \text{(1.3a)}$$

$$[\mu W] = [V][\mu A] \qquad \text{(1.3b)}$$

The sets of Equation (1.1) are used in **power systems,** such as power supplies, electric motors, and power amplifiers, where currents are in the range of amperes. The sets of Equation (1.2) are common in **low-power electronics,** such as home entertainment equipment and personal computer circuits, where currents are in the range of milliamperes. Finally, the sets of Equation (1.3) are used in **micropower electronics,** such as pocket calculators and digital wristwatches, where currents are in the range of microamperes.

Signal Notation

Voltages and currents are usually designated as *quantities* with *subscripts.* The latter may be either *uppercase* or *lowercase,* according to the following convention:

(1) The *instantaneous value* of a signal is represented by *lowercase letters* with *uppercase subscripts,* such as v_S, i_S, v_{AB}, i_{AB}.

(2) *DC signals,* that is, signals that are designed to remain constant with time, are represented by *uppercase letters* with *uppercase subscripts,* such as V_S, I_S, V_{AB}, I_{AB}.

(3) Often a signal consists of a *time-varying* or *ac component* superimposed upon a *constant* or *dc component.* *AC components* are represented by *lowercase letters* with *lowercase subscripts,* such as v_s, i_s, v_{ab}, and i_{ab}. This is illustrated in Figure 1.1 for the case of a voltage signal,

$$\boxed{v_S = V_S + v_s} \qquad \text{(1.4)}$$

In this notation, V_S and v_s are called, respectively, the *dc component* and the *ac component* of the signal, and v_S is called its *total instantaneous value.* For a current signal, we similarly have $i_S = I_S + i_s$.

(4) Special signal values such as *maximum* or *peak values, average values,* and *rms values* are represented by *uppercase letters* with *lowercase subscripts,* such as V_m, I_m, V_{pk}, I_{pk}, V_{av}, I_{av}, V_{rms}, and I_{rms}.

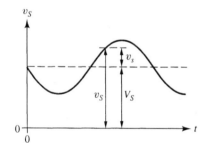

Figure 1.1 Signal notation: $v_S = V_S + v_s$.

1.2 ELECTRIC QUANTITIES

Before embarking upon the study of electric circuits and their laws, let us review the basic concepts of *charge, electric field, voltage, current,* and *power.* A clear understanding of these concepts is critical for developing physical insight into the operation of electric circuits.

Charge

The fundamental quantity of electricity is the **electric charge** which we denote as q. Its SI unit is the *coulomb* (C), named for the French physicist Charles A. de Coulomb (1736–1806). Charges may be either **positive** or **negative.** The most elementary positive charge is that of the proton, whose value is $+1.602 \times 10^{-19}$ C. The most elementary negative charge is that of the **electron,** whose value is -1.602×10^{-19} C. All other charges, whether positive or negative, are integral multiples of these basic charges. For instance the **hole,** a charge carrier found in semiconductor materials of the types used to fabricate diodes and transistors, has a positive charge of $+1.602 \times 10^{-19}$ C.

An atom contains z protons in the nucleus and z electrons in the surrounding electron shells, where z is the atomic number of the element under consideration. Since their positive and negative charges balance out, atoms are **electrically neutral.** However, an atom can be stripped of one or more of its outer-shell electrons. This leaves a corresponding number of protons uncovered, turning the atom into a **positive ion.** Likewise, a **negative ion** is an atom having one or more excess electrons in its outer shell.

To make up a charge of 1 C it takes $1/(1.602 \times 10^{-19}) = 6.24 \times 10^{18}$ elementary charges. This is a mind-boggling number. Luckily, matter consists of atoms so densely packed that to make up a charge this large is a relatively easy task.

Charge is **conservative,** that is, it cannot be created or destroyed. However, it can be *manipulated* in a variety of different ways, and herein lies the foundation of electrical engineering. The most common tool for the manipulation of charge is the **electric field,** which we denote as E. When exposed to an electric field E, a charge experiences a force whose direction is the same as E for positive charges, but opposite to E for negative charges.

Potential Energy

As a consequence of the force exerted by the electric field, a charge possesses **potential energy.** This energy, denoted as w, depends on the magnitude of the charge as well as its location in space. This is similar to a mass exposed to the gravitation of earth. Lifting a mass m to a height h above some reference level, such as sea level, increases its potential energy by the amount $w = mgh$, where g is the gravitational acceleration.

If we let the mass go, it will fall in the direction of decreasing potential energy and acquire kinetic energy in the process. The same can be said of a charge under the effect of an electric field, except that *positive charges* tend to fall in the *same direction* as E, and *negative* charges in the *opposite direction,* or *against* E.

Recall from physics studies that the *reference* or *zero level* of potential energy can be chosen *arbitrarily,* the reason being that only *differences* in potential energy have practical meaning. Just as in the gravitational case it is often convenient to choose sea level as the zero level of potential energy, in the electrical case it has been agreed to regard *earth* as the zero level of potential energy for charges. Home appliances and entertainment equipment are equipped with three-prong power cords for connection to the utility line as well to earth, or **ground.** All points of the equipment that are connected to earth are said to be *grounded,* and any charges residing at these points are thus assumed to have *zero potential energy.* In portable equipment lacking an electric connection to earth, the chassis is often given the role of ground.

Voltage

The rate at which potential energy varies with charge is denoted as v and is called the **electric potential,**

$$v \triangleq \frac{dw}{dq} \tag{1.5}$$

where the symbol "$\triangleq$" stands for "is defined as." Potential can also be viewed as potential energy per unit charge, or *potential energy density*. The SI unit of potential is the *volt* (V), named for the Italian physicist Alessandro Volta (1745–1827). Since the SI unit of energy is the joule (J), it follows that $1 \text{ V} = 1 \text{ J/C}$.

What matters to us, however, is not potential per se, but **potential difference,** or **voltage.** The physical interpretation of Equation (1.5) is as follows: *If a charge dq gives up an amount of energy dw in going from one point to another in space, then we define the voltage between those points as* $v = dw/dq$.

Potential difference or voltage is easily created by means of a battery. The terminal having the *higher* potential of the two is identified by the $+$ sign and is called *positive*. The other terminal is by default called *negative*. Denoting the individual terminal *potentials with respect to ground* as v_P and v_N, a battery yields

$$v_P - v_N = V_S \tag{1.6}$$

where V_S is the voltage rating of the battery. For example, a 9-V radio battery yields $v_P - v_N = V_S = 9$ V, that is, it keeps v_P *higher than* v_N by 9 V, regardless of the individual values of v_N or v_P. If we ground the negative terminal so that $v_N = 0$ V, then the battery will yield $v_P = 9$ V. Conversely, if we ground the positive terminal so that $v_P = 0$, the battery yields $v_N = -9$ V. Likewise, with $v_N = 6$ V the battery yields $v_P = 15$ V, with $v_N = -6$ V it yields $v_P = +3$ V, with $v_P = -5$ V it yields $v_N = -14$ V, and so forth.

Another familiar example is offered by a 12-V car battery, whose negative terminal is designed to be connected to the chassis of the car and, hence, to be at zero potential by definition. We observe that the terms *positive* and *negative* bear no relationship to the individual polarities of v_P and v_N; they only reflect the signs preceding v_P and v_N in Equation (1.6). Clearly, the terms *higher* and *lower* would be more appropriate than the terms *positive* and *negative*.

Relation Between Electric Field and Potential

By establishing a potential difference, a battery generates an **electric field.** Field and potential are related by the important law of physics

$$E = -\text{grad } v \tag{1.7a}$$

If we orient the battery terminals so that the direction of E coincides with the x axis, this relation simplifies as

$$E = -\frac{dv}{dx} \tag{1.7b}$$

where E denotes the magnitude or strength of E, and the negative sign indicates that E points in the direction of *decreasing* potential. The SI unit of electric field is the *volt/meter* (V/m).

The preceding concepts are illustrated in Figure 1.2, where a battery is connected by means of two wires to a bar of conductive material, such as carbon composition, to maintain a prescribed voltage V_S across its extremes. If the material is homogeneous and the bar has uniform cross section, as shown, then the electric field produced inside the bar will be constant throughout. We can thus replace the differentials of Equation (1.7b) with finite differences and write

$$E = -\frac{v(\ell) - v(0)}{\ell - 0} \quad \text{(1.8)}$$

where ℓ is the length of the bar, and $v(\ell)$ and $v(0)$ are, respectively, the voltages at its right and left extremes. Since $v(0) - v(\ell) = V_S$, by battery action, it follows that

$$E = \frac{V_S}{\ell} \quad \text{(1.9)}$$

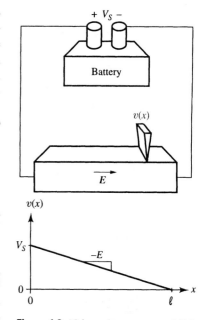

Figure 1.2 Using a battery to establish an electric field inside a homogeneous bar of conductive material.

▶ **Example 1.1**

If a 9-V battery is connected across a 6-cm homogeneous bar of conductive material, what is the strength of the electric field inside the bar?

Solution

Since $V_S = 9$ V and $\ell = 6 \times 10^{-2}$ m, Equation (1.9) yields $E = V_S/\ell = 9/(6 \times 10^{-2}) = 150$ V/m.

Remark You are urged to develop the habit of always expressing your results in their *correct physical units*. Just saying $E = 150$ in our example is not enough. We must also specify V/m. ◀

Equation (1.7b) can be turned around to calculate $v(x)$,

$$v(x) = \int_0^x -E \, d\xi + v(0) \quad \text{(1.10)}$$

where ξ is a dummy integration variable. Using Equation (1.9) yields

$$v(x) = -\frac{V_S}{\ell}x + v(0) \quad \text{(1.11)}$$

This represents a straight line with slope $dv(x)/dx = -(V_S/\ell) = -E$, and with initial value $v(0)$. As we know, this value can be chosen arbitrarily, and in Figure 1.2 we have chosen $v(0) = V_S$ because this yields $v(\ell) = 0$ V and makes things neater. Substituting $v(0) = V_S$ into Equation (1.11) yields

$$v(x) = V_S \left(1 - \frac{x}{\ell}\right) \quad \text{(1.12)}$$

We observe that as we move from the left extreme to the right extreme, the voltage inside the bar varies *linearly* from V_S to 0 V. This can be verified experimentally by tapping off the voltage at any intermediate point by means of a sliding contact, or *wiper,* as shown. For example, with $V_S = 9$ V and with the wiper positioned half-way where $x = \ell/2$, Equation (1.12) predicts $v = 9[1 - (\ell/2)/\ell] = 4.5$ V; with the wiper two-thirds of the way toward the right, where $x = (2/3)\ell$, we obtain $v = 3$ V, and so forth. It is evident that by properly positioning the wiper, we can obtain *any voltage* between V_S and 0 V. This principle forms the basis of slide potentiometers, or *slide pots,* which are used, among other things, to effect volume control in modern audio equipment.

Current

If charges are free to move, exposing them to an electric field will force them to *drift* either along or opposite to E, depending on charge polarity. The resulting stream of charges is called **current.** The rate at which charge crosses some reference plane perpendicular to the stream is denoted as i and is called the **instantaneous current,**

$$i \triangleq \frac{dq}{dt} \qquad (1.13)$$

The SI unit of current is the *ampere* (A), named for the French physicist André M. Ampère (1775–1836). Clearly, 1 A = 1 C/s.

The preceding concepts are illustrated in Figure 1.3 for the case of a bar of *p*-type silicon, a material widely used in the fabrication of integrated circuits (ICs) of the type found in digital wristwatches, personal computers, and other modern electronic equipment. In this material the mobile charges are positive and are called *holes.* Applying a voltage *v across* the bar establishes a field E *inside* the bar which, in turn, causes the holes to drift in the direction of the

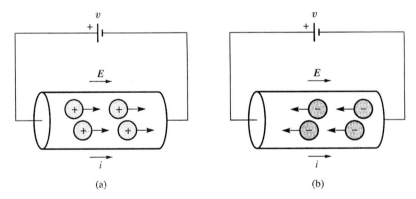

Figure 1.3 Illustrating electric current: (a) *positive* charge flow; (b) *negative* charge flow.

field to produce a current i,

$$v \to \boldsymbol{E} \to i$$

As shown in the figure, we indicate current direction by an arrow alongside the bar. Note also that to simplify our diagram we are representing the battery in terms of its standard circuit symbol.

One can find i by counting the number of holes that pass in 1 s through some reference plane perpendicular to the bar. Multiplying this number by the charge of each hole, or $+1.602 \times 10^{-19}$ C, yields i. This is similar to standing atop a bridge and counting the number of cubic feet of water passing under the bridge in 1 s to find the flow rate of the river below. As an example, suppose we count 2.5×10^{15} holes/s. Then $i = 2.5 \times 10^{15} \times 1.602 \times 10^{-19} = 4.0 \times 10^{-4}$ C/s, or 0.4×10^{-3} A. This is more conveniently expressed as 0.4 mA.

Figure 1.3(b) illustrates the dual situation, namely, that in which the mobile charges are *electrons,* which are negative. This situation is encountered in ordinary conductors like copper as well as in n-type silicon, another widely used material in IC fabrication. Since electrons are negative, the direction of flow is now *opposite* to $\boldsymbol{E}$; however, the inversion in direction is counterbalanced by the inversion in charge polarity, so that the *direction* of i is again the same as $\boldsymbol{E}$.

An alternate way of justifying this is as follows. In Figure 1.3(a), as positive charges drift from left to right, they tend to accumulate positive charge on the negative terminal while removing positive charge from the positive terminal. This is equivalent to accumulating *positive charge* on the *negative terminal,* and *negative charge* on the *positive terminal.* In Figure 1.3(b), as negative charges drift from right to left, they tend to accumulate *negative charge* on the *positive terminal,* and *positive charge* on the *negative terminal.* From the battery viewpoint the two situations are indistinguishable. Hence, the direction of i must be the same in both cases, even though the charges have opposite polarities.

As we proceed, we shall find it convenient to think of current as due *exclusively* to positive charge flow, regardless of the actual physical details. When studying semiconductor devices such as diodes you will find that hole and electron currents may coexist within the same device. Denoting these currents, respectively, as i_P and i_N, the *net current i* is the *additive* combination of the two, or $i = i_P + i_N$. For example, suppose we have 2×10^{15} holes/s flowing from left to right and 3×10^{15} electrons/s flowing from right to left. Then $i = (2 + 3)10^{15} \times 1.602 \times 10^{-19} = 0.8$ mA flowing from left to right. This is the same as if i were due exclusively to $(2 + 3)10^{15}$ holes flowing from left to right.

Equation (1.13) can be turned around to find the amount of charge that passes through the reference plane during the time interval from t_1 to t_2,

$$Q = \int_{t_1}^{t_2} i(t) \, dt \tag{1.14}$$

The ability of a battery to store charge is referred to as the **battery capacity,** and is expressed in *ampere-hours* (Ah), where 1 Ah = (1 C/s) × (3600 s) = 3600 C. Car batteries have capacities on the order of 10^2 Ah.

▶**Example 1.2**

The current entering a certain terminal is $i(t) = 10e^{-10^6 t}$ mA.

(a) Find the total charge entering the terminal between $t = 0$ and $t = 1$ μs.

(b) Assuming $i(t)$ is due to electron motion, express the above charge in terms of the basic electron charge. Are electrons entering or exiting the terminal?

(c) Repeat, but for the interval between $t = 1$ μs and $t = 2$ μs.

Solution

(a) The charge entering the terminal is

$$Q = \int_{t_1}^{t_2} 10e^{-10^6 t} \times 10^{-3}\, dt = 10^{-2} \times \left. \frac{e^{-10^6 t}}{-10^6} \right|_{t_1}^{t_2}$$

$$= 10^{-8} \left(e^{-10^6 t_1} - e^{-10^6 t_2} \right) = 10^{-8} \left(e^0 - e^{-1} \right) = 6.321 \text{ nC}$$

(b) This corresponds to $6.321 \times 10^{-9}/(1.602 \times 10^{-19}) = 3.946 \times 10^{10}$ electrons. Since they are negative, they *exit* the terminal.

(c) Repeating the calculations, we find $Q = 2.325$ nC, or 1.452×10^{10} *exiting* electrons.   ◀

Exercise 1.1   (a) The current entering a certain terminal is $i(t) = 5 \sin 2\pi 10^3 t$ A. Assuming the current is due to hole motion, find the net number of holes entering the terminal (a) between $t = 0$ and $t = 0.5$ ms, and (b) between $t = 0$ and $t = 1$ ms. Compare the two cases, comment.

ANSWER   (a) 9.935×10^{15} holes, (b) zero, because the number of holes entering between 0 and 0.5 ms equals the number exiting between 0.5 ms and 1 ms.

Power

To sustain electric current inside a piece of material takes an expenditure of energy, or work. In the examples of Figure 1.3 this work is performed by the battery. The rate at which energy is expended is denoted as p and is called the **instantaneous power**,

$$p \triangleq \frac{dw}{dt} \tag{1.15}$$

The SI unit of power is the *watt* (W), named for the Scottish inventor James Watt (1736–1819). Clearly, 1 W = 1 J/s.

Writing $dw/dt = (dw/dq) \times (dq/dt)$ and using Equations (1.5) and (1.13) yields

$$p = vi \tag{1.16}$$

In words, this relation states that **whenever a current i flows between two points having a potential difference v, the corresponding instantaneous power is $p = vi$.**

Equation (1.15) can be turned around to find the energy expended over the time interval from t_1 to t_2,

$$W = \int_{t_1}^{t_2} p(t)\, dt \tag{1.17}$$

Since the unit of power is the watt, energy is sometimes expressed in *watt-seconds*. An alternate unit of energy, used especially by electric utility companies, is the *kilowatt-hour* (kWh). This unit represents the amount of energy expended in one hour at the rate of 10^3 J/s, or 10^3 W. Thus, 1 kWh = $(10^3$ J/s$) \times (3600$ s$)$, or

$$1 \text{ kWh} = 3.6 \times 10^6 \text{ J} \tag{1.18}$$

Like energy, power cannot be created or destroyed; however, it can be *converted* from one form to another. A typical example is offered by an ordinary flashlight, which is shown in circuit-diagram form in Figure 1.4, using the standard symbols for the battery and the bulb. As we know, the battery causes current to flow out of its positive terminal, through the bulb and wires, and back into its negative terminal, from where the current returns to the positive terminal to complete the loop. As current flows through the bulb, electric energy is converted to light and heat. Electric energy, in turn, comes from the battery, where it is generated by conversion from chemical energy. Thus, in a flashlight circuit, current serves as a *vehicle* to transfer power from the battery to the bulb. Moreover, we say that the battery **releases** electric power and that the bulb **absorbs** electric power. For this reason, the battery is said to be an **active** circuit element, the bulb a **passive** circuit element.

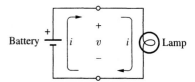

Figure 1.4 In a flashlight circuit, power is transferred from the battery to the bulb.

▶ Example 1.3

A 12-V car battery is connected to a 36-W headlight for 1 hr. Find:

(a) the current passing through the bulb;

(b) the total energy dissipated in the bulb, in J and in kWh;

(c) the total charge that has passed through the bulb, both in coulombs and in terms of the electron charge.

Solution

(a) $i = p/v = 36/12 = 3$ A.

(b) Since p is constant with time, Equation (1.17) yields
$W = p \times (t_2 - t_1) = 36(60 \times 60 - 0) = 129.6$ kJ. This can
also be expressed as $W = (129.6 \times 10^3)/(3.6 \times 10^6) = 0.036$ kWh.

(c) Since i is constant, Equation (1.14) yields $Q = i(t_2 - t_1) = 3(60 \times 60 - 0) = 10.8$ kC. This is equivalent
to $10,800/(1.602 \times 10^{-19}) = 6.742 \times 10^{22}$ electrons. ◀

Exercise 1.2   Assuming the billing rate of your electric utility company
is \$0.10/kWh, estimate the cost of inadvertently leaving your desk light
on all night each night for the entire semester. (*Hint:* To solve this
problem, read the wattage on your lightbulb, and decide how many hours
it is left on each night.)

ANSWER   Mine would cost \$14.40.

It is worth pointing out that electric current can serve not only as a vehicle
to *convert* from one form of energy to another, but also as a vehicle to *transport*
energy over long distances. For example, consider the task of transmitting to
homes and industries the mechanical power produced by turbines at a remote
power plant. Transmission by means of long rotating shafts would be an un-
realistic proposition. Instead, turbine power is first converted by generators to
electrical form and then transported over long distances by high-voltage transmission lines. At the destination, electric motors reconvert electrical power to
mechanical form. You will surely agree that this form of transmission is far
more efficient and reliable, and certainly cheaper.

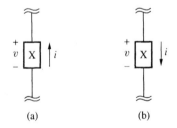

(a)   (b)

Figure 1.5 Relation between
current direction and *voltage
polarity* for a device
(a) *releasing* power, and
(b) *absorbing* power.

The Active and Passive Sign Convention

How can we tell whether a given element in a circuit is absorbing or releasing
power? To this end, we must examine the *direction* of i *through* the element
relative to the *polarity* of v *across* its terminals. Then, we use the following rule:

**Power Rule: If i flows in the direction of *increasing* v, as in the case of a
flashlight battery, then power is being *released*. If i flows in the direction
of *decreasing* v, as in the case of a flashlight bulb, then power is being
absorbed.**

This important rule, summarized in Figure 1.5, holds for any element X, not
just for batteries and bulbs. Taking the *positive* voltage terminal of an element as

the reference terminal for current entry or exit, we can restate the power rule as:

(1) In a *power-releasing* element current *exits* the element via the *positive* terminal. In this case voltage and current are said to conform to the **active sign convention.**

(2) In a *power-absorbing* element current *enters* the element via the *positive* terminal. Voltage and current are now said to conform to the **passive sign convention.**

As we proceed we shall find that elements such as resistors and diodes never release power, indicating that these devices conform, at all times, to the passive sign convention. By contrast, devices such as voltage or current sources, capacitors, and inductors may either release or absorb power, depending on how they are used. For instance, a car battery releases power when connected to the headlights, but it absorbs power when connected to a battery charger. Thus, during a charging cycle current enters the battery via the positive terminal and exits via the negative terminal; during a discharging cycle current exits the battery via the positive terminal and returns via the negative terminal. Keep these important observations in mind because we shall use them over and over again.

1.3   ELECTRIC SIGNALS

Energy and information are the hallmarks of a technological society. Electrical engineering plays a prominent role in the development of suitable means to make energy and information useful to society. At the end of the previous section we pointed out the role of electric quantities such as voltage and current as vehicles for the *conversion, transmission,* and *utilization* of **energy.** Broadly speaking, this is the domain of the branch of engineering dealing with *electrical power* and *electromechanical systems.*

The other most important use of electric quantities is as vehicles to *represent, manipulate, transmit,* and *store* **information.** The disciplines dealing with this aspect are *analog* and *digital electronics, automatic control* and *instrumentation,* and *commmuncations.* Information is often converted to electrical form to take advantage of the great ease with which one can manipulate electric quantities. When used as a vehicle for information, an electric quantity such as a voltage, a current, or an electromagnetic wave is referred to as a **signal.** A common example is offered by the voltage pulses that are used to represent data and programs in computers. The computational power of such systems stems from the ability to process, store, transmit, and receive electric signals rapidly, reliably, and inexpensively.

Signals constitute an important ingredient of circuit theory, and an early exposure to the subject is especially desirable if the study of circuits is accompanied by experimentation in the laboratory, where the student is required to become familiar with signal generators and signal-measuring equipment.

An important feature of a signal is the manner in which it varies with time. Patterns of time variation, or **waveforms,** are usually designed to convey information. In the following we shall examine the most common waveforms and their properties. In accordance with the notation convention described at the end of Section 1.1, we designate the *instantaneous* value of a signal as x_S.

A CONVERSATION WITH
BURKS OAKLEY II
UNIVERSITY OF ILLINOIS—URBANA-CHAMPAIGN

Burks Oakley has recently changed his focus from visual electrophysiology to computer-assisted education.

Burks Oakley II is a professor and the assistant department head in the Department of Electrical and Computer Engineering at the University of Illinois at Urbana-Champaign, as well as course director for the "Introduction to Circuit Analysis" course. He holds a B.S. degree from Northwestern University, and M.S. and Ph.D. degrees in bioengineering from the University of Michigan.

Professor Oakley, what got you interested in electrical engineering?

I was working as a biomedical engineer, designing and using electronic instrumentation to study the electrical properties of living cells. I had to learn more about electrical engineering to conduct my research in biomedical engineering.

What research areas have you worked in?

Until the past few years, most of my work was in visual electrophysiology. I studied the origin of the electroretinogram, or ERG, which is generated by the synchronous electrical activity of neurons in the retina much as the electrocardiogram is generated by cells in the heart or the electroencephalogram by cells in the brain. Ophthalmologists use the ERG to diagnose retinal disease, and I spent quite a few years learning exactly how the various components of the complex ERG waveform are generated.

For the past three years, my main focus has switched to computer-assisted education—how to use computers to help students learn. Because computers have become more powerful while becoming more affordable, and because students start using them so early in school, computers will soon be replacing textbooks as the main teaching tool. Here in

Urbana, the whole campus, including all the dorms, is networked. This coming semester, I will have the students in my class submit their homework to me electronically over the network, from their computer to mine, and my computer will grade the homework

> *"You may be the smartest engineer since Steinmetz, but if you cannot communicate your solution to a problem, who's going to know how smart you are?"*

and record the scores in a spreadsheet. In the near future, I can see students walking into the lecture hall with their notebook computers, connecting to a wireless network, turning in their homework electronically, and then downloading my multi-media presentation right into their computers.

Where is the field of EE headed?

It certainly is getting broader and broader, a revolution made possible by the invention of the transistor in the 1940s and the integrated circuit in the 1960s. Think of all the recent developments that have changed our lives, ranging from consumer products such as VCRs and microwaves to computer networks that enable us to communicate all over the world instantly via electronic mail. Many of these advances are due to the efforts of electrical engineers.

Although one engineer sitting at a workstation may do the design and analysis that used to occupy 25 or 30 engineers, that loss is more than balanced by the numbers of new products that will need the skills of engineers. Cars, for example. Antilock brakes, emission controls, computerized fuel injectors, radios with CD players—all these systems did not exist a few years ago, and they were designed and analyzed by EEs. And that's only one industry. There is now, and will be, just as much need for EEs in many other industries, too.

What advice do you have for today's EE students?

There's so much breadth in the field of electrical engineering: control systems, communications, computers, digital systems, electromagnetics, power, signal processing—the list goes on and on. And so much new knowledge is generated each year. It's impossible for our students to understand everything about all of these areas. The most important thing today's student can learn in college is how to be a lifelong learner. The days are long gone when students took what they learned in four years of college and applied it for the rest of their working days. The half-life of what students learn today is about five years, so the best thing that they can do while in school is learn how to learn on their own.

Engineers today work in interdisciplinary teams, so students should learn how to communicate well and how to work as part of a team. I try to get my students working collaboratively on certain assignments, such as homework, and I even have them solve new problems in lecture by working in groups.

A major complaint that we hear from people in industry is that although they are technically very proficient, our graduate engineers have a difficult time

> *"The most important thing today's student can learn in college is how to be a lifelong learner."*

communicating their ideas. Think about it: You may be the smartest engineer since Steinmetz, but if you cannot communicate your solution to a problem, who's going to know how smart you are? I always recommend that students take a technical writing course while in college, as well as a public speaking course, in order to learn to present their ideas to others. I also think that it is best to take as many social science and humanities courses as possible. All of these courses will help students become well-rounded individuals.

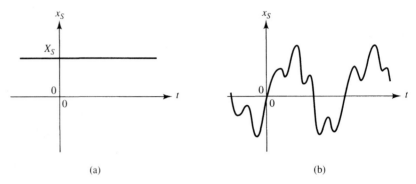

Figure 1.6 (a) *DC* signal and (b) *time-varying* signal.

DC Signals

The most basic signal type is a signal whose value is designed to remain constant with time,

$$x_S = X_S \tag{1.19}$$

where X_S represents the contant value of the signal, which may be positive, negative, or zero. Figure 1.6(a) shows an example of a constant signal. Following the convention of Section 1.1, these signals are denoted with *uppercase letters* and *uppercase subscripts,* such as V_S, I_S, V_{AB}, I_{AB}, and so forth. This allows us to identify the constant nature of a signal just by looking at the notation (provided, of course, the notation is used correctly).

A common example of a constant signal is the current flowing in the flashlight circuit of Figure 1.4. Such a current is referred to as a **direct current,** or **dc** for short. Since a direct current is usually the result of a constant voltage, a constant voltage is likewise referred to as a direct-current voltage, or a **dc voltage.** A familiar example is the 12-V dc voltage produced by a car battery.

One of the most common applications of dc signals is to provide power to electronic and electromechanical devices and systems such as radios, wristwatches, personal computers, dc motors, and so forth. Instruments specifically designed to serve this function are aptly called **dc power supplies.**

Time-Varying Signals

Figure 1.6(b) shows an example of a **time-varying signal.** Such a signal might be, for instance, the voltage produced by a microphone in response to the pressure of a sound wave. Since the time variations of the voltage signal mimic those of the sound pressure, voltage in this case is said to be the *analog* of sound and is thus referred to as an **analog signal.**

Because it converts energy from one form to another, that is, from acoustical to electrical, a microphone is said to be a **transducer.** A loudspeaker is a transducer that converts energy in the opposite direction, that is, from electrical back to acoustical. A home audio system offers an excellent example of why it is desirable to convert sound to an electric signal and then back to sound. It is because of the ease with which such a signal can be (a) **amplified** to produce, via a loudspeaker, a much more powerful sound; (b) **stored** to be played back at a later time, for instance, by means of a compact disc; (c) **transmitted** over long distances, for instance, via radio waves or fiber-optic cables; (d) **processed** to create a variety of special effects, for instance, through the use of the bass and treble controls, graphic equalizers, mixers, and so forth.

Time-varying signals take on a variety of different names, depending on their patterns of time variation as well as the purpose for which they are used.

The Step Function

When we flip the switch of a flashlight, the current through and the voltage across the bulb undergo an abrupt change from zero to some nonzero value. This pattern of time variation is referred to as a **step function.** Taking $t = 0$ as the instant of switch activation, the instantaneous value of a step signal is

$$x_S = 0 \quad \text{for } t < 0 \tag{1.20a}$$

$$x_S = X_m \quad \text{for } t > 0 \tag{1.20b}$$

where X_m is referred to as the **step amplitude.** The step function is depicted in Figure 1.7(a).

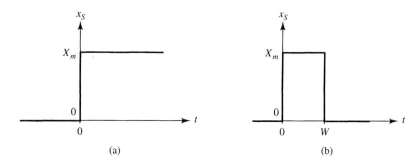

(a)   (b)

Figure 1.7 (a) Step function and (b) pulse.

When we switch the flashlight off, the bulb voltage and current jump back to zero. We now have what is referred to as a *negative-going* step. By contrast, the step of Figure 1.7(a) is also called a *positive-going* step.

The Pulse

If a positive-going step at $t = 0$ is followed by a negative-going step at some later instant $t = W$, the resulting pattern of time variation, depicted in Figure 1.7(b), is called a **pulse.** The parameters X_m and W are called, respectively, the **pulse amplitude** and the **pulse width.**

A sequence of periodically repeated pulses, depicted in Figure 1.8, is refered to as a **pulse train.** Pulses and pulse trains are the kind of signals that are found in digital wristwatches, pocket calculators, and personal computers. Instruments designed to provide pulse trains for laboratory testing are called **pulse generators.**

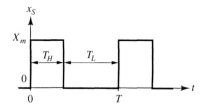

Figure 1.8 Pulse train.

The time intervals T_H and T_L are often called, respectively, **mark** and **space.** The degree of time symmetry of a pulse train is expressed in terms of the **duty cycle** d, which we define as

$$d \triangleq \frac{T_H}{T_L + T_H} \tag{1.21}$$

Multiplying the right-hand side by 100 yields the duty cycle in percentage form. When $T_H = T_L$ we have $d = 0.5$ (or $d = 50\%$), and we say that the pulse train is **symmetric** with respect to time.

Periodic Signals

A signal that repeats itself every T seconds is said to be a **periodic signal** with **period** T. Mathematically, such a signal satisfies the condition

$$x_S(t \pm nT) = x_S(t) \tag{1.22}$$

for any t and $n = 1, 2, 3, \ldots$. The most common periodic signals are those produced by laboratory instruments called **function generators.** They include the **sine, rectangle, triangle,** and **sawtooth** waves shown in Figure 1.9. The signal produced by the human heart and visualized by means of electrocardiographs is an example of an approximately periodic signal.

Periodic voltage waveforms are easily observed with an **oscilloscope,** an instrument that automatically displays voltage versus time on a cathode ray tube (CRT). The oscilloscope can readily be used to measure the period T.

The number of cycles of oscillation completed in one second is denoted as f and is called the **frequency,**

$$f = \frac{1}{T} \tag{1.23}$$

The SI unit of frequency is the *hertz* (Hz), named for the German physicist Heinrich R. Hertz (1857–1894). Clearly, 1 Hz represents one cycle of oscillation per second. Common examples of frequencies are those broadcast by commercial radio stations. AM stations broadcast over the range of 540 kHz to 1600 kHz, and FM stations over the range of 88 MHz to 108 MHz. Recall that $1\text{kHz} = 10^3 \text{ Hz}$ and $1\text{MHz} = 10^6$ Hz.

A dc signal can be regarded as an infinitely slow periodic signal, that is, as a periodic signal having $T = \infty$. Consequently, since $f = 1/T = 1/\infty = 0$, **the frequency of a dc signal is zero.**

Frequency is readily measured with a **frequency counter,** an instrument that utilizes a digital counter and an electronic timepiece to count the number of cycles completed in one second and then displays the result in digital form. Once f is known, the period is readily found as $T = 1/f$.

The degree of time symmetry of a rectangular wave is again expressed in terms of the duty cycle d of Equation (1.21). When $d = 50\%$, the rectangular

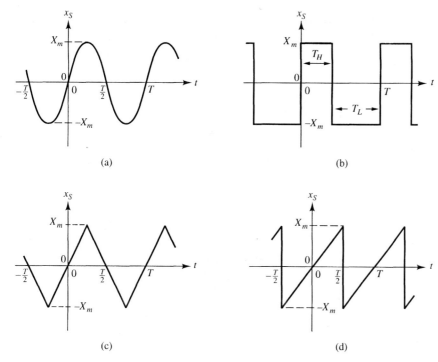

Figure 1.9 Common periodic waveforms: (a) *sine,* (b) *rectangle,* (c) *triangle,* and (d) *sawtooth.*

wave is symmetric with respect to time and is more properly called a **square wave.**

AC Signals

The signal of Figure 1.9(a) alternates *sinusoidally* between the extremes $+X_m$ and $-X_m$. Since a sinusoidal signal tends to produce a sinusoidally **alternating current,** or **ac** for short, we refer to it as an **ac signal.** Thus, while the term *dc* has come to signify *constant over time,* the term *ac* signifies *alternating sinusoidally.*

The quantity X_m is called the **amplitude** or **peak value** of the sinusoidal wave and is sometimes denoted as X_{pk}, so that $X_{pk} = X_m$. The difference between the extremes is called the **peak-to-peak value** and is denoted as X_{pk-pk}. Clearly, $X_{pk-pk} = 2X_{pk} = 2X_m$.

The energetic strength of a signal is characterized in terms of a parameter called the **root-mean-square value,** and denoted as X_{rms}. Though this parameter will be discussed in greater detail in Section 4.4, suffice it to say here that for a sine wave the rms value is related to the peak value as

$$X_{rms} = \frac{X_m}{\sqrt{2}}$$

(1.24)

Note the use of *uppercase letters* with *lowercase subscripts* to designate special values of a signal. This again conforms to the notation convention of Section 1.1.

The most common example of an ac signal is the household voltage, whose rating in North America is 120 V rms, 60 Hz. This means that the peak value is $V_m = V_{rms}\sqrt{2} = 170$ V, so the utility voltage alternates between $+170$ V and -170 V. Moreover, it does so 60 times per second, indicating that the duration of each cycle of oscillation is $T = 1/f = 1/60 = 16.7$ ms.

It is not difficult to see that the *instantaneous value* x_S of a sinusoidal wave can be expressed mathematically as

$$x_S = X_m \sin \frac{2\pi t}{T} \tag{1.25}$$

That the argument of the sine function must be $2\pi t/T$ follows from the fact that as t spans a complete period T, the argument must span 2π radians. Moreover, because the sine function alternates between $+1$ and -1, we must multiply it by X_m to make x_S alternate between $+X_m$ and $-X_m$. By Equation (1.23), we can also write

$$x_S = X_m \sin 2\pi f t \tag{1.26}$$

For example, the household voltage can be expressed as $v_S = 120\sqrt{2} \sin 2\pi 60t$ V $\simeq 170 \sin 377t$ V.

Function generators have provisions to also add a dc component or **offset** V_S to a signal. Thus, when the generator is configured for the sinusoidal waveform, its output can be expressed as

$$v_S = V_S + V_m \sin 2\pi f t \tag{1.27}$$

and it oscillates between the extremes (V_S+V_m) and (V_S-V_m). Separate controls (called, respectively, the *amplitude, offset,* and *frequency* controls) allow for the individual adjustment of V_m, V_S, and f. Clearly, setting V_S to zero yields a purely ac signal, and setting V_m to zero yields a purely dc signal.

▶ **Example 1.4**

Write a mathematical expression for a 1-kHz voltage signal that alternates sinusoidally between (a) -10 V and $+10$ V; (b) -1 V and $+5$ V; (c) -2 V and -1 V.

Solution

We find V_S as the mean between the extremes, and V_m as the difference between the upper extreme and V_S.

(a) $V_S = (-10 + 10)/2 = 0$ and $V_m = 10 - 0 = 10$ V. Thus,

$$v_S = 10 \sin 2\pi 10^3 t \text{ V}$$

(b) $V_S = (-1 + 5)/2 = 2$ V, $V_m = 5 - 2 = 3$ V, and

$$v_S = 2 + 3 \sin 2\pi 10^3 t \text{ V}$$

(c) $V_S = (-2 - 1)/2 = -1.5$ V, $V_m = -1 - (-1.5) = 0.5$ V, and

$$v_S = -1.5 + 0.5 \sin 2\pi 10^3 t \text{ V}$$

◀

Exercise 1.3   The output of a sinusoidal function generator is observed with an oscilloscope and is found to have a peak-to-peak amplitude of 10 V, with the lower extreme being -2 V. Moreover, the signal repeats itself every 100 μs. Write a mathematical expression for such a signal.

ANSWER $v_S = 3 + 5 \sin 2\pi 10^4 t$ V.

Analog and Digital Signals

The distinguishing feature of an analog signal such as that produced by a microphone transducer and then magnified by an audio preamplifier is that it can take on a *continuum of instantaneous values* within a certain range. For this reason, the terms **analog** and **continuous** are used interchangeably when referring to a signal of this sort. An example of an analog signal is depicted in Figure 1.6(b). Electric systems dealing with the generation and processing of these signals form the body of **analog electronics.**

By contrast, the signals found in digital computers, pocket calculators, and digital wristwatches are allowed to assume only a *limited set of values* and are thus referred to as **discrete** or **digital** signals. An example is offered by the pulse train of Figure 1.8, which may assume only *two* values, namely 0 or X_m. Two-valued signals, commonly referred to as **binary signals,** form the domain of **digital electronics,** in which computer electronics plays the dominant role.

A major advantage of representing information in binary form stems from the much greater degree of ease and dependability with which binary signals can be generated, processed, transmitted, and stored as compared to analog signals. The superiority of compact discs over analog tapes and phonograph discs in modern audio systems testifies to this advantage.

Average Value of a Signal

At times it is of great practical interest to know the **average value** of a signal. We define the average value of a time-varying signal $x(t)$ over the time interval t_1 to t_2 as the *steady* or *dc signal* X_{av} that over that time interval would span the *same area* as $x(t)$, or $X_{av} \times (t_2 - t_1) = \int_{t_1}^{t_2} x(t)\, dt$. Solving for X_{av} yields

$$X_{av} = \frac{1}{t_2 - t_1} \int_{t_1}^{t_2} x(t)\, dt \tag{1.28}$$

It is clear that X_{av} depends on the particular waveform $x(t)$ as well as the time interval $t_2 - t_1$.

To understand the physical meaning of the average, suppose $x(t)$ is a current signal $i(t)$. Then, extending the concept of Equation (1.14), we can say that I_{av} is the steady current that over the time interval t_1 to t_2 would transfer the *same amount of charge as* $i(t)$.

Full-Cycle and Half-Cycle Averages

If $x(t)$ is periodic, the time interval $t_2 - t_1$ is usually made to coincide with the period T, and the resulting average value is called the **full-cycle average.** For a time-symmetric waveform such as the sine, triangle, sawtooth, and square wave, the areas spanned by the wave above and below the t-axis during each cycle cancel each other out, yielding a *zero* full-cycle average. However, the full-cycle average of an *asymmetric* wave such as the rectangular wave of Figure 1.9(b) is generally different from zero.

▶ **Example 1.5**

For the rectangular wave of Figure 1.9(b), derive an expression for the full-cycle average in terms of X_m and the duty cycle d.

Solution

The area spanned by x_S during one cycle is $T_H X_m + T_L(-X_m) = (T_H - T_L)X_m$. Hence, $X_{av} = [(T_H - T_L)/(T_L + T_H)]X_m$. Using Equation (1.21), this is readily expressed as

$$X_{av} = (2d - 1)X_m \tag{1.29}$$

indicating that as the duty cycle d is varied from 0%, through 50%, to 100%, X_{av} varies from $-X_m$, through 0, to $+X_m$. ◀

Exercise 1.4   For the pulse train of Figure 1.8, derive an expression for X_{av} in terms of X_m and the duty cycle d.

ANSWER $X_{av} = dX_m$.

In the case of time-symmetric periodic waveforms it is often of interest to know the **half-cycle average,** obtained by letting the time interval $t_2 - t_1$ coincide with the half-period during which $x(t) > 0$. To avoid confusion with the full-wave average, we denote the half-cycle average as $X_{av(hc)}$.

▶ **Example 1.6**

Find the half-cycle average of the sine wave of Figure 1.9(a).

Solution

Letting $t_1 = 0$ and $t_2 = T/2$ in Equation (1.28) yields

$$X_{\mathrm{av(hc)}} = \frac{1}{T/2} \int_0^{T/2} X_m \sin \frac{2\pi t}{T} \, dt = \frac{X_m}{T/2} \left(\frac{-T}{2\pi} \right) \cos \frac{2\pi t}{T} \Bigg|_0^{T/2}$$

The result is

$$\boxed{X_{\mathrm{av(hc)}} = \frac{2}{\pi} X_m} \tag{1.30}$$

or $X_{\mathrm{av(hc)}} = 0.637 X_m$. Physically, we can say that if our signal is an ac current of peak amplitude I_m, the amount of charge it transfers during a positive half-cycle is the same as that transferred by a dc current of value $0.637 I_m$. Figure 1.10 illustrates the relationship among X_m, X_{rms}, and $X_{\mathrm{av(hc)}}$.

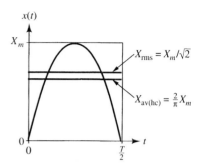

Figure 1.10 Relation among X_m, X_{rms}, and $X_{\mathrm{av(hc)}}$ for an ac signal.

Exercise 1.5 For the triangular and the sawtooth waves of Figure 1.9 derive an expression for $X_{\mathrm{av(hc)}}$ in terms of X_m.

ANSWER $X_{\mathrm{av(hc)}} = \frac{1}{2} X_m$ in both cases.

1.4 ELECTRIC CIRCUITS

An **electric circuit** is a collection of **circuit elements** that have been **connected together** to achieve a specified **goal.** A simple and yet eloquent example is offered by Figure 1.4, where the circuit elements are the battery and the bulb, the interconnections are the wires, and the goal is energy conversion to produce light. Note also that we have represented our circuit in *diagram* form, using standard *circuit symbols* to represent the battery and the bulb, *lines* to represent the leads and connecting wires, and *unfilled circles* to symbolize the points of connection between the leads of the battery and those of the bulb.

The function of a **circuit element** is to ensure a prescribed relationship between the current and voltage at its terminals. An example of such a relationship is Ohm's Law, $v = Ri$. The function of the **interconnections** is to allow circuit elements to share currents and voltages and thus interact with each other to achieve a specified goal.

To facilitate their interconnection, circuit elements are equipped with leads or wires of a good electrical conductor like copper. Ideally, wires pose no resistance to the flow of current, a condition also expressed as:

(1) All points of a wire are at the same potential.

Moreover, wires do not accumulate charge, a condition expressed as:

(2) All current entering one end of a wire exits at the other end.

Practical wires do not satisfy these properties *exactly,* but we shall assume that they do in order to ease our initiation to circuit theory. As you gain experience with practical circuits, you will find that the errors due to nonideal wires can be neglected in most cases of practical interest to the beginner.

Circuit Analysis and Synthesis

Circuit analysis is the process of finding specific voltages and currents in a circuit once its individual elements and their interconnections are known. Conversely, **circuit synthesis** is the process of choosing a set of elements and devising their interconnections to achieve specific voltages and currents in the circuit, such as producing a certain voltage in a part of the circuit in response to a voltage received in another part. Also called **design,** synthesis is usually more difficult than analysis. It involves intuition, creativity, and trial and error. Moreover, the solution may not be unique, and one must know how to choose the one best suited to the particular application. A strong understanding of analysis provides the foundation for intelligent and effective synthesis.

Both analysis and design draw upon two sets of laws:

(1) The **element laws,** such as Ohm's Law, relate the terminal voltages and currents of *individual elements* regardless of how they are connected together to form a circuit.

(2) The **connection laws,** also called *Kirchhoff's Laws* or *circuit laws,* relate the voltages and currents shared at the *interconnections* regardless of the type of elements forming the circuit.

The connection laws are discussed in Section 1.5. The element laws are discussed in Section 1.6 and subsequent sections and chapters, as we introduce new elements. Once we have both sets of laws in hand, we shall be able to analyze moderately complex circuits.

Branches

A circuit is visualized as a network in which each element constitutes a **branch.** The network of Figure 1.11 has six branches, labeled X_1 through X_6. The distinguishing feature of each branch is that at any instant there is some current

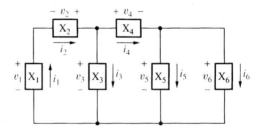

Figure 1.11 Example of an electric circuit with all branch voltages and currents explicitly shown.

through it, called the **branch current,** and some voltage *across* it, called the **branch voltage.**

It is good practice to always *label* the voltages and currents of interest. At times we shall find it convenient to use the letter symbol of the circuit element itself as the subscript. For example, we often denote the current through a resistance R as i_R and the voltage across R as v_R. Likewise, we often denote the current through a voltage source v_S as i_{v_S} and the voltage across a current source i_S as v_{i_S}, and so forth.

Since voltages and currents are *oriented quantities,* besides the labels we must also use arrows to indicate **current directions,** and the $+$ and $-$ signs to indicate **voltage polarities.** Using just the label i without the accompanying arrow, or the label v without the accompanying $+$ and $-$ signs would be meaningless.

Nodes

As the leads of two or more elements are joined together, they form a **node.** The circuit of Figure 1.11 has been redrawn in Figure 1.12 to show that it has four nodes labeled A, B, C, and D. If only two leads converge to a node, as in the case of node A, then we have a **simple node.** If the number of leads is greater than two, we shall emphasize the connections with dots.

The distinguishing feature of a node is that all leads converging to it are at the *same potential* called the **node potential.** Note that the two dots at the bottom of Figure 1.12 are at the same potential because they are connected by an uninterrupted wire. Hence, they are not separate nodes but are part of the *same node* D. To avoid any possible misunderstandings, think of a node as including not only a dot, *but also all other points connected to it by wires.* In Figure 1.12 this has been emphasized by surrounding each node with color.

It is good practice to label all nodes in a circuit *before* you start analyzing it. This will also help identify redundant nodes, as in the case of node D here. When in doubt, mark all points you think are nodes. Then, if any of these points are connected by an uninterrupted wire, they must be taken to form the same node and must thus be labelled with the same symbol. As you draw circuit diagrams, neatness and clarity should be foremost in your mind. It is also important to realize that identical circuits need not look the same on paper. As long as their elements and interconnections are the same, the circuits will also be the same.

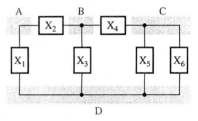

Figure 1.12 Highlighting the nodes of the circuit of Figure 1.11.

Reference Node

Because only potential differences or *voltages* have practical meaning, it is convenient to reference all node potentials in a circuit to the potential of a common node called the **reference** or **datum node.** This node is identified by the symbol $\underline{\underline{\circ}}$ and its potential is zero by definition. When referenced to the datum node, node potentials are simply referred to as **node voltages.**

Given our tendency to visualize potential as height, a logical choice for the reference node is the node at the *bottom* of the circuit diagram, such as node D in Figure 1.12. However, at times it may prove more convenient to designate the node with the *largest* number of connections as the reference node because this may simplify circuit analysis. This is also consistent with the fact that practical circuits include a ground plate to which many elements are connected.

The reader must be careful not to confuse *branch voltages* with *node voltages*. To illustrate the difference and also introduce a simplified voltage notation, consider the circuit of Figure 1.13(a), where the bottom node has been designated as the datum node and the remaining two nodes have been labeled A and B. The potentials of these nodes, denoted v_A and v_B, are referenced to that of the datum node and are thus *node voltages*. By contrast, the *branch voltage* v_{AB} across the element X_2 is *not* a node voltage, because it is not referenced to the datum node. With the polarity shown, this voltage is simply the potential of node A referenced to that of node B, or

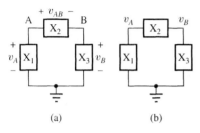

Figure 1.13 (a) Branch voltages and (b) node voltages.

$$v_{AB} = v_A - v_B \tag{1.31}$$

Had we chosen the opposite polarity for this branch voltage, then the proper way of expressing it would have been $v_{BA} = v_B - v_A$. To summarize: v_A, v_B, and v_{AB} are all branch voltages; however, only v_A and v_B are also node voltages.

Once the datum node has been selected, all node voltages are understood to be referenced to this node and the polarity marks can thus be omitted without fear of confusion. In fact we can label a node with its node voltage itself, further reducing the amount of cluttering in our diagrams. This convention is illustrated in Figure 1.13(b). This figure is interpreted by saying that the voltage across X_1 is v_A, positive at the top; that across X_3 is v_B, positive also at the top; and that across X_2 is $v_{AB} = v_A - v_B$ if the polarity is chosen with the positive side at the left, or $v_{BA} = v_B - v_A$ if the positive side is specified at the right. It is important to realize that *no more than one node* in a circuit can be selected as the reference node, and that branch voltages are unaffected by this choice.

► **Example 1.7**

The circuit of Figure 1.14(a) has five branches, labeled X_1 through X_5, and four nodes, labeled A through D. The individual branch voltages have been measured with a voltmeter, and their magnitudes and polarities are as shown. Using the notation of Figure 1.13(b), show all *node voltages* if the datum node is (a) node D; (b) node C.

Solution

(a) With D as the datum node, we have $v_D = 0$ by definition. Moreover, we observe that v_A, v_B, and v_C coincide with the branch voltages across X_1, X_3 and X_5. Hence, $v_A = 1$ V, $v_B = 5$ V, and $v_C = 2$ V. This is shown in Figure 1.14(b), where it is readily seen that in this circuit all node potentials are *higher* than the ground potential.

(b) Changing the datum node from D to C will decrease all node voltages by 2 V. Hence, $v_A = -1$ V, $v_B = 3$ V, $v_C = 0$ V, and $v_D = -2$ V. This is shown in Figure 1.14(c). We observe that the potential at node B is still higher than ground potential;

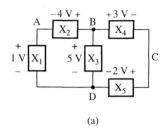

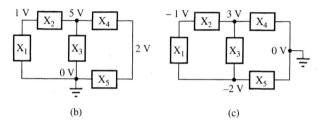

(a)

(b)   (c)

Figure 1.14 (a) Branch voltages; (b), (c) dependency of node voltages upon the choice of the reference node.

however, the potentials at nodes A and D are now *lower* than ground potential.

Exercise 1.6   Repeat Example 1.7, but with the datum node being (a) node B; (b) node A.

ANSWER   (a) $v_A = -4$ V, $v_B = 0$ V, $v_C = -3$ V, $v_D = -5$ V; (b) $v_A = 0$ V, $v_B = 4$ V, $v_C = 1$ V, $v_D = -1$ V.

Loops and Meshes

A **loop** is a closed path such that no node is traversed more than once. A **mesh** is a loop that contains no other loop. Loops and meshes are identified in terms of the branches they traverse. The network of Figure 1.11 has been redrawn in Figure 1.15 to show that it has six loops: $X_1X_2X_3$, $X_3X_4X_5$, X_5X_6, $X_1X_2X_4X_5$, $X_1X_2X_4X_6$, and $X_3X_4X_6$. Of these loops, only the first three are meshes.

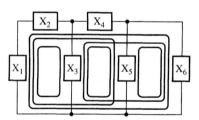

Figure 1.15 Highlighting the loops and meshes of the circuit of Figure 1.11.

Series and Parallel Connections

Two or more circuit elements are said to be connected in **series** if they carry the *same current.* To be in series, two elements must *share a simple node,* as illustrated in Figure 1.16(a). Indeed, since we have stipulated that all current entering one end of a wire exits at the other, it follows that the current leaving X_1 is the same as that entering X_2. Hence, we say that X_1 and X_2 share the same current *i* via the simple node A.

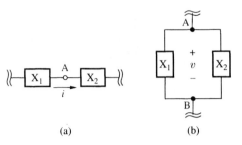

Figure 1.16 Examples of (a) *series* and (b) *parallel* connections.

Looking at the circuit of Figure 1.12, we observe that X_1 and X_2 are in series because they share the simple node A. By contrast X_2 and X_4 are *not* in series because B is not a simple node. For X_2 and X_4 to be in series, X_3 would have to be absent from the circuit.

Two or more elements are said to be connected in **parallel** if they are subjected to the *same voltage*. To be in parallel, the elements must *share the same pair of nodes,* as illustrated in Figure 1.16(b). Indeed, because we have stipulated that all points of a node are at the same potential, it follows that X_1 and X_2 share the same voltage v across the common node pair A and B.

Looking again at the circuit of Figure 1.12, we observe that X_5 and X_6 are in parallel because they share nodes C and D. By contrast, X_3 and X_5 are not in parallel because they only share one node, not a node pair. For X_3 and X_5 to be in parallel, X_4 would have to be a wire in order to coalesce nodes B and C into a single node.

An example of a series circuit is a flashlight, where the battery, the switch, and the bulb carry the same current. An example of a parallel circuit is provided by the various lights and appliances in a home, all of which are subjected to the same voltage, namely, that supplied by the utility company.

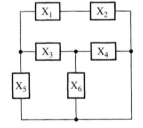

Figure 1.17 Another circuit example.

Exercise 1.7   In the circuit of Figure 1.17: (a) identify all nodes, and state which ones are simple nodes; (b) identify all loops, and state which ones are meshes; (c) state which elements are connected in series and which in parallel.

1.5 KIRCHHOFF'S LAWS

Also referred to as the *circuit laws* or the *connection laws,* **Kirchhoff's laws** (pronounced *kear-koff*), named for the German physicist Gustav Kirchhoff (1824–1887), establish a relation among all the branch currents associated with a node and a relation among all the branch voltages associated with a loop. These laws stem, respectively, from the *charge conservation* and the *energy conservation* principles.

Kirchhoff's Current Law (KCL)

Consider the *branch currents* associated with a *given node n*. At any instant some of these currents will be flowing into the node, others out of the node. These currents obey the following law:

Kirchhoff's Current Law (KCL): At any instant the sum of all currents *entering* a node must equal the sum of all currents *leaving* that node.

Mathematically, KCL is expressed as

$$\sum_n i_{\text{IN}} = \sum_n i_{\text{OUT}} \qquad \textbf{(1.32)}$$

where $\sum_n i_{\text{IN}}$ denotes summation over all currents *entering* and $\sum_n i_{\text{OUT}}$ denotes summation over all currents *leaving* the given node n.

This law is a consequence of the fact that because charge must be conserved, a node cannot accumulate or eliminate charge. Hence, the amount of charge flowing into a node must equal that flowing out. Expressing charge flow in terms of current, therefore, yields KCL. You may find it helpful to liken an electrical node to the junction of two or more water pipes. Clearly, the amount of water entering the junction must equal that leaving it.

To apply KCL successfully, we must first *label* all the branch currents of interest and indicate their *reference directions* by means of arrows. Figure 1.18 shows an example of a properly labeled circuit. By KCL, the following current relations must hold:

- Node A: $i_1 = i_2$
- Node B: $i_2 = i_3 + i_4$
- Node C: $i_4 = i_5 + i_6$
- Node D: $i_3 + i_5 + i_6 = i_1$

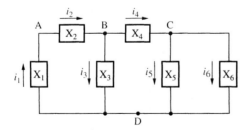

Figure 1.18 Circuit with labeled branch currents.

If a current is unknown, both its magnitude and direction must be found. To this end, we arbitrarily assume a reference *direction* for the unknown current and apply KCL to find its *value*. If this value turns out *positive*, our choice of the reference direction was indeed correct; if it is *negative*, the current actually flows *opposite* to the assumed direction. To make a negative current value come out positive, simply reverse the arrow in the circuit diagram.

▶ **Example 1.8**

In the circuit of Figure 1.18 suppose i_2 and i_3 are given, but i_4 is unknown.

  (a) If at a certain instant $i_2 = 5$ A and $i_3 = 2$ A, what is i_4?

  (b) If at a later instant $i_2 = 6$ A and $i_3 = 7$ A, what is i_4?

Solution

We do not know the direction of i_4, so we arbitrarily assume it flows toward the right as shown.

  (a) By KCL at node B, $i_2 = i_3 + i_4$. This yields

$$i_4 = i_2 - i_3 = 5 - 2 = 3 \text{ A}$$

Since the result is positive, i_4 is indeed flowing toward the right. We express this by saying that the current through the element X_4 is

$$i_{X_4} = 3\text{A} \ (\rightarrow)$$

  (b) Now KCL yields

$$i_4 = i_2 - i_3 = 6 - 7 = -1 \text{ A}$$

Since the result is negative, i_4 now flows toward the left. This makes sense because i_4 must now team up with i_2 to counterbalance i_3. We express this by writing

$$i_{X_4} = 1 \text{ A} \ (\leftarrow)$$

◀

▶ **Example 1.9**

In the circuit of Figure 1.18 let $i_5 = 3$ A and $i_6 = 4$ A. Find i_3 so that $i_2 = 1$ A.

Solution

By KCL at node C, $i_4 = 3 + 4 = 7$ A. By KCL at node B, $i_3 = i_2 - i_4 = 1 - 7 = -6$ A, indicating that the current through X_3 actually flows upward. We express this as $i_{X_3} = 6$ A ($\uparrow$).

◀

Exercise 1.8   If in the circuit of Figure 1.18 $i_1 = 2$ A and $i_5 = 3$ A, find i_6 so that $i_3 = 0$ A. What is i_2?

ANSWER $i_6 = -1$ A, or $i_{X_6} = 1$ A ($\uparrow$); $i_2 = 2$ A, or $i_{X_2} = 2$ A ($\rightarrow$).

Since current labeling is a guessing process, some engineers systematically assume *all* currents associated with a node flow *into* the node; others prefer to assume *all* flow *out* of the node. In either case, after the calculations are done, at least one of the current values will have to come out negative in order to satisfy the charge conservation principle. This also leads to two alternate forms of stating KCL. Rewriting Equation (1.32) as

$$\sum_n i_{\text{IN}} - \sum_n i_{\text{OUT}} = 0 \qquad \text{(1.33)}$$

we can restate KCL by saying that **at any instant the *algebraic sum* of all currents associated with a node must be zero,** where the currents *entering* the node are regarded as *positive,* and those *leaving* the node as *negative.* Applying this form of KCL to Example 1.8 yields $i_2 - i_3 - i_4 = 0$. Substituting the given current values it is readily seen that the results are the same as before.

The other KCL form is obtained by rewriting Equation (1.32) as

$$\sum_n i_{\text{OUT}} - \sum_n i_{\text{IN}} = 0 \qquad \text{(1.34)}$$

In words, **at any instant the *algebraic sum* of all currents associated with a node must be zero,** where the currents *leaving* the node are now regarded as *positive,* and those *entering* the node as *negative.* For instance, applying this form of KCL to Example 1.8 yields $i_3 + i_4 - i_2 = 0$, and you can readily verify that the numerical results are again the same.

It is important to realize that KCL applies not only to nodes but also to entire *portions* of circuits. Consider, for instance, a home appliance connected to the ac power outlet via a three-prong cord. Without knowing the inner workings of the appliance, we can state that because it cannot create or destroy charge, the algebraic sum of the three currents flowing down the cord must, at all times, be zero.

Which KCL form should you use? The first form is physically more intuitive and it offers the advantage that all currents appear in *positive* form, thus reducing the possibility of confusion with signs. By contrast, the other two forms make mathematical bookkeeping more straightforward, a feature especially advantageous in *nodal analysis* (to be addressed in Section 3.2). It does not matter which KCL form you use as long as you are consistent.

Kirchhoff's Voltage Law (KVL)

Consider the *branch voltages* associated with a *given loop* ℓ. As we go around the loop, the voltage across each of its branches may appear either as a *voltage rise* or as a *voltage drop*. For example, if $v_A = 3$ V and $v_B = 7$ V, hopping from node A to node B we experience a 4-V rise. Conversely, if $v_A = 6$ V and $v_B = 1$ V, hopping from A to B we now experience a 5-V drop. The branch voltages around a loop obey the following law:

Kirchhoff's Voltage Law (KVL): At any instant the sum of all voltage *rises* around a loop must equal the sum of all voltage *drops* around that loop.

Mathematically, KVL is expressed as

$$\sum_{\ell} v_{\text{RISE}} = \sum_{\ell} v_{\text{DROP}}$$ (1.35)

where $\sum_{\ell} v_{\text{RISE}}$ denotes summation over all voltage rises and $\sum_{\ell} v_{\text{DROP}}$ denotes summation over all voltage drops around the given loop ℓ.

You may find it convenient to liken a circuit loop to a mountain trail. In this analogy *voltage* corresponds to *elevation,* and *nodes* to *rest stations* along the hike. If you keep separate tabs of all elevation *increases* and elevation *decreases* in going from one rest area to the next, it is clear that by the time you return to the point of departure the two sums must be equal.

To apply KVL successfully, we must first label all the branch voltages of interest and indicate their reference polarities by means of the + and − symbols. Figure 1.19 shows an example of a properly labeled circuit. By KVL, the following holds:

- Loop $X_1X_2X_3$: $v_1 + v_2 = v_3$
- Loop $X_3X_4X_5$: $v_3 = v_4 + v_5$
- Loop X_5X_6: $v_5 = v_6$
- Loop $X_1X_2X_4X_5$: $v_1 + v_2 = v_4 + v_5$
- Loop $X_1X_2X_4X_6$: $v_1 + v_2 = v_4 + v_6$
- Loop $X_3X_4X_6$: $v_3 = v_4 + v_6$

All loops have been traveled in the clockwise direction. Had we chosen the counterclockwise direction, all voltage rises would have appeared as voltage drops and vice versa, leaving the relations unchanged. We conclude that the direction in which we traverse a loop is immaterial as long as we maintain it over the entire loop.

If a voltage is unknown, both its magnitude and polarity must be found. As with currents, we arbitrarily assume a polarity and then use KVL to find the value of the unknown voltage. A positive value indicates a correct choice of the reference polarity. A negative result means that in order to reflect the correct polarity, the + and − signs must be interchanged.

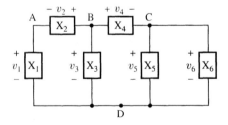

Figure 1.19 Network with labeled branch voltages.

▶ **Example 1.10**

In the circuit of Figure 1.19 suppose v_1 and v_3 are given, but v_2 is unknown.

    (a) If at a certain instant $v_1 = 7$ V and $v_3 = 9$ V, what is v_2?

    (b) If at a later instant $v_1 = 8$ V and $v_3 = 5$ V, what is v_2?

Solution

We do not know the polarity of v_2, so assume it is positive at the right, as shown.

    (a) KVL around loop $X_1 X_2 X_3$ yields $v_1 + v_2 = v_3$, or

$$v_2 = v_3 - v_1 = 9 - 7 = 2 \text{ V}$$

Because the result is positive, the polarity of v_2 is indeed as shown, indicating that v_B is 2 V *higher* than v_A.

    (b) Now KVL yields

$$v_2 = v_3 - v_1 = 5 - 8 = -3 \text{ V}$$

Since the result is negative, the polarity of v_2 is opposite to that shown, indicating that v_B is now 3 V *lower* than v_A. If we place the + sign at A and the − sign at B, then v_2 will be +3 V. ◀

Exercise 1.9   (a) If in the circuit of Figure 1.19 $v_2 = 2$ V and $v_6 = 3$ V, find v_4 so that $v_1 = 0$ V. (b) What is the resulting value and polarity of v_3?

ANSWER   (a) $v_4 = -1$ V; (b) $v_3 = 2$ V with the polarity shown in the figure.

As in the KCL case, there are *two* alternate forms of stating KVL. Rewriting Equation (1.35) as

$$\sum_\ell v_{\text{RISE}} - \sum_\ell v_{\text{DROP}} = 0 \qquad \text{(1.36)}$$

we can restate KVL by saying that **at any instant the *algebraic sum* of all branch voltages around a loop must be zero,** where *voltage rises* are regarded as *positive* and *voltage drops* as *negative*. Applying this form of KVL to Example 1.10 yields $v_1 + v_2 - v_3 = 0$. Clearly, substituting the given numerical values yields the same result as before.

    The other KVL form is obtained by rewriting Equation (1.35) as

$$\sum_\ell v_{\text{DROP}} - \sum_\ell v_{\text{RISE}} = 0 \qquad \text{(1.37)}$$

In words, **at any instant the *algebraic sum* of all branch voltages around a loop must be zero,** where *voltage drops* are now regarded as *positive* and *voltage rises* as *negative*. Applying this form of KVL to Example 1.10 yields $v_3 - v_1 - v_2 = 0$, which again yields the same result.

KVL is a consequence of the fact that the work required to move a charge around a complete loop must, by the energy conservation principle, be zero. This work is obtained by multiplying the given charge by each branch voltage encountered in moving the charge around the loop, and adding all terms together algebraically. Thus, imposing $q\sum_\ell v_{\text{RISE}} - q\sum_\ell v_{\text{DROP}} = 0$ yields Equation (1.36), and hence Equations (1.37) and (1.35). As in the KCL case, it does not matter which of the three KVL forms you use as long as you are consistent.

In applying Kirchhoff's laws it is important to remember that voltages have polarities and currents have directions. Since careless use of signs is one of the most frequent causes of error in circuit analysis, it is important to think of voltage and current values not as plain numbers, but as *numbers preceded by signs*.

Power Conservation

In a circuit, some elements will release power and others will absorb power. Because energy and, hence, power cannot be created or destroyed, **the sum of all *absorbed* powers must, at any instant, equal the sum of all *released* powers,**

$$\sum p_{\text{absorbed}} = \sum p_{\text{released}} \qquad (1.38)$$

As we know, elements conforming to the *active sign* convention *release* power, and elements conforming to the *passive sign* convention *dissipate* power. As we proceed, we shall often exploit Equation (1.38) to check calculations involving branch currents and voltages.

▶ **Example 1.11**

In the circuit of Figure 1.20(a), suppose X_2 releases $p_{X_2} = 20$ W and X_3 absorbs $p_{X_3} = 18$ W. (a) Calculate all branch voltages and currents. (b) Use the power check to verify your calculations.

Solution

(a) Since X_3 absorbs power, its current must flow toward the right. Thus, $i_{X_3} = p_{X_3}/v_{X_3} = 18/6 = 3$ A ($\rightarrow$).

By KCL, $i_{X_2} = i_{X_1} + i_{X_3} = 1 + 3 = 4$ A ($\uparrow$). Since X_2 releases power, its voltage must be positive at the top. Consequently, $v_{X_2} = p_{X_2}/i_{X_2} = 20/4 = 5$ V (+ at the top).

By KCL, $i_{X_4} = i_{X_3} = 3$ A ($\downarrow$). Because in going around the mesh $X_1X_3X_4$ we first encounter a voltage rise $v_{X_2} = 5$ V

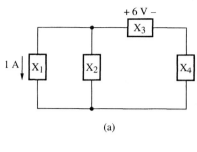

(a)

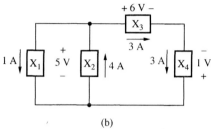

(b)

Figure 1.20 Circuit of Example 1.11

and then a voltage drop $v_{X_3} = 6$ V, it follows that v_{X_4} must, by KVL, be a voltage rise of 1 V, or $v_{X_4} = 1$ (+ at the bottom). The branch voltages and currents are shown in Figure 1.20(b).

(b) Since X_1 conforms to the passive sign convention, it absorbs $p_{X_1} = v_{X_1} i_{X_1} = 5 \times 1 = 5$ W. Since X_4 conforms to the active sign convention, it releases $p_{X_4} = v_{X_4} i_{X_4} = 1 \times 3 = 3$ W.

By Equation (1.38), we must have $p_{X_1} + p_{X_3} = p_{X_2} + p_{X_4}$, or $5 + 18 = 20 + 3$. Thus, the power check is satisfied. ◀

Exercise 1.10   In the circuit of Figure 1.20(a) suppose X_1 releases 2 W and X_4 dissipates 24 W. (a) Calculate all branch voltages and currents. (b) Use the power check to verify your calculations.

ANSWER   (a) $v_{X_1} = v_{X_2} = 2$ V (+ at the bottom); $v_{X_4} = 8$ V (+ at the bottom); $i_{X_2} = 2$ A ($\downarrow$); $i_{X_3} = 3$ A ($\leftarrow$); $i_{X_4} = 3$ A ($\uparrow$). (b) $p_{X_4} = p_{X_1} + p_{X_2} + p_{X_3}$, or $24 = 2 + 4 + 18$.

Exercise 1.11   Discuss the conditions for either X_1 or X_2 under which the circuit of Figure 1.20(a) will yield $p_{X_4} = 0$.

ANSWER $i_{X_2} = 1$ A ($\uparrow$) or $v_{X_1} = v_{X_2} = 6$ V (+ at the top).

1.6 CIRCUIT ELEMENTS

The distinguishing feature of a circuit element is that its electric behavior is described in terms of some **current–voltage relationship** at its terminals. Also called the **element law,** this relationship can be derived *mathematically* through the laws of physics, or it can be determined *experimentally* via point-by-point measurements. In either case, this relationship can be plotted on paper or can be displayed on a cathode ray tube (CRT) for a pictorial visualization of how the element behaves.

The *i–v* Characteristic

An element law can be expressed as

$$i = i(v) \tag{1.39}$$

where v is referred to as the **independent variable** and i as the **dependent variable.** Physically, we regard v as the **cause** and i as the **effect,** since voltage produces an electric field, and the electric field in turn sets charges in motion, $v \rightarrow E \rightarrow i$. Equation (1.39) is called the *i–v* **characteristic** of the element. When plotted in the *i–v* plane, this characteristic will be a curve of any form.

As we proceed, we shall find that a parameter of particular significance is the **slope** of this curve,

$$g \triangleq \frac{di}{dv} \tag{1.40}$$

Since its dimensions are ampere/volts, or siemens, which are the dimensions of conductance, g is called the **dynamic conductance** of the element under consideration. Slope is often expressed as the reciprocal of a parameter denoted as r,

$$\boxed{\frac{1}{r} = \frac{di}{dv}} \tag{1.41}$$

Clearly, $r = 1/g$. Since its dimensions are volt/amperes, or ohms, which are the dimensions of resistance, r is called the **dynamic resistance** of the element. As a general rule, **the slope of an *i–v* curve is always the reciprocal of some resistance called the *dynamic resistance.*** Clearly, the steeper the slope, the smaller the dynamic resistance. Conversely, a large dynamic resistance indicates a mild slope. Figure 1.21(a) shows a circuit to find experimentally the *i–v* characteristic of an unknown element X. The element is subjected to a **test voltage** v, which we represent schematically with a circle and the positive sign to identify voltage polarity. The resulting current i is measured for different values of v, and the data are then plotted point by point to obtain a curve of the type of Figure 1.21(b). If desired, the data can be processed with suitable curve-fitting algorithms to find a best-fit approximation. A test voltage is readily produced with a bench power supply or a signal generator.

A **curve tracer** is an instrument that, upon insertion of the unknown element into an appropriate receptacle, displays its *i–v* characteristic automatically on a CRT. Curve tracers are available either as stand alone instruments, or as plug-in oscilloscope attachments.

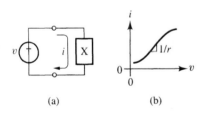

(a)   (b)

Figure 1.21 Circuit arrangement to find experimentally the *i–v* characteristic of a circuit element.

The *v–i* Characteristic

As circuit variables, voltage and current are interchangeable in the sense that at times it may prove more convenient to regard voltage as the cause and current as the effect, at times current as the cause and voltage as the effect. When pursuing the second viewpoint, we express an element law as

$$v = v(i) \tag{1.42}$$

where now i is the independent variable and v is the dependent one. Equation (1.42), referred to as the *v–i* **characteristic,** is obtained by forcing a **test current** i through the element and measuring the resulting voltage v for different values of i. This viewpoint is depicted in Figure 1.22, where the test current has been represented schematically with a circle and an arrow.

Note that the slope of the *v–i* curve is the *reciprocal* of that of the *i–v* curve,

$$r = \frac{1}{g} = \frac{dv}{di} \tag{1.43}$$

In general, the *i–v* or *v–i* characteristic will be a curve of any form and the slope will vary from point to point on the curve. When this is the case, the element is said to be **nonlinear.** Common examples of nonlinear devices are diodes and transistors, the basic ingredients of modern electronic equipment.

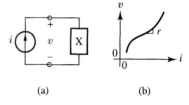

(a)   (b)

Figure 1.22 Circuit arrangement to find experimentally the *v–i* characteristic of a circuit element.

▶ Example 1.12

Shown in Figure 1.23 is the circuit symbol of an element called *rectifier diode,* along with its *i–v* characteristic. Semiconductor theory predicts that for $v \geq 0$ this characteristic can be expressed as

$$i = I_s(e^{v/V_T} - 1) \tag{1.44}$$

where I_s and V_T are device parameters whose values depend on the particular diode sample. Assuming a sample with $I_s = 2 \times 10^{-15}$ A and $V_T = 26$ mV, find (a) the *v–i* characteristic of the rectifier diode; (b) the value of v corresponding to $i = 1$ mA; and (c) the value of r corresponding to $i = 1$ mA. (d) Repeat parts (b) and (c) if $i = 1$ μA.

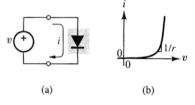

(a)   (b)

Figure 1.23 Rectifier diode and its *i–v* characteristic.

Solution

(a) Rearranging Equation (1.44) yields $\exp(v/V_T) = i/I_s + 1$. Taking the natural logarithm of both sides and solving for v, we obtain the *v–i* characteristic

$$v = V_T \ln(i/I_s + 1) \tag{1.45}$$

(b) By Equation (1.45), $v = 26 \times 10^{-3} \ln[10^{-3}/(2 \times 10^{-15}) + 1] = 0.700$ V.

(c) Taking the derivative of Equation (1.44) yields $g = di/dv = (I_s/V_T) \exp(v/V_T)$. Since $r = 1/g$, this yields

$$r = \frac{V_T}{I_s} e^{-v/V_T} \tag{1.46}$$

Thus, $r = [26 \times 10^{-3}/(2 \times 10^{-15})]\exp[-0.700/(26 \times 10^{-3})] = 26 \; \Omega$.

(d) Similar calculations for $i = 1 \; \mu A$ yield $v = 0.521$ V and $r = 26 \; k\Omega$.

Remark   Note that r varies from one point to another on the i–v curve. ◄

Exercise 1.12   A certain rectifier diode is known to have $V_T = 26$ mV. If applying $i = 1.5$ mA yields $v = 680$ mV, what is the value of I_s for this particular sample?

ANSWER 6.57×10^{-15} A.

► **Example 1.13**

The basic ingredient of digital wristwatches and pocket calculators is the *MOS transistor,* a device having three terminals called *gate G*, *drain D*, and *source S*. Connecting gate to drain as shown in Figure 1.24(a) turns the device into a two-terminal element whose characteristic is

$$i = 0 \quad \text{for } v \leq V_T \tag{1.47a}$$

$$i = \frac{k}{2}(v - V_T)^2 \quad \text{for } v \geq V_T \tag{1.47b}$$

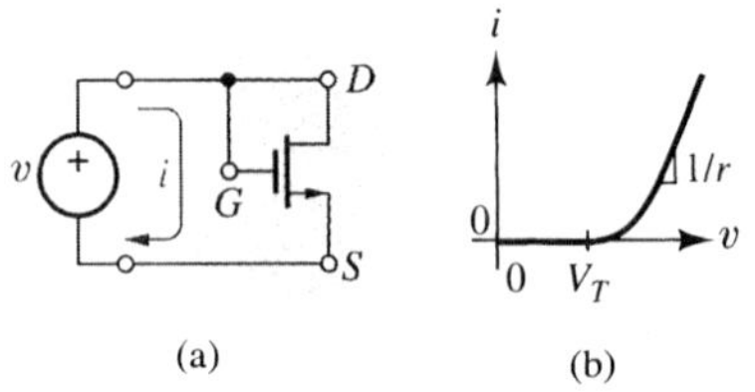

where k and V_T are device parameters that depend on the particular sample. For obvious reasons, this element is said to exhibit *nonlinear resistor* behavior.

If applying $v = 2.5$ V yields $i = 0.4$ mA, and applying $v = 5$ V yields $i = 4.9$ mA, what are the values of k and V_T for this sample?

Solution

Substituting the given data into Equation (1.47b),

$$0.4 \times 10^{-3} = \frac{k}{2}(2.5 - V_T)^2$$

$$4.9 \times 10^{-3} = \frac{k}{2}(5 - V_T)^2$$

This system of two equations in two unknowns is readily solved to obtain $k = 0.8$ mA/V^2 and $V_T = 1.5$ V. ◄

Figure 1.24 Nonlinear MOSFET resistor and its i–v characteristic.

Exercise 1.13   For the nonlinear resistor sample of Example 1.13, predict (a) the value of i corresponding to $v = 6.5$ V and (b) the value of v ($v \geq V_T$) required to achieve $i = 1.6$ mA. (c) Derive an expression for the dynamic conductance and compute it for $v = 2.5$ V and $v = 5$ V. Don't forget to express your results in their correct physical units!

ANSWER (a) 10 mA; (b) 3.5 V; (c) $g = k(v - V_T)$; $g = 0.8$ mS, 2.8 mS.

Straight-Line Characteristics

If the i–v characteristic happens to be a **straight line,** the dynamic resistance and conductance will be *constant* throughout. Following our convention of using uppercase letters to indicate constant quantities, the slope shall now be designated as $1/R$. As shown in Figure 1.25(a), we can replace the differentials with finite differences in Equation (1.41) and find the slope as

$$\frac{1}{R} = \frac{\Delta i}{\Delta v} \qquad \textbf{(1.48)}$$

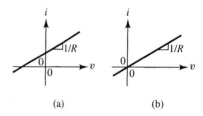

Figure 1.25 Straight-line i–v characteristics: (a) $1/R = \Delta i / \Delta v$; (b) $1/R = i/v$.

If the i–v characteristic is straight and goes through the origin as in Figure 1.25(b), then i and v are said to be **linearly proportional** to each other, and the slope is simply found as

$$\frac{1}{R} = \frac{i}{v} \qquad \textbf{(1.49)}$$

Be aware that this equation holds only if the straight characteristic goes through the origin!

 **Example 1.14**

Shown in Figure 1.26 is the circuit symbol of an element called *zener diode*, along with its i–v characteristic. For $v \leq V_{ZK}$ the characteristic coincides with the v-axis, and for $v \geq V_{ZK}$ it is a straight line with slope $1/R_Z$.

If applying $i = 1$ mA yields $v = 6.0$ V and applying $i = 4$ mA yields $v = 6.3$ V, find V_{ZK} and R_Z.

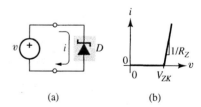

Figure 1.26 Zener diode and its i–v characteristic.

Solution

By Equation (1.48), $R_Z = \Delta v / \Delta i = (6.3 - 6.0)/(4 - 1) = 0.1$ kΩ $= 100$ Ω.

To find V_{ZK} we need an expression for the i–v characteristic for $v \geq V_{ZK}$. Since this characteristic is a straight line with slope $1/R_Z$ and with v-axis intercept at $v = V_{ZK}$, we can write

$$i = \frac{v - V_{ZK}}{R_Z} \tag{1.50}$$

Computing it at the first measurement point, we obtain $1 = (6.0 - V_{ZK})/0.1$, or $V_{ZK} = 5.9$ V.

Exercise 1.14   For the zener diode of Example 1.14, find (a) i if $v = 6.15$ V; (b) v if $i = 1$ mA; (c) the change Δv brought about by a change $\Delta i = 5$ mA.

ANSWER   (a) 2.5 mA; (b) 6.0 V; (c) 0.5 V.

1.7 SOURCES

The function of sources is to initiate voltages and currents in a circuit. As such, they constitute the most basic circuit elements. Because of their ability to release power, sources are said to be **active elements.**

Voltage Sources

A **voltage source,** also called a **voltage generator,** is a circuit element that maintains a prescribed voltage v_S across its terminals *regardless* of the current through it,

$$\boxed{v = v_S} \tag{1.51}$$

Its circuit symbol is a circle with the $+$ sign to identify voltage polarity.

To serve some useful purpose, a voltage source is connected to some external circuit generally referred to as the **load.** This, in turn, will draw some current i from the source, as shown in Figure 1.27(a). It is the function of the source to keep $v = v_S$ regardless of the magnitude and direction of i. The current may be positive, zero, or even negative, in which case i will flow opposite to the reference direction shown in the figure.

Note the absence of i from the source characteristic of Equation (1.51). It takes a load to establish i. The source affects i only indirectly, via the i–v characteristic of the load, $i = i(v_S)$.

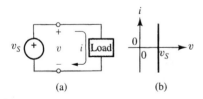

Figure 1.27 Voltage source and its i–v characteristic.

The power of a voltage source is $p = v_S i$. If the current drawn by the load happens to flow into the negative terminal, through the source, and out of the positive terminal, then we have the situation of Figure 1.5(a), and we say that the source is *delivering* power to the load. Such a situation arises, for instance, in the case of a battery powering a light bulb.

Conversely, if i flows into the positive terminal, through the source, and out of the negative terminal, then we conform to Figure 1.5(b), and we say that the source is *receiving* power from the load. Such a situation arises, for instance, when the source is a rechargeable battery that has been connected to a battery charger.

The i–v characteristic of a voltage source is a **vertical line** positioned at $v = v_S$; see Figure 1.27(b). Because its slope is infinite, it follows that **the dynamic resistance of an ideal voltage source is zero.**

Varying v_S will translate the curve right or left, depending on whether v_S is increased or decreased. For $v_S = 0$ the i–v curve coincides with the i axis, where $v = 0$ regardless of the current. This is the characteristic of a **short circuit,** which can thus be regarded as the particular voltage source for which $v_S = 0$. As is shown in Figure 1.28(a), a plain wire of a good conductor is a short circuit.

In a practical voltage source such as a car or a flashlight battery, v does vary with the load current somewhat, indicating that the dynamic resistance r is not exactly zero. For obvious reasons, a voltage source with $r = 0$ is referred to as an **ideal voltage source.** The smaller the value of r relative to the other resistances in the surrounding circuit, the closer the voltage source to ideal.

A good-quality power supply of the type found in an electronics lab provides an excellent approximation to an ideal voltage source. Clever internal circuitry continuously monitors the actual value of v and adjusts itself to varying load current conditions in such a way as to rigorously keep $v = v_S$ at all times.

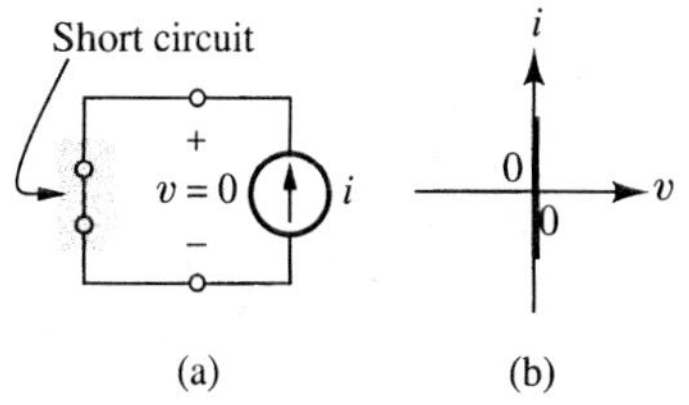

Figure 1.28 A short circuit yields $v = 0$ regardless of i.

Current Sources

A **current source,** also called a **current generator,** is a circuit element that maintains a prescribed current i_S *regardless* of the voltage across its terminals,

$$i = i_S \tag{1.52}$$

Its circuit symbol is a circle with an arrow to indicate current direction.

A current source must always have a load to provide a path for current flow. In response to the current i_S, the load will develop some voltage v, as shown in Figure 1.29(a). It is the function of the source to keep $i = i_S$ regardless of the magnitude and polarity of v. The voltage may be positive, zero, or even negative, in which case the polarity of v will be opposite to that shown in the figure.

Note the absence of v from the source characteristic of Equation (1.52). It takes a load to develop v. The source affects v only indirectly, via the v–i characteristic of the load, $v = v(i_S)$.

With the orientation shown, the current generator is also said to be **sourcing** current to the load via the top terminal and to be **sinking** current from the load via the bottom terminal.

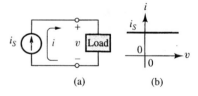

Figure 1.29 Current source and its i–v characteristic.

The power of a current source is $p = v i_S$. If the voltage established by the load is positive, the source will release power; if v is negative, the source will absorb power.

The i–v characteristic of a current source is a **horizontal line** positioned at $i = i_S$; see Figure 1.29(b). Since its slope is zero, it follows that **the dynamic resistance of an ideal current source is infinite.**

Varying i_S will translate the curve up or down, depending on whether i_S is increased or decreased. For $i_S = 0$ the i–v curve coincides with the v axis,

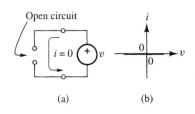

Figure 1.30 An open circuit yields $i = 0$ regardless of v.

where $i = 0$ regardless of the voltage. This is the characteristic of an **open circuit,** which can thus be regarded as the particular current source for which $i_S = 0$. As shown in Figure 1.30(a), an open circuit arises whenever we break a connection to interrupt current flow.

A **switch** is a device that operates as a *short circuit* when it is *closed* and as an *open circuit* when it is *open.* Alternately, we can say that a *closed switch* acts as a 0-V voltage source and an *open switch* acts as a 0-A current source.

Current sources are not as readily available as voltage sources such as batteries and power supplies. However, when studying operational amplifiers in Chapter 6, we shall learn how to construct a current source using these devices as building blocks. In a practical current source, i does vary somewhat with the load voltage, indicating that its dynamic resistance r is not infinite. A current source with $r = \infty$ is referred to as an **ideal current source.** The larger the value of r relative to the other resistances in the surrounding circuit, the closer a current source to ideal.

A Hydraulic Analogy

To appreciate the difference between voltage and current sources, think of a hydraulic analogy in which charge is likened to water, potential to water pressure, and current to water flow,

charge $\leftrightarrow$ water

voltage $\leftrightarrow$ pressure difference

current $\leftrightarrow$ water flow

Then, an electric circuit can be visualized as a system of interconnected pipes in which water pressure and flow are controlled by pumps according to the following analogy:

A *current source* acts as a pump that maintains a prescribed water flow regardless of pressure difference.

A *voltage source* acts as a pump that maintains a prescribed pressure difference regardless of water flow.

Dependent Sources

A source is said to be **dependent** if its value is controlled by some voltage or current located elsewhere in the circuit. A familiar example is an audio amplifier, which delivers at the output jack a voltage v_O that depends on a voltage v_I applied elsewhere to the amplifier, namely, via the input jack. This dependence is of the type $v_0 = Av_I$, where A is called the *gain* of the amplifier. Since v_O is designed to depend on v_I but not on the particular loudspeaker load, the amplifier acts as a dependent voltage source. By contrast, a bench power supply and a waveform generator are examples of **independent** sources because their values do not depend on other voltages or currents in the circuit.

To distinguish dependent from independent sources, we use a diamond-shaped symbol for the former and the conventional circle for the latter. Figure 1.31 shows an example of a circuit containing both an independent

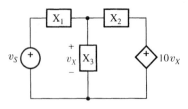

Figure 1.31 Example of a circuit with an independent and a dependent source.

and a dependent source. The value of the dependent source is $10v_X$, where v_X, the controlling signal, is the voltage appearing across the element X_3.

Unless stated to the contrary, we assume that the value of a dependent source is *linearly proportional* to the controlling voltage or current. Figure 1.32 shows the four possible types of linear dependent sources. In each case the controlling variable is a branch voltage v_X or a branch current i_X appearing somewhere else in the circuit, as is shown at the top of the figure. The four types are as follows:

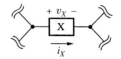

(1) The **voltage-controlled voltage source** (VCVS), which yields

$$v = k_v v_X \qquad \text{(1.53a)}$$

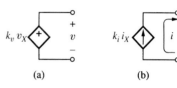

regardless of the load. The constant k_v has the dimensions of V/V, that is, it is a dimensionless number.

(2) The **current-controlled current source** (CCCS), which yields

$$i = k_i i_X \qquad \text{(1.53b)}$$

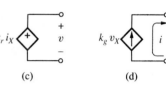

Figure 1.32 The four types of dependent sources: (a) VCVS, (b) CCCS, (c) CCVS, and (d) VCCS. The branch at the top is the *controlling branch.*

regardless of the load. The constant k_i has the dimensions of A/A, that is, it is a dimensionless number.

(3) The **current-controlled voltage source** (CCVS), which yields

$$v = k_r i_X \qquad \text{(1.53c)}$$

regardless of the load. The constant k_r has the dimensions of V/A, that is, the dimensions of *resistance.*

(4) The **voltage-controlled current source** (VCCS), which yields

$$i = k_g v_X \qquad \text{(1.53d)}$$

regardless of the load. The constant k_g has the dimensions of A/V, that is, the dimensions of *conductance.*

You may find it convenient to think of a dependent source as being equipped with an intelligent controller that monitors the controlling signal, v_X or i_X, and continuously adjusts its voltage v if it is a controlled *voltage* source, or its current i if it is a controlled *current* source, in such a way as to ensure the prescribed relationship between the controlled and the controlling signals, regardless of any variations in the controlling signal or in the load. In Chapter 5 we shall see that dependent sources are a necessary ingredient in the modeling of amplifiers and ideal transformers. Until then, to keep things simple, we shall restrict our attention to circuits containing only independent sources.

Voltage Sources in Series

Two or more voltage sources may be connected in series to increase the amount of voltage to be sourced. For instance, connecting two sources as in Figure 1.33(a), the overall voltage is, by KVL,

$$v_S = v_{S1} + v_{S2}$$ **(1.54)**

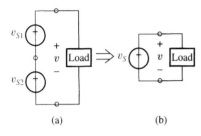

Figure 1.33 Two voltage sources in series are equivalent to a single source $v_S = v_{S1} + v_{S2}$.

indicating that the series combination of two sources has the *same effect* as a single source whose value is the sum of the two. This is expressed by saying that the two sources are **equivalent** to a single source of value $v_S = v_{S1} + v_{S2}$; see Figure 1.33(b). If one of the sources, say v_{S2}, is reversed, then $v_S = v_{S1} - v_{S2}$. A common example of voltage sources in series is a car battery, where six cells of 2 V each are stacked in series to achieve 12 V.

Two voltage sources $v_{S1} \neq v_{S2}$ *must never be connected in parallel* because this would violate KVL, at least as long as the sources are assumed to be ideal. To satisfy KVL, a voltage-dropping device such as a resistor must be interposed between the two unequal sources. This resistor also sets the magnitude of the current flowing from one source to the other. When jumping two car batteries, this function is provided by the internal resistance of the batteries and the resistance of the connecting cables.

Note, in particular, that a *voltage source* $v_S \neq 0$ *must never be short-circuited*, because a short circuit can be regarded as a 0-V voltage source and we have just stated that different voltage sources must never be connected in parallel.

Current Sources in Parallel

Two or more current sources can be connected in parallel to increase the current to be sourced. As shown in Figure 1.34(a), two parallel current sources are equivalent to a single source whose value is, by KCL,

$$i_S = i_{S1} + i_{S2}$$ **(1.55)**

Reversing the direction of the second source, the net current becomes $i_S = i_{S1} - i_{S2}$.

Two current sources $i_{S1} \neq i_{S2}$ *must never be connected in series* because this would violate KCL, at least as long as the sources are assumed to be ideal.

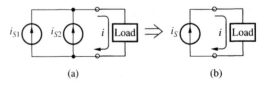

Figure 1.34 Two current sources in parallel are equivalent to a single source $i_S = i_{S1} + i_{S2}$.

In particular, *a current source* $i_S \neq 0$ *must never be left open-circuited,* because this would be tantamount to connecting it in series with a 0-A current source. However, a current source may be short-circuited, in which case i_S will flow through the short circuit and will do so with a voltage drop of 0 V. The forbidden source connections are shown in Figure 1.35.

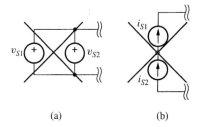

(a)   (b)

Figure 1.35 Unequal *voltage* sources must never be connected in *parallel,* and unequal *current* sources must never be connected in *series.*

▼ SUMMARY

- The physical quantities of interest to us are *charge* q, *electric field* E, *potential* v, *current* i, and *power* p. Potential is not an absolute but a *relative* quantity because only *potential difference* or *voltage* has practical meaning. The zero level of potential is commonly referred to as *ground potential.*

- Applying a *voltage* v across a circuit element produces an *electric field* inside the element, which in turn sets mobile charges in motion to produce a *current* i through the element. The *power* expended to achieve this effect is always $p = vi$.

- Power may be *released* or *absorbed.* An element *releases* power when current inside the element flows from a lower to a higher potential; this is referred to as the *active sign convention.* An element *absorbs* power when current inside the element flows from a higher to a lower potential; this is referred to as the *passive sign convention.*

- A *circuit* is a collection of *circuit elements* that have been *connected* together to serve a *specific purpose.* A circuit is characterized in terms of its *nodes, branches, meshes,* and *loops.* Its variables are its *branch voltages* and *branch currents,* which are generally referred to as *signals.*

- The function of the circuit connections is to allow elements to share node voltages and branch currents. These voltages and currents satisfy the laws of the interconnections, also called *Kirchhoff's laws.*

- KCL states that the sum of the *currents entering* a node must equal the sum of the *currents leaving* it. KVL states that the sum of the *voltage rises* encountered in going around a loop must equal the sum of the *voltage drops.*

- In a circuit the sum of all *dissipated* powers equals the sum of all *released* powers.

- The function of a circuit is to ensure a prescribed i–v (or v–i) relationship at its terminals, called the *element law.* In general, an element law will be nonlinear; however, we are primarily interested in *linear elements* because the equations governing the circuit will then simply be *linear algebraic equations.*

- The slope of an i–v curve is always the reciprocal of a resistance called the *dynamic resistance* of the corresponding element.

- If the i–v characteristic is a straight line, then its slope is constant and it can be found as $1/R = \Delta i / \Delta v$. If the i–v characteristic is a straight line and it goes through the origin, then $1/R = i/v$.

- Two elements are said to be in *series* if they carry the *same current,* in *parallel* if they experience the *same voltage.* To be in series, two elements must share a simple node; to be in parallel, they must share the *same node pair.*

- Currents and voltages in a circuit are initiated by *sources*. A *voltage source* maintains a prescribed *voltage regardless of the load current,* a *current source* maintains a prescribed *current regardless of the load voltage.*

- An ideal voltage source has *zero dynamic resistance;* an ideal current source has *infinite dynamic resistance.*

- A *dependent* source is a special type of source whose value is controlled by a signal somewhere else in the circuit.

- Two *voltage sources* may be connected in *series,* but never in *parallel;* two *current sources* may be connected in *parallel,* but never in *series.*

▼ PROBLEMS

1.1 Units and Notation

1.1   Sketch and label the current signal $i_S = 3[2 \sin 3t - 1]$ A versus t, and mark its dc component I_S and its ac component i_s.

1.2   (a) Sketch and label versus time a voltage signal v_S whose ac and dc components are, respectively, $v_s = 0.5 \sin 2t$ V, and $V_S = 10$ V. (b) Repeat, but with V_S changed to -3 V. (c) Repeat, but with $V_S = 0$ and $v_s = 0.5 \sin 2t$ V. (d) Repeat, but with $V_S = 10$ V and $v_s = 0$.

1.2 Electric Quantities

1.3   Find the energy change experienced by an electron as it flows from the positive terminal of a car battery, through the battery, and out of the negative terminal. Does the energy of the electron increase or decrease? Does the battery release or absorb energy?

1.4   A current of 1 A is flowing down a wire with a cross-sectional area of 1 mm². If the wire contains 10^{23} free electrons per cm³ uniformly distributed throughout the material, what is the average speed with which they drift down the wire?

1.5   The charge entering a certain circuit element is found to be $q = 5te^{-10^3 t}$ C. (a) Find the corresponding current i. (b) Find the instant $t \geq 0$ at which i is minimum, as well as i_{min}.

1.6   A 12-V car battery is being charged by forcing a current of 0.5 A into its positive terminal. Find the energy as well as the charge delivered to the battery in 1 h.

1.7   Assuming the household electric energy costs $0.10/kWh, find: (a) the cost of running a 100-W TV set for 8 h/day, for a week; (b) the cost of running a 25-W bulb continuously for a year.

1.8   If the current out of the positive terminal of an element is $i = 5e^{-10t}$ A and the voltage is $v = 3i$ V, find the energy (absorbed or released?) by the element between 0 and 50 ms.

1.9   If the voltage across an element is $v = 10e^{-5t}$ V and the current entering its positive terminal is $i = 3v$ A, find the energy (absorbed or released?) by the element between 0.1 and 0.5 s.

1.10   Devices X_1 and X_2 share a voltage v and a current i as in Figure P1.10. (a) If $v = 10$ V and $i = 2$ A, what is the power exchanged by the two devices? Which device delivers and which absorbs power? (b) If i is due to electron flow, state whether electrons enter X_1 and leave X_2 via the top or the bottom terminal.

Figure P1.10

1.11   In Figure P1.10 find the power at the interconnection and state whether power is flowing from X_1 to X_2 or from X_2 to X_1 for each of the following cases: (a) $v = 5$ V, $i = 4$ A; (b) $v = 5$ V, $i = -2$ A; (c) $v = -24$ V, $i = 5$ A; (d) $v = -12$ V, $i = -2$ A; (e) $v = 0$ V, $i = 10$ A; (f) $v = 220$ V, $i = 0$ A.

1.12   In Figure P1.10 let $v = 10 \sin 2\pi 10^3 t$ V and $i = 5 \cos 2\pi 10^3 t$ A. (a) Sketch v, i, and p versus t. (b) Identify the time intervals during which power is transferred from X_1 to X_2 and those during which it is transferred from X_2 to X_1. (c) Find the maximum instantaneous power exchanged between X_1 and X_2.

1.13   Repeat Problem 1.12 if $v = 10 \sin 2\pi 10^3 t$ V and $i = 5 \sin(2\pi 10^3 t - \pi/4)$ A.

1.3 Electric Signals

1.14   A pulse-train current alternates betwen 0 and 1 mA with a 25% duty cycle and a frequency of 1 kHz. Sketch its waveform and calculate the net charge transferred between $t = 0$ and $t = 5$ ms.

1.15 (a) Write the mathematical expression for a waveform that alternates *sinusoidally* between -5 mA and 0 mA with a frequency of 1 kHz. (b) Repeat, but for a waveform alternating sinusoidally between -10 V and -15 V with a period of 1 μs.

1.16 (a) Find the average value of $v = 100e^{-10^3 t}$ V between $t_1 = 0$ and $t_2 = 1$ ms. (b) Repeat, but for $t_1 = 0$ and $t_2 = 2$ ms.

1.17 Find the average value of $v = 10e^{-10^3 t}$ V between the instant in which $v = 9$ V to the instant in which $v = 1$ V.

1.18 Find the average value of $i = 20(1 - e^{-500t})$ mA between $t_1 = 1$ ms and $t_2 = 3$ ms.

1.19 Find the average value of $i = 100(1 - e^{-500t})$ mA between the instant in which $i = 0$ mA and the instant in which $i = 50$ mA.

1.20 A signal alternates *abruptly* between -1 V and $+2$ V with a duty cycle d. (a) Sketch the waveform, and find its full-cycle average if $d = 25\%, 50\%, 75\%$. (b) For what value of d is the full-wave average zero?

1.21 Sketch and label the periodic waveform defined as $v(t) = V_m \sin(2\pi t/T)$ for $0 \le t \le T/2$, $v(t) = 0$ for $T/2 \le t \le T$, and $v(t \pm nT) = v(t)$ for all t and $n = 1, 2, 3, \ldots$. This waveform is called a *half-wave rectified* voltage waveform. Find its full-cycle average.

1.22 Sketch the waveform $v(t) = 170 |\sin 120\pi t|$, called a *full-wave rectified* ac voltage waveform. What is its period? What is its full-cycle average?

1.23 Sketch and label the waveform $i(t) = 10 \sin^2 2\pi 10^3 t$ A. Find its period and its full-cycle average.

1.24 Sketch the periodic voltage waveform defined as $v(t) = 4e^{-t}$ V for $0 < t < 1$ s, and $v(t \pm n1) = v(t)$ for all t and $n = 1, 2, 3, \ldots$. Find its full-cycle average.

1.25 Find the full-cycle average of the waveform of Figure P1.25.

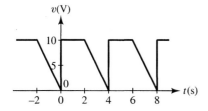

Figure P1.25

1.26 Find the full-cycle average of the waveform of Figure P1.26.

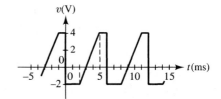

Figure P1.26

1.4 Electric Circuits

1.27 In the circuit of Figure P1.27: (a) identify all nodes and state which ones are simple nodes; (b) identify all loops and state which ones are meshes; (c) identify all elements connected in series and all elements connected in parallel.

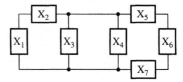

Figure P1.27

1.28 Repeat Problem P1.27 for the circuit of Figure P1.28.

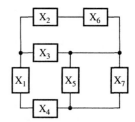

Figure P1.28

1.5 Kirchhoff's Laws

1.29 Given the branch current values shown in Figure P1.29, use KCL to find the magnitudes and directions of all remaining branch currents in the circuit.

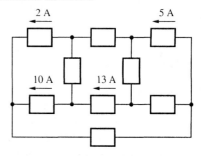

Figure P1.29

1.30   Repeat Problem P1.29 if the branch current at the top left is changed from 2 A (←) to (a) 8 A (←); (b) 2 A (→); (c) 0 A.

1.31   Given the branch current values shown in Figure P1.31, use KCL to find the magnitudes and directions of all remaining branch currents.

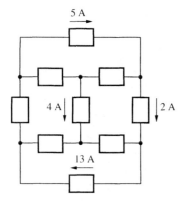

Figure P1.31

1.32   Repeat Problem P1.31 if the branch current at the center is changed from 4 A (↓) to (a) 4 A (↑); (b) 1 A (↓); (c) 0 A.

1.33   Given the branch voltage values shown in Figure P1.33, use KVL to find the magnitudes and polarities of all remaining branch voltages in the circuit.

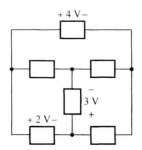

Figure P1.33

1.34   (a) Repeat Problem P1.33 with the polarity of the 4-V branch voltage reversed. (b) Assuming the node at the bottom center is the reference node, show the magnitudes and polarities of all node voltages in the circuit.

1.35   Given the branch voltage values shown in Figure P1.35, use KVL to find the magnitudes and polarities of all remaining branch voltages in the circuit.

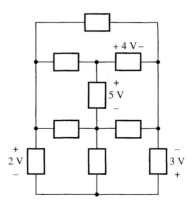

Figure P1.35

1.36   (a) Repeat Problem P1.35 with the polarity of the 2-V branch voltage reversed. (b) Assuming the bottom node is the reference node, show the magnitudes and polarities of all node voltages in the circuit.

1.37   (a) Find all unknown branch voltages and branch currents in the circuit of Figure P1.37. (b) Verify the conservation of power.

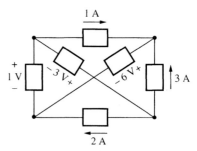

Figure P1.37

1.38   (a) Assuming the branch at the left of Figure P1.38 is absorbing 6 W, find all unknown branch voltages and branch currents. (b) Verify the conservation of power.

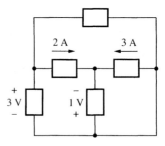

Figure P1.38

1.39   (a) Assuming the branch at the top of Figure P1.38 is releasing 9 W, find all unknown branch voltages and branch currents. (b) Verify the conservation of power.

1.6 Circuit Elements

1.40   Assuming a rectifier diode has $V_T = 26$ mV, what is the value of I_s that will cause it to conduct $i = 1$ mA with $v = 0.700$ V?

1.41   At what current does a rectifier diode exhibit a dynamic resistance of 1 Ω? 1 kΩ? 1 MΩ?

1.42   A zener diode has $V_{ZK} = 10$ V and $R_Z = 150$ Ω. (a) Find v for $i = 10$ mA. What is the corresponding power dissipated by the diode? (b) Find i for $v = 10.6$ V. What is the corresponding power by the diode?

1.43   At $i = 1$ mA a zener diode develops $v = 6.2$ V, and at $v = 7$ V it draws $i = 5$ mA. (a) Find i at $v = 6.5$ V; (b) find v at $i = 3$ mA.

1.44   The nonlinear MOSFET resistor of Figure 1.24 has $V_T = 2$ V and $k = 0.1$ mA/V^2. What is its dynamic resistance at (a) $i = 10$ mA? (b) $i = 1$ mA? (c) $i \rightarrow 0$ mA? (d) $v = 3$ V?

1.45   With $v = 4.5$ V a nonlinear MOSFET resistor draws $i = 0.4$ mA and exhibits a dynamic conductance of 0.4 mA/V. (a) What is its dynamic conductance at $i = 1$ mA? (b) For what value of v does it have a dynamic conductance of 1 mA/V?

1.46   With $i = 0.25$ mA a nonlinear MOSFET resistor exhibits a dynamic resistance of 2 kΩ. (a) What is its dynamic resistance at $i = 1$ mA? (b) For what value of i does it have a dynamic resistance of 10 kΩ?

1.47   An element is known to have an i–v characteristic of the type $i = (v/R) \times (1 - v/2V_p)$ for $0 \leq v \leq V_p$, with R and V_p suitable device parameters. If it is found that $i(0.5$ V$) = 1.75$ mA and $i(1.5$ V$) = 3.75$ mA, find the values of R and V_p, in SI units.

1.48   An element has the i–v characteristic $i = (v/100) \times (1 - v/5)$. What is its dynamic resistance at $v = 0$? 2 V? 2.5 V?

1.7 Sources

1.49   Suppose it is found that the 2-V source of Figure P1.49 is delivering 6 W of power. (a) Find the magnitude and direction of the current i around the loop. (b) Find the power (delivered or absorbed?) by the 6 V source and by the X_1 element.

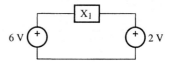

Figure P1.49

1.50   Repeat Problem 1.49, but for the case in which the 2-V source is found to be absorbing 10 W of power.

1.51   Suppose it is found that the 5-A source in Figure P1.51 is delivering 10 W of power. (a) Find the magnitude and polarity of the voltage across X_1. (b) Find the power of the 3 A source and that of X_1. Are these powers delivered or absorbed?

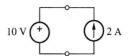

Figure P1.51

1.52   Repeat Problem 1.51 if it is found that the 5-A source is absorbing 6 W.

1.53   (a) Find the power exchanged by the two sources of Figure P1.53. Which source is delivering and which is absorbing power? (b) Repeat, but with the terminals of the current source interchanged, so that its arrow is now pointing downward.

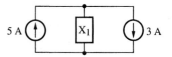

Figure P1.53

1.54   With the switches in the positions shown, the bulb of Figure P1.54 glows. (a) Show two different ways of turning the bulb off. (b) For each way of (a), show two different ways of turning the bulb back on. (c) Suggest an application for this switch arrangement in a home.

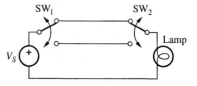

Figure P1.54

RESISTIVE CIRCUITS

H aving familiarized ourselves with the concepts of circuit and circuit element, we are now ready to examine circuits based on the simplest and yet most widely used circuit element: the *linear resistor.* As implied by the name, its distinguishing feature is the ability to *resist* the flow of current in such a way as to maintain a relationship of *linear proportionality* between current and voltage. The most important consequence of this type of relationship is that the equations governing the behavior of resistive circuits are *linear algebraic equations,* the simplest type of equations.

After introducing the concept of resistance and examining its physical foundations, we turn our attention to *series* and *parallel* resistance combinations, voltage and current *dividers,* and other configurations of practical interest such as *bridges* and *ladders.*

The study of resistive circuits offers us the opportunity to introduce the important engineering concepts of **equivalence** and **modeling,** concepts that we exploit to simplify the analysis of more complex networks as well as to facilitate our understanding of practical sources and measurement instrumentation. The section on instrumentation, at the end of the chapter, is intended for students concurrently enrolled in a lab, but it can be omitted without loss of continuity.

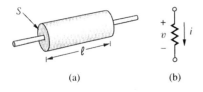

(a)   (b)

Figure 2.1 Resistor and circuit symbol for resistance.

2.1 RESISTANCE

Resistance, denoted as R, represents the **ability to oppose current flow.** Circuit elements specifically designed to provide this function are called **resistors.** As shown in Figure 2.1(a), a resistor consists of a rod of conductive material such as carbon composition, though a variety of other materials and shapes are common. Figure 2.1(b) shows the circuit symbol for resistance, along with the reference polarities for voltage and current, which are those of the **passive sign convention** defined in Section 1.2.

Ohm's Law

When subjected to a test voltage v as in Figure 2.2(a), a resistance R conducts a current i that is *linearly proportional* to the applied voltage,

$$i = \frac{1}{R}v \qquad \text{(2.1)}$$

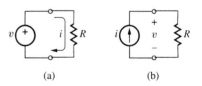

(a)   (b)

Figure 2.2 Resistance subjected to: (a) a test voltage v, and (b) a test current i.

and whose direction is *always out of the source's positive terminal, through the resistance, and into the negative terminal.* Clearly, the larger the value of R the smaller the current for a given applied voltage. The SI unit of resistance is the *ohm* (Ω), named for the German physicist Georg S. Ohm (1787–1854). By Equation (2.1), 1 Ω = 1 V/A. Conversely, when subjected to a test current i as in Figure 2.2(b), a resistance develops a voltage v that is *linearly proportional* to the applied current,

$$v = Ri \qquad \text{(2.2)}$$

and whose polarity is such that the *positive side of* v is that at which current *enters* the resistance, and the *negative side* that at which current *exits* the resistance. Clearly, the smaller the resistance the smaller the voltage drop for a given applied current. A relationship of linear proportionality between voltage and current is referred to as **Ohm's Law,** and the voltage drop developed by a resistance is called an *ohmic voltage drop.*

In applying Ohm's Law we must pay careful attention to the *direction* of i relative to the *polarity* of v, and vice versa. Since the resistance is a dissipative element, it must conform to the *passive sign convention,* that is, current must always flow from a higher to a lower potential. Figure 2.3 depicts a very common situation: a resistance R is connected between nodes A and B of an otherwise arbitrary circuit, and we wish to apply Ohm's Law to it. If the form $i = v/R$ is desired, then we must write

$$i_R = \frac{v_A - v_B}{R} \qquad \text{(2.3)}$$

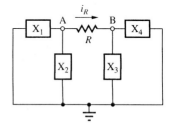

Figure 2.3 Applying Ohm's Law to a resistance embedded in a circuit: $v_A - v_B = Ri_R$, or $i_R = (v_A - v_B)/R$.

In words, **the current equals the *voltage at the tail minus the voltage at the head of the arrow,* divided by the resistance.** Writing it as $i_R = (v_B - v_A)/R$ would be an intolerable mistake! Conversely, if the form $v = Ri$ is desired, then we must write

$$v_A - v_B = Ri_R \qquad \text{(2.4)}$$

In words, **the *voltage at the tail minus the voltage at the head* of the arrow equals the product of the resistance and the current.** Writing $v_B - v_A = Ri_R$ would again be intolerable. If, for some reason, we want the voltage difference to appear as $v_B - v_A$, then we must write $v_B - v_A = -Ri_R$, or $i_R = -(v_B - v_A)/R$. When in doubt, think of a resistor as a water pipe, with water flow likened to current and water pressure to potential. The tendency of water to flow from regions of higher pressure toward regions of lower pressure offers an easy means for relating current direction and voltage polarity.

The *i–v* Characteristic

The $i-v$ characteristic of the resistance is shown in Figure 2.4(a). This is a straight line with slope $1/R$, in agreement with the statement of Section 1.6 that the slope of an $i-v$ curve is always the *reciprocal* of a resistance. The line goes through the origin, and its slope is constant throughout.

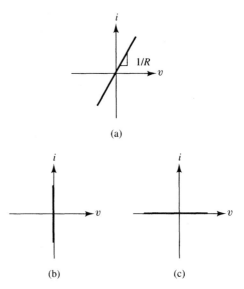

Figure 2.4 The $i-v$ characteristic of (a) a *resistance R,* (b) a *short circuit* $(R = 0)$, and (c) an *open circuit* $(R = \infty)$.

If R is small the slope will be very steep. In the limit $R \to 0$ the line will coalesce with the i axis, where $v = 0$ regardless of i. This is shown in Figure 2.4(b). As we know, this is the characteristic of a **short circuit.**

Conversely, if R is large, the line will lie almost flat. In the limit $R \to \infty$ the line will coalesce with the v axis, where $i = 0$ regardless of v. This is shown in Figure 2.4(c). As we know, this is the characteristic of an **open circuit.** Note that removing a resistance R from a circuit is equivalent to setting $R = \infty$, not $R = 0$! As we know, a **switch** functions as a *short circuit* when *closed,* and as an *open circuit* when *open.* A **fuse** functions as a *short circuit* under normal operating conditions. However, should its current exceed a prescribed limit called the *fuse rating,* the heat resulting from excessive power dissipation will cause the fuse to blow and thus become an *open circuit.*

Conductance

For occasional use it is convenient to work with the reciprocal of resistance

$$G = \frac{1}{R}$$ (2.5)

Aptly called **conductance,** it represents the ability of a circuit element to conduct. Its SI unit is the *siemens* (S), named for the German-born British inventor Karl W. Siemens (1787–1883). Clearly, $1\,S = 1\,A/V = 1\,\Omega^{-1}$. Conductance represents the *slope* of the i–v characteristic. Note that had we chosen to plot v versus i instead of i versus v, then the slope of the resulting $v - i$ characteristic would have been R, or $1/G$.

Power Dissipation

Resistors absorb energy and convert it into heat. In fact, resistance could also be defined as the ability of a circuit element to dissipate power. Ohmic heating is exploited on purpose in home appliances such as ranges, toasters, hair-dryers, irons, and electric heaters. In electronic circuits and instrumentation, however, ohmic heating can lead to intolerable temperature rise and cause damage to the components. This explains the reason for using cooling fans in certain instruments or air conditioning in larger computer installations.

The instantaneous power dissipated by a resistor is

$$p = vi$$ (2.6)

This can be expressed in terms of R and i by writing $p = (Ri)i$ or

$$p = Ri^2$$ (2.7)

Likewise, power can be expressed in terms of R and v by writing $p = v(v/R)$ or

$$p = \frac{v^2}{R}$$ (2.8)

The **wattage rating** of a resistor represents the maximum power that it can safely dissipate before it is damaged due to overheating. This rating depends on the resistive material as well as the resistor's physical size, and is provided by the manufacturer in accordance with the code discussed in Appendix 1. To prevent damage, one must ensure that in any given application the rated wattage is never exceeded.

▶ **Example 2.1**

(a) What is the maximum voltage that can safely be applied across a 10-kΩ, 1/4-W resistor?

(b) What is the maximum current that can be forced through the same?

Solution

(a) By Equation (2.8), $v = \sqrt{Rp}$. Letting $p = p_{max} = 0.25$ W yields $v_{max} = \sqrt{Rp_{max}} = \sqrt{10 \times 10^3 \times 0.25} = 50$ V.

(b) Likewise, Equation (2.7) yields $i_{max} = \sqrt{p_{max}/R} = 5$ mA. ◀

Exercise 2.1   If a resistor dissipates 32 mW when conducting a current of 4 mA, what is its resistance and its voltage?

ANSWER   2 kΩ, 8 V.

• Conduction

Since conduction is at the basis of resistors as well as other devices that you will study in subsequent electronics courses, it is worth examining in detail. In conductive materials such as copper and carbon, atoms occupy fixed positions of an orderly crystal lattice and are so closely spaced that their outer electron shells overlap. The electrons of these shells, called *valence electrons* and no longer belonging to any particular atom, are free to wander from atom to atom throughout the material. A conductor is thus visualized as a periodic array of tightly *bound ions* permeated with a gas of *free valence electrons*.

As a consequence of thermal agitation, ions vibrate about their lattice positions while the free electrons wander around erratically, continuously colliding with the vibrating ions and changing direction. Since electron motion is random in all directions, the net electron flow through any reference plane is, on average, zero.

Consider now the effect of applying a voltage v across a rod of conductive material having length ℓ and cross-sectional area S, as shown in Figure 2.5. This creates an electric field of strength $E = v/\ell$, which in turn accelerates the free electrons toward the left. Their velocity would increase indefinitely, were it not for the collisions with the bound ions, at each of which the electrons lose energy and change direction. Consequently, a steady-state condition is reached in which a net electron *drift* toward the left is superimposed on the random thermal motion. The average drift velocity u is proportional to the strength E of the applied field,

$$u = \mu_n E \qquad (2.9)$$

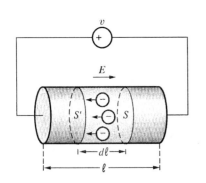

Figure 2.5 Illustrating electron conduction.

The proportionality constant μ_n, aptly called the **mobility,** provides a measure of the ease with which free electrons drift inside a material under the effect of a given field.

To find the current i, we can ignore the random motion and assume the electrons to be uniformly drifting at the velocity u toward the left. Since they are negative, the direction of i will be toward the right. Its magnitude is found by calculating the amount of charge dq crossing the area S in time dt. Consider the electrons that cross S at time t. At time $t + dt$ these electrons will have reached S', whose distance from S is $d\ell = u\, dt$. Clearly, all the electrons that have crossed S between t and $t + dt$ are contained within the volume between S and S'. Their number is thus $dn = nS\, d\ell$, where n is the *free electron density,* that is, the number of free electrons per unit volume, and $S\, d\ell$ is the volume between S and S'. Multiplying dn by the electron charge q yields $dq = nqS\, d\ell = nqSu\, dt$. Substituting $u = \mu_n E = \mu_n v/\ell$ and letting $i = dq/dt$ finally yields

$$i = nq\mu_n \frac{S}{\ell} v \tag{2.10}$$

This confirms that current is *linearly proportional* to the applied voltage. To stress the fact that voltage is the origin of the force that drives current, a battery voltage is also called *electromotive force,* or *emf* for short.

Comparing Equations (2.1) and (2.10) allows us to write

$$\boxed{R = \rho \frac{\ell}{S}} \tag{2.11a}$$

$$\boxed{\rho = \frac{1}{nq\mu_n}} \tag{2.11b}$$

where ρ, called the **resistivity,** represents the intrinsic ability of the given material to resist current flow. The SI units of ρ are $\Omega \times$ m. The reciprocal of ρ, denoted as σ and called the **conductivity,** represents the intrinsic ability of the material to conduct. Clearly,

$$\sigma = nq\mu_n \tag{2.12}$$

indicating that what makes a material a good conductor is a high free electron density n and a high electron mobility μ_n.

Similar considerations hold for the case of positive mobile charges such as holes in p-type silicon. The only difference is that n must be replaced by p, the *free hole density,* and μ_n by μ_p, the *hole mobility.* Thus, the conductivity of p-type silicon is $\sigma = pq\mu_p$. If the material contains both positive and negative mobile charges, then $\sigma = pq\mu_p + nq\mu_n = q(p\mu_p + n\mu_n)$, and $\rho = 1/[q(p\mu_p + n\mu_n)]$.

The reason for expressing R as in Equation (2.11a) is to evidence its dependence on the type of material, or ρ, as well as its physical dimensions ℓ and S. In this respect it is helpful to liken a resistor to a pipe. The amount of fluid flow through the pipe for a given pressure difference between its extremes depends on fluid parameters such as viscosity as well as on the pipe diameter and length. The shorter and thicker the pipe, the greater the amount of flow for a given pressure difference.

▶ **Example 2.2**

A copper wire having $\ell = 10$ cm and $S = 1$ mm² carries a current of 1 A. Assuming $n = 8.5 \times 10^{22}$ cm⁻³ and $\rho = 1.7 \times 10^{-6}$ Ω-cm, find:

(a) the resistance R of the wire;

(b) the voltage v across the wire;

(c) the electric field inside the wire;

(d) the electron drift velocity u.

Solution

(a) $R = \rho(\ell/S) = 1.7 \times 10^{-6} \times 10/(0.1)^2 = 1.7$ mΩ.

(b) $v = Ri = 1.7 \times 10^{-3} \times 1 = 1.7$ mV.

(c) $E = v/\ell = 1.7 \times 10^{-3}/10 = 1.7 \times 10^{-4}$ V/cm.

(d) $\mu_n = 1/nq\rho = 1/(8.5 \times 10^{22} \times 1.602 \times 10^{-19} \times 1.7 \times 10^{-6}) = 43.2$ cm²V⁻¹s⁻¹; $u = \mu_n E = 43.2 \times 1.7 \times 10^{-4} = 7.34 \times 10^{-3}$ cm/s $= 26.44$ cm/hour.   ◀

Exercise 2.2   Pure silicon has both negative and positive mobile charges, and at room temperature $n = p = 1.45 \times 10^{10}$ cm⁻³. Assuming $\mu_n = 1350$ cm²V⁻¹s⁻¹ and $\mu_p = 480$ cm²V⁻¹s⁻¹, find the resistance of a silicon pellet having $\ell = 1$ mm and $S = 1$ cm².

ANSWER   23.52 kΩ.

In general, all practical circuit elements, including capacitors and inductors to be studied later, exhibit a small series resistance and a large parallel resistance. Called *stray* or *parasitic* resistances, they can sometimes affect circuit behavior appreciably. In power devices the series resistance of the leads may be of concern. However, if the leads are fabricated with highly conductive material like copper and are kept sufficiently short and thick, their ohmic drops can often be ignored in comparison with the other voltages in the circuit.

Practical Resistors and Potentiometers

Practical resistors obey Ohm's Law only as long as i and v are confined within certain limits. For obvious reasons, a resistor whose v/i ratio is constant regardless of the magnitudes of v and i is referred to as *ideal*. The closer a practical i–v characteristic to a straight line, the closer the corresponding resistor to ideal.

By proper selection of ρ, S, and ℓ, resistors can be manufactured in a wide variety of values, power ratings, and performance specifications. Discrete resistors usually come in cylindrical shape. In integrated-circuit technology they are fabricated by depositing a thin layer of conducting material on a common insulating material called substrate. Commonly available resistance values range from fractions of ohms to megaohms. See Appendix 1 for the standard resistance values and color codes.

Variable resistors are usually implemented by varying ℓ. A *potentiometer,* or *pot* for short, is a bar of resistive material with a sliding contact, or *wiper,* of the type already encountered in Figure 1.2. The bar is often fabricated in circular shape and the sliding contact is fixed to a shaft to provide rotational control. The circuit symbol for the potentiometer is shown in Figure 2.6(a), where R, the potentiometer rating, represents the resistance between the extremes A and B.

Varying the wiper setting varies its distance from either extreme, thus increasing the resistance between the wiper and one extreme while decreasing that between the wiper and the other extreme. Connecting the wiper to one of the extremes as in Figure 2.6(b) configures the potentiometer as a **variable resistance** between 0 and R. Variable resistance is often represented with the simplified symbol of Figure 2.6(c).

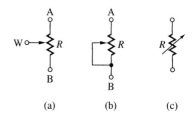

(a)   (b)   (c)

Figure 2.6 (a) Circuit symbol for the *potentiometer;* (b) its connection as a *variable resistance;* (c) circuit symbol for a *variable resistance.*

2.2 SERIES/PARALLEL RESISTANCE COMBINATIONS

Series and parallel resistance combinations appear so frequently, especially as subcircuits, that they are worthy of special consideration. The study of these combinations leads to the important concept of *equivalence,* which we shall often exploit to facilitate the analysis of more complex circuits.

Resistances in Series

The resistances of Figure 2.7(a) share a *simple node* and thus carry the *same current i.* If we go around the loop clockwise, we encounter a voltage rise across the test source and two voltage drops across the resistances. By KVL, the rise must equal the sum of the drops, $v = v_{R_1} + v_{R_2}$. Using Ohm's Law, this can be rewritten as $v = R_1 i + R_2 i = (R_1 + R_2)i$, or

$$v = R_s i \qquad (2.13)$$

$$\boxed{R_s = R_1 + R_2} \qquad (2.14)$$

Equation (2.13) indicates that the composite circuit consisting of two resistances in series exhibits the *same v–i* relationship as a single resistance, aptly called the *equivalent resistance,* whose value is related to the individual values by Equation (2.14).

Note that R_s is *greater* than either R_1 or R_2, indicating that placing resistances in series *increases* the overall resistance. This can be justified physically by assuming R_1 and R_2 to be bars of identical material and cross-sectional area,

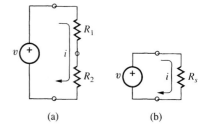

(a)   (b)

Figure 2.7 Two resistances in *series* are equivalent to a single resistance $R_s = R_1 + R_2$.

and lengths ℓ_1 and ℓ_2. Connecting them in series yields a bar of length $\ell_1 + \ell_2$. Its resistance is then $R_s = \rho(\ell_1 + \ell_2)/S = \rho\ell_1/S + \rho\ell_2/S = R_1 + R_2$.

Note that if $R_2 = R_1$, then $R_s = 2R_1$. Moreover, if one of the resistances is much larger than the other—say, at least *an order of magnitude larger*—then the smaller one can be ignored and R_s will essentially coincide with the larger of the two. Thus, if

$$R_1 \gg R_2 \qquad\qquad\qquad \textbf{(2.15a)}$$

(say, $R_1 \geq 10R_2$), then we can write

$$R_s \simeq R_1 \qquad\qquad\qquad \textbf{(2.15b)}$$

where $\simeq$ means "approximately equal to." For example, if $R_1 = 150\ \Omega$ and $R_2 = 10\ \Omega$, then $R_s \simeq 150\ \Omega$. Likewise, if $R_1 = 3\ \text{k}\Omega$ and $R_2 = 100\ \text{k}\Omega$, then $R_s \simeq 100\ \text{k}\Omega$. Keep these simple observations in mind to help develop a feel for how circuits operate.

▶ **Example 2.3**

In the circuit of Figure 2.7(a) let $v = 10\ \text{V}$, $R_1 = 2\ \text{k}\Omega$, and $R_2 = 3\ \text{k}\Omega$. Find R_s, i, v_{R_1}, and v_{R_2}. Hence, use KVL to check the results.

Solution

$R_s = 2 + 3 = 5\ \text{k}\Omega$; $i = 10/5 = 2\ \text{mA}$; $v_{R_1} = 2 \times 2 = 4\ \text{V}$ (+ at the top); $v_{R_2} = 3 \times 2 = 6\ \text{V}$ (+ at the top). As a check, $4\ \text{V} + 6\ \text{V} = 10\ \text{V}$, indicating that KVL is satisfied, as it should be.   ◀

Series resistance combinations are useful when one runs out of specific resistance values. For instance, if you need a 30-kΩ resistance but this value is unavailable, you can synthesize it by connecting a 20-kΩ resistance in series with a 10-kΩ resistance.

This principle is also exploited in resistance adjustments. For instance, if you need a 100.0-kΩ resistance but you only have resistors with a 5% tolerance, you can use a 91-kΩ standard resistance in series with a 20-kΩ potentiometer connected as a variable resistance. The combination of the two will yield a resistance nominally variable from 91 kΩ to 111 kΩ, which can be adjusted to the desired value *exactly*.

The foregoing considerations can easily be generalized to the case of n resistances in series. By Ohm's Law, each resistance develops a voltage drop proportional to the resistance itself, and by KVL the sum of all drops must equal the applied voltage. The overall series resistance is

$$\boxed{R_s = R_1 + R_2 + \cdots + R_n} \qquad\qquad \textbf{(2.16)}$$

If all resistances are equal to R_1, then $R_s = nR_1$.

Resistances in Parallel

The resistances of Figure 2.8(a) share the *same node pair* and thus develop the *same voltage v*. The current from the test source splits between the two resistances, so we can use KCL and write $i = i_{R_1} + i_{R_2}$. By Ohm's Law, this can be rewritten as $i = v/R_1 + v/R_2 = (1/R_1 + 1/R_2)v$, or

$$i = \frac{1}{R_p}v \tag{2.17}$$

$$\boxed{\frac{1}{R_p} = \frac{1}{R_1} + \frac{1}{R_2}} \tag{2.18}$$

Equation (2.17) indicates that the composite device consisting of two resistances in parallel behaves like a single *equivalent resistance* R_p, whose value is related to the individual resistances by Equation (2.18). This is often expressed in short form as

$$\boxed{R_p = R_1 \parallel R_2} \tag{2.19a}$$

Since $1/R_p = (R_1 + R_2)/(R_1 R_2)$, Equation (2.18) can also be expressed as

$$\boxed{R_p = \frac{R_1 R_2}{R_1 + R_2}} \tag{2.19b}$$

However, when using the pocket calculator with the reciprocal function $1/x$, it is preferable to use the form of Equation (2.18).

Note that R_p is *smaller* than either R_1 or R_2, indicating that placing resistances in parallel *decreases* the overall resistance. This makes sense, because current now has more than one path along which to flow. We can also justify this result by assuming R_1 and R_2 to be bars of identical material and length, but cross-sectional areas S_1 and S_2. Connecting them in parallel yields a bar of overall cross-sectional area $S_1 + S_2$, whose resistance R_p is such that $1/R_p = (S_1 + S_2)/(\rho\ell) = S_1/(\rho\ell) + S_2/(\rho\ell) = 1/R_1 + 1/R_2$.

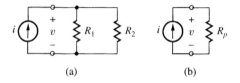

(a)   (b)

Figure 2.8 Two resistances in *parallel* are equivalent to a single resistance R_p such that $1/R_p = 1/R_1 + 1/R_2$.

Note that if $R_2 = R_1$, then $R_p = R_1/2$. Moreover, if one of the resistances is much smaller than the other, say, at least *an order of magnitude smaller,* then the larger one can be ignored and R_p will essentially coincide with the smaller of the two. Thus, if

$$R_1 \ll R_2 \tag{2.20a}$$

(say, $R_1 \leq 0.1R_2$) then we can write

$$R_p \simeq R_1 \tag{2.20b}$$

As an example, if $R_1 = 10\ \Omega$ and $R_2 = 150\ \Omega$, then $R_p \simeq 10\ \Omega$. Likewise, if $R_1 = 100\ \text{k}\Omega$ and $R_2 = 3\ \text{k}\Omega$, then $R_p \simeq 3\ \text{k}\Omega$. You are again encouraged to keep these observations in mind for future use.

▶ **Example 2.4**

What resistance must be placed in parallel with $10\ \Omega$ to achieve $7.5\ \Omega$?

Solution

Let $R_1 = 10\ \Omega$ and let R_2 be the unknown resistance. Then, by Equation (2.18), we have $1/R_2 = 1/R_p - 1/R_1 = 1/7.5 - 1/10 = 1/30$. Thus, the required parallel resistance is $30\ \Omega$. ◀

Exercise 2.3   In the circuit of Figure 2.8(a) let $i = 3$ mA, $R_1 = 2\ \text{k}\Omega$, and $R_2 = 10\ \text{k}\Omega$. Find R_p, v, i_{R_1}, and i_{R_2}. Use KCL to check your results.

ANSWER   1.667 kΩ, 5 V, 2.5 mA ($\downarrow$), 0.5 mA ($\downarrow$).

The preceding considerations can easily be generalized to the case of n resistances in parallel. By Ohm's Law, each resistance conducts a current inversely proportional to the resistance itself, and KCL requires that the sum of all currents equal the applied current. Then, the overall parallel resistance is such that

$$\frac{1}{R_p} = \frac{1}{R_1} + \frac{1}{R_2} + \cdots + \frac{1}{R_n} \tag{2.21}$$

If all resistances are equal to R_1, then $R_p = R_1/n$. Equation (2.21) can be restated in terms of conductances as

$$G_p = G_1 + G_2 + \cdots + G_n \tag{2.22}$$

In a series configuration the *resistances* add up, and in a parallel configuration the *conductances* add up. As a general rule, the formalism will be simpler if we work with *resistances* when dealing with *series* topologies and with *conductances* in *parallel* topologies.

Series/Parallel Resistance Combinations

A resistive network consisting entirely of series and parallel combinations can, by repeated use of Equations (2.16) and (2.21), be reduced to a single **equivalent resistance** R_{eq}—that is, a resistance having the *same i–v characteristic* as the entire network. We can then apply Ohm's Law to this equivalent resistance to find, for instance, the current drawn by the network in response to a given applied voltage or the power dissipated by the network.

In engineering parlance R_{eq} is referred to as the resistance *seen* between the designated terminals of the network. It is customary to indicate the equivalent resistance of a network or of a portion thereof by means of an arrow pointing toward the network itself, or portion thereof. An example will better illustrate.

▶ **Example 2.5**

   (a) Using series/parallel reductions, find the equivalent resistance of the network of Figure 2.9(a).

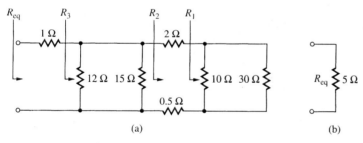

Figure 2.9 A network of series/parallel resistance combinations is equivalent to a single resistance R_{eq}.

   (b) Find the current drawn by the network when its terminals are connected to a source $v_S = 15$ V. Hence, find the power dissipated by the entire network.

Solution

   (a) When applying series/parallel reductions, it is convenient to start from the *branch farthest away* from the terminals and work backward toward the terminals themselves. This leads us through the following steps, illustrated in Figure 2.10:

(1) Since the 10-Ω and 30-Ω resistances in Figure 2.9(a) are in parallel, their combined resistance is $R_1 = 10 \parallel 30 = 7.5 \ \Omega$. The network thus reduces as in Figure 2.10(a).

(2) Since R_1 is in series with the 2-Ω and 0.5-Ω resistances, their combined resistance is $R_2 = 2 + 7.5 + 0.5 = 10 \ \Omega$. The network thus reduces as in Figure 2.10(b).

(3) Since R_2 is in parallel with the 12-Ω and 15-Ω resistances, their combined resistance is found using Equation (2.21). Thus, $1/R_3 = 1/12 + 1/15 + 1/10 = 1/(4 \ \Omega)$, so the network reduces as in Figure 2.10(c).

(4) Finally, since R_3 is in series with the 1-Ω resistance, we have $R_{eq} = 1 + 4 = 5 \ \Omega$.

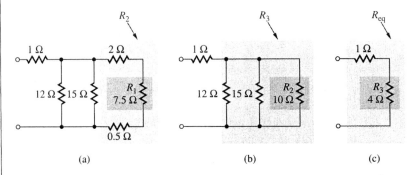

Figure 2.10 Step by step reduction of the network of Figure 2.9(a). Note the use of arrows in the original network to designate intermediate equivalent resistances.

(b) $i = v_S/R_{eq} = 15/5 = 3$ A; $p = R_{eq}i^2 = 5 \times 3^2 = 45$ W. ◀

Exercise 2.4 In the circuit of Figure 1.11, let X_1 through X_5 be resistances with values 1 Ω, 2 Ω, 3 Ω, 4 Ω, and 5 Ω, and let X_6 be a 21-A current source pointing upward. Sketch the circuit. Then, starting from the left and working your way toward the source, use series/parallel resistance reductions to find the voltage across the source.

ANSWER 55 V (positive at the top).

The series/parallel reduction procedure is especially useful when we seek quantities associated with the applied source, such as the resistance seen by the source, the voltage or current at its terminals, or the power supplied by the source.

The Proportionality Analysis Procedure

A resistive network enjoys the property that when connected to a source, all its branch voltages and currents are *linearly proportional* to the source. Though this property will be addressed more systematically in Section 3.4, we exploit it here to find all voltages and currents in a network of series and parallel resistance combinations. To this end, we start from the *branch farthest away* from the source; working our way back toward the source, we express all remaining voltages and currents in terms of the voltage (or current) of this one branch via the repeated use of Ohm's and Kirchhoff's Laws. Once we reach the source, we end up with a relationship between this one branch and the source, which is readily solved for the branch voltage (or current). Finally, we use this result to find all remaining voltages and currents in the circuit.

▶**Example 2.6**

    (a) If the network of Figure 2.9(a) is connected to a 15-V source, find all node voltages and branch currents.

    (b) Verify the value of R_{eq} found in Example 2.5.

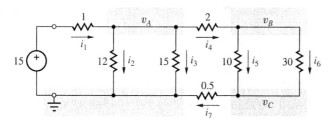

Figure 2.11 Finding all currents and voltages in a network consisting of series/parallel resistance combinations.

Solution

    (a) Sketch the circuit as in Figure 2.11, where the physical units are [V], [Ω], and [A]. Next, label all branch currents and essential nodes, including the reference node, as shown. Then, starting from the branch *farthest away* from the source, express all branch currents and node voltages in terms of i_6 via the repeated use of Ohm's Law, KVL, and KCL:

- Ω: $v_B - v_C = 30i_6$
- Ω: $i_5 = (v_B - v_C)/10 = 3i_6$
- KCL: $i_4 = i_5 + i_6 = 4i_6$
- KCL: $i_7 = i_5 + i_6 = 4i_6$
- KVL: $v_A = 2i_4 + 10i_5 + 0.5i_7 = 40i_6$
- Ω: $i_3 = v_A/15 = (8/3)i_6$
- Ω: $i_2 = v_A/12 = (10/3)i_6$

- KCL: $i_1 = i_2 + i_3 + i_4 = 10i_6$
- KVL: $15 = 1i_1 + 12i_2 = 50i_6$

The last equation yields $i_6 = 15/50$, or

$$i_6 = 0.3 \text{ A}$$

Back-substituting, we find the branch currents as $i_1 = 3$ A, $i_2 = 1$ A, $i_3 = 0.8$ A, $i_4 = 1.2$ A, $i_5 = 0.9$ A, and $i_7 = 1.2$ A. Moreover, using Ohm's Law, we find the node voltages as $v_A = 12i_2 = 12$ V, $v_C = 0.5i_7 = 0.6$ V, and $v_B - v_C = 30i_6$, or $v_B = v_C + 30i_6 = 9.6$ V.

(b) Since the current drawn from the source is $i_1 = 3$ A, the entire network appears to the source as a single resistance $R_{eq} = 15/3 = 5 \ \Omega$. This confirms the result of Example 2.5. ◀

Exercise 2.5 In the circuit of Exercise 2.4 find the voltages across the 1-Ω, 3-Ω, and 5-Ω resistances. What is the overall equivalent resistance seen by the source? (*Hint:* Express all branch currents and voltages in terms of v_1, the voltage across the resistance farthest away from the source.)

ANSWER $v_1 = 5$ V, $v_3 = 15$ V, $v_5 = 55$ V, $R_{eq} = 55/21 \ \Omega$.

▶**Example 2.7**

What happens if the source of Example 2.6 is doubled? Halved? Tripled?

Solution

If the source is changed by a factor k, the last step of Example 2.6 becomes

$$\text{KVL: } k \times 15 = 50i_6$$

which yields

$$i_6 = k \times 0.3 \text{ A}$$

Thus, changing the source by k changes i_6 and, hence, all remaining voltages and currents by the same factor. In particular, doubling ($k = 2$), halving ($k = \frac{1}{2}$), or tripling ($k = 3$) the source will double, halve, or triple all voltages and currents in the circuit. ◀

2.3 VOLTAGE AND CURRENT DIVIDERS

Like series and parallel resistance combinations, the configurations we are about to discuss appear so frequently, either as independent circuits or as subcircuits, that it is worth studying them in proper depth. Then, when we encounter them in the analysis of more complex networks, we can readily apply the formulas of this section to simplify our analytical task.

The Voltage Divider

The circuit of Figure 2.12 consists of two resistances, R_1 and R_2, and two pairs of terminals identified by the unfilled circles and called, respectively, the **input** and **output** terminal pairs. The function of the circuit is to accept an external voltage v_I across the input terminals and develop a voltage v_O across the output terminals. We wish to derive a relationship between v_O and v_I.

Since the two resistances are in *series*, they carry the same current, $i = v_I/(R_1 + R_2)$ (flowing clockwise). The voltage v_I *divides* between the two resistances in amounts that are *linearly proportional* to the resistances themselves, $v_{R_1} = R_1 i$ and $v_{R_2} = R_2 i$. Letting $v_O = v_{R_2} = R_2 i$ and eliminating i yields

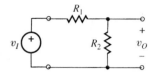

Figure 2.12 Voltage divider.

$$v_O = \frac{R_2}{R_1 + R_2} v_I \tag{2.23}$$

indicating that v_O is a *fraction* of v_I, the value of this fraction being $R_2/(R_1 + R_2)$. Had we chosen to obtain our output across R_1, then the value of the fraction would have been $R_1/(R_1 + R_2)$.

Also called a **voltage attenuator,** the circuit is used whenever it is desired to reduce or attenuate a voltage. Note that the circuit preserves the nature of the input signal: if v_I is dc, so is v_O, and if v_I is ac, so is v_O. Moreover, v_O is positive when v_I is positive, negative when v_I is negative, and zero when v_I is zero. We also say that v_O is **linearly proportional** to v_I. When observed with an oscilloscope, v_O shows the *same waveform as* v_I, differing only in magnitude (because v_O is smaller than v_I).

▶ **Example 2.8**

In the circuit of Figure 2.12 let $R_1 = 10$ kΩ and $R_2 = 20$ kΩ. Find v_O if (a) $v_I = 12$ V; (b) $v_I = 6 \sin 2\pi 10^3 t$ V; (c) $v_I = -3 + 9 \sin 2\pi 10^3 t$ V.

Solution

By Equation (2.23),

$$v_O = \frac{20}{10 + 20} v_I = \frac{2}{3} v_I$$

We thus have (a) $v_O = 2/3(12) = 8$ V; (b) $v_O = 4 \sin 2\pi 10^3 t$ V; (c) $v_O = -2 + 6 \sin 2\pi 10^3 t$ V. ◀

Equation (2.23) reveals the following three important cases:

(1) $R_1 = R_2$. In this case v_I divides *equally* between R_1 and R_2, so $v_O = (1/2)v_I$.

(2) $R_1 \ll R_2$. Now most of v_I is dropped by R_2, so $v_O \simeq v_I$.

(3) $R_1 \gg R_2$. Now most of v_I is dropped by R_1, so $v_O \simeq 0$ V.

Keep these special cases in mind, as they help you develop a better feeling for how differently sized elements participate in the making up of a circuit.

▶ Example 2.9

In the circuit of Figure 2.12 let $v_I = 10$ V. Find v_O for the following resistance pairs: (a) $R_1 = 3\ \Omega$, $R_2 = 2\ \Omega$; (b) $R_1 = 180\ \Omega$, $R_2 = 2\ \Omega$; (c) $R_1 = 1\ \Omega$, $R_2 = 47\ \Omega$.

Solution

(a) By Equation (2.23),

$$v_O = \frac{2}{3+2} 10 = 4 \text{ V}$$

(b) Since $180 \gg 2$, we expect $v_O \ll 10$ V. In fact we have

$$v_O = \frac{2}{180+2} 10 = 0.11 \text{ V}.$$

(c) Since $1 \ll 47$, we expect $v_O \simeq 10$ V. In fact, $v_O = 9.8$ V.   ◀

Gain

When dealing with input-output voltage relationships, it is convenient to work with the ratio v_O/v_I, called the **voltage gain** and expressed in V/V. In the present case the gain is a fraction of 1 V/V. Dividing both the numerator and the denominator by R_2 in Equation (2.23) yields

$$\frac{v_O}{v_I} = \frac{1}{1+R_1/R_2} \tag{2.24}$$

The reason for expressing gain in this form is to evidence its dependence on the *ratio* R_1/R_2, indicating that a given gain can be achieved with a variety of different resistance pairs. For instance, if we need a gain of 1/4 V/V, which requires $R_1/R_2 = 3$, we can use $R_1 = 3\ \Omega$ and $R_2 = 1\ \Omega$, or $R_1 = 30\ \text{k}\Omega$ and $R_2 = 10\ \text{k}\Omega$, or $R_1 = 3.6\ \text{k}\Omega$ and $R_2 = 1.2\ \text{k}\Omega$, and so forth. The choice of a particular pair will affect the current and, hence, the power absorbed by the divider from the input source.

▶ **Example 2.10**

Design a divider circuit which attenuates a 15 V source to 5 V while drawing a current $i = 0.5$ mA.

Solution

With voltages in V and currents in mA, resistances will be in $k\Omega$. By Equation (2.24), $1 + R_1/R_2 = v_I/v_O = 15/5 = 3$ V/V, or

$$R_1 = 2R_2$$

We also want $R_1 + R_2 = v_I/i = 15/0.5$, or

$$R_1 + R_2 = 30 \text{ k}\Omega$$

Substituting the first equation into the second yields $2R_2 + R_2 = 30$, or $R_2 = 10$ kΩ. Back-substituting into the first equation we obtain $R_1 = 20$ kΩ. To summarize, $R_1 = 20$ kΩ and $R_2 = 10$ kΩ.   ◀

Exercise 2.6   Design a voltage divider with a gain of 0.1 V/V and such that it absorbs 1 mW of power when $v_I = 10$ V.

ANSWER $R_1 = 90$ kΩ, $R_2 = 10$ kΩ.

A **variable-gain attenuator** is readily implemented by connecting a potentiometer as in Figure 2.13. Letting R denote the overall potentiometer resistance, the portion of this resistance between the wiper and the bottom lead can be expressed as kR, where k is a fraction that can be varied anywhere over the range $0 \leq k \leq 1$ by varying the wiper position. Applying the voltage divider formula yields $v_O = (kR/R)v_I$, or

$$v_O = kv_I$$

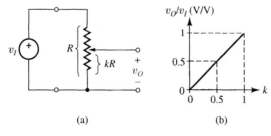

(a)   (b)

Figure 2.13 Variable-gain voltage attenuator.

indicating that the gain coincides with the wiper travel k. Thus, varying the wiper from end to end varies the gain from 0 V/V to 1 V/V and, hence, varies v_O from 0 to v_I. This confirms our observations in connection with Figure 1.2.

A popular application of variable attenuation is *volume control* in audio equipment, where v_I is typically the signal from the preamplifier and v_O is the signal sent to the power amplifier and thence to the loudspeaker. In this case varying the wiper from end to end varies v_O from 0 V, which corresponds to the absence of sound, to v_I, which corresponds to full volume.

A potentiometer can also be used as a **mechanical transducer**, with the wiper voltage providing an electrical representation of the wiper position along a linear or a circular path, depending on potentiometer construction. As such, potentiometers can be used to measure *position, rotation, fluidic level,* and so forth.

The Current Divider

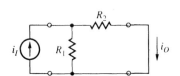

Figure 2.14 Current divider.

The circuit of Figure 2.14 accepts an external current i_I at the input terminals and delivers a current i_O at the output terminals. Since R_1 and R_2 share the same pair of nodes, they are in *parallel* and thus develop the common voltage drop $v = [R_1R_2/(R_1 + R_2)]i_I$ (+ at the top). The current i_I *divides* between the two resistances in amounts *inversely proportional* to the resistances themselves, $i_{R_1} = v/R_1$ ($\downarrow$) and $i_{R_2} = v/R_2$ ($\rightarrow$). Letting $i_O = i_{R_2} = v/R_2$ and eliminating v yields

$$i_O = \frac{R_1}{R_1 + R_2} i_I \qquad \text{(2.25)}$$

indicating that i_O is a *fraction* of i_I. We also observe that i_O is *linearly proportional* to i_I. Note the similarity with the voltage divider formulas, except for the interchanging of R_1 and R_2. The **current gain** of the divider, in A/A, is

$$\frac{i_O}{i_I} = \frac{1}{1 + R_2/R_1} \qquad \text{(2.26)}$$

and it depends exclusively on the *ratio* R_2/R_1.

If $R_1 = R_2$, then i_I splits evenly between the two resistances to yield $i_O = 0.5i_I$. If $R_2 \ll R_1$ most of i_I will flow through R_2, the smaller of the two resistances, and yield $i_O \simeq i_I$. If $R_2 \gg R_1$ most of i_I will flow through R_1 to yield $i_O \simeq 0$.

Exercise 2.7   Design a circuit that attenuates a 1-A current source to 1 mA while developing a voltage $v = 1$ V.

ANSWER $R_1 = 1.001 \ \Omega$, $R_2 = 1 \ \text{k}\Omega$.

Exercise 2.8

(a) Show that if a voltage source v_S drives n resistances in *series* as in Figure 2.15(a), the voltage across the resistance R_k, $k = 1, 2, \ldots, n$, is

$$v_k = \frac{R_k}{R_s} v_S \qquad \text{(2.27)}$$

where $R_s = R_1 + R_2 + \cdots + R_n$.

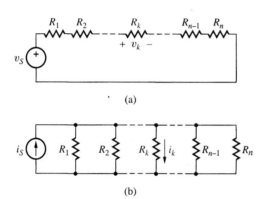

(a)

(b)

Figure 2.15 Generalized voltage and current dividers.

(b) Show that if a current source i_S drives n resistances in *parallel* as in Figure 2.15(b), the current through the resistance R_k, $k = 1, 2, \ldots, n$, is

$$i_k = \frac{R_p}{R_k} i_S \qquad \text{(2.28)}$$

where $R_p = R_1 \parallel R_2 \parallel \cdots \parallel R_n$.

Applying Dividers to Circuit Analysis

Since the voltage and current dividers often appear as subcircuits in more complex networks, the analysis of these networks may benefit from the *judicious* application of the divider formulas. Just keep in mind the following:

(1) For two resistances to form a *voltage divider,* they must be in *series* with each other, that is, they must carry the same *current* and, hence,

share a *simple node.* If one or more other elements are connected to the shared node, we can no longer apply the voltage divider formula, at least not in the simple form of Equation (2.23).

(2) For two resistances to form a *current divider,* they must be in *parallel,* that is, they must share the *same voltage* and, hence, the same *node pair.* If the resistances fail to share both nodes, then we can no longer apply the current divider formula, at least not in the simple form of Equation (2.25).

Some examples will illustrate what we mean by *judicious* application of the divider formulas.

▶ **Example 2.11**

Find v_O in the circuit of Figure 2.16(a).

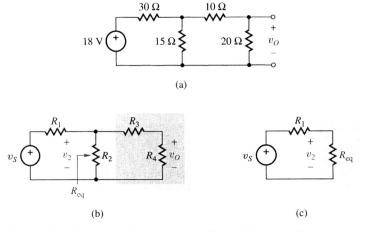

(a)

(b)   (c)

Figure 2.16 (a) Circuit of Example 2.11; (b) and (c) intermediate steps to find, respectively, v_O and v_2.

Solution

For clarity, redraw and label the circuit as in Figure 2.16(b). To find v_O, we observe that the subcircuit consisting of R_3 and R_4 is a voltage divider because its resistances are in series. Letting v_2 denote the voltage across R_2, we thus have $v_O = [R_4/(R_3 + R_4)]v_2 = [20/(10 + 20)]v_2$, or

$$v_O = \frac{2}{3}v_2 \qquad (2.29)$$

To find v_2, one might be tempted to apply the voltage divider formula to R_1 and R_2, but this would be wrong because these resistances share a node with R_3 and are thus not in series. However, if we consider the resistance $R_{eq} = R_2 \parallel (R_3 + R_4) = 15 \parallel (20 + 10) = 10\ \Omega$, then our circuit can be reduced to the equivalent of Figure 2.15(c), which is indeed a voltage divider. Thus, as long as we use R_{eq} instead of R_2, we can still apply the voltage divider formula and write $v_2 = [R_{eq}/(R_1 + R_{eq})]v_S = [10/(30 + 10)]v_S$, or

$$v_2 = \frac{1}{4}v_S \qquad (2.30)$$

Substituting $v_S = 18$ V yields $v_2 = 4.5$ V and, hence, $v_O = 3$ V.   ◀

Exercise 2.9   Suppose in the circuit of Example 2.11 we had mistakenly applied the voltage divider formula to R_1 and R_2 to obtain $v_2 = [R_2/(R_1 + R_2)]v_S = [15/(30 + 15)]18 = 6$ V instead of 4.5 V. Verify that $v_2 = 6$ V is unacceptable because with this value the circuit fails to satisfy KCL at the node common to R_1, R_2, and R_3.

Remark   This exercise shows that it is good practice to always check for the correctness of our results.

▶ **Example 2.12**

Find i_O in the circuit of Figure 2.17(a).

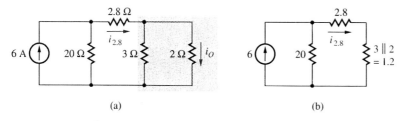

(a)   (b)

Figure 2.17 Circuit of Example 2.12.

Solution

Letting $i_{2.8}$ denote the current through the 2.8-Ω resistance, we can apply the current divider formula to obtain $i_O = 3/(3 + 2)i_{2.8} = 0.6i_{2.8}$.

To find $i_{2.8}$ refer to Figure 2.17(b), where the 3-Ω and 2-Ω resistances have been replaced with their parallel combination $3 \parallel 2 = 1.2\ \Omega$.

Combining this parallel with the 2.8-Ω resistance yields the series combination $2.8 + 1.2 = 4\ \Omega$. We can again apply the current divider formula to obtain

$$i_{2.8} = \frac{20}{20 + 4}6 = 5\ \text{A}$$

Thus, $i_O = 0.6i_{2.8} = 0.6 \times 5 = 3\ \text{A}$.

Exercise 2.10   Find v_O in the circuit of Figure 2.16(a) if the voltage source is replaced with a 3-A current source pointing downward. What is the magnitude and polarity of the voltage across the current source?

ANSWER -20 V; 120 V (positive at the bottom).

2.4 RESISTIVE BRIDGES AND LADDERS

Voltage dividers can be combined in various ways to implement more complex networks. Two popular such networks are the *resistive bridge* and the *resistive ladder*.

The Resistive Bridge

The configuration of Figure 2.18, called a **resistive bridge,** consists of two voltage dividers (R_1 and R_2, and R_3 and R_4) driven by a common source v_S. The two dividers are also referred to as the *bridge arms*. By KVL, the output is $v_O = v_{R_2} - v_{R_4}$. Applying the voltage divider formula in the form of Equation (2.24) twice yields

$$v_O = \left(\frac{1}{1 + R_1/R_2} - \frac{1}{1 + R_3/R_4} \right) v_S \qquad \text{(2.31)}$$

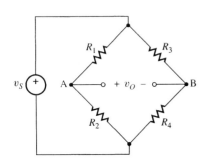

Figure 2.18 Resistive bridge.

Depending on the relative magnitudes of the terms within parentheses, the polarity of v_O may be the same as that of v_S or opposite.

▶ Example 2.13

In the bridge of Figure 2.18 let $v_S = 12$ V and $R_1 = R_2 = R_3 = 2\ \text{k}\Omega$. (a) Find v_O if $R_4 = 1\ \text{k}\Omega$. (b) Repeat, but for $R_4 = 3\ \text{k}\Omega$.

Solution

(a) By Equation (2.31),

$$v_O = \left(\frac{1}{1 + 2/2} - \frac{1}{1 + 2/1} \right) 12 = 2 \text{ V}$$

(b) With $R_4 = 3 \text{ k}\Omega$ we obtain $v_O = -1.2$ V. ◀

Exercise 2.11 In the bridge of Figure 2.18 let $R_1 = R_3 = R_4 = 1 \text{ k}\Omega$. (a) If $v_S = 5$ V, find R_2 so that $v_O = 0.5$ V. (b) If $R_2 = 3 \text{ k}\Omega$, find v_S so that $v_O = 1$ V.

ANSWER 1.5 kΩ, 4 V.

If the two terms within parentheses in Equation (2.31) happen to be identical, then we have $v_O = 0$ V regardless of v_S, and the bridge is said to be a **balanced bridge.** To achieve this condition we need

$$\frac{R_1}{R_2} = \frac{R_3}{R_4} \tag{2.32}$$

In words, **for a bridge to be balanced, the resistances of its arms must be in equal ratios.**

Also called the **Wheatstone bridge,** the circuit finds application in **null measurements,** so called because they are performed by varying one of the elements in the bridge until the output goes to zero. For instance, suppose R_1 and R_2 are fixed resistances, R_3 is a variable resistance R_v such as a precision potentiometer or a precision resistance box, and R_4 is an unknown resistance R_x. If we monitor v_O with an inexpensive voltmeter and vary R_v until a null reading is obtained, then the resistances will satisfy Equation (2.32), which can be rephrased as

$$R_x = \frac{R_2}{R_1} R_v \tag{2.33}$$

Thus, once the ratio R_2/R_1 is known, the value of R_x can be read directly from the calibrated dial of the potentiometer. Note that v_S need not be accurate and that a null reading can be performed with virtually any unsophisticated voltmeter, these being the distinctive advantages of null measurements.

Besides finding application in null measurements, resistive bridges are also used to measure small resistance changes for use with transducers such as *resistive temperature detectors* and *strain gauges*. A *load cell* is a resistive bridge made up of four strain gauges, and it is used in stress measurements.

FAMOUS SCIENTISTS

THOMAS ALVA EDISON

When You're a Genius, Nothing Else Matters

Thomas Alva Edison seems an unlikely American hero: a grammar-school dropout, lazy, frequently fired, insensitive to subordinates, and the world's worst money manager. Add one more adjective to the description, though, and you change the whole picture: genius, a trait Edison was practically bursting with.

Born in 1847 to a school-teacher mother and a father who had just fled a civil rebellion in Canada, Edison grew up mainly in Port Huron, Michigan. He started school at seven but hated it so much that his mother allowed him to quit after only three months and took on the task of teaching him at home. Soon, this "problem child" was reading the great works of literature and science all on his own.

Thomas Edison is shown here with his "Edison effect" lamps. Current will carry between the filament and the plate because there is little or no gas in the tube. (Culver Pictures)

At age 12 he got a job as newsboy on the railroad running from Port Huron to Detroit. He lost that job when a chemistry experiment he was running in the baggage car got out of hand and set the car on fire. At about the same time, he began to lose his hearing, probably a delayed result of a bout with scarlet fever when he was seven but possibly a result of the beating the conductor administered when he found his train aflame.

Edison then sold his newspapers at the Port Huron station, where one day in 1862 he ran through the railyard and scooped the stationmaster's child out of the path of an oncoming train. The grateful papa taught Edison Morse code and gave him a

job as a telegrapher. In an Ontario telegraph office in 1864, one of Edison's jobs was to telegraph to the Toronto office every hour just to keep in contact. Edison rigged up an automatic system that would send the same message to Toronto every 60 minutes. He got fired when a superior caught him asleep while his first invention was dot-dot-dashing away on its own.

This second firing led to several years of drifting from one job to another. During one spell of unemployment, this one in late 1868, Edison read Faraday's marvelous *Experimental Researches in Electricity* and found his calling. His first application of this newly acquired electrical knowledge was to invent and patent an electric vote counter in 1869. It was not a commercial success, and the inventor vowed he would never again "waste time" inventing something that people would not buy.

That same year Edison arrived in New York City and got a job at the Gold Indicator Company, a wire service that monitored gold prices. After making what he thought were major improvements on the ticker machine that carried the quotes, Edison decided to approach his boss and ask for $3,000 for his new version of the machine, a price he chose because he did not have the nerve to ask for the $5,000 he really wanted. Before he could name his price, though, the boss opened the negotiations with, "How does $40,000 sound?"

Edison made so many improvements at Gold Indicator that Western Union soon took over the company. Wanting to work for himself rather than for a large corporation, he quit and formed his own company. Management at Western Union knew a good thing when they saw it, so they made a deal to fund Edison's research and be part owner of any of his telegraph and telephone inventions.

In 1877, Edison got the idea for his most original invention, the phonograph, described in later years as his personal favorite. He was trying to make the receiver key of a telegraph record its message on paper rather than have an operator write it down. Attaching a pointed piece of metal to the diaphragm of a

▼

"Asked once if he wasn't discouraged because his work was not going well, Edison replied, 'Why, I haven't failed, I've just found 10,000 ways that won't work.'"

▲

telephone, he planned to attach this combination to a telegraph key and allow the vibrations of the diaphragm to force the pin to make marks on a cylinder of waxed paper. Then he realized that the human voice could cause the diaphragm to vibrate just as well as a telegraph key. Reverse the setup so that the marked-up cylinder drives the point and diaphragm instead of vice versa and you have a phonograph.

Because he was deaf, the inventor never got to hear his phonograph in action. To get some partial enjoyment from the machine once it was refined and had become a commercial success, he would bite the speaker horn so that the sound could vibrate the bones in his head, allowing him to "hear" the beauty of the music.

By now, Edison, only 30 years old, was rich and famous. He was asked to give a private demonstration of his phonograph to President Rutherford B. Hayes. J. P. Morgan, Cornelius Vanderbilt, and other robber barons bankrolled his Menlo Park, New Jersey, research laboratory, where Edison the insomniac drove a crew to work 18 and 20 hours a day to fulfill his promise to investors of "a minor invention every ten days and a big thing every six months."

Although not as original as his phonograph work, the incandescent lightbulb is the invention we must closely associate with Edison. In both Europe and the United States, inventors had been working on the idea of electric lighting since about 1802, when Sir Humphry Davy used an electric arc to light a London street. Arc lamps, in which an electric current crosses a gap between two carbon rods, gave off a harsh light, however, as well as a foul-smelling gas. Incandescence, the property by which a body heated to a high-enough temperature gives off visible radiation, was the route Edison chose for his light, building

▼

To get some enjoyment from the phonograph he had invented, Edison (who was deaf) "would bite the speaker horn so that the sound could vibrate the bones in his head ...."

▲

on the work of such men as Frederick De Moleyns (who patented a platinum/carbon incandescent bulb in 1841) and Joseph Swan (who made a carbon evacuated bulb in 1860) in England, and J. W. Starr (two incandescent patents in 1845) in the United States.

The problem was how long a filament heated to incandescence would glow before melting. Edison always described himself as a brute-force type of inventor, trying idea after idea with very little appeal to the theories worked out by physicists and mathematicians. Asked once if he wasn't discouraged because his work was not going well, Edison replied, "Why, I haven't failed, I've just found 10,000 ways that won't work." Using this try-everything approach, Edison's workers took almost 16 months to find the

(Continued)

(Continued)
best incandescent filament. On October 19, 1879, the Menlo Park crew got a carbon-covered piece of sewing thread to incandesce in vacuum for 40 hours.

From the 1880s until his death in 1931, Edison worked in several fields but never again attained the great successes of his youth. He connected his phonograph to a camera as a first step on the road to talking pictures, lost a few million dollars on a scheme for mining iron in New Jersey, and spent many years on two ideas very much ahead of their time: an efficient storage battery for electric cars and prefabricated poured-concrete houses that could make a home affordable to everyone. The last major research project Edison undertook was a plan for processing goldenrod and using the extract to make synthetic rubber for the tires of his friend Harvey Firestone.

Edison died of kidney failure brought on by diabetes on October 18, 1931, one day shy of the fifty-second anniversary of his first successful incandescent bulb. In a posthumous footnote of the life of this greatest of America's inventors, the man who died a grammar-school dropout recently became a college graduate. Thomas Edison State College in Trenton, New Jersey, gives college credits to its students for learning acquired outside the classroom, a type of learning Edison surely had an overabundance of. After the college's office of testing and assessment compared Edison's invention record with the curriculum leading to a B.S. degree at Edison State, the college, on October 25, 1992, awarded its favorite son an *earned* bachelor of science degree in applied science and technology with a specialization in electrical technology.

Resistive Ladders

The network of Figure 2.19, consisting of *n resistance pairs* or *stages,* is called an *n-stage* **resistive ladder.** The odd-numbered resistances are referred to as the *series arms,* and the even-numbered resistances as the *shunt arms.* The function of a resistive ladder is to produce *progressively decreasing* shunt-arm voltages and currents as we move from one stage to the next. As such, the ladder forms the basis of a class of attenuators known as *multitapped attenuators.*

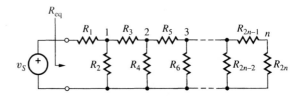

Figure 2.19 An *n*-stage resistive ladder.

The analysis of resistive ladders is facilitated by the fact that they lend themselves to repeated series/parallel combinations. As an example, let us find the equivalent ladder resistance R_{eq} seen by the source. To this end, let us start from R_6 and work backward toward the source. Since R_6 appears in parallel with the equivalent resistance seen looking into the dotted portion of the ladder, their reciprocals may be added together. Their combined resistance appears in series with R_5, so the two may be added together. Proceeding in this manner until we reach the source, there results

$$
R_{\text{eq}} = R_1 + \cfrac{1}{\cfrac{1}{R_2} + \cfrac{1}{R_3 + \cfrac{1}{\cfrac{1}{R_4} + \cfrac{1}{R_5 + \cfrac{1}{\cfrac{1}{R_6} + \cdots}}}}}
$$

(2.34)

This equation, known as a *continued fraction,* is read as well as calculated from the bottom up. Note that in this fraction shunt resistances appear in reciprocal form and series resistances appear in direct form.

▶ Example 2.14

Find the equivalent resistance of a 3-stage resistive ladder with $R_1 = 12$ kΩ, $R_2 = 10$ kΩ, $R_3 = 30$ kΩ, $R_4 = 15$ kΩ, $R_5 = 20$ kΩ, and $R_6 = 10$ kΩ.

Solution

Using Equation (2.34) and calculating it from the bottom up using a pocket calculator yields

$$
R_{\text{eq}} = 12 + \cfrac{1}{\cfrac{1}{10} + \cfrac{1}{30 + \cfrac{1}{\cfrac{1}{15} + \cfrac{1}{20 + 10}}}} = 20 \text{ k}\Omega
$$

◀

Exercise 2.12   A 4-stage ladder consisting of eight identical 10-kΩ resistances is driven by a 5-V source. Find the power dissipated by the ladder.

ANSWER   1.544 mW.

▶ Example 2.15

Find the voltages across the shunt arms of a three-stage ladder with $R_1 = 1$ kΩ, $R_2 = 8$ kΩ, $R_3 = 2$ kΩ, $R_4 = 18$ kΩ, $R_5 = 4$ kΩ, $R_6 = 5$ kΩ, and $v_S = 15$ V.

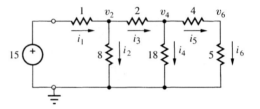

Figure 2.20 Ladder of Example 2.15.

Solution

Sketch the circuit and label all branch currents and essential nodes, including the reference node. The result is shown in Figure 2.20, where the units are [V], [kΩ], and [mA]. Then, apply the *proportionality analysis procedure* introduced at the end of Section 2.2. Thus, starting from the rightmost resistance and working backward toward the source, we obtain:

$$i_6 = v_6/5 = 0.2v_6$$

$$i_5 = i_6 = 0.2v_6$$

$$v_4 = 4i_5 + v_6 = 1.8v_6$$

$$i_4 = v_4/18 = 0.1v_6$$

$$i_3 = i_4 + i_5 = 0.3v_6$$

$$v_2 = 2i_3 + v_4 = 2.4v_6$$

$$i_2 = v_2/8 = 0.3v_6$$

$$i_1 = i_2 + i_3 = 0.6v_6$$

$$15 = 1i_1 + v_2 = 3v_6$$

Clearly, $v_6 = 15/3 = 5$ V, $v_4 = 9$ V, and $v_2 = 12$ V. ◀

Exercise 2.13   A four-stage ladder consisting of eight identical 1-Ω resistances is driven by a source v_S. Find v_S so that the last-stage current is 1 A.

ANSWER   34 V.

• *R-2R* Ladders

The ladder version of Figure 2.21 is called an *n-stage R-2R ladder*. It consists of *n* identical *series* resistances of value *R* and $n+1$ identical *shunt* resistances of value 2*R*. To investigate its properties, we start from the right and work our way back toward the source, one stage at a time.

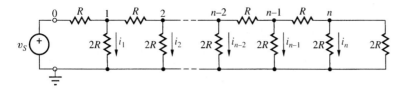

Figure 2.21 *R-2R* ladder.

Referring to the ladder portion shown in Figure 2.22(a), we note that if we stand at the left of node n and look toward the right, we see the equivalent resistance $R_n = (2R) \parallel (2R) = R$. We can thus simplify the rightmost portion of the ladder as in Figure 2.22(b). Then, the voltage at node n is related to that at node $n - 1$ as $v_n = [R_n/(R + R_n)]v_{n-1} = (1/2)v_{n-1}$.

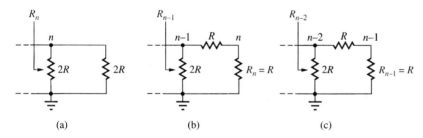

(a) (b) (c)

Figure 2.22 Stage-by-stage analysis of the *R-2R* ladder.

If we now stand at the left of node $n - 1$ in Figure 2.22(b) and look toward the right, we see the equivalent resistance $R_{n-1} = (2R) \parallel (R + R_n) = R$. This leads us to the equivalent circuit of Figure 2.22(c), where we again apply the voltage divider formula to obtain $v_{n-1} = (1/2)v_{n-2}$.

We can readily repeat the preceding line of reasoning for each of the remaining stages, and we can generalize as follows:

(1) If we stand at the left of a node and look toward the right, the equivalent resistance seen between that node and ground is always R, regardless of the node.

(2) The voltage at the node k ($k = 1, 2, \ldots, n$) is related to that at the node $k - 1$ immediately to its left as

$$v_k = \frac{1}{2}v_{k-1} \qquad \text{(2.35)}$$

(3) Letting $i_k = v_k/2R$ and $i_{k-1} = v_{k-1}/2R$ yields a similar relationship between adjacent shunt-arm currents,

$$i_k = \frac{1}{2}i_{k-1} \qquad \text{(2.36)}$$

for $k = 1, 2, \ldots, n$.

These relationships are depicted in Figure 2.23.

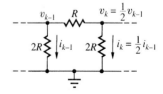

Figure 2.23 As we move from one shunt arm of an *R-2R* ladder to the next on the right, both voltage and current halve.

Because of its ability to provide adjacent currents in a 2-to-1 ratio, the *R-2R* ladder finds wide application in a class of circuits known as *digital-to-analog converters,* or DACs for short. These circuits are used, for instance, in compact disc players to convert the digital codes stored on a disc to the analog signal needed to drive a loudspeaker.

Exercise 2.14   Consider an *R-2R* ladder with $n = 8$, $R = 10$ kΩ, and $v_S = 10.240$ V. Find v_1 through v_8 and i_1 through i_8. What is the total power dissipated by the ladder?

ANSWER   5.12 V, 2.56 V, 1.28 V, 0.64 V, 0.32 V, 0.16 V, 80 mV, 40 mV; 256 μA, 128 μA, 64 μA, 32 μA, 16 μA, 8 μA, 4 μA, 2 μA; 5.243 mW.

2.5 Practical Sources and Loading

A voltage source is designed to supply a prescribed voltage *regardless* of the current drawn by the load. In a practical source, however, it is found that the voltage decreases somewhat as the load current is increased. This is readily verified in the case of a car battery by starting the engine with the headlights on. As we turn the ignition key, the additional current drawn by the starter causes a drop in voltage whose effect is a decrease in light intensity. Likewise, when the home refrigerator goes on, it causes the kitchen lights to dim momentarily as a consequence of the voltage drop due to the current surge in the motor.

Practical Voltage Source Model

Of particular interest is the case when the drop in voltage across the terminals of a voltage source is *linearly proportional* to the load current, for then we can write $v = v_S - \alpha i$, where α is a suitable proportionality constant. Since α must have the dimensions of resistance, the relationship is more properly expressed as

$$v = v_S - R_s i \qquad (2.37)$$

where v is the voltage at the terminals of the source, i is the current drawn by the load, and R_s is called the **internal resistance** of the source. Note that $v = v_S$ only when $i = 0$. Consequently, v_S is called the **open-circuit voltage** of the practical source. This is the voltage that we would measure across the terminals of the practical source in the absence of any load.

Solving Equation (2.37) for i yields the *i–v characteristic* of a practical voltage source,

$$i = \frac{v_S}{R_s} - \frac{1}{R_s}v \qquad (2.38)$$

This represents a straight line with v-axis intercept at $v = v_S$, and with slope $-1/R_s$; see Figure 2.24(a). Note that the negative slope does not imply a negative resistance but is only a consequence of the fact that we choose the reference direction of i *out* of the positive terminal. Clearly, the smaller R_s, the steeper the slope and the closer the source is to ideal.

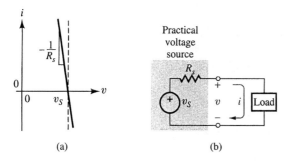

Figure 2.24 Terminal characteristic and circuit model of a practical voltage source.

Equation (2.37) suggests that a practical voltage source behaves as if it consisted of an *ideal* source v_S with a *series* resistance R_s, as depicted in Figure 2.24(b). Indeed, using KVL, we can immediately confirm that $v = v_S - R_s i$. In words, **the voltage available at the terminals equals whatever voltage is produced by the ideal source less the voltage drop across the internal resistance.** Since it appears at the output, R_s is also called the **output resistance** of the source. Thus, *internal resistance, series resistance,* and *output resistance* are equivalent designations for R_s.

The circuit of Figure 2.24(b) is said to be **equivalent** to a practical voltage source because it exhibits the same v–i relation *at its terminals*. Put another way, if we were to encapsulate this circuit with only its terminals protruding, we would be unable, using i–v measurements alone, to tell the difference between this circuit and the practical source it is designed to emulate. However, if we were to open up a practical source, we would be unable to locate either v_S or R_s because they are fictitious elements that we use to mimic practical source behavior. Specifically, we use v_S to account for the voltage produced in the absence of any load, and R_s to account for the voltage drop observed at the terminals when a load is present. The fictitious circuit of Figure 2.24(b) is also said to be a **model** of the practical source.

► **Example 2.16**

The voltage of a supermarket battery is measured with a digital voltmeter, and the reading is 9.250 V. If the battery is connected to a 100-mA bulb, the reading drops to 9.225 V. What are the values of v_S and R_s for this particular battery?

Solution

The open-circuit voltage is $v_S = 9.250$ V. By Equation (2.37) we must have $9.225 = 9.250 - R_s \times 0.1$, or $R_s = 0.25\ \Omega$. Summarizing, $v_S = 9.250$ V and $R_s = 0.25\ \Omega$.

◀

Exercise 2.15   A certain battery is powering three 50-mA bulbs in parallel. If disconnecting one of the bulbs causes the battery voltage to change from 8.910 V to 8.930 V, what are the values of v_S and R_s for this particular battery? (*Hint:* Use Equation (2.37) twice.)

ANSWER   8.97 V, 0.4 Ω.

Although the supermarket battery and the utility power are the first examples of practical sources to come to mind, it must be said that practical sources appear in a variety of other forms. For instance, in a hi-fi system, the phono pickup, the tape head, the microphone, and the radio antenna are examples of practical signal sources, each one characterized by an appropriate open-circuit voltage as well as an internal resistance. Moreover, from the loudspeaker viewpoint, the entire system appears as a voltage source with a suitably small output resistance. As another example, in medical instrumentation, when electrodes are attached to a patient to obtain an electrocardiogram, it is the human body that acts as a voltage source with an appropriate internal resistance!

Practical Current Source Model

Likewise, the current supplied by a practical current source is not strictly independent of the load voltage but usually decreases as the latter is increased. When this decrease is *linearly proportional* to the load voltage, the i–v characteristic of the source can be written as $i = i_S - \beta v$. Since the proportionality constant β must have the dimensions of the reciprocal of resistance, the i–v *characteristic* can be expressed as

$$i = i_S - \frac{1}{R_s} v \qquad\qquad \textbf{(2.39)}$$

where i is the current available at the terminals of the source, v is the voltage developed by the load, and R_s is called the **internal resistance** of the source. Note that $i = i_S$ only when $v = 0$. Consequently, i_S is called the **short-circuit current** of the practical source.

Equation (2.39) represents a straight line with i-axis intercept at $i = i_S$ and with slope $-1/R_s$; see Figure 2.25(a). Clearly, the larger R_s is, the flatter the curve and the closer the source to ideal. We again note that the negative slope is only a consequence of the way in which we have chosen the polarity of v relative to the direction of i.

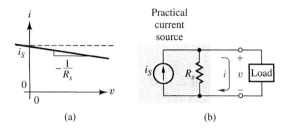

Figure 2.25 Terminal characteristic and circuit model of a practical current source.

Equation (2.39) suggests that a practical current source can be modeled with an *ideal* source i_S and a *parallel* resistance R_s, as depicted in Figure 2.25(b). Indeed, using KCL, it is readily seen that $i = i_S - v/R_s$. In words, **the current available at the terminals equals whatever current is produced by the ideal source less the current diverted through its internal resistance.** Equivalent designations for R_s are *internal resistance, parallel resistance,* and *output resistance* of the practical source. We again note that i_S and R_s are only fictitious elements that we use to model practical current-source behavior.

The foregoing considerations indicate that the concept of resistance need not necessarily always stem from a physical resistor. As it is the case with a practical source, resistance is simply the reciprocal of the slope of its i–v characteristic.

Exercise 2.16   A certain current source has $i_S = 1$ mA and $R_s = 100$ kΩ. Find i for (a) $v = 0$ V; (b) $v = 1$ V; (c) $v = 10$ V; (d) $v = -1$ V.

ANSWER   (a) 1 mA, (b) 0.99 mA, (c) 0.9 mA, (d) 1.01 mA.

The Loading Effect

The fact that connecting a load to a source causes its output to drop in magnitude is referred to as the **loading effect.** Figure 2.26 illustrates this effect for the case of a resistive load R_L.

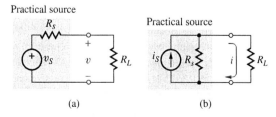

Figure 2.26 Practical voltage and current sources with resistive loading.

In the voltage source case of Figure 2.26(a), the load forms a *voltage divider* with the internal resistance. Thus,

$$v = \frac{R_L}{R_s + R_L} v_S \qquad \text{(2.40)}$$

indicating that the voltage at the load is *less* than the open-circuit voltage, as we anticipated. For obvious reasons, v_S is also called the **unloaded source voltage.** Note that only if $R_s \ll R_L$ can we ignore the voltage loss across R_s and claim that $v \simeq v_S$. In general it can be said that the *smaller R_s is compared to R_L,* the closer the source is to ideal, as we already know. The loading effect can be exploited to indirectly determine the internal resistance of a source, as the following example shows.

▶ **Example 2.17**

The open-circuit measurement of the utility power outlet in a given home yields 120.0 V. After a 12-Ω hot plate is plugged in, the reading drops to 115.0 V. Find R_s for this particular power outlet.

Solution

By Equation (2.40) we must have $115 = [12/(R_s + 12)]120$, or $R_s = 0.522 \ \Omega$.   ◀

In the current source case of Figure 2.26(b), the load forms a *current divider* with the internal resistance,

$$i = \frac{R_s}{R_s + R_L} i_S \qquad \text{(2.41)}$$

indicating that the current at the load is *less* than the short-circuit current (which is also called the **unloaded source current** because of this). Only if $R_s \gg R_L$ can we ignore the current loss through R_s and claim that $i \simeq i_S$. In general, *the larger R_s compared to R_L the closer the source to ideal.* Again, we can exploit the loading effect to determine the parameters of a source.

Exercise 2.17 A practical current source is connected to a load consisting of two 1-kΩ resistances in parallel, and the load voltage is found to be 2.475 V. Removing one of the two 1-kΩ resistances causes the load voltage to rise to 4.902 V. Sketch the circuits corresponding to the two situations; hence, estimate i_S and R_s for this particular source.

ANSWER 5 mA, 50 kΩ.

In general, the loading effect is undesirable because the actual source output varies from load to load. The circuit designer must anticipate load variations and make provisions so that circuit performance is relatively unaffected by these variations. For example, a radio alarm clock that would slow down each time the refrigerator motor goes on would make a very poor product.

• Operating Limits

All practical sources exhibit a *limited range of operation,* indicating that the practical source models developed earlier are valid only within this range. For instance, the utility power outlet will behave as a 120-V source with a suitable series resistance only as long the load current is kept below a certain safety limit. Exceeding this limit will cause the fuses to blow, after which the outlet will cease operating as a source. In the case of a radio or a car battery, the excessive power dissipation due to an overload condition (such as in inadvertent short circuit across its terminals) may result in a significant departure from the designated behavior, not to mention possible damage.

• 2.6 INSTRUMENTATION AND MEASUREMENT

The most basic electrical measurements involve *voltage, current,* and *resistance.* These measurements are performed, respectively, with the **voltmeter,** the **ammeter,** and the **ohmmeter.**

Voltage and Current Measurements

Figure 2.27 shows the basic connections for the measurement of the *voltage across* and the *current through* a given element X of a circuit. The voltmeter, whose symbol is ⓥ, must experience the *same* voltage as the element X. Hence the following rule:

Voltmeter Rule: A voltmeter must be connected *in parallel* with the circuit element under observation.

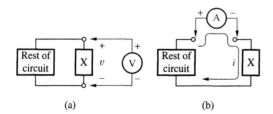

(a)   (b)

Figure 2.27 Connections for (a) *voltage* and (b) *current* measurements.

The ammeter, whose symbol is Ⓐ, must carry the *same current* as the element X, indicating that the circuit must be *broken* to allow for the insertion of the ammeter. Hence the following rule:

Ammeter Rule: An ammeter must be connected *in series* with the circuit element under observation.

To facilitate polarity identification, the positive terminal of a meter is usually color coded in red, and the negative terminal in black. A *positive voltage reading* means that the potential at the red terminal is the higher of the two; a negative reading means that the red terminal potential is the lower of the two. A *positive current reading* means that current enters the instrument via the red terminal and exits via the black terminal; a negative reading means that current enters via the black and exits via the red terminal.

Loading

A reliable measurement must perturb the existing conditions in the circuit as little as possible. Being circuit elements, meters have i–v characteristics of their own. A voltmeter must be capable of accepting any voltage without drawing any current, that is, its characteristic must be $i = 0$ regardless of v. Thus, a voltmeter must appear as an *open circuit* to the circuit under measurement. Conversely, an ammeter must be capable of accepting any current without developing any voltage, that is, its characteristic must be $v = 0$ regardless of i. Thus, an ammeter must appear as a *short circuit* to the circuit under measurement.

These conditions are referred to as *ideal* because practical voltmeters do draw some current and practical ammeters do develop some voltage, however small. The consequence of this is usually a *decrease* in the magnitude of the signal being measured, an effect commonly referred to as **loading.** For obvious reasons, the reading that one would expect with an ideal meter is referred to as the **unloaded** reading, and the reading provided by an actual meter as the **loaded** reading.

Meter deviation from ideality is characterized by means of a resistance R_i called the **internal resistance** or also the **input resistance** of the meter. Its function is to account for the actual terminal characteristic of the meter so that we can predict the amount of loading introduced by the instrument in a given circuit. Clearly, a good voltmeter should have an adequately large R_i (ideally, $R_i = \infty$), and a good ammeter should have an adequately small R_i (ideally $R_i = 0$).

▶ Example 2.18

Figure 2.28(a) shows the voltmeter connection for the measurement of the output of a voltage divider. Let $v_S = 15$ V, $R_1 = 10$ kΩ, and $R_2 = 20$ kΩ.

(a) Predict the ideal reading, that is, the reading that we would expect in the absence of any loading.

(b) Predict the reading obtained with a practical voltmeter having $R_i = 1$ MΩ.

(c) Find the percentage error due to loading.

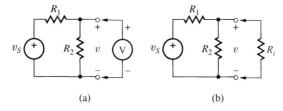

Figure 2.28 Estimating the loading error in voltage measurements.

Solution

(a) Ideally, we would like to obtain the reading

$$v_{\text{unloaded}} = \frac{R_2}{R_1 + R_2} v_S \qquad \text{(2.42)}$$

or

$$v_{\text{unloaded}} = \frac{20}{10 + 20} 15 = 10.000 \text{ V}$$

(b) Connecting the voltmeter to the circuit has the same effect as connecting a resistance R_i in parallel with R_2, as depicted in Figure 2.28(b). We can still apply the voltage divider formula, but we must use $R_2 \parallel R_i$ in place of R_2. The actual reading is now

$$v_{\text{loaded}} = \frac{R_2 \parallel R_i}{R_1 + (R_2 \parallel R_i)} v_S \qquad \text{(2.43)}$$

that is,

$$v_{\text{loaded}} = \frac{20 \parallel 1000}{10 + (20 \parallel 1000)} 15 = 9.934 \text{ V}$$

Clearly, the effect of connecting the meter to the circuit is to cause the output voltage to drop from 10.000 V to 9.934 V. Loading would be absent only if $R_i = \infty$, for then Equation (2.43) would predict $v_{\text{loaded}} = v_{\text{unloaded}}$.

(c) The percentage error due to loading is

$$\varepsilon = 100 \frac{v_{\text{loaded}} - v_{\text{unloaded}}}{v_{\text{unloaded}}} \qquad \text{(2.44)}$$

that is,

$$\varepsilon = 100 \frac{9.934 - 10}{10} = -0.66\%$$

◀

Exercise 2.18   Find the current drawn by the voltmeter in the measurement of Example 2.18.

ANSWER   (a) 0; (b) 9.934 μA.

▶ **Example 2.19**

Find the minimum value of R_i for a loading error magnitude of less than 0.1% in the measurement of Example 2.18.

Solution

Letting $\varepsilon = -0.1$ and $v_{\text{unloaded}} = 10.000$ in Equation (2.44) yields $v_{\text{loaded}} = 9.990$ V. Substituting into Equation (2.43) and solving yields $R_2 \parallel R_i = 19.94$ kΩ. By Equation (2.18) we must have $1/20 + 1/R_i = 1/19.94$, or $R_i = 6.66$ MΩ. Thus, any voltmeter with $R_i \geq 6.66$ MΩ will keep the loading error magnitude below 0.1%.   ◀

Exercise 2.19

(a) Show the ammeter connection for the measurement of the current delivered by the voltage source in the circuit of Figure 2.28(a).

(b) If $v_S = 5$ V, $R_1 = 2$ kΩ, and $R_2 = 3$ kΩ, predict the reading expected ideally as well as the reading obtained with a practical ammeter having $R_i = 200$ Ω. Hence, find the percentage error due to loading.

ANSWER   (b) 1 mA, 0.962 mA, -3.8%.

Multimeters

A **multimeter** is a voltmeter, ammeter, and ohmmeter all in one instrument. The instrument is configured for any of the three measurements by proper connection of the terminals and proper setting of the front-panel controls.

Older multimeter types are based on the *D'Arsonval meter movement,* where a coil is deflected in proportion to the unknown current to provide a reading in *analog* form. Modern multimeters, called **digital multimeters (DMMs),** use microelectronic circuitry to provide readings in *digital form.*

When operating in the voltmeter mode, a DMM is called a **digital voltmeter (DVM).** To minimize errors due to loading, its internal resistance is fairly large, typically from 1 MΩ to 10 MΩ. When operating in the ammeter mode, a DMM is called a **digital ammeter (DAM).** Its deviation from ideality is specified in terms of a voltage called the **burden voltage,** which represents the maximum voltage developed by the instrument (recall that ideally this voltage should be zero). Modern DAMs have burdened voltages on the order of 0.2 V.

DC and AC Multimeter Measurements

Voltmeters and ammeters are equipped with front-panel controls to configure the instrument either for *dc* or *ac* measurements. When measuring a *dc signal,* the instrument must be set on *dc.* Setting it on *ac* will merely yield a zero reading.

When measuring a *time-varying signal,* the instrument may be set either on *dc* or *ac*, depending on the intended measurement.

When a time-varying signal is measured with the multimeter set on *dc,* the instrument yields the **average value** of the signal, also called the **dc component** of the signal. Depending on the case, this value may be positive, negative, or zero.

When a time-varying signal is measured with the multimeter set on *ac,* the instrument yields a non-negative reading. For the case of a **sinusoidal signal,** this reading is designed to coincide with the **rms value.** The peak value is then obtained by multiplying the measured rms value by $\sqrt{2}$. If the signal is *nonsinusoidal,* the instrument still yields a reading, but this is *not necessarily the rms value* of the signal. A suitable correction factor may be needed, depending on the particular waveform. This subject is discussed further in Section 4.4.

▶ **Example 2.20**

Predict the ac and dc readings for the case of a 1-kHz voltage signal v_S that alternates sinusoidally between (a) -2 V and $+8.5$ V; (b) -5 V and $+5$ V; (c) -4 V and -1 V.

Solution

Our signal can be expressed in the form

$$v_S = V_S + V_m \sin 2\pi 10^3 t \qquad (2.45)$$

Retracing the steps of Example 1.4, we have:

(a) $V_S = (-2 + 8.5)/2 = 3.25$ V, and $V_m = 8.5 - 3.25 = 5.25$ V. Thus, the *dc* reading is 3.25 V and the *ac* reading is $5.25/\sqrt{2} = 3.712$ V.

(b) Now $V_S = 0$ V and $V_m = 5$ V, so the *dc* reading is 0 V and the *ac* reading is $5/\sqrt{2} = 3.536$ V.

(c) $V_S = -2.5$ V and $V_m = 1.5$ V, so the *dc* reading is -2.5 V and the *ac* reading is $1.5/\sqrt{2} = 1.061$ V. ◀

Exercise 2.20 The dc and ac readings of a 60-Hz ac current are, respectively, -2.5 A and 3.5 A. Write a mathematical expression for this current, and find its extreme values.

ANSWER $-2.5 + 4.95 \sin 2\pi 60t$ A, 2.45 A, -7.45 A.

Oscilloscopes

An important voltmeter is the **oscilloscope,** especially when the signal to be observed or measured varies rapidly with time. Since it displays amplitude as a function of time, the oscilloscope allows us to observe the actual waveform and to perform amplitude-related measurements like *peak* and *average values,* as well as time-related measurements like *period, duty cycle,* and relative *delays* of different waves. Its front-panel controls allow us to select the vertical sensitivity in volts/division (V/div) and the horizontal sensitivity in seconds/division (s/div).

Like any voltmeter, the oscilloscope must be connected in **parallel** with the element under observation. To minimize loading errors, the internal resistance of an oscilloscope is fairly large, typically on the order of 1 MΩ to 10 MΩ. The instrument can also be configured for indirect current measurements by breaking the circuit, inserting a suitably small series resistance R_s, and observing the voltage drop v across its terminals. The current is then calculated as $i = v/R_s$.

When the oscilloscope is operated in the *dc* mode, the waveform appears offset with respect to the 0 V baseline by the amount V_{av}, where V_{av} is the average value of the waveform determined by internal circuitry over a suitable time interval. The waveform of Figure 2.29(a) has a positive offset of 2 divisions, a peak amplitude of 3 divisions, and a peak-to-peak amplitude of 6 divisions. Were V_{av} negative, the waveform would be offset downward.

When operated in the *ac* mode, the oscilloscope automatically *blocks out the average component,* also called the *dc component,* and it displays only the time-varying component, whose average is thus zero. This is shown in Figure 2.29(b).

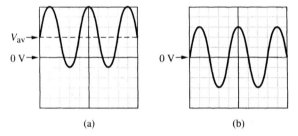

(a)   (b)

Figure 2.29 Illustrating the difference between the *dc* and *ac* modes of operation of the oscilloscope.

▶ Example 2.21

In Figure 2.29(a) let the vertical sensitivity be 0.5 V/div and the horizontal sensitivity be 2 μs/div. Find V_{av}, V_m, T, and f; hence, write a mathematical expression for this signal.

Solution

By inspection, $V_{av} = 2$ div $\times 0.5$ V/div $= 1$ V; $V_m = 3$ div $\times 0.5$ V/div $= 1.5$ V; $T = 4$ div $\times 2$ μs/div $= 8$ μs; $f = 1/T = 1/(8 \times 10^{-6}) = 125$ kHz. Hence,

$$v_S = 1 + 1.5 \sin 785, 398t \text{ V}$$

◀

Exercise 2.21 Referring to the notation of Figure 1.8, show how a 250-Hz, 25% duty-cycle pulse train with $V_m = 4$ V would appear on the screen with the oscilloscope first set in the *dc* mode, then in the *ac* mode. Assume a horizontal sensitivity of 1 ms/div and a vertical sensitivity of 1 V/div. Be neat!

▼ SUMMARY

- Resistance represents the ability of a circuit element to *resist* the flow of current or, equivalently, the ability to *dissipate energy*. The *v–i* characteristic of the resistance is $v = Ri$ and is called *Ohm's Law*.

- When applying Ohm's Law, care must be exercised that voltage and current satisfy the *passive sign convention,* with current flowing in the direction of decreasing potential.

- Series and parallel resistance combinations still behave as resistances. If two resistances R_1 and R_2 are connected in *series,* their equivalent resistance is $R_s = R_1 + R_2$; if they are connected in *parallel,* their equivalent resistance R_p is such that $1/R_p = 1/R_1 + 1/R_2$.

- Using the concept of *equivalent resistance* along with the series and parallel resistance formulas, it is often possible to simplify the analysis of resistive networks via *reduction techniques.*

- The simplest resistive circuits are the *voltage* and *current dividers,* also called *attenuators.* The voltage (or current) divider accepts an *input voltage* (or *current*), and produces an *output voltage* (or *current*) which is a *fraction* of the input. This fraction depends on the *ratio* between the two resistances.

- When voltage or current dividers appear as subcircuits, one can judiciously apply the divider formulas as a means to facilitate circuit analysis.

- Additional resistive circuits of practical interest are the *bridge* and the *ladder.* The bridge finds applications in *null* as well as in *deviation* measurements. Resistive ladders are used to generate progressively smaller voltages and currents.

- In an *R-2R* ladder the shunt branch voltages and currents halve as we move from one shunt branch to the next. As such, the *R-2R* ladder forms the basis of digital-to-analog converters.

- Practical sources depart from the ideal because their terminal characteristics are sensitive to the particular load, a feature referred to as the *loading effect.*

- A practical voltage source can be modeled with an *ideal voltage source* and a *nonzero series resistance;* a practical current source can be modeled with an *ideal current source* and a *noninfinite parallel resistance.*

- In performing laboratory measurements, a *voltmeter* must always be connected in *parallel* and an *ammeter* always in *series* with the element under examination.

- A practical meter tends to load down the circuit under measurement. To minimize loading, the input resistance of a multimeter should be as *large* as possible when used as a *voltmeter,* and as *small* as possible when used as an *ammeter.*

- Resistance is not only the distinctive property of a physical resistor, it also serves the purpose of modeling the terminal characteristics of practical sources as well as of test and measurement equipment.

▼ PROBLEMS

2.1 Resistance

2.1   (a) Use Ohm's Law and KCL to find v_1, i_1, and i_2 in the circuit of Figure P2.1. (b) Verify that the power released by X_1 equals the sum of the powers absorbed by the two resistances.

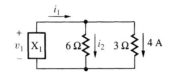

Figure P2.1

2.2   (a) Use Ohm's Law and KVL to find v_1, i_1, and v_2 in the circuit of Figure P2.2. (b) Verify that the power released by X_1 equals the sum of the powers absorbed by the two resistances.

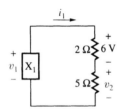

Figure P2.2

2.3   (a) What is the maximum voltage and current that can safely be applied to a 1-kΩ resistor whose power rating is 10 W? 1 W? 1/8 W? (b) Repeat, but for a 10-Ω resistor.

2.4   (a) What is the minimum 1-W resistance that can safely be connected to a car battery?   (b) What is the minimum 1-W resistance that can safely be connected to the household ac power line? (c) If a 2-kΩ resistor can safely carry a current of 10 mA, what is the maximum power that this resistor can safely dissipate and the maximum voltage that it can safely withstand?

2.5   (a) Find the resistance of an aluminum wire 1 m long and 1 mm in diameter. For aluminum, $\rho = 2.8 \times 10^{-6}$ Ω-cm. (b) What is the voltage drop across the wire for a current of 1 A? The strength of the electric field inside the wire?

2.6   A conducting line in an integrated circuit is 1 mm long and has a rectangular cross section of 1 μm × 5 μm. If a current of 10 mA produces a voltage drop of 100 mV across the line, find ρ, in Ω-cm.

2.7   A resistor is fabricated by depositing a thin film of conducting material on a cylinder 10 mm long and 1 mm in diameter. If $\rho = 3 \times 10^{-3}$ Ω-cm, find the film thickness, in μm, needed for $R = 100$ Ω.

2.2 Series/Parallel Resistance Combinations

2.8   Three resistances R_1, R_2, and R_3 are connected in parallel. If $R_1 = 5$ Ω and $R_3 = 20$ Ω, what is the value of R_2 needed to achieve an overall equivalent resistance of 3 Ω? 2 Ω?

2.9   (a) Two resistances, $R_1 = 5$ kΩ±5% and $R_2 = 20$ kΩ± 10%, are connected in *series.* Find the combined resistance, and express it as $R_s \pm x\%$. (b) Repeat, but for the case in which R_1 and R_2 are connected in *parallel,* and express your result as $R_p \pm y\%$.

2.10   (a) Verify the following rule: If you need a resistance R, take a resistance $R_1 = mR$ and a resistance $R_2 = R_1/(m-1)$, $m > 1$, and connect them in parallel. (b) Use this rule to specify five different pairs of standard 5% resistances to achieve 10 kΩ (see Appendix 1 for 5% resistance values).

2.11   In the circuit of Figure P2.11: (a) find R_{eq} if $R = 30$ kΩ; (b) find R for $R_{eq} = 18$ kΩ; (c) find R for $R_{eq} = R$.

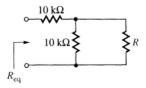

Figure P2.11

2.12   (a) In the circuit of Figure P2.12 find a condition for R in terms of R_1 to ensure $R_{eq} = R$. (b) If $R_1 = 10$ kΩ, what is the value of R for which $R_{eq} = R$?

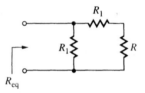

Figure P2.12

2.13 Using only 10-kΩ resistances, devise suitable series/parallel combinations to achieve R_{eq} = (a) 7.5 kΩ; (b) 15 kΩ; (c) 25 kΩ. In each case, try minimizing the number of resistances used.

2.14 (a) Using series/parallel resistance reductions, find the equivalent resistance between terminals A and B in the circuit of Figure P2.14. (b) Repeat, but with C and D shorted together.

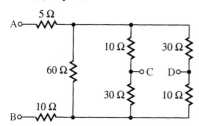

Figure P2.14

2.15 (a) Using series/parallel resistance reductions, find the equivalent resistance seen by the source in the circuit of Figure P2.15. (b) What is the overall dissipated power?

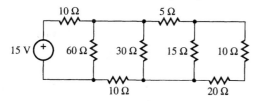

Figure P2.15

2.16 (a) Assuming the bottom terminal of the source in the circuit of Figure P2.15 is grounded, use the proportionality analysis procedure to find all node voltages in the circuit. (b) To what value must the source be changed to ensure 1 A through the 15-Ω resistance?

2.17 (a) Using series/parallel resistance reductions, find the equivalent resistance seen by the source in the circuit of Figure P2.17. (b) What is the voltage across the source?

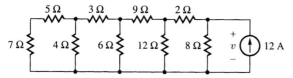

Figure P2.17

2.18 (a) Apply the proportionality analysis procedure to the circuit of Figure P2.17 to find all branch currents. (b) To what value must the source be changed to ensure 9 V across the 9-Ω resistance?

2.19 Find the current delivered by the source in the circuit of Figure P2.19. (*Hint:* Reduce the portions to the left and to the right of node A separately.)

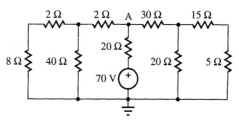

Figure P2.19

2.20 Using the proportionality analysis procedure, find all node voltages in the circuit of Figure P2.19. (*Hint:* Starting from the far left and far right, work your way toward the center and express v_A both in terms of $i_{8\Omega}$ and in terms of $i_{5\Omega}$.)

2.21 (a) Apply the proportionality analysis procedure to find v_1 and v_2 in the circuit of Figure P2.21. (b) What is the net resistance seen by the source? (*Hint:* Express all currents and voltages in terms of the current i through the 4-Ω resistance.)

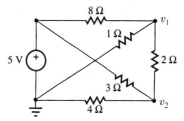

Figure P2. 21

2.22 Suppose in the circuit of Figure P2.14 we connect a 3-V voltage source between terminals C and D, with the positive side at C. Sketch the circuit; then find the voltage between A and B.

2.23 Suppose in the circuit of Figure P2.14 we connect a 0.8-A current source between terminals C and D so that current flows from D, through the source, to C. (a) Sketch the circuit; then find the voltage across the source. (b) Repeat, but with A and B shorted together.

2.3 Voltage and Current Dividers

2.24 In the voltage divider of Figure 2.12 let R_1 = 15 kΩ and R_2 = 3 kΩ. (a) Find v_O if v_I = 12 V. (b) Find v_I for v_O = 1 V.

2.25 A voltage divider is implemented with R_1 = 1 kΩ±5% and R_2 = 3 kΩ ± 5%. Find the range of values for its gain.

2.26 In the voltage divider circuit of Figure 2.12 let v_I = 10 V. (a) If R_1 = 16 kΩ, find R_2 for v_O = 2 V. (b) If R_2 = 2 kΩ, find R_1 for v_O = 4 V. (c) If v_O = 5 V and v_I is known to be releasing 50 mW, find R_1 and R_2.

2.27 Design a voltage divider to obtain a voltage of 5 V from a car battery, under the constraint that the power dissipated by the divider may not exceed 100 mW.

2.28   By properly applying the voltage divider formula twice, find the voltages across the 4-Ω and 2-Ω resistances in Figure P2.28.

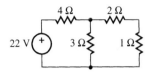

Figure P2.28

2.29   Letting the node at the top of the 3-Ω resistance be the reference node in the circuit of Figure P2.28, find all node voltages, all branch currents, and the power delivered by the source.

2.30   In the voltage divider circuit of Figure 2.12 let $v_I = 220$ V and $v_O = 132$ V. (a) If connecting a 10-kΩ resistance in parallel with R_2 yields $v_O = 60$ V, find R_1 and R_2. (b) Predict v_O if the 10-kΩ resistance is connected in parallel with R_1.

2.31   When connected to a 4-mA source, a current divider divides the current down to 3 mA while dissipating 60 mW. Find R_1 and R_2.

2.32   By properly applying the current divider formula twice, find the currents through the 5-kΩ and 10-kΩ resistances in Figure P2.32.

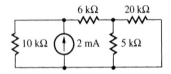

Figure P2.32

2.33   Letting the node at the top terminal of the source be the reference node in the circuit of Figure P2.32, find all node voltages and branch currents. What is the power delivered by the source?

2.34   In the current divider circuit of Figure 2.14 let $i_I = 6$ A and $i_O = 2$ A. (a) If connecting a 10-Ω resistance in series with R_2 yields $i_O = 1.5$ A, find R_1 and R_2. (b) Predict i_O if the 10-Ω resistance is connected in parallel with R_1.

2.35   In the current divider circuit of Figure 2.14 let $i_I = 10$ A and $i_O = 2$ A. If connecting a 2.4-Ω resistance in parallel with the source yields $i_O = 1.5$ A, find R_1 and R_2.

2.36   In the current divider circuit of Figure 2.14 let $i_I = 5$ mA and $i_O = 1$ mA. If connecting a 1-kΩ resistance in series with R_2 yields $i_O = 0.75$ mA, predict the value of i_O if the 1-kΩ resistance is connected in series with R_1.

2.37   The resistances denoted as R_1 and R_2 in Figure P2.37 form an *attenuator pad,* a network often used in audio applications to achieve a prescribed equivalent resistance R_{eq} and gain v_O/v_I. (a) Show that if $R_{eq} = R_L$, then $R_L^2 = 4R_1(R_1 + R_2)$,

and $v_O/v_I = R_2/(2R_1 + R_2 + R_L)$. (b) Specify suitable values for R_1 and R_2 to achieve $R_{eq} = R_L = 600$ Ω with $v_O/v_I = 0.5$ V/V.

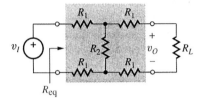

Figure P2.37

2.38   Adding suitable resistances to a linear potentiometer as shown in Figure P2.38 results in *nonlinear* variable-gain voltage division. Let the portion of R between the wiper and the bottom terminal be kR, so that the portion of R between the wiper and the top terminal is $(1 - k)R$, $0 \le k \le 1$. Plot the gain v_O/v_I versus k in k-increments of 0.1 for the following four cases: (a) $R_1 = R_2 = \infty$; (b) $R_1 = \infty$, $R_2 = 0.1R$; (c) $R_1 = 0.1R$, $R_2 = \infty$; (d) $R_1 = R_2 = 0.1R$.

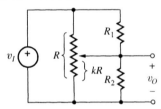

Figure P2.38

2.39   Using a 100-kΩ potentiometer and suitable external resistances, design a voltage divider circuit such that varying the wiper from end to end varies the gain over the range: (a) 0 to 0.75 V/V; (b) 0.2 V/V to 1 V/V; (c) 0.1 V/V to 0.9 V/V.

2.4 Resistive Bridges and Ladders

2.40   In the bridge circuit of Figure 2.18 let $v_S = 18 \times \cos 2\pi 10^3 t$ V, $R_1 = 10$ kΩ, and $R_2 = R_3 = R_4 = 30$ kΩ. (a) Find v_O as a function of time. (b) If a wire is connected between A and B, find the current i_O from A, through the wire, to B as a function of time.

2.41   (a) Given four resistances, 10 kΩ, 30 kΩ, 40 kΩ, and 120 kΩ, show all possible ways of connecting them to form a balanced bridge. (b) Which of the above bridges are less wasteful of energy when powered by a voltage source? By a current source?

2.42   Suppose a 0.4-A current source is connected between terminals A and B of the circuit of Figure P2.14, with current flowing from A, through the source, to B. Draw the circuit; then find the magnitude and polarity of the voltage appearing between A and B, and that between C and D.

2.43 Suppose a 12-V voltage source is connected between nodes A and B of the circuit of Figure P2.14, with the positive side at A. Find the magnitude and polarity of the voltage appearing between C and D, as well as the power supplied by the source.

2.44 (a) Find the voltage v_{AB} that must be applied to the circuit of Figure P2.14 to obtain $v_{CD} = 1$ V. (b) If C and D are shorted together with a wire, find the voltage v_{AB} that must be applied to obtain a current of 1 A flowing from D, through the wire, to C.

2.45 A 5-stage ladder of the type of Figure 2.19 consists of ten 20-kΩ resistances. If the ladder is driven by a 0.25-mA source, what voltage will it develop? How much power will it dissipate?

2.46 (a) Referring to the 5-stage ladder of Figure P2.46, show that in order for the tap voltages to form a *geometric progression* as shown, the resistances must satisfy the conditions

$$R_1 = R(1 - k)/k$$

$$R_2 = R/(1 - k)$$

and that the net resistance seen by the source is $R_{eq} = R/k$. (b) Verify that the case $k = 0.5$ yields the R-$2R$ ladder of Figure 2.21.

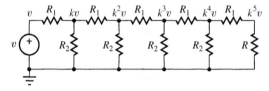

Figure P2.46

2.47 (a) Using the results of Problem 2.46, design a 5-stage ladder having $k = 0.1$ and $R_{eq} = 10$ kΩ. (b) Repeat, but for the case $k = 1/4$ and $R_{eq} = 10$ kΩ.

2.5 Practical Sources and Loading

2.48 With a 1-W load, the voltage at the terminals of a certain voltage source is found to be 10.0 V; without the load, it rises to 10.2 V. Estimate v_S and R_s. Hence, predict the voltage at the terminals of the source for the case of a 22-Ω load.

2.49 A certain transducer is modeled with a source v_S and a series resistance R_s. The voltage at its terminals is measured with a DVM having internal resistance $R_i = 10$ MΩ and is found to be 12.500 mV; adding a 5-MΩ resistance in parallel with the DVM causes the reading to drop to 9.375 mV. Find v_S and R_s.

2.50 (a) If a certain current source supplies $i = 0.46$ A at $v = 10$ V, and $i = 0.4$ A at $v = 25$ V, estimate i_S and R_s. (b) What current would this source supply at $v = 15$ V? At $v = -15$ V?

2.51 The current at the terminals of a certain current source is measured with an ammeter having an internal resistance $R_i = 10$ Ω, and is found to be 11.988 mA; adding a 1.2-kΩ resistance between the source terminals causes the ammeter reading to drop to 11.889 mA. Find i_S and R_s.

2.6 Instrumentation and Measurement

2.52 In the voltage divider circuit of Figure 2.12 let $v_I = 5.000$ V, $R_1 = 200$ kΩ, and $R_2 = 300$ kΩ. (a) What reading will a voltmeter with input resistance $R_i = 1$ MΩ yield when connected across R_1? Across R_2? (b) If an ammeter with $R_i = 10$ kΩ is connected in series with the source, what reading will it yield? How much voltage will it develop?

2.53 In the voltage divider circuit of Figure 2.12 let $v_I = 10.000$ V. If connecting a voltmeter with an input resistance $R_i = 1$ MΩ across R_1 yields the reading $v_{R_1} = 7.407$ V and connecting it across R_2 yields the reading $v_{R_2} = 1.852$ V, find R_1 and R_2.

2.54 In the current divider circuit of Figure 2.14 let $i_I = 2.000$ A, $R_1 = 2$ Ω and $R_2 = 8$ Ω. (a) What reading will an ammeter with input resistance $R_i = 0.1$ Ω yield when connected in series with R_1? In series with R_2? (b) If a voltmeter with $R_i = 100$ kΩ is connected in parallel with the source, what reading will it yield? How much current will it draw?

2.55 In the current divider circuit of Figure 2.14 let $i_I = 2.000$ mA, $R_1 = 2$ kΩ, and $R_2 = 18$ kΩ. If i_I, i_{R_1} and i_{R_2} are measured with an ammeter having internal resistance R_i, what is the maximum allowed value of R_i for loading error magnitudes of less than 0.1%?

2.56 Write a mathematical expression and predict the ac and dc ammeter readings of a current alternating sinusoidally: (a) between $+3$ mA and $+5$ mA with a frequency of 5 kHz; (b) between -3 mA and $+2$ mA with a period of 10 μs; (c) between 0 and $+10$ mA, and such that the net charge transferred in one cycle is 1 nC.

2.57 A sinusoidal voltage waveform is observed with an oscilloscope. Its peak-to-peak amplitude is 8 divisions, and the time interval between two consecutive positive peaks is 4 divisions. If the vertical and horizontal sensitivities are, respectively, 2 V/div and 10 μs/div, find the maximum rate at which the voltage waveform changes with time, in V/s.

2.58 A 10-kHz rectangular wave alternates between -2 V and $+4$ V with a 75% duty cycle. Assuming sensitivities of 1 V/div and 20 μs/div, show how the waveform would appear on the oscilloscope screen with the oscilloscope set first in the dc mode, then in the ac mode.

2.59 (a) Repeat Problem 2.58, but for a 50% duty cycle. (b) Find the duty cycle value for which changing the oscilloscope mode from dc to ac or vice versa leaves the waveform on the screen unaffected.

CIRCUIT ANALYSIS TECHNIQUES

C*ircuit analysis* is the process of finding specific voltages or currents in a circuit. This process is of paramount importance because we must learn to analyze circuits before we can design new ones or adapt existing ones to new needs. In resistive circuits the formulation of the equations governing the voltage and current variables is based solely on Ohm's Law and Kirchhoff's laws.

We have already been exposed to simple instances of circuit analysis in Chapter 2. In the present chapter we approach this topic in a more systematic and general manner. Our goal is not only to teach the mathematical aspects of circuit analysis, but also to help the reader develop a *physical feel* for circuits. To prevent mathematics from obscuring the physical aspect, we shall keep our circuits as simple as possible. Even so, our circuits are of the type and complexity that working engineers see every day. In the next chapter we show that even when dealing with more complex networks it is often possible to break them down into subcircuits of more manageable size and analyze them individually, still using the techniques of the present chapter.

Before embarking upon a systematic approach to analysis, we find it instructive to examine some illustrative circuits from a *synthesis* viewpoint in Section 3.1. We find that synthesis provides the reader with a better feeling for how the element laws and the connection laws interact to make a circuit work.

We then turn to the two most important analysis methods, the *node method* and the *loop method*. The node method allows us to find all *node voltages* in a circuit and the loop method all *mesh currents*. Once either set of variables is known, we can readily find the voltage across or the current through any branch in the circuit. In general, the node and loop methods require the solution of *systems* of simultaneous linear algebraic equations. This solution is carried out using either the *Gaussian elimination method* or *Cramer's rule.* Both methods are reviewed in Appendix 2.

Quite often we are interested in just one variable in a circuit, not in the entire set, and we thus seek ways to expedite our analysis. One important such technique is offered by the *superposition principle,* which exploits a unique property of resistive circuits known as *linearity* to break down a circuit into smaller subcircuits that are easier to analyze. Another technique, known as the *source transformation technique,* exploits the equivalence between a voltage source with a series resistance and a current source with a parallel resistance to effect suitable circuit manipulations that may also help expedite circuit analysis.

Even though these techniques can in principle be applied to the analysis of any network, as circuit complexity increases they become prohibitive both in terms of time and effort. To overcome this obstacle, computer simulation programs have been developed to perform circuit analysis automatically. At the end of the chapter we introduce SPICE, one of the most widely used such programs.

Since linear circuit analysis involves only ordinary algebra, its success relies not on fancy math but on the analyst's ability to clearly understand the problem, formulate its algebraic equations correctly, and be neat and attentive during the algebraic manipulations. A most exasperating aspect of circuit analysis is that a simple sign or magnitude error suffices to invalidate the entire effort of long and tedious calculations! An integral part of the problem-solving process is *checking* for the correctness of the results. In presenting the various methods, we encourage the student to develop the habit of using one method to obtain the solution and another to verify its correctness.

Although our present attention is limited to circuits consisting of resistances and independent sources, we shall see that the techniques of the present and the next chapter are applicable to networks containing other element types such as dependent sources, capacitances, inductances, and even the more sophisticated devices that you will encounter later in the curriculum, such as diodes and transistors. We can say without fear of exaggeration that these two chapters are the cornerstones of your electrical and electronic engineering curriculum. To start out on the right track, you may wish to allocate far more time and attention to this pair of chapters than you would normally do, for once you have mastered the fundamentals of resistive circuit analysis, all subsequent material will flow smoothly, no matter how sophisticated the function of the circuit or of its elements.

3.1 CIRCUIT SOLUTION BY INSPECTION

As elements are connected to form a circuit, each branch current and branch voltage must satisfy two sets of laws, the element laws and the connection laws. This leaves no room for indeterminacy because at any instant each current and voltage must assume a well-defined value and all values must fit together precisely, like the pieces of a jigsaw puzzle. Before embarking upon a systematic approach to circuit analysis, it is instructive to develop a quick and intuitive feeling for this kind of interplay. We devote this section to solving some circuits examples through the repeated use of Ohm's and Kirchhoff's laws. The common thread is the search for suitable branch element values to achieve specific currents or voltages in designated parts of a circuit. This is the kind of task that working engineers face daily.

▶ Example 3.1

In the circuit of Figure 3.1(a) find the value of v_S that will cause the 2-Ω resistance to conduct a current of 3 A flowing downward.

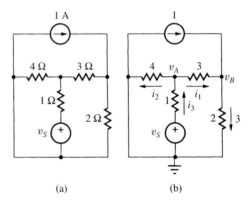

Figure 3.1 Circuit of Example 3.1.

Solution

First, label the essential nodes, select one of them as the reference node, and ground it. The result is shown in Figure 3.1(b), where we have omitted the physical units to further simplify the notation. It is understood that they are [V], [Ω], and [A].

Next, we start from the given 3-A branch current and work our way back toward the unknown source v_S. We shall examine one node and one branch at a time, paying careful attention to *current directions* and *voltage polarities* as we apply Ohm's and Kirchhoff's laws. Once again we stress that **resistances must satisfy the passive sign convention.**

(1) Consider the 2-Ω resistance. Since its current flows downward, we must have, by Ohm's Law, $v_B - 0 = 2 \times 3$, or $v_B = 6$ V.

(2) Consider the node labeled v_B. Since only 1 A is coming in while 3 A are going out, there must be a current i_1 flowing into this node. Indicate this by means of an arrow, as shown. Then, by KCL, we must have $i_1 + 1 = 3$, or $i_1 = 2$ A.

(3) By Ohm's Law, $v_A - v_B = 3i_1$, that is, $v_A = v_B + 3i_1 = 6 + 3 \times 2 = 12$ V.

(4) By Ohm's Law, i_2 must flow toward the left, and $i_2 = (v_A - 0)/4 = 12/4 = 3$ A.

(5) Consider the node labeled v_A. Since both i_1 and i_2 are flowing out of this node, there must be a current i_3 flowing into the node such that $i_3 = i_1 + i_2 = 2 + 3 = 5$ A.

(6) By Ohm's Law, $v_S - v_A = 1i_3$, or $v_S = v_A + 1i_3 = 12 + 1 \times 5 = 17$ V. ◄

Exercise 3.1   Repeat Example 3.1, but with the current source increased from 1 A to 4 A. Be careful with voltage and current polarities as you apply Ohm's Law!

ANSWER $v_S = 2.75$ V.

► **Example 3.2**

In the circuit of Figure 1.11 let X_1 be a 17-V source with the positive terminal at the top; let X_2, X_4, and X_5 be, respectively, 10-Ω, 20-Ω, and 4-Ω resistances; let X_6 be a 2-A source flowing downward. If X_3 is an unknown resistance R, find the value of R that will cause the 2-A source to release 12 W.

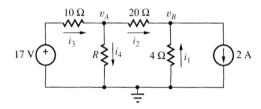

Figure 3.2 Circuit of Example 3.2.

Solution

First draw the circuit as in Figure 3.2, with the elements and the essential nodes properly labeled. Then, proceed one step at a time, as follows:

(1) Recall that for the current source to *release* power, its current must flow in the direction of *increasing* voltage. Thus, v_B must be lower than ground potential, or negative. Since $p = vi$, we must thus have $v_B = -(p/i) = -(12/2) = -6$ V.

(2) By Ohm's Law, i_1 must flow upward, as shown. Thus, $i_1 = (0 - v_B)/4 = [0 - (-6)]/4 = 1.5$ A.

(3) By KCL at the node labeled v_B, $i_2 + i_1 = 2$, or $i_2 = 2 - 1.5 = 0.5$ A.

(4) By Ohm's Law, $v_A - v_B = 20i_2$, or $v_A = v_B + 20i_2 = -6 + 20 \times 0.5 = 4$ V.

(5) By Ohm's Law, i_3 must flow toward the right, as shown, and $i_3 = (17 - 4)/10 = 1.3$ A.

(6) By KCL at the node labeled v_A, $i_3 = i_4 + i_2$, or $i_4 = i_3 - i_2 = 1.3 - 0.5 = 0.8$ A.

(7) By Ohm's Law, $R = v_A/i_4 = 4/0.8 = 5\ \Omega$.   ◀

Exercise 3.2   In the circuit of Figure 3.2 let $R = 5\ \Omega$. To what value must the voltage source be changed in order to cause the current source to absorb 0 W?

ANSWER $v_S = 140$ V.

Exercise 3.3   Referring to the circuit of Figure 3.1(a), find the values of v_S that will cause the 3-Ω resistance to dissipate 12 W. (*Hint:* there are two such values. Why?)

ANSWER   17 V, -12 V.

These circuits may look intimidating at first, but once we start solving them we realize that all it takes is Ohm's and Kirchhoff's laws and some patience to make all currents and voltages fit together properly. The key idea is to follow a step-by-step approach rather than to worry about all currents and voltages simultaneously. It may be reassuring to know that this is how engineers work in practice, even when dealing with more sophisticated tasks such as transistor amplifier design.

Resistive Ladder Design

A similar approach can be used to design resistive ladders meeting given specifications under given constraints. Some examples will better illustrate.

▶ **Example 3.3**

Design a 6-stage resistive ladder that, when driven with a 100-V source, yields the following voltages: 50 V, 20 V, 10 V, 5 V, 2 V, and 1 V. All *series* resistances must be 10 kΩ.

Solution

First redraw the ladder as in Figure 3.3, with the branch currents and designated voltages explicitly labeled. The series resistances, all assumed equal to 10 kΩ, have been labeled as R. Our task is to find the shunt resistance values required to meet the specifications. As usual, we start

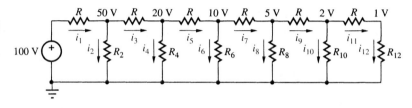

Figure 3.3 Example of resistive ladder design.

from the far right and work our way back toward the source as follows:

(1) By KCL and Ohm's Law, $i_{12} = i_{11} = (2-1)/R = 1/R$. By Ohm's Law, $R_{12} = (1-0)/i_{12} = 1/(1/R) = R = 10$ kΩ.

(2) By KCL and Ohm's Law, $i_{10} = i_9 - i_{11} = (5-2)/10 - (2-1)/10 = 0.2$ mA. Then, Ohm's Law yields $R_{10} = 2/0.2 = 10$ kΩ.

(3) By KCL and Ohm's Law, $i_8 = i_7 - i_9 = (10-5)/10 - (5-2)/10 = 0.2$ mA. By Ohm's Law, $R_8 = 5/0.2 = 25$ kΩ.

(4) By KCL and Ohm's Law, $i_6 = i_5 - i_7 = (20-10)/10 - (10-5)/10 = 0.5$ mA. By Ohm's Law, $R_6 = 10/0.5 = 20$ kΩ.

(5) By KCL and Ohm's Law, $i_4 = i_3 - i_5 = (50-20)/10 - (20-10)/10 = 2$ mA. By Ohm's Law, $R_4 = 20/2 = 10$ kΩ.

(6) By KCL and Ohm's Law, $i_2 = i_1 - i_3 = (100-50)/10 - (50-20)/10 = 2$ mA. By Ohm's Law, $R_2 = 50/2 = 25$ kΩ.

In short, we need $R_2 = R_8 = 25$ kΩ, $R_4 = R_{10} = R_{12} = 10$ kΩ, and $R_6 = 20$ kΩ. ◀

Exercise 3.4 Repeat Example 3.3, but under the constraint that all *shunt* resistances be 20 kΩ.

ANSWER $R_1 = 11.36$ kΩ, $R_3 = 15.79$ kΩ, $R_5 = 11.11$ kΩ, $R_7 = 12.5$ kΩ, $R_9 = R_{11} = 20$ kΩ.

▶ **Example 3.4**

(a) What is the equivalent resistance R_{eq} seen by the source in the circuit of Example 3.3?

(b) Suitably modify the circuit so that $R_{eq} = 100$ kΩ.

Solution

(a) $R_{eq} = v_S/i_1 = 100/[(100 - 50)/10] = 20$ kΩ.

(b) To ensure the same node voltages we must leave all resistance ratios unchanged. This is achieved by multiplying all resistances by the common scale factor

$$\frac{R_{eq(new)}}{R_{eq(old)}} = \frac{100}{20} = 5$$

Thus, $R = 10 \times 5 = 50$ kΩ; $R_2 = R_8 = 25 \times 5 = 125$ kΩ; $R_4 = R_{10} = R_{12} = 10 \times 5 = 50$ kΩ; $R_6 = 20 \times 5 = 100$ kΩ. ◀

Exercise 3.5   Design a 5-stage resistive ladder that, when driven by a 100-V source, yields the following voltages: 10 V, 1 V, 100 mV, 10 mV, and 1 mV. Your ladder must meet the following specifications: (a) the current drawn by the ladder from the source must be 1 mA; (b) the currents through the series resistances must halve as we move from one position to the next toward the right.

ANSWER   90 kΩ, 20 kΩ, 18 kΩ, 4 kΩ, 3.6 kΩ, 800 Ω, 720 Ω, 160 Ω, 144 Ω, 16 Ω.

DC Biasing

Devices like diodes and transistors, the basic ingredients of electronic instrumentation, require prescribed dc currents and voltages to operate properly. Providing these currents and voltages is referred to as **dc biasing.** Resistances play a fundamental role in biasing. As an example, suppose two devices X_1 and X_2 are to be powered by a 15-V source and suppose X_1 requires 2 mA at 5 V, while X_2 requires 0.5 mA at 6 V. Connecting X_1 and X_2 either in series or in parallel with the source would not work because of conflicting voltage and current requirements. To resolve these conflicts, we surround the devices with suitable voltage-dropping, current-setting resistances.

▶ **Example 3.5**

Design a suitable biasing network for the above devices X_1 and X_2.

Solution

Start out with the two devices connected in series; then, as shown in Figure 3.4, use the series resistance R_1 to resolve voltage conflicts and the parallel resistance R_2 to resolve current conflicts.

By KVL, the voltage drop across R_1 is $15 - (6 + 5) = 4$ V, so $R_1 = 4/2 = 2$ kΩ. By KCL, the current through R_2 is $2 - 0.5 = 1.5$ mA, so $R_2 = 6/1.5 = 4$ kΩ. In short, to properly bias X_1 and X_2 we need two resistances: $R_1 = 2$ kΩ and $R_2 = 4$ kΩ. ◀

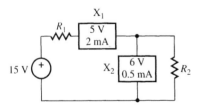

Figure 3.4 Example of resistive biasing.

Exercise 3.6   An alternate way of solving Example 3.5 is to connect X_1 to the 15-V source via a dedicated resistance $R_1 = (15 - 5)/2 = 5\ k\Omega$, and X_2 to the same source via a dedicated resistance $R_2 = (15 - 6)/0.5 = 18\ k\Omega$. Calculate the total power dissipated by the circuit, and verify that this alternative is *more wasteful* of power and, hence, less desirable.

Exercise 3.7   Repeat Example 3.5, but for the case in which the voltage source is 10 V, and under the constraint that the total power dissipated by the circuit be minimized.

3.2 NODAL ANALYSIS

The state of a circuit is known completely once all its branch voltages and currents are known. Two systematic approaches are available to find these variables: the *node method,* discussed in this section, which allows us to find all *node voltages* in a circuit; and the *loop method,* discussed in the next section, which allows us to find all *mesh currents.* Once either set of variables is known, one can readily find all branch voltages and currents in the circuit using the $i-v$ relationships of the individual elements.

To keep things simple, this chapter focuses upon circuits consisting solely of *resistances* and *independent sources.* The more general case of circuits containing *dependent sources* is addressed in Chapter 5.

The Node Method

Nodal analysis allows us to find all unknown *node voltages* in a circuit through the following steps:

(1) Select a *reference* node and ground it. Label all the nodes whose voltages are *unknown,* and use *arrows* with arbitrarily chosen directions to label the currents entering or leaving each of these nodes.

(2) Apply KCL at each *labeled* node, but with labeled currents expressed in terms of *node voltages* via Ohm's Law. In so doing, make sure you relate current directions and voltage polarities correctly, that is, according to the *passive sign convention.*

(3) Solve the resulting set of simultaneous equations for the unknown *node voltages.*

If a *branch voltage* is desired, it can readily be found as the difference between the adjacent node voltages. If a *branch current* is desired, it can be found from the branch voltage via Ohm's Law. If a current comes out negative, this simply means that its direction is opposite to that originally chosen.

Before turning to actual examples, we stress once again the importance of *labeling* the circuit (*reference node,* unknown *node voltages,* and associated *branch currents*) before we start analyzing it. Experience indicates that one of the most frequent reasons beginners get stuck is the failure to adequately label the circuit!

▶ **Example 3.6**

Apply nodal analysis to the circuit of Figure 3.5(a).

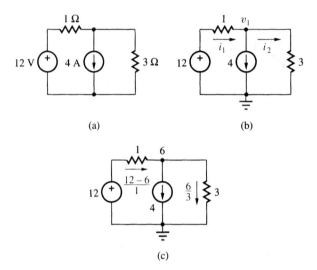

Figure 3.5 Circuit of Example 3.6.

Solution

Ground the bottom node as in Figure 3.5(b). Of the remaining two node voltages, only the one labeled v_1 is unknown, the other being 12 V by voltage source action. We also omit the physical units for simplicity, it being understood that they are [V], [A], and [Ω].

Arbitrarily choosing the current directions as shown, by KCL we have $i_1 = 4 + i_2$. Expressing currents in terms of voltages via Ohm's Law yields

$$\frac{12 - v_1}{1} = 4 + \frac{v_1}{3}$$

Multiplying through by 3 yields $36 - 3v_1 = 12 + v_1$. Collecting yields $(3 + 1)v_1 = 36 - 12$, or

$$v_1 = 6 \text{ V}$$

◀

Exercise 3.8 Repeat Example 3.6, but with the terminals of the current source interchanged so that its current now flows upward.

ANSWER $v_1 = 12$ V.

> ▶ **Example 3.7**

Find the power in each element of the circuit of Example 3.6. Hence, verify that energy is conserved.

Solution

Let us first find the branch currents. Using Ohm's Law we obtain $i_{1\Omega} = (12 - 6)/1 = 6$ A ($\rightarrow$) and $i_{3\Omega} = 6/3 = 2$ A ($\downarrow$). This is shown in Figure 3.5(c). Since the current through the voltage source flows upward, it conforms to the *active* sign convention of Figure 1.5(a). Using $p = vi$, we find that this source is *releasing* $p_{12V} = 12 \times 6 = 72$ W. By contrast, the resistances are *dissipating* power. Using $p = Ri^2$, we find $p_{1\Omega} = 1 \times 6^2 = 36$ W, and $p_{3\Omega} = 3 \times 2^2 = 12$ W. The voltage across the current source is 6 V. Since in this particular case this source conforms to the *passive* sign convention of Figure 1.5(b), it is *absorbing* $p_{4A} = 6 \times 4 = 24$ V. We must have

$$p_{12V} = p_{1\Omega} + p_{3\Omega} + p_{4A}$$

or $72 = 36 + 12 + 24$, thus confirming that energy is conserved. ◀

Exercise 3.9 Find i_1 and i_2 for the circuit of Exercise 3.8. Hence, verify that power is conserved.

ANSWER $i_1 = 0$, $i_2 = 4$ A; $p_{4A} = p_{3\Omega} = 48$ W.

> ▶ **Example 3.8**

Apply nodal analysis to the circuit of Figure 1.11 for the case in which X_1 is a 9-V source with the positive terminal at the top; X_2, X_3, X_4, and X_5 are, respectively, 3-kΩ, 6-kΩ, 4-kΩ, and 2-kΩ resistances; and X_6 is a 5-mA source flowing downward.

Solution

The circuit is shown in Figure 3.6 with all the labels in place and the bottom node as the datum node. Of the three node voltages at the top, only the ones labeled v_1 and v_2 are unknown, the other being 9 V by voltage source action. The SI units are now [V], [mA], and [kΩ].

Applying KCL at the labeled nodes yields, respectively,

$$\frac{9 - v_1}{3} = \frac{v_1}{6} + \frac{v_1 - v_2}{4}$$

$$\frac{v_1 - v_2}{4} = \frac{v_2}{2} + 5$$

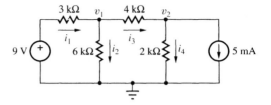

Figure 3.6 Circuit of Example 3.8.

Multiplying through by 12 and 4, respectively, and collecting,

$$3v_1 - v_2 = 12 \tag{3.1}$$

$$v_1 - 3v_2 = 20 \tag{3.2}$$

This system of two equations in two unknowns can be solved either by the *Gaussian elimination method,* or by *Cramer's rule,* as discussed in Appendix 2. We shall illustrate both approaches.

To apply the Gaussian elimination method, multiply Equation (3.1) throughout by -3 and add it pairwise to Equation (3.2). This yields $(-9 + 1)v_1 + (3 - 3)v_2 = (-36 + 20)$, or $v_1 = 2$ V. Back-substituting into Equation (3.1) yields $3 \times 2 - v_2 = 12$, or $v_2 = -6$ V.

Cramer's rule yields

$$v_1 = \frac{\begin{vmatrix} 12 & -1 \\ 20 & -3 \end{vmatrix}}{\begin{vmatrix} 3 & -1 \\ 1 & -3 \end{vmatrix}} = \frac{12(-3) - 20(-1)}{3(-3) - 1(-1)} = 2 \text{ V}$$

$$v_2 = \frac{\begin{vmatrix} 3 & 12 \\ 1 & 20 \end{vmatrix}}{\begin{vmatrix} 3 & -1 \\ 1 & -3 \end{vmatrix}} = \frac{3 \times 20 - 1 \times 12}{3(-3) - 1(-1)} = -6 \text{ V}$$

The fact that v_2 is negative means that the potential at the corresponding node is *lower* than ground potential. ◀

Exercise 3.10   Consider the circuit obtained from that of Figure 3.6 by connecting an additional 1.5-kΩ resistance from the top left node to the top right node. Sketch the modified circuit; hence, find the new values of v_1 and v_2.

ANSWER   4.5 V, 1.5 V.

Checking

How do we know whether our results are correct? When we are solving even moderately complex circuits, the opportunity for mistakes arises in a variety of subtle ways, so we need a reliable check even when the results are obtained with a fancy calculator. A good engineer must not only understand the principles but must also know how to come up with an answer, and one that is correct. Consequently, **the verification of results is just as important as their derivation!**

In the case of nodal analysis, *using the calculated node voltages to find the branch currents and then verifying that they do indeed satisfy KCL* provides a reasonable check. For instance, in Figure 3.5(c), where we found $i_{1\,\Omega} = (12 - 6)/1 = 6$ A $(\rightarrow)$, and $i_{3\,\Omega} = 6/3 = 2$ A $(\downarrow)$, KCL yields $6 = 4 + 2$, which checks.

Suppose, however, that in Example 3.6 we had erroneously written $i_1 = (v_1 - 12)/1$, a common mistake for beginners. Then, the condition $i_1 = 4 + i_2$ would have yielded $(v_1 - 12)/1 = 4 + v_1/3$, whose solution is $v_1 = 24$ V. This, in turn, would have implied $i_{1\,\Omega} = (24 - 12)/1 = 12$ A $(\leftarrow)$, and $i_{3\,\Omega} = 8$ A $(\downarrow)$. With all three currents flowing out of the node, KCL would yield $0 = 12 + 4 + 8$. This untenable result is a warning that an error was committed somewhere. It is our responsibility to go back and recheck our equations to identify the error!

An alternate check is to use $p = vi$ to compute the power of each element and then verify that the *total released power* equals the *total absorbed power,* in the manner of Example 3.7. Though more laborious, this approach has the advantage that it checks *both* the voltages and the currents. Regardless of the approach taken, it is good practice to *always check the validity of the results.*

▶ Example 3.9

Check the results of Example 3.8.

Solution

In Example 3.8 we found $v_1 = 2$ V and $v_2 = -6$ V. Hence, the branch currents are, by Ohm's Law, $i_{3\,k\Omega} = (9 - 2)/3 = 7/3$ mA $(\rightarrow)$, $i_{6\,k\Omega} = 2/6 = 1/3$ mA $(\downarrow)$, $i_{4\,k\Omega} = [2 - (-6)]/4 = 2$ mA $(\rightarrow)$, and $i_{2\,k\Omega} = [0 - (-6)]/2 = 3$ mA $(\uparrow)$.

   KCL at the nodes designated v_1 and v_2 yields, respectively, $7/3 = 1/3 + 2$, and $2 + 3 = 5$. Both check, thus confirming the correctness of our previous results. ◀

Exercise 3.11   Check the results of Exercise 3.10.

▶ Example 3.10

Apply nodal analysis to the circuit of Figure 3.7(a). Check your results.

Solution

Ground the bottom node, label the remaining three node voltages v_1, v_2, and v_3, and assign arbitrary directions to the currents through the

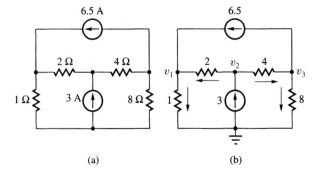

(a)   (b)

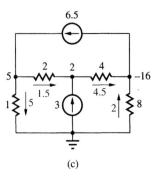

(c)

Figure 3.7 Circuit of Example 3.10 (a) before and (b) after labeling; (c) solved circuit.

resistances. This yields Figure 3.7(b), where the units are [V], [A], and [Ω]. Applying KCL at the three designated nodes, we obtain

$$6.5 + \frac{v_2 - v_1}{2} = \frac{v_1}{1}$$

$$3 = \frac{v_2 - v_1}{2} + \frac{v_2 - v_3}{4}$$

$$\frac{v_2 - v_3}{4} = 6.5 + \frac{v_3}{8}$$

Multiplying through by 2, 4, and 8, respectively, and collecting,

$$v_2 = 3v_1 - 13 \tag{3.3a}$$

$$v_3 = 3v_2 - 2v_1 - 12 \tag{3.3b}$$

$$3v_3 = 2v_2 - 52 \tag{3.3c}$$

Substituting Equation (3.3b) into (3.3c) yields $3(3v_2 - 2v_1 - 12) = 2v_2 - 52$, or

$$7v_2 = 6v_1 - 16 \tag{3.3d}$$

Substituting Equation (3.3a) into (3.3d) and collecting yields $15v_1 = 75$, or $v_1 = 5$ V. Back-substituting into Equation (3.3a) yields $v_2 = 2$ V, and back-substituting into Equation (3.3c) yields $v_3 = -16$ V. In short,

$$v_1 = 5 \text{ V}$$

$$v_2 = 2 \text{ V}$$

$$v_3 = -16 \text{ V}$$

Checking As is shown in Figure 3.7(c), we have, by Ohm's Law, $i_{1\Omega} = 5/1 = 5$ A ($\downarrow$), $i_{2\Omega} = (5 - 2)/2 = 1.5$ A ($\rightarrow$), $i_{4\Omega} = [2 - (-16)]/4 = 4.5$ A ($\rightarrow$), and $i_{8\Omega} = [0 - (-16)]/8 = 2$ A ($\uparrow$).
KCL at the 5-V, 2-V, and −16-V nodes yields, respectively, $6.5 = 5 + 1.5$, $1.5 + 3 = 4.5$, and $4.5 + 2 = 6.5$, all of which check. ◀

Exercise 3.12 Repeat Example 3.10, but with the 6.5-A current source replaced by a 6-Ω resistance.

ANSWER $v_1 = 2.5$ V, $v_2 = 7$ V, $v_3 = 4$ V.

Supernodes

A case requiring special treatment occurs when two labeled nodes are connected by a *voltage source*. By establishing a *constraint* between the corresponding node voltages, the source reduces the number of unknowns by one. However, it also creates a problem because its current is indirectly established by the surrounding elements and, as such, cannot be expressed directly in terms of the source voltage itself.

This difficulty is overcome by regarding the source and its nodes as a *generalized node* or **supernode,** and applying KCL to this generalized node rather than to the two individual nodes. The resulting equation will contain the voltages of both nodes; however, one of them can be eliminated by exploiting the fact that their difference must equal the source voltage itself. Some examples will better illustrate the procedure.

▶ **Example 3.11**

    (a) Apply nodal analysis to the circuit of Figure 3.8(a). Check your results.

    (b) What is the magnitude and direction of the current through the 8-V source? Is the source delivering or absorbing power?

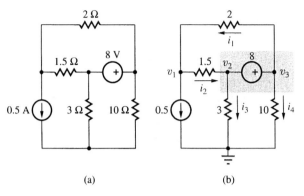

Figure 3.8 Illustrating the supernode technique.

Solution

(a) Surround the voltage source *as well as its nodes* with a dotted line to turn it into a supernode. This is shown in Figure 3.8(b), where the units are [V], [A], and [Ω]. Next, label all essential node voltages and branch currents, as usual. KCL at the node labeled v_1 yields $i_1 = 0.5 + i_2$, or

$$\frac{v_3 - v_1}{2} = 0.5 + \frac{v_1 - v_2}{1.5} \tag{3.4a}$$

Applying KCL at the supernode yields $i_2 = i_3 + i_4 + i_1$, or

$$\frac{v_1 - v_2}{1.5} = \frac{v_2}{3} + \frac{v_3}{10} + \frac{v_3 - v_1}{2} \tag{3.4b}$$

We have three unknowns, but since the 8-V source keeps $v_3 - v_2 = 8$ V, we can substitute

$$v_3 = v_2 + 8 \tag{3.5}$$

into the preceding equations to reduce the number of unknowns to two. After this is done, multiplying Equations (3.4) through, respectively, by 6 and 30, and collecting terms yields

$$v_1 - v_2 = 3$$

$$35v_1 - 48v_2 = 144$$

Solving by the Gaussian elimination method or by Cramer's rule yields $v_1 = 0$ V and $v_2 = -3$ V. Substituting into Equation (3.5) yields $v_3 = 5$ V. To summarize,

$$v_1 = 0 \text{ V}$$

$$v_2 = -3 \text{ V}$$

$$v_3 = 5 \text{ V}$$

Remark   We observe that in this particular case the voltage across the current source is 0 V, confirming that current sources work with any voltage, including zero.

Checking By Ohm's Law, $i_{1.5\,\Omega} = [0 - (-3)]/1.5 = 2$ A ($\rightarrow$), $i_{2\,\Omega} = 5/2 = 2.5$ A ($\leftarrow$), $i_{3\,\Omega} = [0 - (-3)]/3 = 1$ A ($\uparrow$), and $i_{10\,\Omega} = 5/10 = 0.5$ A ($\downarrow$).

KCL at the node labeled v_1 yields $2.5 = 0.5 + 2$, and KCL at the supernode yields $2 + 1 = 0.5 + 2.5$. Both check, confirming that the results are most likely correct.

(b) The current through the 8 V source is $i_{8\text{ V}} = i_{2\,\Omega} + i_{10\,\Omega} = 2.5 + 0.5 = 3$ A ($\rightarrow$). Since it conforms to the *active* sign convention, this source is *delivering* power. ◀

Exercise 3.13

(a) Redraw the circuit of Figure 3.6, but with the 4-kΩ resistance replaced by a 12-V source with the positive terminal at the node labeled v_1. Hence, find v_1 and v_2, and check your results.

(b) What is the power associated with the 12-V source? Is this power released or absorbed?

ANSWER (a) 4 V, -8 V; (b) 12 mW, absorbed.

▶ Example 3.12

Apply nodal analysis to the circuit of Figure 3.9(a).

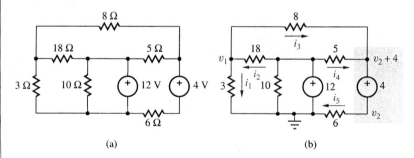

(a)   (b)

Figure 3.9 Nodal analysis using the supernode technique.

Solution

Select the reference node and label all essential node voltages and branch currents as shown in Figure 3.9(b), where the units are [V], [A], and [Ω]. Since the circuit can have only one reference node, the node labeled v_2 becomes an essential node, indicating that the 4-V source forms a supernode. The node at the positive side of the 4-V source is also an essential node. However, instead of labeling it v_3, we directly label it

$v_2 + 4$ in order to minimize the number of unknowns and thus simplify our algebra.

Applying KCL at the node labeled v_1 yields $i_2 = i_1 + i_3$, or

$$\frac{12 - v_1}{18} = \frac{v_1}{3} + \frac{v_1 - (v_2 + 4)}{8}$$

Applying KCL at the supernode yields $i_3 + i_4 = i_5$, or

$$\frac{v_1 - (v_2 + 4)}{8} + \frac{12 - (v_2 + 4)}{5} = \frac{v_2}{6}$$

Multiplying through, respectively, by 72 and 120, and collecting gives

$$37v_1 - 9v_2 = 84$$

$$15v_1 - 59v_2 = -132$$

Solving by the Gaussian elimination method or by Cramer's rule,

$$v_1 = 3 \text{ V}$$

$$v_2 = 3 \text{ V}$$

Clearly, the voltage at the positive side of the 4-V source is $v_2 + 4$, or 7 V.

Checking By Ohm's Law, we have $i_{3\,\Omega} = 3/3 = 1$ A ($\downarrow$), $i_{18\,\Omega} = (12 - 3)/18 = 0.5$ A ($\leftarrow$), $i_{8\,\Omega} = (7 - 3)/8 = 0.5$ A ($\leftarrow$), $i_{5\,\Omega} = (12 - 7)/5 = 1$ A ($\rightarrow$), and $i_{6\,\Omega} = 3/6 = 0.5$ A ($\leftarrow$).

KCL at the node labeled v_1 yields $0.5 + 0.5 = 1$, and KCL at the supernode yields $1 = 0.5 + 0.5$, both of which check.

Remark We observe that the 10-Ω resistance does not intervene in our calculations because it is connected between two *known* node voltages, namely, 12 V and 0 V. All this resistance does is absorb power from the 12-V source, without affecting the remainder of the circuit. ◀

Exercise 3.14 Sketch the circuit of Figure 3.9(a), but with the 8-Ω resistance replaced by a 30-V voltage source, positive at the left. Hence, use nodal analysis to find the voltage across the 3-Ω resistance, and check your result.

ANSWER 19.5 V (positive at the top).

3.3 LOOP ANALYSIS

While nodal analysis allows us to find all unknown *voltages* in a circuit, loop analysis allows us to find all unknown *currents*. We can then use these currents and the v–i relationships of the individual elements to find the branch

voltages. Just as in nodal analysis it was expedient to use *node voltages* as independent variables, even though they do not necessarily coincide with *branch
voltages,* in loop analysis it is expedient to use auxiliary current variables known
as **mesh currents,** even though they do not necessarily coincide with *branch
currents.*

To introduce the mesh current concept, refer to the circuit of Figure 3.10(a),
where the currents through R_1 and R_2 are assumed to flow toward the right. If we
assume the current through R_3 to flow downward, then KCL yields $i_1 = i_{R_3} + i_2$,
or $i_{R_3} = i_1 - i_2$, as is explicitly shown in the figure. The reason for expressing i_{R_3}
in terms of i_1 and i_2 is that this allows us to regard i_1 and i_2 as *circulating* around
the corresponding meshes rather than sinking into or springing from nodes; see
Figure 3.10(b). The advantage of using circulating currents is that they simplify
the application of KVL around a mesh, as we are now about to see.

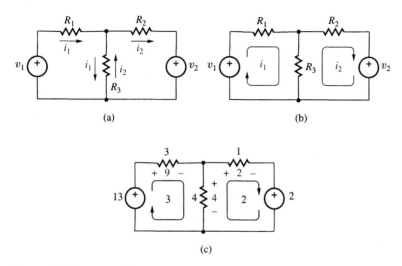

(a)   (b)

(c)

Figure 3.10 Loop analysis: (a) original circuit; (b) labeled circuit; (c) solved
circuit of Example 3.13.

The Loop Method

Loop analysis, applicable to *planar circuits,* or circuits that can be drawn on a
plane with no crossing branches, involves the following steps:

(1) Select a set of meshes such that at least *one mesh* passes through *each
branch.* Label each mesh with the corresponding mesh current, and
assign each mesh current an arbitrarily chosen direction, say, clockwise.

(2) Apply KVL around *each labeled mesh,* but with each branch voltage
expressed in terms of the corresponding mesh currents via Ohm's Law.
Be sure to relate current directions and voltage polarities according to
the *passive sign convention!*

(3) Solve the resulting set of simultaneous equations for the unknown *mesh
currents.*

If a *branch current* is desired, it can readily be found by adding up algebraically all mesh currents passing through that branch. If a *branch voltage* is desired, it can be found from the branch current via Ohm's Law. Before turning to actual examples, we stress once again the importance of properly *labeling* the circuit before we even start analyzing it. Inadequate labeling may easily cause you to get stuck!

▶**Example 3.13**

Apply loop analysis to the circuit of Figure 3.10(b) for the case $v_1 = 13$ V, $v_2 = 2$ V, $R_1 = 3$ Ω, $R_2 = 1$ Ω, and $R_3 = 4$ Ω. What is the current through R_3?

Solution

Our physical units are [V], [A], and [Ω]. Going around the mesh labeled i_1 we encounter a voltage rise v_1 and two voltage drops, $v_{R_1} = R_1 i_1$ and $v_{R_3} = R_3 i_{R_3} = R_3(i_1 - i_2)$. KVL thus yields $v_1 = R_1 i_1 + R_3(i_1 - i_2)$, or

$$13 = 3i_1 + 4(i_1 - i_2)$$

Going around the mesh labeled i_2 we find only voltage drops—$v_{R_3} = R_3(i_2 - i_1)$, $v_{R_2} = R_2 i_2$, and v_2. Applying KVL yields $0 = R_3(i_2 - i_1) + R_2 i_2 + v_2$, or

$$0 = 4(i_2 - i_1) + 1i_2 + 2$$

Collecting terms, we have

$$7i_1 - 4i_2 = 13$$
$$4i_1 - 5i_2 = 2$$

Solving via Gaussian elimination or Cramer's rule yields

$$i_1 = 3 \text{ A}$$
$$i_2 = 2 \text{ A}$$

The branch current through R_3 is $i_{R_3} = i_1 - i_2 = 1$ A ($\downarrow$).   ◀

Exercise 3.15   The circuit of Figure 3.6 has three meshes, with the current around the rightmost mesh being 5 mA and flowing clockwise. Apply the loop method to find the currents through the 6-kΩ and 2-kΩ resistances.

ANSWER   1/3 mA ($\downarrow$), 3 mA ($\uparrow$).

Checking

As usual, we must check our results. This can be done *by using the calculated mesh currents to find the branch currents and, hence, the branch voltages, and then verifying that the latter do indeed satisfy KVL.*

For instance, in Example 3.13 the branch voltages are, by Ohm's Law, $v_{R_1} = R_1 i_1 = 3 \times 3 = 9$ V (positive at the left, abbreviated as + @ left); $v_{R_2} = R_2 i_2 = 1 \times 2 = 2$ V (+ @ left); and $v_{R_3} = R_3 i_{R_3} = 4 \times 1 = 4$ V (+ @ top). This is shown in Figure 3.10(c). KVL around the two meshes yields, respectively, $13 = 9 + 4$ and $4 = 2 + 2$. Both check, indicating that our results are likely to be correct.

Suppose, however, we had erroneously expressed the current through R_3 as $i_{R_3} = i_1 + i_2 = 5$ A, not an uncommon error for beginners. Then, the branch voltages would have been $v_{R_1} = 9$ V, $v_{R_2} = 2$ V, and $v_{R_3} = R_3 i_{R_3} = 4 \times 5 = 20$ V. KVL around the two meshes would have yielded, respectively, $13 = 9 + 20$, and $20 = 2 + 2$. Both identities are untenable, indicating that an error was committed somewhere. It is our responsibility to detect and correct it.

Another way of checking our results is to apply *node analysis* and verify that it yields consistent results with *loop analysis*. Considering the importance of a reliable check, you should never hesitate to try more than one approach!

▶ **Example 3.14**

In Example 3.1 we used Ohm's and Kirchhoff's laws to compute all currents and voltages in order to find v_S in the circuit of Figure 3.1. Now let us follow the inverse approach (namely, assuming $v_S = 17$ V, let us use the loop method to find all branch currents). Then, using Ohm's Law, let us find all branch voltages and check that they obey KVL.

Solution

Redraw the circuit and label the mesh currents as in Figure 3.11(a), where the physical units are [V], [A], and [Ω]. Since the upper mesh current is established by the current source, its value is 1 A, as shown. Going around the mesh labeled i_1 we find only voltage drops. Thus, KVL yields

$$0 = 4(i_1 - 1) + 1(i_1 - i_2) + 17$$

KVL around the mesh labeled i_2 yields

$$17 = 1(i_2 - i_1) + 3(i_2 - 1) + 2i_2$$

Collecting terms, we obtain

$$5i_1 - i_2 = -13$$

$$i_1 - 6i_2 = -20$$

Solving via Gaussian elimination or Cramer's rule yields

$$i_1 = -2 \text{ A}$$
$$i_2 = 3 \text{ A}$$

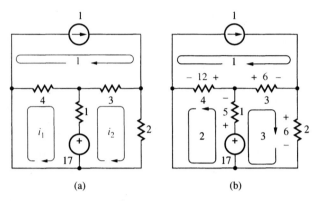

(a)   (b)

Figure 3.11 Circuit of Example 3.14.

Checking   Redrawing the mesh currents as in Figure 3.11(b), we obtain, by Ohm's Law, $v_{4\,\Omega} = 4(2+1) = 12$ V (+ @ right); $v_{1\,\Omega} = 1(2+3) = 5$ V (+ @ bottom); $v_{3\,\Omega} = 3(3-1) = 6$ V (+ @ left); and $v_{2\,\Omega} = 2 \times 3 = 6$ V (+ @ top).

KVL around the bottom meshes yields, respectively, $12 + 5 = 17$, and $17 = 5 + 6 + 6$. Both check, confirming correct results. ◄

Exercise 3.16   Apply the loop method to the circuit of Figure 3.8(a) to find the magnitudes and directions of the currents through the 2-Ω and 10-Ω resistances. Assume the current around the bottom left mesh flows *counterclockwise* to make it coincide with the current of the 0.5-A source. Check your results.

ANSWER   2.5 A ($\leftarrow$), 0.5 A ($\downarrow$).

► Example 3.15

Apply loop analysis to the circuit of Figure 3.12.

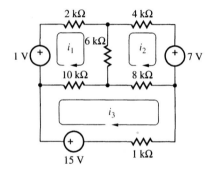

Figure 3.12 Circuit of Example 3.15.

Solution

Applying KVL around the designated meshes yields

$$1 = 2i_1 + 6(i_1 - i_2) + 10(i_1 - i_3)$$

$$0 = 6(i_2 - i_1) + 4i_2 + 7 + 8(i_2 - i_3)$$

$$15 = 10(i_3 - i_1) + 8(i_3 - i_2) + 1i_3$$

Collecting and rearranging,

$$18i_1 - 6i_2 - 10i_3 = 1$$

$$6i_1 - 18i_2 + 8i_3 = 7$$

$$10i_1 + 8i_2 - 19i_3 = -15$$

Solving by Gaussian elimination or Cramer's rule, we obtain $i_1 = 1.5$ mA, $i_2 = 1$ mA, and $i_3 = 2$ mA. ◀

Exercise 3.17 Use KVL to check the results of Example 3.15.

Supermeshes

A case requiring special treatment occurs when two or more unknown mesh currents pass through the same *current source.* By establishing a constraint between these mesh currents, the source reduces the number of unknowns by one. However, it also creates a problem because its voltage, established by the surrounding elements, cannot be expressed directly in terms of the source current.

This difficulty is overcome by altering the choice of the mesh currents so that *only one mesh current passes through the offending current source.* To bypass this source, the remaining meshes will have to be expanded into larger meshes, or **supermeshes.** *Loops* would actually be a better designation; however, the terminology is immaterial as long as we have circulating current paths along which we can apply KVL without encountering unspecifiable branch voltages. Some examples will better illustrate the procedure.

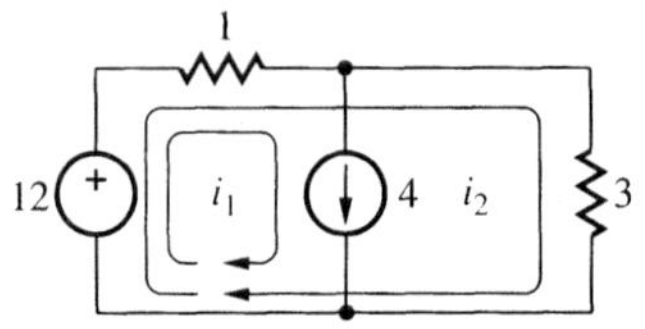

Figure 3.13 Illustrating the supermesh technique.

▶ Example 3.16

Apply loop analysis to the circuit of Figure 3.5.

Solution

Since the circuit contains two meshes sharing the same current source, we let only the first mesh pass through this source and we expand the second mesh into a supermesh or loop to avoid the source. This is shown in Figure 3.13, where the units are [V], [A], and [Ω]. By inspection,

$$i_1 = 4 \text{ A}$$

Moreover, applying KVL around the supermesh yields

$$12 = 1(i_1 + i_2) + 3i_2$$

Solving by the substitution method yields

$$i_2 = 2 \text{ A}$$

Checking   By KCL, $i_{1\Omega} = 4 + 2 = 6$ A ($\rightarrow$), and $i_{3\Omega} = 2$ A ($\downarrow$). By Ohm's Law, $v_{1\Omega} = 1 \times 6 = 6$ V (+ @ left), $v_{3\Omega} = 3 \times 2 = 6$ V (+ @ top). Then, KVL around the loop labeled i_2 yields $12 = 6 + 6$, which checks.   ◀

Exercise 3.18   In the circuit of Figure 3.10(a), let $v_1 = 13$ V, $v_2 = 2$ V, $R_1 = 3$ kΩ, and $R_2 = 1$ kΩ, and let R_3 be replaced by a 1-mA current source flowing upward. Use loop analysis to find the power (delivered or absorbed?) by the 13-V and 2-V sources. Check your results!

ANSWER   32.5 mW, delivered; 7 mW, absorbed.

▶ Example 3.17

Use loop analysis to find the power (released or absorbed?) by each source in the circuit of Figure 3.14(a).

Solution

Expand the upper mesh into a supermesh to avoid passing through the 0.5-mA source twice. This is shown in Figure 3.14(b), where the units are [V], [mA], and [kΩ]. Note also that the current around the bottom right

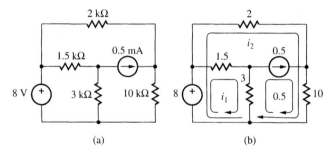

Figure 3.14 Loop analysis using the supermesh technique.

mesh has been directly labeled as 0.5 to avoid introducing an additional variable, which would unnecessarily complicate our algebra.

Applying KVL around the meshes labeled i_1 and i_2 yields

$$8 = 1.5i_1 + 3(i_1 - 0.5)$$

$$8 = 2i_2 + 10(i_2 + 0.5)$$

Solving as usual yields $i_1 = 19/9$ mA and $i_2 = 0.25$ mA.

To find the power of the voltage source we need to know its current $i_{8\text{V}}$. Then, $p_{8\text{V}} = 8 \times i_{8\text{V}}$. By KCL, $i_{8\text{V}} = i_1 + i_2 = 19/9 + 0.25 = 85/36$ mA ($\uparrow$). Since it conforms to the *active* sign convention of Figure 1.5(a), this source is *releasing* $p_{8\text{V}} = 8 \times 85/36 = 170/9$ mW.

To find the power of the current source we need to know the voltage $v_{0.5\text{ mA}}$ across its terminals. Then, $p_{0.5\text{ mA}} = v_{0.5\text{ mA}} \times 0.5$. Using KVL and Ohm's Law, we find $v_{0.5\text{ mA}} = v_{10\Omega} - v_{3\Omega} = 10(0.5 + 0.25) - 3(19/9 - 0.5) = 8/3$ V (+ @ right). Since it conforms to the *active* sign convention, the source is *releasing* $p_{0.5\text{ mA}} = (8/3) \times 0.5 = 4/3$ mW. ◀

Exercise 3.19 Consider the circuit obtained from that of Figure 3.6 by replacing the 6-kΩ resistance with a 5/3-mA current source flowing upward. Use loop analysis to find the power dissipated by the 4-kΩ resistance.

ANSWER 256/9 mW.

Exercise 3.20 Sketch the circuit of Figure 3.14(a), but with the 1.5-kΩ resistance and 0.5-mA source interchanged with each other, the source still pointing toward the right. Then use the loop method to find the voltage across the 3-kΩ resistance and the node method to check your result.

ANSWER 297/74 V (+ @ top).

3.4 LINEARITY AND SUPERPOSITION

A circuit is said to be **linear** if it satisfies the following properties.

(1) **The Scaling Property:** The branch currents and node voltages resulting from a single source in the circuit are *linearly proportional* to the source, indicating that multiplying the value of the source by a given constant multiplies all currents and voltages by the same constant. In particular, setting the source to zero makes all currents and voltages zero. Setting a source to zero is variously referred to as *suppressing, turning off, deactivating,* or *killing* the source.

(2) **The Additive Property:** In a circuit with two or more sources, each branch current and node voltage is the *algebraic sum* of the contributions from each source acting alone (that is, with all other sources set to zero, or suppressed).

All the circuits encountered so far, which consist of sources and resistances, are linear. On the other hand, when studying diodes and transistors, you will see that circuits incorporating these devices generally exhibit a highly nonlinear behavior.

As an example of a linear circuit, let us verify that the voltage v in the circuit of Figure 3.15 satisfies both the above properties. First, let us find the contribution from the voltage source acting alone, that is, the contribution from v_S with $i_S = 0$. Call this contribution v_1. Since $i_S = 0$ represents the i–v characteristic of an *open circuit*, the original network reduces to the subnetwork of Figure 3.16(a), with the current source replaced by an open circuit. Using the voltage divider formula yields

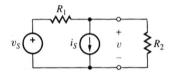

Figure 3.15 A linear circuit with two sources.

$$v_1 = \frac{R_2}{R_1 + R_2} v_S \tag{3.6}$$

indicating that v_1 is linearly proportional to v_S.

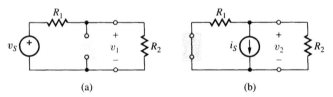

(a)   (b)

Figure 3.16 Subcircuits to find the contributions from each source acting alone.

Next, we find the contribution v_2 from the current source acting alone, that is, the contribution from i_S with $v_S = 0$. Since $v_S = 0$ represents the i–v characteristic of a *short circuit*, we now obtain the subnetwork of Figure 3.16(b), where the voltage source has been replaced with a short circuit. Since i_S is in parallel with R_1 and R_2, Ohm's Law yields $-v_2 = (R_1 \parallel R_2)i_S$, or

$$v_2 = -\frac{R_1 R_2}{R_1 + R_2} i_S \tag{3.7}$$

Clearly, v_2 is linearly proportional to i_S.

Finally, we claim that the voltage v due to sources v_S and i_S acting *simultaneously* is the algebraic sum of their individual contributions,

$$v = v_1 + v_2 \qquad (3.8)$$

To verify this, let us apply nodal analysis to the original circuit of Figure 3.15. Thus, $(v_S - v)/R_1 = i_S + v/R_2$. Collecting terms and using straightforward algebraic manipulation, we can put v in the form

$$v = \frac{R_2}{R_1 + R_2} v_S - \frac{R_1 R_2}{R_1 + R_2} i_S \qquad (3.9)$$

confirming that it satisfies the additive property, $v = v_1 + v_2$.

Exercise 3.21   Show that in the circuit of Figure 3.15 the current flowing through R_1 toward the right is

$$i = \frac{1}{R_1 + R_2} v_S + \frac{R_2}{R_1 + R_2} i_S \qquad (3.10)$$

The Superposition Principle

The preceding considerations indicate that an alternative to the nodal and loop analysis methods is to calculate the contributions from the individual sources acting alone and then to superpose these contributions algebraically. This forms the essence of the **superposition principle.** The advantage of this approach is that it breaks down a complex network into simpler subnetworks, each of which has only one source and is thus easier to solve. The final result is then obtained by adding the individual solutions algebraically. Although we have illustrated the principle for the case of two sources in the circuit, it is readily generalized to an arbitrary number of sources.

While nodal and loop analysis allow us to find *all* voltages or currents in a circuit, the superposition principle is particularly useful when only a *specific* voltage or current is sought, since it may relieve us from having to solve simultaneous equations. Even when nodal or loop analysis is mandatory, we can still use the superposition method to spot-check our results. Another advantage of the superposition approach is that it allows us to compare the relative contributions from the various sources and see which ones affect a given current or voltage the most or the least, once the individual element values are known.

The superposition principle for the case in which the sources in the circuit are only of the *independent* type is summarized as follows:

(1) Label the current or voltage to be found, and indicate its reference direction or polarity explicitly.

(2) Find the contribution from each independent source acting alone, that is, with all other independent sources *set to zero*, or *suppressed*. Note that suppressing a *voltage source* means replacing it with a *short circuit*,

while suppressing a *current source* means replacing it with an *open circuit,*

$$\text{voltage source} \rightarrow \text{short circuit} \qquad \textbf{(3.11a)}$$

$$\text{current source} \rightarrow \text{open circuit} \qquad \textbf{(3.11b)}$$

Note also that a contribution along the chosen reference direction or polarity may be either positive or negative.

(3) Add all contributions *algebraically* (that is, with the signs obtained in step 2) to find the desired current or voltage.

The case when the circuit also contains *dependent* sources requires special care and is discussed in Section 5.2.

Clearly, the superposition principle is a direct consequence of the linearity property. This property is also expressed by saying that in a linear circuit each current and voltage is a *linear combination* of all independent sources in the circuit. With reference to our example of Figure 3.15, this statement is expressed by saying that v must be of the type

$$v = a_1 v_S + a_2 i_S \qquad \textbf{(3.12)}$$

where a_1 and a_2 are suitable coefficients that depend on the resistances but not on the sources. According to Equation (3.9), $a_1 = R_2/(R_1 + R_2)$ and $a_2 = -(R_1 \parallel R_2)$. Note that a_1 is dimensionless because v and v_S have the same dimensions, and a_2 has the dimensions of a resistance because v is a voltage and i_S is a current.

Likewise, the current through R_1 can be expressed as

$$i = a_3 v_S + a_4 i_S \qquad \textbf{(3.13)}$$

where, by Equation (3.10), $a_3 = 1/(R_1 + R_2)$ and $a_4 = R_2/(R_1 + R_2)$. In this case a_3 has the dimensions of a conductance (Ω^{-1}), and a_4 is dimensionless.

Verifying that the calculated coefficients have the *correct physical dimensions* provides a good check for our calculations. For example, suppose in Exercise 3.21 you had erroneously obtained $i = [R_1/(R_1 + R_2)]v_S + [R_2/(R_1 + R_2)]i_S$. Noting that the coefficient of v_S is dimensionless instead of having the dimensions of a conductance indicates that an error was committed somewhere.

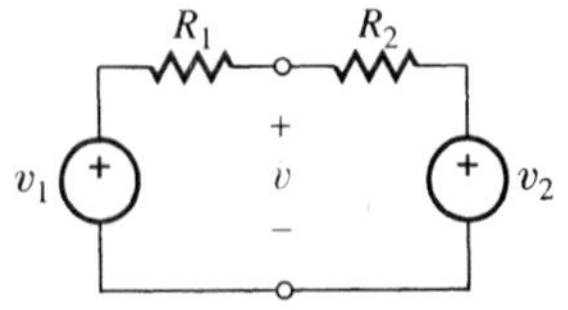

Figure 3.17 Voltage summing circuit.

▶ Example 3.18

(a) Using the superposition principle, derive an expression for v in the circuit of Figure 3.17.

(b) Discuss the special case $R_1 = R_2$.

Solution

(a) Replacing v_2 with a short circuit leaves us with a voltage divider circuit, so the contribution from v_1 is $v_x = [R_2/(R_1 + R_2)]v_1$.

Likewise, suppressing v_1 we find that the contribution from v_2 is $v_y = [R_1/(R_1 + R_2)]v_2$. Letting $v = v_x + v_y$ we obtain

$$v = \frac{R_2}{R_1 + R_2}v_1 + \frac{R_1}{R_1 + R_2}v_2 \qquad (3.14)$$

indicating that the circuit yields a *weighted sum* of v_1 and v_2.

(b) If $R_1 = R_2$, Equation (3.14) simplifies as

$$v = \frac{v_1 + v_2}{2} \qquad (3.15)$$

indicating that now the circuit yields the *mean* of v_1 and v_2. ◀

Exercise 3.22

(a) Applying the superposition principle to the circuit of Figure 3.10(a), show that the voltage across R_3 is

$$v = \frac{v_1}{1 + R_1/R_2 + R_1/R_3} + \frac{v_2}{1 + R_2/R_1 + R_2/R_3} \qquad (3.16)$$

(b) If $R_1 = R_2 = 10 \text{ k}\Omega$, find the value of R_3 that will yield $v = 0.1(v_1 + v_2)$.

ANSWER   (b) 1.25 kΩ.

▶ Example 3.19

Using the superposition principle, find the current i through the voltage source of Figure 3.18.

Solution

Suppressing the sources we obtain the subcircuits of Figure 3.19, where we have omitted the physical units for simplicity, it being understood that they are [V], [A], and [Ω].

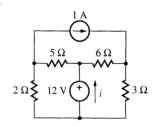

Figure 3.18 Circuit of Example 3.19.

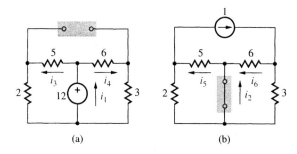

Figure 3.19 Subcircuits arising when applying the superposition principle.

Using KCL in Figure 3.19(a) we obtain $i_1 = i_3 + i_4 = 12/(2 + 5) + 12/(6 + 3) = (12/7 + 4/3)$ A.

Using KCL in Figure 3.19(b) we obtain $i_2 + i_6 = i_5$, or $i_2 = i_5 - i_6$. By the current divider formula, $i_6 = 1 \times 3/(6 + 3) = 1/3$ A, and $i_5 = 1 \times 2/(2 + 5) = 2/7$ A. Thus, $i_2 = (2/7 - 1/3)$ A.

Finally, by the superposition principle, $i = i_1 + i_2 = 12/7 + 4/3 + 2/7 - 1/3 = 3$ A ($\uparrow$). ◀

Exercise 3.23   Using the superposition principle, find the voltage across the current source of Figure 3.18.

ANSWER   4 V (+ @ right).

▶ **Example 3.20**

Using the superposition principle in the circuit of Figure 3.6, find the contributions to v_1 and v_2 due, respectively, to the 9-V source and the 5-mA source.

Solution

Suppressing the current source, we use the voltage divider formula to find the contributions due to the 9-V source,

$$v_{1\,(9\,V)} = \frac{6 \parallel (4 + 2)}{3 + [6 \parallel (4 + 2)]} 9 = 4.5 \text{ V}$$

$$v_{2\,(9\,V)} = \frac{2}{4 + 2} 4.5 = 1.5 \text{ V}$$

Suppressing the voltage source, the contributions due to the 5-mA current source are found via Ohm's Law and the voltage divider formula,

$$v_{2\,(5\,mA)} = -[2 \parallel [4 + (3 \parallel 6)]] 5 = -7.5 \text{ V}$$

$$v_{1\,(5\,mA)} = \frac{3 \parallel 6}{(3 \parallel 6) + 4} (-7.5) = -2.5 \text{ V}$$

Then, $v_1 = 4.5 - 2.5 = 2$ V, and $v_2 = 1.5 - 7.5 = -6$ V. ◀

Exercise 3.24   Using the superposition principle in the circuit of Figure 3.8(a), find the contributions to the voltage across the 2-Ω resistance due to the 0.5-A source and the 8-V source.

ANSWER   3/7 V and 32/7 V (both + @ right).

▶ Example 3.21

Using the superposition principle, find v in the circuit of Figure 3.20(a).

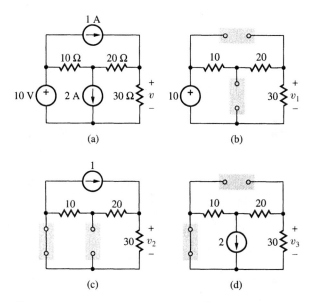

Figure 3.20 Circuits and subcircuits of Example 3.21.

Solution

To find the contribution from one source at a time with the other two sources suppressed, we use the subcircuits of Figure 3.20(b) through (d), where the units are [V], [A], and [Ω].

Applying the voltage divider formula to the circuit of Figure 3.20(b) yields $v_1 = [30/(10 + 20 + 30)]10 = 5$ V.

In the circuit of Figure 3.20(c) we apply the current divider formula to find $i_{30\,\Omega} = [(10 + 20)/(10 + 20 + 30)]1 = 0.5$ A($\downarrow$). Hence, $v_2 = 30 \times 0.5 = 15$ V.

Applying again the current divider formula to the circuit of Figure 3.20(d), we find $i_{30\,\Omega} = [10/(10 + 20 + 30)]2 = 1/3$ A($\uparrow$). Thus, $v_3 = -30 \times 1/3 = -10$ V.

Finally, $v = v_1 + v_2 + v_3 = 5 + 15 - 10 = 10$ V. ◀

Exercise 3.25 Apply the superposition principle to the circuit of Figure 3.20(a) to find the voltage across the 2-A current source.

ANSWER 10/3 V (+ @ bottom).

WHAT CAN I DO?

Naturally you're interested in electricity and electronics, or else why would you be studying electrical engineering, right? What about when school is over, though? What plans have you made for getting your first job? What do EEs actually *do* out there in the real world? It comes as no surprise that a majority are hired into the computer and aerospace industries, but there are many other paths you can take.

Before we go into the details of these jobs, one word of advice. Time and again people in personnel offices say the same thing: "Experience helps." Thus, try to get as much experience as you can while you are still in school. Even if it is far afield from the area you finally decide to go into, your work experience will definitely count.

Precision Engineering. If you are the type of person who sweats the details, the type to whom the thought "it really doesn't have to be *that* good" is like someone scratching fingernails along a chalkboard, precision engineering may be your niche.

In precision engineering, manufacturing is being done at the limits of measurement. Originally the demand for such close work was in the military and in aerospace projects, but today the field takes in medical equipment, scientific instruments, and consumer products. One very familiar consumer example is CD players, a product nonexistent ten years ago.

Lasers are finding applications in a wide range of fields, from telecommunications to medicine. (Larry Mangino/The Image Works)

Patent Law. Despite Jack Kilby's plea to the contrary (Chapter 10), some EEs go on to law school because the EE/law degree combination is a powerful tool much sought after by companies both large and small. Whenever a new idea is conceived in a research laboratory, one of the first steps is down the hall to the company's patent experts so that no money is wasted on already-patented ideas. Naturally these experts must understand the details of the idea in order to research it in patent law. In addition to the private sector, the U.S. Patent Office is always looking for a few good lawyer-engineers.

Cellular Telephones. Nicknamed the "Wild West" of engineering, the cellular telephone industry is technology on the edge. The first cellular phone was sold in 1983 and quickly (and wrongly) tagged as merely a toy for rich folks. Just ten years later, there are more than 10 million in use in the United States alone, and new customers are buying these "toys" at the rate of 7300 per DAY! In this new and explosive field, there are no rule books because none of what needs to be done has ever been done before. Engineers have to be willing to work without the guidance of theory in essentially unstructured jobs and get their hands dirty on every detail from the piece of sand that becomes the chip to the finished phone ready to be turned on.

Security Engineering. Breaches of security are

a fact of modern life, and this sad fact means that the security engineering business is booming. EEs are involved in the protection of U.S. embassies from terrorist attack, in the detection of electronic theft in banks and on Wall Street, in the foiling of would-be highjackers at the airport. Electronic-security systems

▼

*"What do EEs actually do
out there in the real world?"*

▲

for residential and business buildings are becoming more and more common. And, of course, if your political framework allows it, the NSA, CIA, and FBI want YOU!

Public Service. There is a great lack of power plants, telephone networks, and computers in third-world countries. If you feel these things make for a better standard of living and therefore want to help these countries develop twentieth-century technology, you might consider a career in public service. This route is probably not the best bet for brand-new engineers because public service organizations, such as the World Bank and Catholic Relief Services, usually do not hire inexperienced people. These organizations have very limited funds to work with and consequently cannot spend any on training green recruits.

Artificial Intelligence. Most of the people who work in this field refuse to define what the term "artificial intelligence" means, claiming that to do so would needlessly narrow the field's focus. This is a clear indication that here is an extremely new field, ranging wide and moving fast. Industry in the U.S. spends about $300 million annually on AI expert systems (computers that can "think" about a problem the way an expert would). So if words like "hypertext" and "virtual reality" get your brain to churning and your fingers itching to attack a problem, AI may be your field.

The hottest area of AI research these days is neural networks, by which is meant ways of arranging a

computer's power system so that it mimics the way the human brain works, that is, able to process different kinds of information simultaneously. Traffic analysis by AI is almost here, especially on the highways of California. An AI monitor on your dashboard could be told to analyze the data on the *immediate* traffic on every possible road and give you perfect information about which is the best route.

Power Delivery. In the days before electronics, electrical engineers worked mainly in the power industry, designing systems to deliver electrical energy to homes and factories: source to substation and then through a distribution grid of transformers and feeders to the consumer. As electricity use has risen with an ever-increasing population and as the infrastructure has aged, maintaining that delivery system has become an enormous job.

If the power industry had wires that were superconducting at ambient temperatures, we could essentially cut to zero any power losses during transmission. Another application of superconductors is in all the high-tech equipment hospitals use these days. A glance at the front page of almost any newspaper tells us about the crisis in health-care costs,

▼

*"Time and again, people in
personnel offices say the same
thing: 'Experience helps.'"*

▲

and an engineer working with superconductors has a chance to help solve this problem. For instance, a nuclear magnetic resonance machine, used in medicine for magnetic resonance imaging, costs between $1.5 and $2.5 million. Using a high-temperature superconducting magnet in one can cut that cost in half.

On the portable side of power lie all the challenges of better batteries for electric cars, batteries

(Continued)

(Continued)
that will not pollute the environment when they are thrown into the trash, and rechargeable batteries that have a lifetime measured in years rather than in hours. On another branch of the power tree, exciting work is going on in fuel cell technology—at Bell Labs, for instance, where Christopher Dyer recently invented a way of making oxygen-hydrogen cells without the expense of keeping the two gases separate from each other.

If you are one of those people who feel that the problems of our polluted planet are the most pressing ones facing technology in the next century, consider

going into photovoltaic research. Given enough good weather, photovoltaics can power our automobiles, light our homes and offices, run our computers, and perform myriad other tasks that now require the burning of fossil fuels. The only catch has been their very high cost. Texas Instruments has come up with a way of making the cells, which convert sunlight directly to electricity, out of low-purity (and therefore cheap) silicon. Developments like this silicon cell are bringing costs down and giving momentum to this research field, a field that holds out the tantalizing promise of someday freeing us from our petroleum habit.

▶ Example 3.22

In the circuit of Figure 3.15, let $v_S = 12$ V, $R_1 = 1$ kΩ, $i_S = 2$ mA, and $R_2 = 3$ kΩ. Using the superposition principle, find the power dissipated in R_1.

Solution

To find the power we must first find the current i through R_1. Then, $p = R_1 i^2$. By Equation (3.10) we have $i = i_1 + i_2$, where $i_1 = 12/(1 + 3) = 3$ mA and $i_2 = 2 \times 3/(1 + 3) = 1.5$ mA. Thus, $i = 3 + 1.5 = 4.5$ mA, and $p = 1 \times 4.5^2 = 20.25$ mW.   ◀

We observe that the superposition principle is not applicable to resistive power because the latter is a quadratic function and, as such, it is *nonlinear.* Indeed, letting $p_1 = R_1 i_1^2 = 1 \times 3^2 = 9$ mW and $p_2 = R_1 i_2^2 = 1 \times 1.5^2 = 2.25$ in Example 3.22, we immediately see that $p_1 + p_2 \neq p$. However, even though resistive power is a nonlinear function, we can still use the superposition principle to find resistive currents or voltages, which *are* linear, and then use these currents or voltages to find power.

Exercise 3.26   Apply the superposition principle to the circuit of Figure 3.18 to find the power dissipated in the 2-Ω resistance.

ANSWER   2 W.

Concluding Observation

Looking back at the various examples and exercises, we note that the subcircuits arising from the application of the superposition principle can often be solved by inspection, using simple techniques such as *series* or *parallel resistance combinations,* and the *voltage* or *current divider formulas.* Thus, once you have decomposed a circuit into its various subcircuits, learn to *pause a moment* and search for hidden series/parallel combinations or voltage/current dividers which may help you find the solution more easily and quickly. To this end, don't forget that the *voltage divider* formula is applicable only if the resistances are in *series,* and the *current divider* formula only if they are in *parallel.*

3.5 SOURCE TRANSFORMATIONS

The source transformation technique exploits the equivalence between a voltage source in series with a resistance and a current source in parallel with a resistance to effect suitable circuit manipulations to simplify circuit analysis.

To illustrate, consider the situation of Figure 3.21(a), which contains a subcircuit consisting of a *voltage source* v_S in *series* with a resistance R_s. By Ohm's Law we have $i = (v_S - v)/R_s$, or

$$i = \frac{v_S}{R_s} - \frac{v}{R_s} \tag{3.17}$$

Consider now the situation of Figure 3.21(b), which contains a subcircuit consisting of a *current* source of value v_S/R_s in *parallel* with a resistance R_s. By KCL and Ohm's Law we have $i = v_S/R_s - v/R_s$, or again Equation (3.17). Since the two subcircuits yield the same i–v relationship, they are said to be **equivalent.** The operation of the rest of the circuit is unaffected if we replace the *series* subcircuit of Figure 3.21(a) with the *parallel* subcircuit of Figure 3.21(b).

We note that for this transformation to be acceptable, the value of the target source must be related to that of the original source as v_S/R_s and the target parallel resistance must have the same value as the original series resistance. Moreover, the positive terminal of the voltage source must be the same as that corresponding to the arrowhead in the current source.

By similar reasoning, the *parallel* subcircuit of Figure 3.22(a) can be replaced with the *series* subcircuit of Figure 3.22(b) without affecting the operation of the rest of the circuit. To substantiate this claim, apply Ohm's Law and KCL to the circuit of Figure 3.22(a) to obtain

$$i = i_S - \frac{v}{R_s} \tag{3.18}$$

Next, apply Ohm's Law to the circuit of Figure 3.22(b) to obtain $i = (R_s i_S - v)/R_s = i_S - v/R_s$, or again Equation (3.18). We again observe that for the two subcircuits to be equivalent, the value of the target source must be

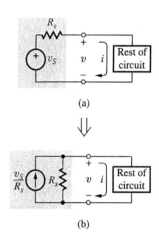

Figure 3.21 Voltage source to current source transformation.

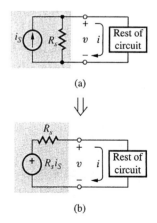

Figure 3.22 Current source to voltage source transformation.

related to that of the original source as $R_s i_S$ and the target series resistance must have the same value as the original parallel resistance. Moreover, the arrowhead inside the current source must face the same terminal as the positive terminal of the voltage source.

It should be clear that the above transformations can be applied only to *source-resistance combinations,* not to ideal sources alone. It is thus fair to say that a practical source, be it a voltage or a current source, admits *two* models.

As an example, a 12-V car battery with a 0.05-Ω internal resistance can be modeled either as a source $v_S = 12$ V with a *series* resistance $R_s = 0.05$ Ω or as a source $i_S = 12/0.05 = 240$ A with a *parallel* resistance $R_s = 0.05$ Ω. We observe that though the two circuits are equivalent at the terminals, they are not necessarily equivalent from an energetic viewpoint. For instance, in the absence of any load, both models yield $v = 12$ V. However, although the voltage source model would in this case dissipate no energy internally, the current source model would dissipate in its internal resistance the power $p = R_s i_S^2 = 0.05 \times 240^2 = 2.88$ kW!

Source transformations can sometimes be exploited to analyze circuits without the need for nodal or loop equations. In applying these transformations, care must be exercised that current direction and voltage polarity be consistent.

▶ Example 3.23

Using source transformations, derive a relationship for v_O in terms of v_S in the ladder network of Figure 3.23.

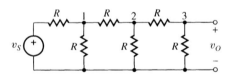

Figure 3.23 Circuit of Example 3.23.

Solution

Start from the left and work your way toward the right, one step at a time. The series combination at the left of node 1 can be converted to a parallel combination as in Figure 3.24(a). Reducing the two parallel resistances yields the combination of Figure 3.24(b), which can then be reconverted to the series form of Figure 3.24(c). Reducing the two series resistances leads, in turn, to the combination of Figure 3.24(d). Repeating the process yields the final circuit of Figure 3.24(h), where we apply the voltage divider formula to obtain $v_O = (v_S/5) \times R/(8R/5 + R)$, or

$$v_O = \frac{1}{13} v_S$$

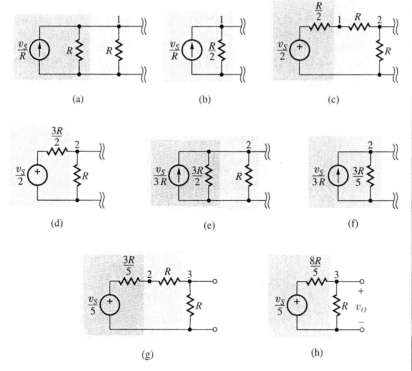

Figure 3.24 Solving the circuit of Figure 3.23 by successive source transformations.

Exercise 3.27   (a) Starting at the left and working your way toward the right, apply suitable source transformations to the circuit of Figure 3.6 to find the *voltage across* the current source. (b) Starting from the right and working your way toward the left, find the *current through* the voltage source.

ANSWER   (a) 6 V (+ @ bottom); (b) 7/3 mA (↑).

▶**Example 3.24**

Apply suitable source transformations to the circuit of Figure 3.25(a) to find the voltage v across the 10-Ω resistance.

Solution

Start at the far sides of the circuit and work your way toward the 10-Ω resistance, one step at a time.

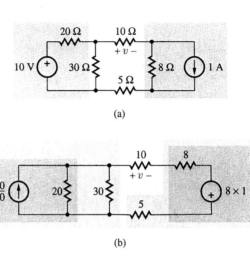

(a)

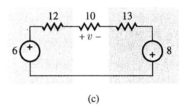

(b)

(c)

Figure 3.25   (a) Circuit of Example 3.24; (b), (c) step-by-step simplification via successive source transformations.

As shown in Figure 3.25(b), the 10-V source and 20-Ω series resistance are transformed to a $10/20 = 0.5$-A source and a 20-Ω parallel resistance. Moreover, the 1-A source and 8-Ω parallel resistance are transformed to an $8 \times 1 = 8$-V source and an 8-Ω series resistance. The positive terminal is at the bottom.

The 20-Ω and 30-Ω resistances are combined into a single $20 \parallel 30 = 12$-Ω resistance. This is then used to convert the 0.5-A source to the $12 \times 0.5 = 6$-V source and 12-Ω series resistance of Figure 3.25(c). Moreover, the 5-Ω and 8-Ω resistances are combined into a single $5 + 8 = 13$-Ω resistance.

It is now a straightforward matter to apply the generalized voltage divider formula to obtain

$$v = \frac{10}{12 + 10 + 13}[6 - (-8)] = 4 \text{ V}$$

◀

Exercise 3.28   Apply suitable source transformations to the circuit of Figure 3.25(a) to find the power dissipated by the 30-Ω resistance.

ANSWER   48 mW.

Exercise 3.29   Apply suitable source transformations to the circuit of Figure 3.6 to find the voltage across the 4-kΩ resistance.

ANSWER   8V (+ @ left).

Analysis Techniques Comparison

So far we have studied four different circuit analysis techniques: the *node* method, the *loop* method, the *superposition principle,* and *source transformations.* Which method is preferable depends on the nature of the circuit as well as the objectives of the problem.

The node or loop methods are generally used when the *complete* state of a circuit is sought. Nodal analysis is generally preferable when the number of unknown node voltages is less than that of unknown mesh currents, and loop analysis is preferable when the number of unknown mesh currents is less than that of unknown node voltages.

Quite often, however, we are only interested in a *specific* voltage or current in the circuit, not in the complete state. In this case then, nodal analysis is preferred when a voltage is sought and loop analysis when a current is sought, all other things being the same. An alternate approach exploits the superposition principle to break down the circuit into simpler subcircuits, which can then be solved via the node or loop methods, not to mention even more direct techniques such as voltage/current division or resistance reductions.

Certain circuits can be solved using exclusively source transformation techniques. Even when circuit topology does not allow this, it is often possible to reduce at least *part* of a circuit via source transformations and then complete the analysis of this simplified circuit using one of the other methods.

As a general rule, before embarking upon the analysis of a circuit, you should develop the habit of pausing a moment to scrutinize the circuit and develop a strategy. Though this may sound forbidding at first, you will find that as you grow in experience you will soon be able to pick the method best suited to the problem at hand. You may find it interesting to know that working engineers describe circuits using almost exclusively nodal equations rather than loop equations. However, for the time being, make sure you become proficient in all four methods, and that you use one method to solve the problem and an alternate method to check the results.

3.6 CIRCUIT ANALYSIS USING SPICE

The techniques of the previous sections allow us to analyze any network. However, as circuit complexity increases, so do the calculations. The amount of time and effort required may soon become prohibitive. When this is the case we resort to special computer programs to perform circuit analysis automatically.

SPICE

A widely used program for the automatic analysis of circuits is the *Simulation Program with Integrated Circuit Emphasis* (SPICE). This program is capable of performing the dc, ac, and transient analysis of circuits containing dependent and independent sources, resistors, capacitors, inductors, mutual inductances, and transmission lines, as well as the four most common semiconductor devices—that is, diodes, BJTs, JFETs, and MOSFETs.

SPICE is available on mainframe computers at most colleges and universities. Check with the computing center of your institution to find how to direct the computer to print out the *SPICE User's Guide,* a document of less than a hundred pages containing the complete description of the program as well as several examples.

SPICE has also been adapted for use with personal computers. One such widely used program is PSpice[TM].[1]   A classroom version of this program is available to students and educators free of charge, and it can be duplicated without restrictions.   Although capable of analyzing only circuits of limited complexity, the classroom version is adequate for most introductory courses in circuits and electronics. All SPICE examples in this book have been run on an inexpensive personal computer using the classroom version of PSpice. To secure your own copy of this version, contact the Microsim Corporation or, better yet, ask around your upperclass peers who may already have it.

In order to run SPICE, we must create an **input file** consisting of (1) **element statements** describing circuit topology and element types and values, and (2) **control statements** or **commands** defining the type of analysis and printout desired. SPICE then analyzes the circuit using numerical techniques the user generally need not be concerned with and presents the results in an **output file,** which can be displayed on the screen or printed out on paper.

Learning to create the input file is a straightforward process that is best illustrated by way of examples. To avoid cramming too much information at once, we shall introduce SPICE gradually, by discussing only the capabilities needed at the time. In this chapter we limit ourselves to the dc analysis of circuits containing independent dc sources and resistances. Additional circuit elements as well as types of analysis will be introduced as we proceed. Although our coverage of SPICE is adequate for the level of this book, you eventually want to secure your own copy of the aforementioned *SPICE User's Guide* for a comprehensive and more systematic description of the program, or you may wish to consult the **Student SPICE Manual** by James S. Kang.[2]

An Illustrative Example

We shall illustrate the use of SPICE to find all node voltages in the circuit of Figure 3.6, which has already been the subject of a number of examples and exercises.

First, we label all nodes 0, 1, 2, 3, ..., *n*, with the reference node being node 0 by definition. This is shown in Figure 3.26.

---

[1]PSpice is a trademark of the Microsim Corporation, 20 Fairbanks, Irvine, CA 92718. Tel. (714) 770-3022. Technical support: (714) 837-0790.

[2]James S. Kang, *Student SPICE Manual,* Saunders College Publishing, Philadelphia, PA, 1994.

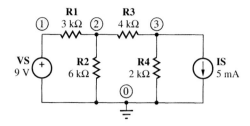

Figure 3.26 Preparing a circuit for analysis via SPICE.

Next, we create a file consisting of element statements and control statements. SPICE requires that the first statement be a **title statement,** and that the last statement be an .END statement. As shown in Table 3.1, we have chosen OUR FIRST SPICE EXAMPLE as the title statement; however, a title like CIRCUIT OF FIGURE 3.26 would have been just as legitimate. As reminders of particular features of our file, we can intersperse it with *comment statements,* which must begin with an asterisk. In Table 3.1 the next to the last statement is a comment statement.

TABLE 3.1 Input File for the Circuit of Figure 3.26

```
OUR FIRST SPICE EXAMPLE
R1 1 2 3K
R2 2 0 6K
R3 2 3 4K
R4 3 0 2K
VS 1 0 DC 9
IS 3 0 DC 5M
* DC ANALYSIS PERFORMED AUTOMATICALLY
.END
```

We are now ready to specify the type of elements forming our circuit, the manner in which they are interconnected, and their values. The element types in our example are **resistors** and **independent dc sources.**

Resistors

The general form of the element statement for a resistance is

RXXX N1 N2 VALUE   **(3.19)**

This statement, illustrated in Figure 3.27(a), contains:

(1) The name of the resistance, an alphanumeric string of up to eight characters which must begin with the letter R. Here, XXX denotes an arbitrary alphanumeric string of up to *seven* characters uniquely identifying the particular resistance.

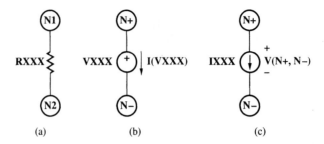

Figure 3.27 SPICE conventions for resistors and independent sources.

(2) The circuit nodes N1 and N2 to which the resistance is connected. Their order is unimportant.

(3) The resistance VALUE, in ohms.

In Table 3.1 we have called our resistances R1, R2, R3, and R4. In subsequent examples we shall use names like RIN, ROUT, REQ, and so forth.

Independent DC Sources

The general forms of the statements for independent dc sources are

VXXX N+ N- DC VALUE   **(3.20a)**

IXXX N+ N- DC VALUE   **(3.20b)**

These statements, illustrated in Figure 3.27(b) and (c), contain:

(1) The name of the source, which must begin with the letter V or I, depending on the source type, and which may contain up to seven additional alphanumeric characters.

(2) The circuit nodes N+ and N- to which the source is connected. For a voltage source, N+ and N- are the positive and negative terminals; for a current source, N+ is the terminal from which the source sinks current and N- the terminal from which it sources current; SPICE always assumes that positive current flows from N+, through the source, to N-. The current through a voltage source is denoted as I(VXXX), the voltage across a current source as V(N+, N-).

(3) The qualifier DC to indicate sources of the dc type.

(4) The source VALUE, in volts or amperes.

In Table 3.1 we have called our sources VS and IS, but names like V1 and I1 or VCC and IEE would have been just as legitimate. Note that in the case of sources the *order* in which the pair of nodes is specified *is important*.

Scale Factors

When expressing element values in terms of powers of ten, we can use either the symbolic form or the exponential form of Table 3.2. Note that M stands for *milli,* not *mega.* For instance, the third and seventh lines in Table 3.1 could also have been expressed as R2 2 0 6E3, and IS 3 0 DC 5E-3.

TABLE 3.2 SPICE Notation for Scale Factors

Value	Symbolic form	Exponential form
10^{-15}	F	1E−15
10^{-12}	P	1E−12
10^{-9}	N	1E−9
10^{-6}	U	1E−6
10^{-3}	M	1E−3
10^{3}	K	1E3
10^{6}	MEG	1E6
10^{9}	G	1E9
10^{12}	T	1E12

Automatic DC Analysis

Having completed the description of the circuit, the next step is to include the control statements that will tell SPICE what type of analysis and printout we want. If these statements are omitted, SPICE will *automatically* perform the dc analysis and print out a table of *all node voltages*. This is what we have chosen to do in Table 3.1, and the comment statement is a remainder of this. If the PSpice version is used, the output file will also contain the *currents through the voltage sources* as well as the total power supplied by these sources and dissipated in the circuit.

Our input file is now ready to be processed by SPICE. The procedure to run SPICE depends on the particular system you are using. Probably the quickest way is to find out from your upperclass peers. Once you have learned to run SPICE, the computer will produce an output listing with the information of Table 3.3.

TABLE 3.3 Output Listing for the SPICE Example of Table 3.1

```
DC ANALYSIS:
NODE   VOLTAGE    NODE   VOLTAGE    NODE    VOLTAGE
(1)    9.0000     (2)    2.0000     (3)     -6.0000

VOLTAGE SOURCE CURRENTS:
NAME             CURRENT
 VS              -2.333E-03

TOTAL POWER DISSIPATION   2.10E-02   WATTS
```

It is a small triumph to realize that the data agree with those derived by hand analysis in Example 3.8, and that the magnitude of the current delivered by the source VS agrees with Exercise 3.27. This current is negative because SPICE assumes positive current to flow from N+, through the source, to N−. The power delivered by VS is, of course, $9 \times 2.333 = 21$ mW.

Try running the example of Table 3.1 on your own system. Once you have succeeded, change the values of some of its elements, rerun the modified file, and verify the results by hand analysis. You are now ready to run your own SPICE files!

Exercise 3.30   Using SPICE, find the node voltages of the circuit of Figure 3.7(a). Then compare them with the results of Example 3.10.

Exercise 3.31   Use SPICE to confirm the results of Example 3.13.

Dummy Voltage Sources

At times we may wish to find the value of a particular branch current in the circuit. Since SPICE calculates only currents through independent voltage sources, we can overcome this limitation by the artifice of inserting a 0-V *dummy voltage source* in series with the given branch. This source acts as an ideal ammeter and it does not perturb the existing conditions in the circuit, yet it allows us to monitor the given branch current.

 ▶**Example 3.25**

Use SPICE to verify that letting $v_S = 17$ V in the circuit of Figure 3.1(a) does indeed cause the 2-Ω resistance to conduct 3 A.

Solution

Inserting a dummy voltage source VD in the given branch, we obtain the circuit of Figure 3.28. Its input file is:

```
USING A DUMMY VOLTAGE SOURCE
VS 1 0 DC 17
R1 1 2 1
R2 2 0 4
R3 2 3 3
IS 0 3 1
R4 3 4 2
* HERE IS THE DUMMY:
VD 4 0 DC 0
.END
```

After SPICE is run, the output listing contains the following:

```
DC ANALYSIS:
NODE VOLTAGE NODE VOLTAGE NODE VOLTAGE NODE VOLTAGE
(1)  17.0000 (2)  12.0000 (3)   6.0000 (4)   0.0000
```

Figure 3.28 Using a dummy voltage source **VD**.

```
VOLTAGE SOURCE CURRENTS:
NAME      CURRENT
 VS     -5.000E+00
 VD      3.000E+00

TOTAL POWER DISSIPATION:  8.50E+01 WATTS
```

This confirms that the current under consideration is indeed 3 A. ◀

Exercise 3.32 Use SPICE with a dummy source to monitor the current through the 2-Ω resistance in Figure 3.18. Then, find the corresponding power and compare with Exercise 3.26.

The `.DC` and `.PRINT DC` Statements

The dc analysis may be performed for a *range* of source values using the `.DC` statement. Its general form is

(3.21)

```
.DC SRCNAME SRCSTART SRCSTOP SRCINCR
```

where `SRCNAME` is the name of the independent voltage or current source to be varied, and `SRCSTART`, `SRCSTOP`, and `SRCINCR` are, respectively, its starting, ending, and incremental values, in volts or amperes. To print the results of this analysis we use the `.PRINT DC` statement, whose general form is

(3.22)

```
.PRINT DC OUTVAR1 OUTVAR2 . . . OUTVAR8
```

where `OUTVAR1` through `OUTVAR8` are the desired current or voltage output variables, for a maximum of eight variables per statement.

The form of a *voltage variable* is `V(N1, N2)`, and it specifies the voltage difference between nodes `N1` and `N2`. If `N2` and the preceding comma are omitted, then ground (`N2 = 0`) is assumed.

The form of a *current variable* is `I(VXXX)`, and it specifies the current flowing through the independent voltage source named `VXXX`. As we know, this current is assumed to flow from the positive terminal, through the source, and out of the negative terminal.

▶ Example 3.26

Referring to the voltage divider circuit of Figure 3.29, use SPICE to find the current delivered by the source as well as the voltages across the resistances for v_I ranging from -5 V to $+5$ V in 1-V increments.

Solution

The input file is:

```
VOLTAGE DIVIDER
VI 1 0
R1 1 2 30K
```

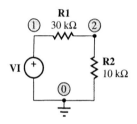

Figure 3.29 Voltage divider circuit.

```
R2 2 0 10K
.DC VI -5 5 1
.PRINT DC I(VI) V(1,2) V(2)
.END
```

After SPICE is run, the output listing contains the following:

VI	I(VI)	V(1,2)	V(2)
-5.000E+00	1.250E-04	-3.750E+00	-1.250E+00
-4.000E+00	1.000E-04	-3.000E+00	-1.000E+00
-3.000E+00	7.500E-05	-2.250E+00	-7.500E-01
-2.000E+00	5.000E-05	-1.500E+00	-5.000E-01
-1.000E+00	2.500E-05	-7.500E-01	-2.500E-01
0.000E+00	0.000E+00	0.000E+00	0.000E+00
1.000E+00	-2.500E-05	7.500E-01	2.500E-01
2.000E+00	-5.000E-05	1.500E+00	5.000E-01
3.000E+00	-7.500E-05	2.250E+00	7.500E-01
4.000E+00	-1.000E-04	3.000E+00	1.000E+00
5.000E+00	-1.250E-04	3.750E+00	1.250E+00

Exercise 3.33 In the current divider circuit of Figure 2.14, let $R_1 = 2$ kΩ and $R_2 = 10$ kΩ. Use SPICE to find the voltage across the source as well as the currents through the resistances for i_I ranging from 0 mA to 10 mA in 0.5-mA increments.

Concluding Remarks

Why should we waste our time with hand analysis when SPICE can handle our circuits much more rapidly and efficiently? To answer this question, we must bear in mind that SPICE, however fast and accurate, is no substitute for the *understanding* of circuit behavior that we can gain only through hand analysis. Even when the level of circuit complexity requires the use of computer power, we still need hand analysis to spot-check the results provided by the machine. A computer will only do what it has been programmed to do, and it is our *responsibility* to ensure that we have written a correct program by checking its results via hand calculations. A *good engineer will always question* what comes out of computers and calculators!

Throughout this book we shall use SPICE as one of various means for checking the results of hand calculations.

▼ Summary

- The state of a circuit is completely known once all its node voltages and branch currents are known. If the circuit consists of sources and resistances, the equations governing its voltages and currents are *algebraic equations* because they are based on Ohm's Law and Kirchhoff's laws.

- The *node method* uses KCL (but with the branch currents expressed in terms of the node voltages) to find all *node voltages* in a circuit.

- The *loop method* uses KVL (but with the branch voltages expressed in terms of mesh currents) to find all *mesh currents* in a circuit.

- In general, both the node and loop methods yield a *system of simultaneous algebraic equations,* which are then solved via the *Gaussian elimination method* or via *Cramer's rule.* Once one set of variables is known, the other set is obtained via the element laws of the individual branches.

- If a branch consists of an ungrounded voltage source, in order to apply the node method we must use the concept of *supernode.* If a branch consists of a current source shared by adjacent meshes, in order to apply the loop method we must use the concept of *supermesh.*

- The *superposition principle* exploits the *linearity* of resistive circuits to break down a circuit into smaller subcircuits, each of which contains just *one independent source* and is thus easier to analyze. We use these subcircuits to find the contributions from the individual sources, and then we add up the contributions algebraically to find a specific voltage or current in the circuit.

- The contribution from a given source in a circuit is found by *suppressing* all other independent sources. Suppressing a *voltage source* means replacing it with a *short circuit;* suppressing a *current source* means replacing it with an *open circuit.*

- A *voltage source* v_S in *series* with a resistance R_s is equivalent to a *current source* v_S/R_s in *parallel* with a resistance R_s. A *current source* i_S in *parallel* with a resistance R_s is equivalent to a *voltage source* $R_s i_S$ in *series* with a resistance R_s.

- The *source transformation technique* exploits the preceding equivalences to effect suitable circuit manipulations to facilitate circuit analysis.

- Circuit analysis can also be performed automatically by computer; however, it is no substitute for the *understanding* that we can gain only through hand analysis. We also need hand analysis to spot-check the results of automatic analysis.

- SPICE, the most popular circuit simulator program, uses *branch statements* to describe the circuit to the simulator and *control statements* to tell the simulator what kind of analysis is to be performed. The program is fed to the simulator via an *input file,* and the results are returned to the user via an *output file.*

- SPICE is specially useful when the level of circuit complexity makes hand analysis prohibitive. You are encouraged to use SPICE to check hand calculations when you solve the exercises and problems of this book.

- An integral part of the problem-solving process is checking for the correctness of the results. A good habit is to analyze a circuit via one method and check the results via another.

▼ PROBLEMS

3.1 Circuit Solution by Inspection

3.1   In the circuit of Figure P3.1 let $v_S = 8$ V. (a) Find i_S such that $i = 2$ mA. (b) Repeat, but with the voltage source terminals interchanged, so that the positive terminal is now at the right.

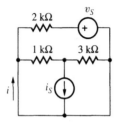

Figure P3.1

3.2   In the circuit of Figure P3.1 let $i_S = 4$ mA. Find v_S so that (a) $i = 1$ mA; (b) $i = 0$ mA; (c) $i = -1$ mA.

3.3   In the circuit of Figure P3.3 let $v_{S2} = 6$ V. For what value of v_{S1} will v_{S2} deliver 6 mW of power? Absorb 6 mW of power?

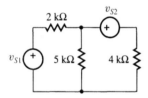

Figure P3.3

3.4   In the circuit of Figure P3.3 let $v_{S1} = 10$ V. Find the two values of v_{S2} that will cause the 5-kΩ resistance to dissipate 20 mW.

3.5   Make up your own problem as follows: In the circuit of Figure 1.17 let X_1 and X_6 be voltage sources and X_2 through X_5 be resistances, with the bottom node as the reference node. You are to come up with suitable component values so that all node voltages and branch currents have integer values of your choice, under the constraint that the component values be also integers.

3.6   Design a resistive ladder that, when driven by a 1000-V source, yields the voltages 100 V, 10 V, 1 V, 100 mV, 10 mV, and 1 mV under the following constraints: (a) the equivalent resistance seen by the source must be 1 MΩ; (b) the current entering each node must split between the subsequent series and shunt resistances in a 3-to-1 ratio.

3.7   Design a circuit to recharge a 1.2-V Ni-Cd battery from a car battery with a current of approximately 10 mA.

3.8   Using a 10-V source, design a resistive network to bias three devices: X_1, requiring 1 mA at 8 V; X_2, requiring 2 mA

at 3 V; and X_3, requiring 3 mA at 4 V. Show at least three designs, and indicate which one is more efficient in terms of energy.

3.9   An *npn bipolar junction transistor* (*npn* BJT) is a three-terminal device that, for biasing purposes, can be modeled with two elements as shown in Figure P3.9. One element draws current I_B at voltage V_{BE}, and the other draws current I_C at voltage V_{CE}. Using a 12-V voltage source and three suitable resistances, design a circuit to bias a BJT at $I_B = 10\mu$A with $V_{BE} = 0.7$ V and at $I_C = 1$ mA with $V_{CE} = 5$ V, under the constraint $V_E = 3$ V.

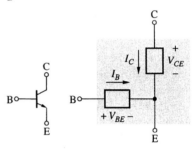

Figure P3.9

3.10   An *n*-channel *metal-oxide-silicon field-effect transistor* (*n*MOSFET) is a three-terminal device that, for biasing purposes, can be modeled with two elements as shown in Figure P3.10. One element requires voltage V_{GS} with a negligible current I_G, the other draws current I_D at voltage V_{DS}. Using a 12-V voltage source and suitable resistances, design a circuit to bias the *n*MOSFET at $V_{GS} = 4$ V with $I_G = 0$ and at $V_{DS} = 5$ V with $I_D = 2$ mA, under the constraint $V_S = 3$ V.

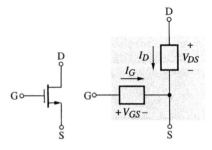

Figure P3.10

3.11   A *light-emitting diode* (LED), shown in Figure P3.11, glows with $I_{AC} \simeq 10$ mA and $V_{AC} \simeq 1.5$ V. If $V_{AC} = 0$ or $I_{AC} = 0$, the LED is off. (a) Using an LED, a 5-V voltage source, a biasing resistor R, and a switch SW, devise a circuit such that the LED glows when SW = CLOSED, and is off when SW = OPEN. (b) Repeat, but so that the LED glows when SW = OPEN, and is off when SW = CLOSED. (c) Which circuit is preferable in terms of saving energy?

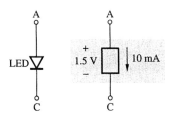

Figure P3.11

3.2 Nodal Analysis

3.12 Apply nodal analysis to the circuit of Figure P3.12. Hence, find all released and absorbed powers, and verify the conservation of power.

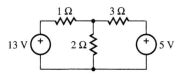

Figure P3.12

3.13 (a) Apply nodal analysis to the circuit of Figure P3.13. Check your results. (b) Find the power (released or absorbed?) by each source.

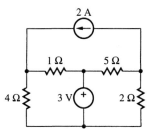

Figure P3.13

3.14 Repeat Problem 3.13, but with direction of the 2-A current source reversed.

3.15 Repeat Problem 3.13, but with the 2-A current source replaced by a 1.8-V voltage source, positive at the right.

3.16 Apply nodal analysis to the circuit of Figure P3.16. Check your results.

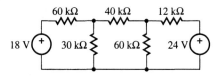

Figure P3.16

3.17 Repeat Problem 3.16, but with the 40-kΩ resistance replaced by a 5-V voltage source, positive at the right.

3.18 Apply nodal analysis to the circuit of Figure P3.18. Check your results.

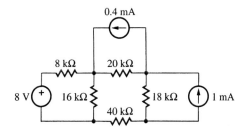

Figure P3.18

3.19 Repeat Problem 3.18, but with the 1-mA current source replaced by a 1-V voltage source, positive at the bottom.

3.20 Apply nodal analysis to the circuit of Figure P3.20 to find the power dissipated by the 2-Ω resistance. Check your result.

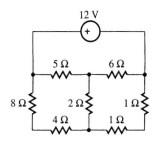

Figure P3.20

3.21 In the circuit of Figure P3.20 replace the 12-V source with a 3-A current source flowing toward the left. Then, use the node method to find the power released by the source.

3.22 Apply nodal analysis to the circuit of Figure P3.22 to find the power (released or absorbed?) by the 3-mA source. Check your result.

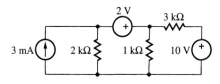

Figure P3.22

3.23 (a) Apply nodal analysis to the circuit of Figure P3.23 to find the power (released or absorbed?) by the 0.4-A source. Check your result. (b) Repeat, but with the direction of the current source reversed.

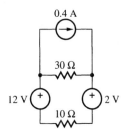

Figure P3.23

3.24   Apply nodal analysis to the circuit of Figure P3.24 to find the power dissipated by the 4-Ω resistance. Check your result.

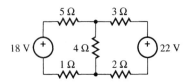

Figure P3.24

3.25   Repeat Problem 3.24 if the circuit includes also a 3-A current source flowing from the node at the positive side of the 22-V source to the node at the positive side of the 18-V source.

3.3 Loop Analysis

3.26   Apply loop analysis to the circuit of Figure P3.26 to find the voltage across the 10-Ω resistance. Check your result.

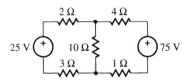

Figure P3.26

3.27   Repeat Problem 3.26 if the circuit includes also a current source (2 A, directed downward) in parallel with the 10-Ω resistance.

3.28   Apply loop analysis to the circuit of Figure P3.28 to find the current through the 2-Ω resistance. Check your result.

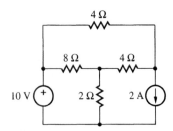

Figure P3.28

3.29   Repeat Problem 3.28, but with the 2-Ω resistance and 2-A current source interchanged with each other.

3.30   Apply loop analysis to the circuit of Figure P3.30 to find the power dissipated by the 4-Ω resistance. Check your result.

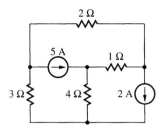

Figure P3.30

3.31   Repeat Problem 3.30, but with an additional 5-Ω resistance in parallel with the 2-A source.

3.32   Apply loop analysis to the circuit of Figure P3.32 to find the magnitude and direction of the current through the 10-V source. Check your result.

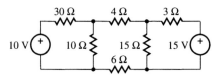

Figure P3.32

3.33   Apply loop analysis to the circuit of Figure P3.16 to find the power dissipated by the 40-kΩ resistance. Check your result.

3.34   Apply loop analysis to the circuit of Figure P3.18 to find the power (released or absorbed?) by the 0.4-mA source. Check your result.

3.35   Apply loop analysis to the circuit of Figure P3.20 to find the equivalent resistance seen by the 12-V source.

3.36   Apply loop analysis to the circuit of Figure P3.22. Check your results.

3.37   Apply loop analysis to the circuit of Figure P3.37. Check your results.

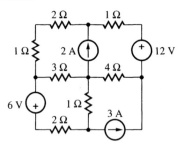

Figure P3.37

3.4 Linearity and Superposition

3.38 Using the superposition principle, find the magnitude and polarity of the voltage across the 3-kΩ resistance in Figure P3.38.

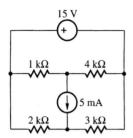

Figure P3.38

3.39 Apply the superposition principle to the circuit of Figure P3.38 to find the power (released or absorbed?) by each source.

3.40 Using the superposition principle, find the voltage across the 4-Ω resistance in Figure P3.24. Check using the loop method.

3.41 Apply the superposition principle to the circuit of Figure P3.28 to find the power dissipated by the 2-Ω resistance. Check using the node method.

3.42 Apply the superposition principle to the circuit of Figure P3.30 to find the power (released or absorbed?) by the 2-A source.

3.43 Using the superposition principle, find the magnitude and polarity of the voltage across the 7-mA source in Figure P3.43.

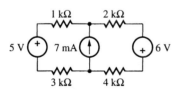

Figure P3.43

3.44 Using the superposition principle, find the power (released or absorbed?) by each voltage source in Figure P3.43.

3.45 Using the superposition principle, find the power dissipated by the 30-Ω resistance in Figure P3.23.

3.46 Using the superposition principle, find the magnitude and direction of the current through the voltage source of Figure P3.46. Which source contributes the most and which the least to this current?

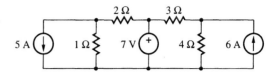

Figure P3.46

3.47 Using the superposition principle, find the power (released or absorbed?) by each current source in Figure P3.46.

3.48 (a) Using the superposition principle, show that the circuit of Figure P3.48 yields

$$v_O = \frac{1}{m+n}(v_1 + v_2 + v_3 + \cdots + v_n)$$

(b) Specify suitable component values to achieve $v_O = (v_1 + v_2 + v_3 + v_4)/4$.

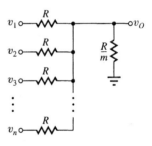

Figure P3.48

3.49 In the circuit of Figure P3.49 find the power (released or absorbed?) by each source using (a) nodal analysis; (b) loop analysis; (c) the superposition principle. Which method do you prefer the most? The least? Why?

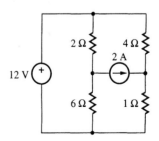

Figure P3.49

3.5 Source Transformations

3.50 Apply source transformations to the circuit of Figure P3.50 to find (a) the magnitude and direction of the current through the 15-V source; (b) the magnitude and polarity of the voltage across the 0.3-A source.

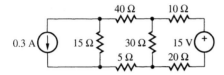

Figure P3.50

3.51 Apply source transformations to the circuit of Figure P3.50 to find (a) the magnitude and direction of the current through the 30-Ω resistance; (b) the magnitude and polarity of the voltage across the 5-Ω resistance.

3.52 Apply source transformations to the circuit of Figure P3.16 to find (a) the voltage across the 40-kΩ resistance; (b) the power (released or absorbed?) by the 18-V source; (c) the power (released or absorbed?) by the 24-V source.

3.53 Using source transformations, find the power (released or absorbed?) by each source in the circuit of Figure P3.22.

3.54 Using source transformations, find the magnitude and direction of the current through the 24-Ω resistance in Figure P3.54.

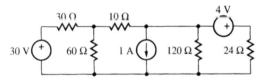

Figure P3.54

3.55 Using source transformations, find the magnitude and direction of the current through the 30-V source in Figure P3.54.

3.56 Using source transformations, find the magnitude and polarity of the voltage across the 12-kΩ resistance in Figure P3.56.

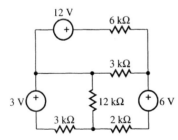

Figure P3.56

3.57 Using source transformations, find the power (released or absorbed?) by the 12-V source in Figure P3.56.

3.6 Circuit Analysis Using SPICE

3.58 Use SPICE to find the voltage across the 2-Ω resistance in the circuit of Figure P3.12.

3.59 Use SPICE to find the voltage across the 10-Ω resistance in the circuit of Figure P3.26.

3.60 Use SPICE to find the voltage across the current source and the current through the voltage source in the circuit of Figure P3.38.

3.61 Use SPICE to find i in the circuit of Figure P3.1 if $v_S = 10$ V and $i_S = 1$ mA.

3.62 For the circuit of Figure P3.50, use SPICE to print the voltage across the current source and the current through the voltage source if the latter is stepped from -15 V to $+15$ V in 5 V increments.

3.63 Given the following SPICE input file,

```
PROBLEM 3.63
V1 1 0 DC 20
R1 1 2 60
R2 2 0 30
R3 2 3 20
I1 0 3 DC 1
R4 3 4 10
V2 0 4 DC 15
.END
```

(a) draw the corresponding circuit, (b) predict the SPICE output, (c) verify your prediction by actually running SPICE.

3.64 Repeat Problem 3.63, but for the following SPICE input file:

```
PROBLEM 3.64
R1 2 0 2K
VS 1 0 DC 3
R2 3 0 4K
IS 2 3 2 M
R3 1 2 5K
R4 1 3 1K
.END
```

CIRCUIT THEOREMS AND POWER CALCULATIONS

I n this chapter we study **one-ports,** a class of circuits equipped with a terminal pair through which signals and energy may enter or leave the circuit. A one-port is uniquely characterized by the current–voltage relationship at its terminals, called the i–v **characteristic.** We are interested in one-ports made up of sources and resistances, because their i–v characteristics are *straight lines,* just like the characteristics of practical voltage and current sources. In fact, we can extend the concept of equivalence and say that in terms of the i–v relationship at its terminals, a linear one-port can be modeled either with an ideal voltage source and a suitable series resistance or with an ideal current source and a suitable parallel resistance. These statements form the essence of the fundamental circuit theorems known, respectively, as **Thévenin's** and **Norton's theorems.**

The ability to model a one-port with just a source and a resistance has far-reaching implications in circuit analysis and design. A complex circuit can often be broken down into separate one-port subcircuits that, on paper at least, can be replaced with their Thévenin or Norton equivalents to facilitate finding the voltages or currents at the common terminal pairs.

In this chapter we first show how to find the Thévenin and Norton equivalents of a one-port. We then apply these equivalents to the analysis of a variety of linear and nonlinear circuits. In the course of nonlinear analysis we introduce *iterative techniques* and *graphical techniques,* which students will find particularly useful in subsequent electronics courses. Though the emphasis of this book is on linear circuits, we feel that a preliminary exposure to nonlinear methods will make the reader better appreciate the usefulness of circuit theorems.

Once we know the voltage and current at the terminals of a one-port, we can also find the power entering or leaving the port. **Power calculations** form an integral part of circuit theory, and this chapter introduces the concepts of *rms value* of a signal, *maximum power transfer,* load and source *matching,* and *efficiency.* An early exposure to these concepts is especially important if the study of circuits is accompanied by experimentation in the laboratory, where the student is required to become familiar with instrumentation and measurement terminology.

We conclude by illustrating the use of SPICE to find Thévenin and Norton equivalents, and also to analyze circuits with nonlinear resistors. The SPICE facilities introduced are the .TF statement and the polynomial approximation of nonlinear resistors. Though no substitute for our understanding of how a circuit operates, SPICE is a valuable tool for checking the results of hand calculations and for analyzing circuits whose complexity would render manual calculations prohibitively long and tedious.

4.1 ONE-PORTS

Just as elements are connected together to form a circuit, circuits can be connected together to form more complex circuits, or networks. This structured approach allows for individual circuit modules to be designed, constructed, analyzed, tested, and repaired separately, thus offering the advantages that we encounter whenever we can break down a complex task into simpler ones. A familiar example is offered by a hi-fi audio system, which consists of various modules connected together by cables: the tuner, the cassette deck, the record player, the compact disc player, and the power amplifier.

If you open up one of these modules, say, the tuner, you will find that it consists of submodules called *printed circuits.* These submodules, in turn, are made up of even smaller modules called *integrated circuits,* or ICs. An IC may consist of tens or even thousands of basic elements such as resistors, capacitors, diodes, and transistors, all fabricated on a common silicon chip.

To interact with each other electrically, circuit modules are equipped with terminals or leads. Clearly, the smallest number of leads a circuit can have in order to share a voltage or a current is *two.* Since it allows us to access the inside of a network, a terminal pair is called a *port.* A network with just one pair of terminals emerging from it is called a *one-port network,* or a **one-port** for short. Note that the ingredients of a one-port are two, the network and its protruding terminals. Since these terminals are connected to a pair of corresponding nodes inside the network, we shall use *node labels* to distinguish one terminal from the other.

The simplest one-ports are the basic circuit elements such as sources and resistances. The ladder of Figure 2.9 is an example of a purely resistive one-port. The practical source models of Figures 2.24 and 2.25 provide additional one-port examples. The bridge circuit of Figure 2.18, which consists of v_S, R_1 through R_4, and the terminal pair denoted as A and B, is yet another example. A more complex one-port is a portable digital multimeter, which consists of sophisticated circuitry inside a case and is equipped with a pair of terminals for connection to the circuit under measurement. The terminals are color coded to allow for their distinction, and they constitute the port of entry to the multimeter, or the input port.

Like a circuit element, a one-port is characterized by a unique relationship between its terminal voltage and current. Called the **terminal characteristic,** it allows us to predict its behavior when the one-port is connected to an external circuit generally referred to as a **load;** see Figure 4.1(a). Clearly, the load itself is a one-port. In light of these considerations, it is fair to say that the concept of a one-port is a generalization of the concept of a circuit element.

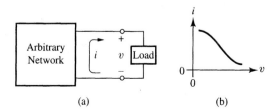

(a)   (b)

Figure 4.1   (a) A one-port terminated on a load; (b) i–v characteristic of an arbitrary one-port.

The choice of the reference direction of i relative to the reference polarity of v is arbitrary. Since in general a one-port may contain sources, it is likely to deliver power to the load, though this need not necessarily be always the case. Hence the reason for choosing the **active sign convention** for our one-port.

In general, the terminal characteristic of a one-port will be a curve of any form; see Figure 4.1(b). However, if the network making up the one-port is *linear,* this characteristic will be a *straight line.* We now wish to substantiate this claim.

The *i*–*v* Characteristics of Linear One-Ports

Figure 4.2(a) shows the circuit arrangement to find the v–i characteristic of a one-port. Note that the test current source is pointing downward to conform to the reference direction of Figure 4.1(a). If the one-port is *linear,* we can apply the superposition principle and state that v must be a linear combination of the independent sources $x_{S1}, x_{S2}, x_{S3}, \ldots, x_{Sn}$ internal to the network, as well as the external source i,

$$v = a_1 x_{S1} + a_2 x_{S2} + a_3 x_{S3} + \cdots + a_n x_{Sn} + a_{n+1} i \qquad \textbf{(4.1)}$$

In this expression x_{S1} through x_{Sn} may be either voltage sources or current sources, and the coefficients a_1 through a_n either are scaling factors or have the dimensions of resistances, depending on whether the corresponding source is, respectively, a voltage or a current source. Clearly, a_{n+1} has the dimensions of a resistance.

To understand the significance of the various terms, consider first the value of v corresponding to $i = 0$. This represents the contributions from the internal sources x_{S1} through x_{Sn} lumped together. Since $i = 0$ implies an open circuit, this value of v is called the **open-circuit voltage** and is denoted as v_{OC}. As

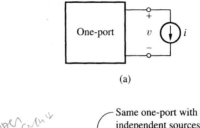

(a)

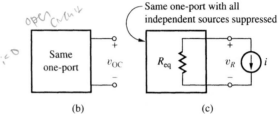

(b)   (c)

Figure 4.2 (a) Circuit arrangement to find the v–i characteristic of a one-port. Subcircuits illustrating the origin of (b) the *open-circuit voltage* v_{OC}, and (c) the *slope* $-1/R_{eq}$.

depicted in Figure 4.2(b), this is simply the voltage that we would measure across the unloaded port with an ideal voltmeter, which acts as an open circuit. Clearly, $v_{OC} = a_1 x_{S1} + a_2 x_{S2} + a_3 x_{S3} + \cdots + a_n x_{Sn}$.

Next, suppose we suppress the internal sources x_{S1} through x_{Sn} so that the network becomes purely resistive and reduces to a single equivalent resistance R_{eq}. Letting v_R denote the resulting value of v, as depicted in Figure 4.2(c), we have $v_R = -R_{eq}i$.

With all sources active, we can apply the superposition principle and write $v = v_{OC} + v_R$, or

$$v = v_{OC} - R_{eq}i \qquad (4.2)$$

It is interesting to note the formal similarity with the practical voltage source characteristic of Equation (2.37). Equation (4.2) is easily turned around to find the *i–v characteristic* of the one-port,

$$i = \frac{v_{OC}}{R_{eq}} - \frac{1}{R_{eq}}v \qquad (4.3)$$

As expected, this represents a *straight line*. As depicted in Figure 4.3, it intercepts the v axis at $v = v_{OC}$, and has slope $-1/R_{eq}$.

Note that the slope is negative only because of our particular choice of the reference direction of i relative to the reference polarity of v in Figure 4.2(a). As mentioned, this choice was made in anticipation of the fact that a one-port with internal sources is likely to *deliver* power when connected to an external load.

Using analogous reasoning, if we terminate the one-port on a test voltage v as in Figure 4.4(a), then i must, by the superposition principle, be a linear

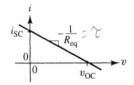

Figure 4.3 The i–v characteristic of a linear one-port.

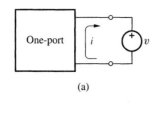

(a)

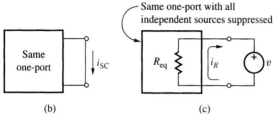

(b)   (c)

Figure 4.4 (a) Circuit arrangement to find the i–v characteristic of a one-port. Subcircuits illustrating the origin of (b) the *short-circuit current* i_{SC}, and (c) the *slope* $-1/R_{eq}$.

combination of the independent sources x_{S1} through x_{Sn} internal to the one-port and the external source v,

$$i = b_1 x_{S1} + b_2 x_{S2} + b_3 x_{S3} + \cdots + b_n x_{Sn} + b_{n+1} v \qquad \textbf{(4.4)}$$

Here b_1 through b_{n+1} are again suitable coefficients.

The value of i corresponding to $v = 0$ represents the contributions from the internal sources x_{S1} through x_{Sn} lumped together. Since $v = 0$ implies a short circuit, this value of i is called the **short-circuit current** and is denoted as i_{SC}. As is shown in Figure 4.4(b), this is simply the current that we would measure at the port with an ideal ammeter, which acts as a short circuit. Clearly, $i_{SC} = b_1 x_{S1} + b_2 x_{S2} + b_3 x_{S3} + \cdots + b_n x_{Sn}$.

If the internal sources are suppressed, the one-port reduces to the same equivalent resistance as before. Letting i_R denote the resulting value of i, we have $i_R = -(1/R_{eq})v$. If all sources are active, we have, by the superposition principle, $i = i_{SC} + i_R$, or

$$i = i_{SC} - \frac{1}{R_{eq}} v \qquad \textbf{(4.5)}$$

This is again a straight line with slope $-1/R_{eq}$ and with i-axis intercept at $i = i_{SC}$. Note again the formal similarity with the practical current source characteristic of Equation (2.39).

Summarizing, the i–v curve of a linear one-port is a **straight line** with v-axis and i-axis intercepts at $v = v_{OC}$ and $i = i_{SC}$, where v_{OC} is the voltage available at the port in the absence of any load, and i_{SC} is the current delivered when the port is short-circuited. Moreover, the slope of the i–v line is $-1/R_{eq}$, where R_{eq} is the **equivalent resistance** of the one-port. Clearly, the i–v line is uniquely defined once two of the parameters are known. We can thus specify it either in terms of its intercepts or in terms of its slope and just one of the intercepts.

Finding R_{eq}: Method 1

Equations (4.3) and (4.5) must be equivalent because they refer to the same one-port. We must thus have $v_{OC}/R_{eq} = i_{SC}$, or $R_{eq} = v_{OC}/i_{SC}$. This forms the basis of the following method to find R_{eq}:

> **Method 1: To find the equivalent resistance of a one-port, find its *open-circuit* voltage v_{OC} and its *short-circuit* current i_{SC}. Then,**

$$R_{eq} = \frac{v_{OC}}{i_{SC}} \tag{4.6}$$

Note that this equation can also be used to find one of the intercepts once R_{eq} and the other intercept are known. For instance, if R_{eq} and v_{OC} are known, then i_{SC} can be found as $i_{SC} = v_{OC}/R_{eq}$. Likewise, if R_{eq} and i_{SC} are known, then $v_{OC} = R_{eq}i_{SC}$.

▶ Example 4.1

Find v_{OC}, i_{SC}, and R_{eq} for a voltage divider circuit consisting of a source $v_S = 12$ V and two resistances, $R_1 = 30$ kΩ and $R_2 = 10$ kΩ.

Solution

Leaving the port open-circuited as in Figure 4.5(a) and applying the voltage divider formula, we obtain

$$v_{OC} = \frac{R_2}{R_1 + R_2} v_S \tag{4.7a}$$

or $v_{OC} = 12 \times 10/(30 + 10) = 3$ V. Short-circuiting the port as in Figure 4.5(b) and noting that $i_{R_2} = 0/R_2 = 0$, we have, by Ohm's Law,

$$i_{SC} = \frac{v_S}{R_1} \tag{4.7b}$$

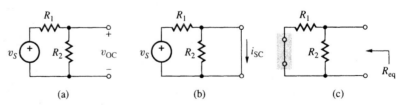

(a)   (b)   (c)

Figure 4.5 Finding v_{OC}, i_{SC}, and R_{eq} for a voltage divider circuit.

or $i_{SC} = 12/30 = 0.4$ mA. Finally, using Method 1 we obtain

$$R_{eq} = \frac{v_{OC}}{i_{SC}} \qquad \text{(4.8)}$$

or $R_{eq} = 3/0.4 = 7.5$ kΩ. As a check, recompute R_{eq} by suppressing the source v_S, as shown in Figure 4.5(c). By inspection, we now find

$$R_{eq} = R_1 \parallel R_2 \qquad \text{(4.9)}$$

or $R_{eq} = 10 \parallel 30 = 7.5$ kΩ, confirming the result via Method 1. ◄

Exercise 4.1   Find v_{OC}, i_{SC}, and R_{eq} for the one-port of Figure 4.6.

ANSWER $v_{OC} = 6$ V (+ @ A), $i_{SC} = 4$ A (A $\rightarrow$ B), and $R_{eq} = 1.5$ Ω.

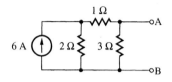

Figure 4.6 Circuit of Exercise 4.1.

►Example 4.2

Find v_{OC}, i_{SC}, and R_{eq} for the one-port of Figure 4.7(a) for the case $v_S = 15$ V and $i_S = 1$ A.

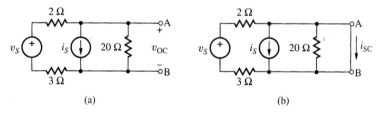

| (a) | (b) |

Figure 4.7   (a) Open-circuited and (b) short-circuited one-port.

Solution

Referring to Figure 4.7(a), we can use the superposition principle along with the voltage divider formula and Ohm's Law to write $v_{OC} = [20/(2 + 20 + 3)]v_S - [(2 + 3) \parallel 20]i_S$, or

$$v_{OC} = 0.8v_S - 4i_S \qquad \text{(4.10a)}$$

Numerically, $v_{OC} = 0.8 \times 15 - 4 \times 1 = 8$ V.

To find i_{SC}, short-circuit the one-port as in Figure 4.7(b). Since the current through the 20-Ω resistance is now zero, we can apply KCL and Ohm's Law to obtain $i_{SC} = v_S/(2 + 3) - i_S$, or

$$i_{SC} = 0.2v_S - i_S \qquad \text{(4.10b)}$$

Numerically, $i_{SC} = 0.2 \times 15 - 1 = 2$ A.

Finally, using Method 1 we obtain

$$R_{eq} = \frac{v_{OC}}{i_{SC}} = \frac{8}{2} = 4 \ \Omega \qquad \text{(4.11)}$$

As a check, recompute R_{eq} by suppressing both sources. Thus, replacing the voltage source with a short circuit and the current source with an open circuit in Figure 4.7(a), we readily find that the resistance between nodes A and B is

$$R_{eq} = 20 \parallel (2 + 3) = 4 \ \Omega \qquad \text{(4.12)}$$

This confirms the result obtained via Method 1.

Exercise 4.2   Find v_{OC}, i_{SC}, and R_{eq} for the circuit of Figure 4.8(a).

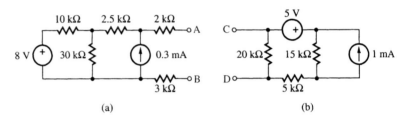

(a)   (b)

Figure 4.8 Additional one-port examples.

ANSWER $v_{OC} = 9$ V (positive at A), $i_{SC} = 0.6$ mA (A $\rightarrow$ B), and $R_{eq} = 15$ kΩ.

Exercise 4.3   Find v_{OC}, i_{SC}, and R_{eq} for the circuit of Figure 4.8(b).

ANSWER $v_{OC} = 5$ V (positive at C), $i_{SC} = 0.5$ mA (C $\rightarrow$ D), and $R_{eq} = 10$ kΩ.

▶ Example 4.3

In Figure 4.9 let $v_S = 15$ V, $R_1 = 1$ kΩ, $R_2 = 4$ kΩ, $R_3 = 3$ kΩ, and $R_4 = 2$ kΩ. Find v_{OC}, i_{SC}, and R_{eq}.

Solution

Leaving the port open-circuited as in Figure 4.9(a) yields, by Equation (2.31),

$$v_{OC} = \left(\frac{1}{1 + R_1/R_2} - \frac{1}{1 + R_3/R_4} \right) v_S \qquad \text{(4.13)}$$

or $v_{OC} = [1/(1 + 1/4) - 1/(1 + 3/2)]15 = 6$ V.

To find R_{eq}, suppress v_S as in Figure 4.9(b). This creates a common node X so that R_1 appears in parallel with R_2, and R_3 in parallel with R_4.

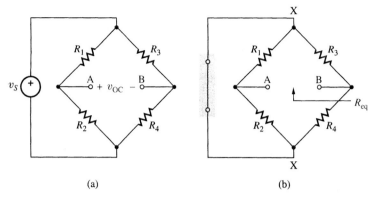

Figure 4.9 Finding v_{OC} and R_{eq} for a resistive bridge circuit.

By inspection, the total resistance encountered in traversing the circuit from node A, through X, to node B is

$$R_{eq} = (R_1 \parallel R_2) + (R_3 \parallel R_4) \tag{4.14}$$

or $R_{eq} = (1 \parallel 4) + (3 \parallel 2) = 2 \text{ k}\Omega$.

Finally, $i_{SC} = v_{OC}/R_{eq} = 6/2 = 3$ mA.

Remark   This example shows that Equation (4.6) can save labor in computing one of the three parameters, i_{SC} in this case.   ◀

Exercise 4.4   Repeat Example 4.3, but with the voltage source v_S replaced by a current source $i_S = 6$ mA pointing upward. Does R_{eq} change? Why?

ANSWER $v_{OC} = 6$ V (positive at A), $i_{SC} = 2.5$ mA (A → B), and $R_{eq} = 2.4 \text{ k}\Omega$.

▶**Example 4.4**

Repeat Example 4.3, but with R_1 changed to: (a) 16 kΩ; (b) 6 kΩ.

Solution

(a) Equation (4.13) yields $v_{OC} = -3$ V, indicating that node A is now *more negative* than node B. Using Equation (4.14), $R_{eq} = (16 \parallel 4) + (3 \parallel 2) = 4.4 \text{ k}\Omega$. Finally, $i_{SC} = v_{OC}/R_{eq} = -3/4.4 = -0.6818$ mA, indicating that the short-circuit current will now flow from node B to node A. Clearly, the i–v line now has *negative intercepts* with the v axis and the i axis.

(b) Equation (4.13) now yields $v_{OC} = 0$ V. Moreover, $R_{eq} = (6 \parallel 4) + (3 \parallel 2) = 3.6$ kΩ, and $i_{SC} = 0/3.6 = 0$ A. Clearly, the i–v line now goes through the *origin*.

Remark   This example shows that v_{OC} and i_{SC} can be negative or even zero.

Exercise 4.5   Find v_{OC}, i_{SC}, and R_{eq} for the one-port of Figure 4.7 for the following cases: (a) $v_S = 15$ V and $i_S = 3$ A; (b) $v_S = 10$ V and $i_S = 3$ A; (c) $v_S = -5$ V and $i_S = 1$ A.

ANSWER $v_{OC} = 0,\ -4$ V, -8 V; $i_{SC} = 0,\ -1$ A, -2 A; $R_{eq} = 4\ \Omega$ throughout.

Finding R_{eq}: Method 2

If the i–v curve happens to go through the origin, then both v_{OC} and i_{SC} are zero and Method 1 cannot be used because it would yield the indeterminate result $R_{eq} = v_{OC}/i_{SC} = 0/0$. An alternate approach is to suppress the independent sources x_{S_1} through x_{S_n} inside the one-port and then use this reduced network to find R_{eq}. In many cases this can be done by simple inspection, using series/parallel reductions as in Equations (4.9), (4.12), and (4.14).

However, there are resistive topologies that do not lend themselves to this simple form of solution. When this is the case, we must use the following more general method, which is simply a restatement of the method of Section 2.1 for finding resistance.

Method 2: To find the equivalent resistance of a linear one-port, first suppress all its *independent* sources. Then apply a *test voltage* v and find the current i out of its positive terminal, as shown in Figure 4.10(a). Finally,

$$R_{eq} = \frac{v}{i} \tag{4.15}$$

Alternately, after suppressing the independent sources, apply a *test current* i and find the resulting voltage v, as shown in Figure 4.10(b). Then, use again $R_{eq} = v/i$.

Recall that to suppress a voltage source we must replace it with a short circuit, and to suppress a current source we must replace it with an open circuit,

$$\text{voltage source} \to \text{short circuit} \tag{4.16a}$$

$$\text{current source} \to \text{open circuit} \tag{4.16b}$$

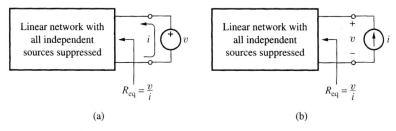

Figure 4.10 R_{eq} can be found by applying either a test voltage v or a test current i.

If the one-port contains also *dependent* sources, these *must be left in the circuit.* This important case will be addressed in great detail in Section 5.2.

We stress once again that the reason for suppressing all the independent sources is to *force the i–v characteristic through the origin* and thus make it possible to obtain its slope as $1/R_{eq} = i/v$, or $R_{eq} = v/i$. We also note that since a one-port with all its independent sources suppressed is *purely resistive,* it is now appropriate to use the *passive sign convention* for R_{eq}, as is shown in Figure 4.10.

It is worth mentioning that the *actual value of the test signal is immaterial.* Since i and v are *linearly proportional* to each other, their ratio is independent of the particular test signal value. Nonetheless, assuming a specific value such as a 1-V test voltage (or a 1-A test current) may simplify the algebra somewhat.

▶ Example 4.5

Find i_{SC}, v_{OC}, and R_{eq} for the circuit of Figure 4.11(a).

Solution

To find i_{SC}, short-circuit the port as in Figure 4.11(b), where the units are [V], [A], and [Ω]. Applying KCL at node A yields $i_1 = i_{SC} + i_2 + 5.8$, or

$$i_{SC} = i_1 - i_2 - 5.8 \text{ A} \qquad (4.17)$$

We observe that shorting nodes A and B together places the 3-Ω resistance in parallel with the 4-Ω resistance and the 8-Ω resistance in parallel with the 1-Ω resistance. Applying the current divider formula, we have

$$i_1 = \frac{4}{3+4}i_5 = \frac{4}{7}i_5 \qquad (4.18a)$$

$$i_2 = \frac{1}{8+1}i_5 = \frac{1}{9}i_5 \qquad (4.18b)$$

Moreover, as i_5 exits the 58-V source via the positive terminal, it sees first the 2-Ω resistance, then the combination (3 ∥ 4) Ω, and finally the combination (8 ∥ 1) Ω, after which it reenters the source via the negative

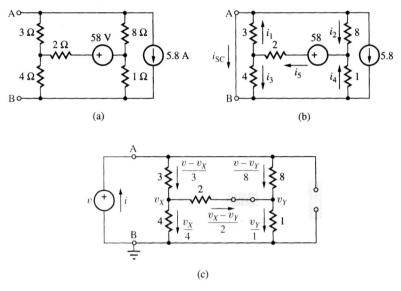

Figure 4.11 Example in which we must use Method 2 to find R_{eq}.

terminal. We can thus apply Ohm's Law and write

$$i_5 = \frac{58}{2 + (3 \parallel 4) + (8 \parallel 1)} = 12.6 \text{ A}$$

Substituting into Equation (4.18) and then into (4.17) yields $i_{SC} = 7.2 - 1.4 - 5.8$, or

$$i_{SC} = 0 \qquad (4.19)$$

Clearly, this is an example in which $v_{OC} = i_{SC} = 0$, so that R_{eq} cannot be found via Method 1.

Let us thus apply Method 2. Suppressing the sources yields the circuit of Figure 4.11(c). This particular network does not lend itself to simple series/parallel reductions, mandating the use of a test signal which we choose to be of the voltage type. For added clarity the expressions for all branch currents have been shown explicitly in color. To find i we apply KCL,

$$i = \frac{v - v_X}{3} + \frac{v - v_Y}{8} \qquad (4.20)$$

We need two additional equations to eliminate v_X and v_Y. These are obtained by applying KCL at the corresponding nodes,

$$\frac{v - v_X}{3} = \frac{v_X}{4} + \frac{v_X - v_Y}{2} \qquad (4.21a)$$

$$\frac{v_X - v_Y}{2} + \frac{v - v_Y}{8} = \frac{v_Y}{1} \qquad (4.21b)$$

These equations are solved for v_X and v_Y in the usual manner to yield $v_X = 0.4v$ and $v_Y = 0.2v$. Substituting into Equation (4.20) we obtain $i = (v - 0.4v)/3 + (v - 0.2v)/8$, or

$$i = \frac{3}{10}v \qquad (4.22)$$

Finally,

$$R_{eq} = \frac{v}{i} = \frac{\not{v}}{(3/10)\,\not{v}} = \frac{10}{3} = 3.333 \ \Omega \qquad \textbf{(4.23)}$$

Remark As expected, v at the numerator and denominator cancel out to yield a result which is independent of the test signal. ◀

Exercise 4.6 Recalculate the equivalent resistance in the circuit of Figure 4.11(c), but using a test current i instead of a test voltage v, and the *loop method* instead of the node method. Clearly, the result must be the same.

▶ **Example 4.6**

Find R_{eq} for the one-port of Figure 4.12(a).

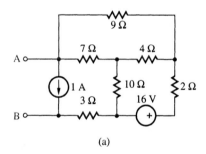

(a)

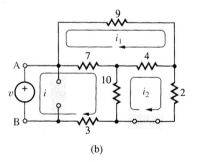

(b)

Figure 4.12 Finding the equivalent resistance of a one-port.

Solution

Let us pause a moment and see which method might be more efficient. Method 1 requires analyzing two circuits, one to find v_{OC} and the other to find i_{SC}. By contrast, Method 2 requires the analysis of only one circuit. Moreover, with its internal sources suppressed, this circuit is likely to be simpler. We thus opt for Method 2.

Let us apply, for instance, a test voltage as in Figure 4.12(b), where the resistances are in [Ω]. We can find the resulting current via *loop analysis*. KVL around the mesh labeled i yields

$$v = 7(i - i_1) + 10(i - i_2) + 3i \tag{4.24}$$

We need two additional equations to eliminate i_1 and i_2. These are obtained by applying KVL around the two remaining meshes,

$$0 = 7(i_1 - i) + 9i_1 + 4(i_1 - i_2) \tag{4.25a}$$

$$0 = 10(i_2 - i) + 4(i_2 - i_1) + 2i_2 \tag{4.25b}$$

These equations can be solved for i_1 and i_2 via Gaussian elimination to yield $i_1 = 0.5i$ and $i_2 = 0.75i$. Substituting into Equation (4.24) and collecting we obtain

$$v = 9i \tag{4.26}$$

Finally,

$$R_{eq} = \frac{v}{i} = \frac{9\,\cancel{i}}{\cancel{i}} = 9 \ \Omega \tag{4.27}$$

Note again the cancellation of i. ◀

Exercise 4.7   Consider the circuit obtained from that of Figure 4.9(a) by breaking the connection between the positive terminal of the source and the node common to R_1 and R_3, and inserting an additional series resistance R_5. Sketch the modified circuit. Hence, assuming $v_S = 15$ V, $R_1 = 1$ kΩ, $R_2 = 4$ kΩ, $R_3 = 3$ kΩ, $R_4 = 2$ kΩ, and $R_5 = 5$ kΩ, find v_{OC}, and then use Method 2 to find R_{eq}, the equivalent resistance between nodes A and B.

ANSWER $v_{OC} = 2$ V (positive at A), $R_{eq} = 34/15$ kΩ.

Exercise 4.8   Repeat Exercise 4.7, but using Method 1 to find R_{eq}.

ANSWER $v_{OC} = 2$ V (positive at A), $i_{SC} = 15/17$ mA (A → B), and $R_{eq} = 34/15$ kΩ.

Remark

To speed up the algebra, some authors suggest using a numerical test signal value, such as $v = 1$ V or $i = 1$ A. However, we prefer to work with *unspecified* values because this requires a *cancellation,* as demonstrated in Equations (4.23) and (4.27), thus providing a *check* for the correctness of our result.

As an example, suppose in Equation (4.26) we had erroneously obtained $v = 9i + 3$. This would have yielded $R_{eq} = v/i = (9i + 3)/i = 9 + 3/i$, which is unacceptable because in a linear circuit R_{eq} must be *independent* of the particular test current i. Had we used a 1-A test current, then our erroneous Equation (4.26) would have yielded $v = 9i + 3 = 9 \times 1 + 3 = 12$ V, or $R_{eq} = v/i = 12/1 = 12 \; \Omega$, with no error warning!

4.2 CIRCUIT THEOREMS

Two one-ports are said to be **equivalent** if they exhibit the same i–v characteristic. If the one-ports are linear, to be equivalent they must exhibit the same values of v_{OC}, i_{SC}, and R_{eq}.

We have already encountered the concept of equivalence in Section 2.2 in connection with series and parallel resistance combinations. For instance, regarding the resistive network of Figure 2.19 as a single equivalent resistance simplifies the task of finding the current drawn from the source. In Section 2.5 we used the equivalence concept to model practical sources. Though they are fictitious circuits, the models allow us to predict source behavior under the effect of loading. Moreover, equivalence provides the basis for the source transformation technique of Section 3.5. Our interest in the equivalence of one-ports stems from the desire to *simplify* analysis and design when different one-ports are connected together to form more complex circuits.

Thévenin's Theorem

Consider the one-port v–i characteristic of Equation (4.2), repeated here for convenience,

$$v = v_{OC} - R_{eq}i \qquad \textbf{(4.28)}$$

Its formal similarity to the v–i characteristic of a practical voltage source, expressed by Equation (2.37), suggests that a linear one-port can be modeled like a practical voltage source, namely, with the *series* combination of an *ideal* voltage source v_{OC} and a resistance R_{eq}. This equivalence, illustrated in Figure 4.13, forms the essence of **Thévenin's Theorem,** named for the French engineer M. L. Thévenin (1857–1926). More specifically, we have:

Thévenin's Theorem: Any linear resistive one-port is equivalent to an *ideal voltage source* v_{OC} in *series* with a resistance R_{eq}, where v_{OC} and R_{eq} are, respectively, the open-circuit voltage and the equivalent resistance of the one-port.

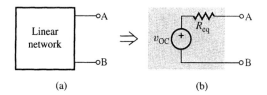

(a) (b)

Figure 4.13 Linear one-port and its Thévenin equivalent.

A CONVERSATION WITH
JACK KILBY

TEXAS INSTRUMENTS

Jack St. Clair Kilby was the coinventor of the integrated circuit while at Texas Instruments. (Courtesy of Texas Instruments)

Engineering was being put to a severe test in the 1950s. The invention of the transistor in 1947 had made miniaturization possible, but a ceiling, a seemingly insurmountable limitation, was already looming: the human skill needed in the manufacture of transistorized devices. Circuit designers could sketch out elaborate systems containing hundreds, thousands of electronic components... in theory, anyway. In practice, each of these components had to be joined perfectly to its neighbors with conducting wire, perfectly because the finished unit must of course be able to conduct electricity. One bad solder joint out of thousands and an entire circuit was useless.

In the industry, the problem was christened "the tyranny of numbers," and, with visions of 100,000-component radar systems and 200,000-component computers dancing in their heads, engineers everywhere were working on it. How could you make and join all these components into a functional circuit without worrying about the mechanical problems of soldering techniques?

Enter Jack Kilby, today, best known as one of the inventors of the integrated circuit.* Kilby had a B.S. degree in EE from the University of Illinois (1947) and a M.S. from the University of Wisconsin (1950), also in EE. His undergraduate work was interrupted

---

*Robert Noyce and Jack Kilby are both given credit for the invention.

by World War II, during which he served in the OSS, precursor of today's CIA. In 1947 he went to work for a Milwaukee electronics company, Centralab, where his job was to develop ceramic thick-film circuits, and he fast made himself an expert on transistor technology. Newly hired by Texas Instruments (TI) in the spring of 1958 and working alone during July because everyone else was on vacation and he had not yet built up any vacation time, Kilby came up with the idea that has allowed electronics fabricators to crash through the size and cost ceiling.

The trick was in what material you used to make your components. From the very beginning of the electric age, resistors had always been made of some inexpensive material having very high resistivity, capacitors of some other inexpensive material having a very high dielectric constant, and so forth. At the time Kilby arrived at Texas Instruments, the company had a very heavy investment in the semiconductor silicon. Kilby figured any idea that was going to fly with management would have to use silicon and so began thinking about this inexpensive material and what it could do: doped one way, it could work in a pinch as a conductor, but certainly not a very good one; doped another way, it could work as a resistor, but again certainly not a very good one. And this was the genius of Jack Kilby: he thought about what would happen when you made a circuit of poor but workable silicon resistors and poor but workable silicon capacitors and poor but workable silicon "everything else." On July 24, he wrote in his lab notebook: "Extreme miniaturization of many electrical circuits could be achieved by making resistors capacitors [sic] and transistors and diodes on a single slice of silicon."

And he was right. In an interview covering his early days and the state of the electronics industry today, Jack Kilby has these thoughts to offer.

What was the world of electronics like before the integrated circuit? In other words, what were the things leading up to your invention?

It was a much smaller field—radio, television, the first computers, that's about all. People could see that there were desirable things not being done because of cost and size. Almost everyone recognized that there had to be a better way than putting things together one piece at a time. To find this better way, a number of programs were started in the mid-1950s. Circuit integration was only one of many ways to solve the size problem.

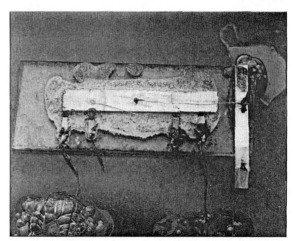

The first integrated circuit, a phase-shift oscillator, was invented by Kilby in 1958. (Courtesy of Texas Instruments)

There were certainly other approaches. A micromodule, which used regular components but with an automated assembly setup, was one way, but it's pretty much dead today. Another tack was the thick- and thin-film approach; there were some components you couldn't make with this method, but it could make great connections and for that reason is still used today in hybrid circuits and some computers.

By the time I got to TI in 1958, lots of people had already been working on the problem. Waiting for everyone to return from vacation that summer, I came

(Continued)

(Continued)

up with what I considered to be merely a tentative approach, this idea of making all the components out of the same semiconductor material. It just worked out, so my tentative approach was the one that stuck. It took several months to build that first circuit—from July 1958 to September—with lots of skilled technicians helping me.

Did all you people at TI recognize right away how important your idea was?

Those of us involved thought it would be important for the existing electronics world. What no one figured on was how very much this idea would *change* that world. Even though the first circuit we built was tested and worked on September 12, we didn't file our patent until the following February. And it wasn't until a year later, February 1960, that I heard about Bob Noyce, when the patent office notified us that a similar patent had been filed—in July of 1959, I believe. Things weren't moving too quickly on any front.

Do you think this failure to instantly see worth is true with all the great ideas?

Yes. To cite just one obvious case, Edison had a very tough job, I am sure. For one thing, the gas companies were big and powerful opponents; for another thing, the first electric light was not that good a product. Edison may have been all enthused about it, but other people surely weren't. Not at the start, at any rate. All ideas have to prove themselves, and all ideas depend on timing. Babbage invented a computer in the middle of the 1800s, but the time for it just was not right.

How does it feel to have invented something so seminal?

Well, my personal feeling is of great satisfaction, of course, but you've got to remember that nothing is ever one person working alone. All ideas have roots in other, earlier ideas. The best you can claim is that you started something big, and that's how I like to think of myself, as having started something nice and big.

It has been written that you insist on being described as an engineer rather than a scientist. Why do you feel so strongly about the distinction?

Because there's such a basic difference in the two functions. A scientist is interested in expanding our knowledge base, and an engineer is interested in applying existing knowledge to new problems. And since that's what I do, solve problems, I'm an engineer.

As an example of the application of this theorem, we can say that the voltage divider circuit of Figure 4.14(a) is equivalent to the series combination of Figure 4.14(b). The expressions for the element values of this equivalent were derived in Example 4.1.

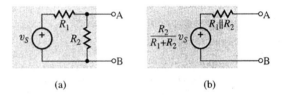

(a)   (b)

Figure 4.14   Voltage divider circuit and its Thévenin equivalent.

Is all your work problem-driven, or have you had research projects that were the result of some question you just began wondering about one day?

Pretty much, it's problem-driven, which goes back to what I just said about why I consider myself an engineer and not a scientist: I love to solve problems. If you're really lucky, sometimes you get to choose your own problem, and in those cases I guess you could say that observation and wondering played a role.

If you had everything to do over, what would you change? What would you be working on if you were twenty-two again?

For the change part, I have to answer nothing of any significance. Sounds awful to say that, but it's true. Things have worked out very well for me over the years.

As for what I'd be working on, there is a whole new set of problems in electronics today. The field is so much broader than the radio, television, and primitive computer of the early days. Because costs have dropped so much—by a factor of about a million—electronics can do so much today that was not feasible thirty years ago. A personal computer in every briefcase, a cellular phone in every car, electronic cars. So many areas. I don't know which one I'd choose.

What are your memories about the circuits course you took in college?

One thing I can tell you is how good the teachers were when I was at Illinois. A great faculty that really helped us over the rough spots. Our teachers knew the mistakes we were going to make before we made them and so could steer us away from trouble. I think we used a circuits book by Knight, and he was one of my teachers.

Back then, electronics was just beginning to come into its own, so "electrical engineering" meant studying power for the most part. We were mostly concerned with the generation and transmission of power from source to consumer, how to design motors, things like that. It really was a completely different world.

As one of the prime players who have made that completely different world, what's your advice to today's EE students?

Electronics is a fascinating field, completely open-ended. Even with all the stupendous advances we've already seen, there's so much more to come. The sense of limitless possibility is exhilarating. So take the courses you need and do whatever else you have to do in school to earn your engineering degree. And don't anybody run off and be a lawyer!

It is intriguing that however complex a one-port may be, its terminal behavior can be mimicked by a circuit as simple as its Thévenin equivalent. Replacing the one-port with its equivalent in the course of our calculations may save considerable time and effort. To illustrate this principle, suppose we wish to find the voltage v in the circuit of Figure 4.15(a) for different values of R, say, for $R = 1\ k\Omega$, $4\ k\Omega$, $6\ k\Omega$, $12\ k\Omega$, and $16\ k\Omega$. One way to proceed is to use nodal or loop analysis to derive a general expression for v as a function of R and then repeatedly substitute the given values of R to find $v(R)$. For example, applying nodal analysis yields

$$\frac{9 - v_A}{3} = \frac{v_A}{6} + \frac{v(R)}{R} \qquad \textbf{(4.29a)}$$

$$\frac{v(R)}{R} = \frac{v_A - v(R)}{2} + 5 \tag{4.29b}$$

where we have used $v_B = v_A - v(R)$. Eliminating v_A and solving for $v(R)$ yields, after suitable algebraic manipulation,

$$v(R) = \frac{R}{4+R} 16 \tag{4.30}$$

It is now a simple matter to substitute the given values of R to obtain, respectively, $v(1) = [1/(4+1)]16 = 3.2$ V, $v(4) = 8$ V, $v(6) = 9.6$ V, $v(12) = 12$ V, and $v(16) = 12.8$ V.

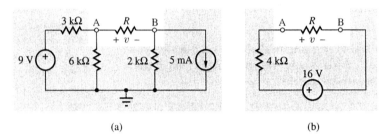

(a)   (b)

Figure 4.15 Replacing the circuit surrounding R with its Thévenin equivalent.

A moment's reflection, however, indicates that Equation (4.30) can be regarded as the voltage divider formula for the case in which our resistance R is connected to the series combination of a 16-V voltage source and a 4-kΩ resistance. This is depicted in Figure 4.15(b). We wonder whether there is a quicker way of reaching this conclusion, without having to solve the system of equations of Equation (4.29). The answer is obtained by applying Thévenin's Theorem to the network *surrounding R*. As depicted in Figure 4.16(a), this network appears to R as a one-port with A and B as its terminals.

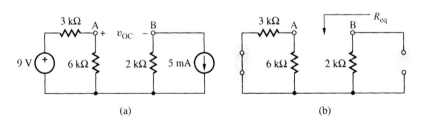

(a)   (b)

Figure 4.16 Finding v_{OC} and R_{eq} for the circuit surrounding R in Figure 4.15(a).

Referring to Figure 4.16(a), v_{OC} is readily found as $v_{OC} = [6/(3+6)]9 - (-2)5 = 16$ V. Likewise, suppressing all sources as in Figure 4.16(b), we readily find $R_{eq} = (3 \parallel 6) + 2 = 4$ kΩ. This confirms the element values in the Thévenin equivalent of Figure 4.15(b).

Even though R is in fact part of the network of Figure 4.15(a), we choose to work with the fictitious equivalent of Figure 4.15(b) merely to *simplify* our analysis. This is legitimized by the fact that the equivalent has the same values of v_{OC} and R_{eq} as the original network. Dealing with the equivalent is much easier because we can obtain Equation (4.30) via the simple voltage divider formula.

From the one-port viewpoint, R is referred to as the *load.* We wish to stress once again that Thévenin's Theorem is applied just to the portion of the circuit of Figure 4.15(a) *surrounding R,* not to the entire circuit!

As we replace a one-port with its equivalent, we must make sure that the *polarity* of the Thévenin source correctly reflects the polarity of v_{OC} in the original port. This is why it is so critical that we label the port terminals *before we even start* analyzing the circuit!

▶ Example 4.7

The bridge circuit of Example 4.3 is connected to a load as in Figure 4.17(a). Find the load voltage and current if the load is

(a) a 1-kΩ resistance;

(b) a 10-V source with the positive terminal at A;

(c) a 2-mA source directed from A to B;

(d) a plain wire.

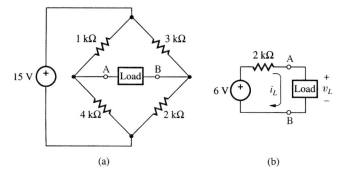

(a) (b)

Figure 4.17 A loaded bridge and its Thévenin equivalent.

Solution

Replace the bridge circuit with its Thévenin equivalent as depicted in Figure 4.17(b). The element values of this equivalent were derived in Example 4.3. In Figure 4.17(b) we have:

(a) $i_L = v_{OC}/(R_{eq} + R_L) = 6/(2 + 1) = 2$ mA; $v_L = R_L i_L = 1 \times 2 = 2$ V.

(b) $v_L = 10$ V; $i_L = (v_{OC} - v_L)/R_{eq} = (6 - 10)/2 = -2$ mA,
indicating that now current flows from B, through the load, to A.

(c) $i_L = 2$ mA; by KVL, $v_L = v_{OC} - R_{eq}i_L = 6 - 2 \times 2 = 2$ V.

(d) $v_L = 0$ V; $i_L = v_{OC}/R_{eq} = 6/2 = 3$ mA.

Exercise 4.9   Using Thévenin's Theorem, find the load voltage and current if the one-port of Figure 4.8(a) is loaded with

(a) a 30-kΩ resistance;

(b) a 12-V source with the positive terminal at A;

(c) a 12-V source with the positive terminal at B;

(d) a 1-mA source pointing downward.

ANSWER   (a) 6 V (positive at A), 0.2 mA (A → load → B);
(b) 0.2 mA (B → load → A); (c) 1.4 mA (A → load → B);
(d) 6 V (positive at B).

Exercise 4.10   Repeat Exercise 4.9, but for the one-port of Figure 4.8(b).

ANSWER   (a) 3.75 V (positive at C), 0.125 mA (C → load → D);
(b) 0.7 mA (D → load → C); (c) 1.7 mA (C → load → D);
(d) 5 V (positive at D).

▶ **Example 4.8**

If the one-ports of Figure 4.8 are connected together, A with C and B with D, so that the composite circuit of Figure 4.18(a) results, find the common voltage v.

Solution

The circuit could in principle be solved via the node or loop methods. However, a much quicker alternative is to replace the one-ports with their individual Thévenin equivalents, whose element values were found in Exercises 4.2 and 4.3. The result is shown in Figure 4.18(b). Then, using the superposition principle, we readily find

$$v = \frac{10}{15 + 10}9 + \frac{15}{15 + 10}5 = 6.6 \text{ V} \qquad \textbf{(4.31)}$$

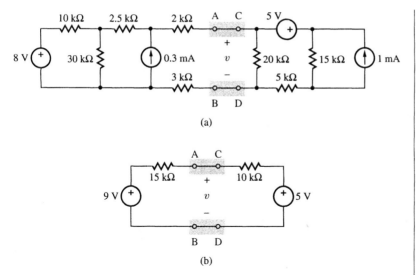

(a)

(b)

Figure 4.18 Using Thévenin's Theorem to find the voltage shared by two one-ports.

Exercise 4.11   Using Thévenin's Theorem, find the magnitude and direction of the current exchanged by the one-ports of Figure 4.8 when they are connected (a) A with C and B with D; (b) A with D and B with C.

ANSWER   (a) 0.16 mA (A → C, and D → B); (b) 0.56 mA (A → D, and C → B).

Norton's Theorem

Let us now turn to the one-port i–v characteristic of Equation (4.5), repeated here for convenience,

$$i = i_{SC} - \frac{1}{R_{eq}}v \qquad (4.32)$$

Its similarity with Equation (2.39) suggests that a linear one-port can also be modeled like a practical current source, namely, with the parallel combination of an ideal current source i_{SC} and a resistance R_{eq}. This equivalence, illustrated in Figure 4.19, forms the essence of **Norton's Theorem**, named for the American scientist Edward L. Norton (born 1898). More specifically, we have:

Norton's Theorem: Any linear resistive one-port is equivalent to an *ideal current source* i_{SC} in *parallel* with a resistance R_{eq}, where i_{SC} and R_{eq} are, respectively, the short-circuit current and the equivalent resistance of the one-port.

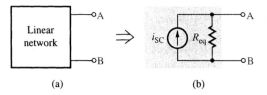

(a)   (b)

Figure 4.19 Linear one-port and its Norton equivalent.

In light of this, a linear one-port admits not one but *two* equivalents, the Thévenin and the Norton equivalents. The value of R_{eq} is the *same* in both equivalents, and it can be found either by Method 1 or by Method 2 of the previous section. Since the Thévenin voltage v_{OC} and the Norton current i_{SC} are related as

$$v_{OC} = R_{eq}i_{SC}$$

(4.33)

once one of the equivalents is known, the other is immediately obtained by a simple source transformation. Like Thévenin's Theorem, Norton's Theorem provides a powerful tool for circuit analysis. It is fair to say that all previous problems in this section could have been solved just as well using Norton's Theorem instead of Thévenin's.

Exercise 4.12   In the one-port of Figure 4.7(a) let $v_S = 20$ V and $i_S = 3$ A. Using Norton's Theorem, find the load voltage and current if the one-port is loaded with

(a) a 2-Ω resistance;

(b) a 4-V source with the positive terminal at A;

(c) a 4-V source with the positive terminal at B;

(d) an open circuit.

ANSWER   (a) 4/3 V (positive at A), 2/3 A (A $\rightarrow$ load $\rightarrow$ B); (b) 0 A; (c) 2 A (A $\rightarrow$ load $\rightarrow$ B); (d) 4 V (positive at A).

▶ **Example 4.9**

In the circuit of Figure 4.20(a) let $R_L = 2$ kΩ. Using Norton's Theorem, find the power delivered to R_L.

Solution

Norton's Theorem must be applied to the circuit *external* to R_L. To find i_{SC}, short-circuit the one-port as in Figure 4.20(c). By KCL, $i_{SC} = i_1 + i_3$.

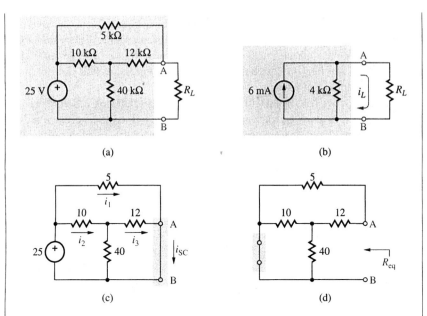

Figure 4.20 Using Norton's Theorem to find the power dissipated in R_L.

By Ohm's Law, $i_1 = 25/5 = 5$ mA. By the current divider formula, $i_3 = i_2 \times 40/(40+12) = (40/52)i_2$. By Ohm's Law, $i_2 = 25/[10+(40 \parallel 12)] = 1.3$ mA. Substituting for i_2 yields $i_3 = (40/52)1.3 = 1$ mA. Finally, $i_{SC} = i_1 + i_3 = 5+1 = 6$ mA.

To find R_{eq}, suppress the 25-V source as in Figure 4.20(d). Then, by inspection, $R_{eq} = 5 \parallel [12 + (10 \parallel 40)] = 4$ kΩ.

Replacing the one-port with its Norton equivalent yields the simplified circuit of Figure 4.20(b). By the current divider formula, the load current is $i_L = 6 \times 4/(4 + 2) = 4$ mA. Finally, the load power is $p_L = R_L i_L^2 = 2 \times 4^2 = 32$ mW.   ◀

Exercise 4.13   Repeat Example 4.9 if the 25-V source is changed to a 9-mA source flowing upwards.

ANSWER   135.1 mW.

Thévenin and Norton Comparison

Although either equivalent can in principle be used to represent a linear one-port, the Thévenin equivalent usually works best with *series* connections, the Norton equivalent with *parallel* connections. Clearly, if R_{eq} happens to be zero, we have no choice but to use the Thévenin equivalent. Likewise, if R_{eq} happens to be infinite, we must use the Norton equivalent.

If R_{eq} happens to be much *smaller* than the external load, the one-port behavior will be closer to that of a voltage source, so the Thévenin model would be more appropriate in this case. Likewise, if R_{eq} is much *greater* than the load, then the Norton equivalent is more appropriate, as the one-port now approximates a current source.

Finally, we wish to point out that Thévenin's and Norton's theorems may also be applied to *portions* of a network to simplify intermediate calculations. Moreover, conversions back and forth between the two equivalents can often simplify circuit analysis significantly.

▶ Example 4.10

Referring once again to the circuit of Figure 4.15(a), repeated in Figure 4.21(a) for convenience, find v by applying suitable Thévenin and Norton manipulations.

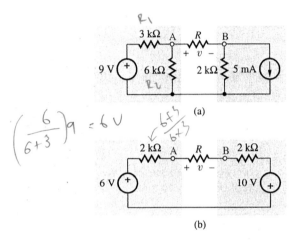

Figure 4.21 An alternate way of solving the circuit of Figure 4.15(a).

Solution

Apply Thévenin's Theorem to the portion of the circuit at the left of R, and a Norton to Thévenin conversion to the portion at the right.

As shown in Figure 4.21(b), the left portion is replaced with a voltage source of value $[6/(3 + 6)]9 = 6$ V and a series resistance of value $3 \parallel 6 = 2$ kΩ. Using a source transformation, the right portion becomes a voltage source of value $-2 \times 5 = -10$ V in series with a 2-kΩ resistance.

It is now straightforward to apply KVL and Ohm's Law to find

$$v(R) = R\frac{6 - (-10)}{2 + R + 2} = \frac{R}{4 + R}16$$

This confirms Equation (4.30), but without having to find the Thévenin equivalent of the entire one-port surrounding *R*.   ◀

Exercise 4.14   In the circuit of Figure 4.21(a), let $R = 4$ kΩ. Find the current through the 6-kΩ resistance by applying Norton's Theorem to the portion of the circuit at its right, and by performing a Thévenin to Norton conversion to the portion at its left.

ANSWER   1/3 mA ($\downarrow$).

CONCLUDING REMARKS

Looking back we note that we have a choice of different analysis techniques: *nodal analysis, loop analysis,* the *superposition principle, source transformations, Thévenin's* and *Norton's theorems.* We are of course interested in minimizing our time and effort as well as reducing the risk of error. Which method is best suited depends on the particular circuit and, to some extent, on personal preference. Until you gain more experience, try being proficient in *all methods,* not just in some. Moreover, try developing the habit of using *one method* to solve your problem and a *different method* to check your results!

• 4.3 NONLINEAR CIRCUIT ELEMENTS

Resistances and sources share the common feature of having i–v characteristics that are *straight lines.* Connecting these elements together yields networks whose terminal characteristics at any pair of nodes are still straight lines. Moreover, the analysis of these networks involves *linear algebraic equations.*

As you proceed through your electronics curriculum, you will find that many circuit elements are inherently nonlinear, with i–v characteristics that are *curves* rather than straight lines. If these characteristics are available in mathematical form, a circuit with one or more nonlinear elements can still be analyzed using Kirchhoff's laws as well as the element laws; however, the resulting equations will in general be *nonlinear.* Moreover, interchanging the terminals of a nonlinear element will in general change its effect in a circuit. For this reason, the terminals of nonlinear elements are usually given separate names, such as *anode* and *cathode* in the case of a diode.

▶ **Example 4.11**

Let the nonlinear element in the circuit of Figure 4.22(a) be the nonlinear resistor discussed in Section 1.6, whose i–v characteristic is

$$i = 0 \quad \text{for } v \leq V_T \tag{4.34a}$$

$$i = \frac{k}{2}(v - V_T)^2 \quad \text{for } v \geq V_T. \tag{4.34b}$$

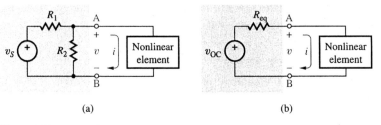

Figure 4.22 (a) Circuit with a nonlinear element. (b) Replacing the linear portion of the circuit with its Thévenin equivalent.

Assuming $k = 0.8$ mA/V^2, $V_T = 1.5$ V, $R_1 = 3$ kΩ, and $R_2 = 2$ kΩ, find the value of v_S that will ensure $v = 4$ V.

Solution

With $v = 4$ V, the current through the nonlinear resistor is, by Equation (4.34b), $i = (0.8/2)(4 - 1.5)^2 = 2.5$ mA. Moreover, by Ohm's Law, $i_{R_2} = v/R_2 = 4/2 = 2$ mA ($\downarrow$); by KCL, $i_{R_1} = i_{R_2} + i = 2 + 2.5 = 4.5$ mA ($\rightarrow$); by Ohm's Law, $v_{R_1} = R_1 i_{R_1} = 3 \times 4.5 = 13.5$ V, positive at the left; by KVL, $v_S = v_{R_1} + v = 13.5 + 4 = 17.5$ V. ◀

Exercise 4.15   In the circuit of Figure 4.22(a) let the nonlinear element be again the nonlinear resistor of Example 4.11. If $v_S = 9$ V and $R_1 = 1000$ Ω, find the value of R_2 that will ensure $i = 1$ mA.

ANSWER   626.4 Ω.

▶ Example 4.12

In the circuit of Figure 4.22(a) let the nonlinear element be again the nonlinear resistor of Example 4.11. If $v_S = 10$ V, $R_1 = 3$ kΩ, and $R_2 = 12$ kΩ, find v and i.

Solution

Though the circuit element is nonlinear, the circuit *surrounding* it is linear, indicating that we can use a Thévenin reduction to simplify our task. The result is shown in Figure 4.22(b), where $v_{OC} = [12/(3 + 12)]10 = 8$ V, and $R_{eq} = 3 \parallel 12 = 2.4$ kΩ. By KVL, $v = v_{OC} - R_{eq}i$. Substituting i as given in Equation (4.34b) yields

$$v = v_{OC} - R_{eq}\frac{k}{2}(v - V_T)^2 \qquad (4.35)$$

Numerically, $v = 8 - 2.4 \times (0.8/2)(v - 1.5)^2$, that is,

$$0.96v^2 - 1.88v - 5.84 = 0 \qquad (4.36)$$

This second-order equation admits the solutions $v_1 = 3.633$ V and $v_2 = -1.675$ V. However, the second solution contradicts Equation (4.34a), indicating that we must discard it. We thus have $v = v_1 = 3.633$ V.

   Substituting into Equation (4.34b) yields $i = 1.820$ mA.   ◀

Exercise 4.16   The nonlinear resistor of Example 4.11 is connected between terminals A and B of the one-port of Figure 4.8(a) in such a way that current flows from A, through the nonlinear resistor, to B. Sketch the circuit and use Thévenin's Theorem to find this current.

ANSWER   0.4308 mA.

Iterative Techniques

Often the equations describing a nonlinear circuit cannot be solved analytically. A case in point is offered by diode and bipolar transistor circuits, whose equations involve linear as well as exponential (or logarithmic) terms. Called *transcendental equations,* they can often be used to find a solution by an *iterative* procedure as follows. Starting with an initial guess for the unknown voltage or current, we use the transcendental equation to find an improved estimate for the unknown variable. The procedure is then repeated, using the improved estimate as a new guess, until the difference between successive estimates becomes suitably small. An example will better illustrate the procedure.

▶ ## Example 4.13

Let the nonlinear element in the circuit of Figure 4.22(a) be the rectifier diode of Section 1.6, whose v–i characteristic is

$$v = V_T \ln\left(\frac{i}{I_s} + 1\right) \qquad (4.37)$$

If $v_S = 10$ V, $R_1 = 3$ kΩ, $R_2 = 12$ kΩ, $V_T = 26$ mV, and $I_s = 2 \times 10^{-15}$ A, find v and i.

Solution

Referring to the equivalent circuit of Figure 4.22(b), we have, by Ohm's Law, $i = (v_{OC} - v)/R_{eq}$. Substituting Equation (4.37) yields

$$i = \frac{v_{OC}}{R_{eq}} - \frac{V_T}{R_{eq}} \ln\left(\frac{i}{I_s} + 1\right) \tag{4.38}$$

Numerically, $i = 8/2400 - (0.026/2400) \ln[i/(2 \times 10^{-15}) + 1]$, or

$$i = \left[3.333 - 0.01083 \ln\left(\frac{i}{2 \times 10^{-15}} + 1\right)\right] 10^{-3} \text{ A} \tag{4.39}$$

This transcendental equation cannot be solved analytically. We therefore start out with a crude guess, say,

$$i_0 = 1 \text{ mA}$$

and we substitute it into Equation (4.39) to find an improved estimate,

$$i_1 = 3.333 - 0.01083 \ln\left(\frac{10^{-3}}{2 \times 10^{-15}} + 1\right) = 3.041 \text{ mA}$$

Repeating the procedure, but using i_1 as our new guess, we find an even better estimate,

$$i_2 = 3.333 - 0.01083 \ln\left(\frac{3.041 \times 10^{-3}}{2 \times 10^{-15}} + 1\right) = 3.030 \text{ mA}$$

Repeating once again yields

$$i_3 = 3.333 - 0.01083 \ln\left(\frac{3.030 \times 10^{-3}}{2 \times 10^{-15}} + 1\right) = 3.030 \text{ mA}$$

Since the difference between the last two estimates is unappreciable within the resolution of four significant digits, which is the resolution we have been using in our calculations, we take $i = 3.030$ mA as our final result.

Substituting into Equation (4.37) yields $v = 0.026 \ln[(3.030 \times 10^{-3})/(2 \times 10^{-15}) + 1] = 0.729$ V. As a check, you can verify that the calculated data do satisfy KCL at node A.   ◀

Exercise 4.17   The diode of Example 4.13 is connected between terminals C and D of the one-port of Figure 4.8(b) in such a way that current flows from C, through the diode, to D. Sketch the circuit. Then, use Thévenin's Theorem and the iterative technique to find this current.

ANSWER   0.432 mA.

Graphical Analysis

The voltage and current developed by a nonlinear element when connected to an arbitrary network as in Figure 4.23(a) can also be found by a graphical technique.

This technique is especially useful when the i–v characteristic of the nonlinear element is available only in graphical form. This characteristic may be provided by the manufacturer in the data sheets, or it may be obtained experimentally by the user with the help of a curve tracer.

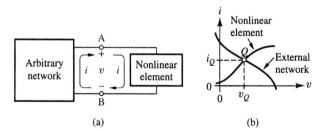

Figure 4.23 Graphical analysis.

Graphical analysis requires that the i–v characteristics of the nonlinear element and the surrounding network be *separately* graphed in the same i–v plane. As is shown in Figure 4.24, each of these characteristics is derived by applying a test voltage v, finding the resulting current i for different values of v, and then plotting the data, point by point. The result is, in general, a pair of curves of any form, as depicted in Figure 4.23(b). Note that in order to graph the curves in the same i–v plane we must express them in terms of the *same* set of i and v variables. In particular, the current *leaving* the network must *enter* the element, and vice versa. It was precisely in anticipation of this constraint that current direction relative to voltage polarity of one-ports in Figure 4.1 was chosen *opposite* to that of individual circuit elements in Figures 1.21 and 1.22.

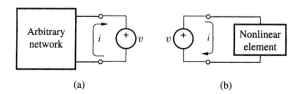

Figure 4.24 Circuit arrangements to separately find the i–v characteristics of a nonlinear element and the surrounding network.

It is now clear that since the nonlinear element and the surrounding network share the same voltage and current, the values of these variables are given by the coordinates v_Q and i_Q of the point Q *common* to both curves. This point is called the **operating point** of the circuit.

Of particular interest is the case in which the nonlinear element is embedded in a *linear network,* for then the latter has a straight-line characteristic, with v-axis intercept at $v = v_{OC}$ and i-axis intercept at $i = i_{SC}$. Graphical analysis for this particular case is summarized as follows:

(1) Graph the $i-v$ characteristic of the nonlinear element.

(2) Draw a straight line with v-axis intercept at $v = v_{OC}$ and i-axis intercept at $i = i_{SC}$, where v_{OC} and i_{SC} are, respectively, the open-circuit voltage and short-circuit current of the network surrounding the nonlinear element.

(3) Locate the intercept Q of the two curves. Then, the values of its coordinates provide the solution for the voltage v and current i shared between the network and the nonlinear element.

▶Example 4.14

In the circuit of Figure 4.25(a) let the nonlinear element be the nonlinear resistor of Example 4.11. Use graphical analysis to estimate v and i.

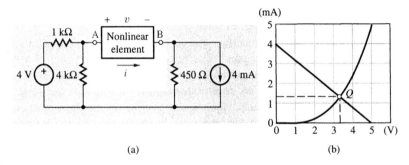

(a)   (b)

Figure 4.25 Graphical analysis of a circuit containing a nonlinear resistor.

Solution

First, graph the $i-v$ characteristic of the nonlinear element. With $k = 0.8$ mA/V^2 and $V_T = 1.5$ V, Equation (4.34) yields $i = 0$ for $v \le 1.5$ V and $i = 0.4(v - 1.5)^2$ mA for $v \ge 1.5$ V. Calculating i for different values of v and plotting the results point by point yields the quadratic curve of Figure 4.25(b).

Next, apply any of the techniques of the previous sections to the one-port surrounding the nonlinear element to find $v_{OC} = 5$ V and $i_{SC} = 4$ mA. This leads to the straight line of Figure 4.25(b).

Finally, by inspection we find that the coordinates of the operating point are, respectively, $v_Q \simeq 3.3$ V and $i_Q \simeq 1.3$ mA.   ◀

Exercise 4.18   In the bridge circuit of Figure 4.17(a), let the value of the source be changed to 5 V and let the load be a rectifier diode such that current flows from A, through the diode, to B. (a) Assuming

$I_s = 2 \times 10^{-15}$ A and $V_T = 0.026$ V, graph the diode i–v characteristic

$$i = I_s \left(e^{v/V_T} - 1 \right) \tag{4.40}$$

(b) Use graphical analysis to estimate the diode voltage and current.

ANSWER   0.69 V, 0.65 mA.

Because of its inherent resolution limitations, the graphical method only provides an *estimate* for the solution. This estimate is often used as the starting point for more refined calculations, such as iterative techniques, or computer simulation techniques, to be discussed in Section 4.5.

Being pictorial in nature, the graphical method offers an alternate viewpoint for understanding circuit behavior. For instance, varying the value of v_{OC} (or i_{SC}) will translate the straight line parallel to itself, thus causing Q to move up or down the nonlinear curve. Conversely, if we manage to translate the nonlinear curve up or down, Q will move up or down the straight line. This feature is encountered in the study of transistor amplifiers.

4.4 POWER CALCULATIONS

Power calculations are an important part of circuit analysis, and an early exposure to the subject is especially desirable if the study of circuits is accompanied by experimentation in the laboratory, where the student is required to become familiar with instrumentation and measurement terminology.

Average Power

If the voltage across or the current through a resistance is of the time-varying type, so is the dissipated power $p(t)$. When dealing with practical power problems, it is of interest to know the **average power** P over a given time interval t_1 to t_2. This is defined as the *steady* or *dc power* P that over that interval would transfer *the same amount of energy* as $p(t)$. Adapting Equation (1.28) to the present situation yields

$$P \times (t_2 - t_1) = \int_{t_1}^{t_2} p(t)\, dt$$

or

$$P \triangleq \frac{1}{t_2 - t_1} \int_{t_1}^{t_2} p(t)\, dt \tag{4.41}$$

If $p(t)$ is periodic, the interval t_1 to t_2 is made to coincide with the period T.

These concepts are illustrated in Figure 4.26 for the case of a resistance subjected to a *sinusoidal* or *ac* voltage

$$v(t) = V_m \sin 2\pi f t \tag{4.42}$$

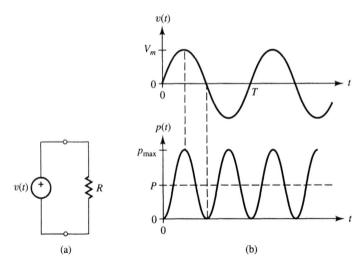

Figure 4.26 AC power dissipated in a resistance.

The instantaneous power is $p(t) = v^2(t)/R = (V_m^2/R) \sin^2 2\pi ft$. By the trigonometric identity $\sin^2 \alpha = (1 - \cos 2\alpha)/2$, we can write

$$p(t) = \frac{1}{2}\frac{V_m^2}{R}[1 - \cos 2\pi(2f)t] \qquad \textbf{(4.43)}$$

As shown in the corresponding graph, this function oscillates between 0 and $p_{max} = V_m^2/R$ at *twice* the frequency of the applied voltage. By inspection, the average power is $P = (1/2)p_{max} = (1/2)V_m^2/R$. This can be written as $P = (V_m/\sqrt{2})^2/R$, or

$$\boxed{P = \frac{V_{rms}^2}{R}} \qquad \textbf{(4.44a)}$$

where, following Equation (1.24), we have set $V_{rms} = V_m/\sqrt{2} = 0.707V_m$. The reason for the designation *rms* will be given shortly. Considering that the peak current I_m is related to the peak voltage V_m as $I_m = V_m/R$, average power can also be expressed as

$$\boxed{P = RI_{rms}^2} \qquad \textbf{(4.44b)}$$

where, using again Equation (1.24), we have set $I_{rms} = I_m/\sqrt{2} = 0.707I_m$. Eliminating R, the above equations yield yet another form for P,

$$\boxed{P = V_{rms}I_{rms}} \qquad \textbf{(4.44c)}$$

Comparing these with Equations (2.6) through (2.8) shows that an *ac voltage* of peak value V_m (or an *ac current* of peak value I_m) causes a resistance to dissipate the *same power* as a *dc voltage* of value $V_{rms} = V_m/\sqrt{2}$ (or a *dc current* of value $I_{rms} = I_m/\sqrt{2}$). For obvious reasons, V_{rms} and I_{rms} are also called the **effective values.**

RMS Values

In general, the effective value of a time-varying voltage $v(t)$ over a certain time interval t_1 to t_2 is defined as the *steady* or *dc voltage* that over that time interval would cause a resistance to dissipate the same amount of energy as $v(t)$. Denoting such a steady voltage as V_{rms} and using Equation (2.8), we write $(V_{rms}^2/R) \times (t_2 - t_1) = \int_{t_1}^{t_2} [v^2(t)/R]\, dt$, or

$$V_{rms} \triangleq \sqrt{\frac{1}{t_2 - t_1} \int_{t_1}^{t_2} v^2(t)\, dt} \qquad \textbf{(4.45a)}$$

Likewise, the rms value of current is

$$I_{rms} \triangleq \sqrt{\frac{1}{t_2 - t_1} \int_{t_1}^{t_2} i^2(t)\, dt} \qquad \textbf{(4.45b)}$$

In words, *the effective value is found by squaring the signal, computing its average or mean, and then taking the square root of the result.* Hence the origin of the designation *rms*. For a periodic signal the interval t_1 to t_2 is made to coincide with the period T.

As an example, consider the *ac* voltage

$$v(t) = V_m \sin \frac{2\pi t}{T}$$

Substituting into Equation (4.45a) and letting $t_1 = 0$ and $t_2 = T$ yields

$$V_{rms}^2 = \frac{V_m^2}{T} \int_0^T \sin^2 \frac{2\pi t}{T}\, dt = \frac{V_m^2}{T} \int_0^T \frac{1}{2}\left[1 - \cos\left(2\frac{2\pi t}{T}\right)\right] dt$$

where we have used the identity $\sin^2 \alpha = (1/2)(1 - \cos 2\alpha)$. Since the cosine integrates to zero, we have $V_{rms}^2 = (V_m^2/T) \times (T/2) = V_m^2/2$, or

$$V_{rms} = \frac{V_m}{\sqrt{2}} \qquad \textbf{(4.46a)}$$

thus providing a justification for Equation (1.24). Likewise, for an **ac current**

$i(t) = I_m \sin(2\pi t/T)$ we have

$$I_{\mathrm{rms}} = \frac{I_m}{\sqrt{2}}$$

(4.46b)

It is important to note that these relations between rms and peak values, which are true for ac signals, in general do not hold for nonsinusoidal signals. To find the correct relationships we must use, case by case, the definitions of Equation (4.45).

▶ **Example 4.15**

Derive an expression for the rms value of a *sawtooth* voltage waveform of the type of Figure 1.9(d).

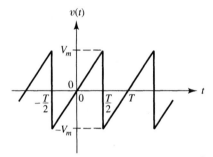

Figure 4.27 Waveform of Example 4.15.

Solution

The waveform is shown in Figure 4.27. Letting $t_1 = -T/2$ and $t_2 = T/2$, it is readily seen that over this time interval the waveform can be expressed as

$$v(t) = \frac{2V_m}{T}t$$

Using Equation (4.45a),

$$V_{\mathrm{rms}} = \sqrt{\frac{1}{T} \int_{-T/2}^{T/2} \left(\frac{2V_m}{T}\right)^2 t^2 \, dt} = \sqrt{\frac{1}{T} \left(\frac{2V_m}{T}\right)^2 \times \left.\frac{t^3}{3}\right|_{-T/2}^{T/2}}$$

Expanding and simplifying yields

$$V_{\mathrm{rms}} = \frac{V_m}{\sqrt{3}}$$

(4.47)

or $V_{\mathrm{rms}} = 0.577 V_m$. Clearly, this relationship is quite different from that of ac signals of Equation (4.46a).   ◀

Exercise 4.19   Derive an expression for the rms value of a *triangular* voltage waveform of the type of Figure 1.9(c).

ANSWER $V_{\text{rms}} = V_m/\sqrt{3}$.

▶ Example 4.16

(a) Given a current *pulse train* of the type of Figure 1.8, derive an expression for I_{rms} in terms of I_m and the duty cycle d.

(b) Find the duty cycle d for which this wave yields the same relationship between I_{rms} and I_m as the sine wave.

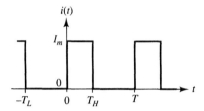

Figure 4.28   Waveform of Example 4.16.

Solution

(a) The waveform is shown in Figure 4.28. Let $t_1 = 0$ and $t_2 = T$. Since $i(t) = I_m$ for $0 \le t \le T_H$ and $i(t) = 0$ for $T_H \le t \le T$, Equation (4.45b) yields $I_{\text{rms}} = \sqrt{(1/T)(I_m^2 \times T_H + 0^2 \times T_L)} = I_m\sqrt{T_H/T}$, or

$$I_{\text{rms}} = I_m\sqrt{d} \qquad \textbf{(4.48)}$$

(b) A pulse train will yield the same relationship as a sine wave only when d is such that $\sqrt{d} = 1/\sqrt{2}$, or $d = 50\%$. ◀

Exercise 4.20   Derive an expression for the rms value of a *rectangular* voltage waveform of the type of Figure 1.9(b). Does this value depend on the duty cycle d? Justify physically.

ANSWER $V_{\text{rms}} = V_m$, regardless of d.

▶ Example 4.17

Find the rms value of the waveform of Figure 4.29(a).

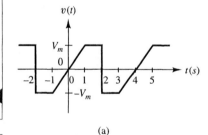

(a)

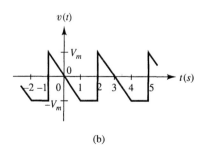

(b)

Figure 4.29   Waveforms of Example 4.17 and Exercise 4.21.

Solution

Given the symmetric nature of the waveform, it is convenient to choose $t_1 = 0$ s and $t_2 = 2$ s. Then,

$$V_{rms} = \sqrt{\frac{1}{2} \int_0^2 v^2(t)\, dt} = \frac{1}{\sqrt{2}} \sqrt{\int_0^1 (V_m t)^2\, dt + \int_1^2 V_m^2\, dt}$$

$$= \frac{1}{\sqrt{2}} \sqrt{V_m^2 \times \frac{t^3}{3}\Big|_0^1 + V_m^2 \times t\Big|_1^2} = V_m \sqrt{2/3}$$

◀

Exercise 4.21   Find the rms value of the waveform of Figure 4.29(b).

ANSWER $V_{rms} = V_m \sqrt{5/9}$.

▶ **Example 4.18**

(a) Find the rms value of a voltage waveform having a *dc component* V_S and an *ac component* of peak amplitude V_m.

(b) If a voltage alternates sinusoidally between 0 V and 1 V, what is its rms value?

Solution

(a) Letting

$$v(t) = V_S + V_m \sin \frac{2\pi t}{T}$$

and $t_2 - t_1 = T$ in Equation (4.45a), we obtain

$$V_{rms} = \sqrt{\frac{1}{T} \int_0^T \left(V_S^2 + 2V_S V_m \sin \frac{2\pi t}{T} + V_m^2 \sin^2 \frac{2\pi t}{T} \right) dt}$$

The first term integrates to $V_S^2 \times T$, the second term integrates to zero, and the third term integrates to $(V_m^2/2) \times T$. Hence,

$$V_{rms} = \sqrt{V_S^2 + V_m^2/2} \qquad\qquad \textbf{(4.49)}$$

(b) A voltage alternating sinusoidally between 0 and 1 V has $V_S = 0.5$ V and $V_m = 0.5$ V. Substituting into Equation (4.49) yields $V_{rms} = \sqrt{0.5^2 + 0.5^2/2} = 0.612$ V.

◀

Exercise 4.22   Sketch and label a voltage waveform alternating between $V_S + V_m$ and $V_S - V_m$ in sawtooth fashion. Hence, find its rms value.

ANSWER $V_{rms} = \sqrt{V_S^2 + V_m^2/3}$.

• AC Multimeters

Practical ac multimeters fall into two categories, *true rms* multimeters and *averaging* multimeters. True rms multimeters include special circuitry to perform the actual calculation of the rms value of the signal. To save the cost of this circuitry, averaging multimeters first synthesize the *half-cycle average* of the signal, which is a relatively easy task, and then multiply this average by an appropriate scale factor to yield a reading that, in the case of a *sinusoidal* or *ac* signal, is the rms value.

The advantage of averaging multimeters is lower cost; the disadvantage is that they provide correct rms readings *only* in the case of purely sinusoidal or ac signals. If the signal is other than sinusoidal, the meter will still provide a reading but this, in general, is *not* the true rms value of the given signal. To obtain the true rms value, the user must multiply the reading by a suitable correction factor, which varies from waveform to waveform.

Maximum Power Transfer

As a practical source is connected to a load, power is dissipated both in the internal resistance of the source and in the load. In the case of a resistively loaded voltage source as in Figure 4.30(a), the power in the load is $p_L = v_L^2/R_L$. By the voltage divider formula, $v_L = [R_L/(R_s + R_L)]v_s$. Eliminating v_L yields

$$p_L = \frac{R_L}{(R_s + R_L)^2} v_S^2 \qquad (4.50)$$

In applications such as audio and communications, where the function of current or voltage signals is to convey energy, it is of interest to *maximize* the power transferred from source to load. Note that p_L vanishes both in the limit $R_L \to 0$ and $R_L \to \infty$, indicating that there must be an intermediate value of R_L for which p_L is maximized. This value is found by differentiating p_L with respect to R_L and setting the result to zero,

$$\frac{dp_L}{dR_L} = \frac{R_s - R_L}{(R_s + R_L)^3} v_S^2 = 0$$

This yields

$$\boxed{R_L = R_s} \qquad (4.51)$$

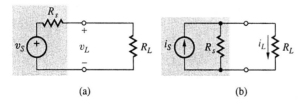

(a)   (b)

Figure 4.30 Power transfer in resistively loaded sources.

indicating that power transfer from a practical voltage source is maximized when the load resistance is made *equal* to the source resistance. When this condition is met, the load is said to be *matched* to the source. Moreover, the terminal voltage is one-half the unloaded voltage. Substituting Equation (4.51) into Equation (4.50) yields the maximum power,

$$p_{L(\max)} = \frac{v_S^2}{4R_s} \qquad \text{(4.52)}$$

Exercise 4.23   Referring to Figure 4.30(b), show that the maximum power transfer from a practical current source occurs when

$$R_L = R_s \qquad \text{(4.53a)}$$

and that its value is

$$p_{L(\max)} = \frac{1}{4}R_s i_S^2 \qquad \text{(4.53b)}$$

▶ **Example 4.19**

A certain radio antenna acts like a 1-mV unloaded source with $R_s = 300 \ \Omega$. What is the power transferred by this antenna to a matched load?

Solution

By Equation (4.52), $p_{L(\max)} = (1 \times 10^{-3})^2/(4 \times 300) = 5/6$ nW.   ◀

In the case of a load connected to an arbitrarily complex linear one-port, power transfer to the load is maximized when the load is made equal to the equivalent resistance R_{eq} of the port. Moreover, adapting Equations (4.52) and (4.53), the maximum power that the port can transfer is $p_{L(\max)} = v_{\text{OC}}^2/(4R_{\text{eq}}) = R_{\text{eq}}i_{\text{SC}}^2/4$, where v_{OC} and i_{SC} are, respectively, the open-circuit voltage and the short-circuit current of the port.

▶ **Example 4.20**

Find the value of R_L in the circuit of Figure 4.31(a) for which power transfer to R_L is maximized. What is the value of $p_{L(\max)}$?

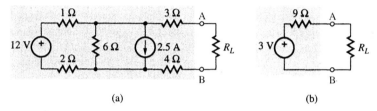

Figure 4.31 Using Thévenin's Theorem to find the maximum power transfer.

Solution

First we must find the Thévenin equivalent of the one-port seen by R_L. To this end, remove R_L from the circuit. Then, using the superposition principle, it is readily seen that

$$v_{OC} = \frac{6}{1+6+2}12 - [(1+2) \parallel 6]2.5 = 3 \text{ V}$$

Moreover, suppressing the sources, we find that

$$R_{eq} = 3 + [6 \parallel (1+2)] + 4 = 9 \ \Omega$$

The circuit of Figure 4.31(a) is thus equivalent to that of Figure 4.31(b). Clearly, power transfer is maximized when $R_L = R_{eq} = 9 \ \Omega$. Moreover, $p_{L(max)} = v_{OC}^2/(4R_{eq}) = 3^2/(4 \times 9) = 0.25$ W.

Exercise 4.24 Repeat Example 4.20, but with the 12-V source replaced by a 1.2-A current source flowing upward.

ANSWER $R_L = 13 \ \Omega$, $p_{L(max)} = 1.17$ W.

Efficiency

The *total* average power P_S *delivered* by a source is the sum of the average power *dissipated* in its internal resistance and the average power P_L *absorbed* by the load. The fraction

$$\eta \triangleq \frac{P_L}{P_S} \qquad \text{(4.54)}$$

is called the **efficiency** of the power-transfer process. Multiplying η by 100 yields the efficiency in percentage form. To conserve power, a designer will strive to maximize the efficiency of the circuit under consideration.

In the case of a **voltage source** with a resistive load R_L, we have $P_S = (R_s + R_L)I_{rms}^2$ and $P_L = R_L I_{rms}^2$, where I_{rms} is the rms value of the current

supplied by the source. This yields

$$\eta = \frac{R_L}{R_s + R_L}$$

(4.55)

indicating that efficiency will approach 1, or 100%, only in the limit $R_s \ll R_L$. Note that with a matched load we have equal power dissipation in the internal resistance and in the load, indicating that *maximum power transfer* is only 50% *efficient*.

By similar reasoning, it is readily seen that the efficiency of a resistively loaded **current source** is

$$\eta = \frac{R_s}{R_s + R_L}$$

(4.56)

indicating that 100% efficiency is approached only in the limit $R_s \gg R_L$.

▶ Example 4.21

Find the efficiency if $R = 5$ kΩ in the circuit of Figure 4.15(a).

Solution

By Equation (4.55), $\eta = 5/(4 + 5) = 0.5556$, or 55.56%.   ◀

 ## 4.5 CIRCUIT ANALYSIS USING SPICE

SPICE can be used to find Thévenin and Norton equivalents as well as to perform nonlinear circuit analysis. As such, SPICE is a very useful tool to check the results of hand calculations.

Finding Thévenin/Norton Equivalents via SPICE

Given an arbitrary one-port, SPICE can be directed to find R_{eq} as well as either v_{OC} or i_{SC} using a special statement called the *transfer function statement*. Its general form is

`.TF OUTVAR INSOURCE`

(4.57)

where `OUTVAR` is the desired voltage or current variable and `INSOURCE` is any of the independent sources in the circuit. For Thévenin equivalents, `OUTVAR` is of the form `V(N1,N2)`, and for Norton equivalents it is of the form `I(VXXX)`. The meaning of the variables `V(N1,N2)` and `I(VXXX)` has already been illustrated in connection with the `.PRINT DC` statement of Equation (3.22).

In response to the `.TF` statement, SPICE performs the dc analysis and also provides what is known as the *small-signal* dc gain from `INSOURCE` to `OUTVAR`, the equivalent resistance seen by `INSOURCE`, and the equivalent resistance seen by `OUTVAR`. The small-signal gain is of particular interest in transistor amplifier analysis, but it does not concern us here, as our primary interest is in the value of `OUTVAR` and the corresponding equivalent resistance.

▶ Example 4.22

Use SPICE to confirm the values of v_{OC} and R_{eq} for the voltage divider circuit of Example 4.1.

Solution

For convenience, the circuit is repeated in Figure 4.32(a). The input file is

```
THEVENIN EQUIVALENT OF THE VOLTAGE DIVIDER
VI 1 0 DC 12
R1 1 2 30K
R2 2 0 10K
.TF V(2) VI
.END
```

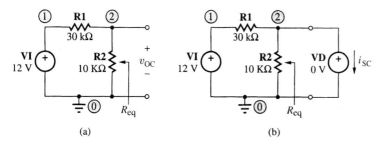

(a)   (b)

Figure 4.32 Using SPICE to find the Thévenin and Norton equivalents of the voltage divider.

After SPICE is run, the output file contains the following dc analysis results:

```
NODE      VOLTAGE    NODE      VOLTAGE
(1)       12.0000    (2)        3.0000
```

along with the following small-signal results,

```
V(2)/VI = 2.500E-01
INPUT RESISTANCE AT VI = 4.000E+04
OUTPUT RESISTANCE AT V(2)  = 7.500E+03
```

Clearly, we have $v_{OC} = $ `V(2)` $= 3$ V, and $R_{eq} = 7.5$ kΩ. ◀

▶ Example 4.23

Use SPICE to confirm the values of i_{SC} and R_{eq} for the voltage divider circuit of Example 4.1.

Solution

As shown in Figure 4.32(b), we insert a 0-V dummy voltage source to create a short and monitor i_{SC}. The input file is

```
NORTON EQUIVALENT OF THE VOLTAGE DIVIDER
VI 1 0 DC 12
R1 1 2 30K
R2 2 0 10K
VD 2 0 DC 0
.TF I(VD) VI
.END
```

After SPICE is run, the output file contains the following:

```
VOLTAGE SOURCE CURRENTS:
NAME        CURRENT
  VI      -4.000E-04
  VD       4.000E-04

SMALL-SIGNAL CHARACTERISTICS:
I(VD)/VI = 3.333E-05
INPUT RESISTANCE AT VI = 3.000E+04
OUTPUT RESISTANCE AT I(VD) = 7.500E+03
```

Clearly, we have $i_{SC} = \text{I(VD)} = 0.4$ ma, and $R_{eq} = 7.5$ kΩ.   ◀

▶ Example 4.24

Use SPICE to find the Thévenin equivalent of the circuit seen by R in Figure 4.15(a).

Solution

Referring to Figure 4.33, the input file is:

```
FINDING THE THEVENIN EQUIVALENT
V1 1 0 DC 9
R1 1 2 3K
R2 2 0 6K
R3 3 0 2K
I1 3 0 DC 5M
.TF V(2,3) V1
.END
```

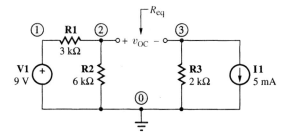

Figure 4.33 Circuit of Example 4.24.

After SPICE is run, the output file contains the following:

```
SMALL-SIGNAL BIAS SOLUTION:
NODE   VOLTAGE   NODE   VOLTAGE   NODE   VOLTAGE
(1)    9.0000    (2)    6.0000    (3)   -10.0000

SMALL-SIGNAL CHARACTERISTICS:
V(2,3)/V1 = 6.667E-01
INPUT RESISTANCE AT V1 = 9.000E+03
OUTPUT RESISTANCE AT V(2,3) = 4.000E+03
```

Clearly, we have $v_{OC} = $ V(2) - V(3) $= 6 - (-10) = 16$ V, and $R_{eq} = 4$ kΩ, thus confirming the equivalent of Figure 4.15(b). ◀

Exercise 4.25   Use SPICE to find the Norton equivalent of the circuit seen by R in Figure 4.15(a).

Exercise 4.26   Use SPICE to find the Thévenin equivalent of the circuit seen by R_L in Figure 4.20(a).

In the particular case in which the one-port circuit contains no independent sources, we are only interested in R_{eq}. This is most readily obtained by driving the port with a test source $i = 1$ A, and letting SPICE find the resulting voltage v. Then, $R_{eq} = v/i = v/1 = v$.

Exercise 4.27   Consider the resistive network obtained from the circuit of Figure 4.11(a) by suppressing all its sources. Sketch it; then use SPICE to find the equivalent resistance seen between nodes A and B.

• Nonlinear Resistors

SPICE allows circuits to contain nonlinear resistors whose i–v characteristics can be expressed in polynomial form as

$$i = p_0 + p_1 v + p_2 v^2 + p_3 v^3 + \cdots$$

where p_0, p_1, p_2, p_3, ... are the coefficients of the polynomial. The general form of the statement for a nonlinear resistor is

```
GXXX N+ N- POLY(1) N+ N- P0 P1 P2 P3 . . .
```
                                                                        (4.58)

This statement contains

   (1) The name of the nonlinear resistor, which must start with the letter G and may contain up to seven additional alphanumeric characters.

   (2) The circuit nodes N+ and N- to which the nonlinear resistor is connected, along with the keyword POLY(1); the order in which these nodes is specified *is important.*

   (3) The polynomial coefficients P0, P1, P2, P3, ..., in A, A/V, A/V^2, etc.

The nonlinear resistor statement is accompanied by the command

```
.NODESET V(N+)=VALUE+ V(N-)=VALUE-
```
                                                                        (4.59)

to provide SPICE with a guess at the initial voltages VALUE+ and VALUE- at nodes N+ and N-. These values can be estimated, for instance, via graphical analysis.

▶ Example 4.25

Use SPICE to find v in the circuit of Example 4.14.

Solution

The circuit is repeated in Figure 4.34 where the nonlinear resistor has been denoted as **G1**. With $k = 0.8$ mA/V^2 and $V_T = 1.5$ V, the i–v characteristic becomes $i = (0.8/2) \times 10^{-3}(v - 1.5)^2$, that is,

$$i = (0.9 - 1.2v + 0.4v^2) \text{ mA} \qquad\qquad \textbf{(4.60)}$$

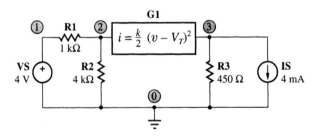

Figure 4.34 Circuit with a nonlinear resistor.

Using the graphical estimates $v_Q \simeq 3.3$ V and $i_Q \simeq 1.3$ mA obtained previously yields V(2) = 2.1 V and V(3) = −1.2 V. The input file is

```
CIRCUIT WITH A NONLINEAR RESISTOR
VS 1 0 DC 4
R1 1 2 1K
R2 2 0 4K
* HERE IS THE NONLINEAR RESISTOR:
G1 2 3 POLY(1) 2 3 0.9M -1.2M 0.4M
.NODESET V(2)=2.1V V(3)=-1.2V
R3 3 0 0.45K
IS 3 0 4M
.DC VS 4 4 1
.PRINT DC V(2,3)
.END
```

After SPICE is run, the output file contains the following:

```
DC ANALYSIS:
   VS            V(2,3)
4.000E+00      3.328E+00
```

Clearly, $v = V(2,3) = 3.328$ V.

Exercise 4.28   Use SPICE to find v and i in the circuit of Exercise 4.16.

▼ SUMMARY

- A *one-port* is a circuit with a *pair of terminals* through which signals or energy can enter or leave the circuit. Since these terminals are connected to internal nodes of the circuit, they must be labeled. Failure to label them may cause confusion in the reference polarities and directions of the port signals.

- The *i–v* characteristic of a *linear one-port* is a *straight line* with intercepts at $v = v_{OC}$ and $i = i_{SC}$, and with slope $-1/R_{eq}$, where v_{OC} is the *open-circuit voltage,* i_{SC} the *short-circuit current,* and R_{eq} the *equivalent resistance* of the one-port.

- v_{OC} and i_{SC} are due to the *internal* sources of the circuit, and they allow us to find the equivalent resistance as $R_{eq} = v_{OC}/i_{SC}$. We call this *Method 1* to find R_{eq}.

- An alternate method is to *suppress all internal independent sources* so as to drive v_{OC} and i_{SC} to zero, apply a *test voltage v* (or a *test current i*), find the *resulting current i* (or the *resulting voltage v*), and finally let $R_{eq} = v/i$. We call this *Method 2* to find R_{eq}.

- *Thévenin's* Theorem states that a linear one-port can be modeled with an ideal voltage source v_{OC} in series with a resistance R_{eq}. *Norton's* Theorem states that the same one-port can be modeled with an ideal current source i_{SC} in parallel with a resistance R_{eq}.

- *Replacing* a one-port with its Thévenin or Norton equivalent in the course of our calculations may simplify circuit analysis tremendously.

- To analyze a circuit containing a nonlinear element, we replace the circuit *surrounding* the element with its Thévenin or Norton equivalent. Then, we analyze the reduced circuit *analytically* if the nonlinear element law is available mathematically, or *graphically* if the element law is available in the form of a curve.

- When a resistance is subjected to an ac signal, the *instantaneous power* alternates at *twice the frequency* of the applied signal.

- The *effective* value of a time-varying signal over a certain time interval is the *steady signal* that over that interval would cause a resistance to dissipate the same amount of power. This value is also called the *rms value*.

- The relationship between the *rms* value X_{rms} and the *peak value* X_m of a periodic signal varies from signal to signal. For an ac signal it is $X_{\text{rms}} = X_m/\sqrt{2}$. However, for other signals it may be quite different.

- The *maximum amount of power transfer* from a source to a load occurs when the load and the source resistance are equal, in which case the load is said to be *matched* to the source.

- SPICE can be used to find the Thévenin or Norton equivalents of a circuit via the `.TF` statement. SPICE can also be used to analyze circuits with nonlinear resistors whose characteristics can be expressed in polynomial form.

▼ PROBLEMS

4.1 One-Ports

4.1 (a) Find v_{OC}, i_{SC}, and R_{eq} for the one-port of Figure P4.1. (b) Repeat, but with the direction of the current source reversed.

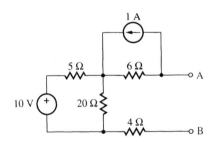

Figure P4.1

4.2 (a) Find v_{OC}, i_{SC}, and R_{eq} for the one-port of Figure P4.2. (b) Repeat, but with the polarity of the voltage source reversed.

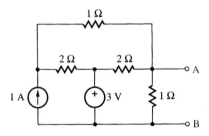

Figure P4.2

4.3 (a) Find v_{OC}, i_{SC}, and R_{eq} for the one-port of Figure P4.3. (b) Repeat, but with the two sources interchanged with each other.

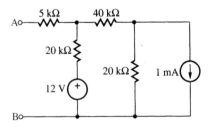

Figure P4.3

4.4 Find v_{OC}, i_{SC}, and R_{eq} for the one-port of Figure P4.4.

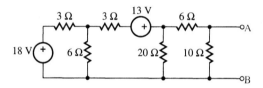

Figure P4.4

4.5 Repeat Problem 4.4, but with an additional 3-Ω resistance connected between node A and the node at the positive side of the 18-V source.

4.6 Find v_{OC}, i_{SC}, and R_{eq} for the one-port of Figure P4.6.

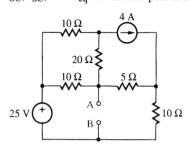

Figure P4.6

4.7 Repeat Problem 4.6, but with the 4-A current source replaced by a voltage source of 100/3 V, positive at the right.

4.8 In the circuit of Figure P4.8 find v_{OC}, R_{eq}, and i_{SC} for the case in which the port terminals are nodes (a) A and B; (b) A and C; (c) A and D; (d) B and C; (e) B and D.

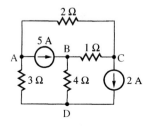

Figure P4.8

4.9 Repeat Problem 4.8, but with the 5-A current source replaced by an 8-V voltage source, positive at the left.

4.10 Find R_{eq} for the one-port of Figure P4.10.

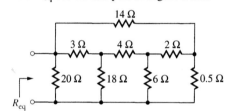

Figure P4.10

4.11 Find R_{eq} for the one-port of Figure P4.11.

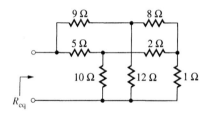

Figure P4.11

4.12 (a) Find the equivalent resistance between C and D in the circuit of Figure P2.14. (b) Repeat if A and B are shorted together.

4.2 Circuit Theorems

4.13 Find v and i in the circuit of Figure P4.13 if the load is (a) a 5-Ω resistance; (b) a 5-V source, positive at the top; (c) a 2/3-A source directed downward; (d) a 5-Ω resistance in series with a 5-V source, positive at the top.

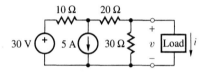

Figure P4.13

4.14 Repeat Problem 4.13, but with the 30-V voltage source changed to 50 V.

4.15 Find the Thévenin and Norton equivalents of the circuit of Figure P4.15. (*Hint:* First find the Norton equivalent of the subcircuit consisting of the voltage source and the surrounding resistive bridge.)

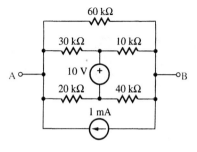

Figure P4.15

4.16 Use repeated applications of Thévenin's Theorem to find the magnitude and direction of the current through the 6-Ω resistance in the circuit of Figure P4.16.

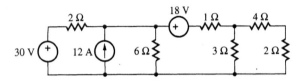

Figure P4.16

4.17   Use repeated applications of Norton's Theorem to find the magnitude and polarity of the voltage across the 12-A current source in the circuit of Figure P4.16.

4.18   If the circuits of Figure P4.18 are connected together, A with E and B with F, find the voltage (magnitude and polarity) and current (magnitude and direction) exchanged by the two circuits.

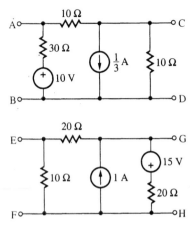

Figure P4.18

4.19   Repeat Problem 4.18 if the connections are C with G and D with H.

4.20   A voltmeter with input resistance R_i is connected between nodes C and D of Figure P4.20 to measure the voltage across the current source. Predict the reading if (a) $R_i = \infty$; (b) $R_i = 1$ MΩ. (*Hint:* Find the Thévenin equivalent of the circuit seen by the voltmeter.)

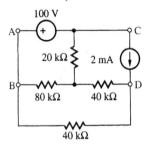

Figure P4.20

4.21   In Figure P4.20 the connection between A and B is opened and an ammeter with input resistance R_i is inserted to measure the current through the voltage source. Predict the

reading if (a) $R_i = 0$; (b) $R_i = 200$ Ω. (*Hint:* Find the Norton equivalent of the circuit seen by the ammeter.)

4.22   A voltmeter with input resistance R_i is connected across the 40-kΩ resistance in Figure P4.3. Predict the reading if (a) $R_i = \infty$; (b) $R_i = 1$ MΩ. (*Hint:* Find the Thévenin equivalent of the circuit seen by the voltmeter.)

4.23   The terminals of an ammeter with internal resistance R_i are connected to nodes A and C of Figure P4.18. Predict the reading if (a) $R_i = 0$; (b) $R_i = 0.2$ Ω. (*Hint:* Find the Norton equivalent of the circuit seen by the ammeter.)

4.24   A linear one-port yields $v = 10$ V when loaded with a resistance $R_L = 100$ kΩ, and $i = 0.25$ mA when loaded with $R_L = 10$ kΩ. Find the Thévenin and Norton equivalents of such a port.

4.3 Nonlinear Circuit Elements

4.25   Find v and i if the nonlinear element in the circuit of Figure P4.25 is a nonlinear resistor with $i = 0$ for $v \leq V_T$, and $i = 0.5k(v - V_T)^2$ for $V \geq V_T$, $k = 0.5$ mA/V^2, $V_T = 3.5$ V. Check your results.

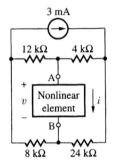

Figure P4.25

4.26   Repeat Problem 4.25, but with the current source replaced by a 20-V voltage source, positive at the right. What happens if the 20-V source is positive at the left?

4.27   A given nonlinear element has $i = (v/100) \times (1 - v/5)$ for $0 \leq v \leq 2.5$ V, with v in volts and i in amperes. Find i and v if this element is connected to a source $v_S = 10$ V with internal resistance $R_s = 1$ kΩ. Check your results.

4.28   (a) Find v and i if the nonlinear element in the circuit of Figure P4.25 is a diode with $i = I_s(e^{v/V_T} - 1)$, $I_s = 2 \times 10^{-15}$ A, $V_T = 0.026$ V. (b) Repeat, but with a 1-kΩ resistance in parallel with the current source.

4.29   A series of measurements at the terminals of an unknown circuit element have provided the following data, in V and mA: $(v, i) = (0, 0), (1, 0.5), (2, 2), (3, 5.5), (4, 7.5), (5, 8)$. Using graphical techniques, estimate i and v if the element is connected to a 6-V source with a 1-kΩ internal resistance.

4.30 The nonlinear element of Problem 4.29 is connected to a source v_S via a 2-kΩ series resistance. What value of v_S will yield $i = 4$ mA through the element? $v = 1.5$ V across the element?

4.4 Power Calculations

4.31 Find the rms value of the waveform of Figure P4.31.

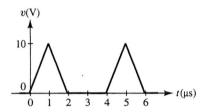

Figure P4.31

4.32 Find the rms value of the waveform of Figure P1.25.

4.33 Find the rms value of the waveform of Figure P4.33.

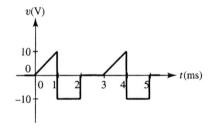

Figure P4.33

4.34 Find the rms value of the waveform of Figure P1.26.

4.35 Find the rms value of the *half-wave rectified* ac voltage waveform, defined as $v(t) = V_m \sin(2\pi t/T)$ for $0 \le t \le T/2$, $v(t) = 0$ for $T/2 \le t \le T$, and $v(t \pm nT) = v(t)$ for all n and t.

4.36 Consider the periodic voltage waveform defined as $v(t) = 4(1 - e^{-t})$ V for $0 < t < 1$ s, and $v(t \pm n1) = v(t)$ for all n and t. Sketch it and find its rms value.

4.37 In household light-dimmers and other appliances, variable power control is achieved by controlling the firing angle θ of an electronic switch, so the ac voltage waveform appears as in Figure P4.37. Find its rms value as a function of θ. Hence, plot the ratio V_{rms}/V_m versus θ for $0 \le \theta \le \pi$ and discuss the cases $\theta = 0$, $\theta = \pi/2$, and $\theta = \pi$.

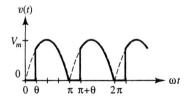

Figure P4.37

4.38 Find the maximum power that a resistive load can absorb from the circuit of Figure P4.38 if the load is connected (a) between nodes A and B; (b) between nodes C and D.

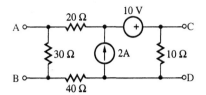

Figure P4.38

4.39 Find the maximum power that a resistive load can absorb from the circuit of Figure P4.38 if the load is connected (a) between nodes A and C; (b) between nodes B and D.

4.40 A linear one-port yields $v = 10$ V when loaded with a resistance $R_L = 10$ kΩ, and $v = 4$ V when loaded with $R_L = 1$ kΩ. (a) Find the maximum power that such a port can deliver to a resistive load, as well as the corresponding load resistance. (b) Find the efficiency in the case of a 5-kΩ load.

4.41 In the circuit of Figure P4.20 the wire connecting A and B is removed and a variable resistance R is inserted. What is the value of R for which it will absorb maximum power from the circuit? What is this power?

4.42 If a variable resistor R is connected between nodes E and G of the circuit of Figure P4.18, what is the value of R for which it will absorb maximum power from the circuit? What is this power?

4.43 (a) In the circuit of Figure P4.43 find a condition for R in terms of R_s and R_L such that the source v_S with resistance R_s is matched with the rest of the circuit. (b) Find R if $R_s = 20$ kΩ and $R_L = 10$ kΩ. (c) Find R if $R_s = 10$ kΩ and $R_L = 20$ kΩ.

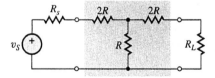

Figure P4.43

4.5 Circuit Analysis Using SPICE

4.44 Use SPICE to find the Thévenin and Norton equivalents of the circuit of Figure P4.4.

4.45 Use SPICE to find the Thévenin and Norton equivalents of the circuit of Figure P4.6.

4.46 Use SPICE to find R_{eq} in the circuit of Figure P4.10.

4.47 Use SPICE to find R_{eq} in the circuit of Figure P4.11.

4.48 Solve Problem 4.25 via SPICE.

4.49 Solve Problem 4.27 via SPICE.

5

TRANSFORMERS AND AMPLIFIERS

Having mastered the tools of circuit analysis, we now wish to apply them to the study of two systems of great practical interest, namely, the **transformer** and the **amplifier,** the workhorses of electrical and electronic engineering. The former finds application especially in the handling of electrical power, the latter in the processing of electronic information. While Chapters 1 through 4 have, of necessity, emphasized analysis, Chapters 5 and 6 shift attention to applications using analysis as a tool for understanding the function of a circuit.

Both transformers and amplifiers belong to a class of circuits known as *two-port circuits,* or **two-ports** for short. One port, called the **input port,** serves as point of *entry* of energy or information coming from a circuit upstream called the **source.** The other port, called the **output port,** serves as point of exit of energy or information for delivery to a circuit downstream called the **load.** The situation is depicted in Figure 5.1.

Perhaps the simplest two-port examples are the voltage and current dividers, whose function is to attenuate a signal coming from an external source. A far more sophisticated two-port example is a hi-fi audio amplifier, which accepts

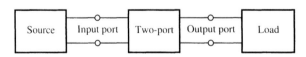

Figure 5.1 Two-port and its typical interconnection.

a weak input signal from a transducer source and magnifies it to produce the powerful output signal required to drive a loudspeaker load.

In light of Section 4.1, both the input and output ports exhibit individual *i–v characteristics,* called, respectively, the **input** and **output characteristics.** If the ports are linear, these characteristics can be modeled in terms of separate Thévenin or Norton equivalents. However, the distinguishing feature of a two-port is the existence of a relationship between its output and input signals, called the **transfer characteristic.** This interdependence between the ports is modeled by means of **dependent sources.**

We begin the chapter by extending the analytical techniques of the previous chapters to circuits containing dependent sources. We find that these sources add a new facet to circuit behavior, the most striking of which is *resistance transformation.* Using just a resistance and a dependent source as a vehicle, we show how the apparent value of the resistance may be increased, decreased, or even made negative! It is precisely because of this extraordinary impact upon circuit behavior that the study of dependent sources has been deferred until now.

Next, we study the *transformer* and its applications to power transmission and resistance matching, followed by the study of the four basic amplifier types, namely, the *voltage, current, transresistance,* and *transconductance* amplifiers.

Finally, we illustrate the use of SPICE to analyze circuits with dependent sources and to find, among others, the Thévenin equivalent of a one-port containing this type of source.

5.1 DEPENDENT SOURCES

In the previous chapters we deliberately avoided dependent sources in order to keep circuits simple and better help develop a feel for their physical operation. It is now time to examine these sources in proper detail, because they are an indispensable ingredient in the modeling of transformers and amplifiers, the central subjects of the present chapter.

Except for the fact that its value depends on a controlling voltage v_X or current i_X *somewhere else* in the circuit, a dependent source functions much like an independent one, providing the designated voltage or the designated current *regardless* of the load. Moreover, a dependent source will either release or absorb power, depending on whether current exits or enters its positive terminal.

In general, the values of the controlling signal and the controlled source are not known, but are circuit variables that must be found by calculation. Some examples will illustrate.

▶ **Example 5.1**

Find i_X and $k_r i_X$ in the circuit of Figure 5.2. Hence, verify the conservation of power.

Solution

The presence of the + symbol inside the diamond indicates that the controlled source is a *voltage* source, and the presence of the symbol i_X

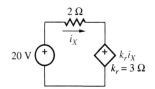

Figure 5.2 Circuit of Example 5.1, containing a dependent source.

next to it indicates that the controlling signal is a *current*. We thus have a CCVS producing the voltage $v_{CCVS} = k_r i_X$, with k_r in V/A, or Ω. To find i_X we apply Ohm's Law,

$$i_X = \frac{20 - 3i_X}{2}$$

Solving, we obtain $i_X = 4$ A, and $k_r i_X = 3 \times 4 = 12$ V.

The resistance is absorbing $p_{2\Omega} = 2 \times i_X^2 = 2 \times 4^2 = 32$ W. Since current *exits* the independent source via the positive terminal and *enters* the dependent source via the positive terminal, the former is releasing $p_{20V} = 20 \times 4 = 80$ W, and the latter is absorbing $p_{CCVS} = 12 \times 4 = 48$ W. Clearly, $32 + 48 = 80$. ◀

If we change the value of any parameter in the circuit, such as the resistance, the parameter k_r, or even the reference direction of i_X, then the controlled variable will also generally change.

Exercise 5.1

(a) Find i_X and $k_r i_X$ if the direction of i_X in Figure 5.2 is reversed so that i_X now flows toward the left, everything else remaining the same. Hence, verify the conservation of power.

(b) Find i_X and $k_r i_X$ if the reference direction of i_X in Figure 5.2 is toward the left, the resistance is raised to 3 Ω, and k_r is lowered to 2 Ω. Hence, verify the conservation of power.

ANSWER   (a) $i_X = 20$ A, $k_r i_X = 60$ V; (b) -20 A, -40 V.

▶ **Example 5.2**

Find v_X and $k_g v_X$ in the circuit of Figure 5.3. Hence, verify the conservation of power.

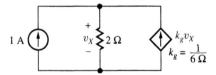

Figure 5.3 Circuit of Example 5.2.

Solution

The arrow inside the diamond indicates that the controlled source is a *current* source, and the presence of the symbol v_X next to it indicates that the controlling signal is a *voltage*. We thus have a VCCS drawing the current $i_{\text{VCCS}} = k_g v_X$, with k_g in A/V, or Ω^{-1}. To find v_X we apply Ohm's Law and KCL,

$$v_X = 2(1 + (1/6)v_X)$$

Solving, we obtain $v_X = 3$ V, and $k_g v_X = (1/6) \times 3 = 0.5$ A.

The resistance is absorbing $p_{2\Omega} = v_X^2/2 = 3^2/2 = 4.5$ W. The sources are releasing, respectively, $p_{1A} = 3 \times 1 = 3$ W, and $p_{\text{VCCS}} = 3 \times 0.5 = 1.5$ W. Clearly, $3 + 1.5 = 4.5$. ◀

Exercise 5.2   Repeat Example 5.2, but with the direction of the VCCS reversed, so that its current now points downward.

ANSWER $v_X = 1.5$ V, $k_r v_X = 0.25$ A.

Resistance Transformation

It is apparent that a dependent source is generally unable to initiate any voltage or current on its own in a circuit. We need an independent source to create the controlling signal; only then will the controlled source respond with its own signal. We then wonder what is the role of a dependent source in a circuit that contains *no* independent sources. In this respect, it is instructive to regard a dependent source as a generalization of the concept of resistance. Whereas a resistance imposes a constraint between the voltage and current at the *same* branch, a dependent source imposes a constraint between *different* branches. It is precisely this interdependence that leads to forms of circuit behavior that may, at times, seem to defy expectations based on our experience with independent sources. Let us illustrate with an example.

We wish to find the equivalent resistance of the circuit of Figure 5.4(a), consisting of just one resistance and one dependent source. The source monitors the voltage v_X applied at the left side of the resistance and drives the right side with a voltage k times as large. We cannot suppress the source to simplify our analysis because this would be tantamount to imposing $k = 0$, and we cannot change the circuit to suit our needs. The proper way to proceed is to apply a test voltage v, find the resulting current i, and obtain the equivalent resistance as $R_{\text{eq}} = v/i$.

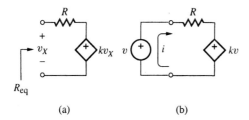

(a)   (b)

Figure 5.4 Finding the equivalent resistance of a circuit containing a dependent source.

Once we apply a test voltage v as in Figure 5.4(b), the dependent source responds with the voltage kv. By Ohm's Law,

$$i = \frac{v - kv}{R} \tag{5.1}$$

or $i = (1 - k)v/R$. Then, $R_{eq} = v/i = v/[(1 - k)v/R]$, or

$$R_{eq} = \frac{R}{1 - k} \tag{5.2}$$

Depending on the value of k, R_{eq} can assume a variety of different values, including infinity and even negative values! All this is due to the presence of the dependent source, which indirectly affects the voltage drop across R and, hence, the value of i, thus leading to unexpected values for the apparent resistance of the circuit. To better appreciate the role of the dependent source let us assume specific element values, say,

$$v = 1 \text{ V}$$

$$R = 1 \text{ } \Omega$$

and examine circuit behavior for different values of k.

(1) $k = 0$. The dependent source behaves as a short circuit so that the current drawn from the test source is $i = v/R = (1 \text{ V})/(1 \text{ } \Omega) = 1$ A. Thus, $R_{eq} = v/i = 1/1 = 1 \text{ } \Omega$, or

$$R_{eq} = R \tag{5.3}$$

as anticipated. Clearly, this is the result that we would obtain with a short in place of the dependent source.

(2) $0 \le k \le 1$. Now the effective voltage drop experienced by R is *less* than the applied test voltage, leading to a decrease in i and, hence, to an increase in the apparent resistance. For instance, with $k = 0.9$, the dependent source responds to the 1-V test voltage with a 0.9-V voltage, so the drop across R is only $1 - 0.9 = 0.1$ V, yielding $i = 0.1/1 = 0.1$ A. Hence, $R_{eq} = v/i = 1/0.1 = 10 \text{ } \Omega$. Making k closer to unity further decreases the effective drop across R and, hence, the value of i, leading to a greater apparent resistance. For instance, with $k = 0.999$, Equation (5.1) indicates an effective drop of only 0.001 V, and Equation (5.2) predicts $R_{eq} = 1000R = 1000 \text{ } \Omega$. In the limit $k \to 1$ we obtain $R_{eq} \to \infty$ because $i \to 0$. In short, for $0 \le k \le 1$ we have

$$R \le R_{eq} \le \infty \tag{5.4}$$

This situation is encountered in a class of amplifiers known as *negative feedback amplifiers,* where it is used to increase the effective resistance of one or both of its ports.

(3) $k \le 0$. Now the dependent voltage is of opposite polarity to the test voltage, thus making the effective drop across R *greater* than the test voltage itself. This causes an increase in the value of i and, hence, a decrease in the apparent resistance. For instance, with $k = -9$, the

dependent voltage is -9 V. The drop across R is $1 - (-9) = 10$ V, so $i = 10/1 = 10$ A and $R_{eq} = v/i = 1/10 = 0.1$ Ω. Making k more negative further increases the effective drop across R and, hence, the value of i, thus reducing the apparent resistance. For instance, with $k = -999$, Equation (5.2) predicts $R_{eq} = R/1000 = 0.001$ Ω, practically a short circuit when compared to R. In short, for $k \leq 0$ we have

$$0 \leq R_{eq} \leq R \qquad (5.5)$$

This situation is again found in *negative feedback amplifiers,* where it is used to decrease the effective resistance of one or both of its ports.

(4) $k > 1$. Now an even more extraordinary effect occurs. Since the dependent voltage is *greater* than the test voltage, current direction is *reversed,* thus resulting in what is known as **negative resistance** behavior. For instance, with $k = 2$, Equation (5.1) yields $i = (1-2)/1 = -1$ A, the negative sign being an indication that i now flows *into* the 1-V test source. Consequently, Equation (5.2) predicts $R_{eq} = -R = -1$ Ω.

Positive and negative resistance behaviors are contrasted in Figure 5.5(a) and (b), where we observe that the test source *supplies* power if the resistance is *positive* but *receives* power if the resistance is *negative.* Even though R itself is positive in Figure 5.5(c), it is the presence of the dependent source that leads to negative resistance behavior. In short, for $k > 1$ we have

$$R_{eq} < 0 \qquad (5.6)$$

This kind of behavior is found in a class of amplifiers known as *positive feedback amplifiers,* and it forms the basis of a class of circuits known as *regenerative circuits,* such as oscillators, Schmitt triggers, and flip-flops.

In spite of its simplicity, the preceding example illustrates a feature that is common to most circuits containing dependent sources: by establishing an interdependence between different branches of a circuit, **a dependent source**

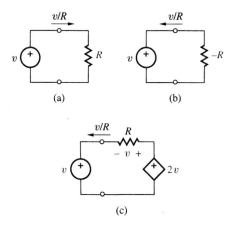

(a)

(b)

(c)

Figure 5.5 (a) Positive resistance, (b) negative resistance, and (c) negative resistance simulation using a positive resistance and a dependent source.

may alter the *apparent value* or even the *apparent nature* of its components, thus creating the conditions for a variety of intriguing effects. As we proceed, we shall see that it is precisely this feature that makes electrical and electronic engineering fascinating and resourceful!

• Transistor Modeling

An important application of dependent sources is the modeling of transistors, the ingredients of electronic amplifiers. One of the most common transistors is the *npn bipolar junction transistor* (*npn* BJT), whose circuit symbol and model are shown in Figure 5.6. As indicated by the model, the transistor accepts an external current i_B at a fixed voltage V_{BE}, multiplies this current by a constant β known as the *current gain,* and draws the magnified current βi_B from the C terminal regardless of the voltage v_{CE}. The values of V_{BE} and β depend on the particular device. Typically, $V_{BE} = 0.7$ V and $\beta = 100$.

To investigate the behavior of a transistor circuit, we simply replace the transistor with its circuit model and then apply the usual techniques of circuit analysis.

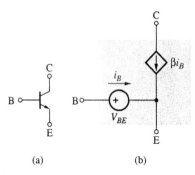

(a)   (b)

Figure 5.6 Circuit symbol and model for the *npn* bipolar junction transistor (*npn* BJT).

▶ **Example 5.3**

In the circuit of Figure 5.7(a) let $R_B = 180$ kΩ, $R_C = 2$ kΩ, and $V_{CC} = 5$ V. Assuming $V_{BE} = 0.7$ V and $\beta = 100$, find v_I for $v_O = 3$ V.

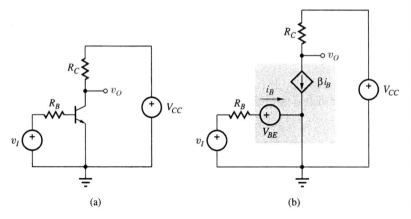

(a)   (b)

Figure 5.7 A transistor circuit and its equivalent.

Solution

Referring to the equivalent circuit of Figure 5.7(b) we have, by KVL and Ohm's Law,

$$v_I = V_{BE} + R_B i_B$$

$$v_O = V_{CC} - R_C \beta i_B$$

Thus, $i_B = (V_{CC} - v_O)/\beta R_C = (5 - 3)/(100 \times 2) = 0.01$ mA, and $v_I = 0.7 + 180 \times 0.01 = 2.5$ V.   ◀

Exercise 5.3 In the circuit of Figure 5.7(a) let $R_B = 100$ kΩ, $R_C = 1$ kΩ, $V_{CC} = 5$ V, and $V_{BE} = 0.7$ V. If it is found that $v_O = 2$ V with $v_I = 3.1$ V, what is the value of β for this BJT sample?

ANSWER 125.

Figure 5.8(a) shows a common circuit to bias a transistor for amplifier operation. We wish to develop a formula for finding I_B in terms of all remaining circuit parameters. Once I_B is known, all other currents and voltages in the circuit can be found by repeated application of Ohm's and Kirchhoff's laws.

To facilitate our task, we replace the transistor by its circuit equivalent, and we also apply Thévenin's Theorem to the portion of the circuit to the left of B. The result is shown in Figure 5.8(b), where we have

$$V_{BB} = \frac{R_2}{R_1 + R_2} V_{CC} \tag{5.7a}$$

$$R_{BB} = R_1 \parallel R_2 \tag{5.7b}$$

The currents *entering* the transistor are I_B and βI_B. By KCL, their sum must equal the current *exiting* the transistor. This allows us to state that the current through R_E is $I_{R_E} = I_B + \beta I_B = (\beta + 1)I_B$, as is shown. Going around the labeled mesh, we encounter a voltage rise and three voltage drops, or $V_{BB} = R_{BB}I_B + V_{BE} + R_E(\beta + 1)I_B$. Solving for I_B yields

$$I_B = \frac{V_{BB} - V_{BE}}{R_{BB} + (\beta + 1)R_E} \tag{5.8}$$

As you progress through the electrical engineering curriculum, you will find yourself using these equations over and over again.

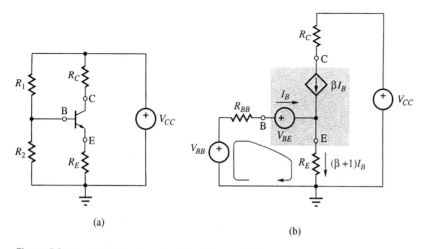

(a)

(b)

Figure 5.8 Transistor biasing circuit and its equivalent.

Lee De Forest

The Father of Radio

Lee De Forest demonstrates an early wireless telephone based on his "Audion" or amplifying diode.

All the electronic devices that fill our lives in these last days of the twentieth century owe their existence to an inventor who worked during the first days of the century: Lee De Forest; for it was De Forest who designed the triode tube—his "Audion"—the first electronic device capable of amplifying a signal.

Born in 1873 in Council Bluffs, Iowa, and growing up in Talladega, Alabama, De Forest described himself as a tinkerer from the time he was ten years old, building an "almost-lifesize" steam locomotive in his backyard in 1882. At 13, he announced his design for a perpetual motion machine: "I, a mere youth of 13 years, by my inventive genius and concentrated thought and study have succeeded where illustrious philosophers in times past have failed . . . ."

As an undergraduate at Yale from 1893 through 1896, De Forest told anyone who would listen that he was going to be like Tesla, Marconi, and Edison, the holy trinity of electricity. After Yale awarded him a doctorate in 1899, he ended up in Chicago, a laborer in the dynamo factory of Western Electric, Ph.D. notwithstanding. When he was promoted to technician in the telephone lab, he worked on his idea for a detector for wireless broadcast signals.

His device, nicknamed the "sponder," worked in 1901. Needing money and publicity to patent it, De Forest persuaded officials of the Publisher's Press Association to let him use his sponder to report on a big yacht race being held in September 1901 in New York Harbor. Marconi's company was under contract to report the race for the Associated Press. In those days, though, the operative word was "broadcast," with the emphasis on the "broad." Neither Marconi nor De Forest bothered to tune his transmitter, and the signal transmitted from each set caused so much interference in the other set that neither man succeeded in reporting any race results. In a high point of journalistic integrity, both press associations, having ballyhooed the wireless angle for weeks, lied and bylined their stories "Received by wireless telegraphy from tug following the yachts." Thus the failure hurt neither inventor, and De Forest was able to persuade investors to stake him in the American De Forest Wireless Telegraphy Company. In 1906, however, the company's directors forced De Forest out, keeping all his wireless patents and leaving him only the supposedly worthless patent on his Audion and $1,000 in "severance pay."

The basic principle of the Audion had been around since 1883, when Edison noticed that a current will carry from a heated filament to a cold receiving plate when both the filament and the plate are in an evacuated tube or in a tube containing gas at low pressure. This "Edison effect" seemed to have no practical application, so no workers devoted any time to it for two decades, until John Fleming, working for the Marconi Company in 1904, figured out that this setup could be used as an ac rectifier *and* as a detector of radio waves. The "Fleming valve" diode was the turn-of-the-century counterpart of the semiconductor diode of the 1940s; however, neither could amplify.

De Forest's genius was to see that a third electrode would control the charge flow and also amplify the signal. A changing voltage applied across this third electrode would cause the current to change,

> *"I, a mere youth of 13 years, by my inventive genius and concentrated thought and study have succeeded where illustrious philosophers in times past have failed...."*

and this changing current passed through an external resistance would generate an amplified copy of the original voltage.

De Forest received a patent for the Audion in January 1907, but until 1912 the applications of his invention were slow in coming. The problem was that no one—De Forest included—realized the importance of vacuum in the tube. His version used a gas-filled tube, and the tubes made in those early years were unreliable because the gas pressure could not be controlled. It was not until the chemist Irving Langmuir demonstrated the importance of a high vacuum that the modern vacuum tube was born.

In 1914, needing funds to keep his foundering Radio Telephone Company going, De Forest sold his Audion patent to AT&T for $50,000. Improved on by H. D. Arnold and other researchers at AT&T, De Forest's triode led to AT&T's first transcontinental telephone network in 1915 and to transatlantic communication soon afterward.

De Forest's life was a long series of patent fights, struggles to raise capital for his inventions, and three failed marriages (his fourth marriage lasted). In

March 1912, he was arrested for having used the mails to defraud. He had mailed out promotional literature to boost sales of stock in his company, and now stockholders, angry about the decline in the stock's worth and nonpayment of dividends, wanted their money back. Also named in the complaint were S. E. Darby (De Forest's patent attorney), and two officers of the company—James Smith and Elmer Burlingame. The trial was held in New York City in November 1913, and the verdicts were guilty for Smith and Burlingame and not guilty for Darby and De Forest. One interesting detail shows how important discoveries can so easily be unrecognized. During the trial, the Audion—the seed of all electronics—was characterized as "a strange device like an incandescent lamp... which ... has proven to be worthless."

In 1914 came litigation with Marconi over the Audion, ending in a stalemate whereby the courts ruled that neither man could manufacture the tubes without the consent of the other, followed by years

> *"...a strange device like an incandescent lamp...which... has proven to be worthless."*

of lawsuits against Edwin Armstrong about who had invented feedback circuitry. This time, the U.S. Supreme Court ruled that De Forest was the correct patent-holder.

The 1920s saw De Forest turn his inventive talents to a talking machine for motion pictures. The result was the 1922 De Forest Phonofilm, but again there were battles with other inventors and investors.

He lived, virtually unknown, from the 1930s on—always struggling to make a comeback. He wrote an autobiography that was published in 1950 but sold poorly. He quietly (and unsuccessfully) campaigned to receive the Nobel Prize in the mid-50s, and died in 1961. As testimony to his pitifully bad business skills, his estate was valued at $1,200.

▶ **Example 5.4**

In the circuit of Figure 5.8(a) let $R_1 = 62$ kΩ, $R_2 = 33$ kΩ, $R_C = 3.0$ kΩ, $R_E = 2.0$ kΩ, and $V_{CC} = 12$ V. Assuming $V_{BE} = 0.7$ V and $\beta = 80$, find V_B, V_C, and V_E.

Solution

Applying Equations (5.7), we find $V_{BB} = 4.168$ V and $R_{BB} = 21.54$ kΩ. Substituting into Equation (5.8) yields $I_B = 18.90$ μA. By Ohm's Law, $V_E = R_E(\beta + 1)I_B = 2 \times 81 \times 0.0189 = 3.061$ V; by KVL, $V_B = V_E + V_{BE} = 3.061 + 0.7 = 3.761$ V; by KVL and Ohm's Law, $V_C = V_{CC} - R_C\beta I_B = 12 - 3 \times 80 \times 0.0189 = 7.465$ V. In short, $V_B = 3.761$ V, $V_C = 7.465$ V, and $V_E = 3.061$ V. ◀

Exercise 5.4   In the circuit of Figure 5.8(a) let $R_1 = 100$ kΩ, $R_E = 3.0$ kΩ, and $V_{CC} = 15$ V. Assuming $V_{BE} = 0.7$ V and $\beta = 115$, find R_2 and R_C to achieve $V_E = 4$ V and $V_C = 10$ V. (*Hint:* Apply KCL at node B.)

ANSWER $R_2 = 51.36$ kΩ, $R_C = 3.783$ kΩ.

5.2 Circuit Analysis with Dependent Sources

The analysis of circuits containing dependent sources relies on the same tools used for circuits with independent sources. However, before embarking upon actual examples, we wish to stress the following:

(1) In general, the values of the controlling signals i_X or v_X are not known but are circuit variables that must be found by calculation, using additional circuit equations as needed.

(2) *Dependent sources cannot be suppressed* to facilitate our analysis because this would invalidate the constraint between controlled source and controlling signal. To find the equivalent resistance of a port, we can still suppress its *independent* sources because their values do not depend on the rest of the circuit. However, *dependent* sources must be left in place in the course of our analysis!

With these observations in mind, we can apply any of the analytical tools of the previous chapters, such as nodal or loop analysis, source transformations, and Thévenin and Norton conversions. Linearity holds also for circuits with dependent sources. However, because these sources cannot be suppressed, little if any is saved by applying the superposition principle, since there will always be at least *two* sources in operation, namely, an unsuppressed independent source and all the dependent sources.

Nodal and Loop Analysis

In circuits with dependent sources, the node-voltage or mesh-current equations must be augmented by the constraint equations stemming from such sources. The following examples will illustrate.

▶ **Example 5.5**

Apply nodal analysis to the circuit of Figure 5.9(a). What are the values of i_X and $k_r i_X$ in this circuit?

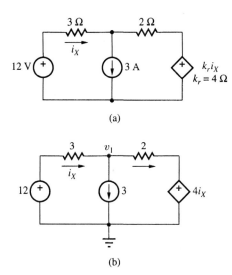

Figure 5.9 Example of nodal analysis with dependent sources.

Solution

Label the circuit as in Figure 5.9(b), where the physical units are [V], [A], and [Ω]. The dependent source is a current-controlled voltage source (CCVS), and the voltage it produces is $v_{CCVS} = 4i_X$. KCL at the node labeled v_1 yields

$$\frac{12 - v_1}{3} = 3 + \frac{v_1 - 4i_X}{2} \tag{5.9}$$

We need an additional equation to eliminate i_X. Using Ohm's Law,

$$i_X = \frac{12 - v_1}{3}$$

Substituting into Equation (5.9), collecting, and solving yields $v_1 = 6$ V. Then, $i_X = (12 - 6)/3 = 2$ A. Summarizing, we have

$$v_1 = 6 \text{ V}$$

$$i_X = 2 \text{ A}$$

The dependent source voltage is $v_{CCVS} = k_r i_X = 4 \times 2 = 8$ V.

Checking   The branch currents are $i_{3\Omega} = 2$ A $(\rightarrow)$ and $i_{2\Omega} = (8-6)/2 = 1$ A $(\leftarrow)$. KCL at node v_1 yields $2 + 1 = 3$. ◀

Exercise 5.5   Consider the circuit obtained from that of Figure 5.9(a) by changing the CCVS to a voltage-controlled voltage source (VCVS) of value $3v_X$, where v_X is the voltage across the 3-A current source, positive at the top. Apply nodal analysis to find the voltage developed by the dependent source. Check your results!

ANSWER $v_{VCVS} = 3v_X = -4.5$ V.

▶ **Example 5.6**

Apply nodal analysis to the circuit of Figure 5.10(a). What is the current supplied by the dependent source?

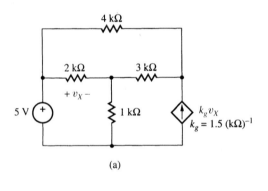

(a)

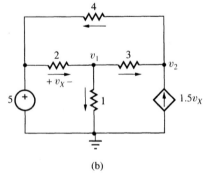

(b)

Figure 5.10 Example of nodal analysis.

Solution

Label the circuit as in Figure 5.10(b), where the physical units are [V], [mA], and [kΩ]. The dependent source is now a voltage-controlled current source (VCCS) supplying the current $i_{VCCS} = k_g v_X$. KCL at the nodes labeled v_1 and v_2 yields

$$\frac{5 - v_1}{2} = \frac{v_1}{1} + \frac{v_1 - v_2}{3} \tag{5.10a}$$

$$\frac{v_1 - v_2}{3} + 1.5v_X = \frac{v_2 - 5}{4} \tag{5.10b}$$

We need an additional equation to eliminate v_X. Using KVL,

$$v_X = 5 - v_1 \tag{5.11}$$

Substituting into Equation (5.10) and collecting yields

$$11v_1 - 2v_2 = 15$$

$$14v_1 + 7v_2 = 105$$

Solving via Gaussian elimination or Cramer's rule, we obtain

$$v_1 = 3 \text{ V}$$

$$v_2 = 9 \text{ V}$$

By Equation (5.11), $v_X = 5 - 3 = 2$ V. Hence, $i_{VCCS} = k_g v_X = 1.5 \times 2 = 3$ mA ($\uparrow$).

Exercise 5.6   Repeat Example 5.6 if the VCCS is changed to a current-controlled current source (CCCS) of value $i_X/16$, where i_X is the current through the 1-kΩ resistance, flowing downward. Check your results!

ANSWER $v_1 = 2$ V, $v_2 = 3.5$ V, $i_{CCCS} = 0.125$ mA.

▶ Example 5.7

Apply loop analysis to the circuit of Figure 5.11(a) to find the power (released or absorbed?) by the voltage source.

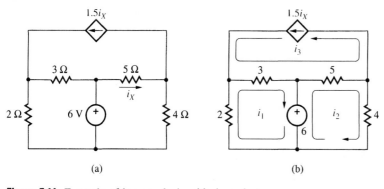

(a)   (b)

Figure 5.11 Example of loop analysis with dependent sources.

Solution

Label the circuit as in Figure 5.11(b), where the units are [V], [A], and [Ω]. KVL around the bottom meshes yields

$$0 = 2i_1 + 3(i_1 + i_3) + 6 \tag{5.12a}$$

$$6 = 5(i_2 + i_3) + 4i_2 \tag{5.12b}$$

where we have, by inspection, $i_3 = 1.5i_X$. We need an additional equation to eliminate i_X. Comparing the two circuits we observe that $i_X = i_2 + i_3$. Hence, $i_3 = 1.5i_X = 1.5(i_2 + i_3)$, or

$$i_3 = -3i_2 \tag{5.12c}$$

Substituting into Equations (5.12a) and (5.12b) and solving yields $i_1 = -3$ A and $i_2 = -1$ A, indicating that both currents actually flow counterclockwise. Moreover, Equation (5.12c) yields $i_3 = 3$ A. In summary,

$$i_1 = -3 \text{ A}$$

$$i_2 = -1 \text{ A}$$

$$i_3 = 3 \text{ A}$$

The current out of the positive terminal of the voltage source is $i_{6\text{V}} = i_2 - i_1 = -1 - (-3) = 2$ A, indicating that this source is *delivering* $p_{6\text{V}} = vi = 6 \times 2 = 12$ W.

Checking We have $i_{2\Omega} = -i_1 = 3$ A ($\downarrow$), so $v_{2\Omega} = 2 \times 3 = 6$ V (positive at the top, or + @ top). We also have $i_{3\Omega} = i_1 + i_3 = -3 + 3 = 0$, so $v_{3\Omega} = 0$. KVL around the bottom-left mesh thus yields $6 + 0 = 6$, which checks.

We have $i_{5\Omega} = i_2 + i_3 = -1 + 3 = 2$ A ($\rightarrow$), so $v_{5\Omega} = 5 \times 2 = 10$ V (+ @ left). Moreover, $i_{4\Omega} = -i_2 = 1$ A ($\uparrow$), so $v_{4\Omega} = 4 \times 1 = 4$ V (+ @ bottom). KVL around the bottom-right mesh thus yields $6 + 4 = 10$, which also checks. ◀

Exercise 5.7 In the circuit of Figure 5.11(a) let the CCCS be changed to a VCCS of value $k_g v_X$, where $k_g = 0.75$ Ω^{-1} and v_X is the voltage across the 4-Ω resistance, positive reference at the top. Use the loop method to find the power (delivered or absorbed?) by the dependent source. Don't forget to check your results!

ANSWER 1.725 W, delivered.

► **Example 5.8**

Apply loop analysis to the circuit of Figure 5.12(a) to find the voltage developed by the dependent source.

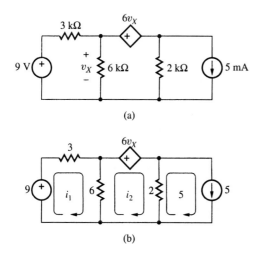

Figure 5.12 Example of loop analysis.

Solution

Label the circuit as in Figure 5.12(b), where the units are [V], [mA], and [kΩ]. KVL around the two leftmost meshes yields

$$9 = 3i_1 + 6(i_1 - i_2) \tag{5.13a}$$

$$0 = 6(i_2 - i_1) + 6v_X + 2(i_2 - 5) \tag{5.13b}$$

We need an additional equation to eliminate v_X. By inspection,

$$v_X = 6(i_1 - i_2)$$

Substituting into Equation (5.13) and solving, $i_1 = 8/3$ mA, and $i_2 = 2.5$ mA. Then, $v_X = 6(8/3 - 2.5) = 1$ V. To summarize,

$$i_1 = 8/3 \text{ mA}$$

$$i_2 = 2.5 \text{ mA}$$

$$v_X = 1 \text{ V}$$

The dependent source voltage is $v_{\text{VCVS}} = 6v_X = 6 \times 1 = 6$ V. ◀

Exercise 5.8   In Figure 5.12(a) let the VCVS be changed to a CCVS of value $k_r i_X$, where $k_r = 5$ kΩ and i_X is the current through the 9-V source flowing upward. Use the loop method to find the power dissipated in the 6-kΩ resistance. Check your results!

ANSWER   1.5 mW.

▶ **Example 5.9**

Use the node method to check that the voltage developed by the dependent source in Figure 5.12(a) is $v_{VCVS} = 6$ V.

Solution

We have $v_{3\ k\Omega} = 9 - v_X$, positive at the left, $v_{6\ k\Omega} = v_X$, positive at the top, and $v_{2\ k\Omega} = v_X - 6v_X = -5v_X$, positive at the top. Applying KCL at the *supernode* encircling the dependent source and its nodes yields

$$\frac{9 - v_X}{3} = \frac{v_X}{6} + \frac{-5v_X}{2} + 5$$

or $v_X = 1$ V. Thus, $v_{VCVS} = 6v_X = 6$ V, as expected.   ◀

Exercise 5.9   Use the node method to check the result of Example 5.7.

Thévenin and Norton Equivalents

Nodal or loop analysis can also be used to find the Thévenin or Norton equivalents of one-ports with dependent sources. Once the open-circuit voltage v_{OC} and the short-circuit current i_{SC} are known, the equivalent resistance is readily obtained via **Method 1,**

$$R_{eq} = \frac{v_{OC}}{i_{SC}} \tag{5.14}$$

It is important to realize that the value of v_X (or i_X) with the port open-circuited is generally *different* from that with the port short-circuited!

If v_{OC} and i_{SC} happen to be zero (as in the case, for instance, of a one-port containing only resistances and dependent sources but no independent sources), then we must use Method 2 which, for circuits with dependent sources, is rephrased as:

> **Method 2: To find the equivalent resistance of a linear one-port, *suppress all independent sources while leaving the dependent sources in place.* Then, apply a test voltage v, find the current i out of the test-source positive terminal, and obtain**

$$R_{eq} = \frac{v}{i} \tag{5.15}$$

> **Alternately, apply a *test current* i, find the ensuing voltage v, and obtain** $R_{eq} = v/i$.

This method is particularly advantageous when only R_{eq} is sought, because it may simplify analysis as compared to Method 1.

Source transformation techniques are applicable also to circuits with dependent sources. However, we must *avoid tampering with the controlling signal,* because the controlled source depends on it!

▶ **Example 5.10**

Find the Thévenin equivalent of the circuit of Figure 5.13(a). To find R_{eq}, use Method 1.

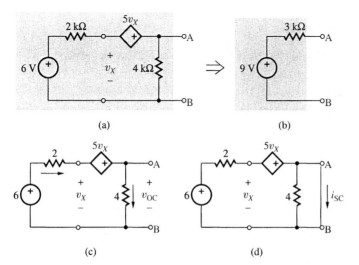

Figure 5.13 Finding the Thévenin equivalent of a circuit with a dependent source.

Solution

Our physical units are [V], [mA], and [kΩ]. To find v_{OC}, refer to the circuit of Figure 5.13(c). Applying KCL at the supernode surrounding the dependent source yields

$$\frac{6 - v_X}{2} = \frac{v_{OC}}{4} \tag{5.16}$$

We need an additional equation to eliminate v_X. Applying KVL, we get

$$v_X + 5v_X = v_{OC}$$

that is, $v_X = v_{OC}/6$. Substituting into Equation (5.16) gives $v_{OC} = 9$ V. Note, incidentally, that $v_X = 9/6 = 1.5$ V.

To find i_{SC}, refer to Figure 5.13(d). Since the current through the 4-kΩ resistance is now zero, we can write, by Ohm's Law,

$$i_{SC} = \frac{6 - v_X}{2} \tag{5.17}$$

To eliminate v_X, we again apply KVL and write

$$v_X + 5v_X = 0$$

that is, $v_X = 0$. Then, Equation (5.17) yields $i_{SC} = 6/2 = 3$ mA.

Finally, $R_{eq} = v_{OC}/i_{SC} = 9/3 = 3$ kΩ. The Thévenin equivalent is shown in Figure 5.13(b).

Remark   Note that v_X with the port open-circuited (1.5 V) is *different* from v_X with the port short-circuited (0 V).

▶ Example 5.11

Recompute R_{eq} for the circuit of Figure 5.13(a) via Method 2.

Solution

Suppress the 6-V independent source, but leave the dependent source in place. Then, apply a test voltage v as in Figure 5.14, and find the current i. Applying KCL at the supernode surrounding the dependent source yields

$$i = \frac{v}{4} + \frac{v_X}{2} \qquad (5.18)$$

To eliminate v_X, apply KVL clockwise around the loop,

$$v_X + 5v_X = v$$

which yields $v_X = v/6$. Substituting into Equation (5.18), we obtain $i = v/4 + (v/6)/2 = v/3$. Hence,

$$R_{eq} = \frac{v}{i} = \frac{\not{v}}{\not{v}/3} = 3 \text{ k}\Omega$$

thus providing a check for the result of Example 5.10.

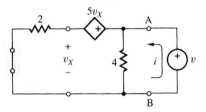

Figure 5.14   Recomputing R_{eq} for the circuit of Figure 5.13(a) via Method 2.

Remark   As expected, v in the numerator and denominator cancel each other out. ◀

Exercise 5.10   Find the Norton equivalent of the circuit of Figure 5.15. To find R_{eq}, use Method 1.

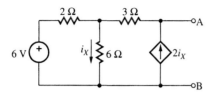

Figure 5.15 Circuit of Exercises 5.10 and 5.11.

ANSWER $i_{SC} = 2$ A, $R_{eq} = 9\ \Omega$.

Exercise 5.11   Find R_{eq} for the circuit of Exercise 5.10 via Method 2.

ANSWER $9\ \Omega$.

▶ **Example 5.12**

    (a) In the circuit of Figure 5.16(a) find the value of R_L for which power transfer to R_L is maximized.

    (b) What is the value of the maximum power?

Solution

    (a) Power transfer is maximized when $R_L = R_{eq}$. Let us find R_{eq} via Method 2. With resistances in Ω, the current of the VCCS can be written as $k_g v_X = (1/5)v_X = 0.2v_X$. Referring to the circuit of Figure 5.16(c) we have, by KCL,

$$i = \frac{v_X}{30} + \frac{v_X - v_1}{20} \tag{5.19}$$

We need to express v_X and v_1 in terms of v. Applying KCL at the node labeled v_1 yields

$$\frac{v_X - v_1}{20} = 0.2v_X + \frac{v_1}{10}$$

which can be solved to yield $v_1 = -v_X$. By KVL, we also have

$$v_X = v$$

Substituting into Equation (5.19) yields $i = v/7.5$. Finally,

$$R_{eq} = \frac{v}{i} = \frac{\not{v}}{\not{v}/7.5} = 7.5\ \Omega \tag{5.20}$$

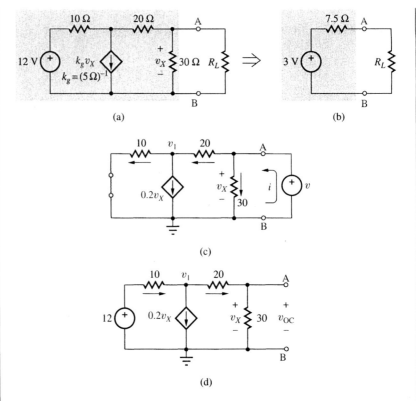

Figure 5.16 Finding maximum power transfer with dependent sources.

(b) Since $p_{L(\max)} = v_{OC}^2/(4R_{eq})$, we need to find v_{OC}. Referring to the circuit of Figure 5.16(d), we observe that $v_{OC} = v_X$. Applying KCL at the node labeled v_1 yields

$$\frac{12 - v_1}{10} = 0.2v_X + \frac{v_1 - v_X}{20} \tag{5.21}$$

By the voltage divider formula,

$$v_X = \frac{30}{20 + 30}v_1$$

or $v_1 = (5/3)v_X$. Substituting into Equation (5.21) and solving, we get $v_X = 3$ V. Hence, $v_{OC} = v_X = 3$ V. Replacing the one-port with its Thévenin equivalent yields the situation of Figure 5.16(b). Finally,

$$p_{L(\max)} = \frac{v_{OC}^2}{4R_{eq}} = \frac{3^2}{4 \times 7.5} = 0.3 \text{ W} \tag{5.22}$$

Exercise 5.12   Repeat Example 5.12 if the VCCS in Figure 5.16(a) is changed to a CCVS of value $k_r i_X = (2\ \Omega)i_X$, positive at the top, where i_X is the current through the 12-V source, flowing upward.

ANSWER $R_L = 12\ \Omega$, $p_{L(\max)} = 0.03$ W.

Concluding Remarks

When looking for Thévenin/Norton equivalents, it is good practice to pause and try developing a strategy to minimize the computational effort and simultaneously reduce the possibility of errors.

(1) First, try anticipating whether it will be easier to find v_{OC} or to find i_{SC}. The answer, of course, depends on the particular circuit as well as personal preference.

(2) Next, try anticipating whether R_{eq} will more easily be found via Method 1 or via Method 2. In the most general case of a one-port containing both *dependent* and *independent* sources, Method 2 is usually preferable because suppressing the independent sources generally results in a *simpler circuit.*

When looking for R_{eq}, the following special cases are worth keeping in mind:

(1) If it is found that $v_{OC} = 0$ (or $i_{SC} = 0$) in spite of the presence of independent sources, then we have no choice but to use Method 2.

(2) If a one-port contains *dependent sources* and resistances but *no independent sources,* then it will automatically yield $v_{OC} = i_{SC} = 0$, again mandating the use of Method 2.

(3) If a one-port is *purely resistive,* as in the case of ladders, see whether you can find R_{eq} via simple series/parallel reductions. If not, you must again use Method 2.

5.3 The Ideal Transformer

As our first example of a practical two-port device we study the **transformer.** As shown in Figure 5.17(a), a transformer consists of two coils wound around a common core of magnetic material to achieve intimate magnetic coupling. The coils are called, respectively, the **primary** and **secondary.** The primary plays the role of the *input port,* the secondary that of the *output port.* Denoting the number of turns in the respective windings as N_1 and N_2, the ratio

$$n = \frac{N_2}{N_1} \tag{5.23}$$

is called the **turns ratio** of the transformer. Figure 5.17(b) shows the circuit symbol of the transformer, along with the reference polarities for voltage and current at both windings.

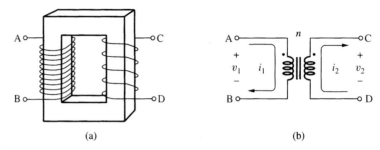

Figure 5.17 The ideal transformer and its circuit symbol.

Figure 5.18 shows a typical transformer connection. Applying a time-varying signal such as an ac signal to the primary generates a *changing flux* in the core. This, in turn, *induces* a similar time-varying signal at the secondary. The induced voltage $v_2(t)$ is related to the applied voltage $v_1(t)$ as

$$v_2 = nv_1 \tag{5.24}$$

It is important to realize that since transformer action is the result of a *changing* magnetic flux, the applied voltage must be of the *time-varying* type. In electric power applications the applied waveform is sinusoidal; in audio applications it is an audio signal; in communication and control it is often a pulse train.

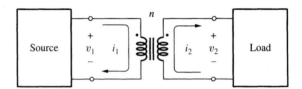

Figure 5.18 Typical transformer connection.

The function of the dots in Figure 5.17(b) is to identify signal polarities. If v_1 and v_2 are chosen with the *positive* references at the *dotted* terminals, then v_1 and v_2 are said to be *in phase* with each other. Otherwise, with the positive reference of one voltage at the dotted terminal of one coil and that of the other at the undotted terminal of the other coil, v_1 and v_2 are *out of phase*. Likewise, the currents are in phase with each other when one *enters* the dotted terminal of one coil and the other *exits* the dotted terminal of the other coil. The dots are omitted when the phase relationship between the two windings is unimportant.

Ideally, a transformer dissipates no energy, so the power absorbed via the primary equals the power released via the secondary. Denoting the primary and secondary currents, respectively, as $i_1(t)$ and $i_2(t)$, this constraint can be expressed as $v_1 i_1 = v_2 i_2$. Substituting $v_2 = nv_1$ and simplifying yields

$$i_1 = ni_2 \tag{5.25}$$

We observe that for current to flow in the primary, the secondary must be terminated on a load. In the absence of any load we have $i_2 = 0$ and hence, by Equation (5.25), $i_1 = 0$ even though $v_1 \neq 0$.

If $n > 1$, the magnitude of the output voltage will be greater than that of the input voltage, and we have a **step-up transformer.** Note, however, that the output current is now less than the input current, owing to the fact that power must be conserved.

Conversely, if $n < 1$, we have a **step-down transformer,** with a smaller output voltage but a greater output current. A common transformer application is to step the utility voltage up or down to meet the specific voltage needs of individual equipment pieces.

Since coupling between the input and output ports is exclusively magnetic, transformers are also used to provide **electrical isolation,** that is, to transmit power or information from one circuit to another without allowing a conducting path to exist between them. Transformers with $n = 1$ are used exclusively for this purpose and are called **isolation transformers.**

Equations (5.24) and (5.25) are referred to as the **ideal transformer** equations to underline the fact that practical transformers satisfy them only on an approximate basis, though modern technology allows the fabrication of certain types of transformers with nearly ideal behavior. These equations can be used to relate the instantaneous values as well as the peak and rms values of the signals at either side of the transformer.

▶ **Example 5.13**

A transformer is to be used to step down the 120-V (rms) utility voltage to an ac voltage of 21 V (rms) for a 10-Ω load. Find n and the peak and rms values of the currents in the two windings.

Solution

By Equation (5.24), $n = v_2/v_1 = 21/120 = 0.175$. By Ohm's Law, $i_2 = v_2/R_L = 21/10 = 2.1$ A (rms), or $i_2 = 2.1\sqrt{2} = 2.97$ A (peak). By Equation (5.25), $i_1 = ni_2 = 0.175 \times 2.1 = 0.367$ A (rms), or $i_1 = 0.175 \times 2.97 = 0.520$ A (peak). ◀

Exercise 5.13 It is desired to dissipate 9 W into a 1-MΩ load using the utility voltage as the power source. If we were to connect the load to the utility voltage *directly,* the power into the load would be only $120^2/10^6 = 14.4$ mW. However, we can achieve our goal by interposing a suitable step-up transformer between the power source and the load. Find the required value of n; hence, find the currents in the two windings.

ANSWER $n = 25$, $i_1 = 75$ mA (rms), $i_2 = 3$ mA (rms).

Circuit Model of the Ideal Transformer

We now wish to develop a circuit model for the ideal transformer. The fact that v_2 depends on the driving source and n, regardless of the load, suggests that the secondary can be modeled with a *dependent voltage source* of value nv_1. Likewise, the fact that i_1 depends on the load and n, regardless of the driving source, suggests that the primary can be modeled with a *dependent current source* of value ni_2. The ideal transformer model is shown in Figure 5.19. When analyzing a circuit containing a transformer, we can replace the latter with this model to better visualize its function in the circuit as well as to facilitate our calculations. When making such a replacement we must be careful that the polarities of the dependent sources relative to the respective dot markings are as shown.

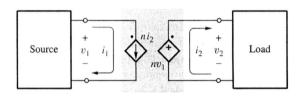

Figure 5.19 Circuit model of the ideal transformer.

Power Transmission

A major application of transformers is in electric power transmission and distribution, where they are used to improve efficiency by reducing transmission line losses. A power transmission system can be modeled with a power source v_s, a transmission line having a net resistance (sum of both sides) R_T, and a load R_L.

In the absence of any transformers, the load and the transmission line would form a voltage divider so that the voltage available at the load would be

$$v_\ell = \frac{R_L}{R_T + R_L} v_s \qquad \text{(5.26a)}$$

The efficiency of this transmission would be $\eta = P_\ell/P_s$, where P_ℓ is the average power absorbed by the load and P_s is that delivered by the source. By Equation (4.56) we would have

$$\eta = \frac{R_L}{R_T + R_L} \qquad \text{(5.26b)}$$

With long transmission lines, R_T can be much greater than R_L, so most voltage and power would be lost in the transmission line, with only a negligible fraction reaching the load.

The culprit is the transmission line current i_{R_T}, which is responsible for the voltage loss $R_T i_{R_T}$ and the power loss $R_T i_{R_T}^2$. These losses can be reduced by making i_{R_T} suitably small. However, to avoid depriving the load of the required power, the decrease in the line current must be counterbalanced by an increase in the line voltage. As depicted in Figure 5.20, these conditions are achieved by using a step-up transformer at the generator end of the line and a step-down transformer at the load end.

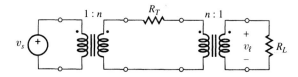

Figure 5.20 Power transmission system.

To analyze the circuit, use Equations (5.24) and (5.25) with n at the step-up side and $1/n$ at the step-down side. Thus, the line current is $i_{R_T} = (1/n)i_\ell$, the line voltage at the generator side is nv_s, and that at the load end is nv_ℓ. Applying Ohm's Law to the line resistance yields $nv_s - nv_\ell = R_T i_{R_T} = R_T i_\ell/n = R_T(v_\ell/R_L)/n$. Collecting and solving for v_ℓ yields, after simple manipulation,

$$v_\ell = \frac{R_L}{R_T/n^2 + R_L} v_s \qquad (5.27a)$$

Comparing this with Equation (5.26a), we note that the effect of using the transformers is to replace the term R_T with the term R_T/n^2. We can thus reuse Equation (5.26b) and write

$$\eta = \frac{R_L}{R_T/n^2 + R_L} \qquad (5.27b)$$

Equation (5.27) indicates that by choosing n sufficiently large to make $R_T/n^2 \ll R_L$, we can make v_ℓ approach v_s and η approach unity. This explains the widespread use of high-voltage transmission lines when power is to be transported over substantial distances.

▶ **Example 5.14**

The power produced by a 120-V (rms), 60-Hz generator is to be transmitted to a 1-Ω load via a 100-mile transmission line having a resistance (sum of both sides) of 0.025 Ω per 1000 feet.

(a) If the voltage at the generator end of the transmission line is 10 kV (rms), find the load voltage and the efficiency.

(b) Compare with the case in which power is transmitted directly, without using any transformers.

Solution

$R_T = [(0.025 \ \Omega)/(10^3 \ \text{ft})] \times (10^2 \ \text{mi}) \times (5280 \ \text{ft/mi}) = 13.2 \ \Omega$.

(a) By Equation (5.24), $n = 10^4/120 = 83.33$. By Equation (5.27), $v_\ell = [1/(13.2/83.33^2 + 1)] \times 120 = 99.81 \times 120 = 119.8$ V (rms) and $\eta = 99.81\%$. These values are quite attractive.

(b) Without transformers we would have, by Equation (5.26), $v_\ell = [1/(13.2 + 1)] \times 120 = 0.0704 \times 120 = 8.45$ V (rms), and $\eta = 7.04\%$, both of which are unacceptably low! ◀

> **Exercise 5.14**   In the transmission system of Example 5.14, find the value of n required to ensure an efficiency of (a) 95%; (b) 99.9%.
>
> **ANSWER**   (a) 15.84, (b) 114.83.

Resistance Transformation

According to Figure 5.19, the primary current i_1 is controlled by the load current i_2. But i_2 depends on v_2, which in turn is controlled by v_1. Consequently, i_1 ultimately depends on v_1. If the secondary is terminated on a resistance R_2, then the ratio v_1/i_1 will also be a resistance, which we shall denote as R_1. We wish to find a relationship between the apparent resistance R_1 and the actual resistance R_2.

To this end we apply a test voltage v as in Figure 5.21(a), find i, and then take the ratio $R_1 = v/i$. By transformer action and Ohm's Law, $i = ni_{R_2} = n(nv/R_2) = n^2 v/R_2$. Letting $R_1 = v/i$ yields

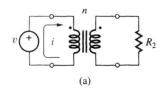

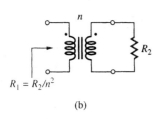

$R_1 = R_2/n^2$

(b)

Figure 5.21 Resistance transformation.

$$R_1 = \frac{R_2}{n^2} \tag{5.28a}$$

indicating that the secondary resistance R_2 appears at the primary as an equivalent resistance $R_1 = R_2/n^2$. R_1 is said to represent the secondary resistance *referred* to the primary, and the transformer is said to perform a **resistance transformation.** This concept is illustrated in Figure 5.21(b).

We observe that resistance transformation applies in both directions. Thus, if the primary is terminated on a resistance R_1, then the equivalent resistance seen looking into the secondary is $R_2 = R_1/(1/n^2)$, or

$$R_2 = n^2 R_1 \tag{5.28b}$$

in agreement with Equation (5.28a), as it should be. We wish to point out that these transformations affect not only resistances but also other circuit elements such as capacitances and inductances, whose ability to oppose current flow is more generally called *impedance*. Impedances are studied in Chapter 11.

Resistance transformation has a significant effect on loading. Consider Figure 5.22(a), where a source v_s with nonzero series resistance R_s is transformer-coupled to a load R_L. As depicted in Figure 5.22(b), the resistance seen looking into the primary is R_L/n^2. Applying the voltage divider formula yields

$$v_1 = \frac{R_L/n^2}{R_s + R_L/n^2} v_s \tag{5.29}$$

Clearly, because of loading, the magnitude of v_1 is *less* than that of v_s. By transformer action, we have $v_2 = nv_1$. Using Equation (5.29) and simplifying yields

$$v_2 = \frac{R_L}{n^2 R_s + R_L} n v_s \tag{5.30}$$

indicating that from the viewpoint of the load, the secondary appears as an unloaded source of value nv_s with a series resistance of value $n^2 R_s$. The situation is depicted in Figure 5.22(c).

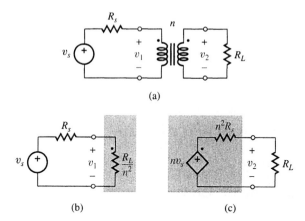

Figure 5.22 (a) Transformer coupling of a source to a load; equivalent circuits at (b) the primary and (c) the secondary.

We define the **source-to-load gain** as the ratio v_ℓ/v_s, in V/V. Thus

$$\frac{v_\ell}{v_s} = n\frac{R_L}{n^2 R_s + R_L} \qquad (5.31)$$

where we have used the fact that $v_\ell = v_2$. Clearly, due to loading, the gain v_ℓ/v_s is *less* than n. Only in the limit $n^2 R_s \ll R_L$ does this gain approach n.

Exercise 5.15   In the circuit of Figure 5.22(a), let $v_s = 30 \sin 2\pi ft$ V, $R_s = 20$ Ω, $n = 5$, and $R_L = 1$ kΩ. Find v_1, v_2, and the source-to-load gain.

ANSWER $20 \sin 2\pi ft$ V, $100 \sin 2\pi ft$ V, 10 V/V.

If n in Figure 5.22(a) is chosen so that

$$\frac{R_L}{n^2} = R_s \qquad (5.32)$$

then the resistance seen looking into the primary is said to be **matched** to that of the source, thus ensuring maximum power transfer from the source to the primary. Since $R_L = n^2 R_s$, there will also be matching at the secondary and, hence, maximum power transfer from the secondary to the load. Clearly, another useful transformer application is to provide **maximum average power transfer** from source to load when $R_s \neq R_L$. For instance, in audio applications, it is often desired to transform the resistance of a loudspeaker load to a value more suited to the power amplifier, or to match the input resistance of a preamplifier to that of the input transducer.

▶ **Example 5.15**

(a) Specify a transformer for maximum average power transfer from a source v_s having $R_s = 900\ \Omega$ to a load $R_L = 100\ \Omega$.

(b) Find expressions for v_1 and v_2 in terms of v_s.

Solution

(a) We want $900 = 100/n^2$. This is met with $n = 1/3$, and the circuit is shown in Figure 5.23.

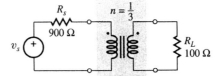

Figure 5.23 Resistance matching circuit of Example 5.15.

(b) Using Equations (5.29) and (5.30) gives $v_1 = [900/(900+900)]v_s = (1/2)v_s$ and $v_2 = nv_1 = (1/3) \times (v_s/2) = (1/6)v_s$.   ◀

Exercise 5.16   (a) Specify a transformer for a maximum power transfer from a source $v_s = 5\sin 2\pi ft$ V with output resistance $R_s = 50\ \Omega$ to a load $R_L = 200\ \Omega$. (b) Find v_1, i_1, v_2, i_2, and the average power transferred to the load.

ANSWER   (a) $n = 2$, (b) $2.5\sin 2\pi ft$ V, $50\sin 2\pi ft$ mA, $5\sin 2\pi ft$ V, $25\sin 2\pi\, ft$ mA, 62.5 mW.

Practical Transformers

Practical transformers depart from the ideal behavior described earlier because of *power losses,* both in the windings (*wire losses*) and in the core (*core losses*), as well as *imperfect coupling* between the secondary and primary windings.

Wire losses result from the nonzero resistance of the windings. Core losses include power dissipation due to magnetic hysteresis and ohmic loss from eddy currents induced in the core by the continuously changing flux. The effect of core losses is that even in the absence of any load there is a current in the primary, called the *magnetization current,* whose function is to supply these losses.

Magnetic coupling is optimized by employing special winding techniques such as wire twisting. In audio and power transformers, eddy current losses are reduced by using laminated cores, which consist of thin metal layers separated by thin insulating layers fabricated perpendicular to the eddy currents. High-frequency transformers reduce eddy current losses by using cores of non-conductive materials known as *ferrites.*

You may find it reassuring to know that in spite of these nonidealities, engineers almost always analyze transformer circuits using the ideal transformer model. Only in the course of a second, more thorough analysis are nonidealities taken into account.

5.4 AMPLIFIERS

The function of an amplifier is to accept a signal from a source and deliver a *magnified replica* of it to a load. Ideally, this function should be independent of the particular source and load to which the amplifier is connected. The most common amplifiers are the *voltage amplifier* and the *current amplifier.*

The transfer characteristic of the **voltage amplifier** is

$$v_O = A_v v_I \qquad\qquad \text{(5.33)}$$

where v_I and v_O are the input and output voltages, and A_v is called the **voltage gain factor,** with dimensions of V/V. Likewise, the transfer characteristic of the **current amplifier** is

$$i_O = A_i i_I \qquad\qquad \text{(5.34)}$$

where i_I and i_O are the input and output currents, and A_i is the **current gain factor,** in A/A.

Gain factors may be positive or negative. If gain is positive, the output has the same polarity as the input; if gain is negative, the output has the opposite polarity. Regardless of gain polarity, the magnifying action by the amplifier affects instantaneous signal values as well as dc and ac values.

▶ **Example 5.16**

    (a) A signal alternating sinusoidally between $+5$ V and -1 V is applied to an amplifier having $A_v = 5$ V/V. What are the dc and ac values of v_I and v_O?

    (b) Repeat if $A_v = -7$ V/V.

Solution

Let $v_I = V_I + V_{im} \sin 2\pi ft$, and $v_O = V_O + V_{om} \sin 2\pi ft$. We have $V_I = [5 + (-1)]/2 = 2$ V, and $V_{im} = 5 - 2 = 3$ V, so that

$$v_I = V_I + v_i = 2 + 3 \sin 2\pi ft \text{ V}$$

The dc value of v_I is 2 V, and the ac value is $3/\sqrt{2} = 2.121$ V.

(a) By amplifier action, $v_O = A_v v_I = 5(2 + 3 \sin 2\pi ft)$, or

$$v_O = V_O + v_o = 10 + 15 \sin 2\pi ft \text{ V}$$

This signal alternates between +25 V and −5. Its dc value is 10 V, and its ac value is $15/\sqrt{2} = 10.61$ V.

(b) Now $v_O = -7(2 + 3 \sin 2\pi ft) = -14 - 21 \sin 2\pi ft$ V, which can be written as

$$v_O = V_O + v_o = -14 + 21 \sin (2\pi ft + 180°) \text{ V}$$

This signal alternates between −35 V and +7 V. Its dc value is −14 V, and its ac value is $21/\sqrt{2} = 14.85$ V. Because of the negative gain, the polarity of V_O is *opposite* to that of V_I, and v_o is 180° *out of phase* with respect to v_i.

◄

Exercise 5.17

(a) The input and output of a voltage amplifier are measured with a DVM. If the dc readings are, respectively, −50 mV and +2 V, find A_v.

(b) If the ac reading of the input is 99 mV, what is the ac reading of the output? What are the values between which the output is alternating?

ANSWER   (a) −40 V/V; (b) 3.96 V, 7.6 V, −3.6 V.

An amplifier with linear terminal characteristics will generally admit a Thévenin or Norton equivalent for each of its ports. The input port usually plays a purely passive role, and we model it with just a resistance, called the **input resistance** R_i. The output port is modeled with an **output resistance** R_o and a **dependent source.** We need this type of source to reflect the fact that the output is controlled by the input.

Voltage Amplifier Model

The model of a voltage amplifier is shown in Figure 5.24, along with the driving source and the load. We wish to find a relationship between the load voltage v_L and the source voltage v_S.

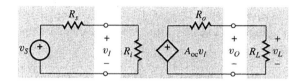

Figure 5.24 Circuit model of a voltage amplifier, with source and load.

Applying the voltage divider formula at the input and output ports,

$$v_I = \frac{R_i}{R_s + R_i} v_S \tag{5.35}$$

and

$$v_O = \frac{R_L}{R_o + R_L} A_{oc} v_I \tag{5.36}$$

We observe that in the absence of any output load ($R_L = \infty$) we would have $v_O = A_{oc} v_I$. Consequently, we refer to A_{oc} as the **open-circuit** or **unloaded voltage gain factor.** Eliminating v_I and using the fact that $v_L = v_O$, we solve for the ratio v_L/v_S, called the **source-to-load voltage gain,**

$$\boxed{\frac{v_L}{v_S} = \frac{R_i}{R_s + R_i} A_{oc} \frac{R_L}{R_o + R_L}} \tag{5.37}$$

This expression contains two voltage-divider terms, reflecting **loading** both at the input and at the output ports, and the term A_{oc}. Clearly, because of loading, we always have $|v_L/v_S| \leq |A_{oc}|$.

▶ Example 5.17

(a) A voltage amplifier with $R_i = 100$ kΩ, $A_{oc} = 100$ V/V, and $R_o = 1$ Ω is driven by a source with $R_s = 20$ kΩ, and drives a load $R_L = 3$ Ω. Calculate the gain v_L/v_S as well as the extent of loading at the input and output ports.

(b) Repeat if $R_s = 10$ kΩ and $R_L = 2$ Ω. Compare.

Solution

(a) Using Equation (5.37), $v_L/v_S = [100/(20 + 100)] \times 100 \times [3/(1 + 3)]$, or

$$\frac{v_L}{v_S} = 0.8333 \times 100 \times 0.75 = 62.5 \text{ V/V}$$

Loading at the input causes the signal to drop to 83.33% of its initial value, and loading at the output causes the amplified signal to drop to 75% of its unloaded value.

(b) We now have $v_L/v_S = 0.9091 \times 100 \times 0.6667 = 60.61$ V/V, indicating less loading at the input but more loading at the output. ◀

Exercise 5.18   A voltage amplifier with $R_i = 20$ kΩ, $A_{oc} = 75$ V/V, and $R_o = 0.25$ Ω is driven by a source with $v_S = 100$ mV and $R_s = 5$ kΩ, and drives a load $R_L = 1$ Ω. Calculate the voltage and current at each port of the amplifier.

ANSWER   Input: 80 mV, 4 μA; output: 4.8 V, 4.8 A.

▶ **Example 5.18**

A 1-mV transducer source with $R_s = 200$ kΩ is to be amplified to produce a 1-V signal across a load $R_L = 8$ Ω. If $R_i = 1$ MΩ and $R_o = 2$ Ω, what is the required unloaded gain?

Solution

We want $v_L/v_S = 1/10^{-3} = 10^3$ V/V. But, according to Equation (5.37), $v_L/v_S = [1/(0.2 + 1)]A_{oc}[8/(2 + 8)] = 0.6667A_{oc}$. Thus, we need $A_{oc} = 10^3/0.6667 = 1500$ V/V. ◀

▶ **Example 5.19**

Loading can be exploited to find the parameters of an amplifier. Thus, suppose a source with $v_S = 200$ mV and $R_s = 10$ kΩ is applied to a voltage amplifier having a load $R_L = 100$ Ω. DVM measurements across the input and output ports yield, respectively, $v_I = 150$ mV and $v_O = 6$ V. Removing R_L raises the output reading to $v_O = 7.5$ V. Find R_i, A_{oc}, and R_o.

Solution

Because of loading at the input we must have $150 = [R_i/(10 + R_i)]200$, or $R_i = 30$ kΩ. Because of loading at the output, we must have $6 = [100/(R_o+100)]7.5$, or $R_o = 25$ Ω. Moreover, $A_{oc} = 7.5/0.150 = 50$ V/V. Summarizing, we have $R_i = 30$ kΩ, $A_{oc} = 50$ V/V, and $R_o = 25$ Ω. ◀

Loading is generally undesirable because whenever we change the source or the load we need to recompute the gain v_L/v_S, not to mention the fact that $|v_L/v_S| \leq |A_{oc}|$. If loading could be eliminated altogether, we would have

$v_L/v_S = A_{oc}$ regardless of R_s and R_L. To minimize loading, the voltage drops across R_s and R_o must be minimized irrespective of R_s and R_L. To meet these objectives we must use a voltage amplifier with $R_i \gg R_s$ and $R_o \ll R_L$. In the limits $R_i = \infty$ and $R_o = 0$ we have what is known as an **ideal voltage amplifier.** Though these limits cannot be achieved exactly, an amplifier designer will strive to approximate them as closely as possible.

Current Amplifier Model

Figure 5.25 shows the model of a current amplifier. Applying the current divider formula twice, we obtain

$$i_I = \frac{R_s}{R_s + R_i} i_S \tag{5.38}$$

and

$$i_O = \frac{R_o}{R_o + R_L} A_{sc} i_I \tag{5.39}$$

Since in the case of a short-circuited output ($R_L = 0$) we would obtain $i_O = A_{sc} i_I$, we refer to A_{sc} as the **short-circuit** or **unloaded current gain factor.** Eliminating i_I and using the fact that $i_L = i_O$, we find the **source-to-load current gain** as

$$\boxed{\frac{i_L}{i_S} = \frac{R_s}{R_s + R_i} A_{sc} \frac{R_o}{R_o + R_L}} \tag{5.40}$$

This gain is *less* than A_{sc}, and it also depends on the particular source and load being used. To minimize loading, a current amplifier should have $R_i \ll R_s$ and $R_o \gg R_L$. In the limiting case of an **ideal current amplifier,** we would have $R_i = 0$, $R_o = \infty$, and $i_L/i_S = A_{sc}$.

Exercise 5.19   A current amplifier with $A_{sc} = 100$ A/A, $R_i = 200\ \Omega$, and $R_o = 50\ k\Omega$ is driven by a source $i_S = 10\ \mu A$ having $R_s = 1\ k\Omega$ and drives a load $R_L = 10\ k\Omega$. (a) Find the load current i_L. (b) To what value must A_{sc} be changed if we want $i_L = 1$ mA?

ANSWER   (a) 25/36 mA, (b) 144 A/A.

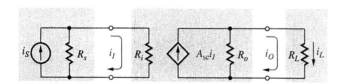

Figure 5.25 Circuit model of a current amplifier.

Transresistance and Transconductance Amps

In voltage and current amplifiers the input and output signals have the same physical dimensions. If the input is a current and the output is a voltage, the amplifier is called a **transresistance amplifier,** and its gain is in V/A. Referring to the circuit model of Figure 5.26, it is readily seen that the source-to-load gain is

$$\frac{v_L}{i_S} = \frac{R_s}{R_s + R_i} A_{\text{oc}} \frac{R_L}{R_o + R_L} \tag{5.41}$$

where the unloaded gain A_{oc} is now in V/A. To avoid loading, a transresistance amplifier should ideally have $R_i = 0$ and $R_o = 0$. Then, $v_L/i_S = A_{\text{oc}}$.

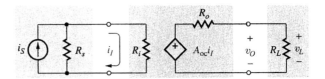

Figure 5.26 Circuit model of a transresistance amplifier.

This type of amplifier finds application as a current-to-voltage converter (*I–V* converter), for instance, to convert the small current supplied by a photomultiplier tube to a voltage for delivery to a low-resistance load. As a rule, when $R_i \ll R_s$ and $R_o \ll R_L$, the amplifier under consideration is more aptly classified as a transresistance type.

If the input is a voltage and the output a current, then we have a **transconductance amplifier,** whose model is shown in Figure 5.27. Its source-to-load gain is

$$\frac{i_L}{v_S} = \frac{R_i}{R_s + R_i} A_{\text{sc}} \frac{R_o}{R_o + R_L} \tag{5.42}$$

where the unloaded gain A_{sc} is now in A/V. To avoid loading, a transconductance amplifier should ideally have $R_i = \infty$ and $R_o = \infty$. This type of amplifier finds application as a voltage-to-current converter (*V–I* converter), and it is also used to model the behavior of transistors. As a rule, when $R_i \gg R_s$ and $R_o \gg R_L$, an amplifier is more aptly classified as a transconductance type.

Table 5.1 summarizes the four amplifier types, along with the corresponding ideal input and output resistance characteristics. Amplifier designers will strive to approximate these characteristics as closely as possible, relative to the expected resistances of the source and the load.

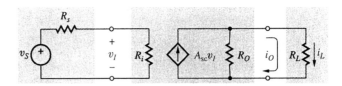

Figure 5.27 Circuit model of a transconductance amplifier.

TABLE 5.1 The Four Amplifier Types and Their Ideal Terminal Characteristics

Input	Output	Amplifier Type	Gain	R_i	R_o
v_I	v_O	voltage	V/V	∞	0
i_I	i_O	current	A/A	0	∞
i_I	v_O	transresistance	V/A	0	0
v_I	i_O	transconductance	A/V	∞	∞

Power Gain

An important feature of amplifiers is that in amplifying signals they also amplify power. A familiar example is offered by a car radio amplifier, which magnifies the small power supplied by the antenna to the much larger power consumed by the loudspeaker. The increased power emerging from the amplifier is drawn from the power supply, which in this case is the car battery.

We define the **source-to-load power gain** of an amplifier as

$$A_p \triangleq \frac{P_L}{P_S} \tag{5.43}$$

where P_S is the average power supplied by the source and P_L that absorbed by the load. Power gain is often expressed in **decibels,** named for the Scottish-American scientist Alexander Graham Bell (1847–1922). Specifically, the decibel value of the power gain A_p is

$$A_{p(\text{dB})} \triangleq 10 \log_{10} A_p \tag{5.44}$$

For instance, if an amplifier delivers a load power $P_L = 20$ W with a source power $P_S = 0.5$ mW, then $A_p = 20/(0.5 \times 10^{-3}) = 4 \times 10^4$, and $A_{p(\text{dB})} = 10 \log_{10}(4 \times 10^4) \simeq 46$ dB.

Referring to Figure 5.24, we note that $P_L = V_{L\,(\text{rms})}^2/R_L$, and $P_S = V_{S\,(\text{rms})}^2/(R_s + R_i)$, where $V_{L\,(\text{rms})}$ and $V_{S\,(\text{rms})}$ are the rms values of v_L and v_S. Thus, $P_L/P_S = (V_{L\,(\text{rms})}/V_{S\,(\text{rms})})^2(R_s + R_i)/R_L$. Using Equation (5.37) and simplifying, we find that the power gain of the *voltage amplifier* is

$$A_p = \left(\frac{R_i}{R_o + R_L} A_{\text{oc}} \right)^2 \frac{R_L}{R_s + R_i} \tag{5.45}$$

Analogous expressions can be derived for the other amplifier types.

Exercise 5.20 Find the source-to-load voltage gain and the power gain for the voltage amplifier of Exercise 5.18.

ANSWER 48 V/V, 77.60 dB.

> **Exercise 5.21**
>
> **(a)** Show that for the current amplifier of Figure 5.25 we have
>
> $$A_p = \frac{R_s}{R_s + R_i} \left(\frac{R_o}{R_o + R_L} A_{sc} \right)^2 \frac{R_L}{R_i} \qquad \textbf{(5.46)}$$
>
> **(b)** Find the power gain of the current amplifier of Exercise 5.19, part (a), in decibels.
>
> **ANSWER**   (b) 54.61 dB.

It is comforting to know that R_i, R_o, and the *unloaded gain* A_{oc} or A_{sc} suffice to fully characterize the terminal behavior of any linear amplifier, no matter how complex its internal circuitry. But then, if all it takes to make an amplifier is a source and two resistances, why do amplifiers such as hi-fi audio amplifiers require so much internal circuitry? To answer this question we must bear in mind that the function of our model is to mimic only the *terminal* behavior of the actual amplifier. Achieving this behavior within the stringent specifications of modern audio equipment can be a formidable task, this being the reason why so much sophisticated circuitry is usually needed.

5.5 CIRCUIT ANALYSIS USING SPICE

SPICE allows circuits to contain linear dependent sources characterized by any of the following equations

$$v = k_v v_X \qquad i = k_g v_X \qquad i = k_i i_X \qquad v = k_r i_X$$

where v and i are the dependent variables, v_X and i_X are the controlling variables, and k_v, k_g, k_i, and k_r are suitable constants.

Voltage-Controlled Sources

The general forms of the element statements for voltage-controlled sources are:

```
EXXX   N+   N-   NC+   NC-   VALUE                    (5.47)
GXXX   N+   N-   NC+   NC-   VALUE                    (5.48)
```

As illustrated in Figure 5.28, these statements contain:

(1) The name of the controlled source, which must begin with the letter E if it is a voltage source (VCVS) and with G if it is a current source (VCCS).

(2) The circuit nodes N+ and N- to which the source is connected. For a VCVS, N+ and N- are, respectively, the positive and negative terminals; for a VCCS, N+ is the node where the source sinks current and N- the node where it sources current. As we know, SPICE assumes positive current to flow from N+ through the source, to N-.

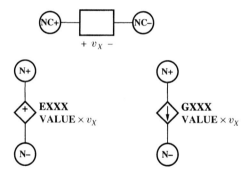

Figure 5.28 SPICE terminology for voltage-controlled sources.

(3) The positive and negative controlling nodes, NC+ and NC-.

(4) The VALUE of the proportionality constant relating the dependent source to the controlling voltage. This value is in V/V for a VCVS, and A/V for a VCCS.

▶ **Example 5.20**

Use SPICE to find all node voltages in the circuit of Figure 5.10(a).

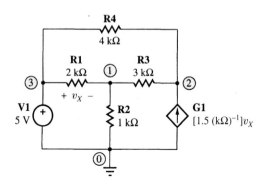

Figure 5.29 Circuit of Example 5.20.

Solution

The circuit is shown in Figure 5.29, and the input file is:

```
CIRCUIT WITH A VCCS
V1 3 0 DC 5
R1 3 1 2K
R2 1 0 1K
R3 1 2 3K
R4 3 2 4K
G1 0 2 3 1 1.5M
.END
```

After SPICE is run, the output file contains:

```
NODE   VOLTAGE   NODE   VOLTAGE   NODE   VOLTAGE
(1)    3.0000    (2)    9.0000    (3)    5.0000
```

Exercise 5.22   Use SPICE to find all node voltages in the circuit of Figure 5.12(a).

Current-Controlled Sources

The general forms of the element statements for current-controlled sources are

$$\text{FXXX} \quad \text{N+} \quad \text{N-} \quad \text{VNAME} \quad \text{VALUE} \tag{5.49}$$

$$\text{HXXX} \quad \text{N+} \quad \text{N-} \quad \text{VNAME} \quad \text{VALUE} \tag{5.50}$$

As illustrated in Figure 5.30, these statements contain

- **(1)** The name of the controlled source, which must begin with the letter F if the source is of the current type (CCCS) and H if it is of the voltage type (CCVS).
- **(2)** The circuit nodes N+ and N- to which the source is connected. For a current source, N+ is the terminal where the source sinks current, N- the terminal where it sources current. For a voltage source, N+ and N- are, respectively, the positive and negative terminals. As we know, SPICE assumes positive current to flow from N+, through the source, to N-.
- **(3)** The name of a voltage source through which the controlling current flows. Since SPICE requires that the controlling current be that of a voltage source, this may require inserting a dummy voltage source in the controlling branch, in the manner discussed in Section 3.6.
- **(4)** The VALUE of the proportionality constant relating the dependent source to the controlling current. This value is in A/A for a CCCS, and in V/A for a CCVS.

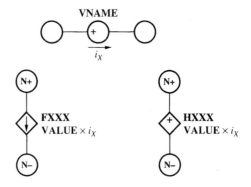

Figure 5.30 SPICE terminology for current-controlled sources.

▶ **Example 5.21**

Using SPICE, find the Thévenin equivalent of the circuit of Figure 5.15.

Solution

The circuit, augmented of the dummy source VD to monitor the controlling current i_X, is shown in Figure 5.31. The input file is:

```
THEVENIN EQUIVALENT WITH DEPENDENT SOURCES
V1 1 0 DC 6
R1 1 2 2
R2 2 4 6
VD 4 0 DC 0
F1 0 3 VD 2
R3 2 3 3
.TF V(3) V1
.END
```

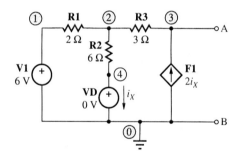

Figure 5.31 Using SPICE to find the Thévenin equivalent of the circuit of Figure 5.15.

After SPICE is run, the output file contains the following:

```
NODE VOLTAGE NODE VOLTAGE NODE VOLTAGE NODE VOLTAGE
(1)   6.0000 (2)   9.0000 (3)  18.0000 (4)   0.0000

OUTPUT RESISTANCE AT V(3)=9.000E+00
```

Clearly, we have $v_{OC} = \text{V}(3) = 18$ V, and $R_{eq} = 9\ \Omega$.   ◀

Exercise 5.23   Use SPICE to check the results of Example 5.5.

The Ideal Transformer

The SPICE analysis of transformer circuits utilizes the ideal transformer model of Figure 5.19. As shown in Figure 5.32, the primary is modeled with a CCCS, the secondary with a VCVS and a 0-V dummy source to sense the control current for the CCCS.

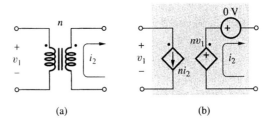

(a)   (b)

Figure 5.32 Ideal transformer and its circuit model suitable for SPICE simulation.

  Example 5.22

Use SPICE to find how a transformer with $n = 5$ reflects a 100-Ω secondary resistance to the primary.

Solution

The circuit, shown in Figure 5.33, uses the dependent sources **F1** and **E2** to model the transformer, the dummy source **VD** to monitor the load current, and the test source **V1**. The input file is

```
RESISTANCE TRANSFORMATION
V1 1 0
F1 1 0 VD 5
E2 3 0 1 0 5
VD 3 2 DC 0
R2 2 0 100
.TF V(2) V1
.END
```

After running SPICE, the output file contains the following:

```
SMALL-SIGNAL CHARACTERISTICS:
V(2)/V1 = 5.000E+00
INPUT RESISTANCE AT V1 = 4.000E+00
OUTPUT RESISTANCE AT V(2) = 0.000E+00
```

Clearly, $R_1 = R_L/n^2 = 100/5^2 = 4\ \Omega$.

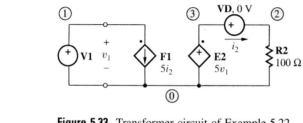

Figure 5.33 Transformer circuit of Example 5.22.

▼ SUMMARY

- A two-port is a circuit equipped with a pair of ports through which signals or energy can enter or leave the circuit. The two-ports considered in this chapter are *transformers* and *amplifiers.*

- A two-port is fully characterized once we know its *terminal* as well as its *transfer characteristics.* The terminal characteristics represent the i–v relationships at the individual ports. The transfer characteristic relates the output to the input.

- To model the dependence of its output upon its input, a two-port requires a *dependent source.*

- The presence of a dependent source may result in unexpected forms of circuit behavior. Using just a resistance and a dependent source as an illustrative example, we see that depending on how we specify the source, the effective value of the resistance may be decreased, increased, or even made negative.

- When analyzing circuits with dependent sources, we still apply the techniques of the previous chapters. However, while independent sources may be suppressed if desired, *dependent sources must always be left in the circuit.* Suppressing them will generally yield wrong results.

- A transformer consists of a *primary* and a *secondary* winding, and is characterized by the *turns ratio n.* The primary and secondary voltages are related as $v_2 = nv_1$ and the currents as $i_1 = ni_2$.

- A transformer is called a *step-up transformer* when $n > 1$, a *step-down transformer* when $n < 1$, and an *isolation transformer* when $n = 1$. Regardless of the value of n, a transformer provides electrical isolation between its primary and secondary.

- A common transformer application is to *improve the efficiency of power transmission* over long transmission lines.

- A transformer effects resistance transformations. A resistance R_2 across the secondary appears at the primary as an equivalent resistance $R_1 = R_2/n^2$. This transformation works in both directions, and it finds application in *resistance matching.*

- There are four types of amplifiers: (a) the *voltage amplifier,* with gain in V/V; (b) the *current amplifier,* with gain in A/A; (c) the *transresistance amplifier,* with gain in V/A; (d) the *transconductance amplifier,* with gain in A/V.

- The *source-to-load* gain is the product of three terms: an input divider term, an unloaded gain factor, and an output divider term. Since the divider terms are less than unity, the source-to-load gain is always *less* than the unloaded gain.

- To avoid loading at the input port, a *voltage-input* amplifier should have $R_i = \infty$, and a *current-input* amplifier $R_i = 0$. To avoid loading at the output port, a *voltage-output* amplifier should have $R_o = 0$, and a *current-output* amplifier $R_o = \infty$.

- The *power gain* of a two-port is the ratio of the average power delivered to the load to that supplied by the source, $A_p = P_L/P_S$. It is often expressed in *decibels* as $A_{p(\text{dB})} \triangleq 10 \log_{10} A_p$.

- Circuits with dependent sources can easily be simulated via SPICE. The name of the source must start with the letter E for a VCVS, G for a VCCS, F for a CCCS, and H for a CCVS.

▼ PROBLEMS

5.1 Dependent Sources

5.1   In the circuit of Figure P5.1 find the two values of k_r (in Ω) for which $p_{1\,\Omega} = 4$ W.

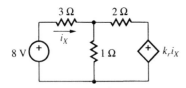

Figure P5.1

5.2   In the circuit of Figure P5.1 find the value of k_r (in Ω) for which $i_{2\,\Omega} = 2$ A ($\leftarrow$). Hence, verify the conservation of power.

5.3   In the circuit of Figure P5.3 verify the conservation of power.

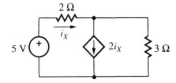

Figure P5.3

5.4   In the circuit of Figure P5.4 find the voltage across the 1-A source. Check your result.

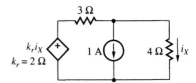

Figure P5.4

5.5   Find R_{eq} in the circuit of Figure P5.5.

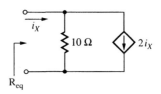

Figure P5.5

5.6   In the BJT circuit of Figure 5.7 let $v_I = 10$ V, $V_{CC} = 10$ V, $R_B = 270$ kΩ, and $R_C = 2.0$ kΩ. Assuming $V_{BE} = 0.7$ V, (a) find v_O if $\beta = 85$; (b) find β for $v_O = 1$ V.

5.7   In the BJT circuit of Figure 5.7 let $V_{CC} = 10$ V, $V_{BE} = 0.7$ V, and $\beta = 100$. Specify R_B and R_C so that $v_O = 5$ V with $v_I = 5$ V. The power dissipated inside the BJT must be 10 mW.

5.8   In the BJT circuit of Figure 5.8 let $R_1 = 30$ kΩ, $R_2 = 15$ kΩ, $R_C = 3.0$ kΩ, $R_E = 1.5$ kΩ, and $V_{CC} = 15$ V. Find V_{CE} if $V_{BE} = 0.7$ V and $\beta = 125$.

5.9   In the BJT of Figure 5.8 let $R_2 = 27$ kΩ, $R_C = 5$ kΩ, and $V_{CC} = 15$ V. If $V_{BE} = 0.7$ V and $\beta = 150$, specify R_1 and R_E to achieve $V_C = 9$ V and $V_B = 4.7$ V.

5.10 In the circuit of Figure P5.10 let $R_1 = R_2 = 100$ kΩ, $R_C = 15$ kΩ, $R_E = 7.5$ kΩ, and $V_{CC} = 12$ V. Assuming $V_{BE} = 0.7$ V and $\beta = 75$, find V_B, V_C, and V_E. (*Hint:* Apply KCL both at B and C.) Check your results.

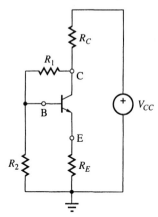

Figure P5.10

5.11 In the circuit of Figure P5.10 let $R_2 = 43$ kΩ, $R_E = 2.0$ kΩ, and $V_{CC} = 10$ V. If $V_{BE} = 0.7$ V and $\beta = 175$, specify R_1 and R_C to achieve $V_C = 7$ V and $V_E = 3$ V.

5.2 Circuit Analysis with Dependent Sources

5.12 Use nodal analysis to find the power dissipated by the 5-kΩ resistance in the circuit of Figure P5.12. Check your results.

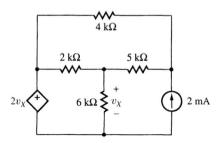

Figure P5.12

5.13 Use loop analysis to find the power dissipated by the 6-kΩ resistance in the circuit of Figure P5.12. Check your results.

5.14 Apply loop analysis to the circuit of Figure P5.14, and check.

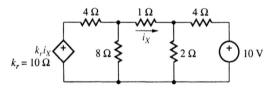

Figure P5.14

5.15 Apply nodal analysis to the circuit of Figure P5.14, and check.

5.16 Use loop analysis to find the power (absorbed or released?) by the dependent source in Figure P5.16. Check.

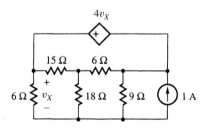

Figure P5.16

5.17 Use nodal analysis to find the voltage across the 1-A source in Figure P5.16. Check.

5.18 Use nodal analysis to find the power dissipated by the 8-kΩ resistance in Figure P5.18. Check.

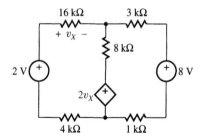

Figure P5.18

5.19 Use loop analysis to find the power (absorbed or released?) by the dependent source in Figure P5.18. Check.

5.20 Use nodal analysis to find i_X in Figure P5.20. Check.

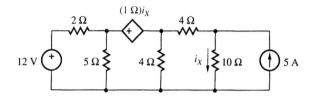

Figure P5.20

5.21 Use loop analysis to find the power (absorbed or released?) by the independent sources of Figure P5.20. Check.

5.22 Find R_{eq} in Figure P5.22. Check.

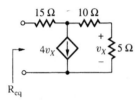

Figure P5.22

5.23 Find R_{eq} in Figure P5.23. Check.

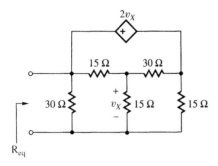

Figure P5.23

5.24 Find R_{eq} in Figure P5.23, but with the controlling and controlled branches interchanged with each other, the reference voltage polarities remaining the same. Check.

5.25 Find R_{eq} in Figure P5.25. Check.

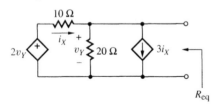

Figure P5.25

5.26 In the circuit of Figure P5.26 find (a) the Thévenin equivalent with respect to terminals A and C; (b) the Norton equivalent with respect to terminals B and C.

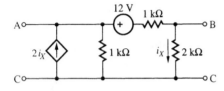

Figure P5.26

5.27 Find the maximum power that the circuit of Figure P5.26 can deliver to a resistive load connected between A and B.

5.28 (a) Find the power dissipated by the 6-Ω resistance in the circuit of Figure P5.16, center. (b) Find the value to which the 6-Ω resistance must be changed if we want it to dissipate as much power as possible. What is this power? (*Hint:* Use Thévenin's or Norton's theorems.)

5.29 Repeat Problem 5.28, but for the upper 4-Ω resistance in the circuit of Figure P5.20.

5.30 Find the maximum power that the circuit of Figure P5.30 can deliver to a resistive load connected between A and B.

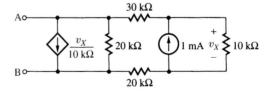

Figure P5.30

5.31 A voltmeter with internal resistance R_i is connected across the 10-kΩ resistance in Figure P5.30. Predict the voltmeter reading with (a) $R_i = \infty$, and (b) $R_i = 100$ kΩ.

5.3 The Ideal Transformer

5.32 In the circuit of Figure 5.22(a) let $v_s = 3$ V (rms), $R_s = 20$ Ω, and $R_L = 1$ kΩ. (a) Find n for a 100-mW average power transfer to the load. (b) Find n for maximum average power transfer to the load. What is this maximum power?

5.33 In the circuit of Figure 5.22(a) let $R_s = 15$ Ω, $n = 20$, and $R_L = 10$ kΩ. (a) Find the rms value of v_s for an average power transfer of 100 mW to R_L. (b) Find the efficiency η of this transfer.

5.34 Find the Thévenin equivalent of the circuit of Figure P5.34 with respect to nodes A and B.

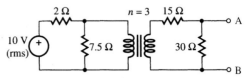

Figure P5.34

5.35 In the circuit of Figure P5.35 let $n_1 = 2$, $n_2 = 5$, and $R = 0$. Find the average power dissipated in the 8-kΩ resistance.

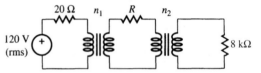

Figure P5.35

5.36 Repeat Problem 5.35, but with $R = 400$ Ω.

5.37 (a) If in the circuit of Figure P5.35 $n_1 = 2$ and $R = 0$, find n_2 to ensure maximum average power transfer to the 8-kΩ load. What is this power? (b) If $n_2 = 5$ and $R = 180$ Ω, find n_1 to ensure maximum power transfer to the 8-kΩ load. What is the maximum power?

5.38 (a) Using the transformer model of Figure 5.19, find R_{eq} in the circuit of Figure P5.38. (b) Repeat if the dot of the secondary is at the bottom.

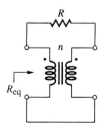

Figure P5.38

5.39 Find the average power dissipated in the 10-Ω resistance of Figure P5.39. (*Hint:* Use the loop method.)

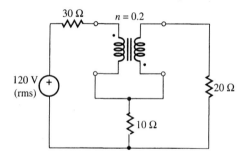

Figure P5.39

5.40 Verify the conservation of power in the circuit of Figure P5.39 if the dot of the secondary is at the top. (*Hint:* Use the loop method.)

5.41 Find the average power dissipated by each resistor in the circuit of Figure P5.41. Then, verify that power is conserved.

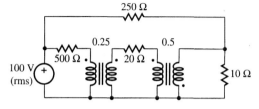

Figure P5.41

5.4 Amplifiers

5.42 A 1.0-V source is applied to a voltage divider consisting of $R_1 = 10$ kΩ and $R_2 = 20$ kΩ. The voltage v_2 developed by

R_2 is fed, in turn, to a voltage amplifier having $R_i = 60$ kΩ, $A_{oc} = 30$ V/V, and $R_o = 600$ Ω, and driving a load $R_L = 900$ Ω. Sketch the circuit, and find v_2 and v_L.

5.43 A source v_S with internal resistance $R_s = 1$ kΩ is applied to a current amplifier having $R_i = 200$ Ω, $A_{sc} = 100$ A/A, and $R_o = 10$ kΩ. Sketch the circuit, and find the gain v_L/v_S for an output load $R_L = 2$ kΩ.

5.44 In the transresistance amplifier of Figure 5.26 let $R_s = 100$ kΩ, $R_i = 15$ kΩ, $A_{oc} = 1000$ V/A, $R_o = 100$ Ω, and $R_L = 500$ Ω. (a) Find the gain v_L/i_S. (b) To what value must the unloaded gain A_{oc} be changed to achieve $v_L/i_S = 1000$ V/A?

5.45 In the transconductance amplifier of Figure 5.27 let $v_S = 150$ mV, $R_s = 100$ Ω, and $R_L = 40$ Ω. DMM readings at the input and output ports yield, respectively, $v_I = 125$ mV and $i_L = 230$ mA. Changing the load to $R_L = 60$ Ω yields $i_L = 220$ mA. Find R_i, A_{sc}, and R_o.

5.46 A transconductance amplifier with $R_i = 2$ kΩ, $A_{sc} = 9$ mA/V, and $R_o = 100$ kΩ is fed by a voltage source v_S having internal resistance $R_s = 1$ kΩ and drives a load $R_L = 20$ kΩ. Sketch the circuit, and find the gain v_L/v_S.

5.47 Shown in Figure P5.47 is a circuit known as the *small-signal model* of a *common-emitter* BJT amplifier. Find the gain v_o/v_S.

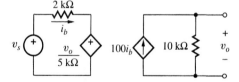

Figure P5.47

5.48 Shown in Figure P5.48 is a circuit known as the *small-signal model* of an *emitter follower* BJT amplifier. (a) Derive expressions for R_i, R_o, and v_o/v_S. (b) Find their numerical values if $R_s = 2$ kΩ, $R_L = 10$ kΩ, $r_\pi = 2.6$ kΩ, and $\beta = 100$.

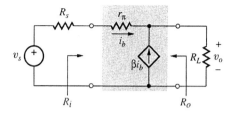

Figure P5.48

5.49 An amplifier having $R_i = 10$ kΩ is fed by a source $i_S = 1$ μA with internal resistance $R_s = 100$ kΩ and yields an output open-circuit voltage of 10 V and short-circuit current of 10 mA. If the output is loaded with $R_L = 5$ kΩ find the voltage gain v_O/v_I, current gain i_O/i_I, and power gain P_L/P_S (in dB).

5.50   A phono cartridge characterized by a 1-V rms voltage and a 1-MΩ series resistance is to drive an 8-Ω loudspeaker. Find the loudspeaker voltage and the power gain, in dB, if the cartridge output is fed to the speaker (a) directly, or (b) via a buffer voltage amplifier having $R_i = 10$ MΩ, $A_{oc} = 1$ V/V, and $R_o = 2$ Ω. Justify physically the difference.

5.51   Figure P5.51 shows two voltage amplifiers connected in *cascade* to achieve a large overall gain. Starting from the load and working your way back toward the source, derive an expression for v_L in terms of v_S. Under what conditions will the source-to-load gain simplify as $v_L/v_S = A_{oc1} \times A_{oc2}$?

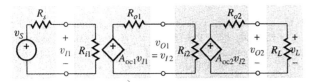

Figure P5.51

5.52   Consider the cascade of a transresistance amplifier having $R_{i1} = 100$ Ω, $A_{oc1} = 0.2$ V/mA, and $R_{o1} = 100$ Ω, and a

transconductance amplifier having $R_{i2} = 1$ kΩ, $A_{sc2} = 100$ mA/V, and $R_{o2} = 100$ kΩ. The cascade is driven by a source i_S with output resistance $R_s = 1$ kΩ and drives a load $R_L = 25$ kΩ. Find the source-to-load gain i_L/i_S and the power gain. Express the latter in dB.

5.5 Circuit Analysis Using SPICE

5.53   Use SPICE to find v_X in the circuit of Figure P5.12.

5.54   Use SPICE to find i_X in the circuit of Figure P5.14.

5.55   Use SPICE to find R_{eq} in the circuit of Figure P5.25.

5.56   Use SPICE to check the Thévenin and Norton equivalents of the circuit of Figure P5.26 as found in Problem 5.26.

5.57   Solve the BJT bias circuit of Problem P5.10 via SPICE.

5.58   Use SPICE to find the Thévenin equivalent of the transformer circuit of Figure P5.34.

5.59   (a) Use SPICE to find R_{eq} in the transformer circuit of Figure P5.38, given that $R = 100$ Ω and $n = 5$. (b) Repeat, but with the secondary dot at the bottom.

OPERATIONAL AMPLIFIERS

T hough amplification might appear as a rather limited form of signal processing, it is nevertheless of fundamental importance because most other signal processing circuits are implemented using amplifiers as basic ingredients. Until a few decades ago, an engineer who needed an amplifier had to design it from scratch, using dozens of assorted components such as resistors, capacitors, diodes, and transistors, not to mention endless slide-rule calculations. Mercifully, it turns out that a considerable portion of this circuitry is *common* to a wide variety of amplifiers. To simplify amplifier design, integrated-circuit manufacturers provide this common circuitry as a ready-made *universal building block.* Then, with the help of only a few external components, the user personalizes this block to the particular application at hand.

Surprisingly, this building block *is itself an amplifier.* Called an **operational amplifier,** or *op amp* for short, it is fabricated on a miniature silicon chip and is made available as an off-the-shelf component. In mass production, its cost may be comparable to that of individual circuit elements such as resistors, transistors, and potentiometers. The combination of low cost, miniature size, high reliability, and easy use makes the op amp the most popular device in analog circuit design.

In this chapter, after introducing op amp models and terminology, we examine the most basic op amp configurations, that is, the *noninverting, inverting, and buffer amplifiers.* In so doing, we introduce another cornerstone concept of electrical engineering, namely, the concept of **negative feedback.** Recognizing that the beginner feels often intimidated by this technology, we develop a friendly **op amp rule** to make our analysis of op amp circuits easy and fast.

We then use this rule to study a variety of op amp circuits that find wide use in electronic instrumentation, such as *summing* and *difference amplifiers, instrumentation amplifiers, I–V* and *V–I converters,* and *current amplifiers.* These circuits are of the type engineers see every day. You can try them out in the lab, or you can even use them as a basis for some hobby or other project.

Our philosophy is to use the op amp also as a vehicle to illustrate intriguing facets of electrical engineering. In the present chapter we use it to implement a *negative resistance,* and we demonstrate its application to the design of current sources. Additional op amp circuits are studied in subsequent chapters as we introduce new elements and concepts.

We conclude by illustrating the use of SPICE to analyze op amp circuits. Computer simulation can be used to check the results of hand calculations as well as to assess the departure of a practical op amp circuit from ideal behavior. The SPICE facility introduced in this chapter is the `.SUBCKT` statement.

6.1 THE OPERATIONAL AMPLIFIER

The operational amplifier is a **high-gain voltage amplifier.** It consists of dozens of circuit components fabricated on a tiny silicon chip and encapsulated in a package with suitable leads or pins for its connection to external components. The most popular package is the 8-pin dual-in-line package of Figure 6.1, called *minidip.* Also shown is the op amp circuit symbol, along with the standard pin designations.

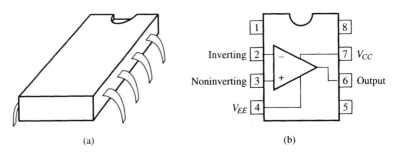

Figure 6.1 A popular op amp package and pin configuration.

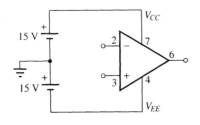

Figure 6.2 Powering the op amp.

In order to function, an op amp needs to be externally powered via two dedicated pins, which, in the case of the minidip package, are pins 7 and 4. The supply voltages are typically $V_{CC} = +15$ V and $V_{EE} = -15$ V, though other values are permissible, as specified in the datasheets. These voltages can be generated with a dual bench power supply or with a pair of batteries.

Op amp powering is shown in Figure 6.2. Its worth noting that the op amp itself does not have a reference or ground pin. The reference node is taken to be the common node of the external power supplies. To avoid cluttering our circuit diagrams, we omit showing the power supply connections; however, remember that they are necessary, and that power must be turned on when an op amp circuit is tried out in the laboratory.

An important limitation of any practical op amp is that it cannot swing its output v_O above V_{CC} or below V_{EE}, or $V_{EE} < v_O < V_{CC}$. If an attempt is made to force v_O outside this range, the op amp will **saturate** and yield either $v_O \simeq V_{CC}$ or $v_O \simeq V_{EE}$. In the following we shall assume that our op amps are operated away from saturation. However, the possibility of saturation must be kept in mind when op amps are tried out in the laboratory.

Op Amp Terminology

Figure 6.3 shows the circuit model of the op amp, consisting of the **input resistance** r_i; the dependent source av_D, where a is the **unloaded voltage gain;** and the **output resistance** r_o. This is the standard amplifier model studied in Chapter 5, except that a is now very large. To give an idea, the 741 op amp, one of the oldest types but still very popular, has typically $a = 200,000$ V/V, though op amp types are now available with a in excess of 10^7 V/V. It is fair to say that what sets op amps apart from other voltage amplifiers is a *high unloaded gain.* As we proceed we shall find that the higher this gain, the better, and that ideally an op amp would have $a \to \infty$.

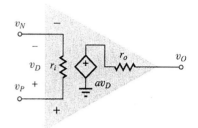

Figure 6.3 Circuit model of the op amp. For the popular 741 op amp, $r_i \simeq 2$ MΩ. $a \simeq 200,000$ V/V, and $r_o \simeq 75$ Ω.

When a load is connected to the output pin, it will generally absorb power. The role of the circuitry internal to the op amp is to take this power from the external supplies V_{CC} and V_{EE} and pass it on to the load. It is certainly a blessing that the whole process can be modeled with a mere dependent source! It is also apparent that without the external supplies, the internal circuitry and, hence, the dependent source would be electrically dead.

It is worth noting that the internal source has been shown grounded even though the op amp has no ground pin provision and the ground reference is established by the supplies externally. Source grounding ensures that our circuit model, viewed as a supernode of a more complex circuit, satisfies KCL.

The input terminals are called **inverting** and **noninverting,** and are identified by the − and + symbols. Their voltages, denoted as v_N and v_P, are referenced to ground, just like the voltage v_O of the output terminal. In the absence of any load the op amp yields

$$v_O = a(v_P - v_N) = av_D \qquad \text{(6.1)}$$

where

$$v_D \triangleq v_P - v_N \qquad \text{(6.2)}$$

is called the **differential input voltage.** Equation (6.2) can be turned around to yield

$$v_D = \frac{v_O}{a} \qquad \text{(6.3)}$$

This allows us to find the input voltage v_D needed to sustain a given output voltage v_O in the absence of any load.

▶ ## Example 6.1

In a certain circuit containing a 741 op amp it is found that $v_P = 1.0$ V and $v_O = 5$ V. Assuming negligible loading, find v_D and v_N.

Solution

We have $v_D = v_O/a = 5/200,000 = 25$ μV. Moreover, by Equation (6.1), $v_N = v_P - v_D = 1.0 - 25 \times 10^{-6} = 0.999975$ V. ◀

Exercise 6.1   In a circuit containing a 741 op amp find (a) v_N if $v_P = 1.0$ V and $v_O = -5$ V, and (b) v_P if $v_N = 3.0$ V and $v_O = -2$ V.

ANSWER   (a) 1.000025 V, (b) 2.99999 V.

The Ideal Op Amp

Equation (6.3) indicates that thanks to its large gain a, an op amp can sustain a nonzero output v_O with a vanishingly small differential input v_D. Indeed, Example 6.1 revealed that in order to sustain $v_O = 5$ V, a 741 op amp needs only $v_D = 25$ μV. To sustain the same output, an op amp with $a = 10^7$ V/V would need an even smaller differential input, namely, $v_D = 5/10^7 = 0.5$ μV! It is apparent that in the limit $a \to \infty$ we obtain $v_D \to 0$. However, the product av_D remains determinate and equal to the given value of v_O.

Our initiation to op amp circuits will be greatly facilitated if we start out with the *idealized* op amp model of Figure 6.4. In this model we let $a \to \infty$. Moreover, to avoid worrying about any loading effects at the input and output ports, we assume $r_i = \infty$, $r_o = 0$, and *zero current flow* at either input terminal. Though these are idealized conditions, current technology makes it possible to realize them to a highly satisfactory degree, so the level of performance of circuits using practical op amps comes quite close to that predicted using the idealized model. This explains the widespread use of op amps in analog design.

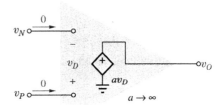

Figure 6.4 Ideal op amp model: $r_i = \infty$, $a \to \infty$, $r_o = 0$, and *no current* flows into or out of either input pin.

6.2 BASIC OP AMP CONFIGURATIONS

In this section we investigate the two most basic op amp circuit topologies, namely, the *inverting* and *noninverting* configurations, and in so doing we introduce a number of important concepts, including negative feedback. Each configuration is analyzed using the op amp model of Figure 6.4 in the limit $a \to \infty$. After acquiring sufficient familiarity with these configurations, we introduce a speedier op amp circuit analysis technique in the next section.

The Noninverting Configuration

The circuit of Figure 6.5(a) consists of an op amp and two external resistances. Since R_2 is connected between the output and one of the inputs, it is called a *feedback resistance.* To facilitate our analysis, we redraw the circuit as in Figure 6.5(b). Using Equation (6.1), along with the voltage divider formula, we have

$$v_O = a(v_P - v_N) = a \left(v_I - \frac{1}{1 + R_2/R_1} v_O \right)$$

Since the voltage divider is feeding the signal $v_O/(1 + R_2/R_1)$ back to the inverting input, it is said to provide **negative feedback** around the op amp. Negative feedback was conceived in 1927 by the American engineer Harold S. Black (1898–1983).

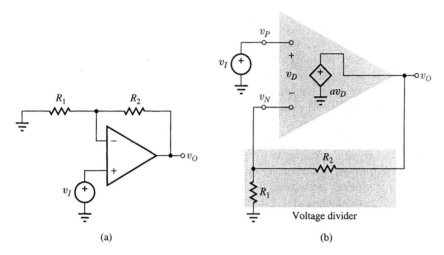

(a)   (b)

Figure 6.5 The noninverting configuration.

Collecting terms we obtain

$$v_O \left(1 + \frac{a}{1 + R_2/R_1} \right) = a v_I$$

If we now define the gain of the op amp circuit as

$$A \triangleq \frac{v_O}{v_I} = \frac{a}{1 + a/(1 + R_2/R_1)} = \left(1 + \frac{R_2}{R_1} \right) \frac{a}{1 + R_2/R_1 + a} \qquad \textbf{(6.4)}$$

then it is readily seen that

$$\lim_{a \to \infty} A = 1 + \frac{R_2}{R_1} \qquad \textbf{(6.5)}$$

This result allows us to make a number of important observations:

(1) The circuit of Figure 6.5(a), made up of an op amp and a voltage divider, is still an amplifier. In fact, it is called a **noninverting amplifier** because its gain A is positive and, hence, v_O has the same polarity as v_I.

(2) The gain A of the op amp circuit is *independent* of the gain a of the op amp. The former is called the **closed-loop gain** because it is obtained by creating a feedback loop around the op amp via an external network. By contrast, a is called the **open-loop gain.**

(3) The gain A is set exclusively by the external components. In fact, the op amp provides the inverse function of the external network. While the divider divides v_O by $1 + R_2/R_1$ to yield v_N, the op amp multiplies v_P by $1 + R_2/R_1$ to yield v_O.

The absence of a from Equation (6.5) is highly welcome because a is a poorly defined parameter: it varies with temperature and time, as well as from one op amp sample to another. However, as long as it is sufficiently large to make $a/(1 + R_2/R_1 + a) \rightarrow 1$ in Equation (6.4), then $A \rightarrow (1 + R_2/R_1)$ regardless of any variations in a. In particular, should the op amp fail, we can simply replace it with one of comparable characteristics without significantly affecting A. We now appreciate why it is desirable to have $a \rightarrow \infty$!

The fact that the closed-loop gain A is set by an external *resistance ratio* makes it easy to customize A to a particular application. If we use resistors of suitable quality, we can make A as accurate as desired. Moreover, we can stabilize it by using resistances that *track* each other with temperature and time in such a way as to ensure a constant ratio. You will agree that the ability to achieve accurate and stable functions using components of lesser individual quality is quite ingenious!

▶ Example 6.2

In the op amp circuit of Figure 6.5(a) specify resistances in the $k\Omega$ range to achieve (a) $A = 2$ V/V, (b) $A = 5$ V/V, and (c) a variable gain over the range 1 V/V $\leq A \leq 10$ V/V by means of a 180-$k\Omega$ potentiometer.

Solution

By Equation (6.5), $R_2 = (A - 1)R_1$.

(a) Let $R_1 = 10$ $k\Omega$. Then, $R_2 = (2 - 1)10 = 10$ $k\Omega$.

(b) Let $R_1 = 30$ $k\Omega$. Then, $R_2 = (5 - 1)30 = 120$ $k\Omega$.

(c) Let R_2 be the 180-$k\Omega$ pot with the wiper tied to one of its extremes, as in Figure 6.6. Then, varying the wiper from end to end will vary R_2 from 0 Ω to 180 $k\Omega$. Since we want A to vary from 1 V/V to 10 V/V, we must have $10 = 1 + 180/R_1$, or $R_1 = 20$ $k\Omega$.   ◀

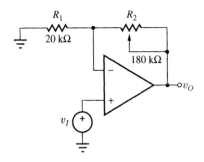

Figure 6.6 Circuit of Example 6.2, (c).

Exercise 6.2 Using a 100-kΩ potentiometer, design a noninverting amplifier whose gain varies from 10 V/V to 100 V/V as the wiper is swept from end to end. (*Hint:* To prevent A from dropping below 10 V/V, you must use a suitable resistance in series with the pot.)

ANSWER $R_1 = 1.111$ kΩ; R_2: 10 kΩ in series with the potentiometer connected as a variable resistance from 0 to 100 kΩ.

To fully characterize the noninverting amplifier we need to know, besides its gain A, its input and output resistances R_i and R_o. These resistances are depicted in Figure 6.7(a). To this end, it is convenient to refer back to Figure 6.5(b). Since no current flows into or out of the noninverting input pin, we have $R_i = \infty$. Moreover, since the output is taken from the ideal dependent source, we have $R_o = 0$. Putting our findings together we end up with the ideal noninverting amplifier equivalent of Figure 6.7(b).

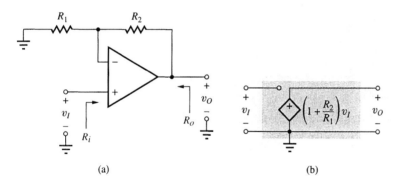

(a)   (b)

Figure 6.7 The ideal noninverting amplifier has $R_i = \infty$, $A = 1 + R_2/R_1$, and $R_o = 0$.

► **Example 6.3**

In the circuit of Figure 6.5(a) let $R_1 = 10$ kΩ, $R_2 = 20$ kΩ, and $a = 10^5$ V/V. Find A, and compare with the ideal case $a \to \infty$.

Solution

By Equation (6.4), $A = (1 + 20/10) \times 10^5/(1 + 20/10 + 10^5) = 2.99991$ V/V. Its percentage deviation from the ideal value $A = 1 + 20/10 = 3$ V/V is $100 \times (2.99991 - 3)/3 = -0.003\%$, a negligible value for all practical purposes! ◄

ROLLS OF HONOR

The professional society you will most probably join when you graduate, the Institute of Electrical and Electronics Engineers (IEEE), has a history going back more than 100 years. By 1884, the year that marks the beginnings of IEEE, electrical theory had borne fruit in the form of numerous applications. Maxwell had formulated his wonderful equations, Faraday had finished his work on electromagnetic induction to clear the way for all types of industrial machinery, and both theory and technology were becoming increasingly complex. It was a boom time for all workers in the field, and what could be more natural than to form a professional group to serve as a forum for exchanging information? The American Institute of Electrical Engineers (AIEE), founded in New York City on May 13, 1884, was the result.

Elihu Thomson, the first recipient of the Edison Medal (1909) (Culver Pictures, Inc.)

Until about thirty years ago, there was a second professional group serving electrical and electronics engineers—the Institute of Radio Engineers (IRE), formed in 1912 as a result of the merger of the Society of Wireless Telegraph Engineers and the Wireless Institute. AIEE and IRE joined forces in 1963, and the merged group is what we know today as IEEE.

IEEE presents numerous awards for outstanding contributions to electrical and electronics engineering. The major ones, usually presented annually, are described here.

The Edison Medal. The oldest IEEE award is the Edison Medal, created by some of Edison's associates in 1904 to celebrate the silver anniversary of the birth of the incandescent lightbulb. In the somewhat flowery words of those associates, "The Edison Medal should, during the centuries to come, serve as an honorable incentive to scientists, engineers, and artisans to maintain by their works the high standard of accomplishment set by the illustrious man whose name and feats shall live while human intelligence continues to inhabit the world."

Elihu Thomson was the first recipient of the Edison Medal, awarded by the AIEE in 1909 on behalf of the medal's creators. One of the first of the great engineers working at General Electric, Thomson held patents for high-frequency generators and transformers, a three-coil generator, a system for incandescent welding, and a watthour meter. He is sometimes called "the father of protective grounding" because of his idea of using a grounding wire attached to the secondary winding of a transformer to prevent shock.

Down through the years since that first medal was awarded, Thomson has been followed by about 80 colleagues. The honor roll contains the most illustrious names in electricity. Some are easily recognizable: George Westinghouse (1911), Alexander Graham Bell (1914), Nikola Tesla (1916), Robert Millikan (1922), Edwin Armstrong (1942)—but most are not: Frank Sprague (1910), John Lieb (1923), Jam Rajchman (1974), and on and on. Unknown to the world at large, but eminent and honored in their own profession.

The Medal of Honor.   The highest award in the IEEE is the Medal of Honor, created by the IRE in 1917 and presented in that year to Edwin Armstrong, generally acknowledged as the inventor of FM radio and

> "... *the Institute of Electrical and Electronics Engineers (IEEE) has a history going back more than 100 years.*"

a pioneer in television. Other recipients include Marconi (1920), De Forest (1922), Kennelly (1932), and Noyce (1978).

The Education Medal.   Engineering educators are honored with this award for "excellence in teaching and ability to inspire students [and] leadership in electrical engineering education."

Some recent recipients are James Gibbons of Stanford (1985), Richard Adler of MIT (1986), Ben Streetman of the University of Texas at Austin (1989), James Meindl of Rensselaer (1990), and Ronald Schafer of the Georgia Institute of Technology (1992).

The Alexander Graham Bell Medal.   Created jointly by AT&T and IEEE in 1976 to commemorate the 100th anniversary of the invention of the telephone, the AGB Medal is given to honor contributions to the field of telecommunications. Among the recipients are Amos Joel, William Keister, and Raymond Ketchledge (1976) for their conception and development of electronic switching systems, David Slepian (1981) for his contribution to communication theory, and Andrew Viterbi (1984) for his work in telecommunications and his inspired teaching.

The Founders Medal.   This honor was established in 1952 by the IRE to mark significant contri-

butions in the "leadership, planning, or administration of affairs of great value to the electrical and electronics engineering profession." It has been presented to such notable managers as David Sarnoff (1953), Elmer Engstrom (1966), William Hewlett and David Packard (1973), and Roland Schmitt (1992).

The Lamme Medal.   Benjamin Lamme was chief engineer of Westinghouse from 1889 to 1924, and this medal was established in 1928 through a bequest from his estate. Lamme invented a motor that Westinghouse used in its move to electrify U.S. railroads with alternating current. In one of the great marketing strategies of the 1890s, Lamme and his boss, George Westinghouse, figured out a way to commercialize the Tesla two-phase induction motor once Westinghouse had purchased the patent from Tesla. The motor would work only in polyphase electrical systems, but none were available in the United States. Lamme and Westinghouse decided to create a lot of hoopla surrounding polyphase generators. Once the country was blanketed with these generators, the sales figures for Tesla's motor automatically shot up. Lamme was one of the principal engineers designing Westinghouse's

> "*The highest award in IEEE is the Medal of Honor...*"

first power plant at Niagara Falls. It was a Lamme-designed 5000-horsepower generator that kicked off power production at the Falls in 1895.

As might be expected, the Lamme Medal is awarded for achievement in the development of electrical apparatus. Some recipients are Rudolf Hellmund (1929, rotating electrical machinery), Henry Warren (1934, electric clocks), S. H. Mortensen (1944, self-starting synchronous motors), and Marion Hines (1983, microwave applications of semiconductor diodes).

The Voltage Follower

Removing R_1 and replacing R_2 with a wire turns the noninverting amplifier into the circuit of Figure 6.8(a). Its closed-loop gain is obtained by letting $R_1 = \infty$ and $R_2 = 0$ in Equation (6.5), and it is

$$\lim_{a \to \infty} A = 1 \text{ V/V} \tag{6.6}$$

Aptly called a **unity-gain noninverting amplifier** or also a **voltage follower,** the circuit admits the equivalent of Figure 6.8(b).

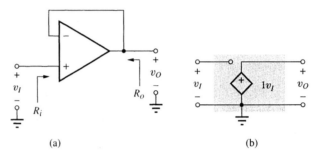

(a)   (b)

Figure 6.8 The voltage follower has $R_i = \infty$, $A = 1$ V/V, and $R_o = 0$.

For voltage purposes the follower acts as a wire shorting the output to the input. However, for current purposes it acts as a *resistance translator* because the resistance seen looking into its input port is infinite, but that looking into its output port is zero. The circuit finds application as a **voltage buffer** when it is desired to eliminate interstage loading.

▶ ## Example 6.4

A grounded source $v_S = 9$ V with internal resistance $R_s = 1$ kΩ is to be fed to a grounded load $R_L = 2$ kΩ. Find the load voltage v_L, the power p_S supplied by the source, and the power p_L dissipated by the load if the source is fed to the load (a) via a plain wire, and (b) via a voltage buffer.

Solution

(a) By connecting the load to the source directly, as in Figure 6.9(a), we create a voltage divider. So

$$v_L = \frac{R_L}{R_s + R_L} v_S = \frac{2}{1 + 2} 9 = 6 \text{ V}$$

Clearly, we have $v_L < v_S$ because of loading. Moreover, $i_{v_S} = i_{R_L} = 9/(1 + 2) = 3$ mA, so $p_S = v_S i_{v_S} = 9 \times 3 = 27$ mW, and $p_L = v_L i_{R_L} = 6 \times 3 = 18.75$ mW.

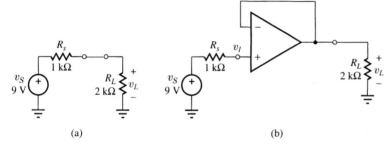

Figure 6.9 Using a voltage buffer to eliminate interstage loading.

(b) If we interpose a voltage buffer between source and load as in Figure 6.9(b), then $i_{v_S} = 0$ because no current flows into the buffer. Hence, $v_I = v_S - R_s \times 0 = v_S$. Applying Equation (6.6) we get $v_L = 1v_I$, or

$$v_L = v_S = 9 \text{ V}$$

Moreover, $p_S = v_S i_{v_S} = 0$, $i_{R_L} = 9/2 = 4.5$ mA, and $p_L = v_L i_{R_L} = 9 \times 4.5 = 40.5$ mW.

Remark   The buffer eliminates loading and also relieves the source from supplying any power to the load. The load is now powered by the op amp, which in turn is powered by the external supplies V_{CC} and V_{EE}. We do not show them in order to reduce cluttering.   ◄

The Inverting Configuration

In the circuit of Figure 6.10 v_I is applied to the inverting input of the op amp via R_1, and R_2 plays again the role of feedback resistance. Since no current flows into or out of the inverting input pin, we must have $i_1 = i_2$, or

$$\frac{v_I - v_N}{R_1} = \frac{v_N - v_O}{R_2}$$

Moreover, by Equation (6.1) we have

$$v_O = a(v_P - v_N) = a(0 - v_N) = -av_N$$

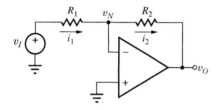

Figure 6.10 The inverting configuration.

Eliminating v_N we obtain, after simple algebra,

$$A \triangleq \frac{v_O}{v_I} = \left(-\frac{R_2}{R_1}\right) \frac{a}{1 + R_2/R_1 + a} \tag{6.7}$$

Clearly,

$$\boxed{\lim_{a \to \infty} A = -\frac{R_2}{R_1}} \tag{6.8}$$

Since its closed-loop gain is *negative,* the circuit is called an **inverting amplifier.** Owing to the fact that v_I is now applied to the inverting input, the polarity of v_O is opposite to that of v_I. We observe that A is set again by an external *resistance ratio.* By proper choice of this ratio, we can set the magnitude of A to any value we wish, including zero. By contrast, the gain of the noninverting configuration is never less than unity.

Exercise 6.3   In the inverting amplifier of Figure 6.10 let $R_1 = 10$ kΩ and let R_2 be a potentiometer configured as a variable resistance from 0 to 100 kΩ. (a) Find the range of values of A if $a = \infty$. (b) Repeat, but with $a = 10^5$ V/V.

ANSWER   (a) -10 V/V $\leq A \leq 0$. (b) -9.9989 V/V $\leq A \leq 0$.

We now wish to find the input and output resistances R_i and R_o, as depicted in Figure 6.11(a). Since the output is taken from the ideal dependent source, we again have $R_o = 0$. To find R_i, we observe that since $v_N = -v_O/a$, for any finite v_O we have

$$\lim_{a \to \infty} v_N = 0 \text{ V} \tag{6.9}$$

In words, the op amp keeps its inverting input node at ground potential regardless of v_I and v_O, this being the reason why this node is called a **virtual ground.** Consequently the resistance seen by the source v_I is simply $R_i = R_1$. The ideal inverting amplifier equivalent is shown in Figure 6.11(b). Compared to the noninverting configuration, which enjoys $R_i = \infty$, the inverting configuration has a finite R_i and will generally cause loading at the input.

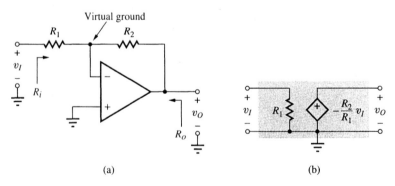

(a)   (b)

Figure 6.11 The ideal inverting amplifier has $R_i = R_1$, $A = -R_2/R_1$, and $R_o = 0$.

▶ Example 6.5

A grounded source $v_S = 2$ V with internal resistance $R_s = 10$ kΩ is connected to an inverting amplifier having $R_1 = 30$ kΩ and $R_2 = 120$ kΩ, so that its unloaded gain is $A = -120/30 = -4$ V/V. Find v_O as well as the amount of input loading.

Solution

The circuit is shown in Figure 6.12. To find v_O we apply Equation (6.8), but with the series $R_s + R_1$ in place of R_1,

$$v_O = -\frac{R_2}{R_s + R_1}v_S = -\frac{120}{10 + 30}2 = -3 \times 2 = -6 \text{ V}$$

To find the amount of input loading we observe that R_s and R_1 form a voltage divider, so

$$v_I = \frac{R_1}{R_s + R_1}v_S = \frac{30}{10 + 30}2 = 0.75 \times 2 = 1.5 \text{ V}$$

indicating a 25% signal loss due to loading.

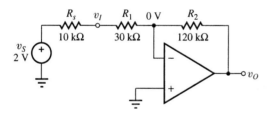

Figure 6.12 Circuit of Example 6.5.

Remark If the source had $R_s = 0$, there would be no loading and we would have $v_O = Av_I = Av_S = -4 \times 2 = -8$ V. However, because of loading, we have $v_O = Av_I = -4 \times 1.5 = -6$ V. ◀

Exercise 6.4 In the circuit of Example 6.5 find the value of R_2 that will compensate for the signal loss due to loading and yield $v_O = -8$ V.

ANSWER 160 kΩ.

Exercise 6.5 Design a circuit which accepts its input from a source having an unknown series resistance R_s, and amplifies it by -100 V/V. Show the final circuit. (*Hint:* Use a *cascade amplifier*, consisting of a noninverting amplifier stage to avoid input loading, followed by an inverting stage to provide polarity inversion.)

6.3 IDEAL OP AMP CIRCUIT ANALYSIS

The analysis of op amp circuits can be speeded up considerably if we exploit the fact that when an ideal op amp is operated with negative feedback, its input port develops *zero voltage* because

$$\lim_{a \to \infty} v_D = \lim_{a \to \infty} \frac{v_O}{a} = 0 \qquad (6.10)$$

for any finite v_O, and it draws *zero current* because this is how op amps are designed. We say that the input port behaves as a **virtual short:** its characteristic is just a point at the origin of the i–v plane. It is important to realize that it is the op amp itself that drives v_D to zero via the negative feedback network. Were we to break the negative feedback path, we would deprive the op amp of its ability to affect its own input and the virtual-short property would no longer hold.

We can visualize the virtual short also from the following alternate viewpoint. Since $v_N = v_P - v_D$, Equation (6.10) yields

$$\lim_{a \to \infty} v_N = \lim_{a \to \infty} \left(v_P - \frac{v_O}{a} \right) = v_P \qquad (6.11)$$

indicating that the voltage at the inverting input of an ideal op amp will *track* that of the noninverting input, regardless of the value of the latter. The foregoing considerations lead to the following rule of op amp behavior:

Op Amp Rule: An ideal op amp with negative feedback will output whatever voltage and current are needed to drive v_D to zero, or, equivalently, to force v_N to track v_P. Moreover, the op amp will do this without drawing any current at either input terminal.

You may find it reassuring to know that this rule is applied routinely by engineers as they analyze existing op amp circuits or design new ones. Though based on the ideal op amp model, this rule provides results that are, in most cases of interest, remarkably close to those provided by real-life op amps.

An Illustrative Example

As an example, let us use the op amp rule to find all node voltages in the circuit of Figure 6.13. To better understand the role of the op amp, consider first its surrounding network, explicitly redrawn in Figure 6.14(a). In the absence of the op amp the circuit carries no current. Hence, $v_P = 0$ and $v_N = 5$ V.

If we now put the op amp in place, it will drive v_O to whatever value is needed to make $v_N = v_P$. To find this special value of v_O, refer to the equivalent circuit of Figure 6.14(b), where we use a broken line to evidence the virtual short established by the op amp. Since the nodes labeled v_N and v_P appear shorted together, the 5-V source delivers the current

$$i = \frac{5}{2 + 3} = 1 \text{ mA} \qquad (6.12)$$

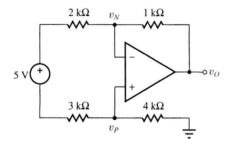

Figure 6.13 An illustrative op amp circuit.

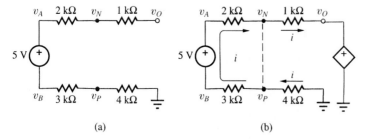

Figure 6.14 Investigating the circuit of Figure 6.13 *before* and *after* connecting the op amp.

as shown. This current, however, cannot flow through the virtual short. It must thus come from the 4-kΩ resistance and then flow into the 1-kΩ resistance. It is now straightforward to use KVL and Ohm's Law to find

$$v_P = -4i = -4 \times 1 = -4 \text{ V}$$

$$v_B = -(3+4)i = -7 \times 1 = -7 \text{ V}$$

$$v_A = v_B + 5 = -7 + 5 = -2 \text{ V}$$

$$v_N = v_P = -4 \text{ V}$$

$$v_O = v_N - 1i = -4 - 1 \times 1 = -5 \text{ V}$$

In summary, once inserted in the circuit, the op amp will cause v_P to change from 0 V to -4 V and v_N to change from 5 V to -4 V in order to make $v_N = v_P$. To achieve this equality, the op amp output pin must swing to $v_O = -5$ V and sink $i_O = 1$ mA. Moreover, the op amp does this without drawing any current at either input terminal.

Exercise 6.6   Using the virtual-short concept, find v_P, v_N, and v_O if the circuit of Figure 6.13 includes also a 6-kΩ resistance between the node at the negative side of the 5-V source and ground. Be sure to check your results.

ANSWER $v_P = v_N = -30/11$ V, $v_O = -185/44$ V.

Exercise 6.7 Use the virtual-short concept to derive the gains of the inverting and noninverting amplifier, $A = -R_2/R_1$ and $A = 1 + R_2/R_1$.

Exercise 6.8

(a) Use the virtual-short concept to find the closed-loop gain of the ideal voltage follower of Figure 6.8(a).

(b) If $v_I = 10.0$ V, find v_D and v_O; hence, compare with the case $a = 10^5$ V/V.

ANSWER (b) 0, 10 V; 100 μV, 9.9999 V.

Gain Polarity Control

The circuit of Figure 6.15 consists of an op amp, three 10-kΩ resistances, and a switch SW. We have two situations:

(1) *SW is closed.* This grounds the noninverting pin, making the op amp act as an inverting amplifier with gain $A = -R_2/R_1 = -10/10$, or

$$A = -1 \text{ V/V} \tag{6.13a}$$

(2) *SW is open.* Since the noninverting input draws no current, we must have $i_{R_3} = 0$, so $v_{R_3} = 0$. By KVL, $v_P = v_I - v_{R_3} = v_I$. By the op amp rule, $v_N = v_P = v_I$. Since the voltage across R_1 is $v_I - v_N = 0$, it follows that $i_{R_1} = 0$. Since the inverting input draws no current, we must have $i_{R_2} = i_{R_1} = 0$, indicating that v_{R_2} must also be zero. Hence, $v_O = v_N - v_{R_2} = v_N = v_I$, indicating that the gain $A = v_O/v_I$ is now

$$A = +1 \text{ V/V} \tag{6.13b}$$

The function of the circuit is summarized by saying that, depending on whether the switch is *open* or *closed*, the circuit acts, respectively, as a unity-gain *noninverting* or *inverting* amplifier.

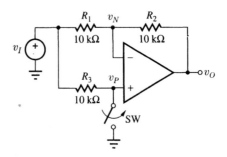

Figure 6.15 Gain polarity control circuit.

Exercise 6.9   Consider the circuit obtained from that of Figure 6.15 by connecting a fourth resistance R_4 between the node labeled v_N and ground.

   (a) Sketch the new circuit, and show that when SW is closed we have $A = -R_2/R_1$, and when SW is open we have $A = 1 + R_2/R_4$.

   (b) Specify standard resistances in the kΩ range so that $A = -10$ V/V when SW is closed, and $A = +10$ V/V when SW is open.

ANSWER   (b) $R_1 = 18$ kΩ, $R_2 = 180$ kΩ, $R_3 = R_4 = 20$ kΩ.

Negative Resistance Converter

In the circuit of Figure 6.16(a) we wish to derive an expression for the equivalent resistance R_{eq} between the noninverting input and ground. To this end, let us apply a test voltage v as in Figure 6.16(b). Since no current flows into the noninverting input, we have

$$i = \frac{v - v_O}{R}$$

Using the noninverting amplifier formula,

$$v_O = \left(1 + \frac{R_2}{R_1}\right) v$$

Eliminating v_O yields, after simplification, $i = -(R_2/R_1)v/R$. Finally, letting $R_{eq} = v/i$ yields

$$R_{eq} = -\frac{R_1}{R_2} R \qquad\qquad \textbf{(6.14)}$$

Clearly, as we apply v, the circuit causes i to flow *into* the positive terminal of the test source rather than out, thus providing **negative resistance** behavior.

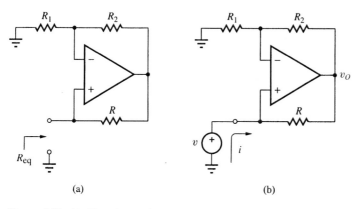

(a)   (b)

Figure 6.16 (a) Negative resistance converter, and (b) test circuit to find R_{eq}.

Examples of applications for this circuit are *V–I converters,* to be studied in Section 6.5, and *noninverting integrators,* to be studied in Section 8.4.

Exercise 6.10   A drawer contains op amps and 10-kΩ resistors. Using only components from the drawer, design a circuit that simulates a −1.25-kΩ negative resistance. Try minimizing the number of components you use. Show your final circuit.

6.4 SUMMING AND DIFFERENCE AMPLIFIERS

Next to amplification, the two most basic operations are signal addition and sub-traction. These functions are provided by two classes of op amp circuits called, respectively, *summing amplifiers* and *difference amplifiers.* In instrumentation and measurement the need often arises for difference amplifiers with infinite input resistance. These are called *instrumentation amplifiers.*

The Summing Amplifier

The circuit of Figure 6.17 has n inputs v_1 through v_n. These voltages supply n currents i_1 through i_n via the corresponding resistances R_1 through R_n. These currents converge at the inverting input node and then, since no current enters or leaves the input pin of the op amp, they flow right into the feedback resistance R_F. We can thus write

$$i_F = i_1 + i_2 + \cdots + i_n \tag{6.15}$$

indicating that the inverting input node acts as a **summing junction** for currents. As we know, the op amp keeps this node at ground potential, so Equation (6.15) can be rephrased as

$$\frac{0 - v_O}{R_F} = \frac{v_1 - 0}{R_1} + \frac{v_2 - 0}{R_2} + \cdots + \frac{v_n - 0}{R_n}$$

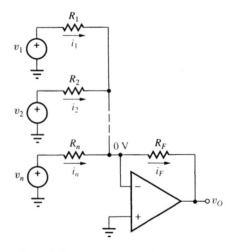

Figure 6.17 Summing amplifier.

Solving for v_O we obtain

$$v_O = -\left(\frac{R_F}{R_1}v_1 + \frac{R_F}{R_2}v_2 + \cdots + \frac{R_F}{R_n}v_n\right) \qquad \textbf{(6.16)}$$

indicating that the circuit yields a *weighted sum* of its inputs, and the weights are set by resistance ratios. Moreover, it is readily seen that the output resistance of the circuit is $R_o = 0$, and that the resistance R_{ik} seen by the driving source v_k coincides with the corresponding series resistance R_k, $k = 1, 2, \ldots, n$.

▶ **Example 6.6**

Using standard 5% resistances in the kΩ range, design a circuit which accepts three inputs, v_1, v_2, and v_3, and yields $v_O = -(2v_1 + 3v_2 + v_3)$.

Solution

Equation (6.16) requires $R_F/R_1 = 2$, $R_F/R_2 = 3$, and $R_F/R_3 = 1$. A set of standard values satisfying these conditions is $R_1 = 15$ kΩ, $R_2 = 10$ kΩ, and $R_3 = R_F = 30$ kΩ.   ◀

Exercise 6.11   (a) Using standard 5% resistances in the kΩ range, design a circuit which accepts two inputs v_1 and v_2 and yields $v_O = -5(v_1 + 2v_2)$. (b) Repeat, but with $v_O = -2(v_1/3 + 5v_2)$.

ANSWER   (a) $R_1 = 20$ kΩ, $R_2 = 10$ kΩ, $R_F = 100$ kΩ; (b) $R_1 = 150$ kΩ, $R_2 = 10$ kΩ, $R_F = 100$ kΩ.

Exercise 6.12   Design a circuit which accepts a source v_{S1} with internal resistance $R_{s1} = 3$ kΩ, and a source v_{S2} with internal resistance $R_{s2} = 10$ kΩ, and yields $v_O = -(20v_{S1} + 10v_{S2})$. Try minimizing the component count. Show the final circuit.

ANSWER $R_1 = 2$ kΩ, $R_2 = 0$, $R_F = 100$ kΩ.

If we make all series resistances equal to R_1, the circuit of Figure 6.17 yields

$$v_O = -\frac{R_F}{R_1}(v_1 + v_2 + \cdots + v_n) \qquad \textbf{(6.17)}$$

and is called a **summing amplifier** because it amplifies the true sum of its inputs. If the feedback resistance is implemented with a potentiometer connected as a variable resistance from 0 to R_F, then we have $v_O = A(v_1 + v_2 + \cdots + v_n)$, with A variable over the range $-R_F/R_1 \le A \le 0$. Summing amplifiers find application in *audio mixing,* where signals representing different sound sources are combined together with different weights to produce a single sound channel. A summing amplifier can also be used to add a dc offset to a signal.

The Difference Amplifier

The op amp of Figure 6.18 is fed by the voltage v_2 at the noninverting side, and by the voltage v_1 at the inverting side. Since the circuit is linear, we anticipate an output of the type

$$v_O = A_2 v_2 - A_1 v_1 \tag{6.18}$$

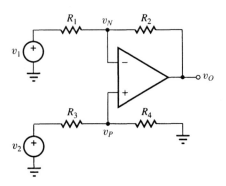

Figure 6.18 Difference amplifier.

To find A_1 we suppress v_2 and calculate the contribution v_{O1} due to v_1 acting alone. It is apparent that with $v_2 = 0$ we also obtain $v_P = 0$, indicating that from the viewpoint of v_1 the circuit acts as an *inverting amplifier.* Thus,

$$v_{O1} = -\frac{R_2}{R_1} v_1 \tag{6.19}$$

so $A_1 = R_2/R_1$. To find A_2 we suppress v_1 and calculate the contribution v_{O2} due to v_2 acting alone. It is apparent that making $v_1 = 0$ turns the circuit into a noninverting amplifier with respect to v_P, so

$$v_{O2} = \left(1 + \frac{R_2}{R_1}\right) v_P = \frac{1 + R_2/R_1}{1 + R_3/R_4} v_2 \tag{6.20}$$

where we have used the voltage divider formula to express v_P in terms of v_2. Thus, $A_2 = (1 + R_2/R_1)/(1 + R_3/R_4)$. Letting $v_O = v_{O1} + v_{O2}$, we can express the output in the insightful form

$$v_O = \frac{R_2}{R_1} \left(\frac{1 + R_1/R_2}{1 + R_3/R_4} v_2 - v_1 \right) \tag{6.21}$$

indicating that the circuit yields a *weighted difference* between its inputs.

We now wish to find the input resistances R_{i1} and R_{i2} seen by the sources v_1 and v_2. Consider R_{i2} first. Since the noninverting input of the op amp draws no current, R_{i2} is found with the help of the partial circuit representation of Figure 6.19(a). Thus,

$$R_{i2} = R_3 + R_4 \qquad \textbf{(6.22a)}$$

To find R_{i1}, we note that the negative feedback action by the op amp can be modeled with an internal VCVS of value $1v_P$ to indicate that v_N is forced to track v_P. This is depicted in Figure 6.19(b). When looking into the inverting input we thus see the dynamic resistance r_{VCVS} of this source, which is zero. Hence, $R_{i1} = R_1 + r_{VCVS}$, or

$$R_{i1} = R_1 \qquad \textbf{(6.22b)}$$

Moreover, the output resistance of the circuit is $R_o = 0$. Since R_{i1} and R_{i2} are finite, they will generally cause input loading if the driving sources have nonzero internal resistances.

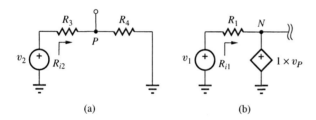

(a)   (b)

Figure 6.19 Partial circuits for the derivation of R_{i1} and R_{i2}.

▶ **Example 6.7**

Design a circuit such that $v_O = 2v_2 - 3v_1$ and $R_{i1} = R_{i2} = 100$ kΩ.

Solution

By Equation (6.22b), $R_1 = R_{i1} = 100$ kΩ. By Equation (6.19), $R_2/R_1 = 3$, or $R_2 = 3R_1 = 3 \times 100 = 300$ kΩ. By Equation (6.22a), $R_3 + R_4 = R_{i2} = 100$ kΩ, so $1 + R_3/R_4 = (R_4 + R_3)/R_4 = 100/R_4$. By Equation (6.20), $(1 + 3/1)/(100/R_4) = 2$, or $R_4 = 50$ kΩ. Then, $R_3 = 100 - R_4 = 50$ kΩ. In summary, $R_1 = 100$ kΩ, $R_2 = 300$ kΩ, and $R_3 = R_4 = 50$ kΩ.   ◀

Exercise 6.13   In the circuit of Figure 6.18 we have $A_2 \leq (A_1 + 1)$. We can achieve $A_2 > (A_1 + 1)$ by connecting a fifth resistance R_5 between the node labeled v_N and ground.

    (a) Sketch the modified circuit and show that A_1, R_{i1}, R_{i2}, and R_o remain unchanged, but $A_2 = (1 + R_2/R_1 + R_2/R_5)/(1 + R_3/R_4)$.

(b) Specify resistances in the kΩ range to achieve $v_O = 3(2v_2 - v_1)$. Try minimizing the component count.

ANSWER (b) $R_1 = 10$ kΩ, $R_2 = 30$ kΩ, $R_3 = 0$, $R_4 = \infty$, $R_5 = 15$ kΩ.

A **difference amplifier** is an amplifier that amplifies the *true difference* of its inputs to yield $v_O = A(v_2 - v_1)$. To achieve this function with the circuit of Figure (6.18) we must impose the condition $1 + R_1/R_2 = 1 + R_3/R_4$, or

$$\frac{R_1}{R_2} = \frac{R_3}{R_4} \tag{6.23}$$

after which Equation (6.21) simplifies as

$$v_O = \frac{R_2}{R_1}(v_2 - v_1) \tag{6.24}$$

Equation (6.23) requires that the resistances form a *balanced bridge*. It is thus fair to say that what it takes to construct a difference amplifier is an op amp and a balanced resistance bridge.

The Instrumentation Amplifier

A common function in measurement and control is the amplification of the voltage difference between two nodes of a circuit, but without disturbing their individual voltages. The difference amplifier of Figure 6.18 does not meet this requirement because its finite input resistances R_{i1} and R_{i2} will load down the individual nodes, altering their voltages as well as the currents entering or leaving them. What is needed is a difference amplifier with $R_{i1} = R_{i2} = \infty$.

One way to meet this requirement is to precede a difference amplifier by two noninverting amplifiers to make the input sources see infinite input resistances. The result is the **instrumentation amplifier** of Figure 6.20(a). Let us derive an expression for v_O in terms of v_1 and v_2.

The op amp denoted as OA_3 is a difference amplifier, so

$$v_O = \frac{R_2}{R_1}(v_{O2} - v_{O1}) \tag{6.25a}$$

where v_{O1} and v_{O2} are, respectively, the outputs of OA_1 and OA_2. Since the inverting inputs of OA_1 and OA_2 draw no current, we can apply Ohm's Law and write $v_{O1} - v_{O2} = (R_3 + R_4 + R_3)i$, or

$$v_{O1} - v_{O2} = (2R_3 + R_4)i \tag{6.25b}$$

where i is the current through the series R_3-R_4-R_3, as shown in the subcircuit of Figure 6.20(b). Considering that OA_1 keeps $v_{N1} = v_{P1} = v_1$, and OA_2 keeps $v_{N2} = v_{P2} = v_2$, the voltage difference across R_4 is $v_1 - v_2$. This is also shown

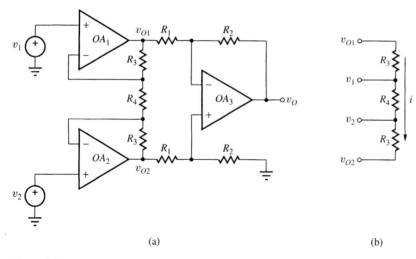

Figure 6.20 (a) Instrumentation amplifier; (b) subcircuit to find a relationship between $(v_{O1} - v_{O2})$ and $(v_1 - v_2)$.

in Figure 6.20(b). Using Ohm's Law we can thus write

$$i = \frac{v_1 - v_2}{R_4} \tag{6.25c}$$

Substituting Equation (6.25c) into (6.25b), and then (6.25b) into (6.25a), we finally obtain

$$\boxed{v_O = \left(1 + \frac{2R_3}{R_4}\right) \frac{R_2}{R_1}(v_2 - v_1)} \tag{6.26}$$

This expression indicates that the gain with which the circuit amplifies the difference $(v_2 - v_1)$ consists of two terms: (a) the term $(1 + 2R_3/R_4)$, representing the gain of the circuit made up of OA_1 and OA_2 and the corresponding resistors (this circuit is referred to as the *first stage*) and (b) the term (R_2/R_1), representing the gain of the difference amplifier based on OA_3 (this is referred to as the *second stage*).

▶ **Example 6.8**

In the circuit of Figure 6.20(a) specify 1% resistances to achieve $v_O = 10^3(v_2 - v_1)$.

Solution

The gain of 10^3 V/V can be factored as the product of two terms in a variety of different ways, such as 1×10^3, 2×500, $\sqrt{10^3} \times \sqrt{10^3}$, 40×25, $10^3 \times 1$, and so forth. Let us arbitrarily implement it as

$10^3 = \sqrt{10^3} \times \sqrt{10^3} = 31.62 \times 31.62$. Let $R_1 = R_4 = 1\ k\Omega$. Then, imposing $1 + 2R_3/1 = 31.62$ yields $R_3 = 15.3\ k\Omega$ (use $15.4\ k\Omega$, 1%). Likewise, imposing $R_2/1 = 31.62$ yields $R_2 = 31.62\ k\Omega$ (use $31.6\ k\Omega$, 1%). In summary, $R_1 = R_4 = 1\ k\Omega$, $R_2 = 31.6\ k\Omega$, and $R_3 = 15.4\ k\Omega$, all 1% resistances. ◄

Exercise 6.14

(a) Repeat Example 6.7, but configuring OA_1 and OA_2 as voltage followers to save three resistances.

(b) Show all node voltages in the circuit if $v_1 = 505.0$ mV and $v_2 = 500.0$ mV.

ANSWER (a) $R_1 = 1\ k\Omega$, $R_2 = 1\ M\Omega$. (b) $v_{O1} = 505.0$ mV, $v_{O2} = 500.0$ mV, $v_{N2} = v_{P2} = 499.5$ mV, $v_O = -5$ V.

6.5 TRANSRESISTANCE, TRANSCONDUCTANCE, AND CURRENT AMPLIFIERS

Even though the op amp is a *voltage* amplifier, with the help of suitable external components it can be configured to operate as any one of the other three amplifier types discussed in Section 5.4.

Transresistance Amplifiers

The **transresistance amplifier,** also called an *I–V* **converter,** is shown in Figure 6.21(a). To find a relationship between v_O and i_I we observe that since no current enters the inverting input pin, i_I must flow right into R. Moreover, since this pin is a virtual ground, we can write $i_I = (0 - v_O)/R$, or

$$v_O = -Ri_I \qquad (6.27)$$

The *transresistance gain* is $-R$, and is expressed in V/A. It is negative because i_I is assumed to flow *into* the circuit. Inverting the reference direction of i_I will of course yield a positive gain. As shown in Figure 6.21(b), the output port behaves as a CCVS. Figure 6.21(c) depicts the transfer characteristic of the circuit.

A relationship of the type of Equation (6.27) could have been obtained by feeding i_I directly to a grounded resistance R without the need for any op amp. However, with the op amp in place the input source sees $R_i = 0$ because it is connected to a virtual ground, and an output load sees $R_o = 0$ because v_O comes from the ideal internal source av_D. According to Table 5.1, these are the ideal terminal characteristics of a transresistance amplifier, indicating that the role of the op amp is to prevent any loading both at the input and at the output.

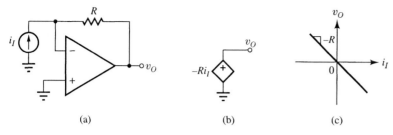

(a)   (b)   (c)

Figure 6.21 (a) Transresistance amplifier, (b) Thévenin equivalent of its output port, and (c) transfer characteristic of the circuit.

▶ Example 6.9

A certain photodiode receives light from a fiber optic cable and, depending on light intensity, it yields a current $0 \le i_D \le 100\ \mu\text{A}$. Design a circuit to convert this current to a voltage $0 \le v_O \le 10$ V for delivery to a grounded load $R_L = 2\ \text{k}\Omega$. What is the source-to-load current gain of the circuit?

Solution

Use the circuit of Figure 6.21 with $R = 10/(100 \times 10^{-6}) = 100\ \text{k}\Omega$. To ensure $v_O \ge 0$, connect the diode so that i_D flows out of the inverting input node, through the diode, and to ground. The current gain is
$i_L/i_D = (v_O/R_L)/i_D = (v_O/i_D)/R_L = R/R_L = 100/2 = 50$ A/A.   ◀

Exercise 6.15   Using resistances not greater than 100 kΩ, design a circuit that accepts two current sources i_1 and i_2 having unknown parallel resistances R_1 and R_2, and delivers $v_O = (1\ \text{V}/\mu\text{A})\ (i_2 - i_1)$ to an unknown load R_L, regardless of R_1, R_2, and R_L. The reference directions of both sources are from ground *into* your circuit. (*Hint:* Use two *I–V* converters and a difference amplifier.)

Figure 6.22 shows an interesting *I–V* converter variant that uses a *T network* to achieve very high transresistance gains without necessarily requiring unrealistically large resistances. To find its transfer characteristic we note that KCL at the node labeled v_1 yields $i_I + i_1 = i_2$, or

$$i_I + \frac{0 - v_1}{R_1} = \frac{v_1 - v_O}{R_2}$$

But, since i_I flows right into R, we have $v_1 = -Ri_I$. Substituting into the above equation and solving for the output yields

$$v_O = -kRi_I \qquad\qquad \text{(6.28a)}$$

$$k = 1 + \frac{R_2}{R} + \frac{R_2}{R_1} \qquad\qquad \text{(6.28b)}$$

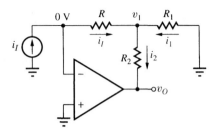

Figure 6.22 *I–V* converter with a *T* network in its feedback loop.

indicating that the presence of R_1 and R_2 effectively increases R by the *multiplicative factor k*.

▶ Example 6.10

Using resistances not greater than 1 MΩ, design an *I–V* converter with a transresistance gain of −0.2 V/nA.

Solution

If we were to use the circuit of Figure 6.21 we would need $R = 0.2/10^{-9} = 200$ MΩ, an unrealistically large resistance. Instead, we use the circuit of Figure 6.22 with $R = 1$ MΩ, and we then exploit its multiplicative action to raise its gain to 200×10^6 V/A. By Equation (6.28b) we want $200 = 1 + R_2/10^6 + R_2/R_1$. To reduce inventory, let R_2 also be 1 MΩ. Then, imposing $200 = 1 + 10^6/10^6 + 10^6/R_1$ yields $R_1 = 5.05$ kΩ. ◀

Exercise 6.16 Repeat Example 6.9, but using resistances not greater than 10 kΩ.

ANSWER Use the circuit of Figure 6.22 with $R = R_2 = 10$ kΩ and $R_1 = 1.25$ kΩ.

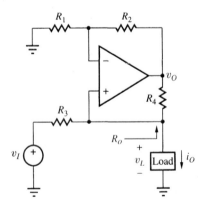

Figure 6.23 Transconductance amplifier.

Transconductance Amplifiers

The role of a **transconductance amplifier,** also called a *V–I* **converter,** is to convert an input voltage v_I to an output current i_O which is *independent* of the particular load appearing at the output. Thus, the load must see an output resistance $R_o = \infty$. A popular *V–I* converter, known as the **Howland circuit** for its inventor, is shown in Figure 6.23. Let us find a relationship between i_O and v_I.

Considering that no current flows into or out of the noninverting pin of the op amp, we have, by KCL,

$$i_O = \frac{v_I - v_L}{R_3} + \frac{v_O - v_L}{R_4}$$

where v_L is the voltage developed by the load in response to the current i_O. The subcircuit made up of R_1, R_2, and the op amp forms a noninverting amplifier, so

$$v_O = \left(1 + \frac{R_2}{R_1}\right) v_L$$

Substituting v_O and rearranging, we get

$$i_O = \frac{v_I}{R_3} - \frac{v_L}{R_4}\left(\frac{R_4}{R_3} - \frac{R_2}{R_1}\right) = \frac{v_I}{R_3} - \frac{v_L}{R_o} \qquad \textbf{(6.29a)}$$

$$R_o = \frac{R_4}{R_4/R_3 - R_2/R_1} \qquad \textbf{(6.29b)}$$

To make i_O independent of the load voltage v_L we impose

$$\boxed{\frac{R_4}{R_3} = \frac{R_2}{R_1}} \qquad \textbf{(6.30a)}$$

or, equivalently,

$$\boxed{R_o = \infty} \qquad \textbf{(6.30b)}$$

after which Equation (6.29a) simplifies as

$$\boxed{i_O = \frac{1}{R_3} v_I} \qquad \textbf{(6.31)}$$

The transconductance gain is $1/R_3$, and is expressed in A/V. As depicted in Figure 6.24, i_O is controlled by v_I but is independent of v_L. Moreover, the circuit *sources current* when $v_I > 0$, and it *sinks* current when $v_I < 0$.

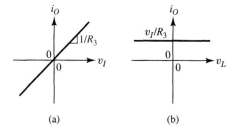

(a)   (b)

Figure 6.24   (a) Transfer characteristic and (b) output characteristic of the *V–I* converter.

In light of Equation (6.30a) it is fair to say that what it takes to make a Howland circuit is an op amp and a *balanced resistance bridge*. The circuit achieves $R_o = \infty$ through the *positive feedback* action provided by R_4.

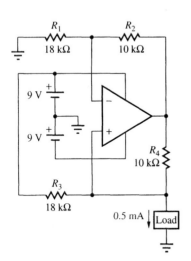

Figure 6.25 Dc current source of Example 6.11.

Exercise 6.17 Consider the circuit obtained from that of Figure 6.23 by lifting the left terminal of R_1 from ground, connecting it to a voltage source v_1, and renaming the source connected to R_3 as v_2. Obtain a relationship between i_O and v_1, v_2, and v_L, and show that if the resistances form a balanced bridge, the circuit provides *difference V–I conversion*: $i_O = (v_2 - v_1)/R_3$.

▶ **Example 6.11**

Design a 0.5-mA source. The circuit is to be powered from ± 9 V batteries.

Solution

Let $v_I = +9$ V. Then, Equation (6.31) requires $R_3 = v_I/i_O = 9/0.5 = 18$ kΩ. A suitable resistance set is $R_1 = R_3 = 18$ kΩ, and $R_2 = R_4 = 10$ kΩ. The circuit is shown in Figure 6.25. ◀

Exercise 6.18 Repeat Example 6.11, but for a 2-mA current *sink* powered from ± 15 V supplies.

ANSWER $v_I = -15$ V, $R_1 = R_3 = 7.5$ kΩ, $R_2 = R_4 = 3$ kΩ.

▶ **Example 6.12**

Use the test-signal method to confirm that the resistance seen by the load in the circuit of Figure 6.25 is infinite.

Solution

Connect the left terminal of R_3 to ground and replace the load with a test source such as $v = 1$ V. Then $i_{R_3} = v/R_3 = 1/18$ mA ($\leftarrow$), $v_O = (1 + R_2/R_1)v = (1 + 10/18)1 = 28/18$ V, and $i_{R_4} = (v_O - v)/R_4 = (28/18 - 1)/10 = 1/18$ mA ($\downarrow$). Since the current coming via R_4 equals that leaving via R_3, the net current out of the test source is $i = 0$. Hence, $R_o = 1/0 = \infty$. ◀

Exercise 6.19   Label all node voltages and branch currents in the circuit of Figure 6.25, and calculate them if the load is (a) a 1-kΩ resistance, (b) a 5-kΩ resistance, (c) a short circuit, (d) a 2-V source, positive at the top, and (e) a 2-V source, positive at the bottom.

There is yet another elegant way of explaining how the circuit achieves $R_o = \infty$. In light of Figure 6.16(a), the Howland circuit can be viewed as consisting of v_I, R_3, and a *negative resistance* toward ground. The situation is depicted in Figure 6.26(a), where we have used Equation (6.14) to find that the value of this resistance is $-(R_1/R_2)R_4$, or $-R_3$, by Equation (6.30a). Applying a source transformation leads to the circuit of Figure 6.26(b), where the resistance seen by the load is $R_o = R_3 \parallel (-R_3) = R_3(-R_3)/(R_3 - R_3) = \infty$, as it should be. This results in the circuit of Figure 6.26(c), where the load current is $i_O = v_I/R_3$, thus confirming Equation (6.31).

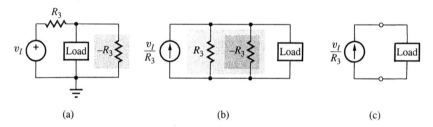

(a)   (b)   (c)

Figure 6.26 Illustrating how the *V–I* converter achieves $R_o = \infty$.

Current Amplifiers

Figure 6.27 shows two ways of configuring the op amp for current-mode operation. In Figure 6.27(a) we have, by Ohm's Law and KVL, $i_O = (v_O - v_P)/R_1 = (v_N - R_2 i_I - v_P)/R_1$. Since the op amp forces v_N to track v_P, we get

$$i_O = -\frac{R_2}{R_1} i_I \qquad (6.32)$$

A negative gain signifies that if we source (or sink) current to the circuit, the latter will in turn sink (or source) current from the load. When the resistances are made equal the circuit is called a **current reverser** or a **current mirror** because it yields $i_O = -i_I$.

Because the op amp inputs appear virtually shorted, the resistances in the circuit of Figure 6.27(b) experience the same voltage drop: $R_1 i_1 = R_2 i_2$. By KCL, $i_O = i_1 + i_2 = i_I + (R_2 i_I)/R_1$, or

$$i_O = \left(1 + \frac{R_2}{R_1}\right) i_I \qquad (6.33)$$

Sourcing (or sinking) current to the circuit causes it to source (or sink) a current $1 + R_2/R_1$ times as large to the load.

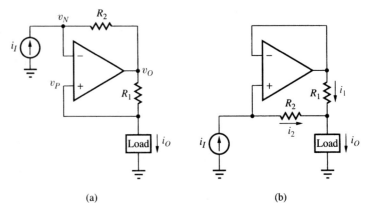

Figure 6.27 Inverting and noninverting current amplifiers.

Exercise 6.20 Using standard 5% resistances, design a current amplifier such that (a) $i_O = 10i_I$ and (b) $i_O = -10i_I$.

ANSWER (a) $R_1 = 2$ kΩ, $R_2 = 18$ kΩ; (b) $R_1 = 1$ kΩ, $R_2 = 10$ kΩ.

▼ 6.6 Op Amp Circuit Analysis Using SPICE

Op amp circuits can easily be simulated via SPICE using either of the op amp models of Figures 6.3 and 6.4.

▶ Example 6.13

Use SPICE to find all node voltages in the circuit of Figure 6.28, whose op amp is assumed ideal.

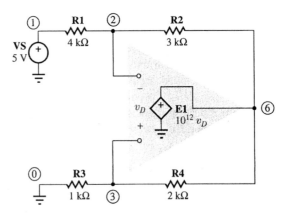

Figure 6.28 Analyzing an op amp circuit via SPICE.

Solution

SPICE does not accept $a = \infty$, so let us arbitrarily use $a = 10^{12}$ V/V, which is a fairly large gain. The input file is

```
A SIMPLE OP AMP CIRCUIT
VS 1 0 DC 5
R1 1 2 4K
R2 2 6 3K
R3 0 3 1K
R4 3 6 2K
*HERE'S THE OP AMP:
E1 6 0 3 2 1.0E12
.END
```

After SPICE is run, the output file contains the following:

```
NODE VOLTAGE NODE VOLTAGE NODE VOLTAGE NODE VOLTAGE
(1)  5.0000    (2) -3.0000   (3) -3.0000   (6) -9.0000
```

You may wish to verify these results by hand calculation. ◀

Exercise 6.21 Use SPICE to find all node voltages in the circuit of Figure 6.13. Do it first using an *ideal* op amp, then using a 741 type for which $r_i = 2$ MΩ, $a = 200,000$ V/V, and $r_o = 75$ Ω. Compare the two cases, and comment.

Transfer Characteristic

SPICE can be directed to find the transfer characteristic of an op amp circuit via the .TF statement. As we know, its general form is

$$.\text{TF}\quad \text{OUTVAR}\quad \text{INSOURCE} \tag{6.34}$$

where OUTVAR is the output signal, which must be either a voltage between a pair of nodes, V(N1,N2), or a current through a voltage source, I(VXXX). Moreover, INSOURCE is the independent input source, either VXXX or IXXX. The inclusion of the .TF statement causes SPICE to compute the gain $A = $ OUTVAR/INSOURCE, as well as the input and output resistances R_i and R_o of the circuit.

▶ **Example 6.14**

Use SPICE to investigate the departure from the ideal of a 741 op amp configured as a noninverting amplifier with $R_1 = 1$ kΩ and $R_2 = 99$ kΩ. Assume the circuit is driven by a 0.1-V source and drives a 2-kΩ load. Comment on the results.

Solution

With the 741 data of Figure 6.3, the circuit appears as in Figure 6.29. The input file is

```
A 741 NONINVERTING AMP
VI 3 0 DC 0.1
R1 0 2 1K
R2 2 6 99K
RL 6 0 2K
*HERE'S THE 741:
RI 2 3 2MEG
E1 1 0 3 2 200K
RO 1 6 75
.TF V(6) VI
.END
```

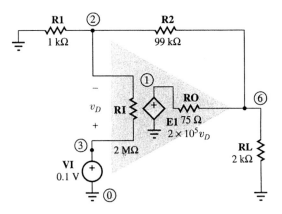

Figure 6.29 Finding the transfer characteristic of a practical op amp circuit.

After SPICE is run, the output file contains

```
NODE VOLTAGE NODE VOLTAGE NODE VOLTAGE NODE VOLTAGE
(1)  10.3770 (2)   0.0999  (3)   0.1000   (6)    9.9948

SMALL-SIGNAL CHARACTERISTICS:
V(6)/VI = 9.995E+01
INPUT RESISTANCE AT VI = 3.855E+09
OUTPUT RESISTANCE AT V(6) = 3.750E-02
```

Following is a comparison of the actual and ideal values, the latter being shown within parentheses: $A = 99.95$ (100) V/V, $R_i = 3.855$ GΩ (∞), $R_o = 37.5$ mΩ (0 Ω). SPICE confirms that, for practical purposes, the departure from the ideal is negligible.

Remark Note that the voltage produced by the dependent source inside the op amp (10.3770 V) is greater than that appearing at the output pin (9.9948 V). This difference is the voltage drop across **RO** due to the current drawn by the load and the external feedback network.

Exercise 6.22   Use SPICE to investigate the departure from the ideal of a 741 op amp configured as an inverting amplifier with $R_1 = 1$ kΩ and $R_2 = 1$ MΩ. Assume the circuit is driven by a 10-mV source and drives a 2-kΩ load. Comment on the results.

Subcircuits

A complex circuit often contains repeated blocks called *subcircuits*. An example is offered by the instrumentation amplifier of Figure 6.20(a) which contains three identical op amps as subcircuits. To avoid repeating the element statements each time a subcircuit is referenced, SPICE allows for a subcircuit to be defined in the input file only once. SPICE then automatically inserts the group of elements forming the given subcircuit wherever the latter is referenced. The group of element statements defining the subcircuit must be preceded by the statement

```
.SUBCKT  SUBNAME  N1  N2 ... NN
```
(6.35)

and must be followed by the statement

```
.ENDS  SUBNAME
```
(6.36)

where `SUBNAME` is the subcircuit name and `N1` through `NN` are the external nodes, which cannot be zero. Subcircuits are referenced by specifying pseudo-elements beginning with the letter `X`, followed by the circuit nodes to be used in expanding the subcircuit, followed by the subcircuit name. The general form of a subcircuit call is

```
XYYY  N1  N2 ... NN  SUBNAME
```
(6.37)

where `YYY` is an arbitrary alphanumeric string. An example will better illustrate the use of this facility.

▶ Example 6.15

Using 741 op amps, design an instrumentation amplifier with $A = 5 \times 20$ V/V. Hence, use SPICE to find v_O if $v_1 = 5.000$ V and $v_2 = 5.010$ V, and the amplifier drives a 2-kΩ load.

Solution

From Figure 6.30, the input file is

```
THREE-OP-AMP INSTRUMENTATION AMPLIFIER
*HERE'S THE OP AMP SUBCIRCUIT
*CONNECTIONS: INVERTING INPUT
*                | NONINVERTING INPUT
*                | | OUTPUT
*                | | |
.SUBCKT OPAMP 2 3 6
RI 2 3 2MEG
```

```
E1 1 0 3 2 200K
RO 1 6 75
.ENDS OPAMP
*HERE'S THE MAIN CIRCUIT:
V1 1 0 DC 5.000
V2 2 0 DC 5.010
R1 3 4 20K
R2 4 5 10K
R3 5 6 20K
*FIRST-STAGE OP AMP REFERENCES:
XOA1 4 1 3 OPAMP
XOA2 5 2 6 OPAMP
R4 3 7 10K
R5 6 8 10K
R6 7 9 200K
R7 8 0 200K
*SECOND-STAGE OP AMP REFERENCE:
XOA3 7 8 9 OPAMP
RL 9 0 2K
.END
```

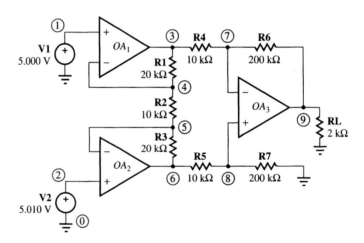

Figure 6.30 SPICE analysis of an instrumentation amplifier.

After SPICE is run, the output file contains the following

```
NODE VOLTAGE NODE VOLTAGE NODE VOLTAGE NODE VOLTAGE
(1)   5.0000   (2)   5.0100 (3)   4.9800 (4)   5.0000
(5)   5.0100   (6)   5.0300 (7)   4.7904 (8)   4.7905
(9)   0.9999
```

You may find it instructive to examine each entry and justify its value, in the manner of Example 6.14.

Exercise 6.23 Using two 741 op amps, design a circuit that accepts two inputs, v_{I1} and v_{I2}, and yields two outputs, $v_{O1} = v_{I1} + v_{I2}$ and $v_{O2} = v_{I1} - v_{I2}$. Use SPICE with the subcircuit facility to verify that your circuit works correctly for $v_{I1} = 3$ V and $v_{I2} = 2$ V.

▼ SUMMARY

- The op amp is a *voltage amplifier* with *extremely high (ideally infinite) gain.* It is used as a *universal building block* in analog circuit design, including the design of other types of amplifiers.

- To function, an op amp needs to be powered. To simplify circuit diagrams we omit showing the power connections. However, power must be turned on when trying out an op amp circuit in the lab.

- An op amp is operated in conjunction with an external network, which is designed to provide *negative feedback* around the op amp, that is, to create an electric path for the op amp output to be reapplied to the inverting input pin.

- Ideally, it is desirable that the op amp have an *infinite open-loop gain* because this will make the *closed-loop gain* depend *exclusively* on the *feedback network.* Then, we can tailor the op amp function to any need by suitable choice of the feedback network, *regardless* of the particular op amp being used.

- The most basic op amp configurations are the *noninverting* and the *inverting amplifiers,* which are implemented with the help of just a pair of external resistances. A particular case of the noninverting amplifier is the *voltage buffer,* which is used as an interstage coupling circuit to *eliminate loading.*

- The analysis of circuits with ideal op amps can be speeded up significantly if we exploit the fact that for *voltage purposes* the input port appears as a *short,* and for *current purposes* it appears as an *open.* This is also referred to as the *virtual-short* concept.

- The virtual-short concept forms the basis of the Op Amp Rule, which states that when operated with negative feedback, *an op amp will output whatever voltage and current are needed to drive its differential input v_D to zero, or, equivalently, to force the inverting input v_N to track the noninverting input v_P.*

- The Op Amp Rule is applied to the analysis of widely used op amp circuits, including *summing and difference amplifiers, instrumentation amplifiers, I–V and V–I converters,* and *current amplifiers.*

- An op amp can be configured to simulate a *negative resistance.* An intriguing application of this type of resistance is to make the effective output resistance of a practical current source go to infinity, thus turning the source into an ideal one.

- Op amp circuits are easily simulated with SPICE. Simulation can be used to check the results of hand calculations, or to assess the departure of a practical op amp circuit from ideal behavior.

▼ PROBLEMS

6.1 The Operational Amplifier

6.1   Given an op amp with gain $a = 10^3$ V/V, find (a) v_N if $v_P = 0$ and $v_O = 3$ V, (b) v_N if $v_P = 1.0$ V and $v_O = -10$ V, (c) v_O if $v_N = 5.002$ V and $v_P = 4.998$ V, and (d) v_P if $v_N = 3.5$ V and $v_O = 3.5$ V.

6.2   An op amp with $r_i = 1$ MΩ, $a = 10^5$ V/V, $r_o = 100$ Ω, and an output load $R_L = 2$ kΩ is part of a circuit where $v_P = 0$ and $v_O = 10.0$ V. Find the voltages across and the currents through r_o and r_i.

6.2 Basic Op Amp Configurations

6.3   (a) Find A for a noninverting amplifier with $R_1 = 10$ kΩ and $R_2 = 20$ kΩ. (b) How does A change if a third resistance $R_3 = 10$ kΩ is connected in series with R_1? In parallel with R_1?

6.4   Repeat Problem 6.3, but for the inverting configuration.

6.5   (a) Design a noninverting amplifier with $A = 5$ V/V under the constraint that the total resistance used be 100 kΩ. (b) Repeat, but for an amplifier of the inverting type.

6.6   (a) A source $v_S = 1$ V is fed to a voltage divider implemented with $R_A = 10$ kΩ and $R_B = 30$ kΩ. The voltage v_B developed by R_B is fed, in turn, to a noninverting amplifier having $R_1 = 10$ kΩ and $R_2 = 30$ kΩ, so that $A = 4$ V/V. Sketch the circuit, and find v_B as well as the amplifier output v_O. (b) Repeat, but for an inverting amplifier with $R_1 = 10$ kΩ and $R_2 = 40$ kΩ, so that now $A = -4$ V/V. Comment on the differences.

6.7   An inverting amplifier is implemented with $R_1 = 10$ kΩ, $R_2 = 100$ kΩ, and an op amp with finite open-loop gain $a = 10^3$ V/V. Sketch and label v_I, v_O, and v_N versus time if v_I is a 1-kHz triangular wave with peak values of ±1 V.

6.8   An inverting amplifier with $R_1 = 10$ kΩ and $R_2 = 1$ MΩ is driven by a source $v_I = 0.1$ V. Find the closed-loop gain A, the percentage deviation of A from the ideal value $-R_2/R_1$, and the inverting input voltage v_N for the cases $a = 10^3$, 10^4, and 10^5 V/V.

6.9   (a) An inverting amplifier is implemented with $R_1 = 1$ kΩ and $R_2 = 100$ kΩ. Find the percentage change in the closed-loop gain A if the open-loop gain a changes from 200,000 V/V to 50,000 V/V. (b) Repeat, but for a noninverting amplifier with $R_1 = 1$ kΩ and $R_2 = 99$ kΩ. (c) Repeat, but for a voltage follower ($R_1 = \infty$, $R_2 = 0$).

6.3 Ideal Op Amp Circuit Analysis

6.10   Find v_N, v_P, and v_O in Figure P6.10. Check your results.

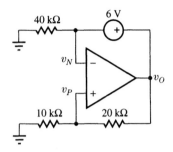

Figure P6.10

6.11   Find v_N, v_P, and v_O in Figure P6.11. Check your results.

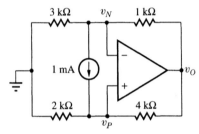

Figure P6.11

6.12   (a) Find v_N, v_P, and v_O in Figure P6.12. (b) Repeat, but with a 5-kΩ resistance connected between the node labeled v_N and ground. Check.

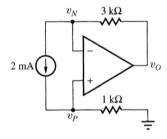

Figure P6.12

6.13   Find v_N, v_P, and v_O in Figure P6.13. Check.

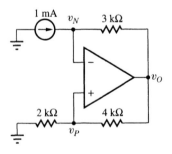

Figure P6.13

6.14 (a) Repeat Problem 6.13, but with a 6-kΩ resistance connected in parallel with the 1-mA source. (b) Repeat Problem 6.13, but with the left terminal of the 1-mA source disconnected from ground and connected instead to the node labeled v_P. Check.

6.15 Find R_{eq} in the circuit of Figure P6.15. Check.

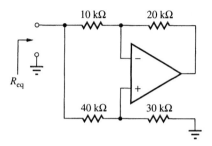

Figure P6.15

6.16 In the circuit of Figure P6.16 find v_O and R_i. Check.

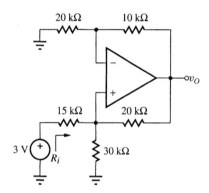

Figure P6.16

6.17 In the circuit of Figure P6.17 let $R_1 = R_2 = 10$ kΩ. Find an expression for the gain $A = v_O/v_I$ in terms of the parameter k, with $0 \le k \le 1$. What is the value of A with the wiper halfway between the top and the bottom? All the way to the top? All the way to the bottom?

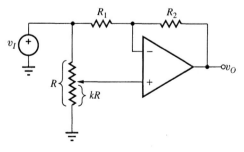

Figure P6.17

6.18 Sketch the circuit of Figure P6.17, but with an additional resistance R_4 between the inverting input pin of the op amp and ground. Then, specify suitable resistances so that the gain $A = v_O/v_I$ varies over the range -100 V/V $\le A \le +100$ V/V as the wiper is varied from end to end. What is the value of A when the wiper is halfway?

6.19 Using a 100-kΩ potentiometer, an op amp, and suitable resistors in the kΩ range, design a circuit to synthesize a negative resistance that can be varied from -1 kΩ to -10 kΩ as the wiper is varied from end to end.

6.20 (a) Show that the circuit of Figure P6.20 has $A = v_O/v_I = -k(R_2/R_1)$ with $k = 1 + R_4/R_2 + R_4/R_3$, and $R_i = R_1$. (b) Specify resistances not larger than 100 kΩ to achieve $A = -200$ V/V and $R_i = 100$ kΩ.

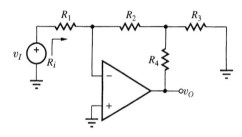

Figure P6.20

6.21 Find the gain v_O/v_I of the circuit of Figure P6.21.

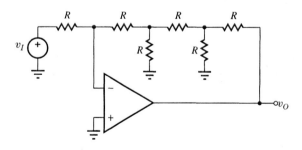

Figure P6.21

6.22 (a) For the circuit of Figure P6.22 show that $A = v_O/v_I = -R_2/R_1$ and $R_i = R_3R_1/(R_3 - R_1)$. (b) Specify standard 1% resistances in the kΩ range to achieve $A = -100$ V/V and approach $R_i = \infty$.

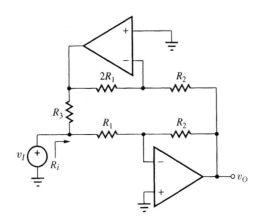

Figure P6.22

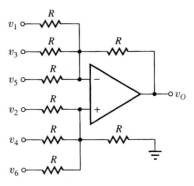

Figure P6.27

6.23   Three students are discussing the input resistance R_i of the noninverting amplifier of Figure 6.7(a). The first student says that since we are looking into the noninverting input, which is an open circuit (oc), we must have $R_i = R_{oc} = \infty$. The second student contends that since the input terminals appear *virtually shorted* (vs) we must have $R_i = R_{vs} + (R_1 \parallel R_2) = 0 + (R_1 \parallel R_2) = (R_1 \parallel R_2)$. The third student claims that since the noninverting input is virtually shorted to the inverting input, which is in turn a virtual ground (vg) node, we must have $R_i = R_{vs} + R_{vg} = 0 + 0 = 0$. Which student is right? How would you refute the other two?

6.24   (a) Show that if the op amp of the illustrative example of Figure 6.13 has finite gain a, then $v_D = -5/(a + 2)$ V. (b) Use this result to recalculate all node voltages if $a = 200,000$ V/V. Compare the results with the ideal case considered in the text and comment.

6.28   (a) Find a relationship between v_O and v_1 through v_3 in the circuit of Figure P6.28. (b) Using standard 5% resistances in the kΩ range, design a circuit to achieve $v_O = 10(v_1 + v_2 + v_3)$. To minimize the component count, use only five resistances. Show the final circuit.

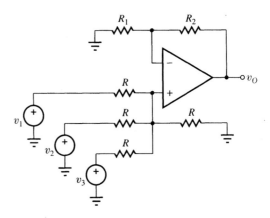

Figure P6.28

6.4 Summing and Difference Amplifiers

6.25   A two-input summing amplifier is implemented with $R_1 = 75$ kΩ, $R_2 = 100$ kΩ, and $R_F = 300$ kΩ. (a) If $v_1 = 5$ V, find v_2 for $v_O = 10$ V; (b) if $v_1 = 5$ V, find v_2 for $v_O = -10$ V; (c) if $v_2 = 3$ V, find v_1 for $v_O = 0$ V.

6.26   Using standard 5% resistances in the kΩ range, design a circuit to yield $v_O = -10(v_1 + 2v_2 + 3v_3 + 4v_4)$.

6.27   Find a relationship between v_O and v_1 through v_6 in the circuit of Figure P6.27.

6.29   In the difference amplifier of Figure 6.18 let $R_1 = 20$ kΩ, $R_2 = 30$ kΩ, $R_3 = 10$ kΩ, and $R_4 = 5$ kΩ. (a) If $v_1 = 2$ V, find v_2 for $v_O = 9$ V. (b) If $v_2 = 5$ V, find v_1 for $v_O = 0$ V.

6.30   (a) Using only one op amp and standard 5% resistances in the kΩ range, design a circuit to achieve $v_O = 10(v_2 - 5v_1)$. (b) Repeat, but for $v_O = 10(5v_2 - v_1)$.

6.31   Show that the difference amplifier of Figure P6.31 yields $v_O = 2(R_2/R_1)(1 + R_3/R_G)(v_2 - v_1)$, so its gain can be made variable by varying the single resistance R_G.

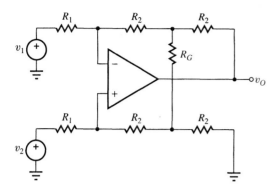

Figure P6.31

6.32 Show that the difference amplifier of Figure P6.32 yields $v_O = (R_2/R_1)(R_G/R_3)(v_2 - v_1)$, so its gain can be made variable by varying the single resistance R_G.

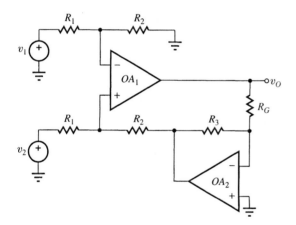

Figure P6.32

6.33 (a) Show that the circuit of Figure P6.33 yields

$$v_O = \left(1 + \frac{R_2}{R_1}\right) \times \left(v_2 - \frac{1 + R_3/R_4}{1 + R_1/R_2} v_1\right)$$

(b) Verify that if $R_3/R_4 = R_1/R_2$, the circuit is an instrumentation amplifier with gain $A = 1 + R_2/R_1$.

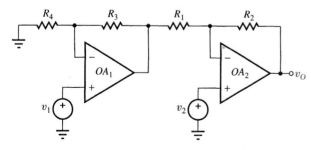

Figure P6.33

6.34 The gain of the circuit of Figure P6.33 can be made variable by connecting a variable resistance R_G between the inverting input nodes of OA_1 and OA_2. Sketch the modified circuit; hence, letting $R_3 = R_1$ and $R_4 = R_2$, show that $v_O = A(v_2 - v_1)$, where $A = 1 + R_2/R_1 + 2R_2/R_G$.

6.35 In the instrumentation amplifier of Problem 6.34 specify standard 1% resistances so that its gain can be varied over the range $10 \leq A \leq 10^3$ V/V by means of a 100-kΩ potentiometer.

6.36 Show that the two-op-amp circuit of Figure P6.36 is an instrumentation amplifier with $v_O = 2(1 + R/R_G)(v_2 - v_1)$, and specify suitable component values to make its gain variable from 10 to 100 V/V by means of a 10-kΩ potentiometer.

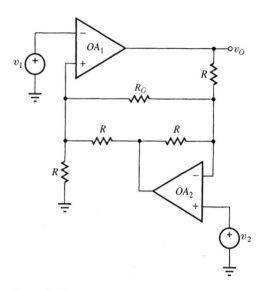

Figure P6.36

6.37 In a certain class of positioning systems the need arises for circuitry to convert from two-axis cartesian coordinate voltages v_x and v_y to three-axis non-orthogonal coordinate voltages v_a, v_b, and v_c. The conversion constraints are $v_x = v_a - v_c$, $v_y = v_b - (v_a + v_c)/2$, and $v_a + v_b + v_c = 0$. Using three op amps and suitable resistances, design a circuit that accepts v_x and v_y and yields v_a, v_b, and v_c under the above constraints.

6.5 Transresistance, Transconductance, and Current Amplifiers

6.38 (a) Show that in Figure P6.38 we have $v_O = -(R_2/R_1) v_I - R_2 i_I$. (b) Specify suitable resistances for gains of -10^2 V/V and -10^6 V/A.

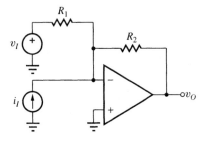

Figure P6.38

6.39 (a) Show that the circuit of Figure P6.39 is a V–I converter with $i_O = v_I/(R_1/k)$, $k = 1 + R_2/R_3$. (b) Assuming $R_1 = 100$ kΩ, specify standard 5% resistances for a gain of 1 mA/V.

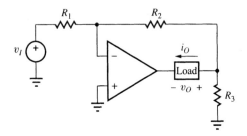

Figure P6.39

6.40 For the circuit of Figure P6.40 obtain an expression of the type $i_O = v_I/R - v_L/R_o$. Hence, verify that if $(R_4 + R_5)/R_3 = R_2/R_1$, the circuit is a V–I converter with $R_o = \infty$ and $R = R_1R_5/R_2$.

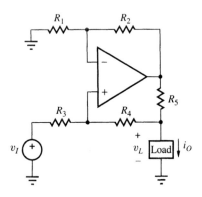

Figure P6.40

6.41 For the circuit of Figure P6.41 obtain an expression of the type $i_O = v_I/R - v_L/R_o$. Hence, verify that if $R_4/R_3 = R_2/R_1$, the circuit is a V–I converter with $R_o = \infty$ and $R = R_1R_5/R_2$.

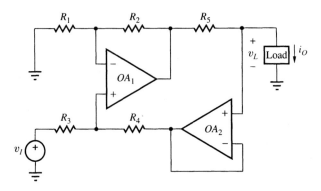

Figure P6.41

6.42 For the circuit of Figure P6.42 obtain an expression of the type $i_O = v_I/R - v_L/R_o$. Hence, verify that if $R_1 + R_2 = R_3$, the circuit is a V–I converter with $R_o = \infty$ and $R = R_2/2$.

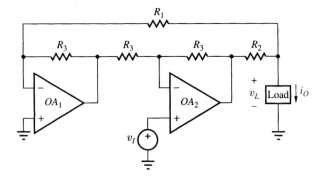

Figure P6.42

6.43 Show that the circuit of Figure P6.43 is a current divider with $i_O = i_I/(1 + R_2/R_1)$, regardless of the load.

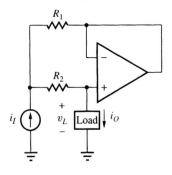

Figure P6.43

6.44 Three students are discussing the resistance R_o seen by the load in the V–I converter of Figure P6.44. The first student notes that at the bottom the load sees the output resistance r_o of the op amp, which is assumed zero, whereas at the top it sees R; consequently, $R_o = R + r_o = R + 0 = R$. The second student observes that at the top the load sees not R but a virtual ground node (vg), so $R_o = R_{vg} + r_o = 0 + 0 = 0$.

The third student reports rumors that $R_o = \infty$. Which student is right? How would you refute the other two? *(Hint:* Use the test-signal method.)

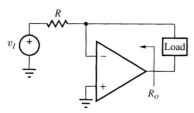

Figure P6.44

▽ 6.6 Op Amp Circuit Analysis Using SPICE

6.45 Use SPICE to find v_N, v_P, and v_O in the circuit of Figure P6.10. Assume the op amp is ideal.

6.46 Use SPICE to investigate the departure from the ideal of a voltage follower implemented with an op amp having $r_i = 250$ kΩ, $r_o = 1$ kΩ, and $a = 10^4$ V/V. Assume the circuit is driven by a 10.0-V source with a 10-kΩ internal resistance, and drives a 2-kΩ load. Comment on the results.

6.47 In the instrumentation amplifier of Figure P6.33 specify suitable resistances for a gain of 1000 V/V. Hence, use the .SUBCKT facility of SPICE to verify the value of v_O if $v_1 = 2.000$ V and $v_2 = 2.004$ V. Assume op amps with $r_i = 1$ MΩ, $r_o = 100$ Ω, and $a = 10^5$ V/V, and a 2-kΩ load.

6.48 In the V–I converter of Figure 6.24(a) let $R_1 = R_2 = R_3 = R_4 = 10$ kΩ, and $v_I = 10$ V. Assuming the op amp is ideal, use SPICE to find the Norton equivalent seen by the load.

6.49 Repeat Problem 6.48, but for the case of a nonideal op amp having $r_i = 100$ kΩ, $r_o = 100$ Ω, and $a = 10^3$ V/V.

6.50 In the I–V converter of Figure 6.23 let $R = 100$ kΩ, $R_1 = 100$ Ω, and $R_2 = 100$ kΩ. Use SPICE to find the equivalent resistance seen by the source i_I if the op amp is the 741 type. Compare with the case of an ideal op amp.

ENERGY STORAGE ELEMENTS

T he circuits examined so far are referred to as *resistive circuits* because the only elements used, besides sources, are resistances. The equations governing these circuits are *algebraic equations* because so are Kirchhoff's laws and Ohm's Law. Moreover, since resistances can only dissipate energy, we need at least one independent source to initiate any voltage or current in the circuit. In the absence of independent sources, all voltages and currents would be zero and the circuit would have no electrical life of its own.

It is now time we turn our attention to the two remaining basic elements, **capacitance** and **inductance.** The first distinguishing feature of these elements is that they exhibit *time-dependent characteristics,* namely, $i = C(dv/dt)$ for capacitance and $v = L(di/dt)$ for inductance. For this reason, capacitances and inductances are said to be **dynamic elements.** By contrast, a resistance is a static element because its $i-v$ characteristic does not involve time. Time dependence adds a new dimension to circuit behavior, allowing for a wider variety of functions as compared to purely resistive circuits.

The second distinguishing feature is that capacitances and inductances can absorb, store, and then release energy, making it possible for a circuit to have an electrical life of its own even in the absence of any sources. For obvious reasons, capacitances and inductances are also referred to as **energy-storage elements.**

The formulation of circuit equations for networks containing capacitances and inductances still relies on the combined use of Kirchhoff's laws and the element laws. However, since the characteristics of these elements depend on time, the resulting equations are no longer plain algebraic equations; they involve time derivatives, or integrals, or both. Generally referred to as **integro-differential**

equations, they are not as straightforward to solve as their algebraic counterparts. In fact, they can be solved analytically only in a limited number of cases. Mercifully, this includes the cases of greatest interest to us. Even when a solution cannot be found analytically, it can be evaluated numerically using a computer, this being the reason why computer simulation such as SPICE plays an indispensable role in the analysis and design of circuits containing energy-storage elements. Though a previous exposure to integro-differential equations is helpful, it is not required because we shall develop solution techniques within the text as they naturally arise.

In the present chapter, after introducing the capacitance and the inductance, we study the **natural response** of the basic *RC* and *RL* circuits, that is, the response provided by the circuit using the energy stored in its capacitance or inductance. This study introduces us to the concept of *root location* in the *s plane,* a powerful concept that we shall explore in greater detail as we proceed.

We then use the *integrating factor method* to investigate how circuits containing these elements react to the application of *dc signals* and *ac signals.* The mathematical level is designed to provide a rigorous understanding of the various response components, namely, the **natural, forced, transient,** and **steady-state** components. Even though the responses to dc and ac signals may seem particular, they nevertheless provide enough insight into the most relevant aspects of circuits containing dynamic elements to allow the designer to predict circuit behavior in most other situations of practical interest.

7.1 CAPACITANCE

Capacitance represents the ability of a circuit element to **store charge in response to voltage.** Circuit elements that are designed to provide this specific function are called **capacitors** or **condensers.**

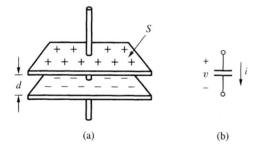

Figure 7.1 Capacitor and circuit symbol for capacitance.

As shown in Figure 7.1(a), a capacitor consists of two conductive plates separated by a thin insulator. Applying a voltage between the plates causes *positive* charge to accumulate on the plate at higher potential and an equal amount of *negative* charge to accumulate on the plate at lower potential. The rate at which the accumulated charge varies with the applied voltage is denoted as C and is called **capacitance,**

$$C \triangleq \frac{dq}{dv} \tag{7.1}$$

Its SI unit is the *farad* (F), named for the English chemist and physicist Michael Faraday (1791–1867). Clearly, 1 F = 1 C/V. Figure 7.1(b) shows the circuit symbol for capacitance, along with the reference polarities for voltage and current.

Recall from basic electricity that capacitance depends on the insulator type and the physical dimensions:

$$C = \varepsilon \frac{S}{d} \tag{7.2}$$

where ε is the *permittivity* of the insulator, S is the *area* of the plates, and d is the *distance* between them. For vacuum space ε takes on the value $\varepsilon_0 = 10^{-9}/(36\pi)$ F/m. Other media are described in terms of the ratio $\varepsilon_r = \varepsilon/\varepsilon_0$, called the *relative permittivity*.

A capacitor storing charge may be likened to a cylindrical tank storing water. The larger its cross-sectional area, the more water the tank can hold. Moreover, the lower the height needed to store a given amount of water, the greater the tank's storage capacity.

In general q may be some arbitrary function of v, indicating that C may itself be a function of v. In this case the capacitance is said to be *nonlinear*.

▶ Example 7.1

In reverse-based *pn* junctions, which form the basis of semiconductor diodes and transistors, q is related to v as

$$q = Q_0 \sqrt{1 + v/V_0}, \text{ and } v > -V_0$$

where Q_0 and V_0 are appropriate *pn* junction parameters. Find the capacitance associated with the *pn* junction.

Solution

Using Equation (7.1) yields

$$C = \frac{Q_0}{2V_0} \frac{1}{\sqrt{1 + v/V_0}}$$

indicating that this type of capacitance depends on the applied voltage v and is thus a nonlinear capacitance.   ◀

Linear Capacitances

Of particular interest is the case in which q is *linearly proportional* to v, for then we must have $q = Cv$ or

$$C = \frac{q}{v} \qquad (7.3)$$

with C *independent of v*. For obvious reasons, this type of capacitance is said to be *linear*. Unless stated otherwise, we shall consider only capacitances of this type.

Equation (7.3) allows us to find the accumulated charge in terms of the applied voltage, or vice versa. For instance, applying 10 V across the terminals of a 1-μF capacitance causes a charge $q = Cv = 10^{-6} \times 10 = +10 \ \mu C$ to accumulate on the plate at higher potential, and a charge $-q = -10 \ \mu C$ to accumulate on the other plate. Even though the *net* charge within the capacitance is always zero, we identify the charge stored in the capacitance as that on the *positive plate*. We thus say that applying 10 V across a 1-μF capacitance results in a stored charge of 10 μC.

The *i–v* Characteristic

To derive the i–v relationship for capacitance, refer to Figure 7.2(a) and suppose the applied voltage is increased by dv. By Equation (7.1), this causes the charge on the top plate to *increase* by $dq = C\,dv$ and the charge on the bottom plate to *decrease* by a similar amount, indicating that current must flow from the source to the top plate and from the bottom plate back to the source. Even

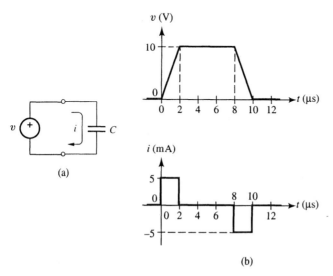

(a)

(b)

Figure 7.2 (a) Voltage-driven capacitance and (b) waveforms of Example 7.3.

though no charges actually cross the insulator gap, it is convenient to regard current as flowing straight through the capacitor because this relieves us from worrying about internal details. This current is found as $i = dq/dt$. Substituting $dq = C\,dv$ yields

$$i = C\frac{dv}{dt} \qquad\qquad (7.4)$$

For obvious reasons, a capacitance is said to perform the operation of **voltage differentiation.**

Equation (7.4) indicates that capacitive current depends not on voltage per se, but on its *rate of change.* This means that to sustain current through a capacitance, the applied voltage *must change.* The more rapidly voltage changes, the larger the current. On the other hand, if voltage is kept constant, no current will flow, no matter how large the voltage. Likewise, if the current through a capacitance is found to be zero, this means that the voltage across it must be constant, not necessarily zero.

This behavior is readily visualized in terms of the aforementioned water tank analogy. Likening voltage to water level and current to water flow, we note that if there is any flow in or out of the tank, then the level must change. Conversely, a stationary level indicates the absence of any flow.

In applying Equation (7.4) careful attention must be paid to the current direction relative to the direction of voltage change. For i to flow in the direction shown, v must *increase.* If v *decreases,* then i will flow in the *opposite* direction. This holds irrespective of the polarity of v. For instance, the current associated with a voltage change from $+3$ V to $+4$ V and that associated with a voltage change from -7 V to -6 V will have the same direction even though the polarities of the voltages themselves are opposite, because in both cases we have a voltage *increase.*

One of the most important applications of capacitors is as *dynamic memories* (DRAMs) in computer systems. In this application a miniature capacitor is charged to some specified voltage and then the connections are broken to prevent current flow. This causes the capacitor to retain its charge and, hence, the specified voltage, thus providing a *memory function.*

Uniform Charge/Discharge

If a capacitance is charged or discharged at a *constant rate,* then dv/dt is constant and so is i, by Equation (7.4). Conversely, if a capacitance is driven with a constant current I, then the rate of voltage change is constant. We can therefore replace the differentials in Equation (7.4) with finite differences and write $I = C\,\Delta v/\Delta t$. Engineers often remember this relation in the form

$$C\,\Delta v = I\,\Delta t \qquad\qquad (7.5)$$

In words, if a capacitance C is charged or discharged with a constant current I, then the voltage change Δv during the time interval Δt is such that $C\,\Delta v = I\,\Delta t$, or **"cee delta vee equals i delta tee."** We stress once again that this relationship holds only if the capacitance current is *constant.*

► **Example 7.2**

How long does it takes for a 1-μF capacitance to charge to 10 V with a constant current of 1 mA?

Solution

$\Delta t = C\,\Delta v/I = 10^{-6} \times 10/10^{-3} = 10$ ms.   ◄

Exercise 7.1

   (a) What current is needed to uniformly charge a 10-pF capacitance to 5 V in 4 ns?

   (b) What capacitance will charge to 10 V in 50 μs with a constant current of 10 mA?

ANSWER   (a) 12.5 mA, (b) 50 nF.

Time Diagrams

Since Equation (7.4) involves time, there is no straightforward way to visualize it in the i–v plane. Rather, we must examine $v(t)$ and $i(t)$ separately, using individual *time diagrams*.

► **Example 7.3**

Sketch and label the current waveform developed by a 1-nF capacitance in response to the applied voltage waveform of Figure 7.2(b), top.

Solution

For $0 \le t \le 2$ μs, v *increases uniformly*. The rate of increase is $\Delta v/\Delta t = (10 - 0)/[(2 - 0) \times 10^{-6}] = 5 \times 10^6$ V/s. During this time interval the capacitance draws a *constant current* $i = C\,\Delta v/\Delta t = 10^{-9} \times 5 \times 10^6 = 5$ mA.

   For 8 μs $\le t \le 10$ μs, v *decreases uniformly* at a rate that is now $\Delta v/\Delta t = (0 - 10)/[(10 - 8) \times 10^{-6}] = -5 \times 10^6$ V/s. The corresponding current is again constant, but with *opposite polarity*, namely, $i = -5$ mA.

   At all other times v is constant, so $i = C\,dv/dt = C \times 0 = 0$. The current waveform is shown in Figure 7.2(b), bottom.

   In short, a voltage increase yields a positive current, a voltage decrease yields a negative current, and a constant voltage (not necessarily zero) yields zero current.   ◄

A CONVERSATION WITH
OSCAR N. GARCIA

NATIONAL SCIENCE FOUNDATION

As Program Director for Interactive Systems at the National Science Foundation, Oscar N. Garcia funds research in interactive artificial intelligence.

I n his new assignment as program director for interactive systems, National Science Foundation, Oscar N. Garcia encourages and promotes research in the field of human-computer interaction, while on leave from his professorship at George Washington University. After earning BS and MS degrees in electrical engineering from North Carolina State University and working in industry and academia for some years, he earned his Ph.D. from the University of Maryland in 1969.

With all your degrees in EE, it seems you never had any doubts about what your major would be. When exactly did you decide to be an electrical engineer?

Probably when I was about ten years old. I was given a toy electric motor for Christmas that had to be assembled and wired, with directions in English—while I was growing up in Cuba, speaking Spanish, of course—and I could only follow the pictures. I took the pieces to an electrical engineering senior student who could read English to see if he could put it together, and he failed to make it work after trying for quite a while. Heartbroken, I went back home and kept working on it while looking at the pictures and finally got it to work perfectly. So I decided that, somehow, I had some talents that qualified me as a potential electrical engineer! And maybe that is how my conviction got started.

Also, my friends told me that mathematics and physics were hard subjects that made engineering the most difficult career at the University of Havana. The challenge appealed to me, so in 1954 I signed up for the EE program. About that time the revolution was starting, there were student strikes and unrest, and the University was closed, so after working for a couple of years I saved enough to come to NC State, with the blessing of my anxious parents.

Your current position has more to do with computers than with EE. When and how did the switch come about?

In 1963, teaching at Old Dominion University, I was asked by my Dean, Harold Lampe, to help bring in the first-ever computer to the school. I had used computers in my MS thesis and had designed some parts for

> *"My advice is graphs, graphs, graphs."*

them in industry, but when the computer came in it was a major task to introduce it to the faculty and the students, and since I was one of the first hired in the EE department, the job fell on my shoulders. Then in 1970 I moved to the University of South Florida in Tampa as an associate professor in EE and continued to work with computers and teach computer and microprocessor courses. I was lucky and received a grant to start a microprocessor laboratory, which was the spark to initiate a growing computer science program followed by a computer science and engineering department.

So you think that there is a natural path from electrical engineering to computer science?

Yes, in particular when you make a mid-point stop at computer engineering. The EE course called digital circuits, for example, which is about how hardware works, started my interest on the software side because I wanted to simulate circuits and systems. As time went on, it became clear that software de-

sign was to become a very important industry, and I became aware of the things that could be attempted in far-out possibilities like understanding human language. Now, most of what I teach are courses in artificial intelligence and speech recognition, so you could say that I have strayed pretty far afield from my original EE days, if you take a narrow view of EE.

From back when you were in classical EE, what do you remember about teaching the circuits course?

That this is the time when students start developing their critical and methodical thinking and analytical problem-solving skills. Maybe the hardest concept to grasp in this course is the idea of transient versus steady-state analysis. If the students can relate these two concepts, and the ideas of frequency- and time-domain analysis, they'll probably do well in the course. A difficult part of this course is to relate the math to the idealized physical situation—what a Laplace transform has to do with the circuit laid out on your lab bench, for instance. My advice here is graphs, graphs, graphs. Drawing graphs always helps you see the relationships you are trying to grasp, so you can never sketch too many of them. The circuits course is not only a good training ground for analytical problem solving, it is also a pillar on which much of the electrical engineering curriculum rests. If the understanding is shaky, then the following courses may be shaky, too, so it is important to have a secure handle on the fundamentals here.

What exactly is Artificial Intelligence or AI?

It is a piece of software, a program, that can do its own reasoning, analogous to the manner in which humans think and learn. In the early days, the cost, low memory capacity, and slow speed of computers prevented users from making them more interactive and "intelligent." All aspects of their operation had to be anticipated and coded. If you missed a part of the syntax of the program—a comma, for example—the program would not run, and you were limited
(Continued)

(Continued)

in the things that you could represent inside the computer. Today, we try to incorporate meaning in the operation of programs, and we have programs capable of understanding natural language, say English or Spanish. When you have a program that handles the record of your checks, for example, you can type a less precise instruction, such as "Find the check for my rent for last June," and the machine goes right to work on it.

Two concepts often associated with classical AI are knowledge-based systems and expert systems. You code into a program knowledge from an expert in a given field and the program can "infer" or reason things out when queried appropriately. The program reasons by drawing inferences: if *this* is true, then *that* is to take place.

The ultimate goal of AI is to have computers do at least as well as humans in their best thinking and interactions, but we are a long way from that goal. My NSF work funds research in interactive AI to try to make human–computer communication easier: computers that understand audible speech, computers that use virtual reality to help you visualize a many-variable problem. I have a graduate student who is hearing-impaired and a good lip-reader. He is teaching the computer how to lip-read.

Are we talking Hal here, the computer in *2001*?

Just about, and not a moment too soon, since 2001 is just around the corner!

Movies aside, how does it work really?

We use a videotape of a person speaking a large number of known sentences in front of a camera. Then, working with the videotape, we find the features of the oral cavity that relate to the sounds and train the computer to identify such sequences of features. The hardest part is to have the computer identify the sequences when they are put together in different sentences, since they are dependent on the surrounding sequences. Also, some sequences that sound differently look alike—"bet" and "pet," for example. The ultimate goal of my student, Alan Goldschen, is to have the lip-reading recognition augment the audible speech recognition.

How will AI affect our daily life years ahead?

I believe there are two parts to the answer here: AI will make it (1) easier for humans to live in a more complex world, and (2) easier for computers and humans to communicate with each other. Imagine having on the dashboard of your car a computer that constantly receives information through radio waves about traffic and weather conditions in the proximity of your location. You get in the car and say, "What is the fastest way to the airport right now?" Wait ten seconds, your best route shows up on the screen, and away you go. Or how about a car with a collision-avoidance radar connected to your computer? We'll never know how many deaths could have been prevented had the drivers made a different split-second decision before the impact. What if all the cars came with radar-equipped computers and could use that nanosecond to make the best decision possible and take over the steering? Surely parents

> *"I have a graduate student who is hearing-impaired.... He is teaching the computer how to lip-read."*

might sleep better when their children are out driving. The aim of the engineer is to make this world a better place to live in. I'd like to think that artificial intelligence is going to have a lot to do with that game plan.

Exercise 7.2   Repeat Example 7.3 if (a) $C = 10$ nF; (b) $C = 1$ nF, but the height of the voltage waveform is 1 V, all time intervals remaining the same; (c) v is uniformly increased from 0 to 10 V over the time interval $0 \le t \le 4$ μs, everything else remaining the same. Be neat and precise as you sketch and label the various waveforms.

The *v–i* Characteristic

Equation (7.4) can be turned around to find the voltage developed by a capacitance in response to an applied current, as in Figure 7.3(a). The result is

$$v(t) = \frac{1}{C} \int_0^t i(\xi)\, d\xi + v(0) \qquad (7.6)$$

where $v(0)$ represents the voltage across the capacitance at $t = 0$. For obvious reasons, a capacitance is said to perform the operation of **current integration.** By Equation (7.3), $v(0)$ is related to the charge $q(0)$ initially stored in the capacitance as

$$v(0) = \frac{q(0)}{C} \qquad (7.7)$$

Of particular interest is the case in which current is *constant,* or $i(t) = I$, for then Equation (7.6) yields

$$v(t) = \frac{I}{C}t + v(0) \qquad (7.8)$$

In words, forcing a *constant current* through a capacitance yields a *linear* voltage waveform, or a *voltage ramp.* The rate at which voltage ramps is $dv/dt = I/C$.

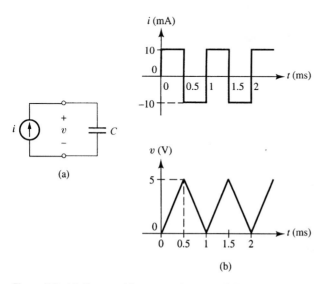

(a)

(b)

Figure 7.3 (a) Current-driven capacitance and (b) waveforms of Example 7.4.

Depending on current polarity, the ramp may be of the *increasing* or *decreasing* type. The capacitance's ability to produce ramps is exploited in the design of *waveform generators* to synthesize the triangle and sawtooth waveforms.

▶ Example 7.4

Sketch and label the voltage waveform developed by a 1-μF capacitance in response to the applied current waveform of Figure 7.3(b), top. Assume $v(0) = 0$ V.

Solution

For $0 \leq t \leq 0.5$ ms, the current is constant. We can use Equation (7.8) and write $v(t) = [(10 \times 10^{-3})/(10^{-6})]t + 0 = 10^4 t$, with t in s. This is an *increasing* voltage ramp because $i > 0$ over this time interval. The rate of increase is $dv/dt = 10^4$ V/s $= 10$ V/ms. At the end of this interval we have $v(0.5$ ms$) = 10^4 \times 0.5 \times 10^{-3} = 5$ V.

For 0.5 ms $\leq t \leq 1$ ms we have $i < 0$, indicating a *decreasing* voltage ramp. Since the magnitude of the current is the same, the rate of decrease is now $dv/dt = -10$ V/ms. Moreover, since the value of v at the beginning of the present time interval is 5 V, the decreasing ramp will be the mirror image of the increasing ramp of the previous interval. Hence, at the end of the present interval we must have $v(1$ ms$) = 0$ V.

Extending analogous reasoning to the subsequent time intervals, it is easily seen that the voltage waveform repeats itself every 1 ms, just as the current waveform does. The voltage response is shown in Figure 7.3(b), bottom.   ◀

The results of Example 7.4 can be generalized by saying that, when driven with a *symmetric square-wave* current, a capacitance will develop a *triangular* voltage waveform having the *same* frequency as the applied waveform.

The triangular wave has a dc component that, for given frequency, capacitance, and current values, depends on the initial condition $v(0)$. In the present case, with $v(0) = 0$ the dc component is $+2.5$ V. Had we imposed $v(0) = -2.5$ V, the dc component would have been 0 V, and the triangular wave would have been centered about the t axis.

Exercise 7.3   If the capacitance of Example 7.4 is changed to 0.5 μF, find the value of $v(0)$ for which the dc component of the triangular waveform will be 0 V.

ANSWER -5 V.

Exercise 7.4   The voltage waveform of Figure 7.4 is of the type used to control the horizontal sweep in TV sets. Sketch and label the current waveform that will cause a 2.2-nF capacitance to develop such a waveform.

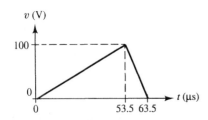

Figure 7.4 Typical TV sweep waveform.

Capacitive Energy

The process of charging a capacitor involves the expenditure of energy. This energy is found by integrating power, which is now $p = vi = vC\,dv/dt$. Thus,

$$w(t) = \int_0^t p(\xi)\,d\xi = \int_0^t vC(dv/d\xi)\,d\xi = \int_0^{v(t)} Cv\,dv$$

For a linear capacitance this yields

$$w(t) = \frac{1}{2}Cv^2(t) \qquad \textbf{(7.9)}$$

For instance, the energy stored in a 1-μF capacitance that has been charged to 10 V is $w = 0.5 \times 10^{-6} \times 10^2 = 50\ \mu$J.

Capacitive energy is stored in the form of *potential energy in the electric field between the plates.* The strength of this field is $E = v/d$, indicating that the thinner the insulator gap, the larger the electric field for a given applied voltage.

Substituting $v = q/C$ into Equation (7.9), we can express capacitive energy in terms of charge as

$$w(t) = \frac{1}{2C}q^2(t) \qquad \textbf{(7.10)}$$

indicating that capacitance can also be viewed as the ability of a device to *store energy in the form of separated charge.*

Unlike resistances, which only absorb energy to dissipate it as heat, capacitances can absorb energy from a circuit, store it, and return it to the circuit at a later time. Because of this, capacitances are said to be *nondissipative* elements.

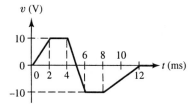

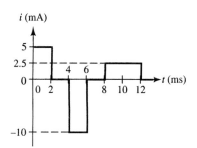

▶ Example 7.5

A 1-μF capacitance is driven with the voltage waveform of Figure 7.5, top. Sketch and label the time diagrams of its current and instantaneous power.

Solution

Since $v(t)$ consists of straight segments, we can again use $i = C\,\Delta v/\Delta t$. With C in μF, Δv in V, and Δt in ms, the units of i will be $[10^{-6}\ \text{F}][\text{V}]/[10^{-3}\ \text{s}] = [10^{-3}\ \text{A}] = [\text{mA}]$. We have the following cases:

(1) 0 ms $\leq t \leq$ 2 ms: $i = C\,\Delta v/\Delta t = 1 \times 10/2 = 5$ mA

(2) 2 ms $\leq t \leq$ 4 ms: $i = 0$ because v is constant

(3) 4 ms $\leq t \leq$ 6 ms: $i = C\,\Delta v/\Delta t = 1 \times (-20)/2 = -10$ mA

(4) 6 ms $\leq t \leq$ 8 ms: $i = 0$ because v is constant

(5) 8 ms $\leq t \leq$ 12 ms: $i = C\,\Delta v/\Delta t = 1 \times 10/4 = 2.5$ mA

The resulting current waveform is shown in Figure 7.5, center.

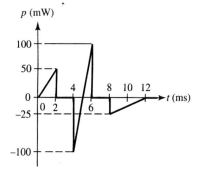

Figure 7.5 Capacitance waveforms of Example 7.5.

Since the instantaneous power is $p(t) = v(t)i(t)$, its time diagram is obtained by multiplying out the voltage and current waveforms point by point. The result is shown in Figure 7.5, bottom, which you are encouraged to verify in detail.

Remark   When $p > 0$ the capacitance *absorbs* energy, and when $p < 0$ it *releases* energy.   ◀

Exercise 7.5   Sketch and label the time diagram of the instantaneous power for the capacitance of Example 7.4.

A Water Tank Analogy

We observe that when v is *positive* and *increasing,* or *negative* and *decreasing,* a capacitance *absorbs* energy because current direction relative to voltage polarity satisfies the *passive sign* convention of Figure 1.5(b), just like a resistance. Conversely, when v is *positive* and *decreasing*, or *negative* and *increasing,* the capacitance releases energy because it conforms to the *active sign* convention of Figure 1.5(a).

Again, the aforementioned water tank analogy may help. As we know, this analogy is based on the following correspondences:

charge ↔ water

voltage ↔ water level

current ↔ water flow

capacitance ↔ water tank

Moreover, to accommodate both positive and negative water levels, we shall assume that the tank extends both above and below sea level, which shall correspond to 0 V.

Now, if the water level inside the tank is *above* sea level and we want to *raise* it further, then we must pump water from sea level up into the tank, thus *expending* energy. Likewise, if the level inside the tank is *below* sea level and we want to *lower* it further, then we must expend energy to pump water out of the tank and up to sea level.

By contrast, if the tank level is *above* sea level and we want to *lower* it, or if it is *below* sea level and we want to *raise* it, then we can let water flow freely from the tank into the sea or from the sea into the tank, indicating that energy is now *released.*

Capacitances in Parallel

If two capacitances are connected in parallel as in Figure 7.6(a), they share the same voltage v. By KCL, their individual currents add up to yield a net current $i = i_{C_1} + i_{C_2} = C_1 \, dv/dt + C_2 \, dv/dt = (C_1 + C_2) \, dv/dt$, that is, $i = C_p \, dv/dt$,

where

$$C_p = C_1 + C_2 \qquad \text{(7.11)}$$

The parallel combination of capacitances behaves as an equivalent capacitance whose value is the *sum* of the individual capacitances; see Figure 7.6(b). This can be justified physically by assuming C_1 and C_2 to be capacitors having the same insulating material and thickness but different areas S_1 and S_2. Connecting their plates in parallel creates an equivalent capacitor with area $S_1 + S_2$. By Equation (7.2), its capacitance is $C_p = \varepsilon(S_1 + S_2)/d = \varepsilon S_1/d + \varepsilon S_2/d = C_1 + C_2$.

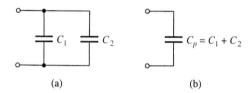

(a)   (b)

Figure 7.6 Two capacitances in parallel are equivalent to a capacitance $C_p = C_1 + C_2$.

Note that if $C_1 = C_2$, then $C_p = 2C_1$. Note also that if one of the capacitances is much larger than the other, the smaller one can be ignored and C_p will essentially coincide with the larger of the two. For instance, if $C_1 \gg C_2$, then $C_p \simeq C_1$.

Capacitances in Series

If two capacitances are connected in series as in Figure 7.7(a), they carry the same current i. By KVL, their individual voltages add up to yield a net voltage $v = v_{C_1} + v_{C_2}$. Using Equation (7.4), we can write $dv/dt = d(v_{C_1} + v_{C_2})/dt = dv_{C_1}/dt + dv_{C_2}/dt = i/C_1 + i/C_2 = i(1/C_1 + 1/C_2)$, that is, $dv/dt = i/C_s$, where

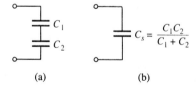

(a)   (b)

Figure 7.7 Two capacitances in series are equivalent to a capacitance $C_s = C_1C_2/(C_1 + C_2)$.

$$\frac{1}{C_s} = \frac{1}{C_1} + \frac{1}{C_2} \qquad \text{(7.12)}$$

The series combination of capacitances behaves as an equivalent capacitance whose *reciprocal* is the *sum* of the *reciprocals* of the individual capacitances; see Figure 7.7(b). To justify physically, assume C_1 and C_2 are capacitors having the same insulating material and area, but different thicknesses d_1 and d_2. Connecting them in series creates an equivalent capacitor with thickness $d_1 + d_2$. By Equation (7.2), its capacitance C_s is such that $1/C_s = (d_1 + d_2)/(\varepsilon S) = d_1/(\varepsilon S) + d_2/(\varepsilon S) = 1/C_1 + 1/C_2$.

C_s is often expressed in the form

$$C_s = \frac{C_1 C_2}{C_1 + C_2} \tag{7.13}$$

Note that if $C_1 = C_2$, then $C_s = C_1/2$. Note also that if one of the capacitances is much smaller than the other, the larger one can be ignored and C_s will essentially coincide with the smaller of the two. For instance, if $C_1 \ll C_2$, then $C_s \simeq C_1$.

It is interesting to note that the formulas for the series and parallel capacitance combinations are similar, respectively, to the formulas for the parallel and series combinations of resistances.

Exercise 7.6   Given a 10-nF capacitance, what capacitance must be placed (a) in *parallel* to achieve 30 nF? (b) In *series* to achieve 2 nF?

ANSWER   20 nF, 2.5 nF.

Practical Capacitors

Real-life capacitors satisfy Equation (7.3) only as long as q and v are confined within suitable limits. Moreover, when its connections are broken, a practical capacitor will not retain its charge indefinitely but will gradually discharge. This phenomenon, referred to as *capacitor leakage,* can be modeled with a parasitic resistance R_p in parallel with a leakage-free capacitance. To cope with leakage in DRAMs, the capacitors are periodically recharged to their intended voltages to prevent loss of information. Aptly enough, this operation is called *memory refreshing.*

By proper selection of ε, S, and d, capacitors can be manufactured in a wide variety of values and performance specifications. One common fabrication technique is to use two sheets of metal foil separated by an insulator, such as paper, and roll them in tubular form to conserve space. In integrated-circuit technology the plates are fabricated by deposition of conductive material on a very small surface and are separated by an extremely thin insulator, in some cases as thin as a few hundred atoms across! Commonly available capacitances range from picofarads to millifarads.

Capacitance arises naturally whenever conducting surfaces such as the wires of a cable, the traces of a printed circuit, or the leads of an electronic device come in close proximity. Called *stray* or *parasitic* capacitance, this unintended capacitance may at times be large enough to influence circuit behavior.

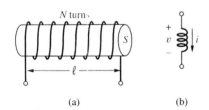

(a)   (b)

Figure 7.8 Inductor and circuit symbol for inductance.

7.2 INDUCTANCE

Inductance represents the ability of a circuit element to **produce magnetic flux linkage in response to current.** Circuit elements specifically designed to provide this function are called **inductors.** As shown in Figure 7.8(a), an inductor consists of a coil of insulated wire wound around a core. Sending current down the wire creates a magnetic field in the core and, hence, a magnetic flux ϕ. If the coil has N turns, the quantity $\lambda = N\phi$ is called the *flux linkage* and is expressed in *weber-turns*. The rate at which λ varies with the applied current is denoted as L and is called the *self-inductance* or simply the **inductance** of the coil,

$$L \triangleq \frac{d\lambda}{di} \tag{7.14}$$

Its SI unit is the *henry* (H), named for the American physicist Joseph Henry (1797–1878). Figure 7.8(b) shows the circuit symbol for inductance, along with the reference polarities for v and i.

Recall from basic magnetism that inductance depends on core material and physical dimensions according to the following generalized expression:

$$L = \mu \frac{N^2 S}{\ell} \tag{7.15}$$

where μ is the *permeability* of the core material, N is the *number* of turns, S is the *cross-sectional* area of the core, and ℓ is the *mean length* of the flux path through the core. For a single-layer solenoid of length ℓ and radius r, the denominator of Equation (7.15) is augmented by the term $0.9r$.

For vacuum space μ takes on the value $\mu_0 = 4\pi 10^{-7}$ H/m. Other media are described in terms of the *relative permeability* $\mu_r = \mu/\mu_0$. When a high inductance value is needed, the core is made of a highly permeable material such as iron or ferrite.

The *v–i* Characteristic

Referring to Figure 7.9(a), suppose the applied current is *increased* by the amount di, so that the flux linkage is increased by $N \, d\phi = L \, di$. By Faraday's Law, this change in flux linkage induces, in turn, a voltage $v = N \, d\phi/dt$ across the coil. Eliminating $N \, d\phi$, we have

$$v = L \frac{di}{dt} \tag{7.16}$$

For obvious reasons, the inductance is said to perform the operation of **current differentiation.**

Equation (7.16) indicates that inductance voltage depends not on current per se but on its *rate of change*. This means that to produce voltage across an inductance, the applied current *must change*. If the current is kept constant, no voltage will be induced, no matter how large the current. Conversely, if it is found that the voltage across an inductance is zero, this means that the current must be constant, not necessarily zero.

Using the current direction arrow as reference, we observe that if i increases, v is positive at the tail of the arrow, indicating that the inductance is *absorbing*

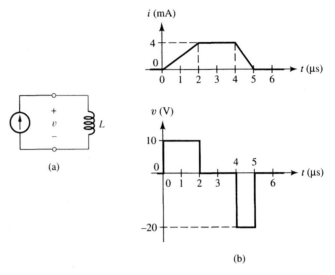

Figure 7.9 (a) Current-driven inductance and (b) waveforms of Example 7.6.

energy; if i decreases, v is positive at the head of the arrow, and the inductance is now *releasing* energy.

▶ Example 7.6

Sketch and label the voltage waveform developed by a 5-mH inductance in response to the applied current waveform of Figure 7.9(b), top.

Solution

For $0 \leq t \leq 2$ μs, i increases at the rate $\Delta i / \Delta t = (4 \text{ mA})/(2 \ \mu\text{s}) = 2 \times 10^3$ A/s. Hence, the inductance develops a constant voltage $v = L \, \Delta i / \Delta t = 5 \times 10^{-3} \times 2 \times 10^3 = 10$ V.

For $4 \ \mu$s $\leq t \leq 5 \ \mu$s, i decreases at the rate $\Delta i / \Delta t = (-4 \text{ mA})/(1 \ \mu\text{s}) = -4 \times 10^3$ A/s and the voltage now developed is $v = 5 \times 10^{-3} \times (-4 \times 10^3) = -20$ V.

At all other times i is constant, so $v = L \, di/dt = L \times 0 = 0$. The current waveform is shown in Figure 7.9(b), bottom.   ◀

Exercise 7.7   A 1-mH inductance is driven with a 100-kHz triangular current waveform having a peak-to-peak amplitude of 10 mA. Sketch and label the applied current waveform and the voltage waveform developed by the inductance. Be neat and precise.

The *i–v* Characteristic

Equation (7.16) can be turned around to find the current carried by an inductance in response to an applied voltage, as shown in Figure 7.10(a). The result is

$$i(t) = \frac{1}{L} \int_0^t v(\xi)\, d\xi + i(0) \qquad \textbf{(7.17)}$$

where $i(0)$ represents the inductance current at $t = 0$. For obvious reasons, an inductance is said to perform the operation of **voltage integration.** If voltage is constant, or $v(t) = V$, we can write

$$i(t) = \frac{V}{L} t + i(0) \qquad \textbf{(7.18)}$$

In words, forcing a *constant voltage* across an inductance yields a *linear* current waveform, or a *current ramp.* The rate at which the current ramps is $di/dt = V/L$. Depending on voltage polarity, the ramp may be of the *increasing* or *decreasing* type.

An important application for inductances is in *switching power supplies,* such as those found in personal computers. Here the inductance is switched periodically between an unregulated input voltage and a regulated output voltage, and the resulting current waveform is of the triangular type.

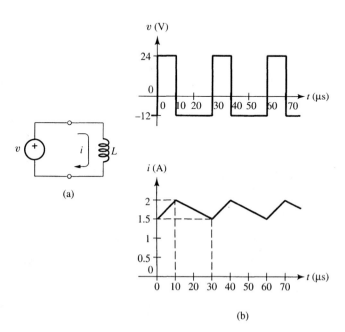

Figure 7.10 (a) Voltage-driven inductance and (b) waveforms of Exercise 7.8.

> **Exercise 7.8**   A certain inductance is used in a switching power supply. If its voltage and current waveforms are as shown in Figure 7.10(b), what is the value of the inductance?
>
> **ANSWER**   480 μH.

The Principle of Duality

You have realized by now that the description of inductance behavior parallels that of capacitance, provided we interchange C with L and i with v. This interchange relationship, called *duality,* stems from the formal similarity between the inductance characteristic $v = L\, di/dt$ and the capacitance characteristic $i = C\, dv/dt$. In words, L does to i what C does to v, and vice versa: C *differentiates* v and *integrates* i; L *differentiates* i and *integrates* v. C responds to an applied *square-wave current* with a *triangular voltage* waveform; L responds to an applied *square-wave voltage* with a *triangular current* waveform. We shall take frequent advantage of the duality principle when studying the transient and ac responses of capacitive and inductive circuits.

Inductive Energy

The process of establishing magnetic flux inside an inductor involves an expenditure of energy. This energy is found by integrating power, which is now $p = vi = iL\, di/dt$. Thus,

$$w(t) = \int_0^t p(\xi)\, d\xi = \int_0^t iL(di/d\xi)\, d\xi = \int_0^{i(t)} Li\, di$$

For a linear inductance this becomes

$$\boxed{w(t) = \frac{1}{2}Li^2(t)} \tag{7.19}$$

It is not surprising that we could have anticipated this result using the duality principle. In fact, replacing C with L and v with i in Equation (7.9) yields Equation (7.19).

Inductive energy is stored in the form of *potential energy in the magnetic field inside the core.* Alternatively, we can view inductance as the *ability of a device to store energy in the form of moving charge, or current.* As current is increased from zero to some nonzero value, energy is absorbed from the external circuit and stored in the inductance. As current is decreased back to zero, the magnetic field collapses and the accumulated energy is returned to the circuit. An ideal inductance is thus *nondissipative.*

Inductances are often used to transfer energy from one circuit to another. In this application, the inductance is first connected to a voltage source to absorb energy, and then is switched to a load to release energy. This forms the principle of a variety of power-handling circuits, such as *voltage converters* and the *switching power supplies* mentioned earlier.

Inductances in Series and in Parallel

If two inductances are connected in series as in Figure 7.11(a), they carry the same current i. By KVL, their individual voltages add up to yield a net voltage $v = v_{L_1} + v_{L_2} = L_1 \, di/dt + L_2 \, di/dt = (L_1 + L_2) \, di/dt$, that is, $v = L_s \, di/dt$, where

$$L_s = L_1 + L_2 \qquad (7.20)$$

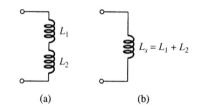

Figure 7.11 Two inductances in series are equivalent to an inductance $L_s = L_1 + L_2$.

The series combination of inductances behaves as an equivalent inductance whose value is the *sum* of the individual inductances, see Figure 7.11(b).

If two inductances are connected in parallel as in Figure 7.12(a), they share the same voltage v. By KCL, their individual currents add up to yield a net current $i = i_{L_1} + i_{L_2}$. Thus, we can write $di/dt = d(i_{L_1} + i_{L_2})/dt = di_{L_1}/dt + di_{L_2}/dt = v/L_1 + v/L_2 = v(1/L_1 + 1/L_2)$, that is, $di/dt = v/L_p$, where

$$\frac{1}{L_p} = \frac{1}{L_1} + \frac{1}{L_2} \qquad (7.21)$$

The parallel combination of inductances behaves as an equivalent inductance whose *reciprocal* is the *sum* of the *reciprocals* of the individual inductances; see Figure 7.12(b). L_p is often expressed in the form

$$L_p = \frac{L_1 L_2}{L_1 + L_2} \qquad (7.22)$$

It is interesting to note that the formulas for the series and parallel combinations of inductances are similar to those for resistances, but are dual to those for capacitances.

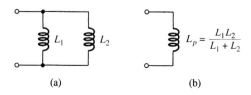

(a)   (b)

Figure 7.12 Two inductances in parallel are equivalent to an inductance $L_p = L_1 L_2/(L_1 + L_2)$.

Practical Inductors

Commonly available inductances range from microhenries to henries. Variable inductances are usually implemented by displacing the core relative to the coil.

Practical inductors satisfy the ideal characteristics discussed earlier only approximately. One of the main limitations stems from the resistance of the winding, whose voltage drop may alter the v–i characteristic significantly. Denoting this unwanted resistance as R_s, we can model a practical inductor with an ideal inductance L in *series* with a resistance R_s, as shown in Figure 7.13. Using

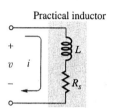

Practical inductor

Figure 7.13 Circuit model of a practical inductor.

KVL, the v–i characteristic of this composite element is then $v = L\,di/dt + R_s i$. Clearly, only in the limit $R_s i \ll L\,di/dt$ can a practical inductor be regarded as ideal.

Of the three basic circuit elements, the inductor is the least ideal. Moreover, its bulkiness and heaviness, and the fact that it does not lend itself to the miniaturization techniques of integrated circuit fabrication, make this element the least popular in modern electronics. Nevertheless, inductors are still widely used in power supply design and high-frequency signal processing.

All circuit elements exhibit a small amount of inductance due to the fact that the current through the element produces magnetic flux, which, in turn, is linked with the element itself. Called *stray* or *parasitic* inductance, this unintended inductance may at times be large enough to influence circuit behavior. Even a plain wire exhibits stray inductance, besides stray resistance and stray capacitance.

Comparison of the Basic Elements

Having completed the study of the three basic circuit elements, we summarize their most important characteristics in Table 7.1.

TABLE 7.1 Comparison of the Basic Elements

Symbol:	R	C	L
Definition:	$R \triangleq \dfrac{dv}{di}$	$C \triangleq \dfrac{dq}{dv}$	$L \triangleq \dfrac{d\lambda}{di}$
Expression:	$R = \rho\dfrac{\ell}{S}$	$C = \varepsilon\dfrac{S}{d}$	$L = \mu\dfrac{N^2 S}{\ell}$
i–v:	$i = \dfrac{1}{R}v$	$i = C\dfrac{dv}{dt}$	$i = \dfrac{1}{L}\displaystyle\int_0^t v\,d\xi + i(0)$
v–i:	$v = Ri$	$v = \dfrac{1}{C}\displaystyle\int_0^t i\,d\xi + v(0)$	$v = L\dfrac{di}{dt}$
p or w:	$p = Ri^2 = \dfrac{v^2}{R}$	$w = \dfrac{1}{2}Cv^2$	$w = \dfrac{1}{2}Li^2$
Series:	$R_s = R_1 + R_2$	$C_s = \dfrac{C_1 C_2}{C_1 + C_2}$	$L_s = L_1 + L_2$
Parallel:	$R_p = \dfrac{R_1 R_2}{R_1 + R_2}$	$C_p = C_1 + C_2$	$L_p = \dfrac{L_1 L_2}{L_1 + L_2}$

7.3 NATURAL RESPONSE OF *RC* AND *RL* CIRCUITS

The analysis of circuits containing energy-storage elements is still based on Kirchhoff's laws and the element laws. However, since these elements exhibit time-dependent i–v characteristics, the resulting circuit equations are no longer plain algebraic equations; they involve time derivatives, or integrals, or both.

The simplest circuits are those consisting of a single energy-storage element embedded in a linear network of sources and resistances. However complex this network may be, we can always replace it with its Thévenin or Norton equivalent to simplify our analysis. After this replacement, the network reduces to either equivalent of Figure 7.14, where for reasons of duality we have chosen to use the Thévenin equivalent in the capacitive case and the Norton equivalent in the inductive case. In either case we wish to find the voltage and current developed by the energy-storage element in terms of the source, the resistance, and the element itself. The manner in which this voltage or current varies with time is referred to as the **time response.**

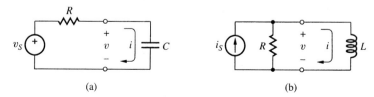

(a)   (b)

Figure 7.14 Basic first-order circuits.

First-Order Differential Equations

In the circuit of Figure 7.14(a) we have, by the capacitance law, $i = C\,dv/dt$. By KVL, we also have $v_S = Ri + v$. Eliminating i we obtain

$$RC\frac{dv}{dt} + v = v_S \tag{7.23}$$

In the circuit of Figure 7.14(b) we have, by the inductance law, $v = L\,di/dt$. By KCL, we also have $i_S = v/R + i$. Eliminating v yields

$$\frac{L}{R}\frac{di}{dt} + i = i_S \tag{7.24}$$

Both equations are of the type

$$\tau\frac{dy(t)}{dt} + y(t) = x(t) \tag{7.25}$$

where $y(t)$, representing either $v(t)$ or $i(t)$, is referred to as the **unknown variable,** and $x(t)$, representing either $v_S(t)$ or $i_S(t)$, is referred to as the **forcing function.** Moreover, we have

$$\tau = RC \tag{7.26}$$

in the *capacitive* case, and

$$\tau = \frac{L}{R} \tag{7.27}$$

in the inductive case. Both RC and L/R have the dimensions of time. Consequently, τ is called the **time constant.**

Since Equation (7.25) contains both the unknown variable and its derivative, it is said to be a **differential equation.** Moreover, it is said to be of the **first order** because this is the order of the highest derivative present. Consequently, the circuits of Figure 7.14, each containing just one energy-storage element, are said to be **first-order circuits.** A circuit containing multiple capacitances (or inductances) is still a first-order circuit if its topology allows for the capacitances (or inductances) to be reduced to a *single* equivalent capacitance (or inductance) through repeated usage of the parallel and series formulas.

Despite its apparent simplicity, Equation (7.25) cannot be solved by purely algebraic manipulations. For instance, rewriting it as $y = x - \tau(dy/dt)$ brings us no closer to the solution because the right-hand side contains the derivative of the unknown as part of the solution itself. Before developing a general solution, in the next section, we study the special but interesting case $x(t) = 0$.

The Source-Free or Natural Response

Letting $x(t) = 0$ in Equation (7.25) yields

$$\tau \frac{dy(t)}{dt} + y(t) = 0 \tag{7.28}$$

Mathematically, this equation is referred to as the **homogeneous differential equation,** and its solution $y(t)$ is referred to as the **homogeneous solution.** Since letting $x(t) = 0$ is equivalent to letting $v_S = 0$ or $i_S = 0$ in the original circuits, this particular solution is physically referred to as the **source-free response.** Lacking any forcing source, the response of the circuit is driven solely by the initial energy of its energy-storage element.

Rewriting Equation (7.28) as $y = -\tau(dy/dt)$, we note that aside from the constant $-\tau$, *the unknown and its derivative must be the same.* You may recall that of all functions encountered in calculus, only the *exponential function* enjoys the unique property that its derivative is still exponential. We thus assume a solution of the type

$$y(t) = Ae^{st} \tag{7.29}$$

where $e = 2.718$ is the base of natural logarithms, and we seek suitable expressions for A and s that will make this solution satisfy Equation (7.28).

To find an expression for s, we substitute Equation (7.29) into Equation (7.28) and obtain $\tau s Ae^{st} + Ae^{st} = 0$, or

$$(\tau s + 1)Ae^{st} = 0$$

Since we are seeking a solution $Ae^{st} \neq 0$, the above equality can hold only if the expression within parentheses vanishes,

$$\tau s + 1 = 0 \tag{7.30}$$

This equation, called the **characteristic equation,** admits the root

$$s = -\frac{1}{\tau} \tag{7.31}$$

Since it has the dimensions of the reciprocal of time, or frequency, the root is variously referred to as the **natural frequency,** the **characteristic frequency,** or the **critical frequency** of the circuit, and it is expressed in *nepers/s* (Np/s). The *neper* (Np) is a dimensionless unit named for the Scottish mathematician John Napier (1550–1617) and used to designate the unit of the exponent of e^{st}, which is a pure number.

Next, we wish to find an expression for A. This is done on the basis of the initial condition $y(0)$ in the circuit, that is, on the basis of the initial voltage $v(0)$ across the capacitance or the initial current $i(0)$ through the inductance. These conditions, in turn, are related to the initial stored energy, which is $w(0) = (1/2)Cv^2(0)$ for the capacitance, and $w(0) = (1/2)Li^2(0)$ for the inductance. Thus, letting $t \to 0$ in Equation (7.29) yields $y(0) = Ae^0$, or

$$A = y(0) \tag{7.32}$$

Substituting Equations (7.31) and (7.32) into (7.29) finally yields

$$y(t) = y(0)e^{-t/\tau} \tag{7.33}$$

As shown in Figure 7.15, $y(t)$ is an **exponentially decaying function** from the initial value $y(0)$ to the final value $y(\infty) = 0$. Since the decay depends only on $y(0)$ and τ, which are peculiar characteristics of the circuit irrespective of any particular forcing function, this solution is also called the **natural response.** Thus, *homogeneous solution, source-free response,* and *natural response* are different terms for the same function of Equation (7.33).

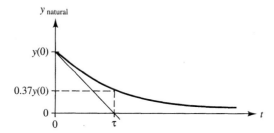

Figure 7.15 The source-free or natural response.

How a circuit manages to produce a nonzero response with a zero forcing function is an intriguing question, but, as stated, this behavior stems from the ability of capacitors and inductors to store energy. It is precisely this energy that allows the circuit to sustain nonzero voltages and currents even in the absence of any forcing source. These voltages and currents will persist until all of the initial energy has been used up by the resistances in the circuit. Contrast this with a purely resistive network where, in the absence of any driving source, each voltage and current in the circuit would at all times be zero.

The Time Constant τ

Mathematically, the time constant τ serves the purpose of making the argument t/τ of the exponential function a *dimensionless number.* Physically, it provides a measure of *how rapidly* the exponential decay takes place. The significance of τ can be visualized in two different ways, as follows:

Evaluating Equation (7.33) at $t = \tau$ yields $y(\tau) = y(0)e^{-1} = 0.37y(0)$, indicating that after τ seconds the natural response has decayed to 37% of its initial value. Equivalently, we can say that after τ seconds the response has accomplished $100 - 37 = 63\%$ of its entire decay. Thus, one way of interpreting the time constant is:

    (1) τ represents the amount of time it takes for the natural response to decay to $1/e$, or to 37% of its initial value.

An alternative interpretation is found by considering the *initial slope* of the response curve. On the one hand, this slope can be found analytically by evaluating the derivative of Equation (7.33) at $t = 0$, that is, $dy(0)/dt = (-1/\tau)y(0)e^0 = -y(0)/\tau$. On the other hand, it can be found geometrically as the ratio of the y-axis to the t-axis intercepts of the tangent to the curve at the origin. Since the y-axis intercept occurs at $y(0)$, it follows that the t-axis intercept must occur at $t = \tau$ in order to make the initial slope equal to the calculated value $-y(0)/\tau$. Thus, an alternate interpretation for the time constant is:

    (2) τ represents the instant at which the tangent to the natural response at the origin intercepts the t axis.

Either of these viewpoints can be exploited to find τ experimentally by observing the natural response with an oscilloscope.

Exercise 7.9   Show that the tangent to $y(t)$ at *any* instant, not necessarily the origin, intersects the t axis τ seconds later.

We observe that the larger the value of τ, the slower the rate of decay because it will take *longer* for the response to decay to 37% or, equivalently, the tangent at the origin will intercept the t axis at a *later* instant. Conversely, a small time-constant results in a rapid decay. This is illustrated in Figure 7.16.

Decay Times

It is often of interest to estimate the amount of time t_ε it takes for the natural response to decay to a given fraction ε of its initial value. By Equation (7.33), t_ε must be such that $\varepsilon y(0) = y(0)\exp(-t_\varepsilon/\tau)$, or $\varepsilon = \exp(-t_\varepsilon/\tau)$. Solving for t_ε yields

$$t_\varepsilon = -\tau \ln \varepsilon \qquad\qquad \textbf{(7.34)}$$

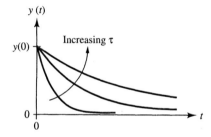

Figure 7.16 The larger the time constant τ, the slower the rate of decay of the natural response.

▶ Example 7.7

Find the time it takes for the natural response to decay to (a) 50%, (b) 1%, and (c) 0.1% of its initial value.

Solution

(a) $t_{0.5} = -\tau \ln 0.5 = 0.69\tau \simeq 0.7\tau$.

(b) $t_{0.01} = -\tau \ln 0.01 = 4.6\tau \simeq 5\tau$.

(c) $t_{0.001} = 6.9\tau \simeq 7\tau$.

◀

Exercise 7.10 Find the time it takes for the natural response to decay from 90% to 10% of its initial value.

ANSWER 2.2τ.

Exercise 7.11 In the circuit of Figure 7.14(b) let $i_S = 0$ and $R = 100\ \Omega$. If it is observed that $i(173.3\ \mu s) = (1/2)i(0) = 50$ mA, find (a) the value of L, and (b) the energy initially stored in the inductance.

ANSWER (a) 25 mH; (b) 125 μJ

Though in theory the response reaches zero only in the limit $t \to \infty$, in practice it is customary to regard the decay as essentially complete after about *five* time constants (5τ), since by this time the response has already dropped below 1% of its initial value, which is negligible in most cases of practical interest.

• The *s* Plane

It is good practice to visualize the root of a characteristic equation as a point in a plane called the *s plane*. Though we shall have more to say about this plane in Chapter 9, for the time being we ignore the vertical axis and use points of the *horizontal axis* to visualize our roots. Negative roots lie on the left portion of this axis, positive roots on the right portion. Moreover, this axis is calibrated in Np/s.

Since the root $s = -1/\tau$ is negative, it lies on the left portion of the axis. Moreover, the *farther away* the root from the origin, the *more rapid* the exponential decay. Conversely, the *closer* the root to the origin, the *slower* the decay. This correspondence is depicted in Figure 7.17(a) and (b).

It is interesting to note that in the limit of a root right *at the origin,* or $1/\tau \to 0$, we have $\tau \to \infty$, indicating an *infinitely slow* decay, as depicted in Figure 7.17(c). In fact, Equation (7.33) predicts $y(t) = y(0)e^{-t/\infty} = y(0)$, that is, a constant natural response. By Equations (7.26) and (7.27), the condition $\tau = \infty$ is achieved by letting $R = \infty$ in the *capacitive* case, and $R = 0$ in the *inductive* case. As we know, when open-circuited, an ideal capacitance will retain its initial voltage indefinitely, so $v(t) = v(0)$ for any $t > 0$; when short-circuited, an ideal inductance will sustain its initial current indefinitely, so

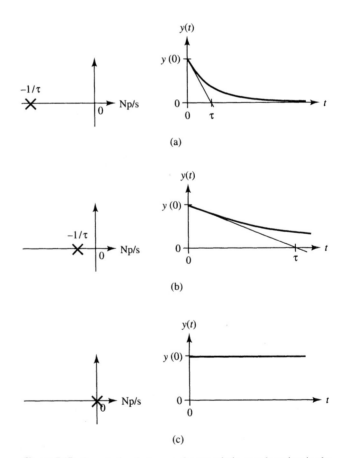

Figure 7.17 Correlation between characteristic root location in the *s* plane, shown at the left, and rate of decay of the natural response, shown at the right: (a) τ small, (b) τ large, (c) $\tau = \infty$.

$i(t) = i(0)$ for any $t > 0$. As we know, the function associated with this type of response is the **memory function.**

▶ **Example 7.8**

In the circuit of Figure 7.14(a) let $v_S = 0$, $R = 10$ kΩ, and $C = 1$ μF, and let the energy initially stored in the capacitance be 50 μJ. (a) Find the location of the root in the s plane. (b) Find $v(t)$.

Solution

    (a) $\tau = RC = 10^4 \times 10^{-6} = 10$ ms. The root is located at
       $s = -1/\tau = -1/0.01 = -100$ Np/s.

    (b) Since $w = (1/2)Cv^2$, the initial voltage is $v(0) = \sqrt{2w(0)/C} = \sqrt{2 \times 50 \times 10^{-6}/10^{-6}} = 10$ V. Then,

$$v(t) = 10e^{-100t} \text{ V}$$

◀

7.4 Response to DC and AC Forcing Functions

Having investigated the particular but important case in which $x(t) = 0$ in Equation (7.25), we now wish to find the solution for the case of an *arbitrary* forcing function $x(t)$. We then turn our attention to the two types of forcing functions of greatest interest, namely, the *dc* and *ac* forcing functions. The level of mathematical detail is designed to provide a clear understanding of the various response components as well as the terminology.

General Solution to the Differential Equation

An elegant method for solving Equation (7.25) is provided by the **integrating-factor approach,** as follows. Multiplying both sides of Equation (7.25) by $(1/\tau) \exp(t/\tau)$ yields

$$\frac{dy(t)}{dt}e^{t/\tau} + \frac{1}{\tau}y(t)e^{t/\tau} = \frac{1}{\tau}x(t)e^{t/\tau}$$

Recognizing that the left-hand side is the derivative of a product, we can write

$$\frac{d}{dt}\left[y(t)e^{t/\tau}\right] = \frac{1}{\tau}x(t)e^{t/\tau}$$

Replacing t with the dummy variable ξ and integrating both sides from 0 to t yields

$$\int_0^t \frac{d}{d\xi}\left[y(\xi)e^{\xi/\tau}\right] d\xi = \int_0^t \frac{1}{\tau}x(\xi)e^{\xi/\tau} d\xi$$

that is,

$$y(t)e^{t/\tau} - y(0)e^0 = \frac{1}{\tau}\int_0^t x(\xi)e^{\xi/\tau} d\xi$$

where $y(0)$ again represents the initial value of y. Finally, solving for $y(t)$ yields the general expression for the solution,

$$y(t) = y(0)e^{-t/\tau} + \frac{1}{\tau}e^{-t/\tau} \int_0^t x(\xi)e^{\xi/\tau} \, d\xi \qquad \text{(7.35)}$$

This expression indicates that the response $y(t)$ consists of *two* components. The first component, which is independent of the forcing function $x(t)$, is the already familiar **natural response.** The second component, which depends on the particular forcing function $x(t)$, is called the **forced response.** For obvious reasons, the sum of the two components is called the **complete response.** To emphasize this, Equation (7.35) is often expressed in the form

$$y(t) = y_{\text{natural}} + y_{\text{forced}} \qquad \text{(7.36)}$$

where

$$y_{\text{natural}} = y(0)e^{-t/\tau} \qquad \text{(7.37)}$$

$$y_{\text{forced}} = \frac{1}{\tau}e^{-t/\tau} \int_0^t x(\xi)e^{\xi/\tau} \, d\xi \qquad \text{(7.38)}$$

It is apparent that the forced response depends on the functional form of $x(\xi)$, that is, on the manner in which v_S or i_S in our RC and RL circuits vary with time. Depending on this form, evaluating the integral of Equation (7.38) analytically may be a formidable task, and numerical methods may have to be used instead. Mercifully, the cases of greatest practical interest are only two, namely, the case of a *constant* or **dc forcing function,** and the case of a *sinusoidal* or **ac forcing function.**

Response to a DC Forcing Function

A **dc forcing function** is a function of the type

$$x(t) = X_S \qquad \text{(7.39)}$$

where X_S is a suitable *constant*. This means that in the case of the RC circuit of Figure 7.14 we have $v_S = V_S$, and in the RL circuit we have $i_S = I_S$. Substituting Equation (7.39) into Equation (7.38) yields

$$y_{\text{forced}} = \frac{1}{\tau}e^{-t/\tau} \int_0^t X_S e^{\xi/\tau} \, d\xi = \frac{X_S}{\tau}e^{-t/\tau} \times \tau e^{\xi/\tau} \Big|_0^t$$

or

$$y_{\text{forced}} = X_S \left(1 - e^{-t/\tau}\right) \qquad \text{(7.40)}$$

Substituting into Equation (7.36) yields the complete response to a dc forcing function,

$$y(t) = y(0)e^{-t/\tau} + X_S\left(1 - e^{-t/\tau}\right)$$

(7.41)

Noting that in the limit $t \rightarrow \infty$ this equation yields $y(\infty) = y(0)e^{-\infty} + X_S(1 - e^{-\infty}) = X_S$, we can also write

$$y(t) = [y(0) - y(\infty)]e^{-t/\tau} + y(\infty)$$

(7.42)

where

$$y(\infty) = X_S$$

(7.43)

As depicted in Figure 7.18, the complete response is an **exponential transition** from the initial value $y(0)$ to the final value $y(\infty)$. This transition is again governed by the time constant τ, which is now visualized as the time it takes for $y(t)$ to accomplish $(100 - 37)\%$, or 63% of the entire transition. Alternatively, τ can be visualized as the time at which the tangent to the curve at the origin intercepts the $y = y(\infty)$ asymptote. Either viewpoint can be exploited to find τ experimentally by observing the response to a dc forcing function with an oscilloscope.

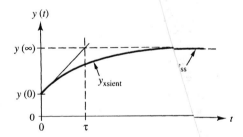

Figure 7.18 Example of a complete response to a dc forcing function.

Figure 7.18 illustrates the complete response for the case $y(\infty) > y(0) > 0$, but analogous diagrams can be constructed for all other possible cases. Just keep in mind that Equation (7.42) holds in general, regardless of whether $y(0)$ and $y(\infty)$ are positive, negative, or zero. Note that if $X_S = 0$, so that $y(\infty) = 0$, the complete response reduces to the natural response.

We shall use Equation (7.42) so often that it is worth committing it to memory. Moreover, when sketching transient responses, it is a good habit to show the tangent at the origin and label its intercept with the $y = y(\infty)$ asymptote explicitly. This makes it easier for the reader to identify the value of the time constant governing the transient.

▶ **Example 7.9**

In the circuit of Figure 7.14(a) let $R = 10$ kΩ, $C = 1$ μF, and $v(0) = 5$ V. Find the response $v(t)$ for $t \geq 0$ if (a) $v_S = 12$ V, (b) $v_S = 1$ V, and (c) $v_S = -3$ V.

Solution

We have $\tau = RC = 10$ ms, or $1/\tau = 100$ Np/s.

    (a) We have $v(\infty) = 12$ V. Using Equation (7.42) yields

$$v(t) = (5 - 12)e^{-100t} + 12 = -7e^{-100t} + 12 \text{ V}$$

    (b) Now $v(\infty) = 1$, so

$$v(t) = 4e^{-100t} + 1 \text{ V}$$

    (c) Now $v(\infty) = -3$ V, so

$$v(t) = 8e^{-100t} - 3 \text{ V}$$

◀

Exercise 7.12   In the circuit of Figure 7.14(b) let $R = 150$ Ω, $L = 30$ mH, and $i(0) = 100$ μA. Find $i(0.1$ ms$)$ if (a) $i_S = 200$ mA, (b) $i_S = 50$ mA, and (c) $i_S = -200$ mA.

ANSWER   (a) 139.3 mA; (b) 80.33 mA; (c) −18.04 mA.

The Transient and DC Steady-State Components

Equation (7.42) reveals that the response to a dc forcing function can be regarded as the sum of two components, namely, an exponentially decaying component with initial magnitude $y(0) - y(\infty)$, called the **transient component,** and a time-independent component of value $y(\infty)$, called the **dc steady-state component** because this is the value to which the complete response will settle once the transient component has died out. To emphasize this, the complete response to a dc forcing function is often expressed as

$$y(t) = y_{\text{xsient}} + y_{\text{ss}}$$

(7.44)

where

$$y_{\text{xsient}} = [y(0) - y(\infty)]e^{-t/\tau}$$

(7.45)

$$y_{\text{ss}} = y(\infty)$$

(7.46)

The two portions of $y(t)$ are shown explicitly in Figure 7.18.

It is interesting to note that y_{xsient} has the *same functional form* as $y_{natural}$ and that it consists of two terms: the natural response itself, $y(0)e^{-t/\tau}$, and the term $y(\infty)e^{-t/\tau}$, which is brought about by the forcing function. We also note that y_{ss} has the *same functional form* as the forcing function, $y_{ss} = y(\infty) = X_S$.

It is common practice to refer to the *complete response* to a dc forcing function as simply the **transient response.** This response, to be studied in detail in Chapters 8 and 9, is of fundamental importance because it allows us to deduce important properties that are *characteristic* of the particular circuit producing it. In fact, the design specifications for circuit systems are often given in terms of transient-response parameters.

Response to an AC Forcing Function

As we shall see in greater detail in Chapter 10, an **ac forcing function** is a sinusoidal function of the type

$$x(t) = X_m \cos \omega t \qquad (7.47)$$

where X_m is called the **amplitude** and ω the **angular frequency** of the ac signal, whose units are *radians/s* (rad/s). This means that in the capacitive circuit of Figure 7.14 we have $v_S = V_m \cos \omega t$, and in the inductive circuit we have $i_S = I_m \cos \omega t$. Substituting Equation (7.47) into Equation (7.38) yields

$$y_{forced} = \frac{X_m}{\tau} e^{-t/\tau} \int_0^t e^{\xi/\tau} \cos \omega \xi \, d\xi \qquad (7.48)$$

To calculate this integral, we use the following formula from the integral tables

$$\int e^{a\xi} \cos b\xi \, d\xi = \frac{e^{a\xi}}{a^2 + b^2} (a \cos b\xi + b \sin b\xi)$$

Substituting into Equation (7.48) with $a = 1/\tau$ and $b = \omega$, we obtain

$$y_{forced} = \frac{X_m}{\tau} e^{-t/\tau} \times \frac{e^{\xi/\tau}}{(1/\tau)^2 + \omega^2} \left(\frac{1}{\tau} \cos \omega \xi + \omega \sin \omega \xi \right) \Big|_0^t$$

or

$$y_{forced} = \frac{X_m}{1 + (\omega\tau)^2} \left(\cos \omega t + \omega\tau \sin \omega t - e^{-t/\tau} \right) \qquad (7.49)$$

Using the trigonometric identity

$$A \cos \alpha + B \sin \alpha = \sqrt{A^2 + B^2} \cos [\alpha - \tan^{-1} (B/A)]$$

with $A = 1$, $B = \omega\tau$, and $\alpha = \omega t$, Equation (7.49) becomes

$$\begin{aligned} y_{forced} = &-\frac{X_m}{1 + (\omega\tau)^2} e^{-t/\tau} \\ &+ \frac{X_m}{\sqrt{1 + (\omega\tau)^2}} \cos (\omega t - \tan^{-1} \omega\tau) \end{aligned} \qquad (7.50)$$

Substituting into Equation (7.36), using also Equation (7.37), and then regrouping, we finally obtain the *complete response*

$$
\begin{aligned}
y(t) = {} & \left(y(0) - \frac{X_m}{1 + (\omega\tau)^2} \right) e^{-t/\tau} \\
& + \frac{X_m}{\sqrt{1 + (\omega\tau)^2}} \cos\left(\omega t - \tan^{-1}\omega\tau\right)
\end{aligned}
\tag{7.51}
$$

An example of a complete response is shown in Figure 7.19.

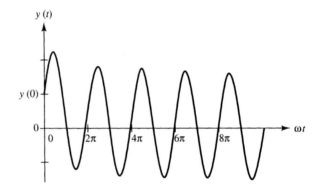

Figure 7.19 Example of a complete response to an ac forcing function.

The Transient and AC Steady-State Components

Equation (7.51) reveals that the response to an ac forcing function consists of two components, namely, an exponentially decaying component, called the **transient component,** and an ac component, called the **ac steady-state component** because this is the value to which the complete response will settle once the transient component has died out. We emphasize this by writing

$$
y(t) = y_{\text{xsient}} + y_{\text{ss}}
\tag{7.52}
$$

where

$$
y_{\text{xsient}} = \left(y(0) - \frac{X_m}{1 + (\omega\tau)^2} \right) e^{-t/\tau}
\tag{7.53}
$$

$$
y_{\text{ss}} = Y_m \cos\left(\omega t + \phi\right)
\tag{7.54}
$$

and,

$$Y_m = \frac{X_m}{\sqrt{1 + (\omega\tau)^2}} \tag{7.55}$$

$$\phi = -\tan^{-1}\omega\tau \tag{7.56}$$

The quantities Y_m and ϕ are called, respectively, the **amplitude** and the **phase angle** of the steady-state component.

Again, we observe that y_{xsient} has the *same functional form* as y_{natural} and that it consists of two terms: the natural response itself, $y(0)e^{-t/\tau}$, and the term $X_m e^{-t/\tau}/(1+\omega^2\tau^2)$, brought about by the forcing function. We also note that y_{ss} has the *same functional form* as the forcing function: they are both sinusoidal, and they have the **same frequency** ω. Moreover, both the amplitude and phase angle of y_{ss} are **frequency dependent** parameters.

It is common practice to refer to the *ac steady-state component* as simply the **ac response.** Like the transient response, the ac response is of fundamental importance because a number of important circuit features can be deduced by observing the ac response. The design specifications for circuit systems are often given in terms of ac-response parameters. The ac response will be studied in detail in Chapters 10 through 12.

▶ Example 7.10

In the *RC* circuit of Figure 7.14(a) let $R = 1$ kΩ, C $= 1$ μF, and $v_S = 10\cos 2\pi 10^3 t$ V. Find the complete response $v(t)$ for $t \geq 0$ if $v(0) = 1$ V.

Solution

We have $\tau = RC = 10^3 \times 10^{-6} = 1$ ms, $\omega\tau = 2\pi 10^3 \times 10^{-3} = 2\pi$, and $X_m = 10$ V. Applying Equation (7.53) yields $v_{\text{xsient}} = [1 - 10/(1 + 2^2\pi^2)]e^{-10^3 t}$, or

$$v_{\text{xsient}} = 0.753 e^{-10^3 t} \text{ V}$$

By Equation (7.55), $Y_m = 10/\sqrt{1 + (2\pi)^2} = 1.572$ V, and $\phi = -\tan^{-1} 2\pi = -80.96°$. Applying Equation (7.54) yields

$$v_{\text{ss}} = 1.572\cos(2\pi 10^3 t - 80.96°) \text{ V}$$

The complete response is thus, by Equation (7.52),

$$v(t) = 0.753 e^{-10^3 t} + 1.572\cos(2\pi 10^3 t - 80.96°) \text{ V}$$

Figure 7.19 shows precisely this response.   ◀

> **Exercise 7.13**   In the *RL* circuit of Figure 7.14(b) let $R = 150 \ \Omega$, $L = 30$ mH, and $i_S = 0.1 \cos 2\pi 10^4 t$ A. Find the complete response $i(t)$ for $t \geq 0$ if $i(0) = 0$.
>
> **ANSWER** $i(t) = -0.629 e^{-5000t} + 7.933 \cos (2\pi 10^4 t - 85.45°)$ mA.

CONCLUDING REMARKS

We conclude with the following observations, which hold regardless of whether the forcing function is of the *dc* or *ac* type:

(1) The *transient* component is *functionally similar* to the *natural response,* that is, it is an *exponentially decaying function.*

(2) The *steady-state* component is *functionally similar* to the *forcing function,* that is, it is of the *dc type* in the case of a dc forcing function, and of the *ac type* in the case of an ac forcing function.

These observations, to which we shall return time and again, provide an important perspective for the material of the subsequent chapters. Keep them in mind.

▼ SUMMARY

- The capacitor and the inductor are *energy-storage* elements. The expression for capacitive energy is $w = (1/2)Cv^2$, and that for inductive energy is $w = (1/2)Li^2$.

- The capacitance law is $i = C \, dv/dt$, and the inductance law is $v = L \, di/dt$. These laws are *dual* of each other. Since their laws are time-dependent, the capacitance and the inductance are also referred to as *dynamic elements.*

- For a capacitance to carry current, its voltage must *change.* If the voltage is kept constant, the current is zero. We also say that the capacitance provides a *memory function.*

- For an inductance to develop voltage, its current must *change.* If the current is kept constant, the voltage is zero.

- The analysis of circuits with energy-storage elements is still based on the combined use of Kirchhoff's laws and the element laws. However, the circuit equations are now *differential equations* as opposed to the *algebraic equations* of purely resistive circuits.

- The manner in which a circuit reacts to the application of a *forcing signal* is called the *response.*

- A circuit with energy-storage elements is capable of producing a response even in the absence of any forcing function. This response is called the *source-free response* or the *natural response* because it is produced by the circuit spontaneously, using the energy available from its energy-storage elements.

- The natural response of a first-order circuit is a *decaying exponential.* The rate of decay is governed by the *time constant* τ. Quantitatively, τ is the amount of time it takes for the natural response to decay to $1/e$ or 37% of its initial value.

- A source-free circuit can be described in terms of its *characteristic equation.* The root s of this equation is related to the time constant τ as $s = -1/\tau$, and it lies on the negative portion of the horizontal axis of the s plane.

- The further away the root is from the origin, the smaller the time constant, and the faster the rate of decay of the natural response. Conversely, the closer the root to the origin, the longer the rate of decay. A root at the origin corresponds to the *memory* function.

- When a forcing function is present, the response consists of two components, the *natural component* and the *forced component.* The natural component is an exponential decay, but the forced component depends on the functional form of the forcing function.

- The two cases of greatest practical interest are the response to a *dc forcing function,* called the *transient response,* and to an *ac forcing function,* called the *ac response.*

- In both cases the *complete response* consists of a *transient component* having the same functional form as the *natural response,* and a *steady state component* having the same functional form as the *forcing function.* The designation steady state refers to the condition achieved once the transient part has died out.

- The ac response has the *same frequency* as the ac forcing function, differing only in *amplitude* and *phase angle.* Moreover, amplitude and phase angle are *frequency dependent.*

▼ PROBLEMS

7.1 Capacitance

7.1 Find the capacitance of two metal plates, each 1 in. × 1 in. in size (1 in. = 2.54 cm) and separated by an air gap 0.1 mm thick. For air, $\varepsilon_r \simeq 1$.

7.2 Find the parasitic capacitance between two traces running parallel to each other on opposite sides of a printed-circuit board. Assume the traces are 2 mm wide and 2 in. long, and the board is 1 mm thick with $\varepsilon_r = 5$.

7.3 (a) A 1-μF parallel-plate capacitor has a 1-mm-thick ceramic dielectric ($\varepsilon_r = 7500$). If the plates are square, find the length of each side of its plates. (b) Repeat for a 1-F capacitance, comment.

7.4 (a) The *parallel* combination of two capacitances, valued 0.2 μF and 0.6 μF and initially discharged, is charged to 10 V. Find the charge and energy stored in each capacitance. (b) Repeat, but for the *series* combination.

7.5 (a) In the circuit of Figure P7.5(a) find a relationship between C and C_1 to make $C_{eq} = C$. (b) Repeat, but for Figure P7.5(b).

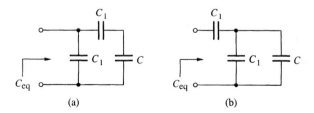

Figure P7.5

7.6 Find all possible capacitance values that can be obtained by interconnecting 2-nF, 3-nF, and 6-nF capacitances.

7.7 (a) Find C_{eq} in Figure P7.7. (b) Repeat if all capacitances are equal to 10 pF.

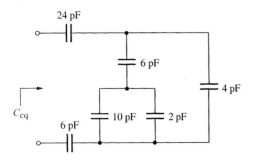

Figure P7.7

7.8 (a) Find C_{eq} in Figure P7.8. (b) Repeat, but with C and D shorted together.

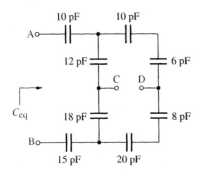

Figure P7.8

7.9 After having been charged separately, two capacitances are connected in series as in Figure P7.9. (a) Find the total stored energy, as well as the fraction of this energy that can be recovered by bringing the terminal voltage v_{AB} back to zero. (b) Repeat if the voltage across the 3-μF capacitance is positive at the bottom.

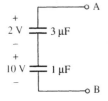

Figure P7.9

7.10 Three capacitances $C_1 = 2$ μF, $C_2 = 3$ μF, and $C_3 = 6$ μF are separately charged to $v_1 = 5$ V, $v_2 = 6$ V, and $v_3 = 3$ V, and then connected in series. Find the voltage

polarity combinations that maximize and minimize the energy recoverable by bringing the voltage across the terminals of the series combination back to zero. What are the maximum and minimum values of this energy?

7.11 (a) Show that a voltage *change* v_i in the circuit of Figure P7.11 produces a voltage *change* v_o such that $v_o = v_i C_1/(C_1 + C_2)$. (b) If $C_2 = 3$ nF, $v_i = 5$ V, and $v_o = 2$ V, find C_1.

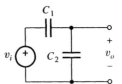

Figure P7.11

7.12 (a) Show that a voltage *change* v_i in the circuit of Figure P7.12 produces a voltage *change* v_o such that $v_o = -(C_1/C_2)v_i$. (b) If $C_1 = 40$ pF and $C_2 = 10$ pF, find v_i for $v_o = +8$ V.

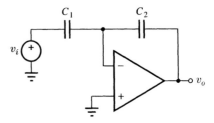

Figure P7.12

7.13 (a) Show that each time the switch of Figure P7.13 is flipped to the right, a net charge $q = C(v_1 - v_2)$ is transferred from v_1 to v_2. (b) Show that flipping the switch back and forth f_0 times per second causes the SW-C combination to act as an effective resistance $R_{eff} = 1/(f_0 C)$. (c) If $C = 10$ pF, what is R_{eff} if $f_0 = 1$ kHz? $f_0 = 1$ MHz?

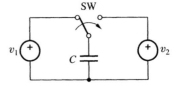

Figure P7.13

7.14 Find the current through a 50-nF capacitance if its voltage is (a) $v = 5(1 - 10^4 t)$ V, (b) $v = 10(1 - e^{-10^4 t})$ V, (c) $v = 4 \sin 10^4 t$, and (d) $v = 15$ V.

7.15 Assuming the initial voltage across a 1-μF capacitance is $v(0) = 5$ V, find v for $t \geq 0$ if the capacitance current is (a) $i = 10^{-5}t$ A, (b) $i = 2e^{-10^5 t}$ A, (c) $i = 0.5 \cos 10^5 t$ A, (d) $i = 5$ μA, and (e) $i = 0$ A.

7.16 The voltage across a 4-nF capacitance has the waveform of Figure P7.16. Sketch and label its current i, instantaneous power p, and stored energy w for $0 \leq t \leq 14$ μs.

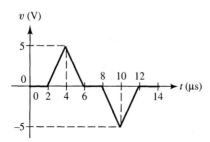

Figure P7.16

7.17 A one-port consists of a 2-nF capacitance in parallel with a 2-kΩ resistance. Sketch and label the current drawn by this one-port in response to the voltage waveform of Figure P7.16.

7.18 Sketch and label, for $0 \leq t \leq 6$ μs, the voltage developed by an initially discharged 5-nF capacitance in response to the current waveform of Figure P7.18.

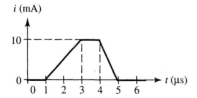

Figure P7.18

7.19 The voltage across a 0.5-μF capacitance is $v = 5 \times \sin 10^3 t$ V. Find the current i, power p, and stored energy w. Hence, plot them versus $10^3 t$ for $0 \leq 10^3 t \leq 2\pi$.

7.2 Inductance

7.20 (a) Find the number of turns wound on an air core 6 mm in diameter and 5 cm in length that are needed to form a 10-μH solenoidal inductor. (b) How many turns must be added to increase the inductance to 15 μH? (c) How does the inductance change if the core is made of iron, for which $\mu_r \simeq 2 \times 10^3$?

7.21 (a) In the circuit of Figure P7.21(a), find a relationship between L and L_1 to make $L_{eq} = L$. (b) Repeat, but for Figure P7.21(b).

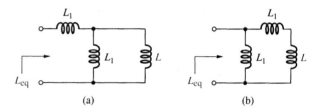

Figure P7.21

7.22 Find L_{eq} in Figure P7.22 if $L = 30$ μH.

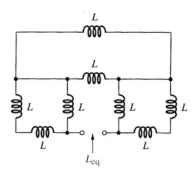

Figure P7.22

7.23 Find the voltage across a 25-mH inductance if its current is (a) $i = 10(1 - 10^2 t)$ A, (b) $i = 5(1 - e^{-10^2 t})$ A, (c) $i = 10$ mA, and (d) $i = 2 \sin 10^2 t$ A.

7.24 Assuming the initial current through a 1-mH inductance is $i(0) = 10$ mA, find i for $t \geq 0$ if the inductance voltage is (a) $v = 10^{-5}t$ V, (b) $v = 2e^{-10^5 t}$ V, (c) $v = 0.5 \cos 10^5 t$ V, (d) $v = 15$ V, and (e) $v = 0$ V.

7.25 The current through a 500-μH inductance has the waveform of Figure P7.25. Sketch and label its voltage v and instantaneous power p for $0 \leq t \leq 6$ μs.

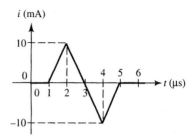

Figure P7.25

7.26 A one-port consists of a 250-μH inductance in series with a 300-Ω resistance. Sketch and label the voltage it develops in response to the current waveform of Figure P7.25.

7.27 Sketch and label, for $0 \leq t \leq 6$ ms, the current drawn by a 2-mH inductance with zero initial stored energy in response to the voltage waveform of Figure P7.27.

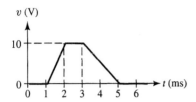

Figure P7.27

7.28 The current through a 1.5 mH inductance is $i = 2 \times \sin 10^6 t$ mA. Find the voltage v, power p, and stored energy w. Hence, plot them versus $10^6 t$ for $0 \leq 10^6 t \leq 2\pi$.

7.29 In Figure P7.29 find i_S if (a) $i_L = 20 \sin (4 \times 10^4 t)$ mA, (b) $i_L = 20 \sin (6 \times 10^4 t)$ mA, and (c) $i_L = 20 \sin (5 \times 10^4 t)$ mA.

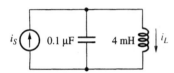

Figure P7.29

7.30 In Figure P7.30 find v_S if (a) $v_C = 10 \sin 10^4 t$ V, (b) $v_C = 10 \sin 10^5 t$ V, and (c) $v_C = 10 \sin 10^6 t$ V.

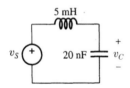

Figure P7.30

7.3 Natural Response of *RC* and *RL* Circuits

7.31 Given the differential equation $5 \times 10^{-3} dy(t)/dt + 2y(t) = 10^{-3} dx(t)/dt - x(t)$, find its characteristic frequency.

7.32 An exponential voltage decay $v(t) = v(0)e^{-t/\tau}$ is observed with an oscilloscope. If it is found that $v(1 \ \mu s) = 12.28$ V and $v(7 \ \mu s) = 3.7$ V, (a) find $v(0)$ and τ, and (b) predict $v(10 \ \mu s)$.

7.33 If it is found that an exponential current $i(t) = i(0)e^{-t/\tau}$ takes 5.545 ms to decay from 8 mA to 2 mA, how long does it take to decay from 6 mA to 4 mA? From 9 mA to 1 mA?

7.4 Response to DC and AC Forcing Functions

7.34 Find the solution to the differential equation $2dy(t)/dt + y(t) = 5e^{-t}$, with the initial condition $y(0) = 10$. Check your result.

7.35 Find the solution to the differential equation $2dy(t)/dt + 3y(t) = x(t)$ if (a) $x(t) = 2te^{-t}$, with the initial condition $y(0) = 5$, and (b) $x(t) = 3e^{-t} \cos 2t$, with the initial condition $y(0) = 1$. Check.

7.36 The *RC* circuit of Figure 7.14(a) consists of a 1-μF capacitance, with an initial stored energy of 450 μJ, and a 1-kΩ resistance. If the circuit is subjected to a dc forcing voltage V_S, and if it is found that $v(2 \ \text{ms}) = 0$, estimate V_S.

7.37 The *RL* circuit of Figure 7.14(b) consists of a 5-mH inductance and a 100-Ω resistance, and is subjected to a dc forcing current of 25 mA. If it is found that $i(40 \ \mu s) = 0$, estimate $i(0)$ as well as the initial stored energy in the inductance.

7.38 For each of the following *complete* voltage responses, identify the *natural, forced, transient,* and *steady-state* components: (a) $v = 5(1 + e^{-t/\tau})$ V; (b) $v = 5(2 - e^{-t/\tau})$ V; (c) $v = 5(3e^{-t/\tau} + 2)$ V.

7.39 For each of the following *complete* current responses, identify the *natural, forced, transient,* and *steady-state* components: (a) $i = 5(2 - 3e^{-t/\tau})$ A; (b) $i = -5(1 - 3e^{-t/\tau})$ A; (c) $i = 10(1 - e^{-t/\tau})$ A; (d) $i = 5e^{-t/\tau}$ A.

7.40 In the *RC* circuit of Figure 7.14(a) let $R = 10$ kΩ, $C = 2$ nF, and $v_S = 10 \cos 10^5 t$ V. Find the complete response $v(t)$ if (a) $v(0) = 5$ V, (b) $v(0) = 2$ V, and (c) $v(0) = 0$.

7.41 In the *RL* circuit of Figure 7.14(b) let $R = 1$ kΩ, $L = 1$ mH, and $i_S = 10 \cos 10^6 t$ mA. Find the complete response $i(t)$ if (a) $i(0) = 0$, (b) the initial stored energy in the inductance is $w(0) = 12.5$ nJ. (Beware that in (b) there are *two* such responses!)

7.42 In the *RL* circuit of Figure 7.14(b) let $R = 100 \ \Omega$, $L = 250 \ \mu$H, and $i_S = I_m \cos 10^6 t$ V. If $i(0) = 2$ mA, find I_m for which the circuit will yield a zero transient component. What is the steady-state component? What is the forced component?

7.43 In the *RC* circuit of Figure 7.14(a) let $R = 20$ kΩ, $C = 1$ nF, and $v_S = 10 \cos \omega t$ V. (a) If $v(0) = 5$ V, find ω for which the circuit will yield a zero transient component. (b) If $\omega = 10^5$ rad/s, find $v(0)$ for which the circuit will yield a zero transient component. (c) If $v(0) = 15$ V, is it possible to adjust ω for a zero transient component?

7.44 If the ac steady-state response component of an *RC* circuit is $3.638 \cos (10^6 t - 75.96°)$ V, find the transient response component, given that $v(0) = 10$ V. What is the forcing function?

7.45 If the transient response component of an *RC* circuit is $3e^{-10^3 t}$ V, find the ac steady-state response component, given that $v(0) = 10$ V and the forcing function has a peak amplitude of 70 V.

7.46 For each of the following *complete* current responses, identify the *natural, forced, transient,* and *steady-state* components: (a) $i = 5[\cos(\omega t + \phi) + e^{-t/\tau}]$ A; (b) $i = 5 \times [\cos(\omega t + \phi) - e^{-t/\tau}]$ A; (c) $i = 5\cos(\omega t + \phi)$ A.

7.47 For each of the following *complete* voltage responses, identify the *natural, forced, transient,* and *steady-state* components: (a) $v = 5e^{-t/\tau}$ V; (b) $v = 5\cos(10^3 t - 11.31°)$ V.

TRANSIENT RESPONSE OF FIRST-ORDER CIRCUITS

In Section 7.4 we found that the response to a dc forcing function consists of a *transient component* and a *dc steady-state component.* The transient component has the same functional form as the *natural response,* which is an exponentially decaying function. The dc steady-state component has the same functional form as the forcing function, which is a constant function. We now wish to apply the mathematical concepts of Sections 7.3 and 7.4 to a variety of circuit configurations, but with greater attention to physical behavior. An engineer must always use physical insight to interpret mathematical results!

In the present chapter we concentrate on first-order circuits, that is, circuits that contain only one energy-storage element or that contain multiple elements, but in such a way that they can be reduced to a single equivalent element via series/parallel combinations. In Chapter 9 we turn our attention to second-order circuits, with special emphasis on circuits that contain one capacitance or one inductance.

We begin by examining in great physical detail how energy builds up and decays in simple capacitive and inductive circuits. In so doing, we develop a set of useful rules to facilitate the transient analysis of more complex networks.

Next, we study the *step* and *pulse responses* of the basic *R-C*, *L-R*, *C-R*, and *R-L* circuits, and observe how these circuits behave in terms of signal distortion as well as passing or blocking the dc component of an input pulse train.

We then turn to capacitive circuits using op amps. After examining the classical *integrator* and *differentiator* configurations, we use the op amp to illustrate another important systems concept, this time the concept of *root location control.* In particular, we illustrate the use of an op amp to create a root in the *right half* of the *s* plane and thus obtain a *diverging response,* a situation that cannot be achieved with a purely passive circuit.

As usual, we conclude the chapter by illustrating the use of SPICE to perform the transient analysis of circuits. The SPICE facilities introduced in this chapter are the .TRAN statement and the PULSE function.

8.1 BASIC *RC* AND *RL* CIRCUITS

In this section we take a closer look at the physical aspects of the transient behavior of basic *RC* and *RL* pairs. However, we must first state two important rules that will greatly facilitate our transient analysis of these circuits.

The Continuity Conditions

Capacitance current and voltage are related as $i_C = C \, dv_C/dt$. The faster the change in v_C, the greater i_C. If v_C were to change *instantaneously,* dv_C/dt would become infinite, and so would i_C. An infinite current, however, would require the existence of an infinite amount of power at the capacitor terminals. Since this is physically impossible, we have the following rule, known as the **voltage continuity rule** for capacitance:

Capacitance Rule 1: The *voltage across* **a capacitance cannot change instantaneously, that is, at any instant** t_0 **we must have** $v_C(t_{0+}) = v_C(t_{0-})$,

where t_{0+} is the instant *right after* t_0, and t_{0-} is the instant *right before* t_0.

Note that the continuity rule applies to v_C, not to i_C; i_C can indeed change instantaneously without violating Rule 1. We can readily visualize this rule using the water tank analogy, where water level is likened to voltage, and water flow to current. To bring about an instantaneous level change would require adding or removing water from the tank in *zero time,* a physically impossible task. Current flow can undergo brusque changes, however, as when we suddenly open a spigot to let water out of the tank.

Turning now to inductance, whose voltage and current are related as $v_L = L \, di_L/dt$, we observe that if i_L were to change *instantaneously,* di_L/dt would become infinite and so would v_L. This is again physically impossible, as it would require the existence of an infinite amount of power at the inductor terminals. Hence, we have the following **current continuity rule** for inductance:

Inductance Rule 1: The *current through* **an inductance cannot change instantaneously, that is, at any instant** t_0 **we must have** $i_L(t_{0+}) = i_L(t_{0-})$.

Inductors are also called *chokes* to reflect their tendency to oppose sudden changes in current. Note that the continuity rule applies to *inductive current,* not to *inductive voltage.* The latter can indeed undergo instantaneous changes without violating Rule 1.

DC Steady-State Behavior

If the *voltage* v_C across a capacitance is known to have stabilized as some constant value V_C, not necessarily zero, the capacitance is said to be in **dc**

steady state. Once in this state, it draws no current from the rest of the circuit because $i_C = C\, dV_C/dt = C \times 0 = 0$. This behavior is summarized as follows:

Capacitance Rule 2: In dc steady state, a capacitance behaves as an *open* circuit, that is, $i_C = 0$.

Likewise, if the current i_L *through* an inductance is known to have stabilized as some constant value I_L, not necessarily zero, the inductance is said to be in **dc steady state,** and the voltage across its terminals is $v_L = L\, dI_L/dt = L \times 0 = 0$. Consequently,

Inductance Rule 2: In dc steady state, an inductance behaves as a *short* circuit, that is, $v_L = 0$.

These rules indicate that if an energy-storage element in a circuit is known to be in dc steady state, we can replace it with an open circuit if it is a capacitance, or a short circuit if it is an inductance, and thus simplify our analysis considerably. Table 8.1 compares capacitance and inductance behavior. Note again the dual behavior of the two elements.

TABLE 8.1   Comparison of *L* and *C* Behavior

Element	C	L
DC steady-state condition	$v_C = $ constant	$i_L = $ constant
DC steady-state behavior	open circuit	short circuit
Cannot change instantaneously	v_C	i_L
Can change instantaneously	i_C	v_L

The *R-C* Circuit

In the circuit of Figure 8.1(a) assume the capacitance is initially discharged so that $q = 0$. Then, by Equation (7.3), we have $v = q/C = 0/C = 0$. Moreover, $i = (0 - v)/R = 0$.

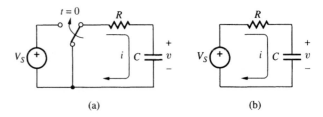

(a)   (b)

Figure 8.1   Flipping a switch to charge a capacitance, and equivalent circuit for $t \ge 0^+$.

At an instant that we arbitrarily choose as $t = 0$ we flip the switch to connect R to V_S, ending up with the situation of Figure 8.1(b). The resulting current $i = (V_S - v)/R$ will charge C toward V_S according to Equation (7.23), which now becomes

$$RC \frac{dv}{dt} + v = V_S \qquad \text{(8.1)}$$

A word of caution is necessary at this point. The instant $t = 0$ is exceptional because the switch voltage is in the process of changing from 0 to V_S and, as such, is not a single-valued function of time. To avoid any ambiguities, we assume Equation (8.1) to begin at $t = 0^+$, where 0^+ denotes the instant *just after $t = 0$*, when the switch voltage has fully attained the value V_S. Likewise, $t = 0^-$ shall denote the instant *just before $t = 0$*, when the switch voltage is still zero. With this in mind, the solution to Equation (8.1) shall also be assumed to begin at $t = 0^+$.

Such a solution is provided by Equation (7.41), but with $t = 0^+$ instead of $t = 0$. Because of the voltage continuity rule, we have $v(0^+) = v(0^-) = 0$, indicating that the natural component in Equation (7.41) is zero. The solution is thus the *forced component*. To summarize, just prior to switch activation we have

$$v(0^-) = 0 \qquad \text{(8.2a)}$$

and thereafter,

$$v(t \geq 0^+) = V_S \left(1 - e^{-t/RC}\right) \qquad \text{(8.2b)}$$

The current is readily found as $i = C\, dv/dt$. Differentiation of the preceding expressions yields

$$i(0^-) = 0 \qquad \text{(8.3a)}$$

$$i(t \geq 0^+) = \frac{V_S}{R} e^{-t/RC} \qquad \text{(8.3b)}$$

The responses are shown in Figure 8.2. It is interesting to note that v is continuous but i is discontinuous at the origin. This does not violate Rule 1, however, as continuity applies to voltage, not to current. Because of its abrupt jump in magnitude, the current waveform is referred to as a **spike.**

Both transients of Figure 8.2 are governed by the time constant $\tau = RC$. Increasing R reduces the charging current, and increasing C increases the amount of charge that needs to be transferred. In either case the transients will be slowed down. Conversely, decreasing R results in faster transients because the charging current is increased. In the limit $R \to 0$, C would try to charge instantaneously.

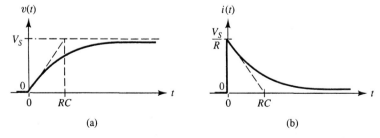

(a)   (b)

Figure 8.2 Voltage and current responses of the circuit of Figure 8.1(a).

But to do so would require an infinitely large current spike from V_S, which we know to be physically impossible. In a practical situation current is limited by the internal resistance of the source and the connecting wires, indicating that charge buildup, however rapid, cannot be instantaneous. The spark that we observe in the laboratory when we connect a capacitor directly across a battery attests to the large current spike involved.

▶ **Example 8.1**

In the circuit of Figure 8.1 let $V_S = 10$ V, $R = 1$ MΩ, and $C = 1$ μF. (a) How long does it take to charge C to 4 V? (b) To what value must we change R for this time to be 0.2 s?

Solution

(a) We have $RC = 10^6 \times 10^{-6} = 1$ s. By Equation (8.2b) we must have $4 = 10(1 - e^{-t/1})$, or $t = \tau \ln (5/3) = 1 \times 0.511 = 0.511$ s.

(b) To decrease the time from 0.511 s to 0.2 s we must decrease R in proportion; $R_{\text{new}} = (0.2/0.511) R_{\text{old}} = 0.392 \times 1$ M$\Omega = 392$ kΩ.   ◀

Exercise 8.1   In the circuit of Figure 8.1, let $V_S = 10$ V and $R = 20$ kΩ. If v is observed with the oscilloscope and it is found that it takes 30.5 μs to rise to 5 V, find C.

ANSWER   2.2 nF.

Suppose that after the capacitance of Figure 8.1(a) has fully charged to V_S, the switch is flipped back down at an instant that we again choose as $t = 0$ for convenience. This is shown in Figure 8.3(a). With the source out of the way, we end up with the *source-free* circuit of Figure 8.3(b).

By the voltage continuity rule we must have $v(0^+) = v(0^-) = V_S$. Consequently, at $t = 0^+$, the top plate holds the charge $+CV_S$ and the bottom

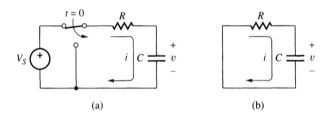

(a)   (b)

Figure 8.3 Flipping a switch to discharge a capacitance, and equivalent circuit for $t \geq 0^+$.

plate the charge $-CV_S$. Moreover, the initial stored energy in the capacitance is $w(0^+) = (1/2)CV_S^2$. Thanks to this energy, C will be able to sustain a nonzero voltage even though V_S has been switched out of the circuit. This voltage, in turn, will cause R to draw current and gradually discharge C. Though the reference direction is still shown clockwise for consistency, the discharge current actually flows counterclockwise, out of the top plate, through the resistance, and back into the bottom plate.

The discharge process is governed by Equation (7.23), which now becomes

$$RC\frac{dv}{dt} + v = 0 \qquad\qquad \textbf{(8.4)}$$

Rewriting as $dv/dt = -v/(RC)$ indicates that v decreases at a rate proportional to v itself. Initially, when v is large, we have a proportionally rapid discharge rate. As the discharge progresses, the rate decreases in proportion, making the discharge slower and slower. This process, similar to emptying a water tank by opening a spigot at the bottom, is an *exponential decay*. In fact, it is the *natural response* predicted by Equation (7.33). We thus have

$$v(0^-) = V_S \qquad\qquad \textbf{(8.5a)}$$

$$v(t \geq 0^+) = V_S e^{-t/RC} \qquad\qquad \textbf{(8.5b)}$$

The current is readily found as $i = C\,dv/dt$. Differentiating, we get

$$i(0^-) = 0 \qquad\qquad \textbf{(8.6a)}$$

$$i(t \geq 0^+) = -\frac{V_S}{R}e^{-t/RC} \qquad\qquad \textbf{(8.6b)}$$

The responses are shown in Figure 8.4.

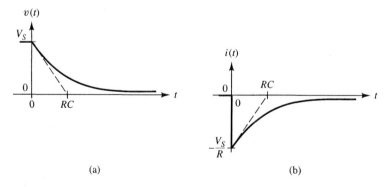

(a)   (b)

Figure 8.4 Voltage and current responses of the circuit of Figure 8.3(a).

The rate of decay is governed by the time constant RC. The larger R or C, the slower the decay. In the limit $R \to \infty$ the decay would become infinitely slow because there would be no current path for C to discharge. Hence, C would retain its initial voltage V_S indefinitely, thus providing a *memory function*.

FAMOUS PHYSICISTS
JOHN BARDEEN

Two-Time Winner

In all the time the Nobel Prize in Physics has been in existence, only one person has received it twice: John Bardeen. In today's slang, we are definitely talking "rocket scientist" here, even though Bardeen's work was never with NASA. His 1972 Prize, shared with Leon Cooper and Robert Schrieffer, was for his work on superconductivity. His 1956 Prize, shared with Walter Brattain and William Shockley, was for the invention of the transistor, which is the work we are most interested in here.

Bardeen was born in Madison, Wisconsin, in 1908 and was a child prodigy. Because of his whiz-kid performance in school, the nine-year-old child was promoted from third grade directly to seventh, and he graduated high school at age 15. He then earned his BS and MS degrees in electrical engineering from the University of Wisconsin. After a few not-very-exciting years as a geophysicist for Gulf Oil, he returned to graduate school, at Princeton this time, and was awarded his doctorate in mathematical physics in 1936. Service in Naval Ordnance during World War II was followed by a position at Bell Labs in 1945.

The scientific world Bardeen entered in 1945 was very different from that of the early workers in electricity (or any other field, for that matter). Research departments were becoming a major part of bigger and bigger corporate behemoths like AT&T and GE,

John Bardeen (shown here), together with William Shockley and Walter Brattain, received a joint Nobel Prize in 1956 for their invention of the transistor. In 1972, Bardeen (together with Leon Cooper and Robert Schrieffer) received a second Nobel for work on superconductivity. (Courtesy of University of Illinois)

and scientific research was done either in these corporate settings or in the academic arena. By the 1950s the days of the lone inventor, struggling to find enough money to buy equipment plus one meal a day, were pretty much over. Figures like Lee De Forest fending off patent suits or Thomas Edison filing them had given way to scientists working as members of teams, taking home regular paychecks, and leaving any legal issues to the corporate lawyers. Thus, Bardeen did not have to track down a warehouse to rent, as Tesla had, or open a laboratory in Newark like Edison. Instead he moved his family to Summit, New Jersey, close by Bell Labs in neighboring Murray Hill, and went to work not as a lone genius but as part of a team. The group he joined was headed by Shockley and was newly formed to study solid-state physics.

Aiming to make an amplifying device using semiconductor material, Shockley had the idea that a positive electrode placed near but not in physical contact with a semiconducting material would pull electrons to the surface of the semiconductor (the "field effect") and thus increase its conducting ability. Such enhanced conductivity was not observed in any of the experiments run by Shockley's team, however, and Bardeen and his colleague Walter Brattain were assigned the task of figuring out why. To test one of their ideas, Bardeen and Brattain designed an experiment that, via the field effect, should have decreased conductivity in the semiconductor. What happened, though, because the gold electrode they were using came into contact with the germanium semiconductor while a voltage was being applied, was that the conductivity *increased:* Bardeen and Brattain had come up with a new device! Their discovery was that an electrode at a semiconductor surface doesn't pull electrons from the interior to the surface but rather pushes positive charge carriers ("holes") from near the surface into the interior, a phenomenon called "hole injection." Seeing that injected holes increased the conductivity at a second electrode placed in contact with the same semiconductor and biased with an opposite voltage, Bardeen and Brattain designed a device in which the two electrodes were very close together. In other words, they made the first injection bipolar transistor, in the form of a point-contact "emitter of holes" and a nearby point-contact "collector of holes."

Thus from Shockley's original quest—an amplifying semiconductor device based on the electric field effect and not realized until much later—was born the hole injection device of Bardeen and Brattain, the prototypical bipolar transistor. The Nobel Prize was awarded to these two men for this most significant discovery of the century and to Shockley for the *pn* junction version of the transistor he developed later.

Once AT&T executives were given a demonstration of how this new device could amplify a signal, in December 1947, work on applications began, a first task being to find a name. The prize went to Bell Labs colleague John Pierce. Because his work at that time was with vacuum tubes, Pierce was used to thinking in terms of transconduction. Realizing that the new invention revolved around the opposite property, transresistance, he started free-associating names that would tie in with "resistor," "thermistor," and so forth, and from "transresistance" came "transistor."

After this epochal work, Bardeen turned the light of his genius on another solid-state phenomenon, superconductivity ("returned," to be more accurate

about it, because he had already been working in this field before his transistor days). He left Bell Labs in 1951 and accepted a professorship in electrical engineering at the University of Illinois, a post he held until his death in 1991.

Bardeen was, to use an old-fashioned word, a gentleman—quiet, modest, and respected and admired by his fellow scientists. (*Too* quiet, some of his graduate students would say, and thus were born the

> *"Bardeen was, to use an old-fashioned word, a gentleman—quiet, modest, and respected and admired by his fellow scientists."*

nicknames "Silent John" and "Whispering John.") He was, in the words of Nick Holonyak, Jr., who was first Bardeen's student at the University of Illinois and then his colleague there, "the nicest man you ever met." One story often told about Bardeen has him going home after work on that day in 1947 when the first successful transistor demonstration was made, sitting down at the kitchen table, and softly saying to his wife, "We discovered something today."

Besides science, Bardeen's great passion was golf. Colleagues report that on the golf course, the quiet, self-effacing Dr. Jekyll scientist was wont to be replaced by, if not so extreme a character as Mr. Hyde, then at least by a more exuberant and expressive individual, liable to be found shouting—"Did you see that? Did you see that shot I just made!"—after a particularly good play. A golf ball containing a Sony transistor radio, presented to Bardeen on his 60th birthday by George Hatoyama, Sony's first research director, was a treasured possession for the remainder of his life.

▶ **Example 8.2**

In the circuit of Figure 8.3 let $V_S = 10$ V, $R = 1$ MΩ, and $C = 1$ μF. Find the time at which the energy in the capacitance has decayed to 50% of its initial value.

Solution

We need to find the time t at which $w(t) = (1/2)w(0^+)$, or $(1/2)Cv^2(t) = 1/2 \times (1/2)CV_S^2$. This yields $v^2(t) = (1/2)V_S^2$, or $v = \sqrt{0.5}V_S$. But, by Equation (7.34) we have $t = -\tau \ln(v/V_S) = -(10^6 \times 10^{-6}) \times \ln\sqrt{0.5} = 0.347$ s.

◀

▶ **Example 8.3**

Referring to Figure 8.3, verify that the energy available from C at the beginning of the transient equals that dissipated by R during the entire transient.

Solution

The initial energy is $w_C(0^+) = (1/2)CV_S^2$. The energy dissipated by R from $t = 0^+$ to $t = \infty$ is

$$w_R = \int_{0^+}^{\infty} p_R(\xi)\,d\xi = \int_{0^+}^{\infty} \frac{v^2(\xi)}{R}\,d\xi = \frac{V_S^2}{R}\int_{0^+}^{\infty} e^{-2\xi/RC}\,d\xi$$

$$= \frac{V_S^2}{R}\left(-\frac{RC}{2}e^{-2\xi/RC}\Big|_{0^+}^{\infty}\right) = \frac{1}{2}CV_S^2$$

that is, $w_R = w_C(0^+)$.

◀

The *C-R* Circuit

Interchanging the roles of the capacitance and resistance in the R-C circuit of Figure 8.1(a) turns it into the C-R circuit of Figure 8.5(a). Though the current response is not affected by this interchange, the voltage response will be different because we are now observing it across R rather than C.

Assume prior to switch activation the circuit is in steady state so that C, by Rule 2, acts as an open circuit. Then, by Ohm's Law, $v(0^-) = Ri(0^-) = R \times 0 = 0$, indicating that the voltage across C must also be zero.

As the switch is flipped up, the voltage of the *left plate* jumps from 0 V to V_S. By Rule 1, the voltage *across* C right after switch activation must still be 0 V. For this to be possible, the voltage of the *right plate* must also jump from 0 V to V_S, so that $v(0^+) = V_S$. With a nonzero voltage drop, R will draw

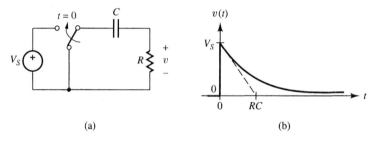

Figure 8.5   *C-R* circuit and its forced response to a dc voltage V_S.

current and cause C to charge exponentially. In summary,

$$v(0^-) = 0 \tag{8.7a}$$

$$v(t \geq 0^+) = V_S e^{-t/RC} \tag{8.7b}$$

This response is shown in Figure 8.5(b). We observe that v now exhibits a discontinuity at $t = 0$. This does not violate Rule 1, however, because this rule applies to the voltage *across* the capacitance, not to the individual plate voltages.

If the switch is left in the up position long enough to allow v to fully decay to 0 V, the capacitance will achieve its dc steady state, in which the voltage at the left plate will be V_S and that at the right plate will be 0 V. Consequently, the final stored energy will be $w_C = (1/2)CV_S^2$.

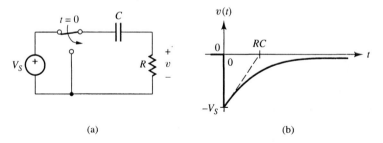

Figure 8.6   *C-R* circuit and its natural response.

Suppose now the switch is flipped back down, at a time that we again choose as $t = 0$, as shown in Figure 8.6(a). This causes the voltage of the *left plate* to jump from V_S to 0 V, and that of the *right* plate to jump from 0 V to $-V_S$, by the continuity rule. Consequently, $v(0^+) = -V_S$, causing R to draw current and discharge C,

$$v(0^-) = 0 \tag{8.8a}$$

$$v(t \geq 0^+) = -V_S e^{-t/RC} \tag{8.8b}$$

This response is shown in Figure 8.6(b). Voltage is again discontinuous at $t = 0$, but without violating Rule 1. We also note the creation of a *negative* voltage transient using a *positive* voltage source. This feature is exploited in the design of certain types of *voltage inverters*.

The *RL* Circuit

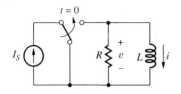

Figure 8.7 Flipping a switch to investigate the forced response of the *RL* circuit.

In the circuit of Figure 8.7 assume the inductance has zero initial stored energy so that, by Equation (7.19), $i(0^-) = 0$. Flipping up the switch connects the source I_S to the *RL* pair, causing current and, hence, energy to build up in the inductance. For $t \geq 0^+$, this process is governed by Equation (7.24),

$$\frac{L}{R}\frac{di}{dt} + i = I_S \tag{8.9}$$

By the current continuity rule we must have $i(0^+) = i(0^-) = 0$. Consequently, the solution to Equation (8.9) is the *forced response*

$$i(0^-) = 0 \tag{8.10a}$$

$$i(t \geq 0^+) = I_S\left(1 - e^{-t/(L/R)}\right) \tag{8.10b}$$

The voltage is found as $v = L\,di/dt$. Differentiation yields

$$v(0^-) = 0 \tag{8.11a}$$

$$v(t \geq 0^+) = RI_S e^{-t/(L/R)} \tag{8.11b}$$

These responses are shown in Figure 8.8. Comparison with the capacitive responses of Figure 8.2 further substantiates the duality principle. We also note that i is continuous at the origin, but v exhibits a spike. This does not violate Rule 1, however, as the latter applies to inductive current, not to voltage.

Both responses are governed by the time constant L/R. Increasing L *slows down* the responses because it takes longer to build up energy in a larger inductance. However, increasing R now *speeds up* the responses. We justify this by noting that the voltage response is scaled by R. Consequently, a larger R results in a larger v and, hence, in a faster *rate* of current buildup, since $di/dt = v/L$.

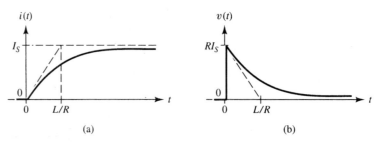

Figure 8.8 Voltage and current responses of the circuit of Figure 8.7.

In the limit $R \rightarrow \infty$, current buildup in the inductance would be instantaneous and it would be accompanied by an infinitely large and infinitely brief voltage spike (a physical impossibility as we know). In a practical circuit this spike will be limited by the finite internal resistance of the source, confirming that current buildup in an inductance, however rapid, cannot be instantaneous.

Exercise 8.3

    (a) In the circuit of Figure 8.7 let $I_S = 1$ mA, $R = 10$ kΩ, and $L = 10$ mH. Assuming zero initial stored energy in the inductance, find the *amplitude* and *duration* of the ensuing voltage spike (take the duration to be about five time constants.)

    (b) Repeat, but for $R = 1$ MΩ. Compare with part (a); comment.

ANSWER   (a) 10 V, 5 μs; (b) 1000 V, 50 ns.

If the switch is left in the up position for a sufficiently long time, all of I_S will eventually be diverted through L and none through R, thus causing v to collapse, by Ohm's Law, to zero. As we know, this is the dc steady state for the inductance. The final stored energy in the inductance is $w_L = (1/2)LI_S^2$.

Suppose now the switch is flipped back down at an instant again chosen as $t = 0$, as depicted in Figure 8.9. Once more, we end up with a *source-free* circuit. By the continuity rule we have $i(0^+) = i(0^-) = I_S$, indicating that after switch activation L will continue to draw current. With the source out of the circuit, this current will have to come from R, so that $v(0^+) = -RI_S$. As a consequence of this voltage, the inductance current will start to decay, producing the *natural response*

$$i(0^-) = I_S \tag{8.12a}$$

$$i(t \geq 0^+) = I_S e^{-t/(L/R)} \tag{8.12b}$$

The voltage is found as $v = L\,di/dt$. Differentiation yields

$$v(0^-) = 0 \tag{8.13a}$$

$$v(t \geq 0^+) = -RI_S e^{-t/(L/R)} \tag{8.13b}$$

Both responses are shown in Figure 8.10.

The responses are governed by the time constant L/R. The smaller R, the slower the decay. In the limit $R \rightarrow 0$ the decay would become infinitely slow because L would have no means to dissipate its energy. Thus, the current I_S would flow indefinitely inside the inductance, and we would have a memory effect. In practice, the nonzero resistance of the winding and connections will cause the inductance current to decay to zero eventually. This is similar to capacitor leakage, except that the parasitic resistance of a practical inductor is a far more serious limitation than is the parallel leakage of a practical capacitor. This explains why inductors are never used for long-term energy storage.

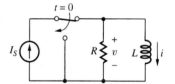

Figure 8.9 Flipping a switch to investigate the natural response of the *RL* circuit.

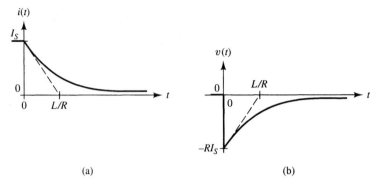

Figure 8.10 Voltage and current responses for the circuit of Figure 8.9.

Conversely, the larger R is, the quicker the decay, and the larger and briefer the voltage spike. If an attempt is made to open-circuit an inductor or a transformer carrying a nonzero current, an arc may develop across the switch whereby the stored energy is dissipated in ionizing the air along the path of the arc. This feature is exploited in automobile ignition systems, where the coil current is periodically interrupted by the distributor to develop an arc across a spark plug.

Exercise 8.4   Referring to Figure 8.9, verify that the energy available from L at the beginning of the transient equals the energy dissipated by R during the entire transient.

Exercise 8.5

    (a) Sketch and label the current through the resistance in the circuit of Figure 8.7. Be neat and precise.

    (b) Repeat, but for the circuit of Figure 8.9.

8.2 TRANSIENTS IN FIRST-ORDER NETWORKS

We now generalize the techniques of the previous section to the case in which an energy-storage element is part of a more complex circuit, or network. Subjecting the network to a sudden change, such as the activation of a switch, will cause the energy-storage element to respond with a transient. This, in turn, will force all other voltage and current variables in the circuit to readjust accordingly, since Kirchhoff's laws must be satisfied at all times.

    If the circuit surrounding the energy-storage element consists of resistances and sources and is thus *linear,* all voltage and current transients in the circuit will be *exponential* and will be governed by the *same* time constant τ. Assuming $t = 0$ as the instant of occurrence of the sudden change, each transient will take on the form of Equation (7.42), namely,

$$y(t \geq 0^+) = [y(0^+) - y(\infty)]e^{-t/\tau} + y(\infty) \qquad \textbf{(8.14)}$$

This expression is uniquely defined once we know $y(0^+)$, $y(\infty)$, and τ. For completeness, we also want to know $y(0^-)$ in order to tell whether there is any discontinuity at $t = 0$. In the following we shall assume that by the time the sudden change is applied, the circuit has had *sufficient time to reach its dc steady state*. Following is a systematic procedure for finding the above parameters:

(1) Find $y(0^-)$, the steady-state value of the response *preceding* the sudden change. This is achieved by examining the circuit *before* the change, when the capacitance acts as an *open circuit* and the inductance as a *short circuit*. Also record the voltage $v_C(0^-)$ *across the capacitance*, or the *current $i_L(0^-)$ through the inductance*, which will be needed in the next step.

(2) Find $y(0^+)$, the value of the response *just after* the sudden change. This is achieved by replacing the capacitance with a *voltage source* of value $v_C(0^+) = v_C(0^-)$, or the inductance with a *current source* of value $i_L(0^+) = i_L(0^-)$. Note that these replacements hold only at the instant $t = 0^+$, after which $v_C(t)$ or $i_C(t)$ will indeed change to reflect energy buildup or decay. Clearly, the equivalent circuit of the present step is only a *snapshot* of our network at $t = 0^+$.

(3) Find $y(\infty)$, the steady-state value of the response in the limit $t \to \infty$. This is achieved by examining the circuit long after the sudden change, when the capacitance will again act as an *open circuit* and the inductance as a *short circuit*.

(4) Find the *equivalent resistance* R_{eq} seen by the energy-storage element *during the transient*. Then use, respectively,

$$\tau = R_{eq}C \qquad \textbf{(8.15a)}$$
$$\tau = L/R_{eq} \qquad \textbf{(8.15b)}$$

R_{eq} is found using Method 2 of Section 5.2, though in particular cases simple inspection may suffice.

In general, the procedure uses *three* equivalent circuits, one for each of the first three steps, and possibly a *fourth* circuit to find R_{eq}. Once we have all parameters in hand, we substitute them into Equation (8.14) to obtain the response as a function of time.

It is often of interest to know how long it takes for a response to swing from a given value $y(t_1)$ to a value $y(t_2)$. By Equation (8.14) we can write $y(t_1) - y(\infty) = [y(0^+) - y(\infty)]e^{-t_1/\tau}$ and $y(t_2) - y(\infty) = [y(0^+) - y(\infty)]e^{-t_2/\tau}$.

Dividing the two equations pairwise yields $[y(t_1) - y(\infty)]/[y(t_2) - y(\infty)] = e^{(t_2 - t_1)/\tau}$. Taking the natural logarithm of both sides and simplifying we obtain

$$t_2 - t_1 = \tau \ln \frac{y(t_1) - y(\infty)}{y(t_2) - y(\infty)}$$ **(8.16)**

Let us now apply the procedure to actual examples.

Capacitive Examples

▶ Example 8.4

Sketch and label $v(t)$ for the circuit of Figure 8.11(a).

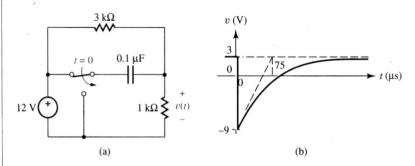

(a)   (b)

Figure 8.11 Circuit of Example 8.4 and its response.

Solution

We need to find $v(0^-)$, $v(0^+)$, $v(\infty)$, and τ. The sequence is illustrated in Figure 8.12, where the units are [V], [mA], and [kΩ].

Figure 8.12(a) depicts the steady state prior to switch activation. With C acting as an open circuit, we can use the voltage divider formula and write $v(0^-) = [1/(3 + 1)]12 = 3$ V. Moreover, using KVL, $v_C(0^-) = 12 - 3 = 9$ V.

Figure 8.12(b) provides a snapshot at $t = 0^+$. By the continuity rule, the voltage *across* the capacitance *just after* switch activation must still be 9 V *regardless* of the external circuit conditions. We model this behavior with a 9-V voltage source, as shown. Then, by KVL, $v(0^+) = -9$ V.

Figure 8.12(c) depicts the steady state after the transient has died out. The voltage divider formula again yields $v(\infty) = 3$ V.

From Figure 8.12(c) we also note that the resistance seen by the capacitance is $R_{eq} = 3 \| 1 = 750$ Ω. Thus, $\tau = RC = 750 \times 10^{-7} = 75$ μs. Substituting into Equation (8.14) yields

$$v(0^-) = 3 \text{ V}$$

$$v(t \geq 0^+) = 3 - 12e^{-t/(75 \ \mu s)} \text{ V}$$

The sketch of v versus t is shown in Figure 8.11(b).

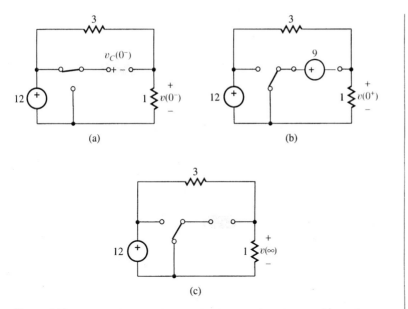

Figure 8.12 Circuit of Figure 8.11(a) for (a) $t = 0^-$, (b) $t = 0^+$, and (c) $t \to \infty$.

Exercise 8.6 Repeat Example 8.4 if the switch, after having been in the *down* position for a sufficiently long time, is flipped *up* at $t = 0$.

ANSWER $v(0^-) = 3$ V, $v(t \geq 0^+) = 3 + 12e^{-t/(75 \ \mu s)}$ V.

▶ **Example 8.5**

Find $v(t)$ in the circuit of Figure 8.13.

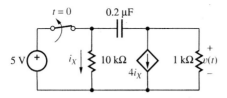

Figure 8.13 Circuit of Example 8.5.

Solution

The sequence is illustrated in Figure 8.14, where the units are [V], [mA], and [kΩ]. In Figure 8.14(a) we have, by Ohm's Law, $v(0^-) = -1 \times 4i_X$ and $i_X = 5/10 = 0.5$ mA. Hence, $v(0^-) = -2$ V. Moreover, KVL yields $v_C(0^-) = 5 - (-2) = 7$ V.

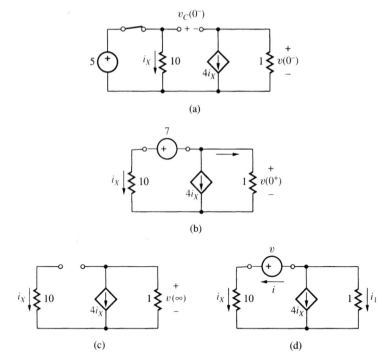

Figure 8.14 Circuit of Figure 8.13 for (a) $t = 0^-$, (b) $t = 0^+$, and (c) $t \to \infty$; (d) finding R_{eq}.

Opening the switch removes the 5-V source from the picture, as shown in the remaining figures. KCL at the supernode surrounding the 7-V source in the snapshot circuit of Figure 8.14(b) yields $0 = i_X + 4i_X + v(0^+)/1$, or $v(0^+) = -5i_X$. KVL around the outer loop clockwise yields $10i_X = 7 + v(0^+)$ or $i_X = [7 + v(0^+)]/10$. Eliminating i_X and solving for $v(0^+)$ yields $v(0^+) = -7/3$ V.

In Figure 8.14(c) it is apparent that $i_X = 0$, so $4i_X = 0$ and, hence, $v(\infty) = 0$.

To find the resistance R_{eq} seen by C during the transient, replace C with a test source as in Figure 8.14(d). KVL around the outer loop clockwise yields $10i_X = v + 1i_1$. KCL at the *supernode* surrounding the test source yields $0 = i_X + 4i_X + i_1$. Eliminating i_1 yields $i_X = v/15$. But, $i = i_X$, so that $i = v/15$. Hence, $R_{eq} = v/i = v/(v/15) = 15$ kΩ. Finally, $\tau = R_{eq}C = 15 \times 10^3 \times 0.2 \times 10^{-6} = 3$ ms. Substituting the calculated data into Equation (8.14),

$$v(0^-) = -2 \text{ V}$$

$$v(t \ge 0^+) = -\frac{7}{3}e^{-t/(3 \text{ ms})} \text{ V}$$

◀

Exercise 8.7   Repeat Example 8.5 if the switch, after having been *open* for a sufficiently long time, is *closed* at $t = 0$. At what time does $v(t)$ reach 0 V?

ANSWER $v(0^-) = 0$, $v(t \geq 0^+) = 7e^{-t/(0.2 \text{ ms})} - 2$ V, 250.6 μs.

▶**Example 8.6**

(a) Sketch and label $i(t)$ in the circuit of Figure 8.15(a).

(b) Find the time at which $i = 0$.

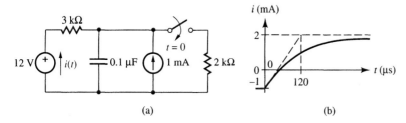

(a)   (b)

Figure 8.15 Circuit of Example 8.6 and its response.

Solution

(a) After gaining sufficient experience we can visualize the intermediate steps mentally.

At $t = 0^-$ the switch is open and C acts as an open. KCL at the upper center node yields $i(0^-) + 1$ mA $= 0$, or $i(0^-) = -1$ mA.

Since the capacitance voltage does not change during switch closure, the voltage across the 3-kΩ resistance also remains unchanged, thus yielding $i(0^+) = i(0^-) = -1$ mA.

For $t \rightarrow \infty$, C acts again as an open. Considering that the switch is now closed, we can apply the superposition principle, along with Ohm's Law and the current divider formula, and write $i(\infty) = 12/(3 + 2) - [2/(3 + 2)]1 = 2$ mA.

The value of R_{eq} during the transient is found by suppressing the sources. Considering that the switch is closed, we have $R_{eq} = 3 \parallel 2 = 1.2$ kΩ. Hence, $\tau = R_{eq}C = 1.2 \times 10^3 \times 0.1 \times 10^{-6} = 120$ μs. The sketch of i versus t is shown in Figure 8.15(b).

(b) The time it takes for $i(t)$ to reach 0 A is, by Equation (8.16), $t = 120 \ln[(-1 - 2)/(0 - 2)] = 48.66$ μs.   ◀

Exercise 8.8   Suppose the switch in the circuit of Figure 8.15(a), after having been *closed* for a sufficiently long time, is *opened* at $t = 0$. Find the voltage $v(t)$ across the current source, positive reference at the top. At what time do we have $v = 10$ V?

ANSWER $v(0^-) = 6$ V, $v(t \geq 0^+) = 15 - 9e^{-t/(0.3 \text{ ms})}$ V, 176.3 μs.

Inductive Examples

▶ Example 8.7

Sketch and label $v(t)$ for the circuit of Figure 8.16(a).

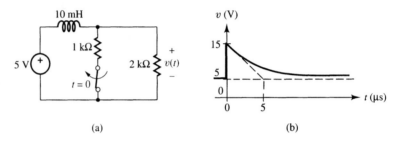

(a)   (b)

Figure 8.16 Circuit of Example 8.7 and its response.

Solution

We need to find $v(0^-)$, $v(0^+)$, $v(\infty)$, and τ. The sequence is illustrated in Figure 8.17, where the units are [V], [mA], and [kΩ].

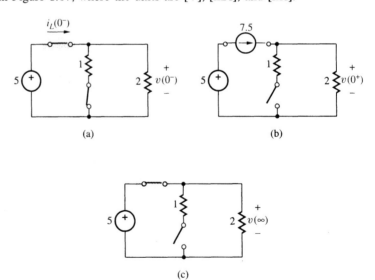

Figure 8.17 Circuit of Figure 8.16(a) for (a) $t = 0^-$, (b) $t = 0^+$, and (c) $t \rightarrow \infty$.

Figure 8.17(a) depicts the steady state prior to switch activation. With L acting as a short circuit, $v(0^-) = 5$ V. Moreover, $i_L(0^-) = 5/1 + 5/2 = 7.5$ mA.

Figure 8.17(b) provides a snapshot at $t = 0^+$, with L replaced by a 7.5-mA source to reflect current continuity. Using Ohm's Law, $v(0^+) = 2 \times 7.5 = 15$ V.

Figure 8.17(c) depicts the steady state after the transient has died out. By inspection, $v(\infty) = 5$ V.

From Figure 8.17(c) we also note that the resistance seen by L during the transient is $R_{eq} = 2$ kΩ. Then, $\tau = L/R_{eq} = 10 \times 10^{-3}/(2 \times 10^3) = 5$ μs. Substituting into Equation (8.14) yields

$$v(0^-) = 5 \text{ V}$$

$$v(t \geq 0^+) = 5 + 10e^{-t/(5 \ \mu s)} \text{ V}$$

The sketch of v versus t is shown in Figure 8.16(b).   ◀

Exercise 8.9   Suppose the switch in the circuit of Figure 8.16(a), after having been *open* for a sufficiently long time, is *closed* at $t = 0$. Find the current $i(t)$ supplied by the 5-V source.

ANSWER $i(0^-) = 2.5$ mA, $i(t \geq 0^+) = 5[1.5 - e^{-t/(15 \ \mu s)}]$ mA.

▶ **Example 8.8**

Sketch and label $v(t)$ in the circuit of Figure 8.18(a); hence, find the instant when $v(t) = 10$ V.

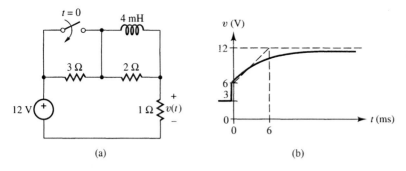

(a)   (b)

Figure 8.18 Circuit of Example 8.8 and its response.

Solution

Refer to Figure 8.19, where the units are [V], [A], and [Ω]. In Figure 8.19(a) we have, by the voltage divider formula, $v(0^-) = [1/(3 + 1)]12 = 3$ V. Also, $i_L(0^-) = 12/(3 + 1) = 3$ A.

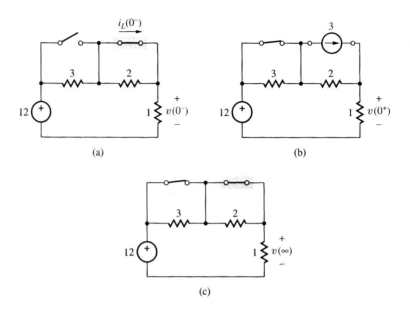

Figure 8.19 Circuit of Figure 8.18(a) for (a) $t = 0^-$, (b) $t = 0^+$, and (c) $t \to \infty$.

In Figure 8.19(b) we apply the superposition principle, along with the voltage divider formula and Ohm's Law, to write $v(0^+) = [1/(2 + 1)]12 + 3 \times (2 \parallel 1) = 6$ V.

In Figure 8.19(c) we have $v(\infty) = 12$ V. Moreover, the resistance seen by L is $R_{eq} = 1 \parallel 2 = 2/3$ Ω. Consequently, $\tau = L/R_{eq} = 4 \times 10^{-3}/(2/3) = 6$ ms. Substituting into Equation (8.14),

$$v(0^-) = 3 \text{ V}$$

$$v(t \geq 0^+) = 12 - 6e^{-t/(6 \text{ ms})} \text{ V}$$

The sketch of v versus t is shown in Figure 8.18(b).

The amount of time it takes for $v(t)$ to reach 10 V is, by Equation (8.16),

$$t = 6\ln\frac{6 - 12}{10 - 12} = 6.592 \text{ ms}$$

Exercise 8.10 Find the power $p(t) = v(t)i(t)$ delivered by the source in the circuit of Figure 8.18(a) if the switch, after having been *closed* for a sufficiently long time, is *opened* at $t = 0$.

ANSWER $p(0^-) = 144$ W, $p(t \geq 0^+) = 36\left(1 + e^{-t/(3 \text{ ms})}\right)$ W.

▶ **Example 8.9**

Find $v(t)$ in the circuit of Figure 8.20.

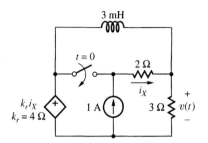

Figure 8.20 Circuit of Example 8.9.

Solution

Refer to the sequence of Figure 8.21, where the units are [V], [A], and [Ω]. In Figure 8.21(a) we have $v(0^-) = 4i_X$, where $i_X = 1$ A. Hence, $v(0^-) = 4$ V. Moreover, KCL at the upper rightmost node yields $i_L(0^-) + i_X = v(0^-)/3$, or $i_L(0^-) = v(0^-)/3 - i_X = 4/3 - 1 = 1/3$ A.

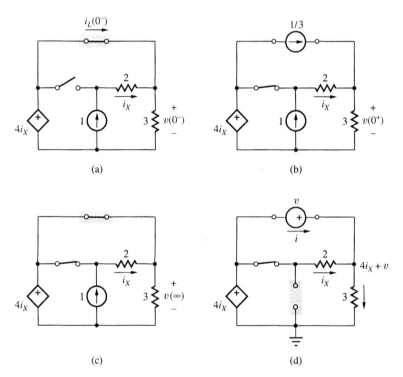

Figure 8.21 Circuit of Figure 8.20 for (a) $t = 0^-$, (b) $t = 0^+$, and (c) $t \to \infty$; (d) finding R_{eq}.

By the continuity rule, the inductance current at $t = 0^+$ is still 1/3 A, as shown in Figure 8.21(b). KCL at the upper rightmost node yields $1/3 + i_X = v(0^+)/3$, where $i_X = [4i_X - v(0^+)]/2$. Eliminating i_X yields $v(0^+) = -2$ V.

In Figure 8.21(c) we have $v(\infty) = 4i_X$, where $i_X = 0$ because the 2-Ω resistance is now short-circuited. Hence, $v(\infty) = 0$.

Suppressing the independent source and applying a test voltage as in Figure 8.21(d), we have, by KCL at the upper rightmost node, $i + i_X = (4i_X + v)/3$. But, by Ohm's Law, $i_X = -v/2$. Eliminating i_X

yields $i = v/6$. Thus, $R_{eq} = v/i = 6\ \Omega$, $\tau = 3 \times 10^{-3}/6 = 0.5$ ms, and

$$v(0^-) = 4\ V$$
$$v(t \geq 0^+) = -2e^{-t/(500\ \mu s)}\ V$$

◀

Exercise 8.11 Repeat Example 8.9, but with the two sources interchanged with each other.

ANSWER $v(0^-) = -6$ V, $v(t \geq 0^+) = -6e^{-t/(500\ \mu s)}$ V.

Exercise 8.12 Referring to Figure 8.13, find $v(t)$ if the capacitance is replaced by a 30-mH inductance.

ANSWER $v(0^-) = -5$ V, $v(t \geq 0^+) = 35e^{-t/(2\ \mu s)}$ V.

Exercise 8.13 Referring to Figure 8.20, find $v(t)$ if the inductance is replaced by a 1-μF capacitance.

ANSWER $v(0^-) = 3$ V, $v(t \geq 0^+) = 1e^{-t/(6\ \mu s)}$ V.

8.3 STEP, PULSE, AND PULSE-TRAIN RESPONSES

The activation of a switch is not the only form of effecting a sudden change in a circuit. Another common form is by means of a *step signal,* that is, a signal that changes *abruptly* from one value to another. As we proceed we shall see that the step response provides precious information about the dynamic characteristics of a circuit.

If a positive-going step is followed at some later instant by a negative-going step of equal amplitude, the resulting signal is the *pulse.* The pulse response is of great interest in computer electronics, where information is represented by means of pulses.

Step Response of *R-C* and *L-R* Circuits

Referring to Figure 8.22, we note that in dc steady state both circuits yield $v_O = v_I$. This is so because in this state C acts as an open, so that $v_O = v_I - Ri = v_I - R \times 0 = v_I$ in the R-C circuit; moreover, L acts as a short, so that $v_O = v_I - v_L = v_I - 0 = v_I$ in the L-R circuit.

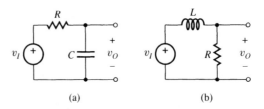

(a)   (b)

Figure 8.22 The *R-C* and *L-R* circuits.

If we apply a *positive-going* step of amplitude V_m, each circuit will yield $v_O(0^-) = 0$ and $v_O(\infty) = V_m$. Moreover, by the continuity rules, $v_O(0^+) = 0$ in both cases. Substituting into Equation (8.14), we have

$$v_O(0^-) = 0 \qquad \text{(8.17a)}$$

$$v_O(t \geq 0^+) = V_m\left(1 - e^{-t/\tau}\right) \qquad \text{(8.17b)}$$

where the two circuits have, respectively,

$$\tau = RC \qquad \text{(8.18a)}$$

$$\tau = L/R \qquad \text{(8.18b)}$$

This response is shown in Figure 8.23(a).

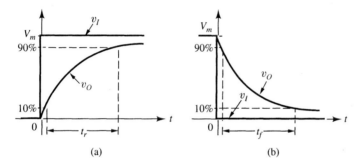

Figure 8.23 Response of the *R-C* and *L-R* circuits to (a) a positive-going step and (b) a negative-going step.

A measure of how rapidly a circuit responds to a positive-going step is the **rise time** t_r, defined as the amount of time it takes for the response to rise from 10% to 90% of its final value. Using Equation (8.16) we get $t_r = \tau \ln\left[(0.1V_m - V_m)/(0.9V_m - V_m)\right] = \tau \ln 9$, or

$$t_r = 2.2\tau \qquad \text{(8.19)}$$

By similar reasoning, the response to a *negative-going step* is

$$v_O(0^-) = V_m \qquad \text{(8.20a)}$$

$$v_O(t \geq 0^+) = V_m e^{-t/\tau} \qquad \text{(8.20b)}$$

and is shown in Figure 8.18(b). This response is characterized by the **fall time** t_f, the time it takes for v_O to fall from 90% to 10% of its initial value. Using Equation (8.16) as before yields

$$t_f = 2.2\tau \qquad \text{(8.21)}$$

▶ Example 8.10

In the circuit of Figure 8.22(a) R is measured with the ohmmeter and found to be 33 kΩ, and t_r is measured with the oscilloscope and found to be 725 μs. Find C.

Solution

By Equations (8.18a) and (8.19), $C = \tau/R = t_r/(2.2R) = 725 \times 10^{-6}/(2.2 \times 33 \times 10^3) = 10$ nF.   ◀

Exercise 8.14   If in the circuit of Figure 8.22(b) $R = 20\ \Omega$ and $t_f = 5.493$ ms, what is the value of L?

ANSWER   50 mH.

Pulse Response of *R-C* and *L-R* Circuits

Figure 8.24 shows the response of the *R-C* and *L-R* circuits to a *single* pulse.

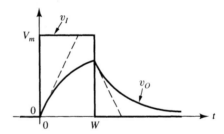

Figure 8.24 Pulse response of the *R-C* and *L-R* circuits.

For $0 \le t \le W$, v_O is an exponential transition from $v(0^+) = 0$ toward $v(\infty) = V_m$,

$$v_O(0 \le t \le W) = V_m(1 - e^{-t/\tau}) \tag{8.22a}$$

For $t \ge W$, that is, for $(t - W) \ge 0$, v_O is an exponential decay with initial value $v_O(W) = V_m(1 - e^{-W/\tau})$ and $v(\infty) = 0$. It is readily seen that this portion of the response can be expressed as $v_O(t \ge W) = v_O(W)e^{-(t-W)/\tau}$. Eliminating $v_O(W)$ and simplifying, we get

$$v_O(t \ge W) = V_m(e^{W/\tau} - 1)e^{-t/\tau} \tag{8.22b}$$

▶ **Example 8.11**

An R-C circuit with $R = 10$ kΩ and $C = 1$ nF is driven with a 5-V, 7.5-μs pulse. Find v_O for (a) $t = 4$ μs, (b) $t = 7.5$ μs, and (c) $t = 12$ μs.

Solution

(a) We have $V_m = 5$ V, $W = 7.5$ μs, and $\tau = 10^4 \times 10^{-9} = 10$ μs. Using Equation (8.22a), $v_O(4$ μs$) = 5(1 - e^{-4/10}) = 1.648$ V.

(b) By Equation (8.22a), $v_O(7.5$ μs$) = 5(1 - e^{-7.5/10}) = 2.638$ V.

(c) By Equation (8.22b), $v_O(12$ μs$) = 5(e^{7.5/10} - 1)e^{-12/10} = 1.682$ V.

◀

Exercise 8.15   An L-R circuit with $L = 50$ mH and $R = 1$ kΩ is driven with a 10-V, 100-μs pulse. Find the two instants at which $v_O(t) = 5$ V.

ANSWER   34.66 μs, 127.4 μs.

Pulse-Train Response of *R-C* and *L-R* Circuits

If an R-C or an L-R circuit is subjected to a pulse train with duty cycle d, the response consists of exponential buildups alternated with exponential decays. After a sufficient number of pulses, the steady-state situation of Figure 8.25 is reached, where the response alternates between V_L and V_H. To find V_L and V_H, we use the following:

$$v_O(0 \leq t \leq dT) = (V_L - V_m)e^{-t/\tau} + V_m$$

$$v_O(dT \leq t \leq T) = V_H e^{-(t-dT)/\tau}$$

Calculation of the first expression at $t = dT$ and the second at $t = T$ yields, respectively,

$$V_H = (V_L - V_m)e^{-dT/\tau} + V_m \qquad \text{(8.23a)}$$

$$V_L = V_H e^{-(T-dT)/\tau} \qquad \text{(8.23b)}$$

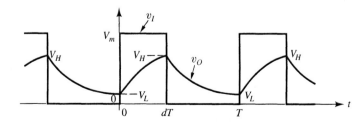

Figure 8.25 Pulse-train response of the *R-C* and *L-R* circuits.

This system of equations is readily solved for the two unknowns,

$$V_L = \frac{e^{dT/\tau} - 1}{e^{T/\tau} - 1} V_m \qquad \text{(8.24a)}$$

$$V_H = \frac{e^{-dT/\tau} - 1}{e^{-T/\tau} - 1} V_m \qquad \text{(8.24b)}$$

Depending on the magnitude of τ relative to T, we have three different possibilities, depicted in Figure 8.26 for the case $d = 50\%$:

(1) $\tau \ll T$. One can readily verify that Equations (8.24) predict $V_L \to 0$ and $V_H \to V_m$, indicating that v_O has sufficient time to essentially reach the extremes of its range. Physically, the pulses and their spacings are long enough to allow for the energy in the capacitance or inductance to *fully* build up or decay within each subcycle. Figure 8.26(a) reveals that, aside from a slight rounding of the edges, the shape of v_O is similar to that of v_I. In the limit $\tau \ll T$ the input pulses are thus transmitted to the output with **negligible distortion.**

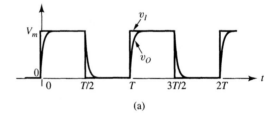

(a)

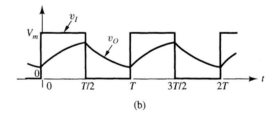

(b)

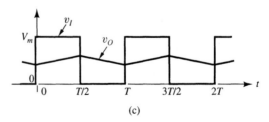

(c)

Figure 8.26 Response of the R-C and L-R circuits to a 50% duty-cycle pulse train for (a) $\tau \ll T$, (b) $\tau = T/2$, and (c) $\tau \gg T$.

(2) $\tau = T/2$. Now v_O has only a limited amount of time to perform its exponential transitions. Letting $T/\tau = 2$ and $d = 0.5$ in Equations (8.24) yield $V_L = 0.269V_m$ and $V_H = 0.731V_m$. As shown in Figure 8.26(b), v_O is a fairly **distorted** version of v_I.

(3) $\tau \gg T$. Because of the long time constant, v_O can now accomplish only the initial portions of its exponential transitions. Since these portions are essentially straight segments, v_O now approximates a *triangle wave* with a small peak-to-peak amplitude. In Chapter 7 we learned that integrating a square wave yields a triangle wave. For this reason the R-C and L-R circuits are, in the limit $\tau \gg T$, referred to as **integrators.**

We also observe that in the limit $T/\tau \to 0$ Equations (8.24) predict $V_L \to V_H \to dV_m$, indicating that v_O now approximates a *dc signal* of value dV_m. Since this coincides with the full-wave average of v_I, the R-C and L-R circuits are, in the limit $\tau \gg T$, also referred to as **averaging filters.**

▶ Example 8.12

In the circuit of Figure 8.22(a), let $R = 10$ kΩ and $C = 0.1$ μF, and let v_I be a 5-V pulse train with a 50% duty cycle. Find the pulse train frequency for which v_O will approach a triangular wave with a peak-to-peak amplitude of 10 mV.

Solution

We have $\tau = RC = 10^4 \times 10^{-7} = 1$ ms. The average value of v_O is 2.5 V, and the peak values of v_O are 2.5 V $\pm$ 5 mV, that is, 2.495 V and 2.505 V. Use Equation (8.23b) to impose

$$2.495 = 2.505e^{-(T-0.5T)/\tau}$$

Solving for T yields $T = 0.008\tau = 0.008 \times 10^{-3} = 8$ μs. Thus, $f = 1/T = 1/(8 \times 10^{-6}) = 125$ kHz. ◀

Exercise 8.16   Repeat Example 8.12, but for a 25% duty cycle.

ANSWER   93.75 kHz.

Step Response of *C-R* and *R-L* Circuits

Referring to Figure 8.27, we observe that in steady state both circuits yield $v_O = 0$, so that we must have $v_O(0^-) = 0$ and $v(\infty) = 0$. Using the continuity rules, it is also seen that, at $t = 0^+$, v_O must experience the *same jump* as v_I.

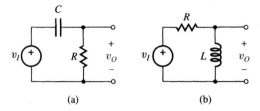

Figure 8.27 The *C-R* and *R-L* circuits.

As shown in Figure 8.28(a), the response to a positive-going step is

$$v_O(0^-) = 0 \tag{8.25a}$$
$$v_O(t \geq 0^+) = V_m e^{-t/\tau} \tag{8.25b}$$

and that to a negative-going step, shown in Figure 8.28(b), is

$$v_O(0^-) = 0 \tag{8.26a}$$
$$v_O(t \geq 0^+) = -V_m e^{-t/\tau} \tag{8.26b}$$

In both cases τ is as given in Equation (8.18). Note the circuit ability to produce a *negative* spike in response to a negative-going input step, even though v_I itself is never negative.

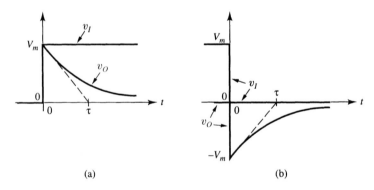

(a)   (b)

Figure 8.28 Response of the *C-R* and *R-L* circuits to (a) a positive-going step and (b) a negative-going step.

Pulse-Train Response of *C-R* and *R-L* Circuits

Figure 8.29 shows the steady-state response of the *C-R* and *R-L* circuits to a 50% duty-cycle pulse train. We observe that the current through C and the voltage across L must *average to zero,* for otherwise there would be unlimited energy buildup in these elements. Consequently, v_O will also average to zero, this being why the *C-R* and *R-L* circuits are also known as **dc blocking circuits.**

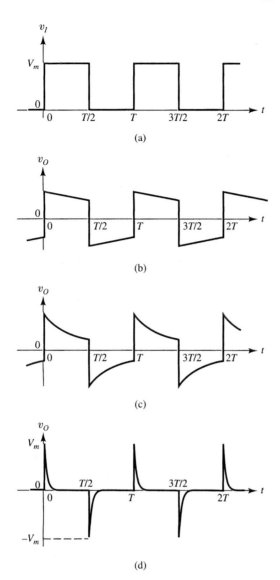

Figure 8.29 (a) Applied pulse train and response of the *C-R* and *R-L* circuits for (b) $\tau \gg T$, (c) $\tau = T/2$, and (d) $\tau \ll T$.

We again identify three cases of interest:

(1) $\tau \gg T$. With a long time constant there is insufficient time for appreciable energy buildup or decay in the capacitance or inductance within each subcycle. The input pulses are thus transmitted to the output with **negligible distortion.**

(2) $\tau = T/2$. The pulses undergo greater **distortion** because there is more time for energy buildup or decay within each subcycle.

(3) $\tau \ll T$. Now there is sufficient time for essentially *complete* energy buildup or decay within each subcycle. Consequently, v_O consists of a

series of *spikes,* the polarity of each spike reflecting the direction of the corresponding transition of v_I. Since the response now approximates the *derivative* of the applied signal, the *C-R* and *R-L* circuits are, in the limit $\tau \ll T$, referred to as **differentiators.**

Exercise 8.17   Referring to the waveforms of Figure 8.29, show that with $d = 0.5$ the peak values of v_O are $\pm V_m/(1 + e^{-T/2\tau})$. Then, discuss the cases $\tau \gg T$, $\tau = T/2$, and $\tau \ll T$.

8.4 FIRST-ORDER OP AMP CIRCUITS

RC circuits provide a fertile area of application for op amps. Thanks to its ability to draw energy from its own power supplies and inject it into the surrounding circuitry, the op amp can be used in ingenious ways to create effects that cannot be achieved with purely passive *RLC* components, let alone with resistors and capacitors but no inductors. The analysis of op amp circuits is based, as usual, on the virtual-short concept.

The Differentiator

In the circuit of Figure 8.30(a) we know that the op amp, via the feedback path provided by the resistance, keeps its inverting input at virtual ground potential. We thus have $i_C = C\, d(v_I - 0)/dt = C\, dv_I/dt$, and $i_R = (0 - v_O)/R = -v_O/R$. Since no current flows into or out of the inverting input pin, we must have

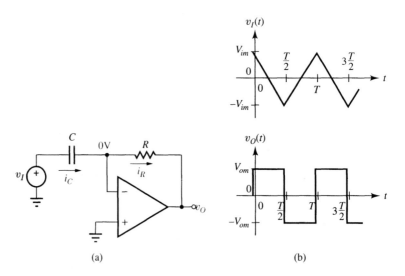

Figure 8.30 Differentiator and input/output waveform example.

$i_C = i_R$. Substituting and rearranging yields the transfer characteristic

$$v_O(t) = -RC \frac{dv_I(t)}{dt}$$ **(8.27)**

In words, the output is proportional to the *time-derivative* of the input. The proportionality constant $-RC$ is negative and has the dimensions of time, or *seconds* (s). The circuit is aptly called a **differentiator.** Figure 8.30(b) shows how it differentiates a triangular input wave to produce a square output wave. Physically, it works as follows:

(1) During the time intervals over which v_I *increases,* i_C flows toward the *right.* By KCL, i_R also flows toward the right, making v_O *negative* as the left terminal of R is at virtual ground potential.

(2) During the times when v_I *decreases,* i_C and, hence, i_R flow toward the *left,* making v_O positive.

Exercise 8.18

(a) Referring to the waveforms of Figure 8.30, show that $V_{om} = 4RCfV_{im}$, where f is the frequency of the input wave.

(b) Repeat, but for an ac input with amplitude V_{im}.

ANSWER   (b) $V_{om} = 2\pi RCfV_{im}$.

▶ Example 8.13

Find the transfer characteristic if the circuit of Figure 8.30(a) also includes a resistance R_p in parallel with C.

Solution

Assuming all currents flow toward the right, we have, by KCL and the op amp rule, $i_{R_p} + i_C = i_R$, or $(v_I - 0)/R_p + C\, d(v_I - 0)/dt = (0 - v_O)/R$. Solving for v_O yields

$$v_O(t) = -\frac{R}{R_p} v_I(t) - RC \frac{dv_I(t)}{dt}$$ ◀

The Integrator

Interchanging R and C in the circuit of Figure 8.30(a) yields the circuit of Figure 8.31(a). KCL and the op-amp rule allow us to write $i_R = i_C$, or $(v_I - 0)/R = C \times d(0 - v_O)/dt$. Solving for dv_O yields $dv_O = -(1/RC)v_I\, dt$. Changing t to the dummy variable ξ, and then integrating both sides from 0 to

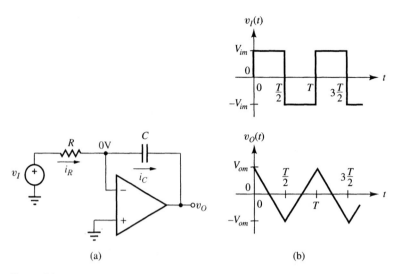

Figure 8.31 Integrator and input/output waveform example.

t, we obtain

$$v_O(t) = -\frac{1}{RC}\int_0^t v_I(\xi)\,d\xi + v_O(0) \tag{8.28}$$

where $v_O(0)$ is the output voltage at $t = 0$. Its value is determined by the charge stored at this time in the capacitance.

Aptly called an **integrator,** the circuit yields an output which is proportional to the *time-integral* of the input. The proportionality constant $-1/RC$ is negative and has the dimensions of the reciprocal of time, or s^{-1}.

Of particular interest is the case in which the input voltage is *constant,* or $v_I(t) = V_I$, for then Equation (8.28) simplifies as

$$v_O(t) = -\frac{V_I}{RC}t + v_O(0) \tag{8.29}$$

In words, applying a *constant voltage* to an integrator yields a *linear* output waveform, or **ramp.** The rate at which the output ramps is $dv_O/dt = -V_I/RC$. With $V_I > 0$ we obtain a *decreasing* ramp, and with $V_I < 0$ we obtain an *increasing* ramp. Thanks to its ability to produce ramps, the integrator finds application in the design of waveform generators to synthesize *triangle* and *sawtooth* waves. Figure 8.30(b) shows how the circuit integrates a square wave to produce a triangular wave. Its physical operation is as follows:

(1) During the times when $v_I > 0$, i_R and, hence, i_C flow toward the *right,* causing the voltage *across* the capacitance, which is $(0 - v_O)$, to *increase.* But, since the left plate is kept at virtual ground potential, v_O must thus *decrease.*

(2) During the times when $v_I < 0$, i_R and, hence i_C flow toward the *left.* This causes the voltage $(0 - v_O)$ *across* the capacitance to *decrease.* Consequently, v_O will now *increase.*

The integrator is the workhorse of op amp RC circuits. Common application areas, besides *signal generation,* are *signal processing* (such as *waveshaping* and *filtering*) and *instrumentation* and *control.*

▶ Example 8.14

An op amp integrator circuit with $R = 100$ kΩ and $C = 0.1$ μF is driven with a constant dc voltage $V_I = 1$ mV. If $v_O(0) = 0$ V, how long will it take for v_O to reach -10 V?

Solution

By Equation (8.29), $v_O(t) = [-10^{-3}/(10^5 \times 10^{-7})]t$, or

$$v_O(t) = -0.1t$$

where v_O is in V and t in s. The time it takes for v_O to reach -10 V is such that $-10 = -0.1t$, or $t = 100$ s. ◀

Exercise 8.19

    (a) Referring to the waveforms of Figure 8.31, show that $V_{om} = V_{im}/(4RCf)$, where f is the input frequency.

    (b) Repeat, but for an ac input with amplitude V_{im}.

ANSWER   (b) $V_{om} = V_{im}/2\pi RCf$.

Exercise 8.20   Develop a differential equation in v_I and v_O if the circuit of Figure 8.31(a) includes also a resistance R_p in parallel with C.

ANSWER $RC\, dv_O(t)/dt + (R/R_p)v_O(t) = -v_I(t)$.

The Noninverting Integrator

The circuit of Figure 8.32 is the V–I converter of Figure 6.23, but with a capacitance as the load. Adapting Equation (6.31) to the present case, the capacitance current is

$$i_C = \frac{v_I}{R_1} \tag{8.30}$$

The op amp acts as a noninverting amplifier with respect to v_C,

$$v_O = \left(1 + \frac{R_2}{R_1}\right) v_C \tag{8.31}$$

Using the capacitance law, we can write $v_I/R_1 = i_C = C\, dv_C/dt = [C/(1 + R_2/R_1)]\, dv_O/dt$, or $dv_O = [(1 + R_2/R_1)/(R_1 C)] \times v_I\, dt$. Changing t to the

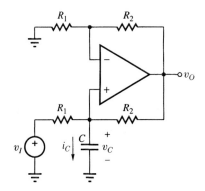

Figure 8.32 The noninverting integrator.

dummy variable ξ and then integrating both sides from 0 to t, we obtain

$$v_O(t) = \frac{1 + R_2/R_1}{R_1 C} \int_0^t v_I(\xi)\,d\xi + v_O(0) \qquad (8.32)$$

The circuit is again an integrator. However, since the constant of proportionality $(1 + R_2/R_1)/(R_1 C)$ is now *positive,* the circuit is called a **noninverting integrator.**

By contrast, the circuit of Figure 8.31(a) is an *inverting integrator.* If v_I is kept constant, the noninverting integrator will yield an *increasing ramp* when $v_I > 0$ and a *decreasing ramp* when $v_I < 0$, just the opposite of the inverting integrator.

Exercise 8.21   Let the circuit of Figure 8.32 have $R_1 = R_2 = 100$ kΩ and $C = 10$ nF. Sketch and label v_I and v_O versus t if v_I is a square wave as in Figure 8.31(b) top. Assume $V_{im} = 1$ V, $T = 2$ ms, and $v_O(0) = 0$ V.

• Creating Divergent Responses

The circuit of Figure 8.33 can be regarded as an R-C circuit terminated on a variable *negative resistance.* Moreover, the switch allows us to exclude the source for the purpose of investigating the *natural response* of the circuit.

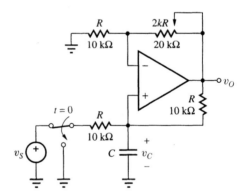

Figure 8.33 Using an op amp to create divergent responses.

We observe that as the wiper is swept from end to end, the variable resistance changes from 0 to 20 kΩ, that is, from 0 to $2R$. Its value can thus be expressed as $2kR$, as shown, where k is a fraction controlled by the wiper position,

$$0 \le k \le 1 \qquad (8.33)$$

By Equation (6.29b), the equivalent resistance R_o seen by C is $R_o = R/(R/R - 2kR/R)$, or

$$R_o = \frac{R}{1 - 2k} \qquad (8.34)$$

As k is varied from 0 to 1, R_o varies from R, through ∞, to $-R$. It is precisely the ability to make R_o *negative* that interests us. To appreciate this feature, let us find a relation between v_O and v_S.

The op amp amplifies the capacitance voltage v_C by $1 + 2kR/R = 1 + 2k$, thus yielding

$$v_O = (1 + 2k)v_C \qquad (8.35)$$

KCL at the noninverting input node and the capacitance law yield

$$\frac{v_S - v_C}{R} = C\frac{dv_C}{dt} + \frac{v_C - v_O}{R} \qquad (8.36)$$

Eliminating v_C, the differential equation governing the circuit is

$$\frac{RC}{1 - 2k}\frac{dv_O}{dt} + v_O = \frac{1 + 2k}{1 - 2k}v_S \qquad (8.37)$$

Since we are only interested in the *natural response,* we first connect the circuit to the source v_S to inject energy into C, then we flip the switch down to exclude the source and see how the circuit does on its own. Thus, for $t \geq 0^+$, the governing equation is

$$\frac{RC}{1 - 2k}\frac{dv_O(t)}{dt} + v_O(t) = 0 \qquad (8.38)$$

This is the familiar homogeneous form of Equation (7.28), but with $\tau = RC/(1 - 2k)$. The root of the *characteristic equation* is $s = -1/\tau$, or

$$s = \frac{2k - 1}{RC} \qquad (8.39)$$

and the solution is $v_O(t \geq 0^+) = v_O(0^+)e^{st}$, or

$$v_O(t \geq 0^+) = v_O(0^+)e^{\frac{2k-1}{RC}t} \qquad (8.40)$$

We identify three important cases:

(1) $k < 0.5$, so that $s < 0$. The location of the root in the s plane is on the *left half* of the horizontal axis. As we know, the natural response is a **decaying exponential,** indicating that for $k < 0.5$, our circuit responds just like an ordinary RC circuit.

(2) $k = 0.5$, so that $s = 0$. The root now lies *right at the origin,* and the natural response is *constant,* $v_O(t \geq 0^+) = v_O(0^+)$. Note that with $k = 0.5$ the resistance ratios are matched, so that our circuit reduces to

the *noninverting integrator* of Figure 8.32. We thus conclude that **the characteristic equation of an integrator admits a root right at the origin of the *s* plane.**

(3) $k > 0.5$, so that $s > 0$. The root now lies on the *right half* of the horizontal axis, and Equation (8.40) predicts an **exponentially diverging** natural response because the coefficient of t in the exponential term is positive. This type of response is depicted in Figure 8.34.

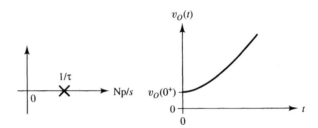

Figure 8.34 A root on the *right half* of the horizontal axis of the *s* plane corresponds to an *exponentially diverging* natural response.

Physically, a diverging natural response stems from the op amp ability to inject into the *RC* circuit *more energy* than *R* can dissipate. It is this additional energy that leads to an ever-increasing voltage buildup across *C*. By contrast, in a purely passive *RC* circuit, energy dissipation by *R* causes *C* to gradually discharge.

▶ **Example 8.15**

In the circuit of Figure 8.33 let $C = 0.1 \ \mu$F. Assuming $v_C(0^+) = 1$ V, find $v_O(t)$ if the wiper is (a) all the way to the left, (b) halfway, and (c) all the way to the right.

Solution

$RC = 10^4 \times 10^{-7} = 1$ ms.

(a) With the wiper to the left, $k = 0$; $v_O(0^+) = (1 + 2k)v_C(0^+) = (1 + 2 \times 0)1 = 1$ V; $s = (2k - 1)/RC = -1/10^{-3}$ Np/s;

$$v_O(t \geq 0^+) = 1e^{-t/(1 \ \text{ms})} \ \text{V}$$

This is an exponentially *decaying* response.

(b) With the wiper halfway we have $k = 0.5$, and $v_O(0^+) = (1 + 2 \times 0.5)1 = 2$ V. The root is $s = (2k - 1)/RC = 0$ Np/s, so that $v_O(t > 0^+) = v(0^+)e^0$, or

$$v_O(t \geq 0^+) = 2 \ \text{V}$$

This is a *constant* response, representing a memory function.

(c) With the wiper to the right, $k = 1$; $v_O(0^+) = (1 + 2 \times 1)1 = 3$ V; $s = (2k - 1)/RC = +1/10^{-3}$ Np/s;

$$v_O(t \geq 0^+) = 3e^{t/(1 \text{ ms})} \text{ V}$$

This is an exponentially *diverging* response.

Exercise 8.22

(a) In the circuit of Figure 8.33 specify C so that, with the wiper set 3/4 of the way to the right, the natural response diverges with a time constant of +10 ms.

(b) If $v_C(0^+) = 1 \, \mu$V, how long does it take for v_O to rise to 10 V?

ANSWER   0.5 μF, 152.0 ms.

• The Root Locus

The effect of k upon the root s of the characteristic equation can be visualized graphically by means of the **root locus,** that is, the trajectory described by s as k is varied from 0 to 1. In our circuit, s moves from $-1/RC$, through 0, to $+1/RC$. The locus is shown in Figure 8.35.

It is interesting to note that the root locus of a *passive RC* circuit extends only up to the origin. By contrast, with the inclusion of an op amp, the locus can be made to spill into the right half of the s plane. It is precisely the ability to control root locations in ways not possible with purely passive circuits that makes op amps especially useful in the design of a variety of important circuits such as *filters* and *oscillators*.

Figure 8.35 Root locus for the circuit of Figure 8.33.

8.5 TRANSIENT ANALYSIS USING SPICE

SPICE has the ability to perform the transient analysis of circuits containing energy-storage elements and to tabulate or plot the response as a function of time. Following is a description of the statements to direct SPICE to perform this type of analysis.

Energy-Storage Elements

The general forms of the statements for the energy-storage elements are

```
CXXX   N+   N-   VALUE   IC=ICOND
```
(8.41a)

```
LXXX   N+   N-   VALUE   IC=ICOND
```
(8.41b)

As depicted in Figure 8.36, these statements contain

(1) The name of the element, which must begin with the letter C if it is a capacitance and the letter L if it is an inductance.

(2) The circuit nodes N+ and N− to which the element is connected.

(3) The element value, in farads or henries.

(4) The initial condition, ICOND. This is the initial voltage $v_C(0^+)$, in volts, across the capacitance, or the initial current $i_L(0^+)$, in amperes, through the inductance. The polarity of $v_C(0^+)$ is positive at N+ and negative at N−. The direction of $i_L(0^+)$ is from N+, through the inductance, to N−. If the initial condition specification is omitted, SPICE will automatically assume it to be zero.

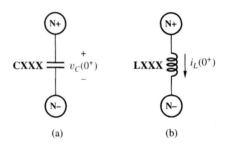

Figure 8.36 SPICE terminology for capacitors and inductors.

As seen in the previous sections, the transient analysis requires the determination of the conditions in the energy-storage elements at the instant in which the transient response is to begin. If prior to this instant the circuit is known to be in dc steady state, these conditions can be found using the dc analysis capabilities of SPICE discussed in Section 3.6.

▶ Example 8.16

Use SPICE to find $v_C(0^-)$ and $v(0^-)$ in the circuit of Figure 8.11(a).

Solution

From Figure 8.37(a), the input file is

```
FINDING THE DC STEADY STATE
V1 1 0 DC 12
R1 1 2 3K
R2 2 0 1K
C 1 2 0.1U
.DC V1 12 12 1
.PRINT DC V(1,2) V(2)
.END
```

After SPICE is run, the output file contains the following:

```
  V1              V(1,2)           V(2)
1.200E+01       9.000E+00        3.000E+00
```

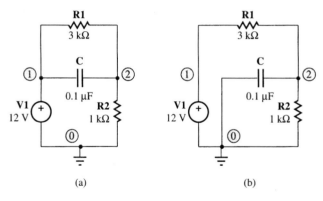

Figure 8.37 SPICE circuits to find the transient response of the circuit of Figure 8.11(a).

We thus have $v_C(0^-) = V(1,2) = 9$ V, and $v(0^-) = V(2) = 3$ V, in agreement with Example 8.4.

Exercise 8.23   Use SPICE to find the dc steady state at $t = 0^-$ for the circuit of Figure 8.13.

Exercise 8.24   Use SPICE to find the dc steady state at $t = 0^-$ for the circuit of Figure 8.16(a).

The `.TRAN` and `.PLOT TRAN` Statements

The transient analysis in SPICE is invoked by means of the `.TRAN` statement, whose general form is

```
.TRAN  TSTEP  TSTOP  TSTART  TMAX  UIC                    (8.42)
```

where `TSTEP` represents the *printing* or *plotting increment,* in seconds; `TSTART`, `TSTOP`, and `TMAX` are, respectively, the start time, stop time, and maximum step for the analysis, also in seconds. If `TSTART` is omitted, SPICE automatically assumes it to be zero. If `TMAX` is omitted, the step defaults to `TSTEP` or to `(TSTOP-TSTART)`/50, whichever is smaller. The optional `UIC` (Use Initial Conditions) keyword directs SPICE to use the initial conditions specified by the `IC=ICOND` options in the element statements of Equation (8.41).

SPICE does not provide the transient response in analytical form. Rather, it gives us the plot of one or more circuit variables. Thus, along with the `.TRAN` statement, we must also use the `.PLOT TRAN` statement, whose general form is

```
.PLOT  TRAN  OUTVAR1  OUTVAR2  ...  OUTVAR8               (8.43)
```

`OUTVAR1` through `OUTVAR8` are the desired voltage or current output variables, for a maximum of eight such variables per statement.

▶ **Example 8.17**

Use SPICE to find the transient response of the circuit of Figure 8.11(a).

Solution

SPICE does not have the ability to open or close switches. However, we can find the transient response using the circuit of Figure 8.37(b), which depicts the situation after switch activation, and we can exploit the initial conditions obtained via the dc analysis of the circuit of Figure 8.37(a), which depicts the situation prior to switch activation. The input file for the circuit of Figure 8.37(b) is

```
PLOTTING THE TRANSIENT RESPONSE
V1 1 0 DC 12
R1 1 2 3K
R2 2 0 1K
C 0 2 0.1U IC=9
.TRAN 10U 400U UIC
.PLOT TRAN V(2)
.END
```

```
  TIME         V(2)
(*)----------  -1.0000E+01  -5.0000E+00   0.0000E+00   5.0000E+00   1.0000E+01
 0.000E+00 -9.000E+00 .  *  -  -  -  -  -  -  -  -  -  -  -  -  -  -  -  -  -  .
 1.000E-05 -7.503E+00 .      *         .            .            .            .
 2.000E-05 -6.198E+00 .          *     .            .            .            .
 3.000E-05 -5.052E+00 .             *  .            .            .            .
 4.000E-05 -4.044E+00 .             .  *            .            .            .
 5.000E-05 -3.158E+00 .             .      *        .            .            .
 6.000E-05 -2.392E+00 .             .         *     .            .            .
 7.000E-05 -1.721E+00 .             .            *  .            .            .
 8.000E-05 -1.130E+00 .             .            .  *            .            .
 9.000E-05 -6.099E-01 .             .            . *             .            .
 1.000E-04 -1.613E-01 .             .            .  *            .            .
 1.100E-04  2.325E-01 .             .            .    *          .            .
 1.200E-04  5.790E-01 .             .            .     *         .            .
 1.300E-04  8.837E-01 .             .            .     *         .            .
 1.400E-04  1.147E+00 .             .            .       *       .            .
 1.500E-04  1.378E+00 .             .            .        *      .            .
 1.600E-04  1.581E+00 .             .            .         *     .            .
 1.700E-04  1.759E+00 .             .            .          *    .            .
 1.800E-04  1.914E+00 .             .            .          *    .            .
 1.900E-04  2.049E+00 .             .            .           *   .            .
 2.000E-04  2.168E+00 .             .            .            *  .            .
 2.100E-04  2.273E+00 .             .            .            *  .            .
 2.200E-04  2.363E+00 .             .            .            *  .            .
 2.300E-04  2.442E+00 .             .            .            *  .            .
 2.400E-04  2.512E+00 .             .            .             * .            .
 2.500E-04  2.574E+00 .             .            .             * .            .
 2.600E-04  2.627E+00 .             .            .             * .            .
 2.700E-04  2.673E+00 .             .            .             * .            .
 2.800E-04  2.714E+00 .             .            .             * .            .
 2.900E-04  2.750E+00 .             .            .             * .            .
 3.000E-04  2.781E+00 .             .            .             * .            .
 3.100E-04  2.808E+00 .             .            .             * .            .
 3.200E-04  2.832E+00 .             .            .             * .            .
 3.300E-04  2.853E+00 .             .            .             * .            .
 3.400E-04  2.872E+00 .             .            .             * .            .
 3.500E-04  2.888E+00 .             .            .              *.            .
 3.600E-04  2.902E+00 .             .            .              *.            .
 3.700E-04  2.914E+00 .             .            .              *.            .
 3.800E-04  2.925E+00 .             .            .              *.            .
 3.900E-04  2.934E+00 .             .            .              *.            .
 4.000E-04  2.942E+00 .             .            .              *.            .
```

Figure 8.38 Plotting the response of the circuit of Figure 8.37(b).

A few remarks are in order. First, note that the left plate of the capacitor is now connected to ground to reflect the new switch position.

Moreover, we express the initial condition as found in Example 8.16 as `IC=9`, and we use the `UIC` keyword in the `.TRAN` statement to direct SPICE to make use of this condition.

For a good visualization of the response, we want `TSTEP` to be much less than the time constant of the circuit and `TSTOP` to be well into the steady-state region. From Example 8.4 we know that $\tau = 75\ \mu s$. A reasonable choice for these parameters is `TSTEP` $= 10\ \mu s$ and `TSTOP` $= 400\ \mu s$. Also, omitting `TSTART` makes it zero by default.

After running SPICE we obtain the listing of Figure 8.38.

Exercise 8.25   Use SPICE to plot the transient response of the circuit of Figure 8.13.

The Graphics Post-Processor

The PSpice™ version of SPICE is equipped with a graphics post-processor which displays the results of SPICE analysis graphically and interactively. Made available also in the classroom version of PSpice, this processor is activated by including the statement `.PROBE` in the input file. Once the PSpice analysis is complete, the Probe post-processor starts up automatically. The types of signals to be displayed and the relative formats are input by the user interactively via self-explanatory menu-driven commands. Once the desired waveform is displayed on the screen, it can be printed out on paper. It is fair to say that the Probe post-processor acts as a form of *software oscilloscope.*

To illustrate the use of this capability, we replace the `.PLOT` statement with the `.PROBE` statement in the input file of Example 8.17, so that it becomes

```
PLOTTING THE TRANSIENT RESPONSE USING PROBE
V1 1 0 DC 12
R1 1 2 3K
R2 2 0 1K
C 0 2 0.1U IC=9
.TRAN 10U 400U UIC
.PROBE
.END
```

After running PSpice, the Probe post-processor comes on automatically and asks for the signals we want to display. We enter `0`, `3`, and `V(2)`, and obtain the printout of Figure 8.39, whose quality is much better than that of Figure 8.38.

Exercise 8.26   Use PSpice along with the Probe post-processor to display the transient response of the circuit of Figure 8.16(a).

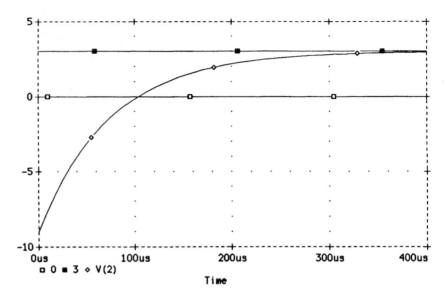

Figure 8.39 Using the Probe post-processor to display the response of the circuit of Figure 8.37(b).

The Pulse Function

To find the pulse response of a circuit, SPICE allows for any independent source to be of the pulse type. The general forms of the statements for pulse sources are

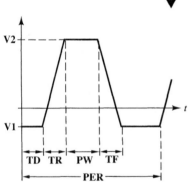

Figure 8.40 SPICE pulse function.

```
VXXX  N+  N-  PULSE  (V1  V2  TD  TR  TF  PW  PER)      (8.44a)

IXXX  N+  N-  PULSE  (I1  I2  TD  TR  TF  PW  PER)      (8.44b)
```

As illustrated in Figure 8.40 for a pulse voltage source, these statements specify trapezoidally shaped pulses that repeat with period PER. V1 (or I1) is the initial value of the source, in volts (or amps), and V2 (or I2) is the value during the pulse, also in volts (or amps). TD is the delay t_d between $t = 0$ and the inception of the pulse, in seconds. TR, TF, PW, and PER are, respectively, the rise time t_r, fall time t_f, pulse width W, and period T, all in seconds.

▶ Example 8.18

Shown in Figure 8.41 are the equivalent circuits of a computer clock generator and its load. Use PSpice to display the source and load waveforms if the source is a 50-MHz pulse train with each pulse having $V_L = 0.2$ V, $V_H = 3.6$ V, $t_r = 3$ ns, $t_f = 2$ ns, $t_d = 0$, and $W = 7.5$ ns.

Solution

The period is $T = 1/(50 \times 10^6) = 20$ ns. Using a step of 0.5 ns to ensure a reasonable resolution, and a window of 50 ns to display several periods,

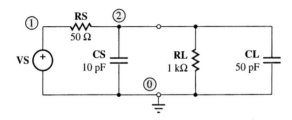

Figure 8.41 Circuit of Example 8.18.

the input file is

```
CLOCK GENERATOR LOADING
VS 1 0 PULSE(0.2 3.6 0 3NS 2NS 7.5NS 20NS)
RS 1 2 50
CS 2 0 10P
RL 2 0 1K
CL 2 0 50P
.TRAN 0.5NS 50NS
.PROBE
.END
```

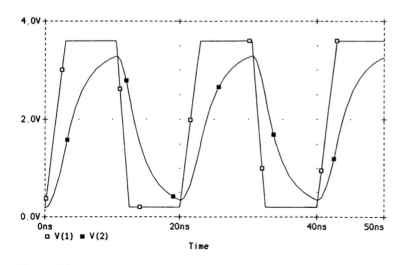

Figure 8.42 Pulse response of the circuit of Figure 8.41.

After directing the Probe post-processor to display `V(1)` and `V(2)` we obtain the traces of Figure 8.42.

We observe that by specifying a pulse width much longer than the transient component of the circuit, the pulse function will approximate a step function. Moreover, if `V1 < V2`, we have a *positive-going step,* and if `V1 > V2`, we have a *negative-going step.*

Exercise 8.27   Use SPICE to confirm the integrator response of Figure 8.31. Specify R and C so that $V_{im} = V_{om} = 5$ V, and $T = 1$ ms.

▼ SUMMARY

- Transients in first-order circuits obey the general expression $y(t) = [y(0) - y(\infty)]e^{-t/\tau} + y(\infty)$.

- The time constant is $\tau = R_{eq}C$ or $\tau = L/R_{eq}$, where R_{eq} is the *equivalent resistance* seen by C or L *during* the transient.

- Transient analysis exploits the fact that the *voltage across* C or the *current through* L *cannot change instantaneously;* moreover, in dc steady state C acts as an *open circuit,* and L as a *short.*

- During a sudden change in the circuit, C acts as a *voltage source* trying to maintain the value it had right before the change, and L acts as a *current source* trying to maintain the value it had right before the change.

- To find how a first-order circuit responds to a sudden change occurring at $t = 0$, we need to examine the circuit at $t = 0^-$, $t = 0^+$, and $t = \infty$. Moreover, we need to find R_{eq} for $t \geq 0^+$.

- The step responses of RC and LR circuits are characterized by the *rise time* and the *fall time.*

- The $R\text{-}C$ and $L\text{-}R$ circuits will pass a voltage pulse train of period T with *negligible distortion* only if $\tau \ll T$. If $\tau \gg T$, the $R\text{-}C$ and $L\text{-}R$ circuits act as *integrators.*

- The $R\text{-}C$ and $L\text{-}R$ circuits *pass* the *dc component* of the input voltage.

- The $C\text{-}R$ and $R\text{-}L$ circuits will pass a voltage pulse train of period T with *negligible distortion* only if $\tau \gg T$. If $\tau \ll T$, the $C\text{-}R$ and $R\text{-}L$ circuits act as *differentiators.*

- The $C\text{-}R$ and $R\text{-}L$ circuits *block out* the *dc component* of the input voltage.

- The most basic first-order op amp circuits are the *differentiator* and the *integrator.* The integrator can be implemented in two versions, the *inverting* and the *noninverting* versions.

- An intriguing application of the op amp is *root location control,* especially the creation of roots in the *right half* of the s plane, a physical impossibility with passive circuits alone.

- A root in the *left half-plane* corresponds to a *decaying* response; a root in the *right half-plane* corresponds to a *diverging* response; a root right at the *origin* corresponds to a *steady* or *constant* response, a function known as the *memory* function.

- SPICE can be directed to perform the transient analysis of a circuit via the .TRAN statement. A useful function in this type of analysis is the PULSE function.

▼ PROBLEMS

8.1 Basic *RC* and *RL* Circuits

8.1 (a) Find the voltage change brought about by subjecting a 1-μF capacitance to a current pulse having an amplitude of 1 mA and a duration of 1 ms. (b) What current pulse amplitude is needed to effect the same voltage change in 1 μs? In 1 ns?

8.2 A 2-nF capacitance with 100 nJ of stored energy is connected in parallel with a 5-kΩ resistance at $t = 0$. How long does it take for the voltage across the parallel combination to decay to 1 V? To 1 mV?

8.3 In the *R-C* circuit of Figure 8.1 let $R = 100$ kΩ and $C = 1$ nF. If it is found that $v(75\ \mu\text{s}) = 2.362$ V, what is the value of v_S? Of $v(100\ \mu\text{s})$?

8.4 Assuming $v_1(0^-) = 20$ V and $v_2(0^-) = 10$ V in the circuit of Figure P8.4, (a) show that $i(t \geq 0^+) = 1.5e^{-t/(16\text{ ms})}$ mA and obtain expressions for $v_1(t \geq 0^+)$ and $v_2(t \geq 0^+)$; (b) find the stored energy at $t = 0^+$ and at $t = \infty$; (c) verify that the difference between the two energies is dissipated by the resistance.

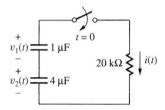

Figure P8.4

8.5 In the circuit of Figure P8.5 the voltage and current for $t \geq 0$ are $v = 10e^{-10^3 t}$ V and $i = 50e^{-10^3 t}$ μA. Find R, C, τ, the initial stored energy, and the energy dissipated by R from $t = 0$ to $t = 0.5$ ms.

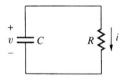

Figure P8.5

8.6 (a) Find the current change brought about by subjecting a 1-mH inductance to a voltage pulse having an amplitude of 1 V and a duration of 1 ms. (b) What voltage pulse amplitude is needed to effect the same change in 1 μs? In 1 ns?

8.7 A source-free *RL* circuit has $L = 10$ mH and $R = 2$ kΩ. If it is found that $v(4\ \mu\text{s}) = 13.48$ V, what is the stored energy in the inductance at $t = 0$? At $t = 10\ \mu$s?

8.8 In the *RL* circuit of Figure 8.9 let $I_S = 1$ A, $R = 20\ \Omega$, and $L = 1$ mH. Sketch and label $v(t)$ and $i(t)$ if the switch is pushed down until the inductance energy decays to 6.25% of its initial value, and then is flipped back up.

8.9 (a) Assuming the switch in the circuit of Figure P8.9 has been open for a long time prior to $t = 0$, find $v(t)$. (b) Repeat, but with the switch closed during $t \leq 0^-$ and opening at $t = 0$. (*Hint:* Use Thévenin's Theorem.)

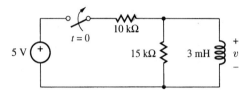

Figure P8.9

8.10 In the circuit of Figure P8.10 the voltage and current for $t \geq 0$ are $v = 10e^{-10^6 t}$ V and $i = 2e^{-10^6 t}$ mA. Find R, L, τ, and the energy dissipated in the resistance from $t = 0$ to $t = 0.5\ \mu$s.

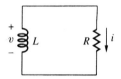

Figure P8.10

8.2 Transients in First-Order Circuits

8.11 (a) Assuming the circuit of Figure P8.11 is in steady state at $t = 0^-$, find $v(t)$. Hence, identify its *natural, forced, transient,* and *steady-state* components. (b) Repeat, but with the switch open for $t \leq 0^-$ and closing at $t = 0$.

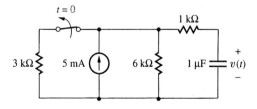

Figure P8.11

8.12 (a) Find the response $v(t)$ for the circuit of Figure P8.12 assuming it is in steady state at $t = 0^-$. Hence, identify its *natural, forced, transient,* and *steady-state* components. (b) Repeat, but with the switch initially closed and opening at $t = 0$.

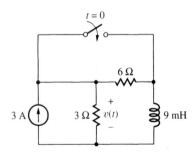

Figure P8.12

8.13 Assuming the circuit of Figure P8.13 is in steady state at $t = 0^-$, find $i(t)$.

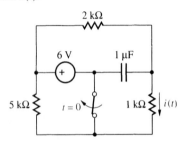

Figure P8.13

8.14 Repeat Problem 8.13, but with the capacitance replaced by a 500-μH inductance, and with the switch open during $t \leq 0^-$, and closing at $t = 0$.

8.15 Assuming the circuit of Figure P8.15 is in steady state at $t = 0^-$, sketch and label $v(t)$.

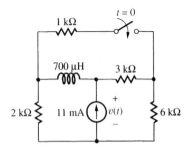

Figure P8.15

8.16 Repeat Problem 8.15, but with the inductance replaced by a 22-nF capacitance and with the switch closed during $t \leq 0^-$ and opening at $t = 0$.

8.17 Assuming the circuit of Figure P8.17 is in steady state at $t = 0^-$, find $v(t)$.

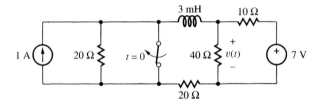

Figure P8.17

8.18 Repeat Problem 8.17, but with the inductance replaced by a 7-μF capacitance and with the switch open during $t \leq 0^-$ and closing at $t = 0$.

8.19 Assuming zero stored energy in the inductance at $t = 0$, sketch and label $v(t)$, $t \geq 0$, in the circuit of Figure P8.19.

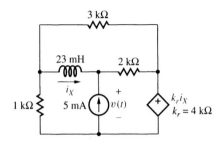

Figure P8.19

8.20 (a) Find k_r in the circuit of Figure P8.20 for a time constant of 20 ms. (b) Assuming zero stored energy in the capacitance of $t = 0$, sketch and label $v(t)$, $t \geq 0$.

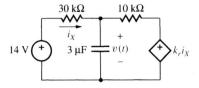

Figure P8.20

8.21 Assuming the circuit of Figure P8.21 is in steady state at $t = 0^-$, find $v_X(t)$.

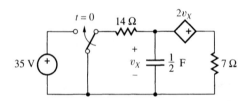

Figure P8.21

8.22 Assuming the circuit of Figure P8.22 is in steady state at $t = 0^-$, find $v(t)$.

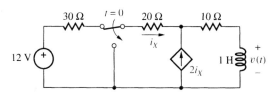

Figure P8.22

8.23 Repeat Problem 8.22, but with the inductance replaced by a 3-μF capacitance and with the switch in the down position for $t \leq 0^-$ and in the up position for $t \geq 0^+$.

8.24 Find $v(t)$ in the circuit of Figure P8.24 if (a) $v(0) = 0$, (b) $v(0) = 6$ V, (c) $v(0) = 12$ V and a 2.4-kΩ resistance is connected in parallel with the dependent source.

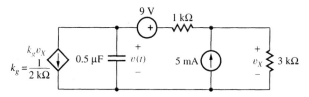

Figure P8.24

8.25 (a) Assuming the circuit of Figure P8.25 is in steady state at $t = 0^-$, find $i(t)$.

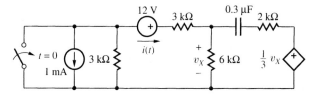

Figure P8.25

8.26 After having been in position A for a long time, the switch of Figure P8.26 is flipped to position B at $t = 0$ and then to position C at $t = 2$ ms. Sketch and label $v(t)$ for -1 ms $\leq t \leq 10$ ms.

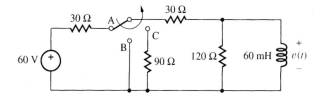

Figure P8.26

8.27 Repeat Problem 8.26, but with the inductance replaced by a 100-μF capacitance.

8.28 After having been closed for a long time, the switches of Figure P8.28 are opened simultaneously at $t = 0$. (a) Find $i(t)$, $t \geq 0^-$. (b) Sketch and label the voltages across the capacitances. What are their values as $t \to \infty$? Justify physically.

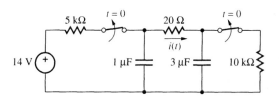

Figure P8.28

8.29 After having been open for a long time, the switches of Figure P8.29 are closed simultaneously at $t = 0$. (a) Find $v(t)$, $t \geq 0^-$. (b) Sketch and label the currents through the inductances. What are their values as $t \to \infty$? Justify physically.

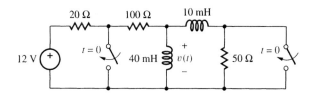

Figure P8.29

8.30 Assuming the circuit of Figure P8.30 is in steady state at $t = 0^-$, sketch and label $v(t)$ for -1 ms $\leq t \leq 5$ ms.

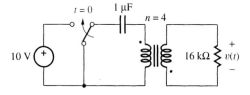

Figure P8.30

8.31 In the circuit of Figure P8.31 the lamp goes on when v reaches 15 V. When on, the lamp acts like a 10-kΩ resistor and discharges the capacitance until v drops to 5 V. At this point the lamp goes off, behaving as an open circuit and thus allowing the capacitance to recharge. Once again v reaches 15 V, the cycle repeats, indicating a flashing effect. (a) Sketch and label v for the first few cycles of oscillation, assuming $v(0) = 5$ V. (b) Find the duration of each cycle and, hence, the flashing frequency. (*Hint:* Use Equation (8.16)).

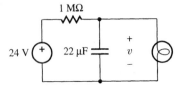

Figure P8.31

8.3 Step, Pulse, and Pulse-Train Responses

8.32   Sketch and label the response v_O of the circuit of Figure P8.32 if (a) v_I is a positive-going step with amplitude $V_m = 9$ V, and (b) v_I is a single pulse with amplitude $V_m = 6$ V and width $W = 3$ μs.

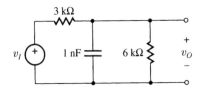

Figure P8.32

8.33   Repeat Problem 8.32, but with the capacitance replaced by a 5-mH inductance.

8.34   In the C-R circuit of Figure 8.27(a) let $R = 20$ kΩ and $C = 2$ nF. Sketch and label both v_I and v_O versus t if v_I has the doublet waveform of Figure P8.34 with $V_m = 10$ V and $W = 50$ μs. Show your waveforms for $-W \leq t \leq 4W$.

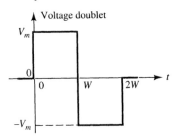

Figure P8.34

8.35   Repeat Problem 8.34, but for the L-R circuit of Figure 8.22(b), with $L = 6$ mH and $R = 200$ Ω.

8.36   Assuming the circuit of Figure P8.36 is in steady state, find $v(t)$ if, at $t = 0$: (a) $i_S = 1$ mA and v_S is suddenly changed from 0 to 9 V; (b) $v_S = 15$ V and i_S is suddenly changed from 0 to 0.5 mA.

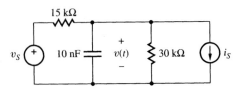

Figure P8.36

8.37   Repeat Problem 8.36, but with the capacitance replaced by a 200-μH inductance.

8.38   In the circuit of Figure P8.19 find $v(t)$ if the current source, after having been at 5 mA for a long time, is suddenly changed to -3 mA at $t = 0$.

8.39   In the circuit of Figure P8.20 let $k_r = 10$ kΩ. Sketch and label i_X versus t if the voltage source, after having been at 14 V for a long time, is suddenly changed to 7 V at $t = 0$.

8.4 First-Order Op Amp Circuits

8.40   (a) An inverting op amp integrator has $R = 100$ kΩ and $C = 1$ nF. Assuming C is initially discharged, sketch and label v_O if v_I is a voltage doublet of the type of Figure P8.34, with $V_m = 0.5$ V and $W = 1$ ms. (b) Repeat, but for a noninverting integrator of the type of Figure 8.32 with $R_1 = R_2 = 100$ kΩ and $C = 1$ nF.

8.41   Find v_O in the circuit of Figure P8.41 if v_I is a 5-V positive-going step and the top plate of C has an initial voltage of 1 V. Then, identify the *natural, forced, transient,* and *steady-state* components of v_O.

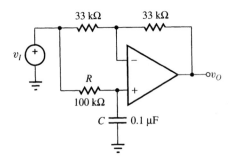

Figure P8.41

8.42   Repeat Problem 8.41, but with R and C interchanged with each other. Assume the capacitance is initially discharged.

8.43 (a) Assuming the capacitance in Figure P8.43 is initially discharged, sketch and label v_O if v_I is a 2-V positive-going step. (b) Repeat, but if v_I is a single, 2-V, 5-ms pulse.

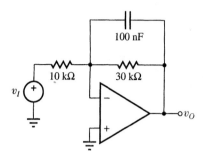

Figure P8.43

8.44 Assuming the circuit of Figure P8.43 is in steady state, sketch and label v_O if v_I is a 1-V, 200-Hz, 50% duty-cycle pulse train. Show v_O both with and without the capacitance.

8.45 Assuming the circuit of Figure P8.45 is initially in steady state, sketch and label v_O versus t if, at $t = 0$, v_I changes abruptly (a) from 0 to 1 V; (b) from 1 V to −1 V.

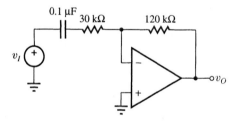

Figure P8.45

8.46 Assuming the left plate of the capacitance of Figure P8.45 has an initial charge of +100 nC, sketch and label v_O versus t for $0^- \leq t \leq 15$ ms if v_I is a voltage doublet of the type of Figure P8.34 with $V_m = 1$ V and $W = 5$ ms.

8.47 For the circuit of Figure P8.47 sketch and label v_O versus t if v_I, after having been at −2 V for a long time, is changed to +1 V at $t = 0$. At what instant is $v_O = 0$ V?

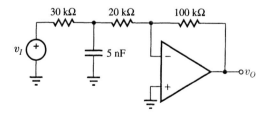

Figure P8.47

8.48 Assuming zero initial energy in the capacitance of Figure P8.48, sketch and label v_O if v_I is a single 5-V, 0.5-ms pulse.

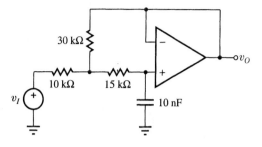

Figure P8.48

8.49 Find v_O for the circuit of Figure P8.49 if v_I is a 10-V positive-going step. Assume an initially discharged capacitance.

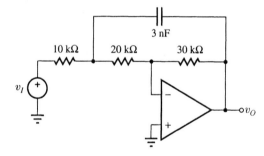

Figure P8.49

8.50 In the circuit of Figure P8.50 find v_O ($t \geq 0$) if $v_C(0) = 1$ mV. How long will it take for v_O to reach 12 V?

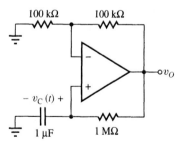

Figure P8.50

8.51 (a) Derive a differential equation relating v_I and v_O in the circuit of Figure P8.41. (b) Repeat with R and C interchanged.

▼ 8.5 Transient Analysis Using SPICE

8.52   Use the Probe post-processor of PSpice to display $v(t)$ in the circuit of Figure P8.11.

8.53   Use the Probe post-processor of PSpice to display $v(t)$ in the circuit of Figure P8.12.

8.54   Use the Probe post-processor of PSpice to display $i(t)$ in the circuit of Figure P8.25.

8.55   Use the Probe post-processor of PSpice to display $v(t)$ in the transformer circuit of Figure P8.30.

8.56   Use the Probe post-processor of PSpice to display v_I and $v_O(t)$ in the op amp circuit of Figure P8.41 if v_I is the voltage doublet of Figure P8.34 with $V_m = 1$ V and $W = 15$ ms. Assume C is initially discharged.

8.57   Use the Probe post-processor of PSpice to verify the converging, constant, and diverging responses of Example 8.15.

TRANSIENT RESPONSE OF SECOND-ORDER CIRCUITS

In this chapter we turn our attention to *second-order* circuits, that is, circuits containing two energy-storage elements that cannot be reduced to a single equivalent element via series/parallel reductions. A second-order circuit may contain two capacitances, two inductances, or one of each. The last case is by far the most interesting because it may result in *oscillatory behavior,* a phenomenon found neither in first-order circuits nor in passive second-order circuits with two energy-storage elements of the same type. This phenomenon stems from the ability of energy to flow back and forth between the capacitance and the inductance, just as energy flows back and forth between the mass and the spring of a mechanical system.

We begin by formulating the differential equations governing the *series RLC* and the *parallel RLC* circuits, and we find that the roots of the characteristic equation and, hence, the natural response, are characterized in terms of two parameters known as the *undamped natural frequency* ω_0 and the *damping ratio* ζ. Varying the damping ratio changes the location of the roots in the s plane as well as the damping characteristics of the response.

Overdamped responses consist of exponentially decaying terms similar to those of first-order circuits; however, *underdamped responses* consist of *decaying oscillations,* a feature unique to higher-order passive circuits with mixed energy-storage element types. When subjected to a step function, an underdamped circuit exhibits *overshoot* and *ringing,* phenomena not possible with first-order circuits. Our mathematical analysis of the response under different damping conditions is of interest also to other fields, such as mechanical engineering and control.

We then turn our attention to second-order op amp circuits and show that using an op amp to provide a controlled amount of *positive feedback,* it is possible to position the roots of the characteristic equation *anywhere* in the *s* plane. This allows us to achieve not only *damped oscillations* using energy-storage elements of the *same* type, but also *diverging oscillations,* a feature not possible with purely passive circuits. Moreover, this kind of behavior is achieved without the use of inductors, which are generally undesirable in modern design. Clearly, in this chapter we are witnessing some of the most intriguing properties of the op amp.

We conclude by illustrating the use of SPICE to display the transient response of second-order circuits.

9.1 NATURAL RESPONSE OF SERIES AND PARALLEL *RLC* CIRCUITS

In this section we study the *source-free* or *natural* response of *linear circuits* containing a *capacitance* and an *inductance* either directly in *series* or in *parallel* with each other. Once the remainder of the circuit is replaced by its Thévenin or Norton equivalent, we are left with the basic configurations of Figure 9.1, known as the **series** *RLC* and the **parallel** *RLC* circuits.

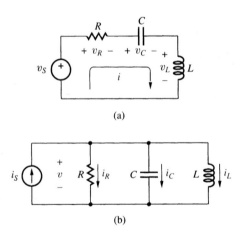

(a)

(b)

Figure 9.1 Basic *second-order RLC* circuits: (a) *series* and (b) *parallel* configurations.

Referring to the series circuit of Figure 9.1(a), we have, by KVL, $v_L + v_R + v_C = v_S$. Differentiating both sides with respect to time yields $dv_L/dt + dv_R/dt + dv_C/dt = dv_S/dt$. Letting $v_L = L\,di/dt$, $v_R = Ri$, and $dv_C/dt = i/C$, and dividing through by L yields

$$\frac{d^2i}{dt^2} + \frac{R}{L}\frac{di}{dt} + \frac{1}{LC}i = \frac{1}{L}\frac{dv_S}{dt} \qquad \textbf{(9.1)}$$

Applying dual reasoning to the parallel circuit of Figure 9.1(b) we have, by KCL, $i_C + i_R + i_L = i_S$. Differentiating both sides yields $di_C/dt + di_R/dt +$

$di_L/dt = di_S/dt$. Letting $i_C = C\,dv/dt$, $i_R = v/R$, and $di_L/dt = v/L$, and dividing through by C yields

$$\frac{d^2v}{dt^2} + \frac{1}{RC}\frac{dv}{dt} + \frac{1}{LC}v = \frac{1}{C}\frac{di_S}{dt} \tag{9.2}$$

Both equations are **second-order differential equations** because this is the order of the highest derivative present. It is not surprising that the inclusion of an additional energy-storage element has increased the order of the circuit and its equation.

The Characteristic Equation

We wish to investigate the **source-free** or **natural response** of both circuits, that is, the response with $v_S = 0$ or $i_S = 0$. Since this makes $dv_S/dt = 0$ or $di_S/dt = 0$, both equations are of the type

$$\frac{d^2y(t)}{dt^2} + 2\zeta\omega_0\frac{dy(t)}{dt} + \omega_0^2 y(t) = 0 \tag{9.3}$$

where $y(t)$, representing either v or i, is the **unknown variable**; ω_0, having the dimensions of *radians/second* (rad/s), is called the **undamped natural frequency**; ζ (zeta) is a dimensionless parameter called the **damping ratio**. The reasons for this form and terminology will become apparent shortly.

The expressions for ω_0 and ζ are found by equating the corresponding coefficients. Thus, letting $\omega_0^2 = 1/LC$ yields

$$\omega_0 = \frac{1}{\sqrt{LC}} \tag{9.4}$$

both for the series and parallel circuits. Moreover, letting $2\zeta\omega_0 = R/L$ yields $\zeta = (R/L)/(2\omega_0)$, or

$$\zeta = \frac{1}{2}R\sqrt{C/L} \tag{9.5}$$

for the *series* circuit; likewise, letting $2\zeta\omega_0 = 1/RC$ yields

$$\zeta = \frac{1}{2}\frac{1}{R\sqrt{C/L}} \tag{9.6}$$

for the *parallel* circuit. Note the duality of the two expressions.

Equation (9.3) states that a linear combination of the unknown function and its first and second derivatives must equal zero. The function and its derivatives must cancel somehow, suggesting an exponential solution of the type

$$y(t) = Ae^{st} \tag{9.7}$$

Let us now seek suitable expressions for A and s that will make this solution work. Substituting into Equation (9.3) yields $s^2 Ae^{st} + 2\zeta\omega_0 s Ae^{st} + \omega_0 Ae^{st} = 0$, or

$$(s^2 + 2\zeta\omega_0 s + \omega_0)Ae^{st} = 0$$

Since we are seeking a solution $Ae^{st} \neq 0$, the expression within parentheses must vanish,

$$s^2 + 2\zeta\omega_0 s + \omega_0^2 = 0 \tag{9.8}$$

This equation is known as the **characteristic equation** because the parameters ω_0 and ζ depend only on the elements R, L, and C and the way they are interconnected to form the circuit, regardless of the voltages or currents. Consequently, we expect it to provide information about the *character* of the natural behavior of the circuit.

The roots of the characteristic equation, variously called the **natural frequencies,** the **characteristic frequencies,** or the **critical frequencies** of the circuit, are readily found as $s_{1,2} = -\zeta\omega_0 \pm \sqrt{\zeta^2\omega_0^2 - \omega_0^2}$, or

$$s_{1,2} = (-\zeta \pm \sqrt{\zeta^2 - 1})\omega_0 \tag{9.9}$$

indicating that the response will actually consist of *two* components, $y_1 = A_1 e^{s_1 t}$ and $y_2 = A_2 e^{s_2 t}$. This is not surprising as we now have *two* energy-storage elements instead of one. If y_1 and y_2 satisfy Equation (9.3), so does their sum, $y = y_1 + y_2$, as you can easily verify by substitution. The most general solution is thus

$$y(t) = A_1 e^{s_1 t} + A_2 e^{s_2 t} \tag{9.10}$$

where A_1 and A_2 are suitable constants to be determined on the basis of the initial conditions for $y(t)$ and its derivative.

While R, L, and C are always real and positive, s_1 and s_2 may be real or complex, depending on whether the *discriminant* $(\zeta^2 - 1)$ in Equation (9.9) is positive or negative. We have the following important cases:

(1) $\zeta > 1$, so that $\zeta^2 - 1 > 0$. In this case the roots are *real, negative,* and *distinct.* For reasons that will become apparent shortly, the corresponding response is said to be **overdamped.**

(2) $0 < \zeta < 1$, so that $\zeta^2 - 1 < 0$. In this case the roots are said to be *complex conjugate,* and the response is said to be **underdamped.**

(3) $\zeta = 1$, so that $\zeta^2 - 1 = 0$. The roots are still real and negative, but they are now *identical.* The response is said to be **critically damped.**

(4) $\zeta = 0$. The roots are said to be *purely imaginary,* and the response is said to be **undamped.**

Equations (9.4) through (9.6) indicate that ω_0 depends only on L and C, while ζ depends also on R.

Varying the Damping Ratio

To investigate the various response types we keep L and C fixed and *vary R* with a potentiometer to achieve different values of ζ. This is shown in Figure 9.2. In either circuit the function of the dc source is to inject energy into the circuit so that the latter can store it in its capacitance and/or inductance, and reuse it to produce the natural response once the source is excluded from the circuit. As usual, we shall assume that the switch has been in the position shown long enough to allow for the circuit to reach its dc steady state, where $i_C = 0$ and $v_L = 0$.

Before considering examples of the various response types, let us find the initial values of $y(t)$ and its derivative because we shall need them later on to calculate the constants A_1 and A_2.

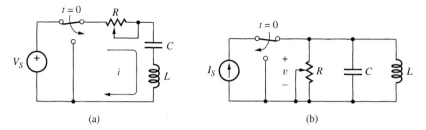

(a)   (b)

Figure 9.2 Investigating the natural response of the *series* and *parallel RLC* configurations for different damping conditions.

▶ **Example 9.1**

In the series RLC circuit of Figure 9.2(a) find the current $i(0^+)$ and its derivative $di(0^+)/dt$ *just after* switch activation.

Solution

During the steady state preceding switch activation, the capacitance acts as an open circuit and the inductance as a short circuit. Hence, $i_L(0^-) = 0$, and $v_C(0^-) = V_S$.

By the inductance and the capacitance continuity rules, $i(0^+) = i_L(0^+) = i_L(0^-) = 0$, and $v_C(0^+) = v_C(0^-) = V_S$. By the inductance law, $di(0^+)/dt = di_L(0^+)/dt = v_L(0^+)/L$. To find $v_L(0^+)$, apply KVL: $v_R(0^+) + v_C(0^+) + v_L(0^+) = 0$, or $v_L(0^+) = -v_R(0^+) - v_C(0^+) = -Ri(0^+) - V_S = -V_S$. Substituting yields $di(0^+)/dt = -V_S/L$. In short, the initial conditions for the given circuit are

$$i(0^+) = 0 \qquad\qquad (9.11a)$$

$$\frac{di(0^+)}{dt} = -\frac{V_S}{L} \qquad\qquad (9.11b) \blacktriangleleft$$

Exercise 9.1   In the parallel RLC circuit of Figure 9.2(b) show that

$$v(0^+) = 0 \qquad \text{(9.12a)}$$

$$\frac{dv(0^+)}{dt} = -\frac{I_S}{C} \qquad \text{(9.12b)}$$

Overdamped Response

If $\zeta > 1$, the roots are **real, negative,** and **distinct,** and they are expressed in *nepers/second* (Np/s), as usual. According to Equation (9.10), the response is the sum of two decaying exponentials having, respectively, A_1 and A_2 as initial values, and $\tau_1 = -1/s_1$ and $\tau_2 = -1/s_2$ as time constants,

$$y(t) = A_1 e^{-t/\tau_1} + A_2 e^{-t/\tau_2} \qquad \text{(9.13)}$$

where, by Equation (9.9),

$$\tau_1 = \frac{1}{\left(\zeta - \sqrt{\zeta^2 - 1}\right)\omega_0} \qquad \text{(9.14a)}$$

$$\tau_2 = \frac{1}{\left(\zeta + \sqrt{\zeta^2 - 1}\right)\omega_0} \qquad \text{(9.14b)}$$

It is good practice to visualize the roots of the characteristic equation as a *pair of points* in a plane of roots called the *s plane,* already introduced in Section 7.3 for first-order circuits. As shown in Figure 9.3(a), a real and negative root pair is represented by a pair of points on the negative portion of the

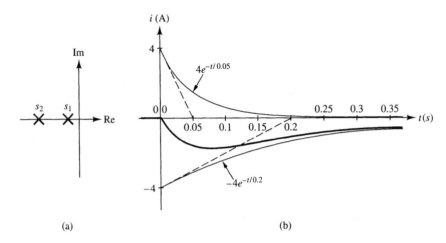

        (a)   (b)

Figure 9.3 *Real, negative roots, and example of an overdamped response.*

horizontal axis at $s_1 = -1/\tau_1$ and $s_2 = -1/\tau_2$. As we know, this axis is calibrated in Np/s.

We now wish to develop expressions for A_1 and A_2 in terms of the initial conditions of the circuit. Evaluating the unknown variable and its derivative at $t = 0^+$ we obtain $y(0^+) = A_1e^0 + A_2e^0 = A_1 + A_2$, and $dy(0^+)/dt = -(1/\tau_1)A_1e^0 - (1/\tau_2)A_2e^0 = -(1/\tau_1)A_1 - (1/\tau_2)A_2$. Solving for A_1 and A_2 yields

$$A_1 = \frac{\tau_1}{\tau_1 - \tau_2}\left(y(0^+) + \tau_2\frac{dy(0^+)}{dt}\right) \tag{9.15a}$$

$$A_2 = \frac{\tau_2}{\tau_2 - \tau_1}\left(y(0^+) + \tau_1\frac{dy(0^+)}{dt}\right) \tag{9.15b}$$

▶ Example 9.2

In the series RLC circuit of Figure 9.2(a) let $V_S = 12$ V, $R = 5\ \Omega$, $C = 50$ mF, and $L = 0.2$ H. Sketch and label $i(t)$.

Solution

For the series configuration we use Equation (9.5) and find $\zeta = 0.5 \times 5 \times \sqrt{50 \times 10^{-3}/0.2} = 1.25$. Since $\zeta > 1$, the response is overdamped. By Equation (9.4), $\omega_0 = 1/\sqrt{0.2 \times 50 \times 10^{-3}} = 10$ rad/s. Substituting into Equation (9.14) we obtain $\tau_1 = 0.2$ s, and $\tau_2 = 0.05$ s.

To find A_1 and A_2, use Equation (9.11), namely, $i(0^+) = 0$ and $di(0^+)/dt = -V_S/L = -12/0.2 = -60$ A/s. Substituting into Equation (9.15) yields $A_1 = [0.2/(0.2 - 0.05)] \times [0 - 0.05 \times 60] = -4$ A, and $A_2 = [0.05/(0.05 - 0.2)] \times [0 - 0.2 \times 60] = 4$ A. Consequently,

$$i(0^-) = 0 \tag{9.16a}$$

$$i(t \geq 0^+) = -4e^{-t/0.2} + 4e^{-t/0.05}\ \text{A} \tag{9.16b}$$

To sketch $i(t)$ for $t \geq 0^+$, we first graph the two exponential decays separately, then we add them up algebraically, point by point. The result is shown in Figure 9.3(b). The fact that the response comes out negative indicates that $i(t)$ actually flows *counterclockwise*. This is not surprising since the polarity of the voltage stored in the capacitance is such that current will flow out of the top plate, through R and L, and back into the bottom plate.   ◀

Exercise 9.2   In the parallel RLC circuit of Figure 9.2(b) let $I_S = 12$ mA, $R = 150\ \Omega$, $L = 5$ mH, and $C = 20$ nF. Derive an expression for $v(t)$. Hence, sketch and label it.

ANSWER $v(0^-) = 0$, $v(t \geq 0^+) = 2.25\left[e^{-t/(3.333\ \mu s)} - e^{-t/(30\ \mu s)}\right]$ V.

▶ Example 9.3

In the overdamped series RLC circuit of Example 9.2 find the voltage across each element as a function of time.

Solution

(a) Referring to the voltage polarities of Figure 9.1(a) we have, by Ohm's Law, $v_R = Ri$. Multiplying both sides of Equations (9.16) by R yields

$$v_R(0^-) = 0$$

$$v_R(t \geq 0^+) = -20e^{-t/0.2} + 20e^{-t/0.05} \text{ V}$$

(b) By the inductance law, $v_L = L\,di/dt$. Differentiating both sides of Equations (9.16) and multiplying by L yields

$$v_L(0^-) = 0$$

$$v_L(t \geq 0^+) = 4e^{-t/0.2} - 16e^{-t/0.05} \text{ V}$$

(c) By the capacitance law, $v_C = (1/C)\int i(t)\,dt$. Integrating both sides of Equations (9.16) with respect to time and dividing by C, we get

$$v_C(0^-) = 12 \text{ V}$$

$$v_C(t \geq 0^+) = 16e^{-t/0.2} - 4e^{-t/0.05} \text{ V}$$

Remark Note that v_R and v_C are continuous at $t = 0$ but v_L jumps from $v_L(0^-) = 0$ to $v_L(0^+) = -16e^{-0} + 4e^{-0} = -12$ V. This jump is required in order to ensure the continuity of v_C. As we know, the continuity rule for inductance applies to i_L but not to v_L. ◀

Exercise 9.3 In the overdamped parallel RLC circuit of Exercise 9.2 find the current through each element as a function of time. Discuss your results.

ANSWER $i_R(0^-) = 0$; $i_R(t \geq 0^+) = 15\left[e^{-t/(3.333\,\mu s)} - e^{-t/(30\,\mu s)}\right]$ mA; $i_C(0^-) = 0$; $i_C(t \geq 0^+) = 1.5e^{-t/(30\,\mu s)} - 13.5e^{-t/(3.333\,\mu s)}$ mA; $i_L(0^-) = 12$ mA; $i_L(t \geq 0^+) = 13.5e^{-t/(30\,\mu s)} - 1.5e^{-t/(3.333\,\mu s)}$ mA.

Underdamped Response

If $0 < \zeta < 1$, the discriminant is negative and can be rewritten as $\sqrt{\zeta^2 - 1} = j\sqrt{1 - \zeta^2}$, where

$$j \triangleq \sqrt{-1} \tag{9.17}$$

is a dimensionless quantity known as the **imaginary unit.** The roots can thus be expressed as

$$s_{1,2} = -\alpha \pm j\omega_d \tag{9.18}$$

where

$$\boxed{\alpha = \zeta\omega_0} \tag{9.19}$$

is called the **damping coefficient,** and

$$\boxed{\omega_d = \omega_0\sqrt{1 - \zeta^2}} \tag{9.20}$$

is called the **damped natural frequency.**

For the root $s_1 = -\alpha + j\omega_d$, ω_d is also referred to as the **imaginary part** of s_1, while $-\alpha$ is, by contrast, referred to as the **real part** of s_1. For the root $s_2 = -\alpha - j\omega_d$ the real part is still $-\alpha$; however, the imaginary part is now $-\omega_d$. Both s_1 and s_2 are examples of **complex variables.** Since their imaginary parts are opposite to each other, s_1 and s_2 are said to form a **complex-conjugate pair.**

Though both α and ω_d have the dimensions of the reciprocal of time, or s^{-1}, we shall continue to express α in *nepers/second* (Np/s), whereas we shall express ω_d in *radians/second* (rad/s) to distinguish between the two. To evidence their complex nature, we shall express s_1 and s_2 in *complex Np/s.* In summary, denoting the physical units of a given quantity x as $[x]$, we have

$$[\alpha] = \text{Np/s}$$

$$[\omega_d] = [\omega_0] = \text{rad/s}$$

$$[s_1] = [s_2] = \text{ complex Np/s}$$

To visualize complex roots in the s plane we need *two* axes: a horizontal axis for plotting real parts, and a vertical axis for imaginary parts. The former is calibrated in Np/s and is called the **real axis;** the latter is calibrated in rad/s and is called the **imaginary axis.** For instance, the root $s_1 = -3 + j4$ complex Np/s is the point of the s plane having an abscissa of -3 Np/s and an ordinate of $+4$ rad/s. Its conjugate $s_2 = -3 - j4$ complex Np/s is the point having an abscissa of -3 Np/s and an ordinate of -4 rad/s. As shown in Figure 9.4(a), two conjugate roots are located symmetrically with respect to the real axis.

We now wish to find the underdamped response. This is obtained by substituting Equation (9.18) into Equation (9.10). However, it is shown in Appendix 3 that $y(t)$ can be put in the more insightful form

$$\boxed{y(t) = Ae^{-\alpha t}\cos(\omega_d t + \theta)} \tag{9.21}$$

where A and θ are the usual initial-condition constants. This function, called a **damped sinusoid,** has **angular frequency** ω_d, **phase angle** θ, and an **exponentially decaying amplitude,** $A^{-\alpha t}$. The rate of decay is governed by the time constant $\tau = 1/\alpha = 1/\zeta\omega_0$. Moreover, by Equation (9.20), the damped

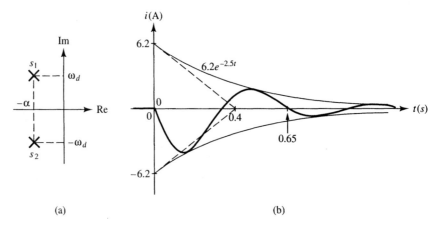

Figure 9.4 *Complex-conjugate roots, and example of an underdamped response.*

frequency ω_d is always *less* than the undamped frequency ω_0. The smaller the value of ζ, the slower the decay and the closer ω_d is to ω_0. We are now better able to appreciate the reason for referring to the roots of the characteristic equation as *frequencies*.

Physically, the oscillations stem from the ability of stored energy to flow back and forth between L and C, whereas the damping stems from energy loss in R. This is similar to a mechanical system consisting of a mass, spring, and damper. If we inject energy into the system by hitting the mass, the system will oscillate at a frequency determined by the mass and spring characteristics. During each cycle, the kinetic energy of the mass is converted to potential energy in the spring and vice versa. At the same time, part of the energy is dissipated into heat because of friction in the damper, thus causing the oscillations to die out. The smaller the friction, the longer the persistence of the oscillations. Clearly, in this analogy the mass and the spring are the energy-storage elements, and the damper simulates dissipation by R.

▶ **Example 9.4**

Repeat Example 9.2 with R lowered to 1 Ω.

Solution

By Equation (9.5), $\zeta = 0.5 \times 1 \times \sqrt{50 \times 10^{-3}/0.2} = 0.25$, indicating an underdamped response. By Equations (9.19) and (9.20), $\alpha = 0.25 \times 10 = 2.5$ Np/s and $\omega_d = 10\sqrt{1 - 0.25^2} = 9.682$ rad/s.

Let us now find A and θ on the basis of the initial conditions of Equation (9.11), namely, $i(0^+) = 0$ and $di(0^+)/dt = -60$ A/s. By Equation (9.21), $i(0^+) = Ae^0 \cos(0 + \theta) = 0$, that is, $A \cos \theta = 0$. Since A cannot be zero, it follows that $\theta = 90°$, indicating that our solution can be expressed as $i(t) = Ae^{-\alpha t} \sin \omega_d t$.

Next, differentiate $i(t)$ and compute it at $t = 0^+$. This yields $di(0^+)/dt = -\alpha A e^0 \sin 0 + \omega_d A e^0 \cos 0 = \omega_d A$, or $A = [di(0^+)/dt]/\omega_d = -60/9.682 = -6.197$ A. Finally,

$$i(0^-) = 0 \qquad (9.22a)$$

$$i(t \geq 0^+) = -6.197 e^{-2.5t} \sin 9.682t \text{ A} \qquad (9.22b)$$

To sketch $i(t)$ for $t \geq 0^+$, we first graph two exponential decays having $\pm A = \pm 6.197$ A as initial values, and $\tau = 1/\alpha = 1/2.5 = 0.4$ s as the common time constant. These curves, shown as thin curves in Figure 9.4(b), are the *envelopes* of the sinusoid. Next, we sketch the sinusoid itself. Expressing it in the more familiar form $\sin \omega_d t = \sin(2\pi t/T_d)$, we find that the period of oscillation T_d is such that $\omega_d = 2\pi/T_d$, or

$$T_d = \frac{2\pi}{\omega_d} \qquad (9.23)$$

In the present case, $T_d = 2\pi/9.682 = 0.6489$ s. As shown, at $t = 0.4$ s the envelopes are down to 37% of their initial values, and at $t = 0.6489$ s the sinusoid completes its first cycle. ◀

Exercise 9.4   Repeat Exercise 9.2, but with $R = 500$ Ω.

ANSWER $v(0^-) = 0$, $v(t \geq 0^+) = -6.928 e^{-50 \times 10^3 t} \sin 86,603t$ V.

Critically Damped Response

Suppose we have an underdamped circuit and we vary R to increase its damping ratio. As ζ approaches unity, the discriminant $(\zeta^2 - 1)$ approaches zero. When ζ crosses unity, the roots change from complex to real because the discriminant crosses zero. Consequently, the response changes from oscillatory to nonoscillatory. Imposing $\zeta = 1$ in Equations (9.5) and (9.6) indicates that this change occurs when R reaches a critical value, R_c, such that

$$R_c = 2\sqrt{L/C} \qquad (9.24)$$

for the *series* circuit, and

$$R_c = \frac{1}{2}\sqrt{L/C} \qquad (9.25)$$

for the *parallel* circuit. Physically, when $R = R_c$, all available energy from the energy-storage elements is dissipated by the resistance within a *single* cycle, thus precluding any further oscillation. A circuit with $\zeta = 1$ is said to be *critically damped* because its response represents the borderline between oscillatory and nonoscillatory responses.

There is a mathematical peculiarity associated with critical damping because the roots are now equal: $s_1 = s_2 = -\omega_0$. Equation (9.10), whose derivation was based on the assumption of distinct roots, is no longer *the* correct solution.

Mathematically it can be shown that repeated roots give rise to natural components of the form te^{st}, along with e^{st}. The general form of a critically damped response is then $y(t) = A_1 e^{-t/\tau} + A_2 t e^{-t/\tau}$, or

$$y(t) = (A_1 + A_2 t)e^{-t/\tau} \tag{9.26a}$$

$$\tau = \frac{1}{\omega_0} \tag{9.26b}$$

where A_1 and A_2 are the usual initial-condition constants. It is easy to verify that this is indeed the correct solution by substituting it into Equation (9.3).

When $\zeta = 1$, the roots are *real, negative,* and *identical.* As shown in Figure 9.5(a), their s-plane representation consists of two coincident points located on the real axis at $-1/\tau$.

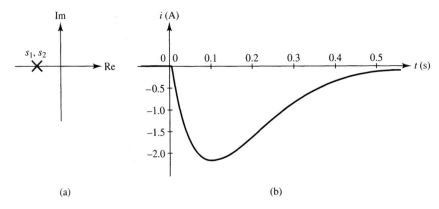

(a)   (b)

Figure 9.5 *Coincident* roots, and example of a *critically damped* response.

▶ Example 9.5

Find the value of R for which the series RLC circuit of Example 9.2 becomes critically damped. Hence, assuming the circuit is in steady state prior to switch activation, sketch and label $i(t)$.

Solution

$R_c = 2\sqrt{L/C} = 2\sqrt{0.2/0.05} = 4\ \Omega;\ \tau = 1/\omega_0 = 1/10 = 0.1$ s.

Next, find A_1 and A_2 using Equation (9.11), that is, $i(0^+) = 0$ and $di(0^+)/dt = -60$ A/s. By Equation (9.26a), $i(0^+) = A_1 e^0 + A_2 0 e^0 = 0$. This yields $A_1 = 0$, so that $i(t) = A_2 t e^{-t/\tau}$. Differentiating and computing at $t = 0^+$ yields $di(0^+)/dt = A_2[e^0 - (1/\tau)0e^0] = -60$ A/s, that is, $A_2 = -60$ A/s. Finally,

$$i(0^-) = 0 \tag{9.27a}$$

$$i(t \geq 0^+) = -60te^{-t/0.1}\ \text{A} \tag{9.27b}$$

As seen in the graph of Figure 9.5(b), the response starts at zero, peaks at $t = \tau$, and then returns asymptotically to zero. ◀

Exercise 9.5   Find the value of R for which the parallel RLC circuit of Exercise 9.2 becomes critically damped. Then, find $v(t)$ if the circuit is in steady state prior to switch activation.

ANSWER $R_c = 250\ \Omega$; $v(0^-) = 0$, $v(t \geq 0^+) = -6 \times 10^5 t e^{-t/(10\ \mu s)}$ V.

A familiar example of a critically damped system is the automobile suspension system, which is so designed to ensure a smooth ride in spite of possible bumps or potholes in the road. The suspension can be checked by suddenly pushing down either the front or the rear end of the car and then observing how it comes back up. Clearly, this is the natural response of the suspension system. Depending on its damping characteristics, a competent mechanic should be able to diagnose the state of the springs and the shock absorbers.

Undamped Response

The decay of an underdamped response is caused by energy loss in the resistance. The smaller this loss, the slower the decay. In the limiting case of zero loss, the response would never die out and the outcome would be a sustained oscillation. In our mechanical analogy this situation corresponds to frictionless oscillation.

Power loss in the resistance is related to the natural response as $p_R = Ri^2$ in the series circuit, and $p_R = v^2/R$ in the parallel circuit. To achieve lossless operation, we must therefore have

$$R = 0 \tag{9.28}$$

for the *series* circuit, and

$$R = \infty \tag{9.29}$$

for the *parallel* circuit. Either condition yields $\zeta = 0$, by Equations (9.5) and (9.6). Letting $\zeta = 0$ in Equations (9.19) through (9.21), we obtain $\alpha = 0$, $\omega_d = \omega_0$, and $y = Ae^0 \cos(\omega_0 t + \theta)$, or

$$\boxed{y(t) = A \cos(\omega_0 t + \theta)} \tag{9.30}$$

where A and θ are the usual initial-condition constants. Since this is an undamped sinusoid, its angular frequency ω_0 is called the **undamped natural frequency.** The period of oscillation is $T_0 = 2\pi/\omega_0$.

When $\zeta = 0$, the roots are **purely imaginary.** As shown in Figure 9.6(a), their s-plane representation consists of two points located symmetrically on the imaginary axis at $\pm\omega_0$.

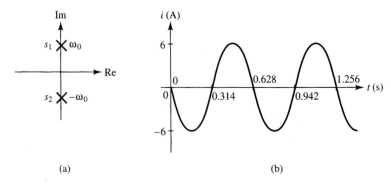

Figure 9.6 *Purely imaginary* roots, and example of an *undamped* response.

▶ **Example 9.6**

Repeat Example 9.2, but with $R = 0$.

Solution

We still have $\omega_0 = 10$ rad/s, so $T = 2\pi/10 = 0.6283$ s. Moreover, $\theta = 90°$, so $i = A \sin \omega_0 t$. Differentiating and computing at $t = 0^+$ yields $di(0^+)/dt = \omega_0 A = -60$ A/s, or $A = -6$ A. We thus have

$$i(0^-) = 0 \tag{9.31a}$$

$$i(t \geq 0^+) = -6 \sin 10t \text{ A} \tag{9.31b}$$

As shown in Figure 9.6(b), this is a sine wave having an amplitude of 6 A and a period of 0.6283 s. ◀

Exercise 9.6 Repeat Exercise 9.2, but with $R = \infty$.

ANSWER $v(0^-) = 0$, $v(t \geq 0^+) = -6 \sin 10^5 t$ V.

▶ **Example 9.7**

For the undamped circuit of Example 9.6 verify that energy flows back and forth between L and C.

Solution

The instantaneous energies of L and C are, respectively, $w_L = (1/2)Li_L^2$ and $w_C = (1/2)Cv_C^2$. For $t \geq 0^+$ we have $i_L = i(t) = -6 \sin 10t$ A, and $v_C = (1/C) \int i(t)\,dt = 12 \cos 10t$ V. Substituting and using the identities $\sin^2 \alpha = (1 - \cos 2\alpha)/2$, and $\cos^2 \alpha = (1 + \cos 2\alpha)/2$ we finally obtain, for $t \geq 0^{+}$,

$$w_L = 1.8(1 - \cos 20t) \text{ J}$$

$$w_C = 1.8(1 + \cos 20t) \text{ J}$$

It is readily seen that both energies alternate between 0 J and 3.6 J but are *out of phase* with each other, indicating that when one reaches its maximum the other reaches its minimum, and vice versa. Moreover, $w_L + w_C = 3.6 \text{ J} =$ constant, confirming the absence of any energy losses. ◀

The underdamped natural frequency ω_d is always *less* than the undamped natural frequency ω_0, and the underdamped period T_d is always *longer* than the undamped period T_0. Increasing ζ decreases ω_d and increases T_d, as per Equations (9.20) and (9.23). As $\zeta \to 1$, we have $\omega_d \to 0$ and $T_d \to \infty$, in agreement with the fact that the response changes from oscillatory to nonoscillatory.

In a practical LC circuit the undamped condition cannot be achieved because the losses in the stray resistances of its elements and interconnections cause the oscillation to eventually die out. However, using additional circuitry to continuously reinject into the system the exact amount of power that is dissipated in its resistances, it is possible to maintain a sustained oscillation, giving the appearance of undamped operation. This principle forms the basis of a certain class of *oscillators*. Typically, an oscillator consists of a timing element, such as an LC pair or a quartz crystal to establish the frequency of oscillation, and an active device such as an amplifier to compensate for energy losses. A familiar analog of an electrical oscillator is a mechanical pocket watch, which uses the energy stored in its spring to compensate for the friction losses of its balance wheel and thus maintain a sustained oscillation.

• The Root Locus

The effect of ζ upon the roots of the characteristic equation can be visualized graphically by means of the **root locus**. This is the system of trajectories described by the roots as ζ is varied over its range of possible values. Figure 9.7 shows the locus for a second-order passive RLC circuit as ζ is varied from $\zeta > 1$ all the way down to $\zeta = 0$.

With $\zeta > 1$, both roots are located on the negative real axis. As ζ decreases, the roots move toward each other on the real axis until they coalesce when the condition $\zeta = 1$ is reached. This corresponds to critical damping.

Decreasing ζ below unity causes the roots to split apart and move along symmetric trajectories toward the imaginary axis, which they reach in the limit $\zeta = 0$. Using the Pythagorean Theorem, the radial distance r of an underdamped root from the origin is, by Equations (9.19) and (9.20), $r = \sqrt{\alpha^2 + \omega_d^2} = \omega_0 = 1/\sqrt{LC} =$ constant, indicating that the trajectories are *circular arcs* with radius ω_0.

As shown in greater detail in Figure 9.8, given a root pair s_1 and s_2 on these arcs, we can state the following:

(1) The abscissa of s_1 (or s_2), called the *real part* of s_1 (or s_2), represents the *negative* of the *damping coefficient*, $-\alpha = -\zeta\omega_0$.

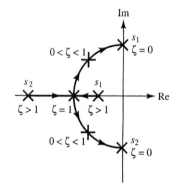

Figure 9.7 Root locus for a second-order passive RLC circuit.

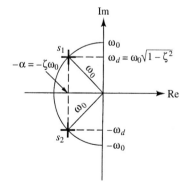

Figure 9.8 The abscissa of s_1 is $-\alpha$, and the ordinate is ω_d.

(2) The ordinate of s_1, also called the *imaginary* part of s_1, represents the *damped frequency* $\omega_d = \omega_0\sqrt{1 - \zeta^2}$.

(3) The *distance* of either root from the origin represents the *undamped frequency* ω_0.

(4) The *damping ratio* is readily found as $\zeta = \sqrt{1 - (\omega_d/\omega_0)^2}$.

9.2 TRANSIENT RESPONSE OF SECOND-ORDER CIRCUITS

Having studied the natural or source-free response of RLC circuits, we now investigate the *transient response,* that is, the response to a dc forcing function $x(t) = X_S$. The equation governing the circuit now takes on the general form

$$\frac{d^2 y(t)}{dt^2} + 2\zeta\omega_0 \frac{dy(t)}{dt} + \omega_0^2 y(t) = \omega_0^2 y(\infty) \qquad \text{(9.32)}$$

The response will generally consist of a *transient component* and a *dc steady-state component,*

$$\boxed{y(t) = y_{\text{xsient}} + y_{\text{ss}}} \qquad \text{(9.33)}$$

The transient component takes on the same functional form as the natural response. Depending on the damping conditions of the circuit, this component will be *overdamped* if $\zeta > 1$, *critically damped* if $\zeta = 1$, and *underdamped* if $0 < \zeta < 1$. As we know, the corresponding functional forms are, respectively,

$$\boxed{\begin{array}{ll} y_{\text{xsient}} = B_1 e^{-t/\tau_1} + B_2 e^{-t/\tau_2} & \text{(9.34a)} \\[2mm] y_{\text{xsient}} = (B_1 + B_2 t)e^{-t/\tau} & \text{(9.34b)} \\[2mm] y_{\text{xsient}} = Be^{-\alpha t}\cos(\omega_d t + \theta) & \text{(9.34c)} \end{array}}$$

where B_1 and B_2, or B and θ, are determined on the basis of the initial value $y(0^+)$ of the function and the initial value $dy(0^+)/dt$ of its derivative. Moreover,

$$\tau_{1,2} = \frac{1}{\left(\zeta \pm \sqrt{\zeta^2 - 1}\right)\omega_0} \qquad \text{(9.35a)}$$

$$\tau = \frac{1}{\omega_0} \qquad \text{(9.35b)}$$

$$\alpha = \zeta\omega_0 \qquad \text{(9.36a)}$$

$$\omega_d = \omega_0\sqrt{1 - \zeta^2} \qquad \text{(9.36b)}$$

The steady-state component represents what is left of $y(t)$ after all transients have died out, that is,

$$\boxed{y_{\text{ss}} = y(\infty)} \qquad \text{(9.37)}$$

The procedure to find how a second-order circuit, assumed to be initially in its dc steady state, responds to a sudden change such as the activation of a switch or the application of a step function is as follows:

(1) Find $y(0^+)$ and $dy(0^+)/dt$ using the continuity rules for capacitance and inductance.

(2) Find $y(\infty)$ using dc steady-state circuit analysis techniques. Recall that in steady state C acts as an open circuit and L as a short circuit.

(3) Find ζ and ω_0. If, with all independent sources suppressed, L and C appear either in *series* or in *parallel* with each other, then $\omega_0 = 1/\sqrt{LC}$ and $\zeta = \frac{1}{2}R_{eq}\sqrt{C/L}$ for the *series* case, and $\zeta = \frac{1}{2}/(R_{eq}\sqrt{C/L})$ for the *parallel* case, where R_{eq} is the equivalent resistance seen by the LC pair. If L and C are neither in series nor in parallel, then we must derive the differential equation governing the circuit and put it in the standard form of Equation (9.3) to find ζ and ω_0.

(4) Based on the value of ζ, select the proper form of Equation (9.34), and find B_1 and B_2, or B and θ, so as to make Equation (9.33) satisfy the initial conditions of Step 1.

Compared with first-order circuits, second-order circuits require the additional initial condition $dy(0^+)/dt$. Moreover, instead of the single circuit parameter τ, we now need the parameter *pair* ζ and ω_0, which may require the tedious task of formulating a differential equation. In Chapters 14 and 16 we shall use, respectively, *network functions* and *Laplace transforms* to find transient responses much more efficiently.

▶ Example 9.8

Assuming the circuit of Figure 9.9 is in steady state prior to switch activation, find $v(t)$.

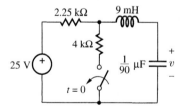

Figure 9.9 Circuit of Example 9.8.

Solution

By the voltage continuity rule, $v(0^+) = v(0^-) = 25$ V. By the capacitance law and the current continuity rule, $dv(0^+)/dt = i_C(0^+)/C = i_L(0^+)/C = i_L(0^-)/C = 0$. In summary,

$$v(0^+) = 25 \text{ V}$$

$$\frac{dv(0^+)}{dt} = 0$$

By inspection, $v(\infty) = [4/(2.25 + 4)]25$, or

$$v(\infty) = 16 \text{ V}$$

Since the resistance and capacitance are in series with each other, we have $\omega_0 = 1/\sqrt{LC} = 1/\sqrt{9 \times 10^{-3}(1/90)10^{-6}} = 100$ krad/s, and $\zeta = (1/2)R_{eq}\sqrt{C/L}$, where R_{eq} is the resistance seen by the LC pair with the source suppressed, or $R_{eq} = 2.25 \parallel 4 = 1.44$ kΩ. Hence, $\zeta = \frac{1}{2} \times 1.44 \times 10^3 \times \sqrt{(1/90)10^{-6}/(9 \times 10^{-3})} = 0.8 < 1$. We thus have an underdamped response with $\alpha = \zeta\omega_0 = 0.8 \times 100 = 80$ kNp/s, and $\omega_d = \sqrt{1 - \zeta^2}\omega_0 = \sqrt{1 - 0.8^2} \times 100 = 60$ krad/s. Its functional form is

$$v(t \geq 0^+) = Be^{-80 \times 10^3 t} \cos(60 \times 10^3 t + \theta) + 16 \text{ V}$$

To find B and θ we observe that

$$v(0^+) = B\cos\theta + 16 = 25 \text{ V}$$

$$\frac{dv(0^+)}{dt} = B(-80 \times 10^3 \cos\theta - 60 \times 10^3 \sin\theta) = 0$$

The second equation yields $\tan\theta = -80/60$, or $\theta = -\tan^{-1}(4/3) = -53.13°$; the first equation yields $B = (25 - 16)/[\cos(-53.13°)] = 15$ V. Consequently,

$$v(t \geq 0^+) = 15e^{-80 \times 10^3 t} \cos(60 \times 10^3 t - 53.13°) + 16 \text{ V}$$

◀

Exercise 9.7   Repeat Example 9.8, but with the switch initially closed and opening at $t = 0$.

ANSWER $v(t \geq 0^+) = 3[e^{-t/(5 \ \mu s)} - 4e^{-t/(20 \ \mu s)}] + 25$ V.

▶ **Example 9.9**

Assuming the circuit of Figure 9.10 is in steady state prior to switch activation, find $v(t)$.

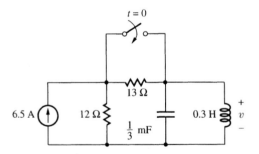

Figure 9.10 Circuit of Example 9.9.

Solution

By the voltage continuity rule, $v(0^+) = v_C(0^+) = v_C(0^-) = v_L(0^-) = 0$. Applying the capacitance law, KCL, the current continuity rule, and the current divider formula, $C\,dv(0^+)/dt = i_C(0^+) = 6.5 - i_L(0^+) =$

$6.5 - i_L(0^-) = 6.5 - [12/(12 + 13)]6.5 = 3.38$ A. Hence, $dv(0^+)/dt = 3.38/(1/3 \times 10^{-3}) = 10,140$ V/s. To summarize,

$$v(0^+) = 0$$

$$\frac{dv(0^+)}{dt} = 10,140 \text{ V/s}$$

By inspection,

$$v(\infty) = 0$$

Since the resistance and capacitance are in parallel with each other, we have $\omega_0 = 1/\sqrt{LC} = 100$ rad/s and $\zeta = 1/(2R_{eq}\sqrt{C/L})$, where R_{eq} is the resistance seen by the LC pair with the source suppressed, or $R_{eq} = 0 + 12 = 12$ Ω. Hence, $\zeta = 1.25 > 1$, indicating an overdamped response. Using Equation (9.35a) we find $\tau_1 = 20$ ms and $\tau_2 = 5$ ms, so

$$v(t \geq 0^+) = B_1 e^{-t/(20 \text{ ms})} + B_2 e^{-t/(5 \text{ ms})}$$

To find B_1 and B_2 we observe that

$$v(0^+) = B_1 + B_2 = 0$$

$$\frac{dv(0^+)}{dt} = \frac{-B_1}{20 \times 10^{-3}} - \frac{B_2}{5 \times 10^{-3}} = 10,140$$

Solving for B_1 and B_2 and substituting gives

$$v(t \geq 0^+) = 67.6(e^{-t/(20 \text{ ms})} - e^{-t/(5 \text{ ms})}) \text{ V}$$

◀

Exercise 9.8   In the circuit of Figure 9.10 find the current flowing downward through the inductance if the switch, after having been closed for a long time, is opened at $t = 0$.

ANSWER $i(t \geq 0^+) = 4.225e^{-60t}\cos(80t - 36.87°) + 3.12$ A.

▶ **Example 9.10**

Assuming the circuit of Figure 9.11 is in steady state prior to switch activation, find $v(t \geq 0^+)$.

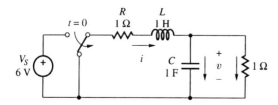

Figure 9.11 Circuit of Example 9.10.

Solution

We have $v(0^+) = v(0^-) = 0$, and $C\,dv(0^+)/dt = i_C(0^+) = i(0^+) - v(0^+)/1 = i(0^-) - 0 = 0$. To summarize,

$$v(0^+) = 0$$

$$\frac{dv(0^+)}{dt} = 0$$

By the voltage divider formula, $v(\infty) = [1/(1+1)]V_S$, or

$$v(\infty) = 3 \text{ V}$$

Suppressing the source reveals that L and C are neither in series nor in parallel with each other. Thus, to find ζ and ω_0, we must derive the differential equation of the circuit. Applying KVL and KCL, along with the element laws, we obtain, for $t \geq 0^+$,

$$V_S = 1i + 1\frac{di}{dt} + v$$

$$i = 1\frac{dv}{dt} + \frac{v}{1}$$

Eliminating i and collecting gives

$$\frac{d^2v}{dt^2} + 2\frac{dv}{dt} + 2v = V_S$$

Comparison with Equation (9.32) yields $\omega_0^2 = 2$ and $2\zeta\omega_0 = 2$, or $\omega_0 = \sqrt{2}$ and $\zeta = 1/\sqrt{2}$. Since $\zeta < 1$, we have an underdamped response with $\alpha = \zeta\omega_0 = 1$ Np/s, and $\omega_d = \sqrt{1 - \zeta^2}\omega_0 = 1$ rad/s. Thus,

$$v(t \geq 0^+) = Be^{-t}\cos(t + \theta) + 3 \text{ V}$$

Imposing the initial conditions to find B and θ yields

$$v(t \geq 0^+) = 3[1 - \sqrt{2}e^{-t}\cos(t - 45°)] \text{ V} \quad \blacktriangleleft$$

Exercise 9.9   Repeat Example 9.10, but with the inductance and capacitance interchanged with each other.

ANSWER $v(t \geq 0) = 3\sqrt{2}e^{-t/2}\cos(t/2 + 45°)$ V.

9.3 STEP RESPONSE OF SECOND-ORDER CIRCUITS

The step response of second-order circuits plays a central role in electrical engineering. Its importance stems from the fact that many circuits of practical interest are of the second order. Even higher-order systems often exhibit a predominantly second-order behavior, this being the reason why design and performance specifications are often given in terms of second-order step-response parameters.

In this section we investigate the step response using the *RLC* circuit of Figure 9.12 as a vehicle. We assume the circuit, after having had sufficient time to reach its dc steady state, is subjected to a voltage step v_I of amplitude V_m, that is

$$v_I = \begin{cases} V_m & \text{for } t > 0 \\ 0 & \text{for } t < 0 \end{cases} \tag{9.38}$$

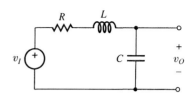

Figure 9.12 Investigating the step response of a series *RLC* circuit.

Using the continuity rules, it is readily seen that the initial conditions are $v_O(0^+) = v_C(0^+) = v_C(0^-) = 0$, or

$$v_O(0^+) = 0 \tag{9.39a}$$

and $dv_O(0^+)/dt = i_C(0^+)/C = i_L(0^+)/C = i_L(0^-)/C = 0/C$, or

$$\frac{dv_O(0^+)}{dt} = 0 \tag{9.39b}$$

Moreover, the steady-state component is readily found as $v_{O(ss)} = v_I(\infty) - Ri_R(\infty) - v_L(\infty) = V_m - R \times 0 - 0$, or

$$v_{O(ss)} = V_m \tag{9.40}$$

As we know, the *undamped natural frequency* ω_0 and the *damping ratio* ζ for a series *RLC* circuit are

$$\omega_0 = \frac{1}{\sqrt{LC}} \tag{9.41a}$$

$$\zeta = \frac{1}{2} R \sqrt{C/L} \tag{9.41b}$$

The step response can be expressed as

$$v_O(t \geq 0^+) = v_{O(\text{xsient})} + V_m \tag{9.42}$$

Depending on the damping conditions, $v_{O(\text{xsient})}$ may take on any one of the three forms of Equation (9.34).

For $\zeta > 1$ the transient component is overdamped, and it consists of two decaying exponentials with time constants

$$\tau_{1,2} = \frac{1}{\left(\zeta \pm \sqrt{\zeta^2 - 1} \right) \omega_0} \tag{9.43}$$

Substituting Equation (9.34a) into Equation (9.42) and imposing the initial conditions of Equation (9.39), we obtain

$$v_O(t \geq 0^+) = V_m \left(1 + \frac{\tau_1}{\tau_2 - \tau_1} e^{-t/\tau_1} + \frac{\tau_2}{\tau_1 - \tau_2} e^{-t/\tau_2} \right) \tag{9.44}$$

A Real Horse Race: The Invention of Television

Every schoolchild can recite the litany of Edison inventing the light bulb, Bell the telephone, Marconi the radio—but who invented television? The answer to this question, in the politically incorrect language of a 1950s magazine article, is that television was definitely "not a one-man invention."

Pioneers in television took one of two roads in designing their systems: mechanical or electronic. In 1884 in Germany, Paul Nipkow patented a television system having a perforated disk that, when rotated in front of an image, divided the image into a series of horizontal lines. This was the first step on the mechanical road to television. The first steps on the electronic road were taken in 1887, when the English physicist William Crookes invented the cathode ray tube (CRT), and in 1897, when Karl Braun of the University of Strasbourg modified the Crookes tube so that its walls would glow when struck by electrons.

In 1911 an English scientist named A. Campbell Swinton described an electronic television system comprised of a Crookes tube as the transmitter and a Braun tube as the receiver. No one working in England at the time, Swinton included, even remotely thought this marvelous system could ever be built, though. At about the same time, a Russian physicist, Boris Rosing, was thinking along the same lines as Swinton. Not so pessimistic as his English counter-parts about the feasibility of assembling such a system, Rosing patented his television tube in 1907. More important as far as the development of tele-

Early televisions were bulky pieces of furniture, as this photograph shows, President Eisenhower is on the screen. (UPI/Bettmann)

vision is concerned, however, is that he passed on his ideas and his enthusiasm to one of his star pupils, Vladimir Zworykin. This young Russian university student, who fled the Revolution and ended up in Pittsburgh working for Westinghouse in 1919, is the first of several inventors who can be called "the father of television."

In the early 1920s progress was being made on both the electronic and the mechanical front. At Westinghouse, Zworykin was perfecting his CRT electronic scanner, while in Schenectady, New York, General Electric was using a mechanical

> *"Every schoolchild can recite the litany of Edison inventing the light bulb, Bell the telephone, Marconi the radio—but who invented television?"*

assemblage of mirrors and rotating perforated disks to send images from a transmitter in the laboratory to receivers in several nearby homes. Meanwhile, out in Utah a high-school student named Philo Farnsworth, the second person entitled to the accolade "father of television," was sketching out for his physics teacher

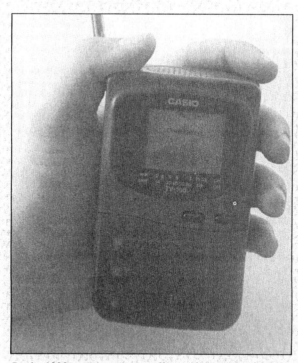

By the 1990s, miniaturization allows televisions to fit in the palm of one's hand—literally. (Michael Newman/PhotoEdit)

how an electronic television system could be built. Farnsworth found a financial angel in the person of George Everson, who bankrolled the boy in a secret laboratory, located first in Los Angeles and later in San Francisco, to work out his electronic television scheme.

Zworykin's work in Pittsburgh bore fruit first, and in December 1923 he filed for a patent on a television camera, billed as "an iconoscope, the first entirely electronic sender." His patent was not granted until December 1938, however, and the intervening unprotected years allowed Farnsworth to get himself firmly established. One possible reason for the problems Zworykin's patent application ran into was lack of support from the people at Westinghouse. After a 1924 demonstration of fuzzy images, Zworykin was sent back to the drawing board, where he stayed for the next six years. When his bosses were no more enthusiastic about his 1930 model, Zworykin decided it was time to quit.

At the same time, a self-taught inventor in Dayton, Ohio, C. Francis Jenkins, working alone with no financial backing, used a mechanical transmitter to send an image across the airwaves from Anacosta, Virginia, to Washington, DC. Setting up his own broadcasting station in Maryland, he began selling television receivers as early as 1928.

The United States was not the only hot place for television research. England had the bug, too. Yet another "father" of the medium is the Scotsman John Logie Baird, who in 1925 used Nipkow's perforated disk to stage London's first television demonstration.

And so the race was on.

1927: Farnsworth applies for a patent on his electronic transmitter, which he calls an "image dissector"; AT&T uses its mechanical setup to broadcast Herbert Hoover's image from Washington to New York and begins broadcasting a signal from New York to the suburbs of New Jersey.

1928: GE broadcasts with a mechanical system in Schenectady; Baird uses his mechanical system to send a signal from England to the United States; Sarnoff tells RCA stockholders that the day of "sight by radio" is just around the corner; Jenkins is selling his receivers to thousands of customers.

1930: Sarnoff hires Zworykin, who tells his new boss it will cost about $100,000 to get his system up and running (the final bill would run to more than $5 million); Farnsworth's patent granted; Zworykin checks up on what Farnsworth is doing and reports that it's nothing RCA should worry about.

1931: Baird is still touting his mechanical system in England; Electric and Musical Industries (EMI) in England and Westinghouse, RCA, AT&T, and Farnsworth in the United States are all taking the electronic path; GE sticks with the mechanical.

1932: *Farnsworth v Zworykin* lawsuit filed, with every judicial level upholding Farnsworth's claim to the first electronic TV patents.

(Continued)

(*Continued*)

1933: Zworykin's improved iconoscope camera allows scenes to be lit with regular levels of light, rather than the extemely bright (and therefore hot) lights necessary for Farnsworth's camera.

1936: England's BBC broadcasts alternate weeks of Baird's mechanical system and EMI's electronic system and lets the public decide; EMI wins.

1939: Its TV system a big hit at the New York World's Fair, RCA begins the first U.S. electronic television service; RCA settles with Farnsworth on the 1932 lawsuit: Sarnoff wants to buy Farnsworth out, but all Farnsworth will agree to is to license RCA to produce television systems using his dissector; it is the first (and last) time RCA is on the paying end of a licensing deal; head RCA patent lawyer Otto Schairer has tears in his eyes when he signs the agreement.

1941: FCC authorizes commercial TV broadcasting; World War II stops all television research.

1946: Uncle Milty enters America's living rooms.

So who *is* the inventor of television? Most historians agree that, on the U.S. front, the joint "winners" were Farnsworth and Zworykin. Farnsworth had most of the ideas in place first, and the patent protection, but his camera—the "image dissector"—needed unbelievably intense lighting; Zworykin's camera needed much less light and is the one that paved the way for the cameras of today.

Jenkins's enterprise was not successful mainly because it was mechanical, and he died in 1934 without ever jumping on the electronic bandwagon. His lack of financial backing up against the research bud-

> "*Pioneers in television took one of two roads in designing their systems: mechanical or electronic.*"

gets of RCA and others probably also had something to do with his oblivion. In Great Britain, Baird is considered the first name in television despite the fact that it is Swinton's electronic system that has endured.

▶ Example 9.11

In the circuit of Figure 9.12 let $R = 3\ \Omega$, $L = 1$ H, and $C = 1$ F. Find the response to a 1-V input step.

Solution

By Equation (9.41) we have $\omega_0 = 1/\sqrt{1 \times 1} = 1$ rad/s and $\zeta = (3/2)\sqrt{1/1} = 1.5 > 1$. By Equation (9.43), $\tau_1 = 1/[(1.5 - \sqrt{1.5^2 - 1})1] = 2.618$ s and $\tau_2 = 1/[(1.5 + \sqrt{1.5^2 - 1})1] = 0.382$ s. Substituting into Equation (9.44) we obtain

$$v_O(t \geq 0^+) = 1 - 1.171e^{-t/2.618} + 0.171e^{-t/0.382} \text{ V}$$

This response is shown in Figure 9.13 as the curve labeled $\zeta = 1.5$.

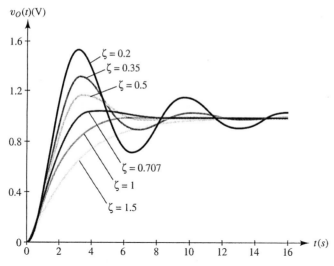

Figure 9.13 Step response of a series RLC circuit with $\omega_0 = 1$ rad/s as a function of the damping ratio ζ.

Exercise 9.10   In the circuit of Figure 9.12 let $L = 10$ mH and $C = 10$ nF. Specify the value of R for an overdamped response with $\zeta = 1.25$. Hence, find the value of the response 10 μs after the application of a 10-V input step.

ANSWER   2.5 kΩ, 2.364 V.

For $\zeta = 1$ the transient component is critically damped with the time constant

$$\tau = \frac{1}{\omega_0} \tag{9.45}$$

Substituting Equation (9.34b) into Equation (9.42) and imposing the initial conditions of Equation (9.39), we obtain

$$v_O(t \geq 0^+) = V_m \left[1 - \left(1 + \frac{t}{\tau} \right) e^{-t/\tau} \right] \tag{9.46}$$

▶ Example 9.12

Find the value of R that makes the circuit of Example 9.11 critically damped. Hence, find the response to a 1-V input step.

Solution

$R = 2\zeta/\sqrt{C/L} = 2 \times 1/\sqrt{1/1} = 2\ \Omega$. Using Equation (9.45), $\tau = 1/1 = 1$ s. Substituting into Equation (9.46) gives

$$v_O(t \geq 0^+) = 1 - (1+t)e^{-t}\ \text{V}$$

This response is shown in Figure 9.13 as the curve labeled $\zeta = 1$.

◀

Exercise 9.11   Repeat Exercise 9.10, but for the case $\zeta = 1$.

ANSWER   2 kΩ, 2.642 V.

It is interesting to compare the step response of a critically damped *RLC* circuit with that of a plain *RC* circuit having the same time constant τ. Recall that the step response of the *RC* circuit is $v_O(t \geq 0^+) = V_m(1 - e^{-t/\tau})$. Since its initial slope is $dv_O(0^+)/dt = V_m/\tau \neq 0$, the *RC* response starts to *rise immediately.* By contrast, the *RLC* response exhibits some *initial delay* because its initial slope is zero, by Equation (9.39b). Physically, this stems from the current-choking action by the inductance.

We also note that because of the presence of the t/τ term in Equation (9.46), the *RLC* response is *slower* than the *RC* response. For instance, at $t = \tau$ the *RC* response has already risen to $v_O(\tau) = V_m(1 - e^{-1}) \simeq 0.63V_m$, while the *RLC* response is still at $v_O(\tau) = V_m[1 - (1 + \tau/\tau)e^{-1}] \simeq 0.26V_m$. Physically, this stems again from the current-choking action by the inductance, which slows down the charging of the capacitance.

For $0 < \zeta < 1$ the transient component is underdamped, and it consists of a decaying sinusoid with

$$\alpha = \zeta\omega_0 \tag{9.47a}$$

$$\omega_d = \omega_0\sqrt{1 - \zeta^2} \tag{9.47b}$$

Substituting Equation (9.34c) into Equation (9.42) and imposing the initial conditions of Equation (9.39), we obtain

$$v_O(t \geq 0^+) = V_m\left(1 - \frac{e^{-\alpha t}}{\sqrt{1 - \zeta^2}}\cos\left(\omega_d t - \sin^{-1}\zeta\right)\right) \tag{9.48}$$

▶ **Example 9.13**

Repeat Example 9.11, but for $R = 1\ \Omega$.

Solution

We now have $\zeta = (1/2)\sqrt{1/1} = 0.5 < 1$, so $\alpha = 0.5 \times 1 = 0.5$ Np/s, and $\omega_d = \sqrt{1 - 0.5^2} \times 1 = 0.8660$ rad/s. Substituting into Equation (9.48) yields

$$v_O(t \geq 0^+) = 1 - 1.155e^{-0.5t}\cos(0.8660t - 30°) \text{ V}$$

This response is shown in Figure 9.13 as the curve labeled $\zeta = 0.5$. ◀

Exercise 9.12   Assuming $C = 1 \mu F$ in the circuit of Figure 9.12, specify suitable values for R and L so that the roots of the characteristic equation are $s_{1,2} = (-4 \pm j3)10^3$ complex Np/s. Hence, find the circuit's response to a 10-V input step.

ANSWER $R = 320 \Omega$, $L = 40$ mH, $v_O(t \geq 0^+) = 10[1 - (5/3) \times e^{-4 \times 10^3 t}\cos(3 \times 10^3 t - 51.13°)]$ V.

For convenience, also shown in Figure 9.13 are the responses corresponding to $\zeta = 1/\sqrt{2}$, $\zeta = 0.35$, and $\zeta = 0.2$. We immediately note that the smaller the value of ζ, the more rapidly the response rises. Moreover, for $\zeta < 1$ we observe two distinctive features:

(1) The underdamped responses rise *above* V_m, a phenomenon referred to as **overshoot.**

(2) After the overshoot, the underdamped responses decay toward the steady-state value V_m in an oscillatory manner, a phenomenon referred to as **ringing.**

It is readily seen that the smaller the value of ζ, the more pronounced the overshoot and the longer the time it takes for the ringing effect to die out.

Overshoot

The maxima and minima of an underdamped response are found by calculating the time derivative of Equation (9.48) and setting the result to zero. The derivative is

$$\frac{dv_O}{dt} = B\left[-\alpha e^{-\alpha t}\cos(\omega_d t + \theta) - \omega_d e^{-\alpha t}\sin(\omega_d t + \theta)\right]$$

where $B = -V_m/\sqrt{1 - \zeta^2}$ and $\theta = -\sin^{-1}\zeta$. Letting $\alpha = \zeta\omega_0 = (-\sin\theta)\omega_0$ and $\omega_d = \omega_0\sqrt{1 - \zeta^2} = (\cos\theta)\omega_0$ and factoring out the common terms we obtain

$$\frac{dv_O}{dt} = B\omega_0 e^{-\alpha t}[\sin\theta\cos(\omega_d t + \theta) - \cos\theta\sin(\omega_d t + \theta)]$$

By a well-known trigonometric identity, the term within brackets reduces to $\sin(\omega_d t + \theta - \theta) = \sin\omega_d t$. Thus

$$\frac{dv_O}{dt} = B\omega_0 e^{-\alpha t}\sin\omega_d t \qquad\qquad (9.49)$$

This derivative vanishes for $t = \infty$ and for $\omega_d t_n = n\pi$, or

$$t_n = \frac{n\pi}{\omega_0\sqrt{1 - \zeta^2}} \qquad n = 0, 1, 2, \dots \qquad (9.50)$$

We are interested in the first maximum, which occurs at

$$t_1 = \frac{\pi}{\omega_0 \sqrt{1 - \zeta^2}}$$

(9.51)

Substituting into Equation (9.48) yields

$$v_O(t_1) = V_m \left[1 - \frac{1}{\sqrt{1 - \zeta^2}} \exp\left(\frac{-\pi\zeta}{\sqrt{1 - \zeta^2}} \right) \cos(\pi - \theta) \right]$$

Since $\cos(\pi - \theta) = -\cos\theta = -\sqrt{1 - \zeta^2}$, we finally obtain

$$v_O(t_1) = V_m \left[1 + \exp\left(\frac{-\pi\zeta}{\sqrt{1 - \zeta^2}} \right) \right]$$

(9.52)

Referring to Figure 9.14, we define the normalized **overshoot** as

$$\text{OS} = \frac{v_O(t_1) - V_m}{V_m}$$

Substituting and simplifying yields

$$\text{OS} = \exp \frac{-\pi\zeta}{\sqrt{1 - \zeta^2}}$$

(9.53)

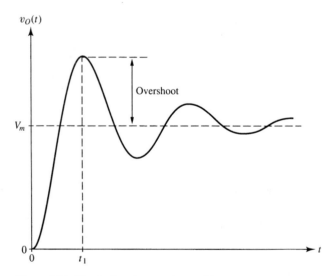

Figure 9.14 Illustrating the overshoot (OS).

Multiplying by 100 yields the overshoot in percentage form. We note that the overshoot depends only on the damping ratio ζ. There is no overshoot for $\zeta \geq 1$. However, for $0 < \zeta < 1$, the smaller the value of ζ, the larger the overshoot. In the limit $\zeta \to 0$ the overshoot approaches 100%.

▶ Example 9.14

Calculate the percentage overshoot of the circuit of Example 9.13 as well as the instant at which the overshoot occurs.

Solution

Using Equation (9.53), OS $= \exp[(-\pi \times 0.5)/\sqrt{1 - 0.5^2}] = 0.163$, or 16.3%. By Equation (9.51), $t_1 = \pi/(1\sqrt{1 - 0.5^2}) = 3.63$ s.   ◀

Exercise 9.13   Calculate the percentage overshoot of the circuit of Exercise 9.12, as well as the instant at which the overshoot occurs.

ANSWER   1.5%, 1.047 ms.

Settling Time

The time it takes for an underdamped response to settle within a given band $V_m(1 \pm \varepsilon)$ is called the **settling time** and is denoted as t_s. As shown in Figure 9.15, this parameter can be estimated as the time it takes for the positive envelope to decay to the value $(1 + \varepsilon)V_m$,

$$V_m\left(1 + \frac{1}{\sqrt{1 - \zeta^2}}e^{-\alpha t_s}\right) \simeq (1 + \varepsilon)V_m$$

Simplifying and rearranging yields $e^{\alpha t_s} \simeq 1/(\varepsilon\sqrt{1 - \zeta^2})$. Taking the natural

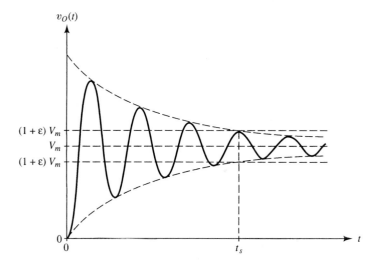

Figure 9.15 Illustrating the settling time t_s.

logarithm of both sides and letting $\alpha = \zeta\omega_0$ finally yields

$$t_s \simeq \frac{-1}{\zeta\omega_0} \ln\left(\varepsilon\sqrt{1-\zeta^2}\right) \qquad\qquad \textbf{(9.54)}$$

It is readily seen that the smaller the value of ε or of ζ, the longer the settling time. For $\zeta \ll 1$, Equation (9.54) can be approximated as $t_s \simeq -(\ln\varepsilon)/\zeta\omega_0$.

▶ **Example 9.15**

(a) Estimate the time it takes for the response of Example 9.13 to settle within 5% of its final value.

(b) Repeat, but for 0.1% of the final value.

Solution

(a) We have $\zeta\omega_0 = 0.5$ Np/s, $\sqrt{1-\zeta^2} = 0.8660$, and $\varepsilon = 0.05$. By Equation (9.54), $t_s = -(1/0.5)\ln(0.05 \times 0.8660) = 6.279$ s.

(b) Likewise, $t_s = -(1/0.5)\ln(0.001 \times 0.8660) = 14.10$ s.   ◀

Exercise 9.14   Find the time it takes for the response of the circuit of Exercise 9.12 to settle within 1 mV of its final value.

ANSWER   2.43 ms.

9.4 SECOND-ORDER OP AMP CIRCUITS

At the end of Section 8.5 we demonstrated the unique ability of first-order op amp circuits to create roots in the **right half of the complex plane,** a feature not possible with purely passive circuits. In this section we investigate second-order op amp circuits and demonstrate that it is possible not only to create roots in the right half of the complex plane but also to create **complex roots without using any inductances,** that is, using only resistances and capacitances as passive elements. Two important advantages accrue from the use of op amps:

(1) The ability to manipulate root locations in the complex plane opens up a wide range of applications for *RC* op amp circuits, such as *filters* and *oscillators.*

(2) The elimination of inductances allows for a high degree of circuit miniaturization, as practical inductors tend to be bulky and heavy. It also results in improved circuit performance, as the inductor is the least ideal of the three basic elements.

To fully appreciate the role of the op amp, we first examine a second-order passive circuit containing resistances and capacitances but no inductances, and we show that the roots of its characteristic equation are always *real* and *negative*. Next, we demonstrate that the inclusion of an op amp to provide a controlled amount of positive feedback allows for the same circuit to realize *complex roots*, and that these roots can be placed anywhere in the complex plane, including the *right half*.

A Second-Order Passive Inductorless Circuit

The passive circuit of Figure 9.16 consists of two cascaded *RC* stages and contains no inductors. To find ζ and ω_0, we first derive the differential equation governing the circuit, then we put it in the standard form of Equation (9.32) to extract ζ and ω_0.

By KCL, $i_1 = i_2 + i_3$ and $i_2 = i_4$. Expressing currents in terms of voltages via Ohm's Law and the capacitance law,

$$\frac{v_S - v_1}{R_1} = \frac{v_1 - v}{R_2} + C_1 \frac{dv_1}{dt}$$

$$\frac{v_1 - v}{R_2} = C_2 \frac{dv}{dt}$$

Eliminating v_1 and rearranging yields,

$$\frac{d^2v}{dt^2} + 2\zeta\omega_0 \frac{dv}{dt} + \omega_0^2 v = \omega_0^2 v_S$$

where

$$\omega_0 = \frac{1}{\sqrt{R_1 R_2 C_1 C_2}} \tag{9.55}$$

$$\zeta = \frac{1}{2} \frac{R_1 C_1 + R_2 C_2 + R_1 C_2}{\sqrt{R_1 R_2 C_1 C_2}} \tag{9.56}$$

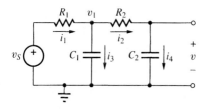

Figure 9.16 A second-order passive inductorless circuit.

As we know, ζ determines the damping characteristics of the circuit. To examine it in closer detail, we introduce the variable

$$x = \frac{R_1 C_1}{R_2 C_2} \tag{9.57}$$

after which ζ can be expressed as

$$\zeta = \frac{1}{2}\left[\left(1 + \frac{R_1}{R_2}\right)\sqrt{x} + \frac{1}{\sqrt{x}}\right] \tag{9.58}$$

The reader can easily verify that

$$\zeta \geq 1 \tag{9.59}$$

and that the *minimum* is reached in the limits $R_1/R_2 \to 0$ and $x \to 1$. Consequently, this circuit admits only **real, negative roots,** so its natural response can never be of the underdamped type.

▶ **Example 9.16**

In the circuit of Figure 9.16 let $C_1 = C_2 = 10$ nF, $R_1 = 40$ kΩ, and $R_2 = 10$ kΩ. Find the roots of its characteristic equation.

Solution

By Equations (9.55) and (9.56),

$$\omega_0 = \frac{1}{\sqrt{4 \times 10^4 \times 10^4 \times (10^{-8})^2}} = 5 \text{ krad/s}$$

$$\zeta = \frac{1}{2}(4 \times 10^4 + 10^4 + 4 \times 10^4)10^{-8} \times \omega_0 = 2.25$$

indicating an overdamped response. The roots are $s_{1,2} = (-\zeta \pm \sqrt{\zeta^2 - 1})\omega_0 = (-2.25 \pm \sqrt{2.25^2 - 1})5000$, or

$$s_1 = -1172 \text{ Np/s}$$

$$s_2 = -21,328 \text{ Np/s}$$

As expected, they are real, negative, and distinct.   ◀

Exercise 9.15   Find $v(t)$ in the circuit of Example 9.16 if v_S is a 2-V positive-going step and both capacitors are initially discharged.

ANSWER $v(t \geq 0^+) = 2 - 2.116e^{-t/(853.1 \ \mu s)} + 0.116e^{-t/(46.89 \ \mu s)}$ V.

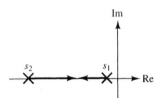

Figure 9.17 Root locus for the passive inductorless circuit of Figure 9.16.

Figure 9.17 shows the root locus of our circuit. It is interesting that the circuit yields **distinct roots** even when $R_1C_1 = R_2C_2$. Physically, this stems from the fact that the R_1C_1 stage is being *loaded* by the R_2C_2 stage. Loading can be *eliminated* by interposing a voltage buffer between the two *RC* stages. If this is done, with $R_1C_1 = R_2C_2$, the roots will be coincident. Alternately, loading can be *minimized* by imposing $R_1 \ll R_2$.

In summary, the natural response of the passive inductorless circuit of Figure 9.16 can never be of the underdamped type. It can be made to approach *critical damping* by letting $R_2 \gg R_1$ to minimize loading, and letting $C_2 = (R_1/R_2)C_1$ to minimize ζ.

▶ **Example 9.17**

   (a) In the circuit of Figure 9.16 specify suitable component values for two real negative roots $s_1 = -500$ Np/s and $s_2 = -2000$ Np/s.

   (b) Modify the component values so that the roots are brought closer together.

Solution

(a) By Equation (9.9) we have

$$\frac{s_2}{s_1} = \frac{-\zeta - \sqrt{\zeta^2 - 1}}{-\zeta + \sqrt{\zeta^2 - 1}} = \frac{-2000}{-500} = 4$$

which yields

$$\zeta = 1.25$$

Then, $\omega_0 = s_1/(-\zeta + \sqrt{\zeta^2 - 1}) = -500/(-1.25 + \sqrt{1.25^2 - 1})$, or

$$\omega_0 = 10^3 \text{ rad/s}$$

Arbitrarily impose $R_1 C_1 = R_2 C_2$ to make $x = 1$, by Equation (9.57). Then, Equation (9.55) requires $R_1 C_1 = R_2 C_2 = 1/\omega_0 = 1$ ms, and Equation (9.58) requires $R_1/R_2 = 2\zeta - 2 = 2 \times 1.25 - 2 = 0.5$.
   Let $C_1 = 10$ nF; then, $R_1 = 10^{-3}/10^{-8} = 100$ kΩ, $R_2 = 100/0.5 = 200$ kΩ, and $C_2 = 10^{-3}/(200 \times 10^3) = 5$ nF. In short, use $R_1 = 100$ kΩ, $R_2 = 200$ kΩ, $C_1 = 10$ nF, and $C_2 = 5$ nF.

(b) To bring the roots closer to each other, impose $R_1 \ll R_2$. For instance, leaving $R_2 = 200$ kΩ and $C_2 = 5$ nF, impose $R_1 = 2$ kΩ, which is two orders of magnitude less than R_2. Then, to retain $x = 1$, impose $C_1 = 10^{-3}/(2 \times 10^3) = 500$ nF. Substituting these values into Equation (9.56) indicates that we now have $\zeta = 1.005$, so that $s_{1,2} = (-\zeta \pm \sqrt{\zeta^2 - 1})\omega_0 = (-1.005 \pm \sqrt{1.005^2 - 1}) \times 10^3$, or $s_1 = -905$ Np/s and $s_2 = -1105$ Np/s. Though not exactly coincident, the roots are close enough to make damping effectively critical. ◀

Exercise 9.16   Modify the component values of Example 9.17, part (b), to bring the roots even closer together at $s_1 = -990$ Np/s and $s_2 = -1010$ Np/s.

ANSWER $R_1 = 20$ Ω, $C_1 = 50$ μF, $R_2 = 200$ kΩ, $C_2 = 5$ nF.

A Second-Order Active Inductorless Circuit

The circuit of Figure 9.18 is similar to that of Figure 9.16 except that the output of the second RC stage is now fed to an amplifier and the amplifier output is then fed back via the upper capacitance. Since the amplifier is of the noninverting type, this modification is said to provide **positive feedback.** The circuit is said to be an **active circuit** because of the presence of the amplifier, with its unique ability to draw energy from its own power supplies and inject it into the surrounding circuitry. By contrast, the circuit of Figure 9.16 is a *purely passive circuit.* To simplify our derivations, we have arbitrarily imposed equal capacitances and 4-to-1 resistances, as shown. We now wish to find ζ and ω_0.

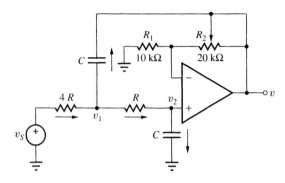

Figure 9.18   A second-order active inductorless circuit.

The transfer characteristic of the amplifier is $v = Kv_2$, where

$$K = 1 + \frac{R_2}{R_1} \qquad (9.60)$$

Since the circuit consists of two RC stages and a gain-of-K amplifier, it is also referred to as a **KRC circuit.** Pursuing a similar approach as for Figure 9.16, we have

$$\frac{v_S - v_1}{4R} = C\frac{d(v_1 - v)}{dt} + \frac{v_1 - v_2}{R}$$

$$\frac{v_1 - v_2}{R} = C\frac{dv_2}{dt}$$

Substituting $v_2 = v/K$, eliminating v_1, and rearranging gives

$$\frac{d^2v}{dt^2} + 2\zeta\omega_0\frac{dv}{dt} + \omega_0^2 v = K\omega_0^2 v_S$$

where

$$\omega_0 = \frac{1}{2RC} \qquad (9.61)$$

$$\zeta = 2.25 - K \qquad (9.62)$$

We observe that now ζ depends on the amplifier gain, indicating that by proper selection of K we can make ζ assume a variety of different values, including *zero* or even *negative* values! With the values of R_1 and R_2 shown, gain is variable over the range 1 V/V $\leq K \leq$ 3 V/V, so ζ can be varied over the range

$$-0.75 \leq \zeta \leq 1.25 \qquad (9.63)$$

Let us now investigate circuit behavior as a function of K.

For 1 V/V $\leq K \leq$ 1.25 V/V we obtain $1 \leq \zeta \leq 1.25$. As for the passive inductorless circuit, the roots are real and negative.

For $K >$ 1.25 V/V we obtain $\zeta < 1$, causing the roots to split apart and become complex conjugate. As we know, this results in oscillatory behavior.

In an *RLC* circuit this behavior stems from the ability of stored energy to flow back and forth between the capacitance and the inductance. In a *KRC* circuit there are no inductances, but it is now the amplifier that acts as a vehicle for energy flow between its own power supplies and the two *RC* stages. This flow takes place via the feedback capacitance.

For $K = 2.25$ V/V we obtain $\zeta = 0$, indicating purely imaginary roots and an *undamped oscillation.* Physically, when K is set to this value, the amount of energy injected via the feedback path matches *exactly* that dissipated by the resistances of the two *RC* stages. In *RLC* circuits the undamped condition is achieved by letting $R \rightarrow 0$ in the series case, or $R \rightarrow \infty$ in the parallel case. In practice, however, neither circuit will meet this condition because of component nonidealities and other parasitic losses, particularly in the inductor. Thus, any oscillation in a passive *RLC* circuit will eventually die out. By contrast, in a *KRC* circuit we can achieve a *sustained oscillation* by adjusting K to 2.25 V/V.

Creating Roots in the Right Half of the *s* Plane

The most striking feature of the *KRC* circuit, however, is its ability to provide *negative values* of ζ, a feature not possible with passive *RLC* circuits. In fact, for $K > 2.25$ V/V we obtain $\zeta < 0$. Since the roots of the characteristic equation are $s_{1,2} = -\zeta\omega_0 \pm j\omega_0\sqrt{1 - \zeta^2}$, with $\zeta < 0$ the real part, $-\zeta\omega_0$ becomes *positive,* indicating that the root locus will now extend into the *right half* of the complex plane. Moreover, the response can be written as

$$v_O(t) = Ae^{|\zeta|\omega_0 t} \cos\left(\omega_0\sqrt{1 - \zeta^2}\,t + \theta\right) \qquad \textbf{(9.64)}$$

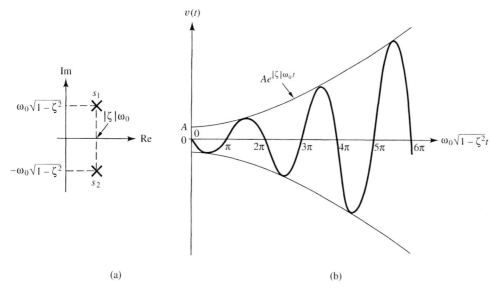

(a)   (b)

Figure 9.19 A complex root pair in the *right half* of the *s* plane corresponds to a *diverging sinusoid.*

where A and θ are the usual initial-conditions constants. This function represents a sinusoid having *angular frequency* $\omega_0\sqrt{1-\zeta^2}$, *phase angle* θ, and an *exponentially increasing amplitude,* $Ae^{|\zeta|\omega_0 t}$. Such a function, shown in Figure 9.19, is referred to as a **diverging sinusoid.** Physically, when $K > 2.25$ V/V, the amount of energy injected by the amplifier during each cycle *exceeds* that dissipated by the resistances of the two RC stages, thus yielding a regenerative effect in which the oscillations grow in magnitude according to a geometric progression.

In a practical circuit the oscillations will grow until some *nonlinearity,* which is always present in some form, limits the output swing and, hence, the amount of energy that the amplifier can reinject into the circuit. Once this condition is achieved, the circuit will provide sustained oscillations. The mechanism for starting oscillations using a negative value of ζ and then exploiting a nonlinearity to achieve $\zeta = 0$ at the desired amplitude is applied to the design of *sinusoidal oscillators.*

• The Root Locus

Figure 9.20 shows the root locus of our illustrative circuit. Its salient features are summarized as follows: for 1 V/V $\leq K \leq 1.25$ V/V the locus lies on the negative real axis; for 1.25 V/V $\leq K \leq 2.25$ V/V the locus consists of circular arcs in the left half of the complex plane; for 2.25 V/V $\leq K \leq 3$ V/V the locus extends into the right half of the complex plane. For $K \geq 1.25$ V/V, the radial distance of either root from the origin is ω_0, indicating that the trajectories are *circular arcs* with radius ω_0.

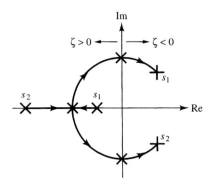

Figure 9.20 Root locus for the circuit of Figure 9.18.

▶ Example 9.18

In the circuit of Figure 9.18 specify suitable component values for $s_{1,2} = +600 \pm j800$ complex Np/s.

Solution

We must have

$$-\zeta\omega_0 = 600 \text{ Np/s}$$

$$\omega_0\sqrt{1-\zeta^2} = 800 \text{ rad/s}$$

Dividing the two equations pairwise yields $-\zeta/\sqrt{1-\zeta^2} = 600/800$, whose solution is $\zeta = -0.6$. Then, $\omega_0 = 600/(-\zeta) = 600/0.6 = 10^3$ rad/s.

To achieve the desired value of ω_0, arbitrarily select $C = 0.1$ μF. Then, Equation (9.61) requires $R = 1/(2\omega_0 C) = 1/(2 \times 10^3 \times 10^{-7}) = 5$ kΩ. We thus use $R = 5$ kΩ and $4R = 20$ kΩ.

To achieve the desired value of ζ we must have, by Equation (9.62), $K = 2.25 - \zeta = 2.25 + 0.6 = 2.85$. Equation (9.60) requires, in turn, that $R_2/R_1 = K - 1 = 1.85$. With $R_1 = 10$ kΩ, R_2 must be set to 18.5 kΩ. ◀

Exercise 9.17 Suppose in the circuit of Figure 9.18 the resistance denoted as $4R$ is changed to R to simplify inventory. Moreover, let R_2 be a 50-kΩ potentiometer. (a) Show that we now have $\omega_0 = 1/RC$ and $\zeta = 1.5 - 0.5K$. (b) Specify R, C, and the wiper settings of R_2 for $\omega_0 = 10^4$ rad/s and $\zeta = 0.5$, 0, and -0.5. In each case, sketch and label the root locations in the s plane.

ANSWER (b) $R = 10$ kΩ, $C = 10$ nF, $R_2 = 10$ kΩ, 20 kΩ, and 30 kΩ.

9.5 TRANSIENT ANALYSIS USING SPICE

The SPICE facilities discussed in Section 8.5 can easily be applied to the analysis of second-order circuits. An example will better illustrate.

▶ Example 9.19

Use PSpice to find the response of a series RLC circuit to a 10-V input step. Use $R = 400$ Ω, $L = 10$ mH, and $C = 0.01$ μF, and observe the response across the capacitance.

Solution

The circuit is shown in Figure 9.21, and the input file is

```
STEP RESPONSE OF SECOND-ORDER CIRCUIT
V1 1 0 PULSE(0 10 10US 1US 1US 250US)
R 1 2 400
L 2 3 10MH IC=0
C 3 0 0.01UF IC=0
.TRAN 5US 250US UIC
.PROBE
.END
```

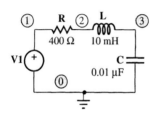

Figure 9.21 Second-order circuit of Example 9.19.

After directing the Probe post-processor to display `V(1)` and `V(3)` we obtain the traces of Figure 9.22.

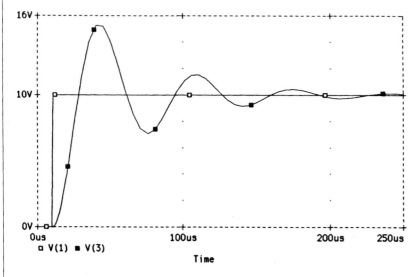

Figure 9.22 Step response of the circuit of Figure 9.21.

Exercise 9.18   Repeat Example 9.19, but for the following resistance values: (a) $R = R_c$, where R_c is the resistance corresponding to critical damping; (b) $R = 10R_c$; (c) $R = 0.1R_c$.

Exercise 9.19   Use PSpice to find the response of the circuit of Example 9.19 to a 5-kHz, 2-V peak-to-peak square wave.

Exercise 9.20   In the second-order op amp circuit of Figure 9.18 let $R = 10$ kΩ, $4R = 40$ kΩ, and $C = 10$ nF. Use PSpice to find the circuit's response to a 1-V input step if the wiper is set halfway.

▼ SUMMARY

- The second-order circuits of greatest interest are the *series RLC* and the *parallel RLC configurations*. Their transient behavior is characterized in terms of the *undamped natural frequency* ω_0 and the *damping ratio* ζ.

- Both circuits have $\omega_0 = 1/\sqrt{LC}$. Moreover, $\zeta = (1/2)R\sqrt{C/L}$ for the *series RLC*, and $\zeta = 1/(2R\sqrt{C/L})$ for the *parallel RLC*. The parameter ζ controls the s-plane locations of the roots of the characteristic equation as well as the nature of the transient response.

- For $\zeta > 1$ the roots are *real, negative,* and *distinct,* and lie on the *negative real axis* of the s plane. The transient response is *overdamped* and consists of two exponentially decaying components.

- For $0 < \zeta < 1$ the roots are *complex conjugate,* and lie in the *left half* of the *s* plane, at locations *symmetric* about the real axis. The transient response is a *damped sinusoid* with *damped frequency* $\omega_d = \omega_0\sqrt{1 - \zeta^2}$, and *damping coefficient* $\alpha = \zeta\omega_0$. Moreover, ω_d represents the *ordinate* of the upper root, and $-\alpha$ its abscissa.

- For $\zeta = 1$ the roots are real and coincident and lie at the same location on the *negative real axis.* The transient response is now said to be *critically damped.*

- For $\zeta = 0$ the roots lie on the imaginary axis, at locations symmetric with respect to the origin. The response is an *undamped sinusoid* with frequency ω_0.

- The *oscillatory behavior* of underdamped and undamped RLC circuits stems from the flow of energy back and forth between L and C.

- The step response of an underdamped circuit exhibits *overshoot* and *ringing.* The smaller ζ is, the greater the overshoot and the longer the ringing.

- The *settling time* is the time it takes for a step response to settle within a given band around its final, steady-state value.

- A second-order passive circuit in which the energy-storage elements are of the same type has always $\zeta \geq 1$, and its transient response can never be underdamped.

- Using an op amp to provide a controlled amount of positive feedback around a second-order passive circuit made up of two RC stages, we can make ζ less than unity, or even negative.

- If ζ is *negative,* the roots lie in the *right half* of the *s* plane. If the root pair is complex, the response is a *diverging sinusoid.*

- The presence of the op amp not only allows for a far greater control over root locations, but it avoids the use of inductors, which are generally undesirable components.

▼ PROBLEMS

9.1 Natural Response of Series and Parallel *RLC* Circuits

9.1 Show that for an *overdamped* circuit with roots s_1 and s_2 we have $\omega_0 = \sqrt{s_1 s_2}$ and $\zeta = |s_1 + s_2|/2\omega_0$.

9.2 (a) A *parallel RLC* circuit has $\zeta = 1.25$ and $\omega_0 = 20$ krad/s. If $R = 80 \ \Omega$, find L and C as well as the roots s_1 and s_2. (b) If the same elements are connected in *series,* how do we change R to ensure the same damping ratio?

9.3 (a) Specify suitable element values for a *parallel RLC* circuit so that $s_1 = -2000$ Np/s and $s_2 = -5000$ Np/s, under the constraint that the initial resistance current, in mA, and the initial resistance voltage, in V, be numerically equal. (b) Repeat, but for a *series RLC* circuit. (*Hint:* Use the results of Problem 9.1.)

9.4 For $t \geq 0^+$, the voltage in the *parallel RLC* circuit of Figure 9.2(b) is $v = 2(5e^{-10t} - e^{-5t})$ V. If $R = 2/3 \ \Omega$, find L and C, as well as their initial stored energies. (*Hint:* Use the results of Problem 9.1.)

9.5 For $t \geq 0^+$, the current in the *series RLC* circuit of Figure 9.2(a) is $i = 7e^{-5t} - 3e^{-t}$ A. If $R = 6 \ \Omega$, find the stored energies in L and C at $t = 0^+$. (*Hint:* Use the results of Problem 9.1.)

9.6 (a) Find the time t_m at which the response of the *overdamped series RLC* circuit of Example 9.2 reaches its greatest magnitude, as well as the magnitude itself. (b) Repeat, but for the *overdamped parallel RLC* circuit of Exercise 9.2.

9.7 Find $v(t \geq 0)$ in the circuit of Figure P9.7, given that $v_C(0) = 9$ V and $i_L(0) = 0.5$ A.

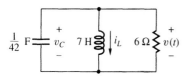

Figure P9.7

9.8 (a) Show that for an *underdamped* circuit with roots $s_{1,2} = -\alpha \pm j\omega_d$, we have $\omega_0 = \sqrt{\alpha^2 + \omega_d^2}$ and $\zeta = \alpha/\sqrt{\alpha^2 + \omega_d^2}$. (b) Specify suitable element values for a *series RLC* circuit with $s_{1,2} = -3000 \pm j4000$ complex Np/s if $C = 0.5\ \mu$F. (c) Repeat, but for a *parallel RLC* circuit with $L = 10$ mH.

9.9 Sketch and label, versus t, the energy $w_C(t)$ stored in the capacitance and the energy $w_L(t)$ stored in the inductance of the *underdamped series RLC* circuit of Example 9.4. Hence, verify the flow of energy back and forth between the two elements. (*Hint:* Use the identity $A\cos\alpha + B\sin\alpha = \sqrt{A^2 + B^2}\cos[\alpha + \tan^{-1}(-B/A)]$.)

9.10 Repeat Problem 9.9, but for the *parallel RLC* circuit of Exercise 9.4.

9.11 (a) Specify suitable element values in a *series RLC* circuit with $R = 10$ kΩ for a coincident root pair $s_{1,2} = -10^5$ Np/s. (b) Repeat, but for a *parallel RLC* circuit.

9.12 (a) Find the voltage across each element in the *critically damped series RLC* circuit of Example 9.5. (b) Find the current through each element in the *critically damped parallel RLC* circuit of Exercise 9.5.

9.13 (a) Find the time t_m at which the response of the *critically damped series RLC* circuit of Example 9.5 reaches its greatest magnitude, as well as the percentage of the stored energy still left in the circuit. (b) Repeat, but for the *critically damped parallel RLC* circuit of Exercise 9.5.

9.14 Assuming the circuit of Figure P9.14 is in steady state at $t = 0^-$, find $v(t \geq 0^+)$.

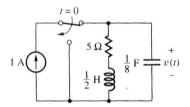

Figure P9.14

9.15 In the circuit of Figure P9.14 find the resistance value that will result in *critical damping*. Hence, find $v(t \geq 0^+)$.

9.16 Assuming the circuit of Figure P9.16 is in steady state at $t = 0^-$, find $i(t \geq 0^+)$.

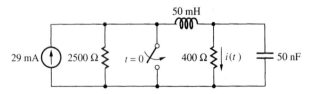

Figure P9.16

9.17 In the circuit of Figure P9.16 increase the 400-Ω resistance until the circuit yields $\zeta = 0.8$. Hence, find $i(t \geq 0^+)$.

9.18 Repeat Problem 9.16 for the circuit of Figure P9.18.

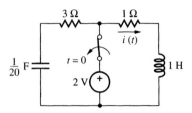

Figure P9.18

9.19 Repeat Problem 9.14 for the circuit of Figure P9.19.

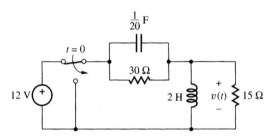

Figure P9.19

9.20 Repeat Problem 9.14 for the circuit of Figure P9.20.

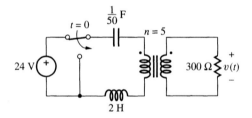

Figure P9.20

9.21 In the circuit of Figure P9.20 find n for $\zeta = 5/3$. Hence, find $v(t \geq 0^+)$ if at $t = 0^-$ the circuit is in steady state.

9.22 Repeat Problem 9.16 for the circuit of Figure P9.22.

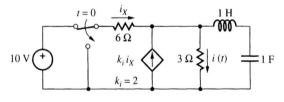

Figure P9.22

9.23 In the circuit of Figure P9.22 find k_i for critical damping. Hence, find $i(t \geq 0^+)$ if the circuit is in steady state at $t = 0^-$.

9.24 (a) In the circuit of Figure P9.24 obtain a differential equation in $v(t)$ for $t \geq 0^+$. Hence, put it in the standard form of Equation (9.3) to show that

$$\omega_0 = \sqrt{\frac{1 + R_1/R_2}{LC}}$$

$$\zeta = \frac{1}{2\sqrt{1 + R_1/R_2}}\left(R_1\sqrt{C/L} + \frac{1}{R_2\sqrt{C/L}}\right)$$

(b) Verify that the preceding expressions tend to the familiar *series RLC* expressions for $R_2 \to \infty$, and the *parallel RLC* expressions for $R_1 \to 0$, as it should be.

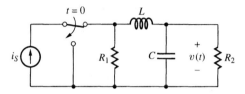

Figure P9.24

9.25 Show that if L and C are interchanged with each other in the circuit of Figure P9.24, ζ remains unchanged and $\omega_0 = 1/\sqrt{(1 + R_1/R_2)LC}$.

9.26 In the circuit of Figure P9.24 let $i_S = 2$ A, $R_1 = 18$ Ω, $R_2 = 6$ Ω, $L = 1$ H, and $C = 1/36$ F. Assuming the circuit is in steady state at $t = 0^-$, obtain the initial conditions $v(0^+)$ and $dv(0^+)/dt$. Hence, using the results of Problem 9.24, find $v(t)$ for $t \geq 0^+$.

9.2 Transient Response of Second-Order Circuits

9.27 Assuming the circuit of Figure P9.27 is in steady state at $t = 0^-$, find $i(t \geq 0^+)$.

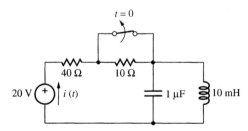

Figure P9.27

9.28 Repeat Problem 9.27, but with the switch initially open and closing at $t = 0$.

9.29 (a) Assuming the circuit of Figure P9.29 is in steady state at $t = 0^-$, find $v(t \geq 0^+)$ if $i_S = 1$ A. (b) Find i_S so that $v_{\text{xsient}} = 0$; what is $v(t \geq 0^+)$? (c) Find i_S so that $v_{ss} = 0$; what is $v(t \geq 0^+)$?

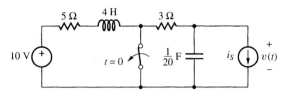

Figure P9.29

9.30 Assuming the circuit of Figure P9.30 is in steady state at $t = 0^-$, use the results of Problem 9.24 to find $v(t \geq 0^+)$.

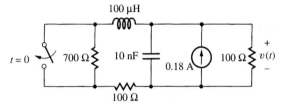

Figure P9.30

9.31 Repeat Problem 9.30, but with the switch originally closed and opening at $t = 0$. How long does it take for $v(t \geq 0^+)$ to come within 1% of its steady-state value?

9.32 Find $i(t \geq 0^+)$ in the circuit of Figure P9.16 if the switch, after having been closed for a long time, is opened at $t = 0$. Hence, estimate the instant $t_1 > 0$ for which $i(t_1) = (1/2)i(\infty)$. (*Hint:* Use the results of Problem 9.24.)

9.33 Assuming the circuit of Figure P9.33 is in steady state at $t = 0^-$, find $v(t \geq 0^+)$.

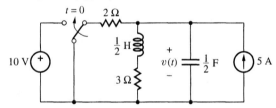

Figure P9.33

9.34 Repeat Problem 9.33, but with the capacitance and inductance interchanged with each other. (*Hint:* Use the results of Problem 9.25.)

9.35 Assuming the circuit of Figure P9.35 is in steady state at $t = 0^-$, find $v(t \geq 0^+)$.

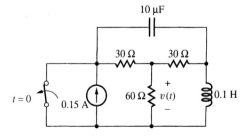

Figure P9.35

9.36   Repeat Problem 9.35, but with the capacitance and inductance interchanged with each other.

9.3 Step Response of Second-Order Circuits

9.37   (a) The step response of a certain *series RLC* circuit exhibits a 30% overshoot and a 1-MHz ringing frequency. If $R = 1$ kΩ, find L and C. (b) Find the overshoot and the ringing frequency if R is changed to 2 kΩ. (c) Find R_c such that for $R \geq R_c$ there is no more ringing.

9.38   (a) In the circuit of Figure P9.38 find $v(t \geq 0^+)$ if i_S, after having been off for a long time, is turned on to 1 A at $t = 0$. (b) Find the percentage overshoot and its instant of occurrence.

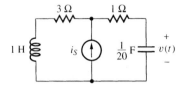

Figure P9.38

9.39   (a) In the circuit of Figure P9.39 find $v_O(t \geq 0^+)$ if i_I, after having been at -1 A for a long time, is changed to $+1$ A at $t = 0$. (b) Find the time it takes for v_O to settle within 0.1 V of its final value.

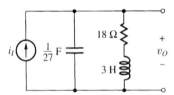

Figure P9.39

9.40   Repeat Problem 9.39, but with the capacitance and inductance interchanged with each other.

9.41   Suppose in the circuit of Figure P9.39 the resistance is lowered to 10.8 Ω. If i_I, after having been at 10 A for a long time, is changed to 5 A at $t = 0$, find the instants at which v_O reaches its maximum and minimum values. What are these values?

9.42   Assuming the circuit of Figure P9.42 is in steady state at $t = 0^-$, find $v_O(t \geq 0^+)$ if v_I is changed from 0 to 12 V at $t = 0$. (*Hint:* Use the results of Problem 9.24.)

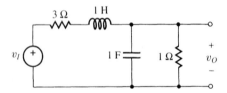

Figure P9.42

9.43   Repeat Problem 9.42, but with the capacitance and inductance interchanged with each other. (*Hint:* Use the results of Problem 9.25.)

9.44   If the circuit in Figure P9.44 is in steady state at $t = 0^-$ and i_S is changed from 0 to 1 mA at $t = 0$, find (a) k_v for a 10% overshoot, (b) the instant at which the overshoot occurs, and (c) the time it takes for v to settle within 10 mV of its final value.

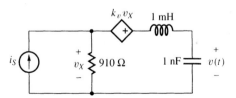

Figure P9.44

9.45   (a) Assuming the circuit of Figure P9.45 is in steady state at $t = 0^-$, find $v(t \geq 0^+)$ if i_S changes from 0 to 9 A at $t = 0$. (b) Repeat, but with the source i_S replaced by a voltage source v_S, positive at the top, changing from 0 to 10 V at $t = 0$.

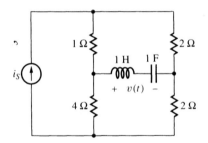

Figure P9.45

9.4 Second-Order Op Amp Circuits

9.46   Find the damping ratio ζ for the circuit of Figure P9.46, and show that $\zeta \geq 1$.

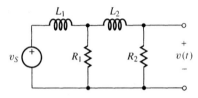

Figure P9.46

9.47   Derive a differential equation relating v_O to v_I in the circuit of Figure P9.47. Hence, find the percentage overshoot of v_O as well as its instant of occurrence if v_I is a 1-V step.

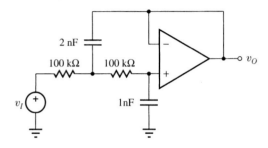

Figure P9.47

9.48 Derive a differential equation relating v_O to v_I in the circuit of Figure P9.48. Hence, assuming zero initial stored energy in its capacitances, find v_O if v_I changes from 0 to 1 V at $t = 0$.

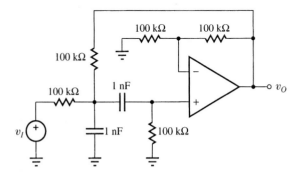

Figure P9.48

9.49 Derive a differential equation relating v_O to v_I in the circuit of Figure P9.49. Hence, find ζ and ω_0.

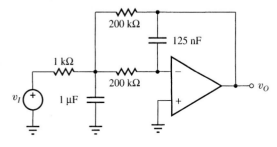

Figure P9.49

9.50 Derive a differential equation relating v_O to v_I in the circuit of Figure P9.50. Hence, find ζ and ω_0.

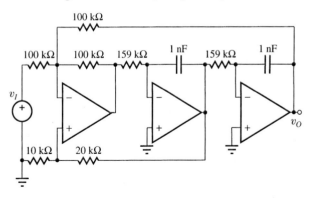

Figure P9.50

⑤ 9.5 Transient Analysis Using SPICE

9.51 Use the Probe post-processor of PSpice to display $v(t)$ in the circuit of Problem 9.7.

9.52 Use the Probe post-processor of PSpice to display $v(t)$ in the circuit of Figure P9.20.

9.53 Use the Probe post-processor of PSpice to display $v(t)$ and graphically estimate the overshoot in the circuit of Figure P9.38 if i_S is a 1-A current step.

9.54 A series *RLC* circuit with $R = 1 \Omega$, $L = 0.2$ H, $C = 0.05$ F, and with zero initial stored energy is driven by a 10-V voltage pulse having a duration of 0.8 s. Use the Probe post-processor of PSpice to display the voltage appearing across each element.

9.55 (a) Use the Probe post-processor of PSpice to display $v_O(t)$ in the op amp circuit of Figure P9.47 if v_I is a 1-V step. Assume both capacitances are initially discharged. (b) Repeat, but with an initial charge of $+2$ nC at the top plate of the 1-nF capacitor.

9.56 Assuming initially discharged capacitances, use the Probe post-processor of PSpice to display $v_O(t)$ in the op amp circuit of Figure P9.50 if v_I is a 1-V step.

AC Response

Having devoted the previous two chapters to the study of the *transient* response, we now turn our attention to the other basic response type, the *ac steady-state* response. As we know, this is the response to a *sinusoidal* forcing function after all transients have died out.

There are many reasons for selecting the **sine wave** as the forcing function. One of these is just a practical one. To begin with, alternators generate electric power in sinusoidal form, and this is also the form in which power is most efficiently transmitted and distributed to households and industries. For reasons of efficiency, sinusoidal waves are also used in commercial radio and television broadcasting. Here, electromagnetic energy is radiated by means of a sinusoidal wave called the carrier. In AM broadcasting the carrier amplitude is changed or modulated according to the information signal to be broadcast; in FM broadcasting the parameter being modulated is the carrier frequency. Finally, in dial-tone telephones information is transmitted in sinusoidal form as, for instance, when computers are linked together via modems. Considering the importance of the utility power and electronic communications in modern life, this reason alone would suffice to justify our study of the ac response.

There is, however, another deeper reason for studying the ac response, namely, because nature itself seems to have a decidedly sinusoidal character. For instance, the natural response of an underdamped second-order circuit is a damped sinusoid, just like the vibration of a guitar string, the motion of a pendulum, the perturbation produced on the surface of a pond by the throwing of a stone, and countless other phenomena. It was perhaps this observation that led the French mathematician Jean Baptiste J. Fourier (1768–1830) to the important discovery that the mathematical functions governing the physical world can be expressed as suitable *sums of purely sinusoidal components*. This conclusion, which the beginner may find hard to accept intuitively, but which can be proven rigorously, has far-reaching implications, for it allows us to find the response to *any* periodic signal, regardless of its waveform. To this end, we first investigate the responses to the signal's individual sinusoidal components; then, using the superposition principle, we add up these partial responses to obtain the overall response. Thus, ac analysis constitutes a far more general analytical tool than one might at first think.

A third reason for studying the ac response is one of mathematical convenience. It stems from the important property that the *derivatives* and *integrals* of a sinusoid are still *sinusoidal*. Since in a linear circuit each branch voltage and current is generally related to the forcing function by a linear differential equation, it follows that a sinusoidal forcing function will induce sinusoidal forced responses throughout the circuit. Thus, the sinusoidal function allows a much easier mathematical analysis than other functional forms.

In Section 7.4 we learned that the response to an ac forcing function consists of two components, the *transient component* and the *ac steady-state component*. The transient component is an exponentially decaying function, and the ac steady-state component is a *sinusoidal function* with the *same frequency* as the forcing function, but differing in *amplitude* and *phase*. Moreover, both the amplitude and phase of the response are *frequency dependent*.

A sufficiently long time after turning on an ac source, the transient component will have died out, leaving only the ac steady-state component. It is precisely upon this component, simply called the **ac response,** that we shall concentrate in the present chapter. This response can be found by formulating and solving the differential equation(s) governing the circuit. Since it deals with sinusoids, which are functions of time, this approach is referred to as **time-domain analysis.** In this chapter we use it to gain a basic understanding of the ac response of the individual circuit elements as well as simple combinations thereof, such as the RC, RL, and RLC circuits. Our goal is to familiarize ourselves with the most salient features of the ac response, particularly the **frequency response.** We also investigate how this response relates to the transient response of the previous chapters, and we do this both for first- and second-order circuits.

Time-domain ac analysis, though feasible for simple circuits, becomes prohibitive as circuit complexity increases. An alternate approach, based on an ingenious technique known as **frequency-domain analysis,** achieves the desired results using *algebraic equations* instead of *differential equations*. Moreover, the formulation of these equations is similar to that of resistive circuits, indicating that we can apply the experience acquired in the earlier part of the book to the analysis of ac circuits as well.

In frequency-domain analysis a sinusoidal signal is represented in terms of a vectorlike quantity known as a **phasor.** We introduce this concept in this chapter to help the student develop the habit of always visualizing an ac signal in terms of its phasor. With this in mind, we shall be better prepared to meet the challenges of the next chapter.

10.1 SINUSOIDS AND PHASORS

Until now we have been expressing an ac signal as $x(t) = X_m \sin(2\pi t/T) = X_m \sin 2\pi f t$, where $x(t)$ is the instantaneous value of the signal, X_m is the amplitude, T is the period, and f is the frequency. As shown in Figure 10.1(a), $x(t)$

alternates between $+X_m$ and $-X_m$, and completes a full cycle of oscillation every T seconds. Moreover, the number of oscillations completed in one second is f.

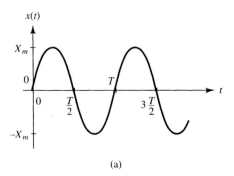

(a)

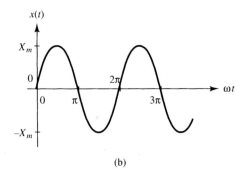

(b)

Figure 10.1 Plot of the sine function: (a) versus time t; (b) versus the angle ωt.

To simplify the notation, it is convenient to introduce the **angular frequency**

$$\omega = 2\pi f = \frac{2\pi}{T} \qquad \textbf{(10.1)}$$

after which our ac signal can be expressed more concisely as $x(t) = X_m \sin \omega t$. Since the argument ωt of the sine function has the dimensions of an angle, its unit is the *radian* (rad). Hence, ω is stated in *radians per second* (rad/s). As shown in Figure 10.1(b), the plot of $x(t)$ versus ωt repeats itself every 2π radians.

By Equation (10.1), T and f are related to ω as

$$f = \frac{\omega}{2\pi} \qquad \textbf{(10.2)}$$

$$T = \frac{2\pi}{\omega} \qquad \textbf{(10.3)}$$

To avoid confusion with the angular frequency ω, f is sometimes called the **cyclical frequency** or the **ordinary frequency.** Recall that ω is stated in rad/s and f in Hz.

Since $\sin(0) = 0$, the sine waveform goes through the origin. When observing an ac signal with the oscilloscope, it is always possible to adjust the trigger control to make the waveform cross the t axis right at the origin. However, when simultaneously observing ac signals at different nodes of a circuit, we find that they generally cross the t axis at different instants. If the observation is made with a dual-trace oscilloscope, we can still adjust the trigger to force one of the waves through the origin, but the other will generally cross the t axis earlier or later, depending on the particular node. Clearly, the form $x(t) = X_m \sin \omega t$ is not general enough because it describes only sinusoids that go through the origin. To represent an ac signal regardless of when it crosses the t axis we must introduce an additional parameter called the *phase angle*.

General Expression for an AC Signal

Following common practice, we shall represent an ac signal in the general form

$$x(t) = X_m \cos(\omega t + \theta) \qquad \text{(10.4)}$$

where

$\quad x(t)$ is the **instantaneous value** of the signal

$\quad X_m$ is its **amplitude** or **peak value**

$\quad \omega$ is the **angular frequency,** in rad/s

$\quad t$ is **time,** in s

$\quad \theta$ is the **phase angle,** in rad or degrees

The reason for switching from the sine to the cosine function is simply one of conformity to prevailing notation. Using the trigonometric identities $\sin \alpha = \cos(\alpha - \pi/2)$ and $\cos \alpha = \sin(\alpha + \pi/2)$, we can easily convert from the sine to the cosine or vice versa, so the choice of one form over the other is purely arbitrary.

To understand the role of the phase angle, recall that the cosine function reaches its maxima when its argument is an integral multiple of 2π, that is, when $\omega t + \theta = \pm 2n\pi$, where n is an arbitrary integer. The maximum corresponding to $n = 0$, referred to as the **central peak,** occurs when $\omega t + \theta = 0$, that is, when

$$\omega t = -\theta \qquad \text{(10.5)}$$

Clearly, with $\theta = 0$, the central peak occurs at $t = 0$. However, when $\theta > 0$, the central peak occurs at a *negative* value of ωt, indicating that the waveform is shifted toward the *left*, as shown in Figure 10.2(a). This shift is referred to as **lead.** Conversely, if $\theta < 0$ the central peak occurs at a *positive* value of ωt. As shown in Figure 10.2(b), the waveform is now shifted toward the *right*, and the shift is now referred to as **lag.**

Given a shift θ along the ωt axis, the corresponding shift along the t axis is, by Equation (10.5), $t = -\theta/\omega$. Letting $\omega = 2\pi/T$, we can express time shift in terms of the period as

$$t = -\frac{\theta}{2\pi} T \qquad \text{(10.6a)}$$

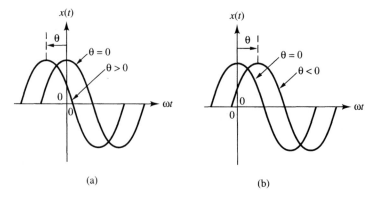

Figure 10.2 Illustrating the effect of the phase angle θ.

where θ is stated in radians. If θ is stated in degrees, then

$$t = -\frac{\theta}{360}T \qquad (10.6b)$$

Of special interest is the case $\theta = \pm\pi$ rad, or $\theta = \pm180°$, which corresponds to a time shift of *half* a period, $T/2$. Using the trigonometric identity $\cos(\omega t \pm \pi) = -\cos\omega t$, we see that subjecting an ac signal to a *phase* shift of π radians, or 180°, or to a *time* shift of $T/2$, has the same effect as multiplying it by -1. Phase reversal is readily achieved by sending the signal through an inverting amplifier.

While θ can be either in radians or in degrees, the term ωt is always in radians because ω is stated in rad/s. When using the calculator to compute the sum $(\omega t + \theta)$, make sure to express ωt and θ in the *same* units, and that the calculator is configured accordingly! To convert from degrees to radians, multiply degrees by $2\pi/360$, or $\pi/180$. To convert from radians to degrees, multiply radians by $180/\pi$.

▶ **Example 10.1**

A 1-kHz sinusoidal voltage reaches its central peak at $t = +0.2$ ms. If its amplitude is 5 V, what is its instantaneous value at (a) $t = 0$? (b) $t = 0.6$ ms?

Solution

By Equation (10.1), $\omega = 2\pi f = 2\pi \times 10^3 = 6{,}283$ rad/s. By Equation (10.5), $\theta = -\omega t = -6{,}283 \times 0.2 \times 10^{-3} = -1.257$ radians $= -1.257 \times 180/\pi = -72°$. We can therefore write

$$v(t) = 5\cos(6{,}283t - 72°) \text{ V}$$

where the term $6{,}283t$ is stated in radians whereas the phase angle is stated in degrees. This mixed notation, though lacking rigor, is quite common in engineering practice. For the given values of t,

    (a) $v(0) = 5\cos(0 - 72°) = 1.545$ V

    (b) $v(0.6 \text{ ms}) = 5\cos(6,283 \times 0.6 \times 10^{-3} - 72°) =$
        $5\cos(3.770 \text{ rad} - 1.257 \text{ rad}) = -4.045$ V.

◀

Exercise 10.1   The instantaneous value of a 1-MHz sinusoidal current is 5 mA at $t = 0$, and -8.660 mA at $t = 0.25$ μs. Find (a) the amplitude and phase angle, and (b) the amount of time shift of the central peak relative to the instant $t = 0$.

ANSWER   (a) 10 mA, 60°; (b) -166.7 ns.

Phase Difference

In ac observations and measurements it is often of interest to know the *phase difference* between two signals having the same frequency. Denoting these signals as

$$x(t) = X_m \cos(\omega t + \theta_x) \qquad \textbf{(10.7a)}$$

$$y(t) = Y_m \cos(\omega t + \theta_y) \qquad \textbf{(10.7b)}$$

we define their **phase difference** as

$$\phi = \theta_y - \theta_x \qquad \textbf{(10.8)}$$

For obvious reasons, ϕ is also said to represent the phase of $y(t)$ *relative* to that of $x(t)$. Note that if $\theta_x = 0$, then $\phi = \theta_y$.

Phase, like voltage, is a relative concept, in the sense that it is always possible to designate a particular signal in a circuit as the *reference signal* and arbitrarily assign it zero phase angle. The phase angle of any other signal in the circuit will then coincide with the phase difference between that signal and the reference signal.

Depending on the value of ϕ, we have several important cases:

(1) $\phi = 0$. The two signals are said to be **in phase** with each other.

(2) $\phi > 0$. In this case we say that $y(t)$ **leads** $x(t)$ by ϕ rad, or by $t = (\phi/2\pi)T$ s.

(3) $\phi < 0$. In this case we say that $y(t)$ **lags** $x(t)$ by ϕ rad, or by $t = (\phi/2\pi)T$ s.

(4) $\phi = \pm\pi/2 = \pm90°$. The two waves are said to be **in quadrature.** When one reaches its maxima or minima, the other goes through zero, and vice versa. Their relative time shift is $T/4$.

(5) $\phi = \pm\pi = \pm180°$. The two waves are said to be in **anti-phase.** When one reaches its maxima, the other reaches its minima, and vice versa. Their relative time shift is $T/2$.

By suitably adding or subtracting integral multiples of 2π rad, or 360°, ϕ can always be constrained within the range $-\pi \leq \phi \leq \pi$, or $-180° \leq \phi \leq 180°$.

For instance, if $y(t)$ *leads* $x(t)$ by 240°, we can also say that $y(t)$ *leads* $x(t)$ by $240 - 360 = -120°$, or that $y(t)$ *lags* $x(t)$ by 120°. This is similar to expressing the time of the day without ever exceeding 30 in the designation of the minutes, as when we express 8:40 as 20 to 9.

Phase difference is readily measured with a *dual-trace oscilloscope.* First, set both channels to the ac mode and adjust the vertical balance control so that the zero-volt baselines of both channels coincide at the center of the screen. Then, apply the signals and measure their relative time *delay* Δt as well as the period T. Finally, obtain the phase difference, in *degrees* or in *radians,* respectively as

$$\phi = 360 \frac{\Delta t}{T} \qquad \textbf{(10.9a)}$$

$$\phi = 2\pi \frac{\Delta t}{T} \qquad \textbf{(10.9b)}$$

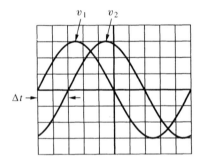

Figure 10.3 Measuring phase difference with the oscilloscope.

▶ Example 10.2

Referring to the oscilloscope display of Figure 10.3, find the phase difference between the two signals.

Solution

By inspection, $\Delta t = 2$ div and $T = 10$ div. By Equation (10.9a), $\phi = 360 \times 2/10 = 72°$. We can thus state that v_1 *leads* v_2 by 72° or, equivalently, that v_2 *lags* v_1 by 72°. ◀

Exercise 10.2 Two signals, $v_1(t) = 8\cos(2\pi 125,000t + \theta_1)$ V and $v_2 = 6\cos(2\pi 125,000t + \theta_1 - 45°)$ V, are to be observed with a dual-trace oscilloscope with a vertical scale setting of 2 V/div and a horizontal setting of 1 μs/div. Show how the two signals would appear on the screen if the oscilloscope is triggered on (a) the positive slope of $v_1(t)$ and (b) the negative slope of v_2.

Phasors

It has been stated that if we apply a sinusoidal signal to a circuit made up of resistances, capacitances, inductances, and possibly dependent sources, all resulting voltages and currents in the circuit will oscillate with the *same frequency* ω as the applied signal, differing only in amplitude and phase angle. The goal of ac analysis, then, is to find the *amplitudes* and *phase angles* of these voltages and currents; ω is of no concern because it is the same throughout. We thus need a notation that focuses on amplitude and phase angle alone. This is provided by the **phasor,** a vectorlike item whose *length* is designed to convey amplitude information and whose *angle* with respect to a reference axis is designed to convey phase-angle information. Specifically, given the ac signal

$$x(t) = X_m \cos(\omega t + \theta) \tag{10.10}$$

the corresponding phasor is denoted as X and is represented in shorthand form as

$$\boxed{X = X_m \underline{/\theta}} \tag{10.11}$$

In words, phasor X has length X_m and angle θ, where X_m is the peak amplitude and θ is the phase angle of the signal. This correspondence, depicted in Figure 10.4, is visualized as

$$x(t) = X_m \cos(\omega t + \theta)$$

$$\updownarrow \quad \updownarrow \qquad\qquad \updownarrow \tag{10.12}$$

$$X = X_m \qquad\qquad \underline{/\theta}$$

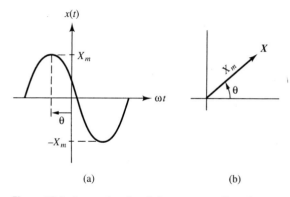

(a)   (b)

Figure 10.4 An ac signal and the corresponding phasor.

The length of phasor X, also called the **modulus** or **magnitude** of X, is *non-negative* by definition, and is denoted as $|X|$,

$$|X| = X_m \tag{10.13a}$$

The angle of phasor X, also called the **argument** of X, is denoted as $\sphericalangle X$ and is always measured *counterclockwise* relative to the positive horizontal axis, as indicated by the arrow,

$$\sphericalangle X = \theta \tag{10.13b}$$

It is important to note that the correspondence between an ac signal and its phasor is bidirectional. Given a phasor X, the corresponding ac signal is

$$x(t) = |X| \cos(\omega t + \sphericalangle X) \tag{10.14}$$

We wish to stress that though phasors look like vectors, they do not obey the rules of vector analysis studied in statics and dynamics courses. The rules of phasor manipulation will be studied in great detail in Section 11.1.

▶ **Example 10.3**

    (a) What is the phasor corresponding to the ac voltage of Example 10.1?

    (b) What is the ac current corresponding to the phasor $I = 0.010\underline{/60°}$ A?

Solution

    (a) Since $v(t) = 5\cos(\omega t - 72°)$ V, we have, by inspection, $V = 5\underline{/-72°}$ V.

    (b) By inspection, $i(t) = 10\cos(\omega t + 60°)$ mA.

        Note that if we wish to find a phasor by mere inspection of its ac signal, the latter must be expressed in the *standard form* of Equation (10.4).

▶ **Example 10.4**

Find the phasors corresponding to (a) $x(t) = 2.5\sin(\omega t + 25°)$ and (b) $x(t) = -30\cos(\omega t + 60°)$.

Solution

    (a) First, convert from the sine to the cosine function, $\sin(\omega t + 25°) = \cos(\omega t + 25° - 90°) = \cos(\omega t - 65°)$. Then, by inspection, $X = 2.5\underline{/-65°}$.

    (b) Because of the negative sign, we must first convert to the standard form. Thus, $x(t) = -30\cos(\omega t + 60°) = 30\cos(\omega t + 60° - 180°) = 30\cos(\omega t - 120°)$. Then, $X = 30\underline{/-120°}$.

Exercise 10.3   The voltage $v_i(t) = 2\cos(\omega t + 45°)$ is applied to an inverting amplifier having gain $A = -5$ V/V. (a) Find the phasor corresponding to the output voltage $v_o(t)$. (b) If the amplifier drives a 2-kΩ load, find the phasor corresponding to the current $i_L(t)$ being sourced to the load.

ANSWER   (a) $V_o = 10\underline{/-135°}$ V; (b) $I_L = 5\underline{/-135°}$ mA.

10.2 AC RESPONSE OF THE BASIC ELEMENTS

Having developed the mathematical tools to handle ac signals, we wish to investigate the ac behavior of the three basic circuit elements, that is, R, L, and C.

In so doing we shall also appreciate the convenience of visualizing ac signals in terms of their phasors. In the following we express ac voltages and currents as

$$v(t) = V_m \cos(\omega t + \theta_v) \qquad \text{(10.15a)}$$

$$i(t) = I_m \cos(\omega t + \theta_i) \qquad \text{(10.15b)}$$

AC Response of the Resistance

Applying the ac voltage of Equation (10.15a) causes a resistance to respond with the current $i(t) = v(t)/R = (V_m/R) \cos(\omega t + \theta_v)$. This current is of the type of Equation (10.15b), with amplitude and phase angle

$$I_m = \frac{V_m}{R} \qquad \text{(10.16a)}$$

$$\theta_i = \theta_v \qquad \text{(10.16b)}$$

In a resistance, the voltage and current amplitudes are related by Ohm's Law, $V_m = RI_m$. Moreover, **in a resistance, voltage and current are always in phase** with each other. Figure 10.5(b) shows the voltage and current wave-forms for the case $\theta_v = 0$. The corresponding phasor diagram is shown in Figure 10.5(c).

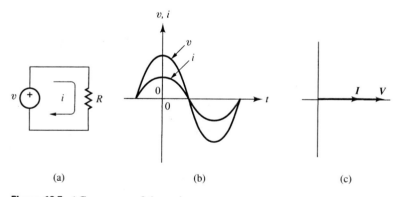

(a)   (b)   (c)

Figure 10.5 AC response of the resistance.

AC Response of the Capacitance

Subjecting a capacitance to an ac voltage as in Figure 10.6(a) results in the current $i = C \, dv/dt = -\omega C V_m \sin(\omega t + \theta_v)$. Using the trigonometric identity $-\sin \alpha = \cos(\alpha + 90°)$, we can write $i(t) = \omega C V_m \cos(\omega t + \theta_v + 90°)$. This is an ac current with amplitude and phase angle

$$I_m = \omega C V_m \qquad \text{(10.17a)}$$

$$\theta_i = \theta_v + 90° \qquad \text{(10.17b)}$$

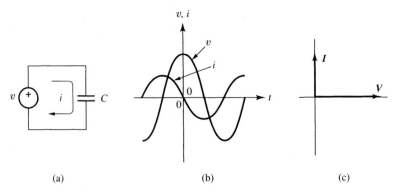

Figure 10.6 AC response of the capacitance.

In a capacitance, the current and voltage amplitudes are still linearly proportional to each other. However, the proportionality constant is now **frequency dependent.** Moreover, **in a capacitance, current always leads voltage by 90°.** This phase relationship is illustrated in Figure 10.6(b) in terms of the waveforms, and in Figure 10.6(c) in terms of the corresponding phasors.

A word of caution is necessary at this point. Considering that the applied voltage is the cause and the resulting current is the effect, the fact that the current *leads* the voltage should not give the impression that the effect is preceding the cause! By the capacitance law, $i(t) = C\, dv(t)/dt$, the instantaneous value of i depends on the rate of change of v at *that specific instant,* not on the previous and certainly not on the subsequent rates! Thus, the current waveform goes through zero when the voltage waveform peaks out, and it peaks out when the voltage waveform reaches its steepest slope, at the instants of its zero crossings. Hence the 90° phase difference between the two waves.

Equation (10.17), derived for the case in which the applied signal is a voltage, also holds for the case of an applied current. The amplitude and phase of the resulting voltage are related to those of the applied current as $V_m = I_m/\omega C$ and $\theta_v = \theta_i - 90°$. We now say that the voltage waveform *lags* the current waveform by 90°.

AC Response of the Inductance

Subjecting an inductance to an ac voltage as in Figure 10.7(a) results in a current i such that $L\, di/dt = v$. Using Equation (10.15b), $L\, di/dt = -\omega L I_m \sin(\omega t + \theta_i)$ $= \omega L I_m \cos(\omega t + \theta_i + 90°)$. Equating this expression to that of v of Equation (10.15a) yields $\omega L I_m = V_m$ and $\theta_i + 90° = \theta_v$, or

$$I_m = \frac{V_m}{\omega L} \tag{10.18a}$$

$$\theta_i = \theta_v - 90° \tag{10.18b}$$

In an inductance, the current and voltage amplitudes are linearly proportional to each other with a **frequency-dependent** proportionality constant. Moreover, **in**

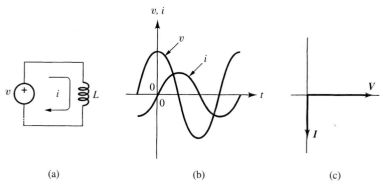

Figure 10.7 AC response of the inductance.

an inductance, current always lags voltage by 90°. These features are depicted in Figure 10.7(b) and (c).

If the applied signal is a current, the resulting voltage is related to the forcing current as $V_m = \omega L I_m$ and $\theta_v = \theta_i + 90°$. We now say that the voltage waveform *leads* the current by 90°.

▶ **Example 10.5**

A circuit element is driven with the household ac voltage. Assuming a phase angle of 0° for the voltage waveform, derive an expression for the current waveform as well as its phasor when the element is (a) a 1-kΩ resistance, (b) a 1-μF capacitance, and (c) a 0.1-H inductance.

Solution

(a) For the household voltage, $V_m = 120\sqrt{2} \simeq 170$ V and $\omega = 2\pi \times 60 = 377$ rad/s. For resistance, $I_m = V_m/R = 170/10^3 = 0.17$ A, and $\theta_i = \theta_v = 0$. Thus,

$$i(t) = 0.17 \cos(377t) \text{ A}$$

$$I = 0.17\underline{/0°} \text{ A}$$

(b) For capacitance, $I_m = \omega C V_m = 377 \times 10^{-6} \times 170 = 64.1$ mA, and $\theta_i = 90° + \theta_v = 90°$. Thus,

$$i(t) = 64.1 \cos(377t + 90°) \text{ mA}$$

$$I = 64.1\underline{/90°} \text{ mA}$$

(c) For inductance, $I_m = V_m/(\omega L) = 170/(377 \times 0.1) = 4.51$ A, and $\theta_i = \theta_v - 90° = -90°$. Thus,

$$i(t) = 4.51 \cos(377t - 90°) \text{ A}$$

$$I = 4.51\underline{/-90°} \text{ A}$$

◀

Exercise 10.4   A circuit element is subjected to the current $i(t) = 10\cos(2\pi 10^3 t + 120°)$ mA. Derive an expression for the voltage waveform as well as its phasor if the element is (a) a 2-kΩ resistance, (b) a 0.5-μF capacitance, and (c) a 10-mH inductance.

ANSWER   (a) $20\underline{/120°}$ V; (b) $3.183\underline{/30°}$ V; (c) $0.628\underline{/-150}$ V.

Limiting Cases

For capacitance, $I_m = \omega C V_m$. In the limit $\omega \to 0$, also called the *low-frequency limit,* we have $I_m \to 0$, indicating that C acts as an *open circuit.* By contrast, since $V_m = I_m/\omega C$, in the limit $\omega \to \infty$, also called the *high-frequency limit,* we have $V_m \to 0$, indicating that C now acts as a *short circuit.* We summarize as

AC Rule for Capacitance: At *low frequencies* **a capacitance acts as an** *open circuit,* **at** *high frequencies* **as a** *short circuit.*

For inductance we have $V_m = \omega L I_m$, indicating that in the *low-frequency* limit L acts as a *short circuit* because $V_m \to 0$. By contrast, since $I_m = V_m/\omega L$, in the *high-frequency* limit, L acts as an *open circuit* because now $I_m \to 0$. As expected, inductive behavior is *dual* to capacitive behavior. In summary,

AC Rule for Inductance: At *low frequencies* **an inductance acts as a** *short circuit,* **at** *high frequencies* **as an** *open circuit.*

These important rules, summarized in Table 10.1, are worth keeping in mind as we start developing a feel for ac circuits.

TABLE 10.1 Limiting AC Behavior for Capacitance and Inductance

Limit	C	L
$\omega \to 0$	open circuit	short circuit
$\omega \to \infty$	short circuit	open circuit

10.3 AC RESPONSE OF FIRST-ORDER CIRCUITS

Having examined the ac behavior of the basic elements, we now investigate the ac response of *combinations* of elements. The simplest such combinations are the series *RC* and *RL* circuits, which we examine in the present section, and the *series* and *parallel RLC* circuits, which we investigate in the next section. Since our analysis deals with sinusoidal functions of time, we refer to it as **time-domain ac analysis.**

We recall from Section 7.4 that the response to an ac forcing function consists of two components: (a) the *transient component,* in the form of a decaying exponential, and (b) the *ac steady-state component,* which is the response after the transient part has died out. In ac analysis we are exclusively interested in the steady-state part, simply called the **ac response.** This response has the *same functional form* as the forcing function, that is, it is sinusoidal with the *same frequency ω.* Moreover, its amplitude and phase angle are *frequency dependent.*

The *C-R* Circuit

In the circuit of Figure 10.8(a) we wish to find the response *i* to the ac voltage *v.* Were *R* present alone, *i* would be in phase with *v;* were *C* present alone, *i* would be leading *v* by 90°. With *R* and *C* present simultaneously, we expect the outcome to be a compromise, with ϕ somewhere in between.

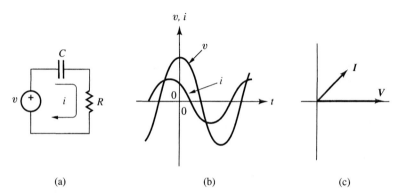

Figure 10.8 AC response of the *C-R* circuit.

Using the capacitance law, along with KVL and Ohm's Law, we have $i = C\,d(v - Ri)/dt$, or

$$RC\frac{di}{dt} + i = C\frac{dv}{dt} \tag{10.19}$$

Assuming a forcing voltage of the type

$$v(t) = V_m \cos \omega t \tag{10.20}$$

we postulate a current response of the type

$$i(t) = I_m \cos(\omega t + \phi) \tag{10.21}$$

and we seek suitable expressions for I_m and ϕ that will make *i* satisfy Equation (10.19).

Substituting Equations (10.20) and (10.21) into Equation (10.19) yields

$$-\omega RC I_m \sin(\omega t + \phi) + I_m \cos(\omega t + \phi) = -\omega C V_m \sin \omega t$$

Using the trigonometric identities $\sin(\omega t + \phi) = \sin \omega t \cos \phi + \cos \omega t \sin \phi$, and $\cos(\omega t + \phi) = \cos \omega t \cos \phi - \sin \omega t \sin \phi$, and collecting terms in $\cos \omega t$ and $\sin \omega t$ we obtain

$$[\cos \phi - \omega RC \sin \phi]I_m \cos \omega t - [(\sin \phi + \omega RC \cos \phi)I_m - \omega C V_m] \sin \omega t = 0$$

In order for this equality to hold at all times, the coefficients of $\cos \omega t$ and $\sin \omega t$ must each vanish, that is,

$$\cos \phi - \omega RC \sin \phi = 0 \qquad \textbf{(10.22a)}$$

$$(\sin \phi + \omega RC \cos \phi)I_m - \omega C V_m = 0 \qquad \textbf{(10.22b)}$$

Using the trigonometric identity $(\sin \phi)/(\cos \phi) = \tan \phi$, Equation (10.22a) yields

$$\tan \phi = \frac{1}{\omega RC} \qquad \textbf{(10.23)}$$

Next, using the trigonometric identities $\cos \phi = 1/\sqrt{1 + \tan^2 \phi}$ and $\sin \phi = (\tan \phi)/\sqrt{1 + \tan^2 \phi}$, along with Equation (10.23), we can readily verify that

$$\sin \phi + \omega RC \cos \phi = \sqrt{1 + (\omega RC)^2}$$

Substituting into Equation (10.22b) and solving for I_m provides the **amplitude relationship** between i and v,

$$\boxed{I_m = \frac{V_m}{R} \frac{\omega RC}{\sqrt{1 + (\omega RC)^2}}} \qquad \textbf{(10.24a)}$$

Moreover, Equation (10.23) yields $\phi = \tan^{-1}(1/\omega RC)$. Using the trigonometric identity $\tan^{-1}(1/\alpha) = 90° - \tan^{-1}\alpha$, we obtain the **phase relationship** between i and v,

$$\boxed{\phi = 90° - \tan^{-1}\omega RC} \qquad \textbf{(10.24b)}$$

These equations confirm the *frequency-dependent* nature of the amplitude and phase relationships. Figures 10.8(b) and 10.8(c) depict these relationships for the special ω value that makes $\omega RC = 1$. Comparison with the resistance and capacitance responses of Figures 10.5 and 10.6 confirms that the C-R circuit yields a response intermediate between those of its individual components.

▶ Example 10.6

In the circuit of Figure 10.8(a) let $V_m = 10$ V, $R = 1$ kΩ, and $C = 1$ nF. Find $v(t)$ and $i(t)$ for (a) $\omega = 10^5$ rad/s, (b) $\omega = 10^6$ rad/s, and (c) $\omega = 10^7$ rad/s.

Solution

(a) We have $V_m/R = 10/1 = 10$ mA, and $\omega RC = 10^5 \times 10^3 \times 10^{-9} = 0.1$ rad. By Equation (10.24), $I_m = 10 \times 0.1/\sqrt{1 + 0.1^2} = 0.9950$ mA and $\phi = 90 - \tan^{-1} 0.1 = 84.29°$. Thus,

$$v = 10 \cos 10^5 t \text{ V}$$

$$i = 0.9950 \cos (10^5 t + 84.29°) \text{ mA}$$

(b) Now $\omega RC = 1$ rad, $I_m = 10 \times 1/\sqrt{1 + 1^2} = 7.071$ mA, and $\phi = 90 - \tan^{-1} 1 = 45°$. Thus,

$$v = 10 \cos 10^6 t \text{ V}$$

$$i = 7.071 \cos (10^6 t + 45°) \text{ mA}$$

(c) $\omega RC = 10$ rad, $I_m = 10 \times 10/\sqrt{1 + 10^2} = 9.950$ mA, and $\phi = 90 - \tan^{-1} 10 = 5.71°$. Thus,

$$v = 10 \cos 10^7 t \text{ V}$$

$$i = 9.950 \cos (10^7 t + 5.71°) \text{ mA} \blacktriangleleft$$

Exercise 10.5   In the circuit of Example 10.6, find (a) the frequency f at which $\phi = 30°$, as well as the corresponding value of I_m and (b) the frequency f at which $I_m = 5$ mA, as well as the corresponding value of ϕ.

ANSWER   (a) 275.7 kHz, 8.660 mA; (b) 91.89 kHz, 60°.

▶ Example 10.7

Let the circuit of Figure 10.8(a) be subjected to a 100-kHz, 12-V peak ac voltage. If it is found that $I_m = 4.695$ mA and $\phi = 38.51°$, find R and C.

Solution

We have $\omega = 2\pi f = 2\pi 10^5$ rad/s. By Equation (10.24b) we must have $38.51° = 90° - \tan^{-1} \omega RC$, or $\omega RC = \tan (90° - 38.51°) = 1.257$ rad. Thus, $RC = 1.257/\omega = 1.257/(2\pi 10^5) = 2 \mu s$.

By Equation (10.24a) we must have $4.695 = (12/R) \times 1.257/\sqrt{1 + 1.257^2}$, or $R = 2$ kΩ. Finally, $C = (2 \mu s)/R = (2 \times 10^{-6})/(2 \times 10^3) = 1$ nF. In summary, $R = 2$ kΩ and $C = 1$ nF. $\blacktriangleleft$

A CONVERSATION WITH
SOCORRO CURIEL

PACIFIC BELL

As a switching equipment engineer for Pacific Bell, Socorro Curiel now manages budgets and human resources instead of transistors and resistors.

Socorro Curiel is currently working as a switching equipment engineer for the Orange-Riverside Business Unit of Pacific Bell in Anaheim, California. She earned a degree in electrical engineering from Loyola Marymount University in Los Angeles in 1988.

Ms. Curiel, what made you decide to study electrical engineering?

Initially, I considered studying computer science in college. However, by the end of my junior year in high school I had already begun to explore other options. Recruiters from the Army came to our school and gave us all the ASVAB aptitude test, and because I scored very high in electronics, one of the recruiters sold me on going to West Point Academy and majoring in avionics. I eventually decided against accepting the scholarship to West Point. I received the nomination from my congressman, but my family talked me out of going 3000 miles away. Instead, I opted to enroll in the electrical engineering program at Loyola Marymount University, which was closer to home.

Are you glad you made that switch?

Definitely. I found the EE courses very challenging, which is probably what I liked best about them. Also, I held a job as a computer operator/programmer during my first two years in college, which included tutoring in computer science. I helped other students learn about mainframe computers and write programs in Pascal and FORTRAN. That early experience convinced me that I would be happier on the engineering side than on the computer side. Consequently, I went to work at Hughes Aircraft Radar Systems Group for the next two years as a student electrical engineer. There I learned the realities of electrical engineering, and then there was no turning back.

Where else have you worked?

I worked with the California Department of Transportation. I designed the highway lighting circuits for some of the on/off ramps that were part of the expansion of the Harbor Freeway in Los Angeles, as well as those for the Century Freeway. This was the first time I got to apply all the theories and laws that I had learned in my circuits courses. The loading information for most of the electronic components came from the manufacturers, but the rest of the design work required working with Kirchhoff's current and voltage laws. I guess my circuits course was the first one that paid off for me.

And then you came to Pacific Bell?

Yes. I'm currently a project manager working with budgets, schedules, and human resources instead of transistors and resistors. It's challenging, but in a different way. The engineering courses I took trained me to handle not only engineering, but all other types of problems as well. Thinking like an engineer means being able to confront *any* problem with the same analytical approach. Whether it's "Which circuit component will work best here?" or "How do we complete this project with x number of dollars and in y amount of time?" or even "Which graduate school should I apply to?" for that matter, I still have to resort to many of the principles I learned in my engineering courses.

What do you predict for the EE industry in the near future?

My bet is that the future is in communications, both personal and global. There are new developments every day in that industry. I plan to move quickly from the installation and implementation end of this business to the front edge, where the research and development happens. For example, the advancement of wireless personal communications

technology—cellular phones, paging systems—is impressive but nowhere near perfection. There are still too many limiting factors. For one thing, these connections still catch too much static. Engineers need to figure out a way to effectively reduce all that noise.

Another problem is that the airwaves can get very crowded, and the FCC enforces strict regulations to control all frequency bands. Since it is inevitable that the number of cellular phones and personal pagers is going to increase, engineers will have to resolve this problem. Multiplexing at extremely high transmission rates is very expensive. Today, only NASA and the federal government use transmission rates in the giga-

> *"…the future is in communications, both personal and global."*

hertz range, but this won't be true for long. Engineers will need to develop a method of using high transmission rates for commercial use that is simple and inexpensive. Today global communication is either by satellite or by underwater fiber-optic cable. Engineers are developing new methods that have higher quality and cost substantially less. These are just a few of the areas in communications that are evolving and are going to explode in the next few years.

What advice do you have for EE students?

Building on what I have just said, I would add that *everything* in the electronics part of our lives is constantly changing and improving. If you're planning to be an engineer, you need to realize that you *can* make significant contributions. Your electrical knowledge in addition to the general problem-solving skills you are learning in your engineering courses can make you a potent agent for change. Don't allow your thinking to conform to the way things are done today. Try always to imagine new and innovative methods of getting the job done.

The High-Pass Function

The manner in which the behavior of a circuit varies as a function of frequency is called the **frequency response.** This response is best visualized by means of **frequency plots,** as follows. If we rewrite Equation (10.24a) as

$$I_m = \frac{V_m}{R} H_m(\omega) \qquad \textbf{(10.25)}$$

then the frequency dependence of the amplitude and phase relationships is uniquely characterized by the pair of functions

$$H_m(\omega) = \frac{\omega/\omega_0}{\sqrt{1 + (\omega/\omega_o)^2}} \qquad \textbf{(10.26a)}$$

$$\phi(\omega) = 90° - \tan^{-1}(\omega/\omega_0) \qquad \textbf{(10.26b)}$$

where we have set

$$\omega_0 = \frac{1}{RC} \qquad \textbf{(10.27)}$$

This quantity, having the same dimensions as ω, or rad/s, is called the **characteristic frequency** of the circuit because it depends only on its components, regardless of the signals.

The plots of $H_m(\omega)$ and $\phi(\omega)$ versus (ω/ω_0), called the **magnitude** and **phase plots,** are shown in Figure 10.9. It is instructive to justify these plots in terms of physical behavior. We do this by considering the following cases:

(1) $\omega/\omega_0 \gg 1$. At high operating frequencies the capacitance is given insufficient time to appreciably charge or discharge within each cycle, indicating that $v_C \to 0$. In fact, at high frequencies C acts as a short circuit compared to R, causing most of the applied voltage to be dropped across R, or $v_R \to v$. Hence, $I_m \to V_m/R$ and i is essentially in phase with v. This is confirmed by Equations (10.26), which predict $H_m \to 1$ and $\phi \to 0$. Clearly, in the high-frequency limit the C-R circuit exhibits **resistive behavior.** This is depicted in Figure 10.10(a).

(2) $\omega/\omega_0 \ll 1$. At low frequencies C has sufficient time to fully charge or discharge within each cycle, so $v_C \to v$. This agrees with the fact that at low frequencies C acts as an open circuit compared to R, or R as a short compared to C. With $v_C \to v$, it follows that $v_R \to 0$, or $I_m \to 0$. Moreover, since $i = C\, dv_C/dt \to C\, dv/dt$, i now leads v by $90°$. This is confirmed by Equations (10.26), which predict $H_m \to 0$ and $\phi \to 90°$. Clearly, in the low-frequency limit the C-R circuit exhibits **capacitive behavior.** This situation is depicted in Figure 10.10(b).

(3) $\omega/\omega_0 = 1$. When the applied frequency ω is made equal to the characteristic frequency ω_0, we expect an intermediate behavior between resistive and capacitive. In fact, Equation (10.26b) predicts $\phi = 45°$. Moreover, Equation (10.26a) predicts $H_m = 1/\sqrt{2} = 0.707$, indicating that we now have $I_m = (V_m/R)/\sqrt{2}$.

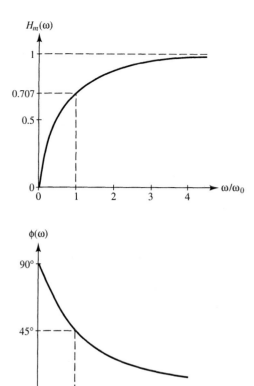

Figure 10.9 Frequency plots of magnitude and phase for the *high-pass function*.

In light of the foregoing discussion it is evident that the C-R circuit provides a substantial amount of current response at high frequencies, but an insignificant amount at low frequencies. For this reason the function $H_m(\omega)$ of Equation (10.26a) is referred to as a **high-pass function.** The frequency selectivity of the C-R circuit is exploited on purpose in the synthesis of *filters.*

The average power dissipated in the resistance is $P = (1/2)RI_m^2$. This power is maximized in the high-frequency limit, where we have $I_m = V_m/R$. Thus,

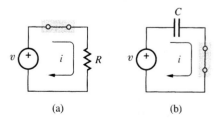

Figure 10.10 Limiting behavior of the C-R circuit for (a) $\omega \rightarrow \infty$, and (b) $\omega \rightarrow 0$.

$P_{max} = (1/2) R(V_m/R)^2$, or

$$P_{max} = \frac{1}{2} \frac{V_m^2}{R} \qquad \text{(10.28a)}$$

By contrast, the average power dissipated at $\omega = \omega_0$, where $I_m = (V_m/R)/\sqrt{2}$, is $P(\omega/\omega_0 = 1) = (1/2) R I_m^2 = (1/2) R(V_m^2/R)/2$, or

$$P(\omega/\omega_0 = 1) = \frac{1}{2} P_{max} \qquad \text{(10.28b)}$$

Consequently, the characteristic frequency ω_0 is also called the **half-power frequency.** We also make the following observations:

(1) Since applied signals whose frequencies lie within the range $\omega_0 < \omega < \infty$ yield responses that are *at least* half as powerful as the strongest response, this range of frequencies is called the **pass band.**

(2) By contrast, signals with frequencies in the range $0 < \omega < \omega_0$ yield responses that are *less* than half as powerful as the strongest response. This range is called the **stop band.**

(3) Since ω_0 marks the edge of the stop band, it is also called the **cutoff frequency.** Thus, *characteristic frequency, half-power frequency,* and *cutoff frequency* are different designations for the same parameter ω_0.

The *L-R* Circuit

In the circuit of Figure 10.11(a) we expect an intermediate response between that of a resistance and that of an inductance. Applying the inductance law, along with KVL and Ohm's Law, we have $v - Ri = L\,di/dt$, or

$$\frac{L}{R} \frac{di}{dt} + i = \frac{1}{R} v \qquad \text{(10.29)}$$

Substituting Equations (10.20) and (10.21) we obtain

$$I_m \cos(\omega t + \phi) - \frac{\omega L}{R} I_m \sin(\omega t + \phi) = \frac{1}{R} V_m \cos \omega t$$

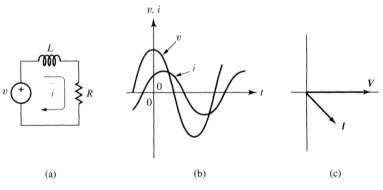

(a)   (b)   (c)

Figure 10.11   AC response of the *L-R* circuit.

Expanding, collecting, and imposing that the coefficients of $\sin \omega t$ and $\cos \omega t$ separately vanish yields

$$\left(\sin \phi + \frac{\omega L}{R} \cos \phi \right) = 0$$

$$\left(\cos \phi + \frac{\omega L}{R} \sin \phi \right) I_m - \frac{V_m}{R} = 0$$

Proceeding as for the *C-R* circuit treated earlier, we obtain the frequency-dependent relationships

$$I_m = \frac{V_m}{R} \frac{1}{\sqrt{1 + (\omega L/R)^2}} \tag{10.30a}$$

$$\phi = -\tan^{-1}(\omega L/R) \tag{10.30b}$$

Figure 10.11(b) and (c) depicts these relationships for $\omega L/R = 1$.

Exercise 10.6   Let the circuit of Figure 10.11(a) be subjected to a 1-MHz, 5-V (peak) ac voltage. If it is found that $I_m = 1.966$ mA and $\phi = -38.15°$, find R and L.

ANSWER $R = 2$ kΩ, $L = 250$ μH.

The Low-Pass Function

If we rewrite Equation (10.30a) as

$$I_m = \frac{V_m}{R} H_m(\omega) \tag{10.31}$$

then the frequency response of the *L-R* circuit is uniquely characterized by the pair of functions

$$H_m(\omega) = \frac{1}{\sqrt{1 + (\omega/\omega_0)^2}} \tag{10.32a}$$

$$\phi(\omega) = -\tan^{-1}(\omega/\omega_0) \tag{10.32b}$$

where

$$\omega_0 = \frac{R}{L} \tag{10.33}$$

is the **characteristic frequency,** in rad/s.

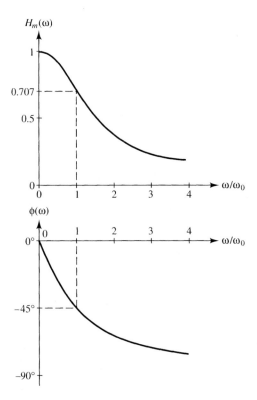

Figure 10.12 Frequency plots of magnitude and phase for the *low-pass function.*

The plots of $H_m(\omega)$ and $\phi(\omega)$ versus (ω/ω_0) are shown in Figure 10.12. We again justify these plots using physical insight.

(1) $\omega/\omega_0 \ll 1$. At low operating frequencies L acts as a short circuit compared to R, making the L-R circuit exhibit **resistive behavior** as depicted in Figure 10.13(a). We thus have $v_R \rightarrow v$, so that $I_m \rightarrow V_m/R$, and i is essentially in phase with v. This agrees with Equations (10.32), which predict $H_m \rightarrow 1$ and $\phi \rightarrow 0°$.

(2) $\omega/\omega_0 \gg 1$. Now L acts as an open circuit compared to R, or R as a short compared to L, making the L-R circuit exhibit **inductive behavior,** as depicted in Figure 10.13(b).

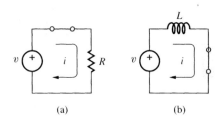

(a)   (b)

Figure 10.13 Limiting behavior of the L-R circuit for (a) $\omega \rightarrow 0$, and (b) $\omega \rightarrow \infty$.

We thus have $v_R \to 0$, so that $I_m \to 0$ and i now lags v by $90°$. Sure enough, Equations (10.32) predict $H_m \to 0$ and $\phi \to -90°$.

(3) $\omega/\omega_0 = 1$. Now the circuit behaves halfway between resistive and inductive, with $\phi = -45°$ and $H_m = 1/\sqrt{2} = 0.707$, or $I_m = (V_m/R)/\sqrt{2}$. When this condition is met the average power dissipated in R is down to *half* its maximum, which now occurs in the low-frequency limit. For this reason ω_0 is also called the **half-power frequency**.

The *L-R* circuit provides a substantial amount of current response at low frequencies, but an insignificant response at high frequencies. The function $H_m(\omega)$ of Equation (10.32a) is aptly referred to as a **low-pass function**. The following definitions hold:

(1) The frequency band $0 < \omega < \omega_0$ is called the **pass band**; applying signals with frequencies within this band yields responses that are *at least* half as powerful as the strongest response.

(2) The frequency band $\omega_0 < \omega < \infty$ is called the **stop band**.

(3) The borderline frequency ω_0 is the **cutoff frequency**.

Relation Between the Transient and AC Responses

Looking back at our studies of first-order circuits, we note, on the one hand, that they can be characterized in terms of their **transient response** via the **time constant** τ, where $\tau = RC$ for capacitive circuits, and $\tau = L/R$ for inductive circuits. On the other hand, they can be characterized in terms of their **ac response** via the **characteristic frequency** ω_0, where $\omega_0 = 1/RC$ for capacitive circuits and $\omega_0 = R/L$ for inductive circuits.

It should be clear by now that the two types of responses are just alternative means for gaining insight into the behavior of the *same* circuit. Once the parameter characterizing one response is known, the other is found via either of the following relations,

$$\omega_0 = 1/\tau \qquad \text{(10.34a)}$$
$$\tau = 1/\omega_0 \qquad \text{(10.34b)}$$

▶ Example 10.8

The transient response of a *C-R* circuit is observed with the oscilloscope and it is found that it takes 1 ms to accomplish 50% of the entire transition. What is the cutoff frequency of this circuit, in Hz?

Solution

By Equation (7.34) we must have $10^{-3} = -\tau \ln 0.5$, or $\tau = 1.443$ ms. Then, the cutoff frequency is $f_0 = \omega_0/2\pi = 1/2\pi\tau = 1/(2\pi \times 1.443 \times 10^{-3}) = 110.3$ Hz. ◀

Exercise 10.7   The ac response of a *C-R* circuit is observed with the oscilloscope, and the frequency of the forcing signal is adjusted until the response is phase-shifted by 30° with respect to the forcing signal. If this frequency is 27.56 kHz, what is the time constant of this circuit?

ANSWER $\tau = 10\ \mu$s.

Concluding Remarks

The time-domain ac analysis of the basic *C-R* and *L-R* circuits requires formulating the differential equation governing the circuit, substituting the expression for the forcing signal and the resulting response, and then performing suitable trigonometric manipulations to find the desired amplitude and phase relationships. This is a tedious process, which soon becomes prohibitive as circuit complexity increases. In the next chapter we present an ingenious technique, known as *phasor analysis,* that allows us to derive the same relationships via a much quicker process based on *algebraic equations* rather than *differential equations.*

10.4 AC Response of Second-Order Circuits

In this section we apply time-domain ac analysis to second-order circuits. Specifically, we study the *series RLC* and the *parallel RLC* configurations, which we already know to be the most interesting examples of this class of circuits.

The Series *RLC* Circuit

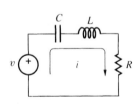

Figure 10.14 The series *RLC* circuit.

As we know, the series *RLC* circuit of Figure 10.14 is governed by Equation (9.1), rewritten here for convenience:

$$LC \frac{d^2i}{dt^2} + RC \frac{di}{dt} + i = C \frac{dv}{dt} \tag{10.35}$$

Assuming an applied voltage of the type

$$v(t) = V_m \cos \omega t \tag{10.36}$$

we postulate a response of the type

$$i(t) = I_m \cos (\omega t + \phi) \tag{10.37}$$

and seek suitable expressions for I_m and ϕ. Substituting into Equation (10.35), we obtain $-\omega^2 LC I_m \cos (\omega t + \phi) - \omega RC I_m \sin (\omega t + \phi) + I_m \cos (\omega t + \phi) = -\omega C V_m \sin \omega t$. Expanding and collecting, we have

$$\left[(1 - \omega^2 LC)\ \cos \phi - \omega RC \sin \phi \right] I_m \cos \omega t -$$
$$\left([(1 - \omega^2 LC) \sin \phi + \omega RC \cos \phi] I_m - \omega C V_m \right) \sin \omega t = 0$$

As we know, the coefficients of $\cos \omega t$ and $\sin \omega t$ must each vanish. Dividing both coefficients by $(1 - \omega^2 LC)$ and equating them to zero gives

$$\cos \phi - \frac{\omega RC}{1 - \omega^2 LC} \sin \phi = 0 \qquad \textbf{(10.38a)}$$

$$\left(\sin \phi + \frac{\omega RC}{1 - \omega^2 LC} \cos \phi \right) I_m - \frac{\omega C V_m}{1 - \omega^2 LC} = 0 \qquad \textbf{(10.38b)}$$

These equations are formally identical to Equation (10.22), provided we make the following substitution in the latter

$$\omega RC \to \frac{\omega RC}{1 - \omega^2 LC} \qquad \textbf{(10.39)}$$

We can thus reuse the results of Equation (10.24) without repeating the tedious trigonometry. Substituting Equation (10.39) into (10.24) yields, after minor manipulations, the frequency-dependent amplitude and phase relationships

$$I_m = \frac{V_m}{R} \frac{\omega RC}{\sqrt{(1 - \omega^2 LC)^2 + (\omega RC)^2}} \qquad \textbf{(10.40a)}$$

$$\phi = \tan^{-1} \frac{1 - \omega^2 LC}{\omega RC} \qquad \textbf{(10.40b)}$$

▶ Example 10.9

(a) In the circuit of Figure 10.14, let $R = 75\ \Omega$, $L = 5$ mH, and $C = 1\ \mu$F. If

$$v(t) = 10 \cos (10^4 t + 60°)\ \text{V}$$

find $i(t)$.

(b) Repeat, but for $\omega = 10^5$ rad/s.

Solution

(a) $\omega RC = 10^4 \times 75 \times 10^{-6} = 0.75$ rad; $1 - \omega^2 LC = 1 - (10^4)^2 \times 5 \times 10^{-3} \times 10^{-6} = 0.5$; $I_m = (10/75) \times 0.75/\sqrt{0.5^2 + 0.75^2} = 110.9$ mA; $\phi = \tan^{-1} (0.5/0.75) = 33.69°$; $\theta_i = \theta_v + \phi = 60 + 33.69 = 93.69°$. Thus,

$$i(t) = 110.9 \cos (10^4 t + 93.69°)\ \text{mA}$$

(b) With $\omega = 10^5$ rad/s we have $\omega RC = 7.5$ rad; $1 - \omega^2 LC = -49$; $I_m = 20.17$ mA; $\phi = -81.30°$. Thus,

$$i(t) = 20.17 \cos (10^5 t - 21.30°)\ \text{mA}$$

◀

The Band-Pass Function

We now wish to investigate the *frequency response* of our circuit. However, before proceeding we wish to put Equation (10.40) in more pleasing and concise form. To this end, consider the term

$$\frac{1 - \omega^2 LC}{\omega RC} = \frac{1}{R}\left(\frac{1}{\omega C} - \omega L\right) = \frac{1}{R\sqrt{C/L}}\left(\frac{1}{\omega\sqrt{LC}} - \omega\sqrt{LC}\right) \qquad \textbf{(10.41)}$$

If we introduce the parameter

$$\omega_0 = \frac{1}{\sqrt{LC}} \qquad \textbf{(10.42)}$$

called the **characteristic frequency,** and the dimensionless number

$$Q = \frac{1}{R\sqrt{C/L}} \qquad \textbf{(10.43)}$$

called the **quality factor of the series *RLC* circuit,** then Equation (10.41) becomes

$$\frac{1 - \omega^2 LC}{\omega RC} = Q\left(\frac{\omega_0}{\omega} - \frac{\omega}{\omega_0}\right) \qquad \textbf{(10.44)}$$

Substituting into Equation (10.40), and rewriting Equation (10.40a) as

$$I_m = \frac{V_m}{R}H_m(\omega) \qquad \textbf{(10.45)}$$

it is readily seen that the frequency dependence of the amplitude and phase relationships is characterized by the function pair

$$H_m(\omega) = \frac{1}{\sqrt{1 + Q^2\left(\dfrac{\omega_0}{\omega} - \dfrac{\omega}{\omega_0}\right)^2}} \qquad \textbf{(10.46a)}$$

$$\phi(\omega) = \tan^{-1} Q\left(\frac{\omega_0}{\omega} - \frac{\omega}{\omega_0}\right) \qquad \textbf{(10.46b)}$$

where frequency appears either as ω/ω_0 or its *reciprocal* ω_0/ω.

Unlike the ac response of a first-order circuit, which is characterized by the single parameter ω_0, the response of a second-order circuit involves *two* parameters, ω_0 and Q. The amplitude and phase plots will thus consist of *families of curves,* each curve corresponding to a different value of Q. These plots are shown in Figure 10.15. For reasons to be discussed in Chapter 14, it is convenient to use a *logarithmic* scale for ω/ω_0. We again find it instructive to justify these plots using physical insight.

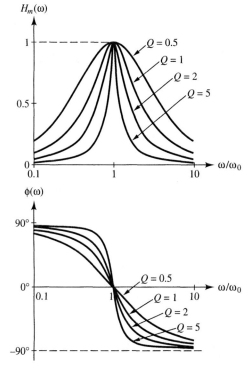

Figure 10.15 Frequency plots of magnitude and phase for the *band-pass function*.

(1) $\omega/\omega_0 \ll 1$. In the low-frequency limit C acts as an open circuit compared to L and R, making $i \rightarrow 0$. Indeed, Equations (10.46) predict $H_m \rightarrow 0$ and $\phi \rightarrow 90°$, indicating **capacitive** behavior at low frequencies. This is depicted in Figure 10.16(a).

(2) $\omega/\omega_0 \gg 1$. In the high-frequency limit L acts as an open circuit compared to C and R, again making $i \rightarrow 0$. We now have $H_m \rightarrow 0$ and $\phi \rightarrow -90°$, indicating **inductive** behavior, as depicted in Figure 10.16(c).

(3) $\omega/\omega_0 = 1$. At this borderline frequency the response exhibits a peak, a phenomenon referred to as **resonance.** Consequently, ω_0 is also called the **resonance frequency.** At this frequency Equations (10.46) predict

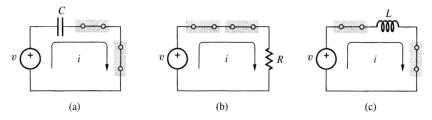

(a)   (b)   (c)

Figure 10.16 Limiting behavior of the series RLC circuit for (a) $\omega \rightarrow 0$, (b) $\omega = \omega_0$, and (c) $\omega \rightarrow \infty$.

$H_m = 1$ and $\phi = 0°$, so $I_m = V_m/R$, by Equation (10.45). The *CL* pair now acts as a short circuit, yielding the equivalent situation of Figure 10.16(b). Evidently, at resonance the *RLC* circuit exhibits **resistive** behavior. Resonance is studied in detail in Chapter 13.

The bell-shaped profile of $H_m(\omega)$ indicates that the series *RLC* circuit provides a substantial amount of current response in the neighborhood of ω_0, but insignificant response at low and high frequencies. Consequently, the function $H_m(\omega)$ of Equation (10.46a) is referred to as a **band-pass** function.

Since $H_m(\omega)$ is maximum for $\omega = \omega_0$, so is I_m and so is the power dissipated in the resistance. Moreover, since $H_m(\omega)$ drops off to zero at either side, two frequencies must exist, $\omega_L < \omega_0$ and $\omega_H > \omega_0$, where power is down to half its maximum value. As depicted in Figure 10.17, these frequencies, aptly called the **half-power frequencies,** must be such that

$$H_m(\omega_L) = H_m(\omega_H) = \frac{H_m(\omega_0)}{\sqrt{2}} = \frac{1}{\sqrt{2}} = 0.707$$

Imposing $H_m(\omega) = 1/\sqrt{2}$ in Equation (10.46a) and squaring both sides gives

$$1 + Q^2 \left(\frac{\omega_0}{\omega} - \frac{\omega}{\omega_0} \right)^2 = 2$$

This second-order equation is readily solved for ω/ω_0 to yield

$$\frac{\omega_L}{\omega_0} = \sqrt{1 + \frac{1}{4Q^2}} - \frac{1}{2Q} \qquad \textbf{(10.47a)}$$

$$\frac{\omega_H}{\omega_0} = \sqrt{1 + \frac{1}{4Q^2}} + \frac{1}{2Q} \qquad \textbf{(10.47b)}$$

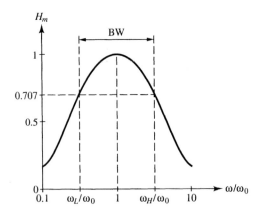

Figure 10.17 Illustrating the *half-power* frequencies ω_L and ω_H, and the *bandwidth* BW.

Multiplying out pairwise and simplifying, we find the interesting property

$$\omega_L \omega_H = \omega_0^2 \qquad \textbf{(10.48)}$$

We now introduce the following definitions:

(1) The frequency ranges $0 < \omega < \omega_L$ and $\omega_H < \omega < \infty$ are called the **stop bands** because applying signals with frequencies within these bands yields responses less than half as powerful as the strongest response. For this reason, ω_L and ω_H are also called the **cutoff frequencies.**

(2) The frequency range $\omega_L < \omega < \omega_H$ is called the **pass band,** and the difference

$$\text{BW} = \omega_H - \omega_L \qquad \textbf{(10.49)}$$

is called the *half-power bandwidth* or simply the **bandwidth** BW of the band-pass function. Subtracting Equation (10.47a) from Equation (10.47b) yields $(\omega_H - \omega_L)/\omega_0 = 1/Q$, or

$$\text{BW} = \frac{\omega_0}{Q} \qquad \textbf{(10.50)}$$

We observe that the higher the Q, the narrower the bandwidth BW for a given resonance frequency ω_0. Clearly, Q gives a measure of the degree of **frequency selectivity** of a band-pass function. A common application of high-Q band-pass circuits is *frequency tuning* in radio receivers.

▶ **Example 10.10**

In the circuit of Figure 10.14 let $R = 50\ \Omega$, $C = 10$ nF, and $L = 10$ mH. Find ω_0, Q, ω_L, ω_H, and BW.

Solution

By Equation (10.42), $\omega_0 = 1/\sqrt{10^{-2} \times 10^{-8}} = 100$ krad/s. By Equation (10.43), $Q = 1/(50\sqrt{10^{-8}/10^{-2}}) = 20$. By Equation (10.47a), $\omega_L/10^5 = \sqrt{1 + 1/(4 \times 20^2)} - 1/(2 \times 20)$, or $\omega_L = 97.531$ krad/s, and $\omega_H = (10^5)^2/(97.531 \times 10^3) = 102.531$ krad/s. Finally, by Equation (10.50), BW $= 10^5/20 = 5$ krad/s.   ◀

Exercise 10.8   Assuming $L = 50$ mH, specify suitable values for R and C in the RLC circuit of Figure 10.14 to implement a band-pass function with $f_0 = 10$ kHz and BW $= 1$ kHz.

ANSWER   314.2 Ω, 5.066 nF.

The Parallel *RLC* Circuit

In the *parallel RLC* circuit of Figure 10.18 we wish to find the voltage response to an ac current. Before committing ourselves to mathematical derivations, let us use physical insight to anticipate the frequency response.

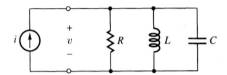

Figure 10.18 The parallel *RLC* circuit.

For $\omega \to 0$, L acts as a short, making $v \to 0$; for $\omega \to \infty$, C acts as a short, again making $v \to 0$. By contrast, at intermediate frequencies, where L and C no longer act as shorts, we expect an appreciable amount of voltage response, or $v \neq 0$. Hence, we anticipate again a *bell-shaped* profile of the type of Figure 10.17.

Turning now to quantitative analysis, we recall that the parallel *RLC* is governed by Equation (9.2), rewritten here for convenience:

$$LC \frac{d^2v}{dt^2} + \frac{L}{R}\frac{dv}{dt} + v = L\frac{di}{dt} \tag{10.51}$$

This equation is formally identical to Equation (10.35), provided we interchange i and v, L and C, and R and $1/R$. This we could have anticipated using *duality* considerations. We can thus reuse the results obtained for the series *RLC* circuit just by effecting the above interchanges. After this is done, the frequency response of the parallel *RLC* is

$$V_m = RI_m \times H_m(\omega) \tag{10.52}$$

$$H_m(\omega) = \frac{\omega L/R}{\sqrt{(1 - \omega^2 LC)^2 + (\omega L/R)^2}} \tag{10.53a}$$

$$\phi(\omega) = \tan^{-1}\frac{1 - \omega^2 LC}{\omega L/R} \tag{10.53b}$$

The functions $H_m(\omega)$ and $\phi(\omega)$ can again be put in the neater forms of Equation (10.46) provided we let

$$\omega_0 = \frac{1}{\sqrt{LC}} \tag{10.54}$$

$$Q = R\sqrt{C/L} \tag{10.55}$$

The expression for the resonance frequency is the same as for the series *RLC* circuit. However, the **quality factor of the parallel *RLC* circuit** is the *reciprocal* of that of the *series RLC* circuit, again reflecting the duality principle. Clearly, Equations (10.47) through (10.50) hold both for the series and the parallel *RLC* circuits, provided we use the appropriate expression for Q.

Exercise 10.9   Repeat Exercise 10.8, but for the parallel *RLC* circuit. Compare with the *series RLC* circuit, comment.

ANSWER   31.42 kΩ, 5.066 nF.

Relation Between the Transient and AC Responses

Looking back at our studies so far, we note, on the one hand, that *RLC* circuits can be characterized in terms of their **transient response,** having the **undamped natural frequency,** $\omega_0 = 1/\sqrt{LC}$, and the **damping ratio** ζ, where $\zeta = (1/2)R\sqrt{C/L}$ for the *series RLC* circuit, and $\zeta = 1/(2R\sqrt{C/L})$ for the *parallel RLC* circuit. On the other hand, *RLC* circuits can be characterized in terms of their **ac response,** having the **characteristic frequency,** $\omega_0 = 1/\sqrt{LC}$, and the **quality factor** Q, where $Q = 1/(R\sqrt{C/L})$ for the *series RLC* circuit, and $Q = R\sqrt{C/L}$ for the *parallel RLC* circuit. The transient and ac responses are just alternative means for gaining insight into the behavior of the *same* circuit. Once the parameters of one response are known, those of the other are found by exploiting the fact that ω_0 is the same for both responses, and Q and ζ are related as

$$Q = 1/2\zeta \qquad \qquad \textbf{(10.56a)}$$

$$\zeta = 1/2Q \qquad \qquad \textbf{(10.56b)}$$

▶ **Example 10.11**

The step response across the capacitor of a *series RLC* circuit is observed with the oscilloscope. If this response has OS = 73% and $f_d = 995$ Hz, find ω_0 and Q.

Solution

Find ζ via Equation (9.53). Thus, letting $73 = 100 \exp\left(-\pi\zeta/\sqrt{1-\zeta^2}\right)$ yields $\zeta = 0.1$. Hence, $Q = 1/2\zeta = 1/0.2 = 5$.
   Find ω_0 via Equation (9.47b). Letting $2\pi \times 995 = \sqrt{1-0.1^2}\,\omega_0$ yields $\omega_0 = 2\pi \times 1000$ rad/s.   ◀

Exercise 10.10   The ac response of a series *RLC* circuit is observed with an oscilloscope, and it is found that it peaks at 100 kHz and has a bandwidth of 25 kHz. If the circuit is now subjected to a 10-V step and the response is observed across the capacitance, predict the maximum value attained by the response, as well the frequency of its damped oscillation.

ANSWER   16.73 V, 99.22 kHz.

▼ SUMMARY

- AC signals play a prominent role in electrical engineering. An ac signal is expressed as $x(t) = X_m \cos(\omega t + \theta)$, where X_m is the *amplitude,* ω the *angular frequency,* and θ the *phase angle.*

- Phase angle represents *angular shift* along the ωt axis. If $\theta > 0$, the shift is toward the *left;* if $\theta < 0$, it is toward the *right.*

- The *ac response* of a circuit is the response to an ac forcing signal after all transients have died out. If the circuit consists of resistances, capacitances, inductances, and possibly dependent sources, all steady-state voltages and currents in the circuit will oscillate with the *same frequency* as the applied source, differing only in amplitude and phase angle.

- The amplitude and phase angle of an ac signal define a vectorlike item called a *phasor.* Given a signal $x(t) = X_m \cos(\omega t + \theta)$, the corresponding phasor is $X = X_m\underline{/\theta}$.

- For resistance we have $I_m = V_m/R$, or $V_m = RI_m$; moreover, voltage and current are always *in phase.*

- For capacitance we have $I_m = \omega C V_m$, or $V_m = (1/\omega C)I_m$; moreover, current always *leads* voltage by $90°$.

- For inductance we have $I_m = V_m/\omega L$, or $V_m = \omega L I_m$; moreover, current always *lags* voltage by $90°$.

- For a *series C-R* circuit we have $I_m = (V_m/R) \times H_m(\omega/\omega_0)$, where $H_m(\omega/\omega_0)$ is the *high-pass function,* and $\omega_0 = 1/RC$ is the *characteristic frequency* of the circuit; moreover, current *leads* voltage by the amount $\phi = 90° - \tan^{-1}(\omega/\omega_0)$, so $0° \leq \phi \leq 90°$.

- For a *series L-R* circuit we have $I_m = (V_m/R) \times H_m(\omega/\omega_0)$, where $H_m(\omega/\omega_0)$ is the *low-pass function* and $\omega_0 = R/L$ is the *characteristic frequency* of the circuit; moreover, current *lags* voltage by the amount $\phi = -\tan^{-1}(\omega/\omega_0)$, so $-90° \leq \phi \leq 0°$.

- The *low-pass function* is $H_m(\omega/\omega_0) = 1/\sqrt{1 + (\omega/\omega_0)^2}$, and the *high-pass function* is $H_m(\omega/\omega_0) = (\omega/\omega_0)/\sqrt{1 + (\omega/\omega_0)^2}$. In either case ω_0 is also called the *half-power frequency* or the *cutoff frequency.*

- For a first-order circuit, the *time constant* τ of the *transient response* and the *characteristic frequency* ω_0 of the *ac response* are related as $\omega_0 = 1/\tau$, or $\tau = 1/\omega_0$.

- For a *series RLC* circuit we have $I_m = (V_m/R) \times H_m(\omega/\omega_0)$, and for a *parallel RLC* circuit we have $V_m = RI_m \times H_m(\omega/\omega_0)$, where $H_m(\omega/\omega_0)$ is the *band-pass* function and ω_0 is the *characteristic frequency* of the circuit.

- The *band-pass* function is $H_m(\omega/\omega_0) = 1/\sqrt{1 + Q^2(\omega_0/\omega - \omega/\omega_0)^2}$, where $\omega_0 = 1/\sqrt{LC}$ is also called the *resonance frequency,* and Q is the *quality factor.* For the *series RLC* circuit, $Q = 1/(R\sqrt{C/L})$; for the *parallel RLC* circuit, $Q = R\sqrt{C/L}$.

- The *bandwidth* of the band-pass function is $\text{BW} = \omega_H - \omega_L$, where ω_L and ω_H are the *half-power,* or *cutoff,* frequencies. We have $Q = \omega_0/\text{BW}$. The higher the Q, the narrower the BW.

- For a second-order circuit, the *undamped natural frequency* of the *transient response* is the same as the *characteristic frequency* of the *ac response.* Moreover, the *damping ratio* ζ and the *quality factor* Q are related as $Q = 1/2\zeta$, or $\zeta = 1/2Q$.

▼ PROBLEMS

10.1 Sinusoids and Phasors

10.1 Given the ac voltage $v(t) = 120\sqrt{2}\cos(10^3\pi t + 30°)$ V, find its amplitude (in V), period (in ms), angular frequency (in rad/s), cyclical frequency (in Hz), and phase angle (in rad). What is v at $t = 0$? At $t = 0.5$ ms? At $t = -0.5$ ms?

10.2 An ac current $i(t)$ has an amplitude of 10 mA and completes one full cycle in 1 μs. If $i(0) = 5$ mA, express $i(t)$ in the standard cosine form, with the phase angle in degrees. Hence, find the instant at which the center peak occurs.

10.3 Given the ac current $i(t) = 10\sqrt{2}\cos(10^5 t + 45°)$ A, find I_m, f, T, θ (in rad), the smallest t ($t > 0$) for which $i = 0$, and the smallest t ($t > 0$) for which $di/dt = 0$.

10.4 At $t = -1$ μs an ac voltage $v(t)$ is zero and going negative; $v(t)$ is next zero and going positive at $t = 3$ μs, and $v(0) = -30$ V. Express $v(t)$ in the standard cosine form.

10.5 Given the ac voltages

$$v_1(t) = 10\sin(2\pi 60t + 45°) \text{ V}$$

$$v_2(t) = 20\cos(2\pi 60t + 30°) \text{ V}$$

$$v_3(t) = -15\sin(2\pi 60t + 75°) \text{ V}$$

sketch their phasors. Then, find the phase angle by which v_1 leads or lags v_2, v_1 leads or lags v_3, and v_2 leads or lags v_3.

10.6 Given the ac voltage $v(t) = 160\cos(2\pi 10^4 t - 135°)$ V, find the expression for the ac voltage obtained by (a) delaying $v(t)$ by 25 μs and (b) advancing $v(t)$ by 50 μs. Show the corresponding phasors.

10.7 Find the phasors corresponding to the following ac signals: (a) $v(t) = 10\cos(2t - 75°)$ V; (b) $i(t) = 10\sin(t - 45°)$ A; (c) $i(t) = -15\cos(10^3 t + 10°)$ mA; (d) $v(t) = 12\cos(10^3 t - \pi/4)$ μV; (e) $i(t) = 100\cos[2\pi(t - 1/16)]$ μA.

10.8 Find the 1-MHz ac signals corresponding to the following phasors: (a) $V_1 = 10\underline{/120°}$ V; (b) $I_2 = 2\underline{/-45°}$ A; (c) $V_3 = 1\underline{/0°}$ V.

10.2 AC Response of the Basic Elements

10.9 (a) A 3-kΩ resistance carries an ac current with an amplitude of 4 mA. If the ac voltage across the resistance has a period of 1 ms and is -6 V at $t = 0$ (using the passive sign convention), find the angular frequency as well as the phase angle of the current. (b) What is the power dissipated by the resistance at $t = 1$ ms? At $t = -0.25$ ms?

10.10 Find the voltage developed in response to the current $i(t) = 5\cos(2\pi 10^3 t + 120°)$ A by (a) a 10-Ω resistance, (b) a 10-μF capacitance, and (c) a 10-mH inductance.

10.11 In response to a certain current, a 1-mH inductance develops the voltage $v(t) = 200\cos(10^5 t + 30°)$ V. What is the voltage developed in response to the same current by a 30-Ω resistance? By a 500-nF capacitance? Express your results in terms of both sinusoids and phasors.

10.12 The household ac voltage is applied to an unknown circuit element, and the resulting current is measured with an appropriate instrument. Let I_{rms} denote the rms value of current and ϕ the phase difference between voltage and current. Assuming the passive sign convention, find the element if (a) $I_{\text{rms}} = 3.183$ A and $\phi = 90°$, (b) $I_{\text{rms}} = 2.262$ A and $\phi = -90°$, and (c) $I_{\text{rms}} = 2$ A and $\phi = 0°$.

10.13 A circuit element is subjected to the ac voltage $v(t) = 100\cos(10^3 t + 30°)$ V. Find the instantaneous power (absorbed or released?) at $t = 0$ if the element is (a) a 1-kΩ resistance, (b) a 10-μF capacitance, and (c) a 25-mH inductance.

10.3 AC Response of First-Order Circuits

10.14 In the C-R circuit of Figure 10.8(a) let $v = 10\cos 10^3 t$ V, $C = 1$ μF, and $R = 500$ Ω. Find the power dissipated by R at $t = 0$.

10.15 In the L-R circuit of Figure 10.11(a) let $v = 10 \times \cos(10^6 t + 90°)$ V, $R = 2$ kΩ, and $L = 3$ mH. Find the power (absorbed or released?) by the inductance at (a) $t = 0$ and (b) $t = 2\ \mu$s.

10.16   If the voltage across and current through the series combination of a resistance and an unknown energy-storage element are

$$v = 200\cos(10^2 t + 60°)\ \text{V}$$

$$i = 5.547\cos(10^2 t + 93.69°)\ \text{mA}$$

find the resistance as well as the unknown element.

10.17   Repeat Problem 10.16, but with

$$i = 3\cos(10^6 t + 30°)\ \text{mA}$$

$$v = 15\cos(10^6 t + 66.87°)\ \text{V}$$

10.18   A voltage source $v_S = 10\cos(10^3 t - 90°)$ V, a resistance $R = 50\ \Omega$, and an inductance $L = 50$ mH are connected in series. Find the maximum instantaneous power delivered by the source, dissipated by R, and absorbed by L. What are the smallest instants ($t > 0$) at which these maxima occur?

10.19   Repeat Problem 10.18, but with the same source connected in series with a 1-kΩ resistance and a 1-μF capacitance.

10.20   In the C-R circuit of Figure 10.8(a) let $v = V_0 \cos\omega t$. Denoting the voltage across C as $v_C = V_1 \cos(\omega t + \phi_1)$, positive reference at the left, plot the ratio V_1/V_0 and ϕ_1 versus frequency. Hence, verify that this is a low-pass function, and justify it physically.

10.21   In the L-R circuit of Figure 10.11(a) let $v = V_0 \cos\omega t$. Denoting the voltage across L as $v_L = V_2 \cos(\omega t + \phi_2)$, positive reference at the left, plot the ratio V_2/V_0 and ϕ_2 versus frequency. Hence, verify that this is a high-pass function, and justify it physically.

10.22   Find $v(t)$ in the circuit of Figure P10.22.

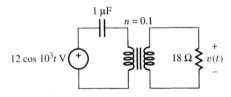

Figure P10.22

10.23   Find $v(t)$ in the circuit of Figure P10.23. Show its phasor.

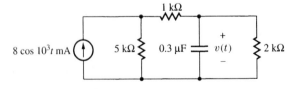

Figure P10.23

10.24   Find $v(t)$ in the circuit of Figure P10.24.

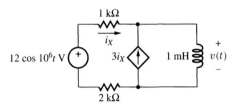

Figure P10.24

10.25   Find $v(t)$ in the circuit of Figure P10.25. Show its phasor.

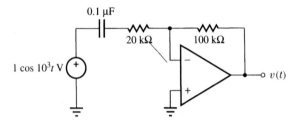

Figure P10.25

10.26   If a C-R circuit responds to a 6-V step with the current $i = 3e^{-t/(10\ \mu s)}$ mA, find the ac voltage needed to ensure the current $i = 3\cos(2 \times 10^5 t)$ mA.

10.27   If the L-R circuit of Figure 10.11(a) responds to the ac voltage $v = 4.717\cos 10^3 t$ V with the current $i = 10 \times \cos(10^3 t - 32°)$ mA, find the current response to a 12-V step.

10.4 AC Response of Second-Order Circuits

10.28   The series RLC circuit of Figure 10.14 has $R = 5\ \Omega$, $L = 250\ \mu$H, and $C = 4$ nF. (a) Find ω_0, Q, BW, ω_L, and ω_H. (b) Find the frequencies ω_1 and ω_2 at which I_m is 50% of its maximum value.

10.29   The parallel RLC circuit of Figure 10.18 has $L = 250\ \mu$H and $C = 4$ nF. Find (a) R for $Q = 100$, (b) BW, and (c) the frequencies ω_1 and ω_2 at which V_m is down to 10% of its maximum value.

10.30   Using $C = 100$ pF, specify R and L in a parallel RLC circuit for $\omega_L = 99.005$ Mrad/s and $\omega_H = 101.005$ Mrad/s.

10.31   In the circuit of Figure P10.31 let $v_i = 10\cos\omega t$ V. Find the value of ω for which the peak amplitude of v_o reaches its maximum value, as well as this value itself.

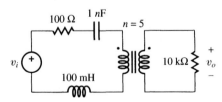

Figure P10.31

10.32 In the series RLC circuit of Figure 10.14 let $v = V_0 \cos \omega t$, and let $v_C = V_1 \cos (\omega t + \phi_1)$ be the voltage across C, positive reference at the left. (a) Find expressions for V_1 and ϕ_1. (b) Plot the ratio V_1/V_0 versus ω if $R = \sqrt{2}\ \Omega$, $L = 1$ H, and $C = 1$ F. Hence, justify the *low-pass* behavior of this circuit.

10.33 In the series RLC circuit of Figure 10.14 let $v = V_0 \cos \omega t$, and let $v_L = V_2 \cos (\omega t + \phi_2)$ be the voltage across L, positive reference at the left. (a) Find expressions for V_2 and ϕ_2. (b) Plot the ratio V_2/V_0 versus ω if $R = \sqrt{2}\ \Omega$, $L = 1$ H, and $C = 1$ F. Hence, justify the *high-pass* behavior of this circuit.

10.34 In the series RLC circuit of Figure 10.14 let $v = V_0 \cos \omega t$, and let $v_3 = v_C + v_L = V_3 \cos (\omega t + \phi_3)$ be the sum of the voltages across the CL pair, positive reference at the left. (a) Find expressions for V_3 and ϕ_3. (b) Plot the ratio V_3/V_0 versus ω if $R = 1\ \Omega$, $L = 1$ H, and $C = 1$ F. Hence, justify the *band-reject* behavior of this circuit.

10.35 If a series RLC circuit responds to a 10-V step with the current $i = [1/(25\sqrt{3})]e^{-10^6 t} \sin 10^6 \sqrt{3} t$ A, find the ac current response to the voltage $v = 10 \cos 10^6 t$ V.

10.36 If a series RLC circuit responds to the ac voltage $v = 100 \cos 250t$ V with the current $i = 4.678 \cos (250t - 79.22°)$ A and the peak amplitude of i is maximized for $\omega = 100$ rad/s, find the current response to a 10-V voltage step.

10.37 If the parallel RLC circuit of Figure 10.18 responds to a 33-mA current step with the voltage $v = 10\sqrt{11}e^{-100t} \times \sin 300\sqrt{11} t$ V, find the current needed to ensure an ac voltage with a peak value of 10 V and in phase with the applied current.

CHAPTER 11

AC Circuit Analysis

11.1 Phasor Algebra

11.2 AC Impedance

11.3 Frequency-Domain Analysis

11.4 Op Amp AC Circuits

11.5 AC Analysis Using SPICE

The objective of *ac analysis* is to find the *steady-state response* $y(t)$ *to an ac forcing function* $x(t)$. Chapter 10 focused on *time-domain ac analysis,* so called because it deals with functions of time. This type of analysis requires the formulation and solution of *differential equations* and is illustrated diagrammatically in the left column of Table 11.1. Though it can in principle be applied to the study of any circuit, it is tedious and lengthy, requiring extravagant trigonometric manipulations. As circuit complexity increases, this approach becomes prohibitive both in terms of time and effort, and we need a more efficient method to obtain the ac response of a circuit.

TABLE 11.1 Time and Frequency Domain Analysis

Time Domain	Frequency Domain
$x(t)$	$X(\omega)$
↓	↓
Differential eqns.	Algebraic eqns.
↓	↓
$y(t)$	$Y(\omega)$

The answer is provided by **phasor analysis,** an ingenious technique developed by the German-born American mathematician and engineer Charles P. Steinmetz (1865–1923). Instead of dealing with sinusoidal functions of time, this technique deals directly with their *phasors,* and it achieves the goals of ac analysis via *algebraic equations* rather than differential equations. Moreover, it uses the circuit theorems and techniques of resistive circuits, indicating that we can extend the experience acquired in earlier chapters to the analysis of ac circuits as well. Also called **frequency-domain ac analysis** because phasors are, in general, functions of frequency, phasor analysis is one of the most important tools of electrical engineering, and (though referred to by different names) it is applied also in other branches of science and engineering. Frequency-domain analysis is illustrated diagrammatically in the right column of Table 11.1, where $X(\omega)$ and $Y(\omega)$ are the phasors of $x(t)$ and $y(t)$. This type of analysis is based on complex variables. Though a previous exposure to this topic is helpful, it is not required because we develop the analysis from the beginning and only to the extent that fulfills our needs. The properties of complex variables are summarized in Appendix 4.

The chapter begins with the study of how *time-domain* operations upon *sinusoids* translate into *frequency-domain* operations upon their *phasors*. Next, we introduce the important concept of **impedance.** This is a generalization of the concept of resistance, which reduces ac circuit analysis to a procedure much like resistive circuit analysis. In fact, we illustrate its application to ac voltage dividers, the node and loop methods, Thévenin and Norton transformations, and so forth. The circuit examples considered are of the types engineers see every day.

AC circuits provide a fertile area of application for op amps. After investigating basic circuits such as integrators, differentiators, and low-pass and high-pass circuits with gain, we illustrate another interesting op amp ability, namely, *impedance transformation.* We do this by discussing the *capacitance multiplier* and the *inductance simulator.*

We conclude by illustrating the use of SPICE to check the results of ac analysis by hand. SPICE can be directed to plot or tabulate the *frequency response* via the `.AC` statement. Alternately, it can be directed to tabulate or plot the *response as a function of time* via the `SIN` function and the `.TRAN` statements.

11.1 PHASOR ALGEBRA

In this section we develop the necessary tools for the mathematical representation and manipulation of phasors. Though phasors resemble vectors, they do not necessarily obey the same rules. We recall from Section 10.1 that given a sinusoidal function of time $x(t) = X_m \cos(\omega t + \theta)$, the corresponding phasor is $X(\omega) = X_m \underline{/\theta}$. When analyzing circuits in the time domain, the most common operations that we perform upon sinusoids are (a) multiplication or division by a constant, or *magnitude scaling,* an operation arising whenever we apply Ohm's Law or divider and amplifier formulas; (b) *differentiation* and *integration,* arising when we apply the capacitance and inductance laws; and (c) *addition* and *subtraction*, arising when we apply Kirchhoff's laws. We now wish to express these operations in phasor form. Namely, given a *time-domain (t-domain) operation* upon a sinusoid $x(t)$ to obtain a new sinusoid $y(t)$,

$$x(t) \xrightarrow[\text{operation}]{t\text{-domain}} y(t)$$

we wish to find the corresponding *phasor operation* upon $X(\omega)$, also called *frequency-domain (ω-domain) operation,* that will yield the correct phasor $Y(\omega)$,

$$X(\omega) \xrightarrow[\text{operation}]{\omega\text{-domain}} Y(\omega)$$

where $X(\omega)$ and $Y(\omega)$ are the phasors of $x(t)$ and $y(t)$.

Magnitude Scaling

Consider the sinusoid $x(t) = X_m \cos(\omega t + \theta)$ having the phasor

$$X = X_m\underline{/\theta} \tag{11.1}$$

Multiplying $x(t)$ by a *positive* constant k yields $kx(t) = kX_m \cos(\omega t + \theta)$. This is a sinusoid with the same frequency and phase angle, but amplitude kX_m. Denoting its phasor as kX, we have

$$\boxed{kX = kX_m\underline{/\theta}} \tag{11.2}$$

Similar considerations hold when $x(t)$ is being *divided* by a constant k, since dividing by k is equivalent to multiplying by $1/k$. This can be summarized as

Phasor Rule 1: Multiplying (or dividing) a sinusoid by a positive factor k is equivalent to *multiplying* (or dividing) its phasor by the same factor.

As an example, let us express Ohm's Law $v(t) = Ri(t)$ in phasor form. Denoting the phasors of $v(t)$ and $i(t)$ as V and I, Rule 1 states that we can simply write $V = RI$. As another example, consider the voltage divider formula, $v_o(t) = [R_2/(R_1 + R_2)]v_i(t)$. Denoting the phasors of $v_i(t)$ and $v_o(t)$ as V_i and V_o, Rule 1 states that $V_o = [R_2/(R_1 + R_2)]V_i$.

Polarity Inversion

Inverting an ac signal is equivalent to phase-shifting it by $180°$. Indeed, if the original signal is $x(t) = X_m \cos(\omega t + \theta)$, then $-x(t) = -X_m \cos(\omega t + \theta) = X_m \cos(\omega t + \theta \pm 180°)$, where $\pm 180°$ means that we can either add or subtract $180°$, the result being the same. Denoting the phasor of $-x(t)$ as $-X$, we have

$$\boxed{-X = X_m\underline{/\theta \pm 180°}} \tag{11.3}$$

The phasor $-X$ can be thought of as the result of subjecting X to a $180°$ **rotation.** This rotation is symbolized by the placement of the **operator** $-$ in front of X. Just as placing the operator $\sqrt{}$ in front of 9 yields 3, placing the operator $-$ in front of a phasor signifies a $180°$ rotation. Clearly, applying the $-$ operator twice yields the original phasor, $-(-X) = X$. Moreover, multiplying a sinusoid by a *negative* constant k is equivalent to first scaling its phasor by $|k|$, and then rotating it by $180°$.

The *j* Operator

A 180° rotation can also be viewed as the result of two consecutive 90° rotations. Let the symbol *j* denote the operator yielding a 90° *counterclockwise rotation* so that we have, by definition,

$$jX = X_m \underline{/\theta + 90°} \qquad \textbf{(11.4)}$$

Applying the *j* operator twice yields $j(jX) = j(X_m\underline{/\theta + 90°}) = X_m \underline{/\theta + 180°} = -X$, that is, $j(jX) = -X$. For this identity to hold, we must in effect have $j^2 = -1$, or

$$j = \sqrt{-1} \qquad \textbf{(11.5)}$$

indicating that the *j* operator is just the **imaginary unit** $\sqrt{-1}$.

Applying the *j* operator three times yields $j(j(jX)) = j(j^2X) = j(-X) = -jX = X_m\underline{/\theta + 270°}$. Applying it four times yields again the original phasor, $j^4X = j^2(j^2X) = -(-X) = X$. These operations are depicted in Figure 11.1(a).

At times we shall find it convenient to use the *reciprocal operator,* $1/j$. To find its effect, multiply numerator and denominator by *j* so that $1/j = j/j^2 = j/(-1) = -j$. Thus, $(1/j)X = -jX = -(jX) = -(X_m\underline{/\theta + 90°}) = X_m\underline{/\theta + 90° - 180°}$, or

$$\frac{1}{j}X = X_m\underline{/\theta - 90°} \qquad \textbf{(11.6)}$$

Likewise, we have $(1/j^2)X = X_m\underline{/\theta - 180°}$, $(1/j^3)X = X_m\underline{/\theta - 270°}$, and $(1/j^4)X = X$. It is apparent that dividing a phasor by *j* rotates it clockwise by 90°. This is depicted in Figure 11.1(b). Summarizing, we can say that **multiplying a phasor by *j* rotates it counterclockwise by 90° and dividing a phasor by *j* rotates it clockwise by 90°.**

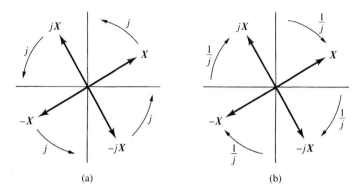

(a)   (b)

Figure 11.1 Geometric interpretation of the *j* and $1/j$ operators.

The j and $1/j$ operators are very important concepts because they provide *algebraic* interpretations for *geometric* operations. We thus expect phasor manipulations to be *algebraic,* in contrast with manipulations of sinusoids, which are of the integro-differential type. Moreover, since this algebra involves both *real* quantities, such as magnitudes and arguments, and *imaginary* quantities, such as j and $1/j$, it is called *complex algebra.*

Differentiation

Given the sinusoid $x(t) = X_m \cos(\omega t + \theta)$, whose phasor is $X = X_m\underline{/\theta}$, the derivative

$$x_{\text{der}}(t) \triangleq \frac{dx(t)}{dt} \tag{11.7}$$

is $x_{\text{der}} = -\omega X_m \sin(\omega t + \theta) = \omega X_m \cos(\omega t + \theta + 90°)$. The corresponding phasor is $X_{\text{der}} = \omega X_m\underline{/\theta + 90°}$, which can be viewed as the result of *scaling* X by ω to obtain the phasor ωX and then applying the operator j to *rotate* ωX counterclockwise by 90°. We can thus express X_{der} in terms of X in the concise form

$$\boxed{X_{\text{der}} = j\omega X} \tag{11.8}$$

In words,

Phasor Rule 2: Differentiating a sinusoid is equivalent to multiplying its phasor by $j\omega$.

As an example of the application of this rule, let us express the capacitance law $i(t) = C\,dv(t)/dt$ in phasor form. Denoting the phasors of $v(t)$ and $i(t)$ as V and I, Rule 2 states that we can simply write $I = j\omega CV$. Likewise, the phasor form of the inductance law, $v(t) = L\,di(t)/dt$, is $V = j\omega LI$. These examples clearly show how *differential* operations upon sinusoids translate into *algebraic* operations upon the corresponding phasors.

Integration

Consider the indefinite integral

$$x_{\text{int}}(t) \triangleq \int x(t)\,dt \tag{11.9}$$

or, $x_{\text{int}} = \int X_m \cos(\omega t + \theta)\,dt = (1/\omega)X_m \sin(\omega t + \theta) = (1/\omega)X_m \cos(\omega t + \theta - 90°)$. Its phasor is $X_{\text{int}} = (1/\omega)X_m\underline{/\theta - 90°}$, and it can be viewed as the result of *scaling* X by $1/\omega$ to obtain the phasor $(1/\omega)X$ and then applying the operator $1/j$ to rotate $(1/\omega)X$ clockwise by 90°. We can thus express X_{der} in terms of X concisely as

$$\boxed{X_{\text{int}} = \frac{1}{j\omega}X} \tag{11.10}$$

In words,

Phasor Rule 3: **Integrating a sinusoid is equivalent to *dividing* its phasor by $j\omega$.**

As an example of the application of this rule, the phasor form of the capacitance law $v(t) = (1/C) \int i(t)\, dt$ is $V = (1/j\omega C)I$. Likewise, the phasor form of the inductance law $i(t) = (1/L) \int v(t)\, dt$ is $I = (1/j\omega L)V$. Clearly, *integration operations* upon sinusoids turn into *algebraic operations* upon the corresponding phasors.

The Complex Plane

The presence of the j operator allows us to express a phasor as a **complex variable** and to visualize it in a two-dimensional space called the **complex plane.** The horizontal axis of this plane is called the **real axis** because its points represent **real numbers.** Positive numbers lie on its right portion, negative numbers on its left portion, and zero at the origin. A positive real number such as 3 can be expressed as $3\underline{/0°}$. A negative real number such as -4 can be expressed as $4\underline{/180°}$.

Subjecting a positive real number such as $3\underline{/0°}$ to a 90° counterclockwise rotation yields $j(3\underline{/0°}) = 3\underline{/90°}$, which lies on the upper portion of the vertical axis. This is depicted in Figure 11.2. Conversely, subjecting a negative real number such as $4\underline{/180°}$ to a 90° counterclockwise rotation yields $j(4\underline{/180°}) = 4\underline{/-90°}$, which lies on the lower portion of the vertical axis. This is also depicted in Figure 11.2. Of course, the numbers $3\underline{/90°}$ and $4\underline{/-90°}$ could also have been obtained by applying a 90° clockwise rotation to, respectively, the numbers $3\underline{/180°}$ and $4\underline{/0°}$. Since $j(3\underline{/0°}) = j3$ and $j(4\underline{/180°}) = j(-4) = -j4$, both numbers are said to be **imaginary numbers** because they are multiples of $j = \sqrt{-1}$. Moreover, since both numbers lie on the vertical axis, the latter is aptly called the **imaginary axis.**

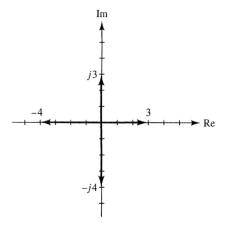

Figure 11.2 The complex plane.

Rectangular Coordinates

To develop a complex-plane representation for a phasor, consider once again the sinusoid $x(t) = X_m \cos(\omega t + \theta)$. Expanding gives $x(t) = X_m(\cos \omega t \cos \theta - \sin \omega t \sin \theta)$, which can be rearranged as

$$x(t) = (X_m \cos \theta) \cos \omega t + (X_m \sin \theta) \cos(\omega t + 90°)$$

Letting

$$X_r = X_m \cos \theta \tag{11.11a}$$
$$X_i = X_m \sin \theta \tag{11.11b}$$

allows us to write

$$x(t) = X_r \cos(\omega t + 0°) + X_i \cos(\omega t + 90°) \tag{11.12}$$

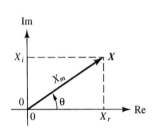

Figure 11.3 Rectangular representation of a complex variable X.

indicating that an arbitrary sinusoid can be expressed as the *sum* of two sinusoidal components in *quadrature* with each other. These components have, respectively, X_r and X_i as amplitudes, and $0°$ and $90°$ as phase angles. Their phasors are thus $\mathbf{X}_r = X_r/\underline{0°} = X_r$ and $\mathbf{X}_i = X_i/\underline{90°} = jX_i$. This is shown in Figure 11.3. Regarding $\mathbf{X}_r$ and $\mathbf{X}_i$ as vectorlike components of $\mathbf{X}$, we can write $\mathbf{X} = \mathbf{X}_r + \mathbf{X}_i$, or

$$\mathbf{X} = X_r + jX_i \tag{11.13}$$

This is called the **rectangular** or **cartesian form** of representing $\mathbf{X}$, and X_r and X_i are called the **rectangular coordinates** of $\mathbf{X}$. By contrast, X_m and θ are called the **polar coordinates.** As we shall see shortly, the rectangular form proves especially useful in the addition and subtraction of phasors.

The coordinates X_r and X_i are also called, respectively, the **real part** and the **imaginary part** of $\mathbf{X}$, and are denoted as

$$X_r = \text{Re}[\mathbf{X}] \tag{11.14a}$$
$$X_i = \text{Im}[\mathbf{X}] \tag{11.14b}$$

Both parts are *real* quantities, and each may be positive, negative, or zero. Note that the imaginary component is X_i, *not* jX_i, as j is not part of the definition of the imaginary part. Of course, there is nothing unreal or mysterious about Im $[\mathbf{X}]$. The designation *imaginary* is used simply to distinguish the vertical from the horizontal component. In the rectangular representation of Equation (11.13), the vertical component is readily identified by the presence of the operator j.

If $X_i = 0$, then $\mathbf{X}$ is said to be **purely real** because it reduces to $\mathbf{X} = X_r$. Conversely, if $X_r = 0$, then $\mathbf{X}$ is said to be **purely imaginary** because it reduces to $\mathbf{X} = jX_i$. Since in general a phasor has both a real and an imaginary part, we refer to it as a **complex variable.**

It is interesting to note that the j and $1/j$ operators themselves admit vectorlike representations. This can be seen by writing $j = 0 + j1$, that is,

$$j = 1\underline{/90°} \qquad \textbf{(11.15a)}$$

and $1/j = j/j^2 = -j = 0 - j1$, that is,

$$\frac{1}{j} = 1\underline{/-90°} \qquad \textbf{(11.15b)}$$

▶ **Example 11.1**

Express the following complex variables in rectangular form:
(a) $X_1 = 7\underline{/45°}$; (b) $X_2 = 5\underline{/120°}$; (c) $X_3 = 1.5\underline{/-30°}$;
(d) $X_4 = \sqrt{2}\underline{/180°}$; (e) $X_5 = 7.5\underline{/-90°}$.

Solution

(a) By Equation (11.11), $X_r = 7\cos 45° = 4.95$ and $X_i = 7\sin 45° = 4.95$. Thus, $X_1 = 4.95 + j4.95$.

By a similar procedure, we obtain (b) $X_2 = -2.5 + j4.33$,
(c) $X_3 = 1.3 - j0.75$, (d) $X_4 = -\sqrt{2}$, a purely real variable, and
(e) $X_5 = -j7.5$, a purely imaginary variable. ◀

Exercise 11.1 Find the *rectangular* representation of the phasors corresponding to the following sinusoids: (a) $x(t) = -10\cos(\omega t + 45°)$; (b) $x(t) = 5\sin(\omega t - 24°)$.

ANSWER $-7.071 - j7.071$; $-2.034 - j4.568$.

Polar Coordinates

Equation (11.11) allows us to find the rectangular coordinates X_r and X_i in terms of the polar coordinates X_m and θ. Conversely, given the rectangular coordinates, the polar coordinates are readily found, with the help of Figure 11.3, as

$$X_m = \sqrt{X_r^2 + X_i^2} \qquad \textbf{(11.16a)}$$

$$\theta = \tan^{-1}\frac{X_i}{X_r} \qquad \textbf{(11.16b)}$$

When computing θ with the calculator, care must be exercised to ensure that the result belongs to the correct quadrant. Entering Equation (11.16b) into the calculator yields the correct result only if $X_r > 0$. If $X_r < 0$ this equation must be modified as

$$\theta = \tan^{-1}\frac{X_i}{X_r} \pm 180° \qquad \textbf{(11.16c)}$$

where the sign is chosen so as to ensure $-180° \leq \theta \leq 180°$. As a precaution, check the following list of rules to ensure that the calculated value of θ comes out in the correct quadrant:

$$\text{If} \quad X_r > 0 \quad \text{and} \quad X_i > 0 \quad \text{then} \quad 0° < \theta < 90° \qquad \textbf{(11.17a)}$$

$$\text{If} \quad X_r < 0 \quad \text{and} \quad X_i > 0 \quad \text{then} \quad 90° < \theta < 180° \qquad \textbf{(11.17b)}$$

$$\text{If} \quad X_r < 0 \quad \text{and} \quad X_i < 0 \quad \text{then} \quad -180° < \theta < -90° \qquad \textbf{(11.17c)}$$

$$\text{If} \quad X_r > 0 \quad \text{and} \quad X_i < 0 \quad \text{then} \quad -90° < \theta < 0° \qquad \textbf{(11.17d)}$$

▶ **Example 11.2**

Express the following complex variables in the form $X = X_m\underline{/\theta}$:
(a) $X_1 = 3 - j4$; (b) $X_2 = -2 - j1$; (c) $X_3 = -1.5 + j3$; (d) $X_4 = -j20$.

Solution

(a) By Equation (11.16a), $X_m = \sqrt{3^2 + (-4)^2} = 5$. Since $X_r = 3 > 0$, use Equation (11.16b) to find $\theta = \tan^{-1}(-4/3) = -53.13°$. This checks with Equation (11.17d). Thus, $X_1 = 5\underline{/-53.13°}$.

(b) $X_m = \sqrt{(-2)^2 + (-1)^2} = 2.236$. Since $X_r = -2 < 0$, use Equation (11.16c) to find $\theta = \tan^{-1}[-1/(-2)] \pm 180° = 26.57° - 180° = -153.43°$. This checks with Equation (11.17c). Thus, $X_2 = 2.236\underline{/-153.43°}$.

(c) $X_m = 3.354$, $\theta = \tan^{-1}[3/(-1.5)] \pm 180° = -63.43° + 180° = 116.57°$. This checks with Equation (11.17b). Thus, $X_3 = 3.354\underline{/116.57°}$.

(d) X_4 is purely imaginary, $X_4 = 20\underline{/-90°}$. ◀

Exercise 11.2   Find the sinusoids whose phasors are (a) $X_1 = 2(1 - j1.5)$, (b) $X_2 = -3(2 + j1)$, (c) $X_3 = j5$, and (d) $X_4 = -3$.

ANSWER   (a) $\sqrt{13}\cos(\omega t - 56.31°)$; (b) $\sqrt{45}\cos(\omega t - 153.43°)$; (c) $5\cos(\omega t + 90°)$; (d) $3\cos(\omega t + 180°)$.

Addition and Subtraction

When applying Kirchhoff's laws, we add up signals algebraically to form expressions such as $x(t) \pm y(t)$. Let

$$x(t) = X_m \cos(\omega t + \theta_x)$$

$$y(t) = Y_m \cos(\omega t + \theta_y)$$

and let the corresponding phasors be $X = X_r + jX_i$ and $Y = Y_r + jY_i$. We wish to show that the combination $x(t) \pm y(t)$ is still a sinusoid with frequency ω. Moreover, we wish to find the corresponding phasor, which we shall denote as

$X \pm Y$. Expanding and regrouping gives

$$x(t) \pm y(t) = X_m (\cos \omega t \cos \theta_x - \sin \omega t \sin \theta_x)$$

$$\pm Y_m (\cos \omega t \cos \theta_y - \sin \omega t \sin \theta_y)$$

$$= (X_m \cos \theta_x \pm Y_m \cos \theta_y) \cos \omega t$$

$$+ (X_m \sin \theta_x \pm Y_m \sin \theta_y) \cos (\omega t + 90°)$$

that is,

$$x(t) \pm y(t) = (X_r \pm Y_r) \cos \omega t + (X_i \pm Y_i) \cos (\omega t + 90°) \qquad \textbf{(11.18)}$$

It was shown earlier that the combination of two sinusoids in quadrature is still a sinusoid with the same frequency. It is also apparent that the phasor of this combination is

$$\boxed{X \pm Y = (X_r \pm Y_r) + j(X_i \pm Y_i)} \qquad \textbf{(11.19)}$$

This can be summarized as follows:

Phasor Rule 4: **Adding or subtracting two sinusoids having the same frequency is equivalent to pairwise adding or subtracting the real and imaginary parts of their phasors.**

▶ Example 11.3

The voltages at nodes A and B of a certain circuit are

$$v_A(t) = 5 \cos (\omega t - 30°) \text{ V}$$

$$v_B(t) = 3 \cos (\omega t + 45°) \text{ V}$$

Find an expression for their difference, $v_{AB}(t) = v_A(t) - v_B(t)$.

Solution

Work with phasors rather than sinusoids! By Rule 4, the phasor of $v_{AB}(t)$ is $\mathbf{V}_{AB} = 5\underline{/-30°} - 3\underline{/45°}$. Using Equation (11.11),

$$\mathbf{V}_{AB} = [5 \cos (-30°) + j5 \sin (-30°)] - [3 \cos 45° + j3 \sin 45°]$$

$$= (4.33 - j2.50) - (2.12 + j2.12) = 2.21 - j4.62$$

Next, use Equation (11.16) to obtain $|\mathbf{V}_{AB}| = \sqrt{2.21^2 + 4.62^2} = 5.12$ V, and $\sphericalangle \mathbf{V}_{AB} = \tan^{-1}(-4.62/2.21) = -64.45°$. Finally,

$$v_{AB}(t) = 5.12 \cos (\omega t - 64.45°) \text{ V} \qquad \blacktriangleleft$$

Exercise 11.3 Two currents, i_1 and i_2, are entering a node, and the current i_3 is leaving it. If $i_2(t) = 1.5 \cos (\omega t + 15°)$ mA and $i_3(t) = 2.5 \cos (\omega t - 90°)$ mA, find $i_1(t)$.

ANSWER $3.231 \cos (\omega t - 116.64°)$ mA.

CHARLES PROTEUS STEINMETZ

The Wizard of Schenectady

The electrical-mathematical genius known as Charles Proteus Steinmetz was born Karl August Rudolph Steinmetz in 1865 in Germany. The name change was the result of an unexpected change in plans. Steinmetz was two days away from receiving a doctorate degree from the University of Breslau and looking forward to life as a university professor in Germany when he found out that he was about to be arrested for publishing a Socialist newspaper. Taking nothing with him, pretending to be on a one-day jaunt in the mountains, he escaped across the Austrian border and made his way to Zurich. A chance friendship with a Zurich student who had wealthy relatives in the United States, and Steinmetz was sailing into New York Harbor on June 1, 1889.

To honor his adopted country, he changed the "Karl" to "Charles" and dropped the "August Rudolph." "Proteus" was a nickname from university days, after the Greek legendary Proteus, the old man of the sea capable of changing himself into any form he liked. Because Steinmetz's brilliant mind could veer in any direction in debates at school, his friends dubbed him Proteus.

Once in the United States, Steinmetz found work at the Yonkers, New York, firm of Eickenmeyer and Osterheld. His first assignment was to design a new electric motor for streetcars. He knew an ac motor

This photograph of Charles Steinmetz was taken shortly after he arrived in the U.S. in 1889. He had barely escaped arrest in Germany for publishing a Socialist newspaper. At his first job, Steinmetz designed the first ac motor to compensate for hysteresis losses. (The Bettmann Archive)

would be better than a dc one, but in those days ac-motor production was a hit-or-miss proposition because no one could figure out the best design. The problem was the overheating caused by hysteresis. Steinmetz spent months testing the hysteresis properties of every known magnetic substance and finally derived an empirical formula for calculating hysteresis in any motor. After his "Law of Hysteresis" was published in 1891, Steinmetz became world-famous as the man who made the first ac motor to compensate for hysteresis losses.

In 1892, General Electric bought out Eickenmeyer and Osterheld, and in February 1893 Steinmetz found himself on a train for Boston, headed to work at the Lynn, Massachusetts, plant of his new employer. There he worked out the second great problem of his career: a mathematical treatment of transformers. As you probably know, electricity cannot economically be transported any great distance unless it is stepped up to very high voltages for the journey and then stepped down again at the end. Sitting down with his logarithm tables and ever-present cigars, Steinmetz worked out all the equations needed for calculating alternating currents so that dependable transformers could be built. Known as Steinmetz's "General Number" or sometimes his "Complex Number," these equations added to his fame. Not yet 30 years old, he had solved two of the salient problems of the electrical industry!

Steinmetz was a marvelously colorful character, if a bit mean-spirited in some of his practical jokes. For his cactus collection, he had a desert conservatory built next to his house, a conservatory inhabited over the years by crows (John and Mary; he had them stuffed when they died), a Gila monster (starved to death after refusing to eat anything once it came to live in Schenectady), and an alligator (which escaped, putting the whole city into a panic until it was caught;

> *"He was founder of ...the Society for the Equalization of Engineers' Salaries ...poker night for his GE colleagues."*

it, too, went on a hunger strike and died when the well-meaning Steinmetz forced too big a hunk of meat down its throat and it suffocated). He was founder of what sounds like a holdover from his Socialist days in college—the Society for the Equalization of Engineers' Salaries—but which was merely poker night for his GE colleagues. The day of any formal party at his home, he would replace the regular lighting in the ladies' powder room with a mercury vapor lamp and howl in delight anytime an unsuspecting guest was heard crying in dismay at her purple appearance.

For 10 years beginning in 1902, Steinmetz was head of the electrical engineering department at Union College in Schenectady. In 1913, however, he was named the first consulting engineer at GE, a post created to free him from paperwork and allow him complete freedom in his laboratory. So he resigned his teaching position and devoted himself full-time (i.e., 18 hours a day) to research.

Steinmetz was an environmentalist years before the term was coined. In 1903 he was telling his audiences that we would run out of coal in only a couple of hundred years and we had better start exploiting "white coal"—hydroelectric power. (Perhaps "environmentalist" is not quite the right word, however, since his plan for using white coal was to harness Niagara Falls. In answer to the hue and cry raised by people opposed to the idea of Niagara as a feeble trickle, he proposed harnessing it only on weekdays and pulling the plug to restore the majesty of the full flow every weekend!) He warned that the rich soil of the U.S. Midwest would someday be as depleted as the Saudi desert and described how electricity could be used for soil nitrification. Another plan was to use electricity to convert sewage to fertilizer. These ideas sounded hilarious in his day, but read any current ecology text and you'll find these procedures described as some of our last best hopes for saving the planet.

Unlike some other early giants of electricity such as Tesla, and De Forest, Steinmetz had a happy and rewarding life. He was respected as a genius while he was still alive to enjoy it, and his colorful personality made him the Steven Hawking of his day and the darling of features editors at newspapers all over

> *"...his plan was to harness Niagara Falls. In answer to the hue and cry raised ..., he proposed harnessing it only on weekdays and pulling the plug to restore the majesty of the full flow every weekend!"*

the country. In 1923, a train journey to California turned into a series of standing-room-only stops in every major city.

Back home after this trip, the highlight of which was a meeting with his favorite movie star, Douglas Fairbanks, Sr., the Wizard of Schenectady died quietly in his home on October 26, 1923.

Graphical Addition and Subtraction

The sum and difference of two complex variables can also be found graphically, using the familiar *parallelogram rule* of vector addition. As depicted in Figure 11.4(a), the sum $X + Y$ is readily found as the diagonal of the parallelogram having X and Y as sides. To find the difference $X - Y$, first rotate Y by 180° to obtain $-Y$, as shown in Figure 11.4(b). Then, obtain $X - Y$ as the diagonal of the parallelogram with X and $-Y$ as sides.

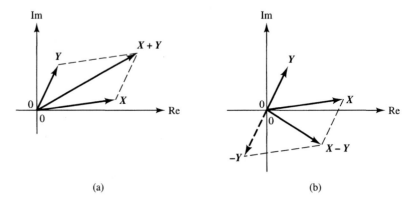

(a)   (b)

Figure 11.4 Geometric interpretation of addition and subtraction of two complex variables X and Y.

Exercise 11.4

  (a) Use the parallelogram method to find the sum and the difference of phasors X_1 and X_2 of Exercise 11.2.

  (b) Repeat, but for phasors X_3 and X_4 of the same exercise.

Useful Approximations

Using the parallelogram rule as a visual aid, it is easily seen that if one of the two complex variables has a much greater length than the other, say, at least *an order of magnitude greater,* then the sum or the difference of the two will essentially coincide with the variable having the greater length. Mathematically, if

$$|X| \gg |Y| \tag{11.20}$$

(say, $|X| \geq 10|Y|$), then we can write

$$X + Y \simeq X \tag{11.21a}$$

$$X - Y \simeq X \tag{11.21b}$$

In particular, given a complex variable $X = X_r + jX_i$, if its real part is much bigger than its imaginary part, or

$$|X_r| \gg |X_i| \tag{11.22a}$$

then we can approximate

$$X \simeq X_r \qquad\qquad \textbf{(11.22b)}$$

For instance, if $X = 15 + j1$, then $X \simeq 15$; if $X = -5 + j0.4$, then $X \simeq -5$. Conversely, if the imaginary part is much bigger, or

$$|X_r| \ll |X_i| \qquad\qquad \textbf{(11.23a)}$$

then we can approximate

$$X \simeq jX_i \qquad\qquad \textbf{(11.23b)}$$

For instance, if $X = 0.1 - j3$, then $X \simeq -j3$; if $X = 5 + j100$, then $X \simeq j100$. These important approximations are worth keeping in mind because they may help speed up our calculations significantly and may also facilitate our physical understanding of ac circuits.

Exponential Form

By Equations (11.11) and (11.13), we can write $X = X_r + jX_i = X_m(\cos\theta + j\sin\theta)$. Euler's identity, discussed in Appendix 3, states that $\cos\theta + j\sin\theta = e^{j\theta}$. Consequently, X can also be expressed in the form

$$\boxed{X = X_m e^{j\theta}} \qquad\qquad \textbf{(11.24)}$$

Aptly called the **exponential form,** it proves especially convenient in complex-number multiplications and divisions, which we shall study shortly. By now we have learned three different ways of representing a complex variable:

(1) The *shorthand* or *polar* form, $X = X_m\underline{/\theta}$
(2) The *rectangular* form, $X = X_r + jX_i$
(3) The *exponential* form, $X = X_m e^{j\theta}$.

Conversion from one form to another is accomplished via Equations (11.11) and (11.16). Note that the j and $1/j$ operators admit the polar forms $j = 1\underline{/90°} = 1e^{j\pi/2} = e^{j\pi/2}$, and $1/j = e^{-j\pi/2}$.

Exercise 11.5   Let k be an operator yielding a 45° counterclockwise rotation. That is, if $X = X_m\underline{/\theta}$, then $kX \triangleq X_m\underline{/\theta + 45°}$. Obtain a relationship between k and j; hence, find the polar as well as the rectangular representations of k.

ANSWER $k = \sqrt{j} = 1\underline{/45°} = 0.7071 + j0.7071$.

Multiplication and Division

As we proceed, we shall often need to multiply or divide complex variables. Given two complex variables $X = X_m e^{j\theta_x}$ and $Y = Y_m e^{j\theta_y}$, their **product** is

$XY = \left(X_m e^{j\theta_x}\right) \times \left(Y_m e^{j\theta_y}\right) = X_m Y_m \times e^{j(\theta_x + \theta_y)}$, that is,

$$XY = X_m Y_m \,\underline{/\theta_x + \theta_y} \qquad\qquad\text{(11.25)}$$

This shows the product of two complex variables is still a complex variable. Moreover,

Phasor Rule 5: The *modulus of a product* equals the *product of the moduli*, and the *argument of a product* equals the *sum of the arguments*.

The multiplication of X by Y can be visualized geometrically as the result of first *scaling* X by Y_m, to obtain $Y_m X$, and then *rotating* $Y_m X$ by θ_y.

The **ratio** of two complex variables is $X/Y = \left(X_m e^{j\theta_x}\right)/\left(Y_m e^{j\theta_y}\right) = (X_m/Y_m)e^{j(\theta_x - \theta_y)}$, that is,

$$\frac{X}{Y} = \frac{X_m}{Y_m} \,\underline{/\theta_x - \theta_y} \qquad\qquad\text{(11.26)}$$

This is still a complex variable. Moreover:

Phasor Rule 6: The *modulus of a ratio* equals the *ratio of the moduli*, and the *argument of a ratio* equals the *difference of the arguments*.

The division of X by Y can be visualized geometrically as the result of first *scaling* X by $1/Y_m$ to obtain $(1/Y_m)X$, and then *rotating* $(1/Y_m)X$ by $-\theta_y$.

As we shall see in the following sections, Rules 5 and 6 prove especially useful when stating circuit element laws in phasor form or when manipulating ac impedances and ac transfer functions. We stress that though phasors look like vectors, the rules of *phasor* multiplication and division are *different* from the rules of *vector* multiplication and division that you learned in vector analysis!

▶ Example 11.4

Find the product as well as the ratio of the following complex-variable pairs:

     (a) $X = 6\underline{/120°}$ and $Y = 2\underline{/-90°}$

     (b) $X = 4 - j3$ and $Y = 1 + j2$

Solution

     (a) By Equation (11.25), $XY = (6 \times 2)\underline{/120° - 90°} = 12\underline{/30°}$. By Equation (11.26), $X/Y = (6/2)\underline{/120° - (-90°)} = 3\underline{/210°}$. This is more conveniently expressed as $X/Y = 3\underline{/210° - 360°} = 3\underline{/-150°}$.

(b) First, convert from rectangular to polar form using Equation (11.16),

$$X = \sqrt{4^2 + 3^2}/\tan^{-1}(-3/4) = 5/-36.9°$$

$$Y = \sqrt{1^2 + 2^2}/\tan^{-1}(2/1) = 2.24/63.4°$$

Then, $XY = (5 \times 2.24)/-36.9° + 63.4° = 11.2/26.5°$ and $X/Y = (5/2.24)/-36.9° - 63.4° = 2.24/-100.3°$.

Exercise 11.6   Given the complex numbers $X = 10/60°$ and $Y = -2 + j3$, find the polar representation of (a) the sum $X + Y$, (b) the product XY, (c) the difference $X-Y$, (d) the ratio X/Y, (e) the difference $Y-X$, and (f) the ratio Y/X.

ANSWER   (a) $12.04/75.57°$; (b) $36.06/-176.31°$; (c) $9.002/38.96°$; (d) $2.773/-63.69°$; (e) $9.002/-141.04°$; (f) $0.3606/63.69°$.

Complex Conjugate

Given a complex variable $X = X_r + jX_i = X_m/\theta$, its **complex conjugate** is defined as

$$X^* \triangleq X_r - jX_i = X_m/-\theta \qquad \textbf{(11.27)}$$

As shown in Figure 11.5, X^* is obtained graphically by reflecting X about the real axis.

An interesting complex-variable property is that the product XX^* is always real: $XX^* = (X_m/\theta)(X_m/-\theta) = X_m^2/\theta - \theta = X_m^2/0°$, or

$$XX^* = X_m^2 = X_r^2 + X_i^2 \qquad \textbf{(11.28)}$$

This feature is used to **rationalize** the denominator of a complex ratio, that is, to convert the denominator to a real number.

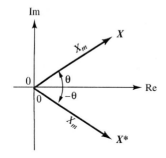

Figure 11.5 A complex variable X and its *complex conjugate X^**.

▶**Example 11.5**

Obtain the rectangular form for the ratio

$$X = \frac{4 + j3}{1 + j2}$$

Solution

One way of doing it is to convert the numerator and denominator to polar form, take their ratio, and then convert the polar result back to rectangular form. A quicker method is to **multiply the numerator and denominator by the complex conjugate of the denominator.** This will eliminate any j from the denominator, thus expediting the transformation to rectangular form:

$$X = \frac{4+j3}{1+j2} \times \frac{1-j2}{1-j2} = \frac{4-j8+j3+6}{1^2+2^2} = \frac{10-j5}{5} = 2 - j1 \blacktriangleleft$$

The rationalization of the denominator is especially useful when it is desired to find the **reciprocal** of a complex variable X. Expressing Equation (11.28) as $XX^* = |X|^2$, the reciprocal is found as

$$\frac{1}{X} = \frac{X^*}{|X|^2} \tag{11.29a}$$

or, in rectangular form, as

$$\frac{1}{X_r + jX_i} = \frac{X_r - jX_i}{X_r^2 + X_i^2} \tag{11.29b}$$

In words, the *reciprocal* of a complex variable is found by taking its *complex conjugate* and *dividing* this by its *modulus squared.* Modern scientific calculators provide this function as well as most other complex-variable operations automatically.

Given a complex variable X and its complex conjugate X^*, it is readily seen that the following additional properties hold:

$$X + X^* = 2 \operatorname{Re}[X] \tag{11.30a}$$

$$X - X^* = 2 \operatorname{Im}[X] \tag{11.30b}$$

or $\operatorname{Re}[X] = (X + X^*)/2$, and $\operatorname{Im}[X] = (X - X^*)/2$.

• Using Phasors to Solve Differential Equations

We conclude by demonstrating the use of phasors to solve *linear differential equations* with *time-independent coefficients* for the case of ac forcing functions. For instance, consider the equation

$$\tau \frac{dy(t)}{dt} + y(t) = x(t) \tag{11.31}$$

where $x(t) = X_m \cos(\omega t + \theta_x)$. Let the phasors of $x(t)$ and $y(t)$ be, respectively, $X = X_m\underline{/\theta_x}$ and $Y = Y_m\underline{/\theta_y}$. Then, by Rules 1 and 2, the phasor of $\tau\, dy/dt$ is $\tau(j\omega Y)$, or $j\omega\tau Y$. Moreover, by Rule 4, adding sinusoidal terms is equivalent to adding the corresponding phasors. Hence, Equation (11.31) becomes, in phasor form,

$$j\omega\tau Y + Y = X \tag{11.32}$$

Note how a *differential* equation in $y(t)$ and $x(t)$ has become an *algebraic* equation in Y and X. Collecting, we get $(j\omega\tau + 1)Y = X$, or

$$Y = \frac{X}{1 + j\omega\tau} \tag{11.33}$$

Application of Rule 6 finally yields $Y_m = |X|/|1 + j\omega\tau|$, or

$$Y_m = \frac{X_m}{\sqrt{1 + (\omega\tau)^2}} \tag{11.34a}$$

and $\angle Y = \angle X - \angle(1 + j\omega\tau)$, or

$$\theta_y = \theta_x - \tan^{-1}\omega\tau \tag{11.34b}$$

Once again we stress that the solution of linear differential equations via phasor algebra is applicable only if the forcing function is of the *ac* type!

Exercise 11.7 Consider the differential equation

$$\tau\frac{dy(t)}{dt} + y(t) = k\tau\frac{dx(t)}{dt}$$

where τ is a time constant, k is a suitable scale factor, and $x(t) = X_m\cos(\omega t + \theta_x)$. Letting $y(t) = Y_m\cos(\omega t + \theta_y)$, express this differential equation in phasor form; hence, obtain a relationship between Y_m and X_m, and a relationship between θ_y and θ_x.

ANSWER $Y_m = kX_m\omega\tau/\sqrt{1 + (\omega\tau)^2}$, $\theta_y = 90° - \tan^{-1}\omega\tau + \theta_x$.

11.2 AC IMPEDANCE

Having learned how to represent and manipulate phasors, we are now ready to express the laws of the basic elements in phasor form. This will lead us to the important concept of *ac impedance*.

Generalized Ohm's Law

For **resistance,** the constitutive law is $v = Ri$. If we denote the phasors of v and i as V and I, Phasor Rule 1 of the previous section states that the phasor of Ri is RI. We can thus write

$$V = RI \tag{11.35}$$

indicating that Ohm's Law holds also in phasor form. For **inductance,** the constitutive law is $v = L\,di/dt$. By Phasor Rules 1 and 2, the phasor of $L\,di/dt$ is $L(j\omega I)$, indicating that

$$V = j\omega LI \tag{11.36}$$

For **capacitance,** the constitutive law is $v = (1/C)\int i\,dt$. Using Phasor Rules 1 and 3, we can express this law in phasor form as $V = (1/C) \times (1/j\omega)I$, or

$$V = \frac{1}{j\omega C}I \tag{11.37}$$

All three equations are of the type

$$V = ZI \qquad (11.38)$$

where Z, a frequency-dependent complex quantity having the dimensions of *ohms* (Ω), is called the **ac impedance.** Equation (11.38) is referred to as the *generalized Ohm's Law for ac circuits.* It is apparent that the constitutive laws, which are *integro-differential* equations when expressed in the time domain, become *algebraic* equations when expressed in the frequency domain.

Impedance and Admittance

Just as R represents the ability of a circuit element to oppose current flow and is defined as $R = v/i$, Z represents the **ability of a circuit element to oppose the flow of ac current** and is defined as

$$Z = \frac{V}{I} \qquad (11.39)$$

In light of Equations (11.35) through (11.37), the impedances of the individual circuit elements take on the forms

$$Z_R = R \qquad (11.40)$$

$$Z_L = j\omega L \qquad (11.41)$$

$$Z_C = \frac{1}{j\omega C} = -j\frac{1}{\omega C} \qquad (11.42)$$

Whereas Z_R is real and frequency independent, Z_C and Z_L contain (a) ω to reflect *frequency dependency,* and (b) j to reflect the 90° *phase shift* between voltage and current. The three impedance forms are depicted in Figure 11.6, which emphasizes the $\pm 90°$ phase shift between voltage and current in the energy-storage elements.

At times it is convenient to work with the reciprocal of impedance, or **admittance,**

$$Y = \frac{1}{Z} \qquad (11.43)$$

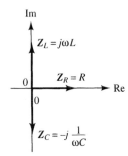

Figure 11.6 Illustrating the three basic forms of impedance.

It can be obtained as $Y = I/V$, and is expressed in Ω^{-1}, or *siemens* (S). Clearly, $Y_R = 1/R = G$, $Y_C = j\omega C$, and $Y_L = 1/j\omega L = -j/\omega L$.

▶ Example 11.6

Find the impedance of a 1-kΩ resistance, a 1-mH inductance, and a 1-μF capacitance at (a) 1 kHz and (b) 100 kHz.

Solution

(a) $\mathbf{Z}_R = 1 \text{ k}\Omega$; $\mathbf{Z}_L = j\omega L = j2\pi f L = j2\pi 10^3 \times 10^{-3} = j6.283 \ \Omega$; $\mathbf{Z}_C = -j/\omega C = -j/2\pi f C = -j/(2\pi 10^3 \times 10^{-6}) = -j159.2 \ \Omega$.

(b) $\mathbf{Z}_R = 1 \text{ k}\Omega$; $\mathbf{Z}_L = j2\pi 10^5 \times 10^{-3} = j628.3 \ \Omega$; $\mathbf{Z}_C = -j/(2\pi 10^5 \times 10^{-6}) = -j1.592 \ \Omega$.

Remark   Increasing frequency increases $\mathbf{Z}_L$ and decreases $\mathbf{Z}_C$ in proportion. However, it does not affect $\mathbf{Z}_R$. ◀

Exercise 11.8

(a) Find the capacitance needed to achieve an impedance of $-j1 \text{ k}\Omega$ at 1 MHz.

(b) Find the frequency, in Hz, at which a 25-mH inductance provides an impedance of $j1 \text{ k}\Omega$.

ANSWER $C = 159.2 \text{ pF}$, $f = 6.366 \text{ kHz}$.

Limiting Cases for the Element Impedances

An engineer must be able to predict the behavior of an ac circuit in the low-frequency or the high-frequency limits using simple inspection. This task is facilitated by the knowledge of how capacitive and inductive impedances behave at either extreme. Using Equations (11.41) and (11.42), it is readily seen that

$$\lim_{\omega \to 0} \mathbf{Z}_C = \infty \tag{11.44}$$

$$\lim_{\omega \to 0} \mathbf{Z}_L = 0 \tag{11.45}$$

$$\lim_{\omega \to \infty} \mathbf{Z}_C = 0 \tag{11.46}$$

$$\lim_{\omega \to \infty} \mathbf{Z}_L = \infty \tag{11.47}$$

An infinite impedance allows no ac current to flow regardless of the applied voltage, and thus behaves as an *open circuit*. A zero impedance develops zero ac voltage regardless of the applied current, and thus behaves as a *short circuit*. In summary, **at sufficiently low operating frequencies, capacitances act as open circuits, and inductances as short circuits. At sufficiently high operating frequencies, capacitances act as short circuits, and inductances as open circuits.** These rules, worth remembering, confirm our physical considerations of the previous chapter.

Series/Parallel Impedance Combinations

Just as it is advantageous to find the equivalent resistance of series and parallel resistance combinations, we now wish to find the equivalent impedance of series and parallel combinations of *RLC* elements.

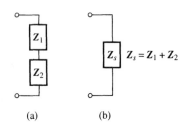

Figure 11.7 Two impedances in *series* are equivalent to a single impedance $Z_p = Z_1 + Z_2$.

When two elements with impedances Z_1 and Z_2 are connected in **series** as in Figure 11.7(a), they carry the same current i and develop a net voltage v which is the sum of the individual voltages, $v = v_1 + v_2$. Phasor Rule 4 allows us to put this in phasor form as $V = V_1 + V_2$. Dividing through by I yields $V/I = V_1/I + V_2/I$, or

$$Z_s = Z_1 + Z_2 \qquad (11.48)$$

where $Z_1 = V_1/I$ and $Z_2 = V_2/I$ are the individual impedances, and $Z_s = V/I$ is the *equivalent impedance* of their series combination shown in Figure 11.7(b). We observe that even though series elements carry the same current, their individual voltages will generally differ not only in *amplitude,* but also in *phase angle.*

Likewise, two elements connected in **parallel** as in Figure 11.8(a) share the same voltage v and conduct a net current i that is the sum of the individual currents, $i = i_1 + i_2$. In phasor form, $I = I_1 + I_2$. Dividing through by V yields $I/V = I_1/V + I_2/V$, or

$$\frac{1}{Z_p} = \frac{1}{Z_1} + \frac{1}{Z_2} \qquad (11.49a)$$

where $Z_1 = V/I_1$ and $Z_2 = V/I_2$ are the impedances of the individual elements, and $Z_p = V/I$ is the *equivalent impedance* of their parallel combination shown in Figure 11.8(b). This is often expressed in shorthand form as $Z_p = Z_1 \parallel Z_2$. Note that even though parallel elements share the same voltage, their individual currents generally differ both in *amplitude* and *phase angle.*

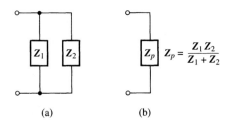

Figure 11.8 Two impedances in *parallel* are equivalent to a single impedance $Z_p = Z_1 \parallel Z_2 = Z_1Z_2/(Z_1 + Z_2)$.

We can also express Equation (11.49a) in terms of admittances as

$$Y_p = Y_1 + Y_2 \qquad (11.49b)$$

Clearly, when dealing with *parallel combinations* it is more convenient to work with *admittances,* and when dealing with *series combinations* it is more convenient to work with *impedances.*

► **Example 11.7**

    (a) If a 1-kΩ resistance and a 0.1-μF capacitance are connected in *series,* what is the corresponding impedance at 1 kHz? Express the result both in rectangular and in polar form.

    (b) Repeat, but with R and C connected in *parallel.*

Solution

    (a) By Equation (11.48), $\mathbf{Z}_s = R - j/\omega C = 10^3 - j/(2\pi \times 10^3 \times 0.1 \times 10^{-6})$, or $\mathbf{Z}_s = 1000 - j1592\ \Omega$. Using Equation (11.16) to convert from rectangular to polar form yields

$$\mathbf{Z}_s = 1000 - j1592 = 1879\underline{/-57.86°}\ \Omega$$

    (b) By Equation (11.49b), $\mathbf{Y}_p = 1/R + j\omega C = 1/1000 + j/1592\ \Omega^{-1}$. To obtain $\mathbf{Z}_p$, find the reciprocal of $\mathbf{Y}_p$ via Equation (11.29b). Thus, $\mathbf{Z}_p = (1/1000 - j/1592)/(1/1000^2 + 1/1592^2) = 717.0 - j450.5\ \Omega$. Converting to polar form gives

$$\mathbf{Z}_p = 717.0 - j450.5 = 846.7\underline{/-32.14°}\ \Omega$$

◄

Exercise 11.9   Repeat Example 11.7, but for a 50-Ω resistance and a 10-mH inductance.

ANSWER $\mathbf{Z}_s = 50 + j62.83 = 80.30\underline{/51.49°}\ \Omega$; $\mathbf{Z}_p = 30.61 + j24.36 = 39.12\underline{/38.51°}\ \Omega$.

► **Example 11.8**

    (a) What resistance and inductance must be connected in *series* to obtain $\mathbf{Z} = 300 + j400\ \Omega$ at $\omega = 10^5$ rad/s? At $\omega = 10^6$ rad/s?

    (b) Repeat, but with the elements connected in *parallel.*

Solution

    (a) Impose $\mathbf{Z}_R + \mathbf{Z}_L = \mathbf{Z}$, or

$$R + j\omega L = 300 + j400\ \Omega$$

Equating the real and imaginary parts pairwise yields $R = 300\ \Omega$ and $\omega L = 400\ \Omega$. Thus, for $\omega = 10^5$ rad/s we need $L = 400/10^5 = 4$ mH, and for $\omega = 10^6$ rad/s we need $L = 0.4$ mH.

(b) Impose $1/\mathbf{Z}_R + 1/\mathbf{Z}_L = 1/\mathbf{Z}$, or $1/R + 1/j\omega L = 1/(300 + j400)$. Using Equation (11.29) to rationalize $1/\mathbf{Z}$, we obtain

$$\frac{1}{R} - j\frac{1}{\omega L} = (1.2 - j1.6)10^{-3}\ \Omega^{-1}$$

Equating the real and imaginary parts yields $1/R = 1.2 \times 10^{-3}$, or $R = 833.3\ \Omega$, and $\omega L = 1/(1.6 \times 10^{-3}) = 625\ \Omega$. Then, for $\omega = 10^5$ rad/s, $L = 625/10^5 = 6.25$ mH; for $\omega = 10^6$ rad/s, $L = 625\ \mu$H.

Remark The required value of L depends both on topology and frequency, that of R only on topology. ◀

Exercise 11.10 Repeat Example 11.8 if the elements are a resistance and a capacitance, and $\mathbf{Z} = 300 - j400\ \Omega$.

ANSWER (a) $R = 300\ \Omega$, $C = 25$ nF, 2.5 nF; (b) $R = 833.3\ \Omega$, $C = 16$ nF, 1.6 nF.

AC Resistance and Reactance

Being a complex quantity, $\mathbf{Z}$ can be expressed either in polar or in rectangular form,

$$\mathbf{Z}(j\omega) = Z_m(\omega)\underline{/\phi(\omega)} \tag{11.50}$$

$$\mathbf{Z}(j\omega) = R(\omega) + jX(\omega) \tag{11.51}$$

where $Z_m(\omega)$ and $\phi(\omega)$ are the **magnitude** and the **phase angle** of $\mathbf{Z}$, and $R(\omega)$ and $X(\omega)$ are called the **ac resistance** and the **ac reactance.** All four quantities are *real*. Moreover, $Z_m(\omega)$, $R(\omega)$, and $X(\omega)$ are in Ω and $\phi(\omega)$ is in rad or in degrees. The complex-plane representation of $\mathbf{Z}$ is shown in Figure 11.9. As we know, rectangular coordinates are converted to polar ones via

$$Z_m = \sqrt{R^2 + X^2} \tag{11.52a}$$

$$\phi = \tan^{-1}\frac{X}{R} \tag{11.52b}$$

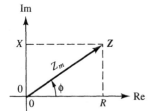

Figure 11.9 Complex-plane representation of impedance.

and polar coordinates are converted to rectangular ones via

$$R = Z_m \cos\phi \tag{11.53a}$$

$$X = Z_m \sin\phi \tag{11.53b}$$

The ac resistance $R(\omega)$, representing the **real part** of **Z**, may or may not involve ω, depending on the particular circuit. Moreover, it may or may not coincide with the ohmic resistance of the circuit. For instance, in part (a) of Example 11.7, the ac resistance was 1000 Ω, the same as the ohmic resistance; however, in part (b) it was 717.0 Ω, even though the ohmic resistance was still 1000 Ω.

The reactance $X(\omega)$, representing the **imaginary part** of **Z**, must always involve ω to reflect the presence of energy-storage elements in the circuit. For this reason, capacitances and inductances are also referred to as **reactive elements.** The two basic reactances are those of the inductance and capacitance themselves,

$$X_L = \omega L \tag{11.54}$$

$$X_C = -\frac{1}{\omega C} \tag{11.55}$$

Note the *opposite* signs and the *reciprocal* frequency dependence, another reflection of the duality principle.

It can be proved that a passive ac one-port—that is, a one-port containing only resistances, inductances, and capacitances but no sources—has always $R(\omega) \geq 0$, indicating that its impedance will lie only in the *first* or *fourth* quadrants of the complex plane. We identify three important cases, depicted in Figure 11.10:

(1) $X(\omega) > 0$: **Z** lies in the first quadrant and is said to be an **inductive impedance.** Since $\phi > 0$, current *lags* voltage.

(2) $X(\omega) < 0$: **Z** lies in the fourth quadrant and is said to be a **capacitive impedance.** Since $\phi < 0$, current *leads* voltage.

(3) $X(\omega) = 0$: **Z** is real; $\mathbf{Z} = R(\omega) + j0 = R(\omega)\underline{/0°}$ and is said to be **purely resistive.** Even though the circuit contains reactive elements, current and voltage are *in phase* with each other. We refer to this phenomenon as **unity power-factor resonance.**

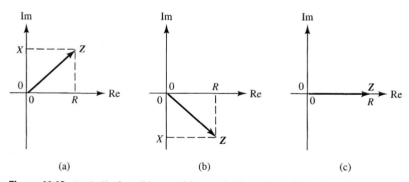

(a)   (b)   (c)

Figure 11.10 (a) Inductive, (b) capacitive, and (c) purely resistive impedance.

▶ Example 11.9

In the ac port of Figure 11.11, let $R_1 = 2$ kΩ, $R_2 = 1$ kΩ, $C = 0.1$ μF, and $L = 50$ mH. Find **Z** and state its type if (a) $\omega = 10^4$ rad/s, (b) $\omega = 10^5$ rad/s, and (c) $\omega = 2 \times 10^4$ rad/s.

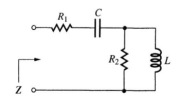

Figure 11.11 An example of an ac port.

Solution

(a) We have $\mathbf{Z} = \mathbf{Z}_{R_1} + \mathbf{Z}_C + (\mathbf{Z}_{R_2} \| \mathbf{Z}_L)$, or

$$\mathbf{Z} = R_1 - j\frac{1}{\omega C} + \frac{R_2 \times j\omega L}{R_2 + j\omega L} = R_1 - j\frac{1}{\omega C} + \omega R_2 L \frac{\omega L + jR_2}{R_2^2 + (\omega L)^2}$$

Substituting the given values yields $\mathbf{Z} = 2000 - j1000 + 200 + j400$ Ω, or

$$\mathbf{Z} = 2200 - j600 = 2280\underline{/-15.26°} \ \Omega$$

This is a *capacitive* impedance with $R = 2200$ Ω, $X = -600$ Ω, $Z_m = 2280$ Ω, and $\phi = -15.26°$. We also observe that the ac resistance is different from the ohmic resistance of the circuit.

(b) For $\omega = 10^5$ rad/s we obtain, by similar calculations,

$$\mathbf{Z} = 2962 + j92.31 = 2963\underline{/1.79°} \ \Omega$$

indicating that $\mathbf{Z}$ is now *inductive*.

(c) For $\omega = 2 \times 10^4$ rad/s, $\mathbf{Z} = 2000 - j500 + 500 + j500$ Ω, or

$$\mathbf{Z} = 2500 + j0 \ \Omega$$

indicating a *resonant* ac one-port.

Remark As ω is increased from 10^4 to 10^5 rad/s, this port changes from capacitive to resistive to inductive. ◀

Exercise 11.11 Consider the one-port of Example 11.9, but with L and C interchanged with each other. Find $\mathbf{Z}$ if (a) $\omega = 0.5 \times 10^4$ rad/s, (b) $\omega = 2 \times 10^4$ rad/s, and (c) $\omega = 10^4$ rad/s, and verify that $\mathbf{Z}$ is, respectively, capacitive, inductive, and resistive.

ANSWER (a) $\mathbf{Z} = 2800 - j150$ Ω; (b) $\mathbf{Z} = 2200 + j600$ Ω; (c) $\mathbf{Z} = 2500 + j0$ Ω.

▶ Example 11.10

If the one-port of Example 11.9 is subjected to the voltage

$$v(t) = 10\cos(10^4 t + 30°) \ \text{V}$$

positive reference at the top, find the current drawn by the port.

Solution

In Example 11.9 it was found that at $\omega = 10^4$ rad/s the port impedance is $\mathbf{Z} = 2280\underline{/-15.26°}$ Ω. By Ohm's Law,

$$I = \frac{V}{Z} = \frac{10\underline{/30°}}{2280\underline{/-15.26°}}$$

Phasor Rule 6 of Section 11.1 states that the modulus and phase of a ratio are, respectively, the ratio of the moduli and the difference of the phases. Thus, $|\boldsymbol{I}| = |\boldsymbol{V}|/|\boldsymbol{I}| = 10/2280 = 4.386$ mA, and $\measuredangle \boldsymbol{I} = \measuredangle \boldsymbol{V} - \measuredangle \boldsymbol{Z} = 30 - (-15.26) = 45.26°$. The current entering the port at the positive side of v is thus

$$i(t) = 4.386 \cos{(10^4 t + 45.26°)} \text{ mA}$$

Exercise 11.12   In the one-port of Figure 11.11 let $R_1 = 150$ Ω, $C = 10$ μF, $R_2 = 50$ Ω, and $L = 25$ mH. If the port is subjected to the current $i(t) = 0.1 \cos{(10^3 t + 45°)}$ A, find the voltage $v(t)$ developed by the port. Assume the passive sign convention.

ANSWER $v(t) = 17.89 \cos{(10^3 t + 18.43°)}$ V.

AC Conductance and Susceptance

Since admittance is the reciprocal of impedance, we have

$$\boldsymbol{Y}(\omega) = Z_m^{-1}(\omega)\underline{/-\phi(\omega)} \tag{11.56}$$

In rectangular form, admittance is expressed as

$$\boldsymbol{Y}(\omega) = G(\omega) + jB(\omega) \tag{11.57}$$

where $G(\omega)$ is called the **ac conductance** and $B(\omega)$ the **ac susceptance.** Both G and B are expressed in Ω^{-1}, or *siemens.* Clearly, we have $\boldsymbol{Y}_C = j\omega C$, $B_C = \omega C$, $\boldsymbol{Y}_L = 1/j\omega L = -j/\omega L$, and $B_L = -1/\omega L$.

Even though $\boldsymbol{Y}$ and $\boldsymbol{Z}$ are reciprocal of each other, in general neither R and G nor X and B are. Using Equation (11.29b), it is easily seen that once R and X are known, G and B can be found as

$$G = \frac{R}{R^2 + X^2} \qquad B = \frac{-X}{R^2 + X^2} \tag{11.58}$$

Conversely, once G and B are known, R and X can be found as

$$R = \frac{G}{G^2 + B^2} \qquad X = \frac{-B}{G^2 + B^2} \tag{11.59}$$

Exercise 11.13   Assuming $\omega = 10^5$ rad/s, (a) find R and X for the case of a 20-Ω resistance and a 1-μF capacitance connected in *parallel.* (b) Find G and B if the same elements are connected in *series.*

ANSWER $R + jX = 4 - j8$ Ω, $G + jB = 0.04 + j0.02$ Ω^{-1}.

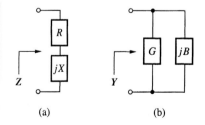

(a)   (b)

Figure 11.12 Impedance and admittance, and their real and imaginary parts.

The preceding considerations indicate that a given impedance can be realized either as the *series* combination of R and X or as the *parallel* combination of G and B. The equivalence of the two topologies is depicted in Figure 11.12, and conversion from one topology to the other is readily done via Equations (11.58) and (11.59).

▶ Example 11.11

Obtain simple (a) *series* and (b) *parallel* equivalents for the one-port of Example 11.9(a).

Solution

(a) In Example 11.9(a) it was found that for $\omega = 10^4$ rad/s the port is *capacitive* and $Z = 2200 - j600$ Ω. A simple *series* equivalent is thus $R_s - j/\omega C_s = 2200 - j600$ Ω. Then, $R_s = 2200$ Ω and $C_s = 1/(600\omega) = 1/(600 \times 10^4) = 166.7$ nF.

(b) A simple *parallel* equivalent is expressed as $1/R_p + j\omega C_p = G + jB$. By Equation (11.58), $G = 2200/[2200^2 + (-600)^2] = 4.231 \times 10^{-4}$ Ω^{-1}, and $B = -(-600)/[2200^2 + (-600)^2] = 1.154 \times 10^{-4}$ Ω^{-1}. Then, $R_p = 1/G = 2364$ Ω, and $C_p = B/\omega = 11.54$ nF.

The two equivalents are shown in Figure 11.13. We stress that this equivalence holds only for $\omega = 10^4$ rad/s.

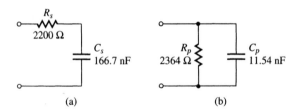

(a)   (b)

Figure 11.13 Simple series and parallel equivalents for the one port of Example 11.11 at $\omega = 10^4$ rad/s.

Exercise 11.14 Obtain simple (a) *series* and (b) *parallel* equivalents for the one-port of Example 11.9(b).

ANSWER (a) 2962 Ω, 923.1 μH; (b) 2964 Ω, 0.9511 H.

• *RL, RC,* and *LC* Pairs

Series and parallel combinations of element pairs occur so often that it is worth deriving expressions for their impedances, and investigating their low- and high-frequency behavior using *physical insight*. This investigation is facilitated by the observation that if one impedance is much larger than the other, say, if

$$|\mathbf{Z}_1| \gg |\mathbf{Z}_2| \tag{11.60}$$

then their series and parallel combinations can be approximated as

$$\mathbf{Z}_1 + \mathbf{Z}_2 \simeq \mathbf{Z}_1 \qquad \text{(11.61a)}$$

$$\mathbf{Z}_1 \parallel \mathbf{Z}_2 \simeq \mathbf{Z}_2 \qquad \text{(11.61b)}$$

The results are summarized in Table 11.2. We discuss only the first row and leave the others as an exercise for the reader.

TABLE 11.2 Behavior of the Impedances of Basic Element Pairs

$\mathbf{Z}$	$\mathbf{Z}(j\omega)$	ω_0	$\mathbf{Z}_{\omega \to 0}$	$\mathbf{Z}_{\omega = \omega_0}$	$\mathbf{Z}_{\omega \to \infty}$
$\mathbf{Z}_R + \mathbf{Z}_C$	$R\left(\dfrac{1 + j\omega RC}{j\omega RC}\right)$	$\dfrac{1}{RC}$	$\mathbf{Z}_C$	$R\sqrt{2}\underline{/-45°}$	$\mathbf{Z}_R$
$\mathbf{Z}_R \parallel \mathbf{Z}_C$	$R\left(\dfrac{1}{1 + j\omega RC}\right)$	$\dfrac{1}{RC}$	$\mathbf{Z}_R$	$\dfrac{R}{\sqrt{2}}\underline{/-45°}$	$\mathbf{Z}_C$
$\mathbf{Z}_R + \mathbf{Z}_L$	$R\left(1 + j\omega\dfrac{L}{R}\right)$	$\dfrac{R}{L}$	$\mathbf{Z}_R$	$R\sqrt{2}\underline{/45°}$	$\mathbf{Z}_L$
$\mathbf{Z}_R \parallel \mathbf{Z}_L$	$R\left(\dfrac{j\omega L/R}{1 + j\omega L/R}\right)$	$\dfrac{R}{L}$	$\mathbf{Z}_L$	$\dfrac{R}{\sqrt{2}}\underline{/45°}$	$\mathbf{Z}_R$
$\mathbf{Z}_L + \mathbf{Z}_C$	$\dfrac{1 - \omega^2 LC}{j\omega C}$	$\dfrac{1}{\sqrt{LC}}$	$\mathbf{Z}_C$	0	$\mathbf{Z}_L$
$\mathbf{Z}_L \parallel \mathbf{Z}_C$	$\dfrac{j\omega L}{1 - \omega^2 LC}$	$\dfrac{1}{\sqrt{LC}}$	$\mathbf{Z}_L$	∞	$\mathbf{Z}_C$

Thus, the impedance of a resistance in series with a capacitance is $\mathbf{Z} = \mathbf{Z}_R + \mathbf{Z}_C$, which can be written as $\mathbf{Z} = R + 1/j\omega C = R(1 + j\omega RC)/(j\omega RC)$. We have three significant cases:

(1) $\omega \to 0$. At low frequencies, where $|\mathbf{Z}_C| \gg \mathbf{Z}_R$, we can approximate $\mathbf{Z} \simeq \mathbf{Z}_C$, indicating *capacitive behavior.*

(2) $\omega \to \infty$. At high frequencies, where $|\mathbf{Z}_C| \ll \mathbf{Z}_R$, we have $\mathbf{Z} \simeq \mathbf{Z}_R$, indicating *resistive* behavior.

(3) The borderline between the two cases occurs at the special frequency ω_0 that makes $|\mathbf{Z}_C| = \mathbf{Z}_R$, or $1/(\omega_0 C) = R$. This is the **characteristic frequency** $\omega_0 = 1/RC$. The impedance at this frequency is $\mathbf{Z}(j\omega_0) = R(1 + j1)/(j1) = R\sqrt{2}\underline{/-45°}$.

The reader is encouraged to verify all remaining rows in detail. In each row, the borderline frequency ω_0 is found as the frequency at which the individual impedances making up $\mathbf{Z}$ achieve *equal* magnitudes. In this respect we find particularly interesting the LC combinations of the last two rows, because at this frequency their individual impedances are *opposite* to each other, or $\mathbf{Z}_L(\omega_0) = -\mathbf{Z}_C(\omega_0)$, as one can readily verify. Consequently, for $\omega = \omega_0$ a *series LC* pair acts as a *short circuit,* and a *parallel LC* pair as an *open circuit.* These interesting phenomena form the basis of *resonance,* which is studied in Chapter 13.

11.3 FREQUENCY-DOMAIN ANALYSIS

We are finally ready to analyze ac circuits by working *directly* with phasors. In the previous chapter we saw that ac analysis in the time domain relies on the use of Kirchhoff's laws and the element laws. Phasor Rule 4 of Section 11.1 states that adding ac signals is equivalent to adding their phasors. The time-domain form of KCL of Equation (1.32) is thus restated in phasor form as

$$\sum_n I_{\text{in}} = \sum_n I_{\text{out}} \qquad\qquad \textbf{(11.62)}$$

where $\sum_n$ denotes summation over the phasors of all currents respectively entering and leaving a given node n. Likewise, the phasor form of KVL of Equation (1.35) is

$$\sum_\ell V_{\text{rise}} = \sum_\ell V_{\text{drop}} \qquad\qquad \textbf{(11.63)}$$

where $\sum_\ell$ denotes summation over the phasors of, respectively, all voltage rises and all voltage drops around a given loop ℓ. We have also seen that the constitutive laws of the basic elements, when expressed in terms of phasors, take on the common form of the generalized Ohm's Law,

$$V = ZI \qquad\qquad \textbf{(11.64)}$$

The formal similarity between the phasor laws for ac circuits and the ordinary laws for resistive circuits indicates that ac analysis via phasors can be carried out using the very techniques developed in Chapters 1 through 6 for resistive circuits. This much simpler form of ac analysis, called **frequency-domain analysis,** involves the following steps:

(1) Redraw the circuit, but with ac signals replaced by their phasors, and the R, L, and C elements replaced by their respective impedances, Z_R, Z_L, and Z_C. The outcome is called the *frequency-domain representation* of the circuit.

(2) Analyze this circuit as if it were purely resistive, except that its elements are now *impedances* and the signals are *phasors*. Depending on the case, this analysis may require *series* or *parallel impedance reductions,* the *voltage* or *current divider formulas,* the *proportionality analysis procedure, node or loop methods, Thévenin* or *Norton reductions,* the *op amp rule,* and so forth. The outcome is one or more *algebraic equations* involving *phasors* and having coefficients that are functions of the *complex impedances* Z_R, Z_C, and Z_L.

(3) Solve for the unknown phasors or phasor ratios.

(4) In the expressions for the unknowns replace Z_R with R, Z_L with $j\omega L$, and Z_C with $1/j\omega C$. The outcome is, in general, one or more

frequency-dependent complex quantities. Depending on the problem statement, the final task may be the derivation of expressions for the modulus or the argument of each unknown, their calculation at a given frequency, and so forth.

Let us illustrate the procedure by actual examples.

AC Dividers

In the R-C circuit of Figure 11.14(a) let $v_i = V_{im} \cos \omega t$ and $v_o = V_{om} \cos (\omega t + \phi)$. We wish to find V_{om}/V_{im}, and ϕ. Replacing ac signals with phasors, and circuit elements with impedances, leads to the frequency-domain circuit of Figure 11.14(b). Considering that phasors and impedances are *complex* quantities, this circuit is only a conceptualization. The circuit that we would actually try out in the laboratory is clearly that of Figure 11.14(a); however, to facilitate our analysis, we work with its mathematical counterpart of Figure 11.14(b). As long as it yields the correct results, its complex nature need not intimidate us!

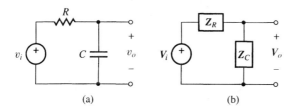

(a)   (b)

Figure 11.14 The R-C circuit and its frequency-domain form.

The circuit of Figure 11.14(b) is an ac voltage divider, so

$$V_o = \frac{Z_C}{Z_R + Z_C} V_i \qquad \text{(11.65)}$$

As anticipated, this is an algebraic equation with $Z_C/(Z_R + Z_C)$ as its coefficient. Next, substitute $Z_R = R$ and $Z_C = 1/j\omega C$ to obtain

$$V_o = \frac{1/j\omega C}{R + 1/j\omega C} V_i$$

Now simplify by multiplying the numerator and denominator by $j\omega C$,

$$V_o = \frac{V_i}{1 + j\omega RC} \qquad \text{(11.66)}$$

As expected, our unknown V_o is a frequency-dependent complex quantity. Our final step is the calculation of $|V_o|$ and $\angle V_o$.

To find $|V_o|$, use the fact that the modulus of a ratio equals the ratio of the moduli: $|V_o| = |V_i|/|1 + j\omega RC|$, or

$$V_{om} = \frac{1}{\sqrt{1 + (\omega RC)^2}} V_{im} \qquad \text{(11.67a)}$$

To find $\sphericalangle V_o$, use the fact that the phase of a ratio equals the difference of the phases: $\sphericalangle V_o = \sphericalangle V_i - \sphericalangle (1 + j\omega RC) = 0° - \tan^{-1} \omega RC$, or

$$\phi = -\tan^{-1} \omega RC \tag{11.67b}$$

Comparing with Equation (10.32) indicates that the *R-C* circuit is a **low-pass filter.** We justify this physically by noting that at sufficiently *low frequencies,* where $|Z_C| \gg Z_R$, Equation (11.65) predicts $V_o \to V_i$. Low-frequency input signals make it to the output port with negligible loss. By contrast, at sufficiently *high frequencies,* where $|Z_C| \ll Z_R$, Equation (11.65) predicts $V_o \to (1/j\omega C)V_i \to 0$, indicating that high-frequency signals are cut off because C shunts them to ground. We also observe that in the high-frequency limit the *R-C* circuit approximates an **integrator,** thus confirming our time-domain findings of Section 8.3.

The *borderline frequency* is that special frequency ω_0 that makes $|Z_C(\omega_0)| = Z_R$, that is, $1/(\omega_0 C) = R$, or

$$\omega_0 = \frac{1}{RC} \tag{11.68}$$

Clearly, this is the *cutoff frequency* of the filter.

Exercise 11.15 Show that the *L-R* circuit of Figure 11.15(a) is also a *low-pass* filter. Use physical reasoning to show that its cutoff frequency is $\omega_0 = R/L$.

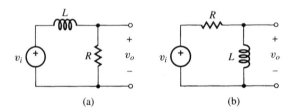

(a)   (b)

Figure 11.15 *L-R* and *R-L* circuits of Exercises 11.15 and 11.16.

Turning now to the *C-R* circuit of Figure 11.16(a), we redraw it in the frequency-domain form of Figure 11.16(b) and again apply the ac voltage divider formula to obtain

$$V_o = \frac{Z_R}{Z_R + Z_C} V_i \tag{11.69}$$

Letting $Z_R = R$ and $Z_C = 1/j\omega C$ and simplifying yields

$$V_o = \frac{j\omega RC}{1 + j\omega RC} V_i \tag{11.70}$$

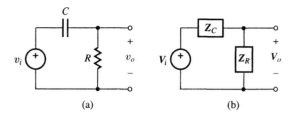

Figure 11.16 The C-R circuit and its frequency-domain form.

Proceeding as before, we find

$$V_{om} = \frac{\omega RC}{\sqrt{1 + (\omega RC)^2}} V_{im} \tag{11.71a}$$

$$\phi = 90° - \tan^{-1} \omega RC \tag{11.71b}$$

Comparison with Equation (10.26) reveals that the C-R circuit is a **high-pass filter.** Physically, at sufficiently *high frequencies,* where $|Z_C| \ll Z_R$, Equation (11.69) predicts $V_o \to V_i$, indicating that high-frequency input signals make it to the output port with negligible loss. By contrast, at sufficiently *low frequencies,* where $|Z_C| \gg Z_R$, Equation (11.69) predicts $V_o \to j\omega RCV_i \to 0$, indicating that low-frequency signals are cut off because of the open-circuit action by C. We also note that in the low-frequency limit the C-R circuit approximates a **differentiator,** thus confirming our time-domain findings of Section 8.3.

The *borderline frequency* is that special frequency ω_0 that makes $|Z_C(\omega_0)| = Z_R$, that is, $1/(\omega_0 C) = R$, or $\omega_0 = 1/RC$. This is the *cutoff frequency* of the filter.

Exercise 11.16   Show that the R-L circuit of Figure 11.15(b) is a *high-pass filter.* Use physical reasoning to show that its cutoff frequency is $\omega_0 = R/L$.

▶ **Example 11.12**

In the circuit of Figure 11.17(a) let $C = 1 \ \mu$F, $R = 300 \ \Omega$, and $L = 0.25$ H. Find i if

$$i_s = 0.1 \cos 10^3 t \ \text{A}$$

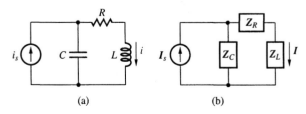

(a)   (b)

Figure 11.17 Circuit of Example 11.12.

Solution

Redraw the circuit in the frequency-domain form of Figure 11.17(b) and apply the current divider formula to obtain

$$I = \frac{Z_C}{Z_C + Z_R + Z_L} I_s$$

At the given operating frequency we have $Z_C = -j/(10^3 \times 10^{-6}) = -j10^3 \ \Omega$, $Z_R = 300 \ \Omega$, and $Z_L = j10^3 \times 0.25 = j250 \ \Omega$, so

$$I = \frac{-j1000}{-j1000 + 300 + j250} I_s = \frac{-j1000}{300 - j750} I_s$$

By Equations (11.25) and (11.26), $|I| = \left(1000/\sqrt{300^2 + 750^2}\right) \times 0.1 = 0.1238$ A and $\sphericalangle I = -90 - \tan^{-1}(-750/300) = -21.80°$. Consequently,

$$i = 0.1238 \cos(10^3 t - 21.80°) \text{ A}$$

◄

Exercise 11.17   In the circuit of Example 11.12 find the voltage across C.

ANSWER $48.34 \cos(10^3 t + 18°)$ V, positive reference at top.

The Proportionality Analysis Procedure

The proportionality analysis procedure, introduced in Section 2.2 for resistive networks, can readily be extended to ac networks such as *impedance ladders*. As we know, a distinguishing feature of a ladder network is that all its branch voltages and currents can be expressed in terms of the voltage or current of the *branch farthest away* from the applied source.

► Example 11.13

In the impedance ladder of Figure 11.18 let Z_1, Z_3, and Z_5 be 20-Ω resistances and Z_2, Z_4, and Z_6 be 0.1-μF capacitances. Find V_2, V_4, and V_6 if $V_s = 10\underline{/0°}$ V and $\omega = 10^6$ rad/s.

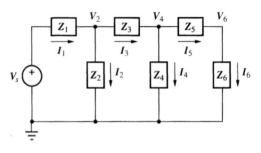

Figure 11.18 Impedance ladder.

Solution

Since $-j/\omega C = -j/(10^6 \times 10^{-7}) = -j10 \ \Omega$, we have

$$Z_1 = Z_3 = Z_5 = 20 \ \Omega$$

$$Z_2 = Z_4 = Z_6 = -j10 \ \Omega$$

Starting with I_6 and repeatedly applying Ohm's and Kirchhoff's laws gives

Ω: $\quad V_6 = Z_6 I_6 = -j10 I_6$

KCL: $\quad I_5 = I_6$

KVL: $\quad V_4 = Z_5 I_5 + V_6 = 20 I_6 - j10 I_6 = (20 - j10) I_6$

Ω: $\quad I_4 = V_4/Z_4 = [(20 - j10)/(-j10)] I_6 = (1 + j2) I_6$

KCL: $\quad I_3 = I_4 + I_5 = (1 + j2 + 1) I_6 = (2 + j2) I_6$

KVL: $\quad V_2 = Z_3 I_3 + V_4 = 20(2 + j2) I_6 + (20 - j10) I_6 = (60 + j30) I_6$

Ω: $\quad I_2 = V_2/Z_2 = [(60 + j30)/(-j10)] I_6 = (-3 + j6) I_6$

KCL: $\quad I_1 = I_2 + I_3 = (-3 + j6 + 2 + j2) I_6 = (-1 + j8) I_6$

KVL: $\quad V_s = Z_1 I_1 + V_2 = 20(-1 + j8) I_6 + (60 + j30) I_6 = (40 + j190) I_6$

Solving for I_6 yields

$$I_6 = \frac{10\underline{/0°}}{40 + j190} = \frac{10}{\sqrt{40^2 + 190^2}}\underline{/0° - \tan^{-1}(190/40)}$$

$$= 51.50\underline{/-78.11°} \ \text{mA}$$

Back-substituting,

$$V_2 = \left[\sqrt{60^2 + 30^2}\underline{/\tan^{-1}(30/60)} \right] 51.50 \times 10^{-3}\underline{/-78.11°}$$

$$= 3.455\underline{/-51.55°} \ \text{V}$$

Likewise, $V_4 = 1.152\underline{/-104.68°}$ V, and $V_6 = 0.5150\underline{/-168.11°}$ V. ◀

Exercise 11.18 In the impedance ladder of Figure 11.18 let Z_1, Z_3, and Z_5 be 200-μH inductances and Z_2, Z_4, and Z_6 be 2-nF capacitances. Find V_2, V_4, and V_6 if $V_s = 10\underline{/0°}$ and $\omega = 10^6$ rad/s.

ANSWER $0.6024\underline{/0°}$ V, $9.036\underline{/180°}$ V, $15.06\underline{/180°}$ V.

Nodal and Loop Analysis

The superiority of phasor analysis over time-domain analysis is even more apparent when circuit complexity is such as to require the use of nodal or loop analysis. If this analysis were performed in the time domain, it would require formulating and solving a *system of differential equations*. In the frequency domain, however, it only requires solving a *system of algebraic equations*.

▶ **Example 11.14**

Find v_1 and v_2 in the circuit of Figure 11.19(a).

Solution

At $\omega = 10$ rad/s the impedances are as shown in Figure 11.19(b), where the units are [V], [A], and [Ω]. Before proceeding, it is convenient to combine parallel elements as shown in the simpler circuit of Figure 11.19(c). Applying KCL at the nodes labeled V_1 and V_2 we obtain, respectively,

$$2 = \frac{V_1}{1 - j1} + \frac{V_1 - V_2}{-j2}$$

$$\frac{V_1 - V_2}{-j2} = \frac{V_2}{1} + \frac{V_2 - 5}{j1}$$

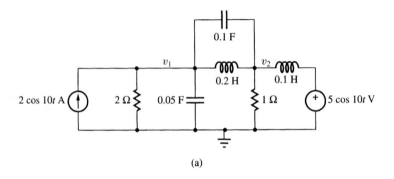

(a)

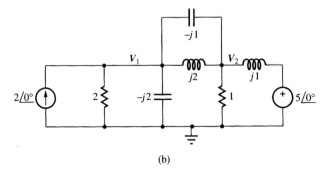

(b)

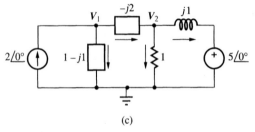

(c)

Figure 11.19 AC circuit analysis via the node method.

Expanding and collecting, we get

$$(1 - j3)V_1 + (-1 + j1)V_2 = -4 - j4$$

$$V_1 + (1 + j2)V_2 = 10$$

This system of two equations in two unknowns can be solved via Cramer's rule or Gaussian elimination. For instance, multiplying the second equation by $-(1 - j3)$ throughout, and then adding it to the first equation pairwise, we obtain

$$-(1 - j3)(1 + j2)V_2 + (-1 + j1)V_2 = -(1 - j3)10 - 4 - j4$$

Collecting, simplifying, and solving gives

$$V_2 = \frac{7 - j13}{4 - j1} = 3.581\underline{/-47.66°} \text{ V}$$

Back-substituting gives

$$V_1 = 10 - (1 + j2)V_2 = \frac{7 - j11}{4 - j1} = \sqrt{10}\underline{/-43.49°} \text{ V}$$

Hence,

$$v_1 = \sqrt{10}\cos(10t - 43.49°) \text{ V}$$

$$v_2 = 3.581\cos(10t - 47.66°) \text{ V}$$

▶ Example 11.15

The circuit of Figure 11.20 is already expressed in frequency-domain form. Find V_1 and V_2 via nodal analysis.

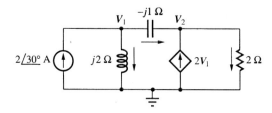

Figure 11.20 Frequency-domain nodal analysis.

Solution

Applying KCL at the designated nodes yields

$$2\underline{/30°} = \frac{V_1}{j2} + \frac{V_1 - V_2}{-j1}$$

$$2V_1 + \frac{V_1 - V_2}{-j1} = \frac{V_2}{2}$$

Expanding and collecting gives

$$j0.5V_1 - jV_2 = 2\underline{/30°}$$

$$(2 + j1)V_1 - (0.5 + j1)V_2 = 0$$

Multiplying the first equation by j throughout yields $-0.5V_1 + V_2 = j(2\underline{/30°}) = 2\underline{/120°}$, or

$$V_2 = 0.5V_1 + 2\underline{/120°}$$

Substituting into the second equation, we get

$$(2 + j1)V_1 - (0.5 + j1)(0.5V_1 + 2\underline{/120°}) = 0$$

Collecting terms yields $(1.75 + j0.5)V_1 = (0.5 + j1)(2\underline{/120°})$, or

$$V_1 = \frac{(0.5 + j1)2\underline{/120°}}{1.75 + j0.5} = 1.229\underline{/167.49°} \text{ V}$$

Back substituting, we have

$$V_2 = 0.5 \times 1.229\underline{/167.49°} + 2\underline{/120°} = 2.457\underline{/130.62°} \text{ V}$$

◀

Exercise 11.19   Redraw the circuit of Figure 11.20, but with the inductance and the capacitance interchanged with each other. Hence, find V_1 and V_2 via nodal analysis.

ANSWER $1.050\underline{/-83.20°}$ V, $3.063\underline{/-52.23°}$ V.

▶ Example 11.16

The circuit of Figure 11.21 is already expressed in frequency-domain form. Find I_1 and I_2 via loop analysis.

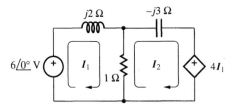

Figure 11.21 Frequency-domain loop analysis.

Solution

Applying KVL around the designated meshes yields

$$6\underline{/0°} = j2I_1 + 1(I_1 - I_2)$$

$$0 = 1(I_2 - I_1) - j3I_2 + 4I_1$$

Expanding and collecting,

$$(1 + j2)I_1 - I_2 = 6$$

$$3I_1 + (1 - j3)I_2 = 0$$

Multiplying the first equation throughout by $(1 - j3)$ and then adding it to the second equation pairwise yields $(1 - j3)(1 + j2)I_1 + 3I_1 = (1 - j3)6$, or $I_1 = 6(1 - j3)/(10 - j1) = 1.888\underline{/-65.85°}$ A. Back substituting yields $I_2 = -18/(10 - j1) = 1.791\underline{/-174.29°}$ A. ◀

Exercise 11.20   Solve the circuit of Figure 11.19(c) via the loop method. Check.

Exercise 11.21   Solve the circuit of Figure 11.20 via the loop method.

Exercise 11.22   Solve the circuit of Figure 11.21 via the node method.

Finding the Equivalent Impedance of an AC Port

The equivalent impedance of an ac port plays a central role in loading and power calculations. As in the resistive case, the equivalent impedance can often be found by inspection, via series/parallel reductions. However, if the circuit topology does not lend itself to this approach, or if the circuit contains dependent sources, then we must use the phasor form of Method 2 of Section 4.1, which is now rephrased as follows:

To find the equivalent impedance Z_{eq} of an ac port, suppress all independent sources, apply a *test phasor V* and find the *resulting phasor I*. Then,

$$Z_{eq} = \frac{V}{I} \qquad \textbf{(11.72)}$$

Alternately, suppress all independent sources, apply a *test phasor I*, find the *resulting phasor V*, and let $Z_{eq} = V/I$.

This method is illustrated in Figure 11.22. Once again we recall that suppressing

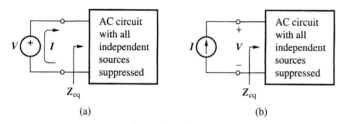

(a)   (b)

Figure 11.22 Finding the equivalent impedance Z_{eq} of an ac port using (a) a *test voltage* or (b) a *test current*.

an *independent voltage source* means replacing it with a *short circuit,* and suppressing an *independent current source* means replacing it with an *open circuit.* However, *dependent* sources must be left in the circuit!

▶ **Example 11.17**

Find **Z** in the circuit of Figure 11.23(a) and show its complex-plane representation.

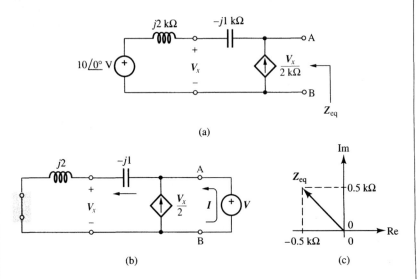

Figure 11.23 Steps to find the impedance of an ac port.

Solution

Suppressing the independent voltage source and applying a test phasor **V** we obtain the circuit of Figure 11.23(b), where all impedances are in kΩ. Applying KCL at node A yields

$$I + \frac{V_x}{2} = \frac{V - V_x}{-j1}$$

By the voltage divider formula we have

$$V_x = \frac{j2}{j2 - j1}V = 2V$$

Eliminating V_x yields $I + 2V/2 = (V - 2V)/(-j1)$, or

$$I = (-1 - j1)V$$

Finally,

$$Z_{eq} = \frac{V}{I} = \frac{\not V}{(-1 - j1)\not V} = \frac{-1 + j1}{1^2 + 1^2} = -0.5 + j0.5 \text{ k}\Omega$$

The complex plane representation of Z_{eq} is shown in Figure 11.23(c).

Remark   It is interesting to observe that in this example the ac resistance is different from zero, even though there are no ohmic resistances in the circuit. Moreover, this resistance is negative. With a dependent source in the circuit you never can tell what you are going to get until you solve it! ◀

Exercise 11.23   Referring to the circuit of Figure 11.23(a), find the impedance seen by the independent source.

ANSWER $-2 + j2$ kΩ.

Thévenin and Norton Equivalents

Thévenin's and Norton's theorems hold in the frequency domain, as do the source transformation techniques illustrated in Section 3.5 for resistive circuits.

▶ **Example 11.18**

Find the (a) Thévenin and (b) Norton equivalents of the circuit of Figure 11.23(a).

Solution

(a) The phasor V_{oc} of the open-circuit voltage between nodes A and B in Figure 11.23(a) is, by KVL and Ohm's Law,

$$V_{oc} = V_x + (-j1)\frac{V_x}{2} = (1 - j0.5)V_x$$

We need an additional equation to eliminate V_x. Again offered by KVL and Ohm's Law, this is

$$V_x = 10\underline{/0°} + (j2)\frac{V_x}{2} = 10 + jV_x$$

Collecting terms yields $V_x(1 - j1) = 10$, or

$$V_x = \frac{10}{1 - j1} = 5 + j5 \text{ V}$$

Substituting into the expression for V_{oc} we obtain

$$V_{oc} = (1 - j0.5)(5 + j5) = 7.906\underline{/18.43°} \text{ V}$$

In Example 11.17 we found $Z_{eq} = -0.5 + j0.5$ kΩ, or

$$Z_{eq} = 0.7071\underline{/135°} \text{ kΩ}$$

Consequently, the Thévenin equivalent appears as in Figure 11.24(a).

(b) Applying a source transformation, we obtain

$$I_{sc} = \frac{V_{oc}}{Z_{eq}} = \frac{7.906\underline{/18.43°}}{0.7071\underline{/135°}} = 11.18\underline{/-116.57°} \text{ mA}$$

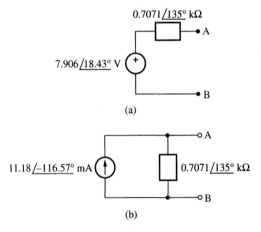

Figure 11.24 Thévenin and Norton equivalents of the circuit of Figure 11.23(a).

The Norton equivalent is shown in Figure 11.24(b). ◀

Exercise 11.24   In the *RC* circuit of Figure 11.14, let $R = 1$ kΩ, $C = 1$ μF, and $v_i = 10 \cos \omega t$ V. Find its Thévenin and Norton equivalents for (a) $\omega = 500$ rad/s and (b) $\omega = 2$ krad/s.

ANSWER   (a) $V_{oc} = 8.944\underline{/-26.60°}$ V, $Z_{eq} = 0.8944\underline{/-26.60°}$ kΩ, $I_{sc} = 10\underline{/0°}$ mA; (b) $V_{oc} = 4.472\underline{/-63.43°}$ V, $Z_{eq} = 0.4472\underline{/-63.43°}$ kΩ, $I_{sc} = 10\underline{/0°}$ mA.

CONCLUDING REMARKS

Compared with *time-domain* ac analysis, *frequency-domain* ac analysis involves *algebraic equations* as opposed to *differential equations,* and requires no trigonometric acrobatics. This ought to convince you of the unquestionable superiority of the frequency-domain method, even though it requires learning phasor algebra. This superiority becomes more apparent when we analyze more complex circuits.

Compared with purely resistive circuits, ac circuits generally require longer computations because of the *complex* nature of their impedances. But, considering the advantages over time-domain techniques, this price is well worth paying!

11.4 OP AMP AC CIRCUITS

The analysis techniques of the last section can also be applied to ac circuits containing op amps. Once the circuit is redrawn in frequency-domain form, its analysis proceeds on the basis of the virtual-short concept, which is rephrased in ac form as follows:

Op Amp Rule for AC Circuits: An op amp will swing its output phasor V_o to whatever value is needed to drive its difference input phasor V_d to zero or,

equivalently, to force V_n to track V_p, where V_n and V_p are the phasors of the inverting and noninverting inputs, and $V_d = V_p - V_n$.

Some examples will illustrate the use of this rule.

Integrators

The circuit of Figure 11.25(a) has already been investigated in Section 8.4 from the *time-domain* viewpoint. It was shown that it yields $v_o = -(1/RC) \int_0^t v_i(\xi) \, d\xi + v_o(0)$ and, hence, that it functions as an *integrator.* We now wish to reexamine it from the *frequency-domain* viewpoint. To this end, we redraw it as in Figure 11.25(b), and observe that the phasors of the resistance and capacitance currents must satisfy $I_R = I_C$. By the virtual-ground concept and the generalized Ohm's Law, $(V_i - 0)/Z_R = (0 - V_o)/Z_C$, or $V_o/V_i = -Z_C/Z_R$. This is the familiar *inverting amplifier gain formula,* but with impedances instead of resistances. Letting $Z_R = R$ and $Z_C = 1/j\omega C$ yields

$$V_o = -\frac{1}{j\omega RC} V_i \qquad (11.73)$$

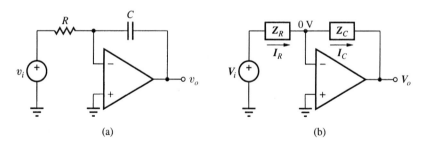

Figure 11.25 The inverting integrator and its frequency-domain representation.

To understand the function of the circuit, rewrite the equation as $V_o = (-1/RC) \times V_i/j\omega$. Phasor Rule 3 of Section 11.1 states that dividing a phasor by $j\omega$ is equivalent to *integrating* the corresponding ac signal. Consequently, the ac function of the circuit is proportional to **integration,** thus confirming our time-domain findings. The constant of proportionality is $-1/RC$, whose dimensions are s^{-1}.

▶ Example 11.19

In the integrator of Figure 11.25 let $R = 159$ kΩ, $C = 1$ nF, $V_i = V_{im}\underline{/0°}$ and $V_o = V_{om}\underline{/\phi}$. If $V_{im} = 1$ V, find V_{om} and ϕ for (a) $f = 100$ Hz, (b) $f = 1$ kHz, and (c) $f = 10$ kHz.

Solution

Multiplying the numerator and denominator of Equation (11.73) by j yields $V_o = (j/\omega RC)V_i$, indicating that the circuit introduces a 90° phase lead. Thus, $\phi = 90°$, regardless of f.

By Phasor Rule 6 we have $V_{om} = V_{im}/\omega RC = 1/(2\pi f \times 159 \times 10^3 \times 10^{-9}) = 10^3/f$. Calculating at the given values of f gives (a) $V_{om} = 1 \times 10^3/100 = 10$ V, (b) $V_{om} = 1$ V, and (c) $V_{om} = 0.1$ V. ◀

Exercise 11.25   In the integrator of Figure 11.25(a) let $R = 159$ kΩ and let the input be a 1-kHz ac wave with a peak amplitude of 2.5 V. Specify C for an output peak amplitude of 5 V.

ANSWER   0.5 nF.

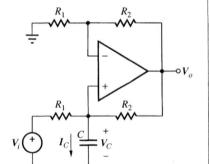

Figure 11.26 The noninverting integrator.

▶ Example 11.20

Derive a relationship between the output and input phasors of the *noninverting integrator* of Figure 8.32.

Solution

The circuit is repeated in frequency-domain form in Figure 11.26. Phasor Rule 1 of Section 11.1 allows us to reuse Equations (8.30) and (8.31) in phasor form as $I_C = V_i/R_1$ and $V_o = (1 + R_2/R_1)V_C$. But, $V_C = (1/j\omega C)I_C$. Eliminating V_C and I_C yields

$$V_o = \frac{1 + R_2/R_1}{j\omega R_1 C} V_i \qquad (11.74)$$

The fact that V_o is proportional to $V_i/j\omega$ with the constant of proportionality $(1 + R_2/R_1)/(R_1 C) > 0$ confirms the designation *noninverting integrator* for the circuit. ◀

The Differentiator

Interchanging R and C in the circuit of Figure 11.25(a) yields the circuit of Figure 11.27(a). Referring to its frequency-domain representation of Figure 11.27(b), we can directly apply the phasor form of the *inverting gain formula*, $V_o/V_i = -Z_R/Z_C = -R/(1/j\omega C)$, or

$$\boxed{V_o = -j\omega RCV_i} \qquad (11.75)$$

Rewriting the equation as $V_o = (-RC) \times j\omega V_i$ and recalling that multiplying a phasor by $j\omega$ is equivalent to *differentiating* the corresponding ac signal, we conclude that the ac function of the circuit is proportional to **differentiation.** The constant of proportionality is $-RC$, whose dimensions are s.

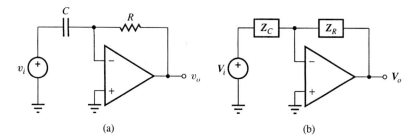

(a)   (b)

Figure 11.27 The differentiator and its frequency-domain form.

Exercise 11.26 Repeat Example 11.19, but for the differentiator of Figure 11.27(a). Hence, compare with the example.

ANSWER $\phi = -90°$ regardless of f; $V_{om} = 0.1$ V, 1 V, 10 V.

Low-Pass Circuit with Gain

To find the function provided by the circuit of Figure 11.28, we first derive a phasor relationship between V_o and V_i. Applying again the inverting amplifier gain formula yields $V_o/V_i = -Z_p/Z_{R_1}$, where $Z_p = Z_C \parallel Z_{R_2} = R_2/(1 + j\omega R_2 C)$, and $Z_{R_1} = R_1$. The result is

$$V_o = -\frac{R_2}{R_1}\frac{1}{1 + j\omega R_2 C}V_i \tag{11.76}$$

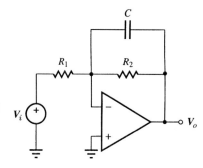

Figure 11.28 Low-pass circuit with gain.

The input-output amplitude and phase relationships are thus

$$V_{om} = \frac{R_2}{R_1} \times \frac{1}{\sqrt{1 + (\omega R_2 C)^2}} \times V_{im} \tag{11.77a}$$

$$\phi = 180° - \tan^{-1}\omega R_2 C \tag{11.77b}$$

The first term at the right-hand side of Equation (11.77a) represents *gain,* and the second term is the *low-pass* function discussed in Section 10.3. We conclude that our circuit provides a **low-pass** function with **gain.** The gain magnitude is R_2/R_1, and the characteristic frequency of the low-pass function is $\omega_0 = 1/R_2 C$.

▶ **Example 11.21**

In the circuit of Figure 11.28 let $R_1 = 10$ kΩ, $R_2 = 100$ kΩ, and $C = 1$ nF. Find $v_o(t)$ if

$$v_i(t) = 1 \cos(2\pi 10^3 t + 45°) \text{ V}$$

Solution

We have $R_2/R_1 = 100/10 = 10$ and $\omega R_2 C = 2\pi 10^3 \times 10^5 \times 10^{-9} = 0.6283$ rad. Thus,

$$V_o = \frac{-10}{1 + j0.6283} V_i$$

Moreover, $|V_o| = (10/\sqrt{1 + 0.6283^2})1 = 8.467$ V, and $\sphericalangle V_o = -180 - \tan^{-1} 0.6283 + 45 = -167.14°$. Finally,

$$v_o(t) = 8.467 \cos(2\pi 10^3 t - 167.14°) \text{ V}$$

◀

▶ **Example 11.22**

In the circuit of Figure 11.28 let $C = 1$ nF. Specify R_1 and R_2 for a characteristic frequency of 10 kHz and a gain magnitude of 2.

Solution

Imposing $\omega_0 = 1/R_2 C$ yields $R_2 = 1/\omega_0 C = 1/(2\pi 10^4 \times 10^{-9}) = 15.9$ kΩ. Imposing $R_2/R_1 = 2$ yields $R_1 = R_2/2 = 15.9/2 = 7.96$ kΩ.

◀

High-Pass Circuit with Gain

For the circuit of Figure 11.29 we have $V_o/V_i = -Z_{R_2}/Z_s$, where $Z_s = Z_C + Z_{R_1} = R_1(1 + j\omega R_1 C)/j\omega R_1 C$, and $Z_{R_2} = R_2$. Consequently,

$$V_o = -\frac{R_2}{R_1} \frac{j\omega R_1 C}{1 + j\omega R_1 C} V_i \tag{11.78}$$

Moreover,

$$V_{om} = \frac{R_2}{R_1} \times \frac{\omega R_1 C}{\sqrt{1 + (\omega R_1 C)^2}} \times V_{im} \tag{11.79a}$$

$$\phi = -90° - \tan^{-1} \omega R_1 C \tag{11.79b}$$

The first term at the right-hand side of Equation (11.79a) represents *gain,* and the second term is the *high-pass* function discussed in Section 10.3. Our circuit provides a **high-pass** function with **gain** R_2/R_1, and with characteristic frequency $\omega_0 = 1/R_1 C$.

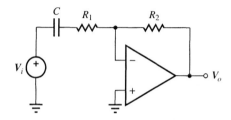

Figure 11.29 High-pass circuit with gain.

Exercise 11.27   Find $v_o(t)$ in Figure 11.29 if $R_1 = 10$ kΩ, $R_2 = 30$ kΩ, $C = 22$ nF, and $v_i(t) = 2\cos(2\pi 10^3 t - 30°)$ V.

ANSWER $v_o(t) = 4.861\cos(2\pi 10^3 t - 174.12°)$ V.

Exercise 11.28

   (a) In the circuit of Figure 11.29 let $C = 0.1$ μF. Specify R_1 and R_2 for a characteristic frequency of 100 Hz and a gain magnitude of 10.

   (b) Find $v_i(t)$ for $v_o(t) = 5\cos 400t$ V.

ANSWER   (a) $R_1 = 15.9$ kΩ, $R_2 = 159$ kΩ; (b) $v_i(t) = 0.5154 \times \cos(400t + 165.96°)$ V.

• Capacitance Multiplication

We claim that the one-port op amp circuit of Figure 11.30(a) behaves as a capacitance C_{eq}. To substantiate this claim as well as to derive an expression for C_{eq}, we apply a test voltage V as in Figure 11.30(b), we find the resulting current I, and we then evaluate the one-port impedance as $Z_{eq} = V/I$. Clearly, for Z_{eq} to be capacitive, it must be of the type $Z_{eq} = 1/j\omega C_{eq}$.

Since OA_1 operates as a voltage follower, we have $V_1 = V$. OA_2 operates as an inverting amplifier to yield $V_2 = -(R_2/R_1)V_1 = -(R_2/R_1)V$. Since no current flows into or out of the noninverting input of OA_1, we can apply Ohm's Law and write

$$I = \frac{V - V_2}{Z_C} = \frac{V + (R_2/R_1)V}{Z_C} = \frac{(1 + R_2/R_1)V}{Z_C} \qquad \textbf{(11.80)}$$

Thus, $Z_{eq} = V/I = Z_C/(1 + R_2/R_1) = 1/[j\omega C(1 + R_2/R_1)]$. This can be put in the form

$$Z_{eq} = \frac{1}{j\omega C_{eq}} \qquad \textbf{(11.81a)}$$

$$C_{eq} = \left(1 + \frac{R_2}{R_1}\right)C \qquad \textbf{(11.81b)}$$

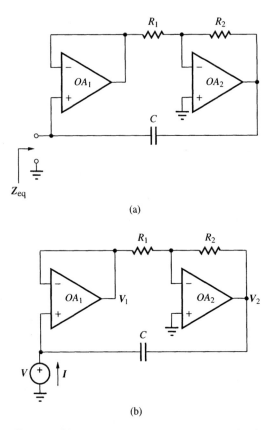

Figure 11.30 Capacitance multiplier and test circuit to find Z_{eq}.

For obvious reasons, the circuit is referred to as a **capacitance multiplier.** It finds application especially in integrated-circuit technology where, for reasons of miniaturization, only very small capacitors can practically be fabricated. When larger capacitances are needed, a multiplier is used. This capacitance multiplication effect is referred to as the **Miller effect** for capacitances.

▶Example 11.23

In the circuit of Figure 11.30 let $C = 10$ pF, $R_1 = 100$ Ω, and $R_2 = 100$ kΩ. Find C_{eq}.

Solution

$C_{eq} = (1 + 10^5/10^2)10 \times 10^{-12} \simeq 10$ nF, indicating that the effective capacitance is 1000 times as large as the physical capacitance.   ◀

• Inductance Simulation

We claim that the one-port op amp circuit of Figure 11.31(a) behaves as an inductance. To substantiate this, apply a test voltage V as in Figure 11.31(b),

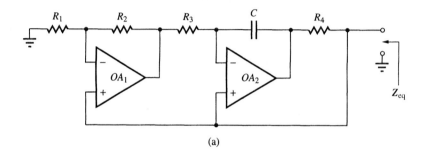

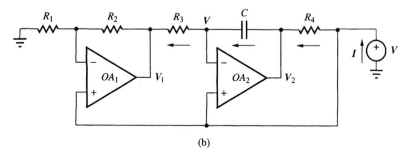

Figure 11.31 Inductance simulator and test circuit to find Z_{eq}.

find *I*, and then let $Z_{eq} = V/I$. Clearly, for Z_{eq} to be inductive, it must be of the type $Z_{eq} = j\omega L_{eq}$.

By Ohm's Law and the op amp input current constraint,

$$I = \frac{V - V_2}{R_4} \tag{11.82}$$

Since OA_2 keeps its inverting input pin at *V*, KCL yields,

$$\frac{V_2 - V}{1/j\omega C} = \frac{V - V_1}{R_3} \tag{11.83}$$

Since OA_1 operates as a noninverting amplifier, we have

$$V_1 = \left(1 + \frac{R_2}{R_1}\right) V \tag{11.84}$$

Substituting Equation (11.84) into Equation (11.83), solving for the difference $V_2 - V$, and then substituting it into Equation (11.82), we get

$$I = \frac{V}{j\omega R_1 R_3 R_4 C / R_2}$$

Thus, $Z_{eq} = V/I = j\omega R_1 R_3 R_4 C / R_2$, which can be put in the form

$$Z_{eq} = j\omega L_{eq} \tag{11.85a}$$

$$L_{eq} = \frac{R_1 R_3 R_4}{R_2} C \tag{11.85b}$$

For obvious reasons the circuit is referred to as an **inductance simulator.** Not only does it use no physical inductors, it can also simulate fairly large inductances using miniature components. The ability of op amp circuits to simulate inductances finds application in the synthesis of an important class of circuits known as *active filters.*

▶ Example 11.24

In the circuit of Figure 11.31(a) let $C = 0.1$ μF and $R_1 = R_2 = R_3 = R_4 = 100$ kΩ. Find L_{eq}.

Solution

$$L_{eq} = [(10^5)^3/10^5] \times 10^{-7} = 1 \text{ kH}.$$

Remark   If such a large inductance were to be implemented with a coil, its bulk and weight would far exceed those of the simulator circuit!   ◀

Looking back at the examples of this section, we observe that op amps in ac circuits can be used to provide **gain** as well as to effect **impedance changes.** As you progress through your electronics curriculum, you will have countless opportunities to observe how these features are applied to the design of a variety of circuits of practical interest.

11.5 AC ANALYSIS USING SPICE

SPICE has the ability to perform ac analysis and to tabulate or plot the response as a function of frequency. Following is a description of the necessary SPICE statements.

Independent AC Sources

The general forms of the statements for independent ac sources are

```
VXXX   N+   N-   AC   ACMAG   ACPHASE
```
  (11.86a)

```
IXXX   N+   N-   AC   ACMAG   ACPHASE
```
  (11.86b)

As for dc sources, these statements contain the name of the source, which must begin with the letter V or I, and the circuit nodes N+ and N- to which the source is connected. However, the qualifier AC indicates that we now have a source of the ac type. Its amplitude and phase angle are specified by ACMAG, in V or A, and by ACPHASE, in degrees. If ACMAG is omitted, a value of unity is assumed, and if ACPHASE is omitted, a value of zero is assumed.

The .AC Statement

The ac analysis in SPICE is invoked by means of the .AC statement, which has one of the following forms:

```
.AC   LIN   NP   FSTART   FSTOP
```
  (11.87a)

```
.AC   OCT   NO   FSTART   FSTOP
```
(11.87b)

```
.AC   DEC   ND   FSTART   FSTOP
```
(11.87c)

In each statement, `FSTART` and `FSTOP` specify the lower and upper limits of the frequency range over which ac analysis is to be performed. If analysis at a single frequency is desired, then `FSTART` and `FSTOP` are made both equal to that frequency. The keywords `LIN`, `OCT`, and `DEC` specify the type of frequency sweep as follows:

- `LIN`: Frequency is swept *linearly,* and `NP` is the number of frequency points in the entire sweep.
- `OCT`: Frequency is swept logarithmically by *octaves,* and `NO` is the number of frequency points per octave.
- `DEC`: Frequency is swept logarithmically by *decades,* and `ND` is the number of frequency points per decade.

The `.PRINT AC` **and** `.PLOT AC` **Statements**

SPICE does not provide the ac response in analytical form. Rather, it tabulates or plots circuit variables at the specified frequency points. Thus, along with the `.AC` statement, we must also use the `.PRINT AC` or the `.PLOT AC` statements, whose forms are

```
.PRINT   AC   OUTVAR1   OUTVAR2 ... OUTVAR8
```
(11.88a)

```
.PLOT   AC   OUTVAR1   OUTVAR2 ... OUTVAR8
```
(11.88b)

In these statements, `OUTVAR1` through `OUTVAR8` are the desired voltage or current output variables, for a maximum of eight such variables per statement. As in the case of dc analysis, the form of a *voltage* variable is `V(N1,N2)`, and it specifies the voltage difference between nodes `N1` and `N2`. If `N2` and the preceding comma are omitted, then ground ($N2 = 0$) is assumed. Likewise, the form of a *current* variable is `I(VXXX)`, and it specifies the current flowing through the independent voltage source named `VXXX`.

For ac analysis, five additional output variables can be specified by replacing the letter `V` or the letter `I` by

- `VR` or `IR`   real part
- `VI` or `II`   imaginary part
- `VM` or `IM`   magnitude
- `VP` or `IP`   phase
- `VDB` or `IDB` $20 \log_{10}$ (`VM` or `IM`)

Some examples will better illustrate the use of these statements.

▶ Example 11.25

Use SPICE to verify the results of Example 11.9(a).

Solution

The circuit is redrawn in Figure 11.32. To find the impedance Z, apply the test current $I = 1\underline{/0°}$ A. Then, $Z = V/I = Z = V/(1\underline{/0°}) = V$, indicating that the variables of interest can be obtained from SPICE as $R(\omega) = $ VR(1), $X(\omega) = $ VI(1), $Z_m = $ VM(1), and $\phi = $ VP(1). They are to be calculated at one frequency point, namely, at $f = \omega/2\pi = 10^4/2\pi = 1591.55$ Hz. The input file is

```
FINDING THE IMPEDANCE OF A ONE-PORT
I 0 1 AC 1 0
R1 1 2 2K
C 2 3 0.1U
R2 3 0 1K
L 3 0 50M
* DUMMY RESISTANCE TO SATISFY SPICE
RDUMMY 1 0 1.0E15
.AC LIN 1 1591.55 1591.55
.PRINT AC VR(1) VI(1) VM(1) VP(1)
.END
```

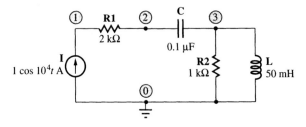

Figure 11.32 Circuit of Example 11.25.

After SPICE is run, the output file contains the following:

```
 FREQ        VR(1)        VI(1)        VM(1)        VP(1)
1.592E+03 2.200E+03 -6.000E+02 2.280E+03 -1.526E+01
```

Clearly, these data match the results of Example 11.9(a).

Exercise 11.29   Use SPICE to verify the results of Exercise 11.11.

Exercise 11.30   Use SPICE to find the *admittance,* the ac *conductance,* and the *susceptance* of the circuit of Example 11.9(a). (*Hint:* Apply a test voltage $V = 1\underline{/0°}$ V.)

▶ Example 11.26

Use SPICE to verify the results of Example 11.15.

Solution

Since the original circuit is in frequency-domain form, it must be transformed into a time-domain form in order to be acceptable to SPICE. Thus, arbitrarily assume $\omega = 1$ rad/s so that $f = \omega/2\pi = 0.159155$ Hz. Then, since $j\omega L = j2$ Ω and $1/j\omega C = -j1$ Ω, it follows that with $\omega = 1$ rad/s we must have $L = 2$ H and $C = 1$ F. The circuit is shown in Figure 11.33. The input file is

```
EXAMPLE OF NODAL ANALYSIS
IS 0 1 AC 2 30
L 1 0 2
C 1 2 1
G1 0 2 1 0 2
R 2 0 2
.AC LIN 1 0.159155 0.159155
.PRINT AC VM(1) VP(1) VM(2) VP(2)
.END
```

After SPICE is run, the output file contains the following:

```
 FREQ       VM(1)      VP(1)      VM(2)      VP(2)
1.592E-01 1.229E+00 1.675E+02 2.457E+00 1.306E+02
```

Clearly, these results agree with those of Example 11.15.

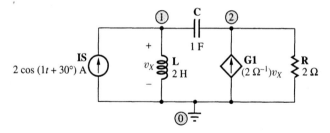

Figure 11.33 Circuit of Example 11.26.

Exercise 11.31   Use SPICE to verify the results of Exercise 11.18.

▶ Example 11.27

Use SPICE to verify the results of the integrator circuit of Example 11.19. Assume a 741 op amp.

Solution

The input file is

```
A 741 INTEGRATOR
VI 1 0 AC 1 0
```

```
R 1 2 159K
C 2 6 1N
*741 OP AMP:
RD 2 0 2MEG
E1 4 0 0 2 2.0E5
RO 4 6 75
.AC DEC 1 100 10K
.PRINT AC VM(6) VP(6)
.END
```

After SPICE is run, the output file contains the following:

```
 FREQ          VM(6)          VP(6)
1.000E+02     1.001E+01     9.000E+01
1.000E+03     1.001E+00     9.000E+01
1.000E+04     1.001E-01     9.000E+01
```

This confirms the results of Example 11.19. ◀

Exercise 11.32   Use SPICE to verify the results of Example 11.21. Assume the op amp is the 741 type.

▶ **Example 11.28**

A series RLC circuit with $R = 400 \; \Omega$, $L = 10$ mH, and $C = 10$ nF, is driven by a source $v(t) = 10 \cos \omega t$ V. Use PSpice to obtain the decade frequency plot of the amplitude of the current $i(t)$ over the range $0.1\omega_0 \leq \omega \leq 10\omega_0$.

Solution

The circuit is shown in Figure 11.34. The resonance frequency is $f_0 = \omega_0/2\pi = 1/2\pi \sqrt{LC} = 1/2\pi \sqrt{10^{-2} \times 10^{-8}} = 15.9$ kHz. We shall therefore use FSTART = 1.6 kHz and FSTOP = 160 kHz. Moreover, we shall use both the .PLOT and the .PROBE statements to compare the

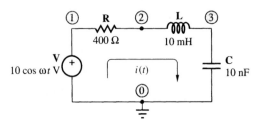

Figure 11.34 Circuit of Example 11.28.

quality of the plots. The input file is

```
SERIES RLC CIRCUIT
V 1 0 AC 10
R 1 2 400
L 2 3 10M
C 3 0 10N
.AC DEC 10 1.6K 160K
.PLOT AC IM(V)
.PROBE
.END
```

After running PSpice, we obtain the plot of Figure 11.35. Using the Probe post-processor and directing it to plot `IM(V)`, the magnitude of the current through the test source, we obtain the trace of Figure 11.36.

```
   FREQ       IM(V)
(*)----------   1.0000E-03    1.0000E-02    1.0000E-01    1.0000E+00    1.0000E+01
                - - - - - - - - - - - - - - - - - - - - - - - - - - - - - - - -
   1.600E+03   1.015E-03 .*                .             .             .
   2.014E+03   1.285E-03 . *               .             .             .
   2.536E+03   1.631E-03 .  *              .             .             .
   3.192E+03   2.083E-03 .    *            .             .             .
   4.019E+03   2.682E-03 .     *           .             .             .
   5.060E+03   3.502E-03 .       *         .             .             .
   6.370E+03   4.681E-03 .        *        .             .             .
   8.019E+03   6.519E-03 .          *      .             .             .
   1.010E+04   9.769E-03 .            *    .             .             .
   1.271E+04   1.653E-02 .              *  .             .             .
   1.600E+04   2.499E-02 .                *.             .             .
   2.014E+04   1.609E-02 .              *  .             .             .
   2.536E+04   9.567E-03 .            *    .             .             .
   3.192E+04   6.412E-03 .         *       .             .             .
   4.019E+04   4.616E-03 .        *        .             .             .
   5.060E+04   3.457E-03 .       *         .             .             .
   6.370E+04   2.650E-03 .     *           .             .             .
   8.019E+04   2.059E-03 .    *            .             .             .
   1.010E+05   1.613E-03 .  *              .             .             .
   1.271E+05   1.271E-03 . *               .             .             .
   1.600E+05   1.004E-03 .*                .             .             .
                - - - - - - - - - - - - - - - - - - - - - - - - - - - - - - - -
```

Figure 11.35 Output listing for the series *RLC* circuit of Figure 11.34.

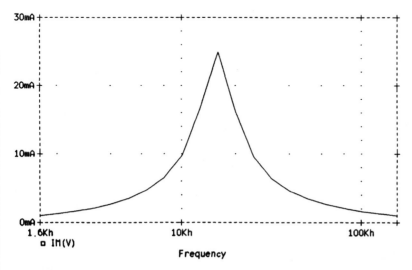

Figure 11.36 Using the Probe post-processor to display the frequency response of the series *RLC* circuit of Figure 11.34.

Exercise 11.33   In the circuit of Figure 11.17(a), let $i_s(t) = 0.1 \cos \omega t$ A, $C = 1$ μF, $R = 300$ Ω, and $L = 0.25$ H. Use PSpice to obtain the decade frequency plots of the amplitude and phase angle of $i(t)$ over the range $200 \leq \omega \leq 20{,}000$ rad/s.

▼ SUMMARY

- AC analysis via phasors, also called *frequency-domain analysis,* is much simpler and quicker than ac analysis in the time domain. Phasors are *complex variables,* and the mathematics required to represent and manipulate phasors is called *complex algebra.*

- A complex variable can be represented in the *shorthand form,* the *rectangular form,* or the *exponential form,* $X = X_m\underline{/\theta} = X_r + jX_i = X_m e^{j\theta}$, where X_m is the *modulus,* θ is the *phase,* and X_r and X_i are the *real* and *imaginary parts.*

- Conversion from one form to the other is done via $X_r = X_m \cos \theta$ and $X_i = X_m \sin \theta$ or via $X_m = \sqrt{X_r^2 + X_i^2}$ and $\theta = \tan^{-1}(X_i/X_r)$.

- The *sum* (or *difference*) of two complex variables is the variable whose *real part* is the *sum* (or *difference*) of the *individual real parts,* and whose *imaginary part* is the *sum* (or *difference*) of the *individual imaginary parts.*

- The *product* (or *ratio*) of two complex variables is the variable whose *modulus* is the *product* (or *ratio*) of the *individual moduli,* and whose *phase* is the *sum* (or *difference*) of the *individual phases.*

- The *time-domain* operations of *differentiation* and *integration* correspond, respectively, to the *frequency-domain* operations of *multiplication* by $j\omega$ and *division* by $j\omega$.

- The phasor forms of Kirchhoff's laws are

    KCL: $\sum_n I_{\text{in}} = \sum_n I_{\text{out}}$

    KVL: $\sum_\ell V_{\text{rise}} = \sum_\ell V_{\text{drop}}$

- The phasor form of Ohm's Law is $V = ZI$, where Z is called the *impedance.* The impedances of the basic elements are $Z_R = R$, $Z_C = 1/j\omega C = -j/\omega C$, and $Z_L = j\omega L$.

- *Series impedances* combine as $Z_s = Z_1 + Z_2$, *parallel impedances* as $1/Z_p = 1/Z_1 + 1/Z_2$.

- To analyze an ac circuit we redraw it in frequency-domain form by replacing its *signals* with their *phasors* and its *circuit elements* with their *impedances.* We then analyze the circuit using the techniques developed in Chapters 2 through 6 for resistive circuits.

- The *voltage* and *current divider formulas,* the *proportionality analysis procedure, the node* and *loop methods, Thévenin's* and *Norton's theorems, source transformation techniques,* and inverting and noninverting *op amp gain formulas* hold in the frequency domain as well.

- The equivalent impedance **Z** of an ac port is found by suppressing its independent sources, applying a *test phasor* **V** (or a *test phasor* **I**), determining the *resulting phasor* **I** (or *phasor* **V**), and then letting $\mathbf{Z} = \mathbf{V}/\mathbf{I}$.

- In a frequency-domain circuit an op amp will output whatever voltage phasor is needed to ensure $\mathbf{V}_d \to 0$ or, equivalently, to ensure $\mathbf{V}_n \to \mathbf{V}_p$, where $\mathbf{V}_n$ and $\mathbf{V}_p$ are the voltage phasors at the inverting and noninverting inputs, and $\mathbf{V}_d = \mathbf{V}_p - \mathbf{V}_n$.

- Important ac op amp circuits are the *integrator,* the *differentiator,* and the *low-pass* and *high-pass filters* with *gain.* Op amps can also be used to effect impedance changes. Two important examples are *capacitance multiplication* and *inductance simulation.*

- SPICE can be directed to plot or tabulate the *frequency response* via the `.AC` statement.

▼ PROBLEMS

11.1 Phasor Algebra

11.1 If a sinusoid with $\omega = 5$ rad/s has the phasor $\mathbf{V} = 10/\underline{15°}$ V, what is the phasor of its time derivative? Of its time integral? Of the time derivative of its time derivative? Of the time integral of its time integral?

11.2 If the phasors of the time derivative and time integral of $x(t)$ are, respectively, $10^3/\underline{60°}$ and $0.1/\underline{-120°}$, find $x(t)$.

11.3 Assuming $\omega = 5$ rad/s and using only frequency-domain techniques, (a) find $x(t)$ if $\mathbf{X} = -j6$; (b) find $y(t)$ and $dy(t)/dt$ if $\mathbf{Y} = -5(2 - j5)$; (c) find $z(t)$ and $\int z(t)\,dt$ if $\mathbf{Z} = e^{j135°}/(2 + j1)$.

11.4 Using phasor additions/subtractions, find (a) the sinusoid that must be added to $x(t) = 10\cos(10^3 t - 45°)$ to obtain $y(t) = 5\cos(10^3 t + 60°)$; and (b) the sinusoid that must be subtracted from $x(t) = 2\cos(10^6 t - 12°)$ to obtain $y(t) = 2\cos(10^6 t + 90°)$.

11.5 Using phasor additions/subtractions, combine the following functions into a single sinusoid: (a) $x(t) = 50\cos\omega t + 30\cos(\omega t + 120°)$; (b) $y(t) = 20\sin 10t - 15\cos(10t + 36°)$; (c) $z(t) = 2\cos(2\pi 10^3 t - 30°) - \cos(2\pi 10^3 t + 60°) + 3\cos(2\pi 10^3 t - 90°)$.

11.6 Find the exponential form of the following complex variables: (a) $3 + j4$; (b) $4 - j3$; (c) $-1 + j1$; (d) $-6 - j8$; (e) 1; (f) -2; (g) $j3$; (h) $-j4$.

11.7 Find the polar as well as the rectangular representation of the operators k_1, k_2, k_3, and k_4, whose effects upon a phasor $\mathbf{A} = A_m/\underline{\theta}$ are defined as (a) $k_1\mathbf{A} = A_m/\underline{\theta + 120°}$, (b) $k_2\mathbf{A} = A_m/\underline{\theta - 120°}$, (c) $k_3\mathbf{A} = \sqrt{3}A_m/\underline{\theta + 30°}$, and (d) $k_4\mathbf{A} = \sqrt{3}A_m/\underline{\theta - 30°}$.

11.8 Find the rectangular form of the following complex variables: (a) $2e^{j30°}$; (b) $10^3 e^{-j60°}$; (c) $e^{-j120°}$; (d) $0.5e^{j135°}$; (e) $10^{-3}e^{j189°}$; (f) $(1/8)e^{-j180°}$; (g) $5e^{j90°}$; (h) $2e^{-j90°}$.

11.9 Given that $\mathbf{A} = 2 - \sqrt{12}e^{-j60°} + \sqrt{3}$, and $\mathbf{B} = e^{j45°}/(2 - j1)$, find $\mathbf{A} + \mathbf{B}$, $\mathbf{AB}$, $\mathbf{A} - \mathbf{B}$, and $\mathbf{A}/\mathbf{B}$. Express your results both in polar and rectangular form.

11.10 Given that $\mathbf{Z} = 3 + j2$, find both the polar and rectangular form of $\mathbf{Z}^*$, $1/\mathbf{Z}$, $\mathbf{Z} + \mathbf{Z}^*$, $\mathbf{Z} - \mathbf{Z}^*$, $\mathbf{ZZ}^*$, $\mathbf{Z}/\mathbf{Z}^*$, and $\mathbf{Z}^*/\mathbf{Z}$.

11.11 Find the real part of $20e^{j(\omega t + 3\pi/4)}/(3 - j4)$, and express it in the standard cosine form, with the phase angle in degrees.

11.12 (a) Given the differential equation $d^2 y/dt^2 + 2\,dy/dt + 9y = 2(dx/dt - 5x)$, express it in phasor form, with $\mathbf{X}$ and $\mathbf{Y}$ being the phasors of x and y. (b) Obtain relationships between $|\mathbf{Y}|$ and $|\mathbf{X}|$, and between $\angle \mathbf{Y}$ and $\angle \mathbf{X}$ if $\omega = 2$ rad/s.

11.13 Given that two phasors satisfy $\mathbf{Y}/\mathbf{X} = (j\omega - 1)/(2 - \omega^2 + j5\omega)$, find a differential equation relating the corresponding ac signals $y(t)$ and $x(t)$. Hence, find $y(t)$ if $x(t) = 10\cos(5t - 30°)$.

11.2 AC Impedance

11.14 Find the polar and rectangular forms of the impedance exhibited by (a) a 100-mH inductance at 100 Hz, (b) a 10-μF capacitance at 100 rad/s, (c) a 1-pF capacitance at 1 GHz, (d) a 100-μH inductance at 500π krad/s, and (e) a 1-kΩ resistance at $2\pi 60$ rad/s.

11.15 (a) Find the inductance needed to achieve an impedance of $j10$ Ω at 100 kHz. What is its reactance at 1 kHz? (b) Find the capacitance needed to achieve an impedance of

$-j5 \, \Omega$ at 10 kHz. What is its reactance at 1 MHz? (c) Find the resistance needed to achieve an impedance of 100 Ω at 60 Hz.

11.16   (a) Find the range of frequencies over which a 1-μF capacitance exhibits an impedance Z_C such that $1 \, \Omega \leq |Z_C| \leq 1$ kΩ. (b) What is the range of values achieved over the same frequency range by the magnitude $|Z_L|$ of the impedance of a 1-mH inductance?

11.17   (a) Find the impedance Z_1 that must be connected in parallel with $Z_2 = 3 - j4 \, \Omega$ to achieve $Z_p = 4 - j3 \, \Omega$. Is Z_1 capacitive or inductive? (b) Find the impedance Z_3 (capacitive or inductive?) that must be connected in series with $Z_4 = 10/60° \, \Omega$ to achieve $Z_s = 10 \, \Omega$.

11.18   (a) In the circuit of Figure 11.11 derive a general expression for $Z = R(\omega) + jX(\omega)$. (b) Show that $X(\omega)$ vanishes for $\omega = \infty$ and $\omega = 1/\sqrt{LC - L^2/R_2^2}$. (c) Verify that the corresponding expressions for $R(\omega)$ reduce to $R_1 + R_2$ and $R_1 + L/(R_2 C)$.

11.19   Consider the circuit of Figure 11.11, but with the capacitance and inductance interchanged with each other. (a) Derive an expression for $Z = R(\omega) + jX(\omega)$. (b) Find the frequencies for which $X(\omega)$ vanishes, and the corresponding expressions for $R(\omega)$.

11.20   In the circuit of Figure P11.20 find the impedance between (a) nodes A and D, (b) nodes B and D, and (c) nodes C and D.

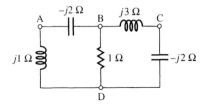

Figure P11.20

11.21   The frequency of the ac source in Figure P11.21 is adjusted until i is in phase with v. If the peak value of v is 10 V, find that frequency as well as the peak value of i at that frequency.

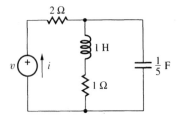

Figure P11.21

11.22   A source $V = 10/0°$ V drives an unknown impedance made up of a resistance in series with an unknown reactance. If at $\omega = 1$ krad/s the current is $I = 3 - j4$ mA, what is I at $\omega = 2$ krad/s? At $\omega = 0.5$ krad/s?

11.23   At 60 Hz the individual impedances of a series RLC circuit are $Z_R = 3 \, \Omega$, $Z_L = j3 \, \Omega$, and $Z_C = -j4 \, \Omega$. Find the current $i(t)$ drawn by the circuit in response to the applied voltage $v(t) = 15\cos(2\pi 120t + 30°)$ V.

11.24   Assuming $\omega = 10$ krad/s in the circuit of Figure P11.24, find the impedance Z between terminals A and B if terminals C and D are (a) open-circuited, and (b) short-circuited.

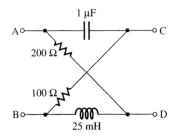

Figure P11.24

11.25   (a) In the circuit of Figure P11.24 find an expression for the admittance $Y(\omega)$ between terminals A and D. (b) Find the low- and high-frequency limits of Y, and justify them physically. (c) Find the frequency ω at which $|Y|$ is maximum, and the corresponding polar value of Y.

11.26   In the circuit of Figure P11.26 let $Z = R + jX$ and $Y = G + jB$ denote, respectively, the impedance and admittance between terminals A and B. Find the frequencies at which (a) $R = 3$ kΩ, (b) $X = -2$ kΩ, (c) $G = 1/(2 \, \text{k}\Omega)$, and (d) $B = 1/(3 \, \text{k}\Omega)$.

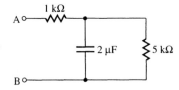

Figure P11.26

11.27   (a) Obtain simple series and parallel equivalents of $Z = 5/60°$ kΩ at $\omega = 1$ Mrad/s. (b) Obtain simple series and parallel equivalents of $Z = 10/-30°$ kΩ at $\omega = 100$ krad/s.

11.28   (a) Given a parallel RLC circuit with $L = 10$ mH, find R and C so that the impedance of the circuit at $\omega = 10$ krad/s is $Z_p = 500/0° \, \Omega$. (b) Find the two frequencies at which $|Z_p| = 250 \, \Omega$.

11.29   If $\omega = 100$ krad/s in the circuit of Figure P11.29, find a simple series equivalent for $\mathbf{Z}_s$ to cause $\mathbf{V}$ to be in phase with $\mathbf{I}$, and such that $|\mathbf{V}|/|\mathbf{I}| = 10$ A.

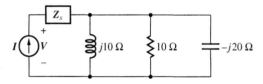

Figure P11.29

11.30   (a) Show that the impedance seen looking into the primary of the transformer of Figure 11.30 is $\mathbf{Z}_1 = \mathbf{Z}_2/n^2$. (b) If $n = 10$ and the secondary is terminated on a resistance $R_2 = 10$ kΩ in parallel with a capacitance $C_2 = 1$ μF, find the equivalent resistance R_1 and capacitance C_1 seen looking into the primary.

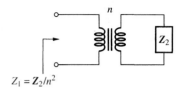

$$Z_1 = \mathbf{Z}_2/n^2$$

Figure P11.30

11.31   Find $\mathbf{Y}$, G, and B in the circuit of Figure P11.31.

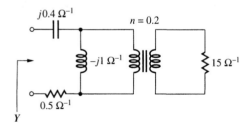

Figure P11.31

11.3 Frequency-Domain Analysis

11.32   In the circuit of Figure P11.32 find v_i for $v_o = 3\cos(2\pi 10^5 t + 20°)$ V.

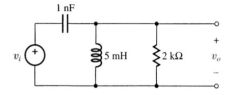

Figure P11.32

11.33   (a) In the circuit of Figure P11.33 find i_i for $i_o = 2.5\cos(2\pi 10^5 t - 15°)$ mA. (b) Repeat, but with the inductance and capacitance interchanged with each other.

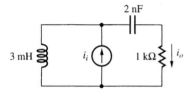

Figure P11.33

11.34   Using the proportionality analysis procedure, find v in the circuit of Figure P11.34 if $i = 12\cos 10^3 t$ A.

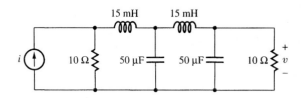

Figure P11.34

11.35   Repeat Problem 11.34, but using source transformations.

11.36   Using the proportionality analysis procedure, find $\mathbf{I}$ in the circuit of Figure P11.36.

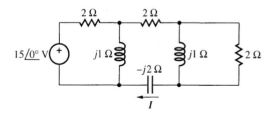

Figure P11.36

11.37   Using the proportionality analysis procedure, find the phasors of the voltages across the 1-Ω resistances in Figure P11.37.

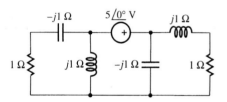

Figure P11.37

11.38 (a) If $v_1 = 10\cos 10^5 t$ V and $v_2 = 20\sin 10^5 t$ V in the circuit of Figure P11.38, find v via the node method. (b) Check your result using source transformations.

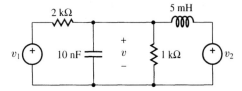

Figure P11.38

11.39 If $v_1 = 20\sin 10^5 t$ V and $v_2 = 10\cos 10^5 t$ V in the circuit of Figure P11.38, find v via the loop method. (b) Check your result using the superposition principle.

11.40 Apply nodal analysis to the circuit of Figure P11.40 to find the current supplied by the source.

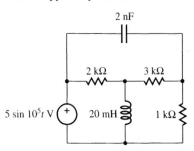

Figure P11.40

11.41 (a) Find I_x in Figure P11.41 using the node method. (b) Check your result using the loop method.

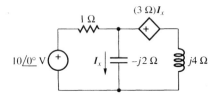

Figure P11.41

11.42 Apply nodal analysis to the circuit of Figure P11.42 to find the voltage across the current source.

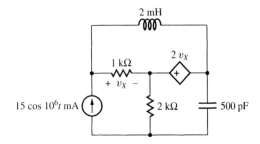

Figure P11.42

11.43 Repeat Problem 11.42, but using loop analysis.

11.44 Use the node method to find the voltage across the current source in the circuit of Figure P11.44.

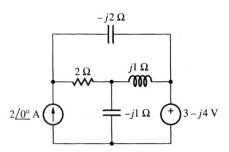

Figure P11.44

11.45 Use the loop method to find the current through the voltage source in the circuit of Figure P11.44.

11.46 In the circuit of Figure P11.46 use the loop method to find i_x. Use the node method to check your result.

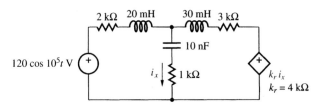

Figure P11.46

11.47 In the circuit of Figure P11.47 find the phasor of the current through the voltage source and the phasor of the voltage across the current source.

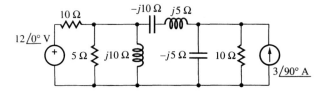

Figure P11.47

11.48 Find I_1 and I_2 in the circuit of Figure P11.48.

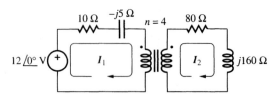

Figure P11.48

11.49 Find **Z** at $\omega = 10^5$ rad/s in the circuit of Figure P11.49. Is **Z** capacitive or inductive?

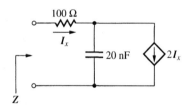

Figure P11.49

11.50 (a) Find **Z** at $\omega = 10^5$ rad/s in the circuit of Figure P11.50. Is **Z** capacitive or inductive? (b) Repeat, but with the inductance and capacitance interchanged with each other.

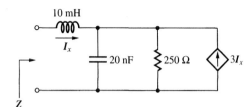

Figure P11.50

11.51 If in the circuit of Figure P11.20 a source $V = 5\underline{/0°}$ V is connected between nodes A and C, positive at C, find the current I out of the source's positive terminal.

11.52 Obtain a relationship between V_o and V_i in the circuit of Figure P11.52, and verify that (a) $|V_o| = |V_i|$ regardless of ω, and that (b) sweeping ω from 0 to ∞ causes the phase difference $\measuredangle V_o - \measuredangle V_i$ to change from 0 to $-180°$. Can you justify this physically?

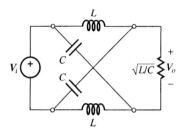

Figure P11.52

11.53 Find the Thévenin and Norton equivalents of the circuit of Figure P11.53 with respect to terminals A and B.

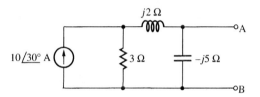

Figure P11.53

11.54 Find the Norton equivalent of the circuit of Figure P11.54 with respect to terminals A and B.

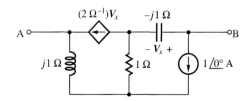

Figure P11.54

11.55 Find the Thévenin equivalent of the circuit of Figure P11.55 with respect to terminals A and B.

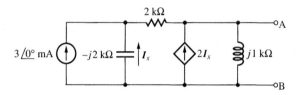

Figure P11.55

11.56 Find the Norton equivalent of the circuit of Figure P11.56 with respect to terminals A and B.

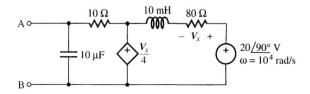

Figure P11.56

11.57 Find i_x in the circuit of Figure P11.46 if the 1-kΩ resistance is changed to (a) a 2-kΩ resistance, (b) a 25-mH inductance, and (c) a 20-nF capacitance. (*Hint:* Replace the circuit external to the 1-kΩ resistance with its Thévenin equivalent.)

11.58 (a) Find the Thévenin and Norton equivalents of an active ac port, given that it yields the terminal voltage $v = 2\cos(10^3 t - 15°)$ V when terminated on a 1-kΩ resistance, and $v = 3\cos(10^3 t - 105°)$ V when terminated on a 1-μF capacitance. (b) What is the terminal current if the port is terminated on a 1-H inductance?

11.59 (a) Show that if

$$Z_1/Z_2 = Z_3/Z_4$$

in the *impedance bridge* of Figure P11.59, then the bridge is *balanced*, that is, it yields $V_o = 0$ regardless of V_s. (b) If at $\omega = 10^4$ rad/s $Z_1 = 1 + j2$ kΩ, $Z_2 = 2 - j1$ kΩ, and $Z_3 = 1 + j1$ kΩ, find simple series and parallel equivalents for Z_4 to balance the bridge.

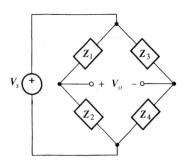

Figure P11.59

11.60 The bridge of Figure P11.60, known as a *capacitance comparison bridge,* is used to find an unknown capacitance C_x having stray series resistance R_x, via null measurements.

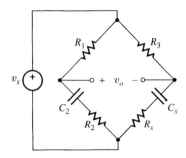

Figure P11.60

(a) Using the balance condition of Problem 11.59, show that the bridge is balanced for all ω if $C_x = R_1 C_2/R_3$ and $R_x = R_2 R_3/R_1$.

(b) Suppose v_s is a 1-kHz ac source, v_o is monitored with earphones, $C_2 = 5.000$ nF, $R_3 = 2.000$ kΩ, and R_1 and R_2 are varied until the sound disappears. If R_1 and R_2 are found to be, respectively, 1.234 kΩ and 0 Ω, what are the values of C_x and R_x?

11.61 The bridge of Figure P11.61, known as the *Maxwell bridge,* is used to find an unknown inductance L_x having stray series resistance R_x, via null measurements.

(a) Using the balance condition of Problem 11.59, show that the bridge is balanced for all ω if $L_x = R_2 R_3 C_1$ and $R_x = R_2 R_3/R_1$.

(b) Suppose $C_1 = 10.00$ nF, $R_3 = 100.0$ Ω, and R_1 and R_2 are varied until a null is detected. If R_1 and R_2 are found to be, respectively, 6789 Ω and 2345 Ω, what are the values of R_x and L_x?

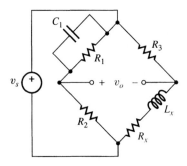

Figure P11.61

11.62 The bridge of Figure P11.62, known as the *Hay bridge,* is used to find an unknown inductance L_x having stray series resistance R_x, via null measurements.

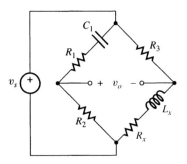

Figure P11.62

(a) Using the balance conditions of Problem 11.59, show that the bridge is balanced at a given frequency ω if

$$L_x = \frac{R_2 R_3 C_1}{1 + (\omega R_1 C_1)^2} \quad \text{and} \quad R_x = \frac{\omega^2 R_1 R_2 R_3 C_1^2}{1 + (\omega R_1 C_1)^2}$$

(b) If at 100 kHz the bridge is balanced with $R_1 = 25.33$ Ω, $C_1 = 100.0$ pF, and $R_2 = R_3 = 1.000$ kΩ, find R_x and L_x.

11.63 (a) Using the balance condition of Problem 11.59, find whether an impedance bridge with $Z_1 = Z_3 = 1$ kΩ, $Z_2 =$

$j1$ kΩ, and $Z_4 = -j1$ kΩ balances. (b) Using the balance conditions of Problem 11.61(a), find whether a Maxwell bridge with $R_1 = 4$ kΩ, $C_1 = 1.5$ μF, $R_2 = 8$ kΩ, $R_3 = 1$ kΩ, $R_x = 2$ kΩ, and $L_x = 12$ H balances at $\omega = 1$ krad/s.

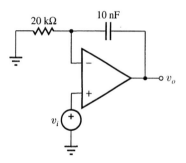

Figure P11.66

11.4 Op Amp AC Circuits

11.64 (a) Show that the circuit of Figure P11.64 is an inverting integrator with $V_o = [-1/(j\omega kRC)]V_i$, where $k = 1 + R_2/R_1 + R_2/R$. (b) Using a 0.1-μF capacitance, select resistances not greater than 100 kΩ to yield $V_o = -(1/j\omega)V_i$.

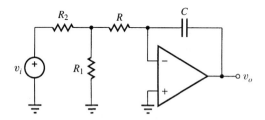

Figure P11.64

11.65 (a) Show that the circuit of Figure P11.65 is a non-inverting integrator with $V_o = [1/(j\omega RC)]V_i$. (b) Specify component values so that $|V_o| = |V_i|$ for $\omega = 1$ krad/s.

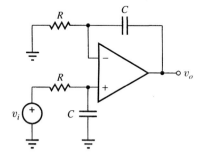

Figure P11.65

11.66 (a) In the circuit of Figure P11.66 find $v_o(t)$ if $v_i(t) = 2\cos(10^4 t + 90°)$ V. (b) Repeat, but with the capacitance and resistance interchanged with each other.

11.67 (a) Show that the circuit of Figure P11.67 is a noninverting differentiator with $V_o = j\omega 2RCV_i$. (b) Specify components for $|V_o| = 10|V_i|$ at $\omega = 1$ krad/s.

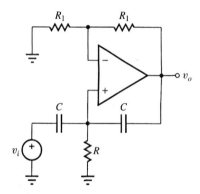

Figure P11.67

11.68 (a) Show that the circuit of Figure P11.68 is an inductance simulator with $Z = j\omega L_{eq}$, where $L_{eq} = R_1 R_2 R_4 C/R_3$. (b) If $C = 1$ nF, specify suitable resistances to synthesize a 10-mH inductance.

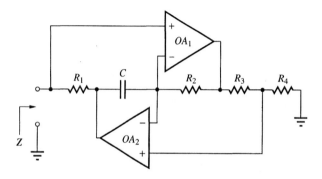

Figure P11.68

11.69   In the circuit of Figure P11.69 find $v_o(t)$ if $v_i(t) =$ 10 cos ωt V, and $\omega =$ (a) 0.1 krad/s, (b) 1 krad/s, (c) 10 krad/s.

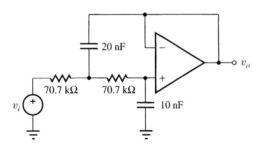

Figure P11.69

11.70   In the circuit of Figure P11.70 find $v_o(t)$ if $v_i(t) =$ 10 cos ωt V, and $\omega =$ (a) 1 krad/s, (b) 10 krad/s, (c) 100 krad/s.

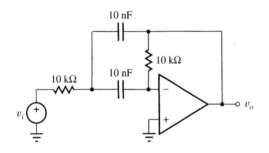

Figure P11.70

11.71   Derive a relationship between the output and input phasors V_o and V_i of the circuit of Figure P11.71. Hence, show

that $|V_o| = |V_i| \times H_m$, where H_m is the *band-pass* function introduced in Section 8.4. What are the values of ω_0 and Q?

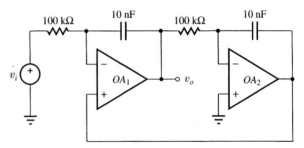

Figure P11.71

⚡ 11.5 AC Analysis Using SPICE

11.72   Solve Problem 11.44 via SPICE.

11.73   Use SPICE to find Z in the circuit of Figure P11.50. Have your result printed both in polar and rectangular form.

11.74   In Figure P11.34 let $i(t) = 1 \cos 2\pi f t$ A. Use the Probe post-processor to display the decade frequency plot of the amplitude and phase angle of $v(t)$ over the range 10 Hz $\leq f \leq$ 100 kHz. Hence, justify the claim that the circuit is a low-pass filter.

11.75   Solve Problem 11.55 via SPICE.

11.76   Solve Problem 11.66 via SPICE.

11.77   Solve Problem 11.70 via SPICE, and obtain the frequency plot of the amplitude of $v_o(t)$ for $10^3 \leq \omega \leq 10^5$ rad/s.

AC Power and Three-Phase Systems

In this chapter we apply phasor techniques to the study of *ac power* and *three-phase* (3-ϕ) systems, the most widely used power generation and distribution systems. Our interest in ac power stems from a variety of reasons. First, electrical power is most efficiently and economically generated, transmitted, and distributed in *ac form*. Second, all electrical and electronic devices and systems have *power ratings* that must not be exceeded if satisfactory operation is desired. Finally, power is a *precious resource,* and whether we are designers or users of electrical equipment, it is our responsibility to strive for an efficient utilization of this resource. After all, ac power forms the basis of our monthly bill from the electric utility company. Depending on the application, the amount of power involved may range from a few picowatts (1 pW $= 10^{-12}$ W) as in telecommunication signals, to several gigawatts (1 GW $= 10^9$ W) as in large power-generating plants. This is an amazingly wide range of values!

We begin by examining the transfer of power from an ac source to a load impedance and find that the *average power,* which is the type of power of greatest practical interest, depends not only on the magnitudes of the voltage and current but also on their *phase difference*. Moreover, power transfer is maximized when the load impedance is made equal to the *complex conjugate* of the impedance of the source, a condition referred to as *ac matching.*

A distinguishing feature of a load containing reactive elements is that in addition to the irreversible flow of energy from the source to the load, where it is converted from electrical to nonelectrical form such as heat or mechanical energy, there is also a flow of energy *back and forth* between the source and the reactive elements of the load. This phenomenon, stemming from the ability of reactive elements to store as well as release energy, may unduly tax the power generation and distribution systems, thus calling for means of easing the capacity demands upon these systems. This is achieved through a technique known as *power-factor correction.*

A major drawback of power sources such as the common household utility power, also called a *single-phase* (1-ϕ) source, is the *pulsating* nature of the instantaneous power flow from source to load. When it comes to generating, transmitting, and utilizing large banks of power, 3-ϕ systems are decidedly superior because of their ability to ensure a *constant* power flow. Consequently, 3-ϕ motors or generators experience a *uniform torque* and thus operate more smoothly and efficiently.

In this chapter, after reviewing the concept of *instantaneous power,* we develop the concepts of *average power, reactive power, power factor, maximum power transfer, ac matching,* and *power-factor correction.* We also introduce the concept of *complex power* as a useful analytical tool in ac power calculations.

We then turn to 3-ϕ *systems* and investigate *Y-connected* and Δ-*connected* sources and loads, as well as transformations from one type of connection to the other. The 3-ϕ systems of greatest practical interest are the *balanced* Y-Y and Y-Δ *systems,* and we show how their analysis can be reduced to that of a 1-ϕ equivalent.

Finally, *power calculations* and *power measurements* in 1-ϕ and 3-ϕ systems are demonstrated via a variety of actual examples. We conclude by illustrating the use of SPICE to analyze 3-ϕ systems.

12.1 AC POWER

Applying an ac signal to a circuit containing resistive as well as reactive elements as in Figure 12.1 results in power transfer from the source to the circuit, henceforth referred to as the *load.* Let the applied signal be a current, whose phase angle we assume to be zero for simplicity,

$$i(t) = I_m \cos \omega t \qquad \text{(12.1a)}$$

The resulting voltage is then

$$v(t) = V_m \cos (\omega t + \phi) \qquad \text{(12.1b)}$$

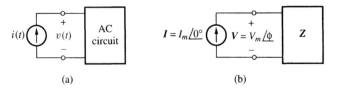

(a)   (b)

Figure 12.1 AC source driving a load: (a) time domain and (b) frequency-domain representation.

where ϕ represents the *phase angle of voltage relative to that of current*.   If $\phi > 0$, v *leads* i (or i *lags* v), indicating an *inductive* load.   Conversely, if $\phi < 0$, v *lags* i (or i *leads* v), and the load is *capacitive*.

An example of ac current and voltage waveforms is shown in Figure 12.2, top. The instantaneous power is $p(t) = v(t)i(t)$. Its diagram can be obtained by multiplying out the voltage and current waveforms point by point, as shown at the bottom. The alternations in the polarity of $p(t)$ indicate that the load *absorbs* energy when $p(t) > 0$ and *returns* energy to the source when $p(t) < 0$,

$$p(t) > 0 \Longleftrightarrow \text{load } \textit{absorbs} \text{ energy}$$

$$p(t) < 0 \Longleftrightarrow \text{load } \textit{releases} \text{ energy}$$

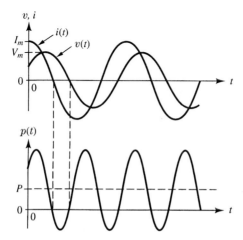

Figure 12.2 Instantaneous current, voltage, and power.

Even though both $i(t)$ and $v(t)$ average to zero, $p(t)$ is in general *shifted upward*, indicating that the load, *on average, absorbs* more power than it returns to the source. From symmetry, the **average ac power** P is represented by the dashed line *halfway* between the upper and lower peaks of $p(t)$. We now wish to find P.

The Power Factor

By Equation (12.1), $p(t) = v(t)i(t) = V_m \cos (\omega t + \phi) \times I_m \cos \omega t$. Using the trigonometric identity $\cos \alpha \cos \beta = \frac{1}{2}[\cos (\alpha + \beta) + \cos (\alpha - \beta)]$, we get

$$p(t) = \frac{1}{2}V_m I_m \cos \phi + \frac{1}{2}V_m I_m \cos (2\omega t + \phi) \qquad \textbf{(12.2)}$$

This expression indicates that the instantaneous power consists of two components: (a) a *time-independent* component having the value $(1/2)V_m I_m \cos\phi$, and (b) a *time-dependent* component having amplitude $(1/2)V_m I_m$ and oscillating at *twice* the applied frequency. Since the time-dependent component of $p(t)$ averages to zero, it follows that the *time average* P of $p(t)$ coincides with the constant term,

$$P = \frac{1}{2}V_m I_m \cos\phi \qquad \text{(12.3a)}$$

where P is in *watts* (W). Since for ac signals $V_{rms} = V_m/\sqrt{2}$ and $I_{rms} = I_m/\sqrt{2}$, we can also use the form

$$P = V_{rms} I_{rms} \cos\phi \qquad \text{(12.3b)}$$

We identify three important cases:

(1) $\phi = 0°$. This condition arises when v and i are *in phase* with each other, as in the case of a *purely resistive* load. Since $\cos 0° = 1$, Equation (12.2) reduces to

$$p(t) = \frac{1}{2}V_m I_m (1 + \cos 2\omega t) \qquad \text{(12.4)}$$

This function alternates between 0 and $V_m I_m$ and is thus always positive, indicating that energy transfer is at all times *from the source to the load.* Moreover, since $\cos\phi$ is maximized, so is P. This maximum, obtained by letting $\cos\phi = 1$ in Equation (12.3), is denoted as S and is called the **apparent power,**

$$S = \frac{1}{2}V_m I_m = V_{rms} I_{rms} \qquad \text{(12.5)}$$

Even though it has the same dimensions as P, S is expressed in *volt-amperes* (VA) to distinguish it from P.

(2) $\phi = \pm 90°$. This condition arises when the circuit is *purely reactive.* Since $\cos\pm90° = 0$, Equation (12.2) reduces to

$$p(t) = \frac{1}{2}V_m I_m \cos(2\omega t \pm 90°) \qquad \text{(12.6)}$$

which averages to zero. This confirms that *purely reactive loads dissipate no power.* The energy absorbed during a positive alternation of $p(t)$ is returned to the source during the subsequent negative alternation.

(3) More generally an ac circuit will be partly resistive and partly reactive, indicating that $\cos\phi$ will generally lie between zero and unity. By Equation (12.3) we thus expect

$$0 \leq P \leq S \qquad \text{(12.7)}$$

The term $\cos\phi$, called the **power factor** (pf),

$$pf = \cos\phi \qquad\qquad (12.8)$$

is, in general, *frequency dependent* because so is ϕ. Hence, the average ac power P is also frequency dependent.

We observe that since $\cos(-\phi) = \cos\phi$, the power factor is the *same* whether current lags or leads voltage by the given phase angle. To resolve any ambiguity, we use the designations *leading pf* and *lagging pf,* the terms *leading* and *lagging* referring to the *phase of current relative to voltage.* Clearly,

$$leading\ pf \Longleftrightarrow capacitive\ load$$

$$lagging\ pf \Longleftrightarrow inductive\ load$$

▶ Example 12.1

Find the apparent power as well as the average power delivered by the source in Figure 12.3. What is the *pf* of this circuit? Is it of the leading or lagging type?

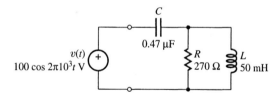

Figure 12.3 Circuit of Example 12.1.

Solution

We need to find V_m, I_m, and ϕ. By inspection, $V = 100\underline{/0°}$ V. By Ohm's Law, $I = V/Z$, where $Z = Z_C + Z_R \parallel Z_L$, or

$$Z = -j\frac{1}{\omega C} + \frac{j\omega L R}{R + j\omega L}$$

$$= \frac{-j1}{2\pi 10^3 \times 0.47 \times 10^{-6}} + \frac{j2\pi 10^3 \times 0.05 \times 270}{270 + j2\pi 10^3 \times 0.05}$$

$$= -j338.6 + \frac{j84,823}{270 + j314} = -j338.6 + j84,823\frac{270 - j314}{270^2 + 314^2}$$

$$= 155.3 - j205.2 = 257.3\underline{/-52.88°}\ \Omega$$

indicating a *capacitive load.* Then,

$$I = \frac{V}{Z} = \frac{100\underline{/0°}}{257.3\underline{/-52.88°}} = 0.3866\underline{/52.88°}\ A$$

indicating a *leading pf.* Finally, we have

$$S = \frac{1}{2} \times 100 \times 0.3866 = 19.43 \text{ VA}$$

$$pf = \cos 52.88° = 0.6035, \text{ leading}$$

$$P = 19.43 \times 0.6035 = 11.72 \text{ W}$$

Remark   The source delivers an apparent power of 19.43 VA; however, only 11.72 W, or 60.35% of it is dissipated as heat.   ◀

Exercise 12.1   Repeat Example 12.1, but with the capacitance and the inductance interchanged with each other.

ANSWER   20.3 VA, 13.6 W, 0.67, lagging.

Power and Impedance

Additional insight can be gained by referring to the frequency-domain representation of Figure 12.1(b). Since $\phi = \sphericalangle V - \sphericalangle I$ and $Z = V/I$, it follows that

$$\phi = \sphericalangle Z \tag{12.9}$$

indicating that Equation (12.3) can also be written as

$$P = V_{\text{rms}} I_{\text{rms}} \cos \sphericalangle Z \tag{12.10}$$

An alternate expression for P is obtained by letting $V_{\text{rms}} = |Z| \times I_{\text{rms}}$, so that $P = |Z| I_{\text{rms}}^2 \cos \sphericalangle Z$. But, $|Z| \cos \sphericalangle Z = R(\omega)$, so

$$\boxed{P = R(\omega) I_{\text{rms}}^2} \tag{12.11}$$

The similarity with the expression for ohmic power of Equation (2.7) suggests the following physical interpretation for ac resistance:

$R(\omega)$ **represents the ability of a circuit to dissipate power in response to ac current.**

Recall from Section 11.2 that $R(\omega)$ does not necessarily coincide with the ohmic resistance of the circuit. Moreover, the ability to dissipate ac power depends not only on the elements and the way they are interconnected but also on the operating frequency ω.

We can obtain yet another useful expression for P as follows: By Equation (12.11), $I_{\text{rms}}^2 = P/R(\omega)$. Squaring both sides of Equation (12.10) yields

$P^2 = (V_{\text{rms}} \cos \phi)^2 P / R(\omega)$, or

$$P = \frac{(V_{\text{rms}} \cos \phi)^2}{R(\omega)} \tag{12.12}$$

▶ **Example 12.2**

An ac circuit is connected to the 120-V (rms), 60-Hz household ac line and dissipates an average power of 1 kW with a leading *pf* of 0.75. Find simple *series* and *parallel* equivalents for the circuit.

Solution

With a leading *pf*, voltage is lagging and, hence, ϕ is *negative,* indicating a *capacitive* impedance. We have

$$\phi = -\cos^{-1} pf = -\cos^{-1} 0.75 = -41.41°$$

Moreover, letting $\mathbf{Z} = R + jX$, we have

$$R = \frac{(V_{\text{rms}} \cos \phi)^2}{P} = \frac{(120 \times 0.75)^2}{10^3} = 8.1 \ \Omega$$

$$X = R \tan \phi = 8.1 \tan (-41.41°) = -7.144 \ \Omega$$

The elements R_s and C_s of the *series* equivalent must be such that $R_s = R = 8.1 \ \Omega$ and $-1/\omega C_s = X$, or $C_s = -1/\omega X = -1/[2\pi 60(-7.144)] = 371.3 \ \mu F$.

For the parallel case, it is better to work with the admittance $\mathbf{Y} = G + jB$, where, as discussed in Section 11.2,

$$G = \frac{R}{R^2 + X^2} = \frac{8.1}{8.1^2 + 7.144^2} = 69.44 \times 10^{-3} \ \Omega^{-1}$$

$$B = \frac{-X}{R^2 + X^2} = \frac{-(-7.144)}{8.1^2 + 7.144^2} = 61.24 \times 10^{-3} \ \Omega^{-1}$$

The elements R_p and C_p of the *parallel* equivalent must be such that $R_p = 1/G$, or $R_p = 1/(69.44 \times 10^{-3}) = 14.40 \ \Omega$ and $\omega C_p = B$, or $C_p = B/\omega = 61.24 \times 10^{-3}/377 = 162.5 \ \mu F$. The two equivalents are shown in Figure 12.4.

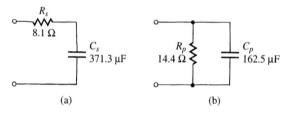

(a)   (b)

Figure 12.4 Circuit equivalents for Example 12.2.

Exercise 12.2   Repeat Example 12.2, but for a *lagging pf.*

ANSWER   8.1 Ω in series with 18.9 mH; 14.4 Ω in parallel with 43.3 mH.

▶Example 12.3

A *series RL* circuit is connected to the household ac line. If the voltage across the inductance is 60 V (rms) and the average power dissipated by the circuit is 25 W, what are the values of R, L, and *pf*?

Solution

By the ac voltage divider formula, we must have

$$\frac{60}{120} = \left| \frac{Z_L}{Z_R + Z_L} \right| = \frac{\omega L}{\sqrt{R^2 + (\omega L)^2}}$$

Squaring both sides yields $0.25 = (\omega L)^2/[R^2 + (\omega L)^2]$, or $0.25R^2 + 0.25(\omega L)^2 = (\omega L)^2$. Collecting terms and taking the square root of both sides yields $\omega L = R/\sqrt{3}$.

We also know that for the series RL combination we have $Z = R + j\omega L$, so $\phi = \tan^{-1}(\omega L/R) = \tan^{-1}(1/\sqrt{3}) = 30°$. Hence, by Equation (12.12), we must have $25 = (120 \cos 30°)^2/R$, or $R = 432 \ \Omega$. Finally, $L = R/\sqrt{3}\omega = 432/(\sqrt{3} \times 2\pi 60) = 0.6616$ H. Clearly, $pf = \cos 30° = 0.8660$.   ◀

Exercise 12.3   A *series RC* circuit is connected to the household ac line. If the voltage across the resistance is 30 V (rms) and the average power dissipated by the circuit is 50 W, what are the values of R, C, and *pf*?

ANSWER   18 Ω, 38.05 μF, 0.25, leading.

Maximum Power Transfer

In our study of resistive circuits, in Section 4.4, we found that power transfer is maximized when the load is made *equal* to the source resistance, a condition also referred to as *matching*. We now wish to examine power transfer for the case of an *active ac port* driving a *load impedance*

$$Z_L = R_L + jX_L \qquad \text{(12.13)}$$

Note that subscript L now identifies parameters associated with the *load.* Given the present context, there should be no reason for confusion with inductance! Once we replace the active port with its Thévenin equivalent, consisting of a voltage source V_{oc} and a series impedance

$$Z_{eq} = R_{eq} + jX_{eq} \qquad \textbf{(12.14)}$$

we end up with the situation of Figure 12.5. We wish to find suitable conditions for the load parameters R_L and X_L that will maximize power transfer to the load.

By Equation (12.11), the load power is $P_L = R_L I_{rms}^2 = (1/2)R_L I_m^2$, where $I_m = |V_{oc}/(Z_{eq} + Z_L)|$. Using Equations (12.13) and (12.14) and applying the rules of Section 11.1 yields

$$P_L = \frac{1}{2}R_L \frac{|V_{oc}|^2}{(R_{eq} + R_L)^2 + (X_{eq} + X_L)^2} \qquad \textbf{(12.15)}$$

On average, reactances absorb zero power and thus any nonzero value of $(X_{eq} + X_L)^2$ would only reduce P_L. We can easily eliminate this term altogether by imposing

$$X_L = -X_{eq} \qquad \textbf{(12.16a)}$$

After this is done, we are left with the problem of finding the condition for R_L that will maximize $P_L = \frac{1}{2}|V_{oc}|^2 R_L/(R_{eq} + R_L)^2$. But this problem has already been solved in Section 4.4, where it was shown that P_L is maximized when

$$R_L = R_{eq} \qquad \textbf{(12.16b)}$$

Equation (12.16) is expressed concisely as $R_L + jX_L = R_{eq} - jX_{eq}$, or

$$Z_L = Z_{eq}^* \qquad \textbf{(12.17)}$$

In words, *power transfer from an active ac port to a load is maximized when the load impedance is made equal to the complex conjugate of the equivalent impedance of the port.* When this condition is met, the load is said to be **matched** to the active ac port. Physically, the equivalent reactance of the ac port and that of the load cancel each other out, in effect forming a *short circuit.* In light of the observations at the end of Section 11.2, we can state that the ac port and its matched load form a *resonant* circuit.

The maximum average power is $P_{L(max)} = (1/2)|V_{oc}|^2 R_{eq}/(R_{eq} + R_{eq})^2$, or

$$P_{L(max)} = \frac{1}{8}\frac{|V_{oc}|^2}{R_{eq}} \qquad \textbf{(12.18)}$$

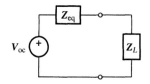

Figure 12.5 Circuit to investigate power transfer.

▶ Example 12.4

(a) In the circuit of Figure 12.6(a) find the load that will receive the maximum power.

(b) What is this power?

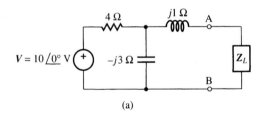

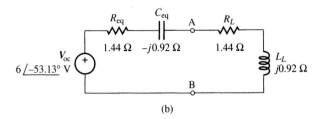

(b)

Figure 12.6 Circuit of Example 12.4 and its Thévenin equivalent with a matched load.

Solution

(a) First, find the Thévenin equivalent of the ac port exclusive of the load. Using the ac voltage divider formula,

$$V_{oc} = \frac{-j3}{4 - j3} 10\underline{/0°} = 6\underline{/-53.13°} \text{ V}$$

With the source suppressed, the impedance Z_{eq} seen by the load is

$$Z_{eq} = j1 + [4 \parallel (-j3)] = j1 + \frac{4(-j3)}{4 - j3} = j1 - j12\frac{4 + j3}{4^2 + 3^2}$$

$$= j1 - j1.92 + 1.44 = 1.44 - j0.92 \ \Omega$$

Figure 12.6(b) shows the Thévenin equivalent as well as the load required for maximum power transfer, $Z_L = 1.44 + j0.92 \ \Omega$.

(b) By Equation (12.18), $P_{L(max)} = (1/8) \times 6^2/1.44 = 3.125$ W. ◀

Exercise 12.4   Repeat Example 12.4, but with the capacitance and inductance interchanged with each other.

ANSWER $Z_L = (4 + j35)/17 \ \Omega$, $P_{L(max)} = 3.125$ W.

12.2 COMPLEX POWER

Average power is of great importance because it provides the basis for consumer billing by the utility company. However, from the viewpoint of power generation and transmission, average power alone is not enough to fully characterize the capacity requirements and efficiency of the system. There are additional parameters that we must take into consideration.

Real and Reactive Power

Let us start with the expression for the *instantaneous power* of Equation (12.2). Using the trigonometric identity $\cos(\alpha + \beta) = \cos\alpha\cos\beta - \sin\alpha\sin\beta$, along with $\frac{1}{2}V_m I_m = V_{rms}I_{rms}$, we obtain

$$p(t) = V_{rms}I_{rms}[\cos\phi + \cos 2\omega t \cos\phi - \sin 2\omega t \sin\phi]$$

where ϕ is the *phase difference between voltage and current.* It is readily seen that if we introduce the quantities

$$P = V_{rms}I_{rms}\cos\phi \qquad \text{(12.19)}$$
$$Q = V_{rms}I_{rms}\sin\phi \qquad \text{(12.20)}$$

then $p(t)$ can be decomposed into two separate components,

$$p(t) = p_R(t) - p_X(t) \qquad \text{(12.21)}$$

where

$$p_R(t) = P(1 + \cos 2\omega t) \qquad \text{(12.22a)}$$

$$p_X(t) = Q\sin 2\omega t \qquad \text{(12.22b)}$$

As illustrated in Figure 12.7, the component $p_R(t)$ alternates between 0 and $2P$ and averages to P. To understand its significance, rewrite Equation (12.19) as $P = [(V_{rms}/I_{rms}) \times \cos\phi]I_{rms}^2 = (Z_m\cos\phi)I_{rms}^2$, or

$$P = R(\omega)I_{rms}^2 \qquad \text{(12.23a)}$$

Clearly, $p_R(t)$ represents the instantaneous power absorbed by the *resistive component* $R(\omega)$ of the load. Aptly called the *instantaneous real power, $p_R(t)$* is irreversibly converted from electrical to nonelectrical form. Accordingly, the average power P is also referred to as the **real power.** As we know, the SI unit of P is the *watt* (W).

The component $p_X(t)$ alternates between $+Q$ and $-Q$ and averages to zero; consequently, it does not intervene in the irreversible power-transfer process. To understand its significance, rewrite Equation (12.20) as $Q = [(V_{rms}/I_{rms}) \times \sin\phi]I_{rms}^2 = (Z_m\sin\phi)I_{rms}^2$, or

$$Q = X(\omega)I_{rms}^2 \qquad \text{(12.23b)}$$

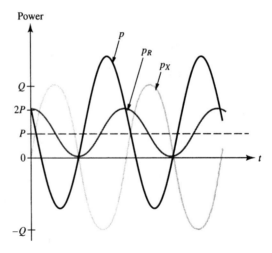

Figure 12.7 The *instantaneous power* $p(t)$, along with its *real* and *reactive* components $p_R(t)$ and $p_X(t)$.

Q accounts for the *energy exchange* between the source and the *reactive component* $X(\omega)$ of the load. During the half-cycle when $p_X > 0$, energy is stored in the electric or magnetic fields of the reactive elements, and during the half-cycle when $p_X < 0$, energy is returned to the source. For obvious reasons, Q is referred to as the **reactive power.** Even though it has the same dimensions as P, we express Q in *volt-amperes reactive* (VAR) to distinguish it from P.

Recall that in inductive loads voltage leads current and thus $\phi > 0$; conversely, in capacitive loads voltage lags current and thus $\phi < 0$. We consequently have, by Equation (12.20),

$$Q > 0 \qquad \text{for } \textit{inductive} \text{ loads}$$
$$Q < 0 \qquad \text{for } \textit{capacitive} \text{ loads}$$

Power engineers recognize this difference by saying that inductive loads *consume* and capacitive loads *produce* reactive power. Using Equations (12.19) and (12.20), it is readily seen that once P and ϕ or P and *pf* are known, Q is found as

$$Q = P \tan \phi = P \tan (\pm \cos^{-1} pf) \qquad \textbf{(12.24)}$$

where we use the $+$ sign if the load is *inductive (lagging pf)*, and the $-$ sign if the load is *capacitive (leading pf)*.

▶ **Example 12.5**

If the household ac voltage is applied to a load $\mathbf{Z} = 100 + j300$ Ω, find S, P, and Q. Hence, assuming $\sphericalangle \mathbf{I} = 0$, find $p(t)$, $p_R(t)$, and $p_X(t)$.

Solution

$2\omega = 2 \times 2\pi 60 = 754$ rad/s. Moreover, $V_{\text{rms}} = 120$ V and $I_{\text{rms}} = V_{\text{rms}}/\sqrt{R^2 + X^2} = 120/\sqrt{100^2 + 300^2} = 0.3795$ A. Thus,

$$S = V_{\text{rms}} I_{\text{rms}} = 120 \times 0.3795 = 45.54 \text{ VA}$$

$$P = RI_{rms}^2 = 100 \times 0.3795^2 = 14.40 \text{ W}$$

$$Q = XI_{rms}^2 = 300 \times 0.3795^2 = 43.20 \text{ VAR}$$

$$p(t) = p_R(t) - p_X(t) = 14.40(1 + \cos 754t) - 43.20 \sin 754t \text{ VA} \quad \blacktriangleleft$$

Exercise 12.5   Repeat Example 12.5, but with a load made up of a 25-Ω resistance in parallel with a 100-μF capacitance.

ANSWER $S = 791.4$ VA, $P = 576.0$ W, $Q = -542.9$ VAR.

Power-Factor Correction

Figure 12.7 shows that even though $p_X(t)$ does not contribute to average power, it nevertheless places an additional burden upon the source, which, besides supplying the rated real power, must also contend with reactive power exchanges. This burden is especially undesirable in the generation and distribution of industrial power. Indeed, rewriting Equation (12.3b) as

$$I_{rms} = \frac{P}{V_{rms} \times pf} \tag{12.25}$$

reveals that the current needed to supply a specified ac power at the rated line voltage is *inversely proportional* to the power factor of the load. Consequently, lower power-factors require that *greater current-carrying capacity* be built into the system. Moreover, the larger currents will result in *greater ohmic losses* in the power-line resistance, thus lowering efficiency.

As an example, suppose $\phi = \pm 45°$; then $pf = \cos 45° = 1/\sqrt{2} = 0.707$. Using Equation (12.25), it is readily seen that

$$I_{rms(pf=0.707)} = \sqrt{2} I_{rms(pf=1)} = 1.414 I_{rms(pf=1)}$$

indicating that the current capacity of the system will have to be increased by 41.4% in order to accommodate the reactive power exchanges that go with the lower power factor. Moreover, denoting the resistance of the power line as R_{line}, we have

$$R_{line} I_{rms(pf=0.707)}^2 = 2 \times R_{line} I_{rms(pf=1)}^2$$

indicating a doubling in line-power loss due to the reactive exchanges.

It is apparent that in order to optimize power generation and distribution, it is desirable to have the power factor approach unity, or $\phi \to 0$. When this condition is met, we have $Q \to 0$, so $p_X(t) \to 0$ and $p(t) \to p_R(t)$. Such a condition is met by connecting a suitable reactive element at the load end of the line so that reactive power is now exchanged between the load and this element, rather than between the load and the source. This procedure is referred to as **power-factor correction.** Though the correcting element may be connected either in series or in parallel, the latter is generally preferred because it spares the element from having to carry the average power absorbed by the load. Moreover, the parallel connection requires no wire cutting.

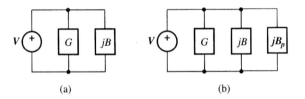

Figure 12.8 A load before and after power-factor correction.

Since we are dealing with parallel connections, our calculations are facilitated if we work with admittances. Thus, let us consider the load of Figure 12.8(a), having the admittance

$$Y = G + jB \qquad \text{(12.26)}$$

Using Equation (12.23), along with Equation (11.59), it is readily seen that the real and reactive power of this load can be expressed as

$$P = GV_{\text{rms}}^2 \qquad \text{(12.27a)}$$

$$Q = -BV_{\text{rms}}^2 \qquad \text{(12.27b)}$$

We now wish to find an element with susceptance B_p that, when connected in parallel with the load as in Figure 12.8(b), will raise the power factor of the composite admittance $Y + jB_p = G + j(B + B_p)$ to unity. This is equivalent to imposing $j(B + B_p) = 0$, or

$$B_p = -B \qquad \text{(12.28)}$$

Once this condition is met, the composite load as seen by the source becomes *purely resistive,* $Y + jB_p = G$. Hence, $Q = 0$, by Equation (12.27b). No longer having to contend with reactive exchanges, the source is now supplying only real power, $p(t) = p_R(t)$.

We can also justify this physically by recalling from Section 11.2 that the parallel combination of two *equal* but *opposite* reactances (or susceptances) acts as an *open circuit,* a phenomenon known as *resonance.* Though no current, and hence no power, enters or leaves the combination, the individual impedances do exchange current and power between themselves. It is thus fair to say that to raise its power factor to unity, the load must be made *resonant.*

Most industrial loads are inductive and thus have *lagging* power factors. These factors are corrected by suitable banks of capacitors. Homes have power factors approaching 1 and thus do not require any correction.

▶ Example 12.6

Find a suitable reactive element that, when connected in parallel with the load of Example 12.5, will yield $pf = 1$.

Solution

With $\mathbf{Z} = 100 + j300$ Ω we have $pf = \cos \sphericalangle \mathbf{Z} = \cos (\tan^{-1} 300/100) = 0.316$, lagging. To achieve $pf = 1$ we need a parallel capacitance C_p such that $\omega C_p = B_p = -B$ where, by Equation (11.58), $B = -300/(100^2 + 300^2) = -0.003$ Ω^{-1}. Thus, $C_p = -B/\omega = -(-0.003)/377 = 7.958$ μF.

◀

Exercise 12.6   If in Example 12.6 we use a parallel capacitance of 5 μF instead of 7.958 μF, what is the resulting power factor? What if we use 10 μF?

ANSWER   0.668 lagging, 0.792 leading.

In practical situations a *pf* suitably *close* to unity is often sufficient. Moreover, the specifications are usually given in terms of the load impedance rather than the load admittance. Thus, given a load $\mathbf{Z} = R + jX$, we wish to find a reactance X_p that, when connected in parallel with $\mathbf{Z}$, will raise the power factor of the composite load $jX_p \parallel \mathbf{Z}$ to a new value pf_{new}. Expanding the expression for the composite load and observing that $pf_{\text{new}} = \cos [\sphericalangle (jX_P \parallel \mathbf{Z})]$, we readily find the desired reactance as

$$X_p = \frac{R^2 + X^2}{R \tan (\pm \cos^{-1} pf_{\text{new}}) - X} \qquad \textbf{(12.29)}$$

where we use the $+$ sign if pf_{new} is *lagging,* and the $-$ sign if it is *leading.* An alternate useful form is obtained by multiplying numerator and denominator by I_{rms}^2. Recognizing that $(R^2 + X^2)I_{\text{rms}}^2 = V_{\text{rms}}^2$, and using also Equations (12.23) and (12.24), we obtain

$$X_p = \frac{V_{\text{rms}}^2/P}{\tan (\pm \cos^{-1} pf_{\text{new}}) - \tan (\pm \cos^{-1} pf_{\text{old}})} \qquad \textbf{(12.30)}$$

where P is the average ac power, and pf_{old} and pf_{new} are, respectively, the values of the power factor before and after correction.

▶ **Example 12.7**

An industrial plant is powered from a 220-V (rms), 60-Hz source, and it consumes 50 kW with $pf = 0.75$, lagging. Find a suitable bank of parallel capacitors to ensure $pf = 0.95$, lagging. What current must these capacitors be able to withstand?

Solution

By Equation (12.30),

$$X_p = \frac{220^2/(50 \times 10^3)}{\tan(\cos^{-1} 0.95) - \tan(\cos^{-1} 0.75)} = -1.750 \; \Omega$$

We thus need a capacitor bank C_p such that $-1/\omega C_p = X_p$, or $C_p = -1/\omega X_p = -1/[2\pi 60(-1.750)] = 1516 \; \mu F$. The load and the correcting capacitor bank are shown in Figure 12.9.

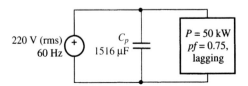

220 V (rms) 60 Hz C_p 1516 μF $P = 50$ kW $pf = 0.75,$ lagging

Figure 12.9 Power-factor correction for Example 12.7.

By Ohm's Law, the current through the capacitor bank is $I_{rms} = V_{rms}/|\mathbf{Z}_{C_p}| = 220/1.750 = 125.7$ A. This requirement must be kept in mind when specifying the capacitor bank. ◀

Exercise 12.7   Find a suitable reactive element that, when connected in parallel with the load of Example 12.5, will yield $pf = 0.95$, lagging. What is the current through this element?

ANSWER $C_p = 7.086 \; \mu F$, $I_{rms} = 0.3206$ A.

Complex Power

Squaring both sides of Equations (12.19) and (12.20) and adding terms pairwise yields $P^2 + Q^2 = (V_{rms}I_{rms})^2(\cos^2 \phi + \sin^2 \phi) = S^2$, where $S = V_{rms}I_{rms}$ is the *apparent* power introduced in the previous section. This Pythagorean relationship suggests that P and Q can be regarded as the *real* and *imaginary* parts of a complex variable S called the **complex power** and having length S and phase angle ϕ,

$$S = P + jQ = S\underline{/\phi} \qquad \text{(12.31)}$$

This relationship is conveniently visualized in terms of a right triangle called the **power triangle** and having S as the hypotenuse, P as the horizontal side, and Q as the vertical side.

As illustrated in Figure 12.10, the power triangle has, in turn, the same shape as the **impedance triangle.** In fact, combining Equation (12.31) and

Equation (12.23), we can write $S = (R + jX)I_{rms}^2$, or

$$S = ZI_{rms}^2 \qquad \text{(12.32)}$$

indicating that the power triangle can be obtained from the impedance triangle simply by multiplying each of its sides by I_{rms}^2.

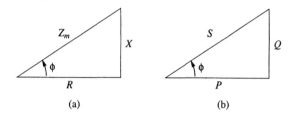

Figure 12.10 Impedance triangle and power triangle.

Once again we emphasize that X and Q are positive for inductive loads and negative for capacitive loads. Hence, Z and S lie in the first quadrant if the load is inductive, and the fourth quadrant if the load is capacitive. The terminology and physical units are summarized as follows:

S *Apparent power* [VA]
P *Real power* [W]
Q *Reactive power* [VAR]

There is yet another insightful way of expressing the complex power of a load. Denoting the phasors of the voltage across the load and the current through the load as

$$V_{rms} = V_{rms}\underline{/\theta_v} \qquad \text{(12.33a)}$$

$$I_{rms} = I_{rms}\underline{/\theta_i} \qquad \text{(12.33b)}$$

and exploiting the fact that $I_{rms}^2 = I_{rms}I_{rms}^*$, Equation (12.32) yields $S = ZI_{rms}^2 = ZI_{rms}I_{rms}^*$. But, by Ohm's Law, $ZI_{rms} = V_{rms}$. Hence,

$$S = V_{rms}I_{rms}^* \qquad \text{(12.34)}$$

The concept of complex power relates P, Q, and S, as well as V_{rms} and I_{rms} in a concise mathematical form, and it thus constitutes a useful analytical tool. Moreover, given an ac source driving a network of interconnected impedances, the following holds:

AC Power Conservation Principle: The complex, real, and reactive power of the source equals, respectively, the sums of the complex, real, and reactive powers of the individual impedances.

Be aware, however, that the *apparent* power of the source is generally different from the sum of the individual apparent powers!

▶ **Example 12.8**

In the circuit of Figure 12.11 a source V_s feeds two industrial loads via a power line whose impedance Z_{line} is as shown. At the load end of the line a parallel capacitance bank C_p is used to correct the power factor.

   (a) Assuming $C_p = 0$, find the source voltage and current capacity required to ensure 220 V (rms) across the load. Hence, find the efficiency of the system.

   (b) Specify a suitable value for C_p to ensure $pf = 0.95$, lagging. Hence, recompute the source capacity and the efficiency, and compare with the uncorrected case.

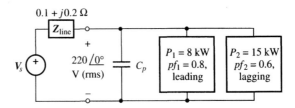

Figure 12.11   Circuit of Example 12.8.

Solution

   (a) At the load we have, by Equation (12.24),

$$Q_1 = P_1 \tan \phi_1 = 8 \tan(-\cos^{-1} 0.8) = -6 \text{ kVAR}$$

$$Q_2 = P_2 \tan \phi_2 = 15 \tan(\cos^{-1} 0.6) = 20 \text{ kVAR}$$

Hence, by the ac power conservation principle,

$$S_{load} = S_1 + S_2 = (8 - j6) + (15 + j20) = 23 + j14 \text{ kVA}$$

Moreover, the current is, by Equation (12.34),

$$I_{rms} = (I^*_{rms})^* = (S_{load}/V_{rms})^* = (23 - j14)10^3/220$$

$$= 104.5 - j63.64 = 122.4\underline{/-31.33°} \text{ A}$$

At the source we have, by KVL,

$$V_{s(rms)} = 220\underline{/0°} + Z_{line}I_{rms}$$

$$= 220 + (0.1 + j0.2)(104.5 - j63.64)$$

$$= 243.2 + j14.53 \text{ V} = 243.6\underline{/3.42°} \text{ V}$$

Moreover, by Equation (12.34),

$$S_{source} = V_{s(rms)}I^*_{rms} = (243.6\underline{/3.42°}) \times (122.4\underline{/31.33°})$$

$$= 29.82\underline{/34.75°} = 24.50 + j17.00 \text{ kVA}$$

Finally, the efficiency of the system is

$$\eta = \frac{P_{\text{load}}}{P_{\text{source}}} = \frac{23}{24.50} = 93.9\%$$

(b) Before power-factor correction, $pf = \cos(\tan^{-1} 14/23) = 0.854$. To raise pf to 0.95 we need $C_p = -1/\omega X_p$, where X_p is found via Equation (12.30). Thus,

$$C_p = -\frac{\tan(\cos^{-1} 0.95) - \tan(\cos^{-1} 0.854)}{377 \times 220^2/(23 \times 10^3)} = 353 \ \mu\text{F}$$

With C_p in place, the complex power of the load becomes, by Equation (12.24),

$$S_{\text{load}} = 23 + j23 \tan(\cos^{-1} 0.95) = 23 + j7.56 \ \text{kVA}$$

Repeating analogous calculations we find $I_{\text{rms}} = 110.0\underline{/-18.20°}$ A, $V_{s(\text{rms})} = 238.0\underline{/4.21°}$ V, $S_{\text{source}} = 24.21 + j9.98$ kVA, and $\eta = 95\%$.

In summary, with C_p in place, the required source capacity is *decreased* from 243.6 V, 122.4 A to 238.0 V, 110.0 A (rms); moreover, system efficiency is *increased* from 93.9% to 95%. ◀

Exercise 12.8   In the circuit of Example 12.8 find C_p so that the pf is raised to unity. Hence, find the improvement in source capacity and efficiency compared with the case $C_p = 0$.

ANSWER   768 μF, 231.4 V (rms), 104.5 A (rms), and 95.5%.

AC Power Measurements

Using a common multimeter, we can measure V_{rms} and I_{rms} at the terminals of an ac load and then calculate their product. The result is the *apparent power* $S = V_{\text{rms}} I_{\text{rms}}$, in VA. If we want the *real* or *average power* P, whose unit is the W, we must multiply S by the *power factor*. Instruments specifically designed to measure average power are called *wattmeters*.

As shown in Figure 12.12(a), the heart of a traditional wattmeter is an *electrodynamometer movement* with a current-sensing coil (*i-coil*), and a voltage sensing coil (*v-coil*). The *i*-coil is wound on a pivoting structure held in place by a spring. When both coils are energized, a torque is developed that twists the pivoting structure against the spring to produce a deflection proportional to the product $v(t)i(t)$. Though ac signals produce pulsating torques, the mechanical inertia of the system provides an averaging effect, resulting in a steady deflection angle that is thus proportional to the *time average* of the product $v(t)i(t)$. As we know, this is the *average power* $P = V_{\text{rms}} I_{\text{rms}} \cos\phi$, indicating that the deflection angle can directly be calibrated in W. The circuit symbol of the wattmeter is shown in Figure 12.12(b).

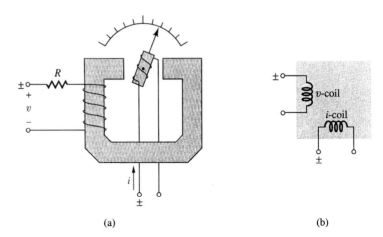

(a)   (b)

Figure 12.12 A wattmeter and its circuit symbol.

To perform power measurements, the *i*-coil must be connected in *series* with the load and the *v*-coil must be connected in *parallel*. To avoid perturbing the circuit under measurement, the *i*-coil is a *low-impedance* coil (ideally, a short circuit, like an ideal *ammeter*), and the *v*-coil is a *high-impedance* coil (ideally an open circuit, like an ideal *voltmeter*). The *v*-coil current is obtained from the applied voltage via a suitably large series resistance *R*. Moreover, each coil has a terminal labeled ± for polarity identification. The wattmeter will deflect *upscale* when the ± terminal of the *i*-coil faces the source and the ± terminal of the *v*-coil is connected to the same line as the *i*-coil.

▶ **Example 12.9**

In the circuit of Figure 12.13 let $R_1 = 10$ Ω, $R_2 = 20$ Ω, and $L = 50$ mH. If v_s is the household ac voltage, what is the wattmeter reading?

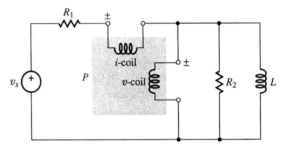

Figure 12.13 Circuit of Example 12.9.

Solution

The load seen by the wattmeter is $Z = (R_2 \parallel j\omega L) = 20 \parallel (j2\pi 60 \times 0.05) = 9.408 + j9.982 \ \Omega$. The rms current through the wattmeter is related to the rms voltage at the source as $I_{rms} = V_{rms}/|R_1 + Z| = 120/|10 + 9.408 + j9.982| = 5.498$ A. The wattmeter reading is thus $P = \text{Re}[Z] \times I_{rms}^2 = 19.41 \times 5.498^2 = 586.7$ W. ◀

Exercise 12.9 If a load consisting of a 15-Ω resistance in series with an unknown inductance is connected to the household ac voltage and dissipates 612 W, what is the value of the inductance?

ANSWER   30 mH.

12.3 THREE-PHASE SYSTEMS

The ac systems we have studied so far are *single-phase* (1-ϕ) systems because they are powered by a single source. A distinctive feature of 1-ϕ systems is that the pulsating voltages and currents go to zero twice within each cycle. When a 1-ϕ source such as the 60-Hz household voltage is applied to an electric motor, the torque disappears 120 times a second. Were it not for the mechanical inertia of the motor, a jerky operation might result. Much can be gained in terms of efficiency and cost if rotating machinery is made to operate with *uniform* torque. This condition is achieved by using *balanced three-phase* (3-ϕ) *power systems*.

To develop a quick appreciation for the advantages of 3-ϕ systems, refer to the simple but informative diagram of Figure 12.14. The power generator consists of three sources having the *same amplitude* and *frequency*, but differing from each other in phase angle by *one-third of a period*, or 120°,

$$v_a = V_m \cos(\omega t + 0°) \qquad \text{(12.35a)}$$
$$v_b = V_m \cos(\omega t - 120°) \qquad \text{(12.35b)}$$
$$v_c = V_m \cos(\omega t + 120°) \qquad \text{(12.35c)}$$

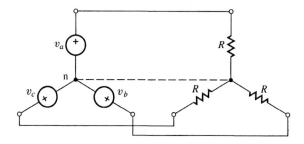

Figure 12.14 Simplified balanced 3-ϕ system.

These waveforms are depicted in Figure 12.15, top. The generator is connected to a *balanced* resistive load, so called because it consists of three *identical* resistances. The individual power components are $p_a = v_a^2/R$, $p_b = v_b^2/R$, and $p_c = v_c^2/R$, and are shown in Figure 12.15, bottom. Each component oscillates between 0 and V_m^2/R and averages to $(1/2)V_m^2/R$. Adding these components point by point reveals that the *total instantaneous power* is at all times *constant* and equal to three times the individual averages,

$$p = p_a + p_b + p_c = 1.5\frac{V_m^2}{R}$$

(12.36)

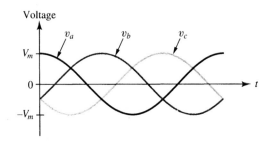

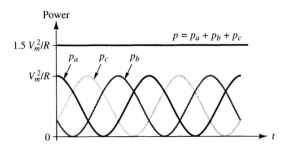

Figure 12.15 Voltage (top) and power (bottom) waveforms for the 3-ϕ system of Figure 12.14.

Power is constant also for other types of balanced loads, such as 3-ϕ electric motors. Consequently, 3-ϕ motors or generators experience a *uniform* torque and thus operate more smoothly and efficiently, just like a multicylinder engine compared to a single-cylinder one. It also turns out that for a given frame size, 3-ϕ motors or generators offer greater power capacity than 1-ϕ types. Moreover, thanks to the inherent sharing of conductors in a 3-ϕ system, the transmission of power requires fewer conductors than three separate 1-ϕ systems handling the same amount of power. For these reasons, virtually all bulk electric power is generated, transmitted, and consumed in 3-ϕ form, though 1-ϕ operation is still made available to small users such as households. We thus find it appropriate to study 3-ϕ systems in detail.

Three-Phase Sources

A 3-ϕ generator consists of a *stator* with three separate windings symmetrically distributed around its periphery, and a *rotor* electromagnet driven at synchronous speed by a steam or gas turbine, a hydraulic turbine, or a diesel engine. As it rotates, the electromagnet induces in each winding a sinusoidal voltage called the **phase voltage.** The three voltages are equal in frequency and amplitude, but differ in phase angle by 120°, and are said to form a **balanced voltage set.** There are two different ways of interconnecting the windings of the generator to form a balanced 3-ϕ source: the **Y (wye)** configuration of Figure 12.16(a), and the **Δ (delta)** configuration shown in Figure 12.18(a).

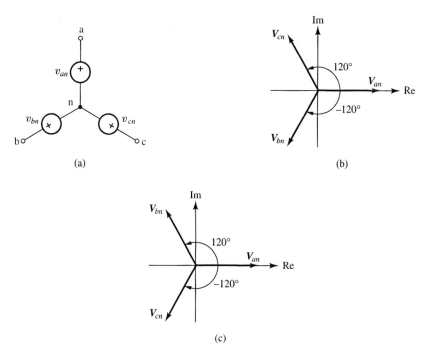

Figure 12.16 (a) Y-connected source and its phasor diagrams for the (b) *abc* (or *positive*), and (c) *acb* (or *negative)* phase sequence.

In the Y configuration the three windings share a common node *n* called the *neutral* node. Using the letters *a*, *b*, and *c* to identify the individual phases, and assuming a zero phase angle for the *a-phase* voltage, we can express the individual phase voltages as

$$
\begin{aligned}
\mathbf{V}_{an} &= V_\phi \underline{/0°} & \text{(12.37a)} \\
\mathbf{V}_{bn} &= V_\phi \underline{/-120°} & \text{(12.37b)} \\
\mathbf{V}_{cn} &= V_\phi \underline{/+120°} & \text{(12.37c)}
\end{aligned}
$$

A few observations about the notation are in order. First, we use *double subscripts* to convey reference-polarity information. For instance, V_{an} shall mean the *voltage difference between nodes a and n,* with the *first* subscript being the node corresponding to the $+$ sign and the *second* subscript being the node corresponding to the $-$ sign. Second, to simplify power calculations we elect to work with *rms values* rather than peak values, and we denote the rms value common to the three phases as

$$V_\phi = \frac{V_m}{\sqrt{2}}$$

(12.38)

where V_m is the peak value. With this notation in mind, the time-domain representation of the individual phase voltages becomes

$$v_{an} = \sqrt{2}V_\phi \cos(\omega t + 0°)$$ (12.39a)

$$v_{bn} = \sqrt{2}V_\phi \cos(\omega t - 120°)$$ (12.39b)

$$v_{cn} = \sqrt{2}V_\phi \cos(\omega t + 120°)$$ (12.39c)

The phasor diagram for this voltage set is shown in Figure 12.16(b). This representation is referred to as the *positive* or *abc* **phase sequence.** By contrast, the representation of Figure 12.16(c) is referred to as the *negative* or *acb phase sequence.* The order of the phase sequence is important in parallel connections, where circuits must share the same sequence to be considered in parallel. Clearly, one set is readily converted to the other by relabeling the phases.

Applying the concept of *geometric phasor addition* to either diagram of Figure 12.16, it is readily seen that the phasors associated with a balanced source always add up to zero,

$$V_{an} + V_{bn} + V_{cn} = 0$$

(12.40a)

Translated to the time domain, this yields

$$v_{an} + v_{bn} + v_{cn} = 0$$

(12.40b)

that is, the *sum of the instantaneous phase voltages of a balanced source is always zero.*

Since the terminals of a 3-ϕ source are connected to a load via power lines, these terminals are also referred to as the *lines.* In addition to the phase voltages, which in a Y-connected source are the line-to-neutral voltages, it is of interest to know the line-to-line voltages. Denoted as V_{ab}, V_{bc}, and V_{ca}, and simply called the **line voltages,** they are readily found as

$$V_{ab} = V_{an} - V_{bn} = V_\phi\underline{/0°} - V_\phi\underline{/-120°} = \sqrt{3}V_\phi\underline{/30°}$$

$$V_{bc} = V_{bn} - V_{cn} = V_\phi\underline{/-120°} - V_\phi\underline{/120°} = \sqrt{3}V_\phi\underline{/-90°}$$

$$V_{ca} = V_{cn} - V_{an} = V_\phi\underline{/120°} - V_\phi\underline{/0°} = \sqrt{3}V_\phi\underline{/150°}$$

These results allow us to express the *line voltages* in terms of the *phase voltages* more concisely as

$$V_{ab} = \left(\sqrt{3}\underline{/30^\circ} \right) V_{an} \tag{12.41a}$$

$$V_{bc} = \left(\sqrt{3}\underline{/30^\circ} \right) V_{bn} \tag{12.41b}$$

$$V_{ca} = \left(\sqrt{3}\underline{/30^\circ} \right) V_{cn} \tag{12.41c}$$

These relationships hold for the *positive* phase sequence. It can easily be shown that for a *negative* phase sequence the line voltages, V_{ab}, V_{ca}, and V_{bc}, are obtained by multiplying the corresponding phase voltages, V_{an}, V_{cn}, and V_{bn}, by $\sqrt{3}\underline{/-30^\circ}$.

The line voltages can also be obtained from the phase voltages via *geometric phasor subtraction*, as depicted in Figure 12.17(a) for the positive phase sequence. Applying, in turn, *geometric phasor addition* to the line voltages, it is readily seen that they add up to zero and thus form a *balanced voltage set*,

$$V_{ab} + V_{bc} + V_{ca} = 0 \tag{12.42}$$

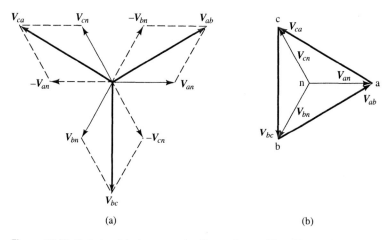

(a)   (b)

Figure 12.17 Relationship between the *line voltages* V_{ab}, V_{bc}, and V_{ca}, and the *phase voltages* V_{an}, V_{bn}, and V_{cn} for a positive phase sequence.

The relationship between the line and phase voltages is conveniently visualized in terms of the *equilateral triangle* of Figure 12.17(b), which has *a*, *b*, and *c* as the vertices, and *n* as the center. We observe that the *sides* are the *line voltages,* and the *lines* from the vertices to the center are the *phase voltages.* We summarize the relationship between the line and the phase voltages of a

Y-connected source as follows:

(1) The amplitude of a *line* voltage is $\sqrt{3} = 1.732$ times that of a *phase* voltage. For instance, if the phase voltage is 120 V (rms), then the line voltage will be $\sqrt{3} \times 120 \simeq 208$ V (rms).

(2) The line voltages form a *balanced set.*

(3) For a *positive phase sequence,* the line voltage set *leads* the phase voltage set by 30°; for a *negative sequence* it lags by 30°.

▶ Example 12.10

If a balanced Y-connected source yields $V_{bn} = 110\underline{/60°}$ V, find V_{ca} for the case of a positive phase sequence.

Solution

By Equation (12.41c), $V_{ca} = (\sqrt{3}\underline{/30°})V_{cn}$. Figure 12.17(b) indicates that for a positive phase sequence V_{cn} can be obtained from V_{bn} via a 120° *clockwise* rotation, that is, by *subtracting* 120° from the argument of V_{bn}. Thus, $V_{cn} = 110\underline{/60° - 120°} = 110\underline{/-60°}$ V, so that $V_{ca} = (\sqrt{3}\underline{/30°})(110\underline{/-60°}) = 190.5\underline{/-30°}$ V.   ◀

Exercise 12.10   If $V_{bc} = 240\underline{/45°}$ V, find V_{an}, V_{bn}, and V_{cn} for the case of a positive phase sequence.

ANSWER $138.6\underline{/135°}$ V, $138.6\underline{/15°}$ V, $138.6\underline{/-105°}$ V.

In principle, a set of line voltages can also be produced by using the *phase voltages* of a Δ-connected source, as depicted in Figure 12.18(a). A Δ-connected source is implemented by connecting adjacent windings of the generator so as to form a closed loop. This configuration is seldom used in practice,

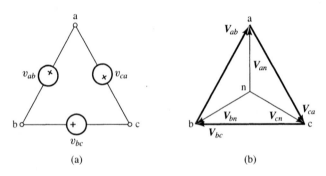

(a)   (b)

Figure 12.18 Δ-connected source and its phasor diagram.

however, because any imbalance in the voltage set would result in undesirable internal currents around the delta loop. It is nevertheless of interest in circuit theory. Note that the Δ-connected source does not have a neutral. However, as the phasor diagram of Figure 12.18(b) shows, a Δ-connected source can be regarded as *equivalent* to a Y-connected source whose *phase voltages* are given by the *lines* from the vertices to the center of the equilateral triangle having the *phase voltages* of the Δ-connected source as its sides.

Three-Phase Loads

As shown in Figure 12.19, a 3-ϕ load consists of three *impedance legs*. Depending on how these impedances are interconnected, we can have the **wye (Y-connected)** load or the **delta (Δ-connected)** load. Note the use of capital letters to denote the line and neutral terminals of the load. As with sources, we can have two phase sequences: *ABC* (or *positive*), and *ACB* (or *negative*).

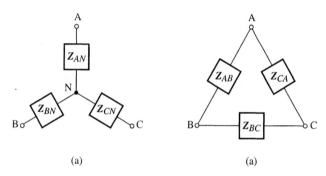

(a)   (a)

Figure 12.19 Y-connected and Δ-connected loads.

Regardless of the load type, the voltages across the legs and the currents through the legs are called the **phase voltages** and **phase currents** of the load. Thus, for a positive phase sequence, the phase voltages of the Y-connected load are V_{AN}, V_{BN}, and V_{CN}, and the corresponding phase currents are I_{AN}, I_{BN}, and I_{CN}; for the Δ-connected load, the phase voltages are V_{AB}, V_{BC}, and V_{CA}, and the phase currents are I_{AB}, I_{BC}, and I_{CA}.

A load is said to be **balanced** if its three impedance legs are identical, that is, if

$$Z_{AN} = Z_{BN} = Z_{CN} = Z_Y \qquad \text{(12.43a)}$$

for the *Y-connected* load and

$$Z_{AB} = Z_{BC} = Z_{CA} = Z_\Delta \qquad \text{(12.43b)}$$

for the *Δ-connected* load.

Δ-Y and Y-Δ Transformations

To simplify the analysis of 3-ϕ circuits, it is often desirable to convert from one load type to the other. We now wish to develop the appropriate transformation equations. In order for a Y connection and a Δ connection to be equivalent, they must exhibit the same terminal behavior. Referring, respectively, to the terminal pairs AB, BC, and CA of Figure 12.19, we must thus have

$$Z_{AN} + Z_{BN} = Z_{AB} \parallel (Z_{BC} + Z_{CA}) = \frac{Z_{AB}(Z_{BC} + Z_{CA})}{Z_{AB} + Z_{BC} + Z_{CA}}$$

$$Z_{BN} + Z_{CN} = Z_{BC} \parallel (Z_{CA} + Z_{AB}) = \frac{Z_{BC}(Z_{CA} + Z_{AB})}{Z_{BC} + Z_{CA} + Z_{AB}}$$

$$Z_{CN} + Z_{AN} = Z_{CA} \parallel (Z_{AB} + Z_{BC}) = \frac{Z_{CA}(Z_{AB} + Z_{BC})}{Z_{CA} + Z_{AB} + Z_{BC}}$$

This system of three equations can be solved for Z_{AN}, Z_{BN}, and Z_{CN} to yield the equations governing the **Δ-Y transformation,**

$$Z_{AN} = \frac{Z_{CA}Z_{AB}}{Z_{AB} + Z_{BC} + Z_{CA}} \tag{12.44a}$$

$$Z_{BN} = \frac{Z_{AB}Z_{BC}}{Z_{AB} + Z_{BC} + Z_{CA}} \tag{12.44b}$$

$$Z_{CN} = \frac{Z_{BC}Z_{CA}}{Z_{AB} + Z_{BC} + Z_{CA}} \tag{12.44c}$$

Conversely, we can solve for Z_{AB}, Z_{BC}, and Z_{CA} to obtain the equations governing the **Y-Δ transformation,**

$$Z_{AB} = \frac{Z_{AN}Z_{BN} + Z_{BN}Z_{CN} + Z_{CN}Z_{AN}}{Z_{CN}} \tag{12.45a}$$

$$Z_{BC} = \frac{Z_{AN}Z_{BN} + Z_{BN}Z_{CN} + Z_{CN}Z_{AN}}{Z_{AN}} \tag{12.45b}$$

$$Z_{CA} = \frac{Z_{AN}Z_{BN} + Z_{BN}Z_{CN} + Z_{CN}Z_{AN}}{Z_{BN}} \tag{12.45c}$$

If the loads are *balanced*, these equations simplify as

$$Z_{AN} = Z_{BN} = Z_{CN} = Z_Y = \frac{1}{3}Z_\Delta \tag{12.46}$$

$$Z_{AB} = Z_{BC} = Z_{CA} = Z_\Delta = 3Z_Y \tag{12.47}$$

▶ **Example 12.11**

Figure 12.20 shows a composite load made up of a Δ-connected load and a Y-connected load in parallel with each other. Assuming $Z_\Delta = 20 + j25$ Ω, and $Z_Y = 5 + j10$ Ω, find a Δ-equivalent as well as a Y-equivalent for the composite load.

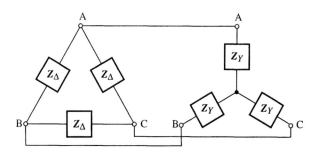

Figure 12.20 Composite load for Example 12.11.

Solution

Performing a Y-Δ transformation on the Y-portion of the load yields a Δ-portion having impedances $3Z_Y = 15 + j30$ Ω. We thus end up with two Δ-connected loads in parallel. The phase impedances of the composite Δ-equivalent are thus

$$Z_{\Delta(\text{composite})} = (20 + j25) \parallel (15 + j30) = \frac{(20 + j25)(15 + j30)}{20 + j25 + 15 + j30}$$

$$= 16.47\underline{/57.25°} = 8.910 + j13.85 \text{ Ω}$$

Performing a Δ-Y transformation gives the phase impedances of the composite Y-equivalent as

$$Z_{Y(\text{composite})} = \frac{Z_{\Delta(\text{composite})}}{3} = 5.491\underline{/57.25°} = 2.970 + j4.618 \text{ Ω}$$

◀

Exercise 12.11 If in Figure 12.20 $Z_Y = 10 + j20$ Ω, find Z_Δ so that the phase impedances of the composite Δ-equivalent load are $Z_{\Delta(\text{composite})} = 50\underline{/0°}$ Ω.

ANSWER $37.5 - j37.5$ Ω.

When a 3-ϕ source and load are connected together via a power line to make up a power system, we have four possible arrangements, depending on the source and load types: Y-Y, Y-Δ, Δ-Δ, and Δ-Y. Since Δ-connected sources are seldom used in practice, the systems of interest to us are the Y-Y and Y-Δ arrangements, to be studied in the next section.

New Words for Mr. Webster's Dictionary: AC versus DC

The guillotine used for executions in France ever since the French Revolution was named after the inventor of that grim machine, Dr. Joseph-Ignace Guillotin. Using that eponymous event as his precedent, Thomas Edison once proposed the noun 'westinghouse' as the official name of the electric chair and the verb 'to westinghouse' for the process we know as electrocution. It was another maneuver in the great financial struggle known as the ac/dc battle.

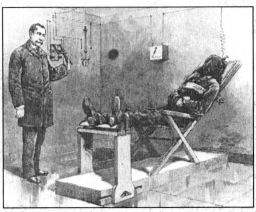

Execution by electrocution using ac (rather than dc) electricity was first used in 1890 by New York state as a humane alternative to hanging and the guillotine. But this evidence of the lethal power of ac electricity failed to override its advantages of thinner copper wires and greater transportability. (University of Illinois)

the system profitable. If every building in the country was to have electricity, there would have to be a generating station literally on every corner. Once the transformer was invented, however, alternating current could be stepped up to thousands of volts at a central generating station, transported long distances over high-tension wires, and then stepped down to more usable voltages at the point of use.

As his first step into the world of electricity, Westinghouse bought the rights to the transformer patent in 1885. Then, while Edison was electrifying New York City with

Like all other down-and-dirty fights in the business world, this one came down to money. By the end of the 1880s, Edison was totally invested in the direct-current system he had developed in the years since he came up with his first working lightbulb in 1879. Out in Pittsburgh, though, was another clever and very rich inventor, George Westinghouse (1846–1914), who had made his fortune by inventing air brakes for trains. Once he got interested in electricity and particularly in the challenge of bringing this marvelous new power into the American home, Westinghouse decided that alternating current was the way to go.

The problem with direct current in the early days was that it could not efficiently be transported any great distances. It was generated at the 120 volts the consumer needed, and any long-distance movement through wiring would mean too much loss to make

direct current, Westinghouse was establishing ac systems in other areas of the country. Slowly he made inroads into Edison's territory, but one big problem still had to be solved: The only thing alternating current could do so far was light; it could not run the motors of American industry because a motor that could run on alternating current did not exist. Surely no one was going to want two systems—Westinghouse's ac for lighting and Edison's dc for everything else—and surely no one was ready to give up this wonderful new invention called the electric motor. No, if he was going to topple Edison from his lofty position as king of the electricity hill, Westinghouse needed a motor that ran on alternating current. Enter Nikola Tesla, inventor of the ac induction motor in 1888. One million dollars changes hands and Westinghouse has the motor he needs to take on Edison in New York.

This was only part of Edison's troubles. The financial picture was beginning to favor alternating current because some stock manipulators (as busy in the 1880s as they were going to be in the 1980s) had driven the price of copper from nine cents a pound all the way up to twenty cents. DC needs fat, therefore expensive, copper wires, but ac can be carried on skinny, therefore much cheaper, wires. Edison's whole empire was threatened. All along he had been proclaiming the safety of dc and the danger of ac. All those thousands of volts would be killing people left and right, whereas there was no danger whatsoever in his mere 120 volts. Self-anointed electrical experts were hired by Edison to educate the public about the dangers of alternating current.

The dc forces acquired a powerful ally: the State of New York, which was investigating "more humane" ways of execution. Partly because of Edison's efforts, the state opted for electricity as its

> ▼
> *"Edison once proposed the noun 'westinghouse' as the official name of the electric chair and the verb 'to westinghouse' for the process we know as electrocution."*
> ▲

new capital-punishment tool, but now the question became, which kind of electricity, ac or dc? Direct-current champion H. P. Brown, a self-promoter hired by Edison in 1888, knew exactly what the answer should be and decided to show the state the beauty of alternating current for this lethal purpose. In December 1888, he called a press conference in Edison's laboratory and used alternating current to kill several large animals, including a cow and a horse.

The state was convinced and carried out its first electrocution on August 6, 1890. To be helpful to the journalists covering this history-making event, the Edison camp suggested that the newly built electric chair be called a "westinghouse" in honor of that great electrical pioneer whose alternating current made possible such a way of leaving this world. To further honor this pioneer, the process should be referred to as "westinghousing" the criminal, as in, "This murderer is scheduled to be westinghoused next month."

Despite Edison's bid to expand the English language by a couple of words, the expressions never took hold and bad press for Westinghouse never materialized. The price of copper stayed high, favoring

> ▼
> *"The problem with direct current in the early days was that it could not efficiently be transported any great distances."*
> ▲

the thinner wires of ac over the thicker ones needed for dc. With the help of Tesla's great mind, Westinghouse continued his quest to electrify the entire nation with alternating current. When he won the contract to light the great Chicago Exposition (where a twenty-year-old named Lee De Forest worked the summer before entering Yale and fell in love with the world of electricity) in 1893, he essentially won the war. In the fall of that year, The Westinghouse Electric Company was awarded the contract to build the world's first hydroelectric plant, at Niagara Falls, and supply electricity—ac, of course—to the city of Buffalo. And that is why the computer this story was typed into and the lamp you read it by are powered by alternating-current electricity.

12.4 Y-Y AND Y-Δ SYSTEMS

Among the four possible types of balanced 3-ϕ systems, the Y-Y and Y-Δ arrangements are the ones of greatest practical interest. Their analysis is based on standard phasor algebra. In fact, we shall shortly see that thanks to the balance property, the analysis of a 3-ϕ system can be reduced to that of a 1-ϕ equivalent.

The Y-Y System

Connecting a Y-source to a Y-load yields the 3-ϕ system of Figure 12.21. The generator is modeled in terms of three Thévenin equivalents, consisting of the per-phase sources V_{sa}, V_{sb}, and V_{sc}, and the series impedances Z_{sa}, Z_{sb}, and Z_{sc}, which account for the internal impedances of the phase windings. The source is connected to the load via three wires or lines, referred to as **hot wires** and having impedances Z_{aA}, Z_{bB}, and Z_{cC}. These impedances account for the distributed impedances of the actual lines as well as the winding impedances of any step-up and step-down transformers that may be present at the generator and load ends of the transmission line. As discussed in Section 5.3, these transformers serve to improve power transmission efficiency. A fourth conductor, referred to as **neutral** and having impedance Z_{Nn}, connects the neutral nodes of the source and the load, and it may or may not be present, as we shall see shortly. Finally, the per-phase impedances of the load are Z_{AN}, Z_{BN}, and Z_{CN}.

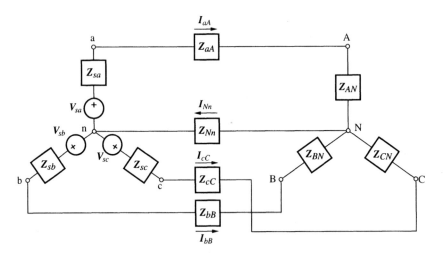

Figure 12.21 Balanced 3-ϕ Y-Y system.

To analyze the system it is convenient to designate the neutral node of the source as the *reference node* for the entire circuit,

$$V_n = 0 \tag{12.48}$$

Applying KCL at node N we have

$$\boldsymbol{I}_{Nn} = \boldsymbol{I}_{aA} + \boldsymbol{I}_{bB} + \boldsymbol{I}_{cC} \tag{12.49}$$

or

$$\frac{\boldsymbol{V}_N}{\boldsymbol{Z}_{Nn}} = \frac{\boldsymbol{V}_{sa} - \boldsymbol{V}_N}{\boldsymbol{Z}_{sa} + \boldsymbol{Z}_{aA} + \boldsymbol{Z}_{AN}} + \frac{\boldsymbol{V}_{sb} - \boldsymbol{V}_N}{\boldsymbol{Z}_{sb} + \boldsymbol{Z}_{bB} + \boldsymbol{Z}_{BN}} + \frac{\boldsymbol{V}_{sc} - \boldsymbol{V}_N}{\boldsymbol{Z}_{sc} + \boldsymbol{Z}_{cC} + \boldsymbol{Z}_{CN}}$$

We again note the use of double-subscript notation to indicate current direction. By this notation, current is defined to flow from the node corresponding to the first subscript to the second. Thus, $\boldsymbol{I}_{Nn}$ flows from N to n. In a *balanced* Y-Y system the following conditions hold:

$$\boldsymbol{V}_{sa} + \boldsymbol{V}_{sb} + \boldsymbol{V}_{sc} = 0 \tag{12.50}$$

$$\boldsymbol{Z}_{sa} = \boldsymbol{Z}_{sb} = \boldsymbol{Z}_{sc} = \boldsymbol{Z}_{\text{winding}} \tag{12.51a}$$

$$\boldsymbol{Z}_{aA} = \boldsymbol{Z}_{bB} = \boldsymbol{Z}_{cC} = \boldsymbol{Z}_{\text{line}} \tag{12.51b}$$

$$\boldsymbol{Z}_{AN} = \boldsymbol{Z}_{BN} = \boldsymbol{Z}_{CN} = \boldsymbol{Z}_Y \tag{12.51c}$$

If we introduce the net **per-phase impedance**

$$\boldsymbol{Z}_\phi = \boldsymbol{Z}_{\text{winding}} + \boldsymbol{Z}_{\text{line}} + \boldsymbol{Z}_Y \tag{12.52}$$

then the equation governing our circuit simplifies as

$$\frac{\boldsymbol{V}_N}{\boldsymbol{Z}_{Nn}} = \frac{\boldsymbol{V}_{sa} + \boldsymbol{V}_{sb} + \boldsymbol{V}_{sc}}{\boldsymbol{Z}_\phi} - 3\frac{\boldsymbol{V}_N}{\boldsymbol{Z}_\phi} = 0 - 3\frac{\boldsymbol{V}_N}{\boldsymbol{Z}_\phi}$$

or

$$\boldsymbol{V}_N \left(\frac{1}{\boldsymbol{Z}_{Nn}} + \frac{3}{\boldsymbol{Z}_\phi} \right) = 0$$

For this condition to hold regardless of $\boldsymbol{Z}_{Nn}$ and $\boldsymbol{Z}_\phi$, we must have

$$\boxed{\boldsymbol{V}_N = 0} \tag{12.53a}$$

Consequently, since $\boldsymbol{I}_{Nn} = \boldsymbol{V}_N / \boldsymbol{Z}_{Nn}$, we also obtain

$$\boxed{\boldsymbol{I}_{Nn} = 0} \tag{12.53b}$$

In words, in a balanced Y-Y system there is *no voltage difference* between the neutral nodes of the source and the load, and *no current* through the neutral connector. In fact, we can eliminate this connector altogether without affecting the operation of the system, thus saving the cost of the wire! This is why the neutral wire is shown as a broken line in Figure 12.14. If the system is not perfectly balanced, some current will flow through the neutral wire, but we expect it to be much smaller than the hot-wire currents, justifying the use of a smaller and cheaper neutral wire.

The **phase currents** in the Y-Y system are

$$I_{aA} = \frac{V_{sa}}{Z_{sa} + Z_{aA} + Z_{AN}} = \frac{V_{sa}}{Z_\phi} \quad \text{(12.54a)}$$

$$I_{bB} = \frac{V_{sb}}{Z_{sb} + Z_{bB} + Z_{BN}} = \frac{V_{sb}}{Z_\phi} \quad \text{(12.54b)}$$

$$I_{cC} = \frac{V_{sc}}{Z_{sc} + Z_{cC} + Z_{CN}} = \frac{V_{sc}}{Z_\phi} \quad \text{(12.54c)}$$

In light of Equation (12.50), the phase currents form a balanced set,

$$I_{aA} + I_{bB} + I_{cC} = 0 \quad \text{(12.55)}$$

a result that we could also have anticipated by letting $I_{Nn} = 0$ in Equation (12.49).

It is apparent that thanks to the balance property, the analysis of Y-Y systems can be simplified significantly. In fact, we need to carry out the detailed analysis of *only one phase,* typically the *a*-phase, using the simplified equivalent of Figure 12.22. Then, we adapt the results of the *a*-phase to the other two phases by simply shifting the *a*-phase voltages and currents by 120°. The shifting rule is as follows:

(1) For a *positive* phase sequence, rotate the *a*-phase phasors by 120° *clockwise* to obtain the *b*-phase phasors, and by 120° *counterclockwise* to obtain those of the *c*-phase. Recall that to rotate a phasor *clockwise* we *subtract* 120° from its argument, and to rotate it *counterclockwise* we add 120°.

(2) For a *negative* phase sequence, rotate the *a-phase* phasors by 120° *counterclockwise* (add 120°) to obtain the *b*-phase phasors, and by 120° *clockwise* (subtract 120°) to obtain the *c*-phase.

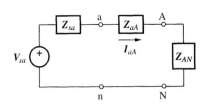

Figure 12.22 Single-phase equivalent of the Y-Y system.

We observe that since there is no voltage difference between nodes *n* and *N*, the neutral wire in the equivalent of Figure 12.22 has been replaced by a perfect conductor. In this 1-ϕ model the current I_{aA} returns via the neutral wire; however, when the three phases are considered simultaneously, the three currents on this wire will, by Equation (12.55), add up to zero, thus justifying the use of a perfect conductor in our model.

▶ **Example 12.12**

A balanced, positive-phase sequence, Y-Y system consists of a generator with an internal per-phase voltage of 120 V (rms) and winding impedance of $0.1 + j0.2$ Ω; a load with a per-phase impedance of $15 + j10$ Ω; and a power line having a per-phase impedance of $0.5 + j1$ Ω. Find

(a) The line currents I_{aA}, I_{bB}, and I_{cC}
(b) The phase voltages V_{AN}, V_{BN}, and V_{CN} at the load
(c) The phase voltages V_{an}, V_{bn}, and V_{cn} at the source
(d) The line voltages V_{AB}, V_{BC}, and V_{CA} at the load
(e) The line voltages V_{ab}, V_{bc}, and V_{ca} at the source

Solution

By Equation (12.52) we have

$$Z_\phi = 0.1 + j0.2 + 0.5 + j1 + 15 + j10 = 15.6 + j11.2 \text{ Ω}$$

(a) Assuming $\measuredangle V_{sa} = 0°$ in Figure 12.22, we have, by Equation (12.54a),

$$I_{aA} = \frac{120/0°}{15.6 + j11.2} = 6.249/-35.68° \text{ A (rms)}$$

Since the sequence is positive, we subtract and add $120°$ to obtain, respectively, $I_{bB} = 6.249/-155.68°$ A (rms) and $I_{cC} = 6.249/84.32°$ A (rms).

(b) By Ohm's Law,

$$V_{AN} = Z_{AN}I_{aA} = (15 + j10)6.249/-35.68° = 112.6/-1.99° \text{ V (rms)}$$

Subtracting and adding $120°$ yields, respectively, $V_{BN} = 112.6/-121.99°$ V (rms) and $V_{CN} = 112.6/118.01°$ V (rms).

(c) Applying KVL and Ohm's Law gives

$$V_{an} = 120/0° - (0.1 + j0.2)6.249/-35.68° = 118.8/-0.31° \text{ V (rms)}$$

Likewise, $V_{bn} = 118.8/-120.31°$ V (rms) and $V_{cn} = 118.8/119.69°$ V (rms).

(d) Applying Equation (12.41a), we get

$$V_{AB} = (\sqrt{3}/30°)V_{AN}$$

$$= (\sqrt{3}/30°)112.6/-1.99°$$

$$= 195.0/28.01° \text{ V (rms)}$$

Likewise, $V_{BC} = 195.0/-91.99°$ V (rms) and $V_{CA} = 195.0/148.01°$ V (rms).

(e) Again applying Equation (12.41a) gives

$$V_{ab} = (\sqrt{3}/30°)V_{an}$$

$$= (\sqrt{3}/30°)118.8/-0.31° = 205.8/29.69° \text{ V (rms)}$$

Likewise, $V_{bc} = 205.8/-90.31°$ V (rms) and $V_{ca} = 205.8/149.69°$ V (rms).

Exercise 12.12   A balanced, positive-phase sequence, Y-Y system has $Z_{\text{winding}} = 0.2 + j0.5\ \Omega$, $Z_{\text{line}} = 1 + j1.5\ \Omega$, and $Z_Y = 20 + j15\ \Omega$. If it is desired to have $V_{AN} = 120\underline{/0^\circ}$ V (rms), what must be the phase voltages V_{sa}, V_{sb}, and V_{sc} at the source?

ANSWER $130.4\underline{/1.86^\circ}$ V, $130.4\underline{/-118.14^\circ}$ V, $130.4\underline{/121.86^\circ}$ V, all (rms).

The Y-Δ System

Connecting a Y-source to a Δ-load yields the 3-ϕ system of Figure 12.23. Since Δ-loads do not have a neutral node, the neutral wire is now absent. Practical 3-ϕ loads are more likely to be found Δ-connected than Y-connected to take advantage of the ease with which, at least for the unbalanced case, individual legs may be added to or removed from a single phase.

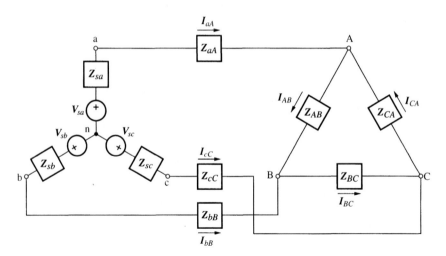

Figure 12.23 Balanced 3-ϕ Y-Δ system.

Using the terminology introduced in the previous section, we observe that the *line voltages* of the Y-connected source are the *phase voltages* of the Δ-connected load. However, the *line currents* of the source and the *phase currents* of the load are no longer coincident. In fact, KCL at the nodes of the Δ-load yields

$$I_{aA} = I_{AB} - I_{CA}$$

$$I_{bB} = I_{BC} - I_{AB}$$

$$I_{cC} = I_{CA} - I_{BC}$$

In a balanced system both the load and line currents form balanced sets. In particular, we can use the following identities:

$$I_{CA} = (1\underline{/120^\circ})I_{AB} \qquad I_{AB} = (1\underline{/120^\circ})I_{BC} \qquad I_{BC} = (1\underline{/120^\circ})I_{CA}$$

Substituting into the previous equations and expanding, we obtain the following relationships between the line currents and the phase currents of a Δ-load:

$$I_{aA} = \left(\sqrt{3}\underline{/-30^\circ}\right)I_{AB} \tag{12.56a}$$

$$I_{bB} = \left(\sqrt{3}\underline{/-30^\circ}\right)I_{BC} \tag{12.56b}$$

$$I_{cC} = \left(\sqrt{3}\underline{/-30^\circ}\right)I_{CA} \tag{12.56c}$$

In words, in a Δ-load the *line currents* are $\sqrt{3} = 1.732$ times the *phase currents*. Moreover, the line current set *lags* the phase current set by 30° for a *positive* phase sequence, and it *leads* by 30° for a *negative* phase sequence.

▶ Example 12.13

Assuming a positive phase sequence, find I_{CA} if $I_{bB} = 12\underline{/90^\circ}$ A (rms).

Solution

We have $I_{CA} = I_{cC}/(\sqrt{3}\underline{/-30^\circ})$, where $I_{cC} = (1\underline{/-120^\circ})I_{bB} = (1\underline{/-120^\circ}) \times (12\underline{/90^\circ}) = 12\underline{/-30^\circ}$ A (rms). Thus, $I_{CA} = (12\underline{/-30^\circ})/(\sqrt{3}\underline{/-30^\circ}) = 6.928\underline{/0^\circ}$ A (rms). ◀

Exercise 12.13   Assuming a positive phase sequence, find I_{aA} if $I_{BC} = 25\underline{/60^\circ}$ A (rms).

ANSWER $43.3\underline{/150^\circ}$ A (rms).

The analysis of the Y-Δ system can be made similar to that of the Y-Y system if we transform the Δ-connected load into a Y-connected equivalent. For a balanced load, this transformation is

$$Z_Y = \frac{1}{3}Z_\Delta \tag{12.57}$$

Once this transformation has been performed, we can use the 1-ϕ equivalent of Figure 12.24 to analyze our circuit. Note again the use of the perfect conductor nN, even though the Δ-connected load has no neutral node.

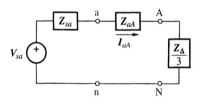

Figure 12.24 Single-phase equivalent of the Y-Δ system.

▶ **Example 12.14**

A balanced, positive-phase sequence, Y-Δ system consists of a generator with an internal per-phase voltage of 120 V (rms) and winding impedance of $0.2 + j0.3$ Ω; a load with a per-phase impedance of $90 + j60$ Ω; and a power line having a per-phase impedance of $1 + j2$ Ω. Find

(a) The line currents I_{aA}, I_{bB}, and I_{cC}

(b) The phase currents I_{AB}, I_{BC}, and I_{CA} at the load

(c) The phase voltages V_{AB}, V_{BC}, and V_{CA} at the load

Solution

By Equation (12.57) the per-phase impedance of the equivalent Y-load is $Z_Y = (1/3)(90 + j60) = 30 + j20$ Ω, so that $Z_\phi = 0.2 + j0.3 + 1 + j2 + 30 + j20 = 31.2 + j22.3$ Ω.

(a) Assuming $\angle V_{sa} = 0°$ in Figure 12.24, we have

$$I_{aA} = \frac{120\underline{/0°}}{31.2 + j22.3} = 3.129\underline{/-35.56°} \text{ A (rms)}$$

Since the sequence is positive we have $I_{bB} = 3.129\underline{/-155.56°}$ A (rms) and $I_{cC} = 3.129\underline{/84.44°}$ A (rms).

(b) By Equation (12.56a),

$$I_{AB} = I_{aA}/(\sqrt{3}\underline{/-30°}) = (3.129\underline{/-35.56°})/(\sqrt{3}\underline{/-30°})$$
$$= 1.807\underline{/-5.56°} \text{ A (rms)}$$

Likewise, $I_{BC} = 1.807\underline{/-125.56°}$ A (rms) and $I_{CA} = 1.807\underline{/114.44°}$ A (rms).

(c) By Ohm's Law,

$$V_{AB} = Z_{AB}I_{AB}$$
$$= (90 + j60)(1.807\underline{/-5.56°}) = 195.5\underline{/28.14°} \text{ V (rms)}$$
$$V_{BC} = 195.5\underline{/-91.86°} \text{ V(rms) and}$$
$$V_{CA} = 195.5\underline{/148.14°} \text{ V(rms).}$$ ◀

Exercise 12.14   Measurements at a balanced Δ-connected load yield $|V_{BC}| = 220$ V (rms) and $|I_{aA}| = 10$ A (rms). Moreover, it is found that I_{aA} leads V_{BC} by 60°. What is the load impedance?

ANSWER $33 + j19.05$ Ω.

12.5 POWER IN THREE-PHASE SYSTEMS

Since in a balanced system the same-set voltages and currents have equal magnitudes, we can simplify our notation if we denote the rms *phase* voltages and rms *phase* currents at the *load* as V_ϕ and I_ϕ, and the rms *line* voltages and rms *line* currents at the *load* as V_L and I_L. Then, we have

$$V_\phi = |V_{AN}| = |V_{BN}| = |V_{CN}| \qquad \textbf{(12.58a)}$$

$$I_\phi = |I_{AN}| = |I_{BN}| = |I_{CN}| \qquad \textbf{(12.58b)}$$

for a *Y-connected* load, and

$$V_\phi = |V_{AB}| = |V_{BC}| = |V_{CA}| \qquad \textbf{(12.59a)}$$

$$I_\phi = |I_{AB}| = |I_{BC}| = |I_{CA}| \qquad \textbf{(12.59b)}$$

for a Δ-*connected* load. Moreover, we can write

$$V_L = |V_{AB}| = |V_{BC}| = |V_{CA}| = \sqrt{3}\,V_\phi \qquad \textbf{(12.60a)}$$

$$I_L = |I_{AN}| = |I_{BN}| = |I_{CN}| = I_\phi \qquad \textbf{(12.60b)}$$

for a *Y-connected* load, and

$$V_L = |V_{AB}| = |V_{BC}| = |V_{CA}| = V_\phi \qquad \textbf{(12.61a)}$$

$$I_L = \sqrt{3}|I_{AB}| = \sqrt{3}|I_{BC}| = \sqrt{3}|I_{CA}| = \sqrt{3}\,I_\phi \qquad \textbf{(12.61b)}$$

for a Δ-*connected* load.

Instantaneous Power

The *total instantaneous power* delivered to a load in a 3-ϕ system is

$$p_T = v_{AB}i_{AB} + v_{BC}i_{BC} + v_{CA}i_{CA} \qquad \textbf{(12.62)}$$

where we are assuming a Δ-connected load with phase voltages v_{AB}, v_{BC}, and v_{CA}, and phase currents i_{AB}, i_{BC}, and i_{CA}. Though we are considering a Δ-connected load, the results we are about to derive apply just as well to Y-connected loads. For a balanced, positive-phase-sequence system we can write

$$i_{AB} = \sqrt{2}I_\phi \cos \omega t$$

$$i_{BC} = \sqrt{2}I_\phi \cos (\omega t - 120°)$$

$$i_{CA} = \sqrt{2}I_\phi \cos (\omega t + 120°)$$

$$v_{AB} = \sqrt{2}V_\phi \cos (\omega t + \phi)$$

$$v_{BC} = \sqrt{2}V_\phi \cos (\omega t + \phi - 120°)$$

$$v_{CA} = \sqrt{2}V_\phi \cos (\omega t + \phi + 120°)$$

where we have taken the phase angle of i_{AB} as the reference. Substituting into Equation (12.62) and exploiting the fact that $v_{AB} + v_{BC} + v_{CA} = 0$, one can readily show (see Problem 12.59) that

$$p_T = 3V_\phi I_\phi \cos \phi \qquad \text{(12.63)}$$

that is, p_T is invariant with time! As we know, this property is particularly useful when the load is a 3-ϕ motor, where it results in a constant torque and, hence, it reduces vibration.

Complex, Real, and Reactive Power

The **per-phase complex power** S_ϕ associated with the load is

$$S_\phi = P_\phi + jQ_\phi \qquad \text{(12.64)}$$

where P_ϕ is the **per-phase real power** and Q_ϕ is the **per-phase reactive power**. By Equations (12.19) and (12.20), these components can be expressed either in terms of *phase* quantities or in terms of *line* quantities as

$$P_\phi = V_\phi I_\phi \cos \phi_\phi = \frac{V_L I_L}{\sqrt{3}} \cos \phi_\phi \qquad \text{(12.65a)}$$

$$Q_\phi = V_\phi I_\phi \sin \phi_\phi = \frac{V_L I_L}{\sqrt{3}} \sin \phi_\phi \qquad \text{(12.65b)}$$

where ϕ_ϕ is the *phase difference* between the voltage and current of the same phase or of the same line. Clearly, the *per-phase apparent power* is $S_\phi = V_\phi I_\phi = V_L I_L/\sqrt{3}$.

The **total complex power** S_T absorbed by the load is the sum of the three per-phase components,

$$S_T = 3S_\phi = P_T + jQ_T \qquad \text{(12.66)}$$

where P_T is the **total real power** and Q_T the **total reactive power**,

$$P_T = 3V_\phi I_\phi \cos \phi_\phi = \sqrt{3}V_L I_L \cos \phi_\phi \qquad \text{(12.67a)}$$

$$Q_T = 3V_\phi I_\phi \sin \phi_\phi = \sqrt{3}V_L I_L \sin \phi_\phi \qquad \text{(12.67b)}$$

Clearly, the *total apparent power* is $S_T = 3V_\phi I_\phi = \sqrt{3}V_L I_L$. Comparison with Equation (12.63) indicates that the total instantaneous power p_T coincides with the *total real power* P_T, something you might have expected. Let us illustrate via actual examples.

▶ Example 12.15

For the Y-Y system of Example 12.12,

(a) Find the *total complex power* absorbed by the load, the lines, and the generator windings.

(b) Find the *total complex power* delivered by the source.

(c) Verify that $S_{T(\text{delivered})} = S_{T(\text{absorbed})}$.

(d) Find the *efficiency* of the system.

Solution

(a) Using the results of Example 12.12, we have, at the load, $V_\phi = 112.6$ V (rms), $I_\phi = 6.249$ A (rms), and $\phi_\phi = \tan^{-1}(10/15) = 33.69°$. Hence,

$$S_{T(\text{load})} = P_T + j\,Q_T = 3V_\phi I_\phi (\cos\phi_\phi + j\sin\phi_\phi)$$
$$= 3 \times 112.6 \times 6.249 \times (\cos 33.69° + j\sin 33.69°)$$
$$= 1757.0 + j1170.6 \text{ VA}$$

Using the data of Example 12.12, along with Equation (12.32) we have, on the line,

$$S_{T(\text{line})} = 3(R + j\,X)I_\phi^2 = 3(0.5 + j1)6.249^2 = 58.6 + j117.1 \text{ VA}$$

Likewise in the generator windings we have

$$S_{T(\text{windings})} = 3(R + j\,X)I_\phi^2$$
$$= 3(0.1 + j0.2)6.249^2 = 11.7 + j23.4 \text{ VA}$$

(b) At the source we have $V_\phi = 120.0$ V (rms), $I_\phi = 6.249$ A (rms), and $\phi_\phi = 0 - (-35.68°) = 35.68°$, so

$$S_{T(\text{source})} = P_T + j\,Q_T = 3V_\phi I_\phi (\cos\phi_\phi + j\sin\phi_\phi)$$
$$= 3 \times 120 \times 6.249 (\cos 35.68° + j\sin 35.68°)$$
$$= 1827.3 + j1311.1 \text{ VA}$$

(c) We have

$$S_{T(\text{absorbed})} = S_{T(\text{load})} + S_{T(\text{line})} + S_{T(\text{windings})}$$
$$= 1757.0 + j1170.6 + 58.6 + j117.1 + 11.7 + j23.4$$
$$= 1827.3 + j1311.1 \text{ VA}$$

$$S_{T(\text{delivered})} = S_{T(\text{source})} = 1827.3 + j1311.1 \text{ VA}$$

thus confirming the conservation of complex power.

(d) The efficiency of the system is

$$\eta = \frac{P_{T(\text{load})}}{P_{T(\text{source})}} = \frac{1757.0}{1827.3} = 96.2\%$$ ◀

Exercise 12.15 Verify the conservation of the *total complex power* for the Y-Δ system of Example 12.14.

ANSWER $S_{T(\text{delivered})} = S_{T(\text{absorbed})} = 917 + j655$ VA.

▶ **Example 12.16**

The two balanced systems of Figure 12.25 are interconnected via lines having per-phase impedance $\mathbf{Z}_{\text{line}} = 1 + j2 \ \Omega$. If $\mathbf{V}_{ab} = 10\underline{/0°}$ kV (rms) and $\mathbf{V}_{AB} = 10\underline{/6°}$ kV (rms), (a) which system is the source and which is the load? What is the power supplied by the source and the power absorbed by the load? (b) Verify the conservation of power.

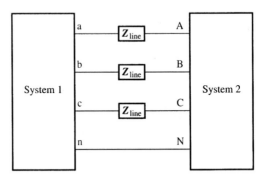

Figure 12.25 3-ϕ system of Example 12.16.

Solution

(a) Assuming a Y-Y system, we have, by Ohm's Law and Equation (12.41)

$$\mathbf{I}_{aA} = \frac{\mathbf{V}_{an} - \mathbf{V}_{AN}}{\mathbf{Z}_{\text{line}}} = \frac{\left(10^4/\sqrt{3}\right)\underline{/-30°} - \left(10^4/\sqrt{3}\right)\underline{/-24°}}{1 + j2}$$

$$= 270.3\underline{/179.56°} \text{ A (rms)}$$

By Equation (12.67a), the power absorbed by System 2 is

$$P_2 = 3\,|\mathbf{V}_{AN}| \times |\mathbf{I}_{aA}|\cos\left(\sphericalangle \mathbf{V}_{AN} - \sphericalangle \mathbf{I}_{aA}\right)$$

$$= 3(10^4/\sqrt{3})270.3\cos\left(-24° - 179.56°\right)$$

$$= -4.291 \text{ MW}$$

Since P_2 is negative, System 2 is actually *delivering* power and is thus the source. The power absorbed by System 1 is

$$P_1 = 3\,|\mathbf{V}_{an}| \times |\mathbf{I}_{Aa}|\cos\left(\sphericalangle \mathbf{V}_{an} - \sphericalangle \mathbf{I}_{Aa}\right)$$

$$= 3(10^4/\sqrt{3})270.3\cos\left[-30° - (179.56° - 180°)\right]$$

$$= 4.072 \text{ MW}$$

Its positive value confirms that System 1 is indeed the load.

(b) The power lost in the line is $P_{\text{line}} = 3R_{\text{line}}|I_{aA}|^2 = 3 \times 1 \times 270.3^2 = 0.219$ MW. Since $4.072 + 0.219 = 4.291$, the power absorbed by System 1 plus that lost in the line equals that delivered by System 2, as it should be.   ◀

Power Measurements in 3-ϕ Systems

The power measurement arrangement for 1-ϕ systems can easily be extended, at least in principle, to 3-ϕ systems. If the system is *balanced,* we need only measure the power of one of its phases and then multiply the wattmeter reading by 3. In practice, however, the neutral terminal may be inaccessible in the case of Y-connected loads, and it is definitely absent in the case of Δ-connected loads, so 3-ϕ phase measurements must be performed on the *lines.* This goal is achieved by the **two-wattmeter method** depicted in Figure 12.26, which uses a wattmeter pair to measure power flow on two of the lines, with the third line acting as common reference line for the v-coils. We now show that this method allows for the indirect measurement of P_T, Q_T, and *pf.*

We begin by observing that the phase impedances of the load are

$$\mathbf{Z}_{AB} = \mathbf{Z}_{BC} = \mathbf{Z}_{CA} = \mathbf{Z}_\phi = Z_\phi\underline{/\phi_\phi}$$

Though the load shown is of the Δ type, the results we are about to derive apply just as well to Y-type loads, as one can always transform one load type to the other. In the following we assume a positive phase sequence and denote the magnitudes of the phase voltages at the load as V_L,

$$|V_{AB}| = |V_{BC}| = |V_{CA}| = V_L$$

and the magnitudes of the line currents as I_L

$$|I_{aA}| = |I_{bB}| = |I_{cC}| = I_L$$

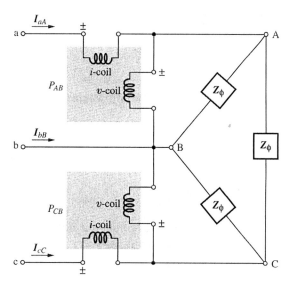

Figure 12.26 Two-wattmeter method for 3-ϕ power measurements.

The reading provided by the top wattmeter is

$$P_{AB} = |V_{AB}| \times |I_{aA}| \cos(\angle V_{AB} - \angle I_{aA})$$

$$= V_L I_L \cos[\angle(Z_\phi I_{AB}) - \angle I_{aA}] = V_L I_L \cos(\phi_\phi + \angle I_{AB} - \angle I_{aA})$$

By Equation (12.56a), $\angle I_{aA} = \angle I_{AB} - 30°$. Hence, $\phi_\phi + \angle I_{AB} - \angle I_{aA} = \phi_\phi + 30°$, so

$$P_{AB} = V_L I_L \cos(\phi_\phi + 30°) \qquad (12.68)$$

The reading provided by the bottom wattmeter is

$$P_{CB} = |V_{CB}| \times |I_{cC}| \cos(\angle V_{CB} - \angle I_{cC})$$

$$= V_L I_L \cos(\angle Z_\phi I_{CB} - \angle I_{cC}) = V_L I_L \cos(\phi_\phi + \angle I_{CB} - \angle I_{cC})$$

Since we have a balanced current set, we can write $\angle I_{CB} = \angle I_{BC} \pm 180° = \angle I_{CA} + 120° \pm 180° = \angle I_{CA} - 60°$. Moreover, Equation (12.56c) yields $\angle I_{cC} = \angle I_{CA} - 30°$. Hence, $\phi_\phi + \angle I_{CB} - \angle I_{cC} = \phi_\phi + \angle I_{CA} - 60° - (\angle I_{CA} - 30°) = \phi_\phi - 30°$, so that

$$P_{CB} = V_L I_L \cos(\phi_\phi - 30°) \qquad (12.69)$$

The *sum* of the two wattmeter readings is

$$P_{AB} + P_{CB} = V_L I_L [\cos(\phi_\phi + 30°) + \cos(\phi_\phi - 30°)]$$

Using a well-known trigonometric identity, the term within brackets reduces to $2\cos\phi_\phi \cos 30° = \sqrt{3}\cos\phi_\phi$. Hence,

$$P_{AB} + P_{CB} = \sqrt{3}V_L I_L \cos\phi_\phi$$

Comparing with Equation (12.67a), we see that the sum of the wattmeter readings is simply the *total real power*,

$$\boxed{P_T = P_{AB} + P_{CB}} \qquad (12.70)$$

Next, consider the *difference* between the two wattmeter readings,

$$P_{CB} - P_{AB} = V_L I_L [\cos(\phi_\phi - 30°) - \cos(\phi_\phi + 30°)]$$

Using again a well-known trigonometric identity, this becomes $P_{CB} - P_{AB} = 2V_L I_L \sin\phi_\phi \sin 30°$, or

$$P_{CB} - P_{AB} = V_L I_L \sin\phi_\phi$$

Comparing with Equation (12.67b), we see that the *difference* of the wattmeter readings is proportional to the *total reactive power*,

$$\boxed{Q_T = \sqrt{3}(P_{CB} - P_{AB})} \qquad (12.71)$$

Finally, using the fact that $\phi_\phi = \tan^{-1}(Q_T/P_T)$, we can use the two-wattmeter method to find also the *power factor* of the load,

$$pf = \cos\left(\tan^{-1}\sqrt{3}\frac{P_{CB} - P_{AB}}{P_{CB} + P_{AB}}\right) \qquad \textbf{(12.72)}$$

These equations allow us to make the following observations:

(1) If $P_{CB} = P_{AB}$, the load is *resistive*.
(2) If $P_{CB} > P_{AB}$, the load is *inductive*.
(3) If $P_{CB} < P_{AB}$, the load is *capacitive*.

Once again we emphasize that these results and observations, derived for balanced systems, hold whether the load is Y-connected or Δ-connected.

▶ Example 12.17

In the balanced system of Figure 12.26 let $V_L = 220$ V (rms). Predict the wattmeter readings as well as P_T and Q_T for the following phase impedances:

(a) $\mathbf{Z}_\phi = 30 + j40\ \Omega$
(b) $\mathbf{Z}_\phi = 30 - j40\ \Omega$
(c) $\mathbf{Z}_\phi = 50\ \Omega$

Solution

(a) We have $I_L = V_L/Z_\phi = 220/\sqrt{30^2 + 40^2} = 4.4$ A (rms), and $\phi_\phi = \tan^{-1}(40/30) = 53.13°$. Hence,

$$P_{AB} = 220 \times 4.4 \times \cos(53.13° + 30°) = 115.8\ \text{W}$$

$$P_{CB} = 220 \times 4.4 \times \cos(53.13° - 30°) = 890.2\ \text{W}$$

$$P_T = 115.8 = 1006\ \text{W}$$

$$Q_T = \sqrt{3}(890.2 - 115.8) = 1341.3\ \text{VAR}$$

(b) We still have $I_L = 4.4$ A (rms); however, now $\phi_\phi = -53.13°$. Thus,

$$P_{AB} = 220 \times 4.4 \times \cos(-53.13° + 30°) = 890.2\ \text{W}$$

$$P_{CB} = 220 \times 4.4 \times \cos(53.13° - 30°) = 115.8\ \text{W}$$

$$P_T = 890.2 + 115.8 = 1006\ \text{W}$$

$$Q_T = \sqrt{3}(115.8 - 890.2) = -1341.3\ \text{VAR}$$

(c) We still have $I_L = 4.4$ A (rms); however, now $\phi_\phi = 0°$. Thus,

$$P_{AB} = 220 \times 4.4 \times \cos(0° + 30°) = 838.3 \text{ W}$$

$$P_{CB} = 220 \times 4.4 \times \cos(0° - 30°) = 838.3 \text{ W}$$

$$P_T = 2 \times 838.3 = 1676.6 \text{ W}$$

$$Q_T = 0 \text{ VAR}$$

◀

Exercise 12.16   In the circuit of Figure 12.26 let $V_L = 220$ V (rms). Predict the wattmeter readings for the following phase impedances:

(a) $\mathbf{Z}_\phi = 100(1 + j\sqrt{3})$ Ω

(b) $\mathbf{Z}_\phi = 100\underline{/70°}$ Ω

(c) $\mathbf{Z}_\phi = j100$ Ω

ANSWER   (a) 0 W, 209.6 W; (b) −84.05 W, 370.8 W; (c) −242 W, 242 W.

▶ Example 12.18

Let the wattmeter readings in the balanced system of Figure 12.26 be $P_{AB} = 1000$ W and $P_{CB} = 200$ W. If the line voltage is $V_L = 220$ V (rms), find

(a) The per-phase average power P_ϕ

(b) The per-phase reactive power Q_ϕ

(c) The power factor *pf*

(d) The phase impedance $\mathbf{Z}_\phi$. Is it inductive or capacitive?

Solution

(a) $P_\phi = (P_{AB} + P_{CB})/3 = (1000 + 200)/3 = 400$ W.

(b) $Q_\phi = \sqrt{3}(P_{CB} - P_{AB})/3 = (200 - 1000)/\sqrt{3} = -461.9$ VAR.

(c) $pf = \cos[\tan^{-1}(Q_\phi/P_\phi)] = \cos[\tan^{-1}(-461.9/400)] = 0.655$, leading.

(d) Using Equation (12.12) we obtain $R_\phi = (V_L \cos\phi_\phi)^2/P_\phi = (220 \times 0.655)^2/400 = 51.86$ Ω. Moreover, by the similarity between the impedance triangle and the power triangle, we can write $X_\phi/R_\phi = Q_\phi/P_\phi$, or $X_\phi = R_\phi Q_\phi/P_\phi = 51.86 \times (-461.9/400) = -59.88$ Ω. Hence, $\mathbf{Z}_\phi = 51.86 - j59.88$ Ω, indicating a capacitive phase impedance.

◀

Exercise 12.17   In the balanced system of Figure 12.26 let $Z_\phi = 10 + j10 \ \Omega$. If the top wattmeter reading is $P_{AB} = 600$ W, find the bottom wattmeter reading P_{CB}, as well as the line voltage V_L and line current I_L.

ANSWER   2239 W, 137.6 V (rms), and 16.85 A (rms).

12.6 SPICE ANALYSIS OF 3-φ SYSTEMS

The ac SPICE facilities introduced in the previous chapter can readily be extended to the analysis of three-phase systems. In fact, SPICE can be used not only to verify calculations for *balanced systems,* but also to investigate *unbalanced systems,* whose hand analysis is usually more laborious.

▶**Example 12.19**

Use SPICE to verify the results of Example 12.12. Assume the operating frequency is 60 Hz.

Solution

To make the circuit acceptable to SPICE we must express its impedances in terms of actual R, L, and C element values. The data of Example 12.12 indicate that all impedances are *inductive.* Since $\omega = 2\pi 60 = 377$ rad/s, the source, line, and load inductances are, respectively, $0.2/377 = 0.5305$ mH, $1/377 = 2.653$ mH, and $10/377 = 26.53$ mH. For the circuit representation of Figure 12.27, the input file is

```
BALANCED Y-Y SYSTEM
VSA 11 0 AC 120 0
VSB 21 0 AC 120 -120
VSC 31 0 AC 120 120
RSA 11 12 0.1
RSB 21 22 0.1
RSC 31 32 0.1
LSA 12 13 0.5305M
LSB 22 23 0.5305M
LSC 32 33 0.5305M
RAA 13 14 0.5
RBB 23 24 0.5
RCC 33 34 0.5
LAA 14 15 2.653M
LBB 24 25 2.653M
LCC 34 35 2.653M
RAN 15 16 15
RBN 25 26 15
RCN 35 36 15
```

```
LAN 16 1 26.53M
LBN 26 1 26.53M
LCN 36 1 26.53M
.AC LIN 1 60 60
.PRINT AC IM(VSA) VM(15,1) VM(13) VM(15,25) VM(13,23)
.PRINT AC IP(VSA) VP(15,1) VP(13) VP(15,25) VP(13,23)
.PRINT AC IM(VSB) VM(25,1) VM(23) VM(25,35) VM(23,33)
.PRINT AC IP(VSB) VP(25,1) VP(23) VP(25,35) VP(23,33)
.PRINT AC IM(VSC) VM(35,1) VM(33) VM(35,15) VM(33,13)
.PRINT AC IP(VSC) VP(35,1) VP(33) VP(35,15) VP(33,13)
.PRINT AC VM(1)
.END
```

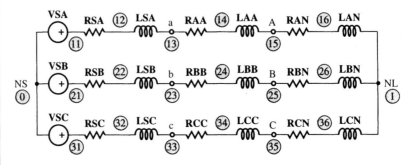

Figure 12.27 Balanced Y-Y system labeled for SPICE analysis.

The last .PRINT statement is used to obtain the voltage at the load neutral which, for a balanced system, is expected to be zero. After SPICE is run, the output file contains:

```
 IM(VSA)      VM(15,1)    VM(13)      VM(15,25)   VM(13,23)
6.248E+00    1.126E+02   1.188E+02   1.951E+02   2.057E+02

 IP(VSA)      VP(15,1)    VP(13)      VP(15,25)   VP(13,23)
1.443E+02    -1.986E+00  -3.139E-01  2.801E+01   2.969E+01

 IM(VSB)      VM(25,1)    VM(23)      VM(25,35)   VM(23,33)
6.248E+00    1.126E+02   1.188E+02   1.951E+02   2.057E+02

 IP(VSB)      VP(25,1)    VP(23)      VP(25,35)   VP(23,33)
2.432E+01    -1.220E+02  -1.203E+02  -9.199E+01  -9.031E+01

 IM(VSC)      VM(35,1)    VM(33)      VM(35,15)   VM(33,13)
6.248E+00    1.126E+02   1.188E+02   1.951E+02   2.057E+02

 IP(VSC)      VP(35,1)    VP(33)      VP(35,15)   VP(33,13)
-9.568E+01   1.180E+02   1.197E+02   1.480E+02   1.497E+02

 VM(1)
1.421E-14
```

These data agree with the results of hand calculations, except for an apparent discrepancy in the phase angles of the line currents, which in

Example 12.12 were assumed to flow out of the sources. As we know, SPICE assumes a voltage source current that flows from the positive terminal, through the source, and to the negative terminal, indicating a 180° discrepancy. For instance, in the preceding listing we have IP(VSA) = 1.443E+02 = 144.3°. Then, $\angle I_{aA} = $ IP(VSA) $- 180° = 144.3 - 180 = -35.7°$, in full agreement with Example 12.12. ◀

Exercise 12.18   Use SPICE to verify the results of Example 12.14, dealing with a balanced Y-Δ system. Assume the operating frequency is 60 Hz.

▶ **Example 12.20**

Use SPICE to find the load voltages in the Y-Y system of Figure 12.27 for the case of slightly *unbalanced* load impedances, $Z_{AN} = 16 + j11$ Ω, $Z_{BN} = 14 + j9$ Ω, and $Z_{CN} = 17 + j8$ Ω.

Solution

The input file is now

```
UNBALANCED Y-Y SYSTEM
VSA 11 0 AC 120 0
VSB 21 0 AC 120 -120
VSC 31 0 AC 120 120
RSA 11 12 0.1
RSB 21 22 0.1
RSC 31 32 0.1
LSA 12 13 0.5305M
LSB 22 23 0.5305M
LSC 32 33 0.5305M
RAA 13 14 0.5
RBB 23 24 0.5
RCC 33 34 0.5
LAA 14 15 2.653M
LBB 24 25 2.653M
LCC 34 35 2.653M
*UNBALANCED LOAD
RAN 15 16 16
RBN 25 26 14
RCN 35 36 17
LAN 16 1 29.18M
LBN 26 1 23.87M
LCN 36 1 21.22M
.AC LIN 1 60 60
.PRINT AC VM(15) VM(25) VM(35) VM(1)
.END
```

After SPICE is run, the output file contains

```
VM(15)      VM(25)      VM(35)      VM(1)
1.129E+02 1.129E+02 1.130E+02 1.089E+01
```

Compared with Example 12.19, we observe that the line voltages have changed from 112.7 V to 112.9 V, 112.9 V, and 113.0 V, and the voltage of the load neutral with respect to the source neutral is now 10.89 V instead of 0 V. ◀

Exercise 12.19   Use SPICE to investigate the behavior of the Y-Δ system of Exercise 12.18 for the case of heavily unbalanced load impedances, $Z_{AB} = 90 + j60$ Ω, $Z_{BC} = 50 - j40$ Ω, and $Z_{CA} = 70$ Ω. Assume the operating frequency is 60 Hz.

▼ SUMMARY

- The *instantaneous ac power* delivered by a source to a load is $p(t) = (1/2)V_m I_m \cos\phi + (1/2)V_m I_m \cos(2\omega t + \phi)$, where V_m and I_m are the peak voltage and current amplitudes, and ϕ is the phase angle of voltage relative to current.

- The *average ac power* is $P = (1/2)V_m I_m \cos\phi = V_{rms}I_{rms}\cos\phi$. The term $\cos\phi$ is called the *power factor pf*.

- The average power delivered to an impedance $Z = R(\omega) + jX(\omega)$ is $P = R(\omega)I_{rms}^2$, indicating that the ac resistance $R(\omega)$ represents the ability of an impedance to dissipate power.

- The average power delivered to an impedance $Z = Z_m\underline{/\phi}$ can also be expressed in the alternate form $P = (V_{rms}\cos\phi)^2/R(\omega)$.

- Given an ac port with open-circuit voltage V_{oc} and equivalent impedance Z_{eq}, and given a load Z_L, power transfer from the port to the load is *maximized* when $Z_L = Z_{eq}^*$, that is, when the load is made equal to the *complex conjugate* of the port impedance. We then have $P_{(max)} = |V_{oc}|^2/8R_{eq}$.

- *Complex power* is defined as $S = P + jQ = S\underline{/\phi}$, where $S = V_{rms}I_{rms}$ is called the *apparent power* and is expressed in *volt-amperes* (VA); $P = S\cos\phi$ is called the *real power* and is expressed in *watts* (W); and $Q = S\sin\phi$ is called the *reactive power* and is expressed in *volt-amperes reactive* (VAR). These relationships are easily visualized in terms of the *ac power triangle*.

- We have $P = R(\omega)I_{rms}^2$, $Q = X(\omega)I_{rms}^2$, and $S = Z(\omega)I_{rms}^2$. Moreover, the complex, real, and reactive powers of a source equal the sums of the complex, real, and reactive powers of the individual impedances comprising the load.

- *Power-factor correction* is the process of adding suitable reactive elements either in parallel or in series with a load to make the power factor of the composite load approach unity.

• The generation, transmission, and utilization of large banks of electric power is accomplished by means of *three-phase* (3-ϕ) *systems* because of their ability to ensure *continuous* power flow as opposed to the *pulsating* power flow of single-phase (1-ϕ) systems.

• A balanced 3-ϕ source produces three sinusoidal voltages having equal amplitude and frequency, and differing in phase angle by 120°. At any instant, the sum of the three voltages is zero.

• Depending on the manner in which its windings are interconnected, a 3-ϕ source may be of the Y or Δ type. Likewise, a 3-ϕ load may be of the Y or Δ type. A source or a load of one type can always be converted to the other via suitable transformations.

• The most common 3-ϕ systems are Y-Y and Y-Δ. The Y-Y system may have a neutral line. If the system is balanced, this line carries zero current. Y-Δ systems have no neutral line. Balanced 3-ϕ systems may be analyzed on a per-phase basis using a simple 1-ϕ equivalent.

• Power is measured by means of the *wattmeter*. The *i*-coil, acting as an ammeter, is connected in *series* with the load; the *v*-coil, acting as a voltmeter, is connected in *parallel*.

• Power measurements in balanced 3-ϕ systems via the *two-wattmeter method* use *line* voltage and current measurements to indirectly obtain P_T, Q_T, and *pf*.

• SPICE provides a powerful tool for the analysis of 3-ϕ systems, especially when they are *unbalanced*.

▼ PROBLEMS

12.1 AC Power

12.1   A source $v_s = 15 \cos 10^3 t$ V, a resistance $R = 10\ \Omega$, an inductance $L = 25$ mH, and a capacitance $C = 100\ \mu$F are connected in series. Calculate the instantaneous power (absorbed or delivered?) by each element at $t = 0$. Hence, verify the conservation of power.

12.2   Verify that the average power supplied by the source in Figure P12.2 equals the sum of the average powers dissipated by the resistances.

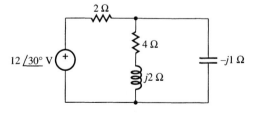

Figure P12.2

12.3   (a) Find S, P, and *pf* for the circuit of Figure P12.3. Is *pf* leading or lagging? (b) Repeat, but with the capacitance removed from the circuit. Comment on your results.

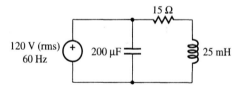

Figure P12.3

12.4   Find the average power absorbed by each impedance in Figure P12.4. Hence, verify that the total absorbed power equals that released by the source.

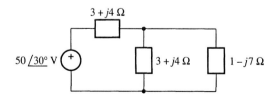

Figure P12.4

12.5 Verify the conservation of average power in the circuit of Figure P12.5.

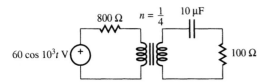

Figure P12.5

12.6 Find the average power dissipated by the resistance of Figure P12.6 if $v_s = 120 \cos 10^3 t$ V.

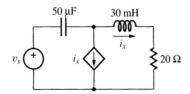

Figure P12.6

12.7 A source $v_s = 120 \cos 2\pi 400 t$ V drives a load that dissipates an average power of 100 W with a lagging power factor of 0.85. Find simple series and parallel equivalents for the load.

12.8 Find the average power dissipated by the 2-kΩ load in Figure P12.8.

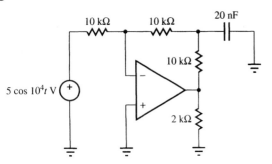

Figure P12.8

12.9 A source $v_s = 100 \cos \pi 100 t$ V with a series impedance of $60 + j80$ Ω drives a 60-Ω load. (a) Find the average power absorbed by the load. (b) Specify a reactive element that, when connected in series with the load, will maximize its average power. Compare this power with that of part (a), and comment.

12.10 If an active ac port delivers the current $i = \cos 10^3 t$ A when terminated on a 10-Ω resistance, and $i = 2 \cos 10^3 t$ A when terminated on a 100-μF capacitance, find the load that will absorb the maximum average power from this ac port, and find this power.

12.11 If an active ac port yields $V = 120\underline{/0°}$ V when open-circuited, and $V = 60\underline{/16.26°}$ V when terminated on a load

of $30 + j40$ Ω, find the load that will absorb the maximum average power from this ac port, and find this power.

12.12 In the circuit of Figure P12.12 find the maximum average power that can be transferred to Z_L as well as Z_L itself.

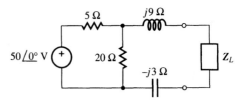

Figure P12.12

12.13 Repeat Problem 12.12 for the circuit of Figure P12.13.

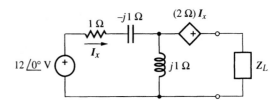

Figure P12.13

12.14 Repeat Problem 12.12 for the circuit of Figure P12.14.

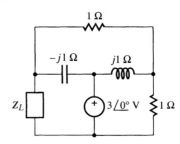

Figure P12.14

12.15 Show that the average power P_L transferred to a load Z_L for a given $\measuredangle Z_L$, not necessarily equal to $-\measuredangle Z_{eq}$, is maximized when $|Z_L|$ is made equal to $|Z_{eq}|$. (*Hint:* Derive an expression for P_L in terms of $Z_L = |Z_L|(\cos \measuredangle Z_L + j \sin \measuredangle Z_L)$ and impose $dP_L/d|Z_L| = 0$.)

12.16 (a) Using the results of Problem 12.15, find a purely resistive load that will dissipate as much average power as possible in the circuit of Figure P12.12. What is this power? (b) Find n for which the transformer of Figure P12.5 will transfer as much average power as possible. What is this power?

12.2 Complex Power

12.17 (a) Show that the maximum and minimum values of the instantaneous power are $p_{max} = P + \sqrt{P^2 + Q^2}$ and $p_{min} = P - \sqrt{P^2 + Q^2}$. (b) Find p_{max} and p_{min} for the household ac line driving a load that consists of a 10-Ω resistance in series with a 35-mH inductance.

12.18 If the instantaneous power associated with a load is $p(t) = 10 + 10\cos 2\pi 100t + 17.32 \sin 2\pi 100t$ VA, find (a) the source frequency, in Hz, (b) P, (c) Q, (d) S, and (e) the maximum and the minimum instantaneous power.

12.19 Let the voltage and current at the terminals of a certain ac port be $v = 120\cos(\omega t + 30°)$ V, and $i = 5\sin \omega t$ A. Assuming the passive sign convention, find the real power and the reactive power. Is the port delivering or absorbing average power? Reactive power?

12.20 In the arrangement of Figure P12.20 find P and Q, and state whether average power flow and reactive power flow is from System 1 to System 2 or vice versa if

(a) $v = 120\cos(\omega t + 60°)$ V, $i = 2\cos \omega t$ A
(b) $v = 60\cos(\omega t + 30°)$ V, $i = 3\cos(\omega t + 90°)$ A
(c) $v = 240\cos \omega t$ V, $i = 15\cos(\omega t + 150°)$ A
(d) $v = 10\cos(\omega t + 60°)$ V, $i = 1\sin(\omega t + 60°)$ A

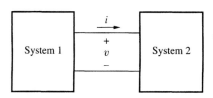

Figure P12.20

12.21 (a) Find the real, reactive, and complex power delivered by the source in Figure P12.21. (b) What reactive element must be connected in parallel with the source to make its power factor 1?

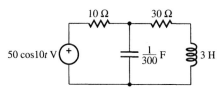

Figure P12.21

12.22 A source $v_s = 120\sqrt{2}\cos 377t$ V supplies a load with the complex power $S = 300 + j400$ VA. What element must be connected in parallel with the load to ensure (a) $pf = 1$? (b) $pf = 0.9$ lagging?

12.23 A source $v_s = 120\sqrt{2}\cos 2\pi 50t$ V drives a series *RLC* circuit having $R = 8$ Ω, $L = 50$ mH, and $C = 500$ μF. What reactive element must be connected directly across the source to achieve $pf = 0.95$, lagging? $pf = 0.95$, leading?

12.24 A load absorbs 10 kW at 220 V (rms) and $pf = 0.8$, lagging. Find the impedance as well as the complex power of the load.

12.25 A source V_s supplies ac power to a load Z_L via a series impedance $Z_s = 5 + j2$ Ω. Find V_s if the load absorbs 1 kW at $V_L = 220/0°$ V (rms) and $pf = 0.8$, leading.

12.26 A source $V_s = 100/0°$ V (rms) supplies ac power to a load $Z_L = 28 + j25$ Ω via a distribution line with impedance $Z_{line} = 2 + j5$ Ω. (a) Find the voltage V_L and current I_L at the load. (b) Verify the conservation of ac power.

12.27 A source supplies ac power to an inductive load via a distribution line with impedance $Z_{line} = 0.1 + j0.3$ Ω. If the load requires 10 kW at $V_L = 220/0°$ V (rms) and the power loss on the line is 0.5 kW, find the complex power delivered by the source.

12.28 Assuming a source frequency of 60 Hz in Problem 12.27, find the capacitance bank that must be connected in parallel with the load to raise its power factor to 0.9, lagging. What is the load voltage after power-factor correction?

12.29 Three loads are in parallel. Load 1 absorbs 100 kW and 40 kVAR; Load 2 absorbs 25 kVA at $pf = 0.9$, leading; Load 3 absorbs 80 kW at $pf = 1$. Find the equivalent impedance and power factor of the parallel combination if its voltage is 240 V (rms).

12.30 Three loads are in parallel and absorb, respectively, $S_1 = 16 + j20$ VA, $S_2 = 12 - j10$ VA, and $S_3 = 5 + j15$ VA from a $120/0°$ V (rms) source. Find the current supplied by the source as well as the overall power factor seen by the source.

12.31 Two loads are in parallel and absorb a total of 5 kW at 120 V (rms), 60 Hz, and $pf = 0.8$, lagging. If Load 1 is known to absorb 2 kW at $pf = 0.7$, lagging, find (a) the power factor of Load 2 and (b) the parallel reactive element necessary to raise the combined power factor to 0.9, lagging.

12.32 Two loads are in parallel and are powered by an ac source V_s via a distribution line having impedance $Z_{line} = 0.1 + j0.25$ Ω. The common load voltage is $V_L = 220/0°$ V (rms). Using complex power calculations, find V_s if Load 1 absorbs 20 kW at $pf = 0.75$, lagging, and Load 2 absorbs 50 kW at $pf = 0.8$, lagging.

12.33 Predict the wattmeter reading in Figure P12.33.

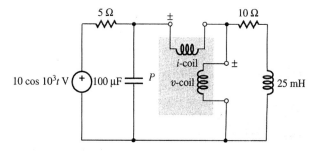

Figure P12.33

12.34   (a) Predict the wattmeter reading in Figure P12.34. (b) Repeat if the voltage is changed to $12\underline{/0°}$ V and the current source to $6\underline{/60°}$ A. Comment.

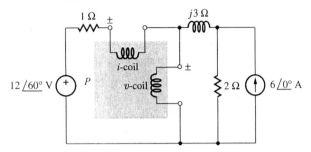

Figure P12.34

12.35   Predict the wattmeter reading in Figure P12.35.

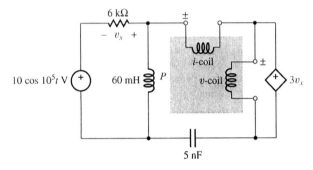

Figure P12.35

12.3 Three-Phase Systems

12.36   Find the phase sequence of each of the following sets:
(a)

$$v_a = 120\cos(\omega t + 75°) \text{ V}$$

$$v_b = 120\sin(\omega t + 45°) \text{ V}$$

$$v_c = 120\cos(\omega t - 165°) \text{ V}$$

(b)

$$v_a = 208\cos(\omega t + 60°) \text{ V}$$

$$v_b = -208\cos\omega t \text{ V}$$

$$v_c = 208\cos(\omega t - 60°) \text{ V}$$

12.37   Given a balanced Y-connected source with a positive phase sequence, find (a) the line voltages if $V_{an} = 120\underline{/90°}$ V (rms) and (b) the phase voltages if $V_{ab} = 208\underline{/60°}$ V (rms).

12.38   Given a balanced Y-connected load, (a) find V_{BC} if $V_{AN} = 120\underline{/60°}$ V (rms) and the phase sequence is positive and (b) V_{AB} if $V_{CN} = 240\underline{/135°}$ V (rms) and the phase sequence is negative.

12.39   Find R_{eq} in the circuit of Figure P12.39(a). To avoid node or loop equations, apply a Y-Δ transformation to the 2-Ω, 6-Ω, and 3-Ω resistances to obtain the equivalent of Figure P12.39(b), then apply series/parallel reductions.

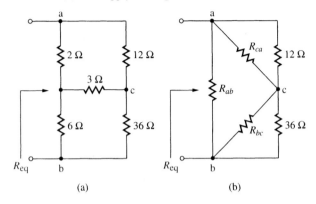

Figure P12.39

12.40   Find R_{eq} in the circuit of Figure P12.40(a). To avoid node or loop equations, apply a Δ-Y transformation to the 2-Ω, 3-Ω, and 24-Ω resistances to obtain the equivalent of Figure P12.40(b), then apply series/parallel reductions.

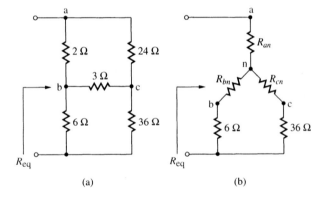

Figure P12.40

12.41   Find Z in the circuit of Figure P12.41 using a Y-Δ transformation.

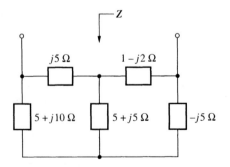

Figure P12.41

12.42 Find $\mathbf{Z}$ in the circuit of Figure P12.41 using a Δ-Y transformation.

12.43 (a) Find the Δ-equivalent of a Y-connected load having $\mathbf{Z}_{AN} = 10\ \Omega$, $\mathbf{Z}_{BN} = j2\ \Omega$, and $\mathbf{Z}_{CN} = -j2\ \Omega$. (b) Find the Y-equivalent of a Δ-connected load having $\mathbf{Z}_{AB} = 10 + j10\ \Omega$, $\mathbf{Z}_{BC} = 10\ \Omega$, and $\mathbf{Z}_{CA} = -j10\ \Omega$.

12.44 If in Figure 12.20 $\mathbf{Z}_\Delta = 18 + j6\ \Omega$, find $\mathbf{Z}_Y$ so that $\mathbf{Z}_{AB} = \mathbf{Z}_{BC} = \mathbf{Z}_{CA} = 4 + j2\ \Omega$. What is the Y-equivalent of the composite load?

12.45 Three balanced loads are connected in parallel. The first is a Δ-connected load with $\mathbf{Z}_\Delta = j20\ \Omega$; the second is a Y-connected load with $\mathbf{Z}_Y = 20\ \Omega$; the third is a Δ-connected load with $\mathbf{Z}_\Delta = -j30\ \Omega$. Find a Y-equivalent as well as a Δ-equivalent for the composite load.

12.46 Two balanced Δ-connected loads are in parallel and are subjected to a balanced voltage set with a per-phase magnitude of 220 V (rms). If the first load draws 20 kW per phase at $pf = 0.8$, lagging, and the second load draws 10 kW per phase at $pf = 0.9$, leading, find a Δ-equivalent as well as a Y-equivalent for the composite load.

12.4 Y-Y and Y-Δ Systems

12.47 A balanced Y-connected load with a per-phase impedance of $30 + j20\ \Omega$ is powered by a balanced source via a distribution line having a per-phase impedance of $0.5 + j1\ \Omega$. If the line voltage at the load is 400 V (rms) find (a) the line current and (b) the line voltage at the source end of the line.

12.48 A balanced, positive-phase-sequence Y-connected source with $\mathbf{V}_{an} = 120\underline{/0^\circ}$ V (rms) supplies power to a balanced Y-connected load. If the per-phase load impedance is $20 + j10\ \Omega$ and the per-phase impedance of the distribution line is $1 + j2\ \Omega$, find (a) the line currents and (b) the phase voltages at the load.

12.49 A balanced Y-connected source with $\mathbf{V}_{an} = 120\underline{/0^\circ}$ V (rms) supplies power to a balanced Y-connected load with a per-phase impedance of $8 + j4\ \Omega$. If $\mathbf{V}_{AN} = 110\underline{/-1.50^\circ}$ V (rms), find the per-phase impedance of the power-distribution line.

12.50 The power-distribution line of a balanced Y-Y system has a per-phase impedance of $0.6 + j0.3\ \Omega$. If the per-phase load impedance is $10 + j15\ \Omega$ and the per-phase power loss on the line is 60 W, find the per-phase power absorbed by the load as well as the per-phase complex power supplied by the source.

12.51 A balanced, positive-phase-sequence Y-connected source with $\mathbf{V}_{an} = 120\underline{/0^\circ}$ V (rms) is connected by four perfect conductor lines to an unbalanced Y-connected load having $\mathbf{Z}_{AN} = 10\ \Omega$, $\mathbf{Z}_{BN} = 6 + j8\ \Omega$, and $\mathbf{Z}_{CN} = 8 - j6\ \Omega$. Find the current on the neutral line.

12.52 If the neutral line in the Y-Y system of Problem 12.51 is removed, find $\mathbf{V}_{Nn}$, the voltage of the load neutral node referenced to that of the source neutral node.

12.53 A balanced Δ-connected load with a phase impedance of $50 - j20\ \Omega$ is powered by a balanced, positive-phase-sequence source via a distribution line having a per-phase impedance of $0.5 + j1\ \Omega$. If $\mathbf{V}_{AB} = 208\underline{/0^\circ}$ V (rms), find (a) the line currents and (b) the line voltages at the source end of the line.

12.54 A balanced Δ-connected load with $\mathbf{Z}_\Delta = 30 + j15\ \Omega$ is in parallel with a balanced Y-connected load having $\mathbf{Z}_Y = 15\ \Omega$ and $\mathbf{V}_{AN} = 240\underline{/0^\circ}$ V (rms). The composite load is powered by a balanced, negative-phase-sequence source via a distribution line having a per-phase impedance of $1 + j2\ \Omega$. Find (a) the line currents, (b) the line voltages at the source end of the line, (c) the phase currents in the Δ-connected load, and (d) the phase currents in the Y-connected load.

12.55 A balanced, negative-phase-sequence Y-Δ system has a per-phase distribution-line impedance of $1 + j0.5\ \Omega$ and a per-phase load impedance of $20 + j10\ \Omega$. If $\mathbf{V}_{an} = 120\underline{/0^\circ}$ V (rms), find (a) the line currents and (b) the phase voltages at the load.

12.56 In a balanced, negative-phase-sequence Y-Δ system each phase has a winding impedance of $0.5\underline{/60^\circ}\ \Omega$, a distribution-line impedance of $0.5 + j0.3\ \Omega$ and a load impedance of $10 + j8\ \Omega$. If $\mathbf{I}_{AB} = 10\underline{/30^\circ}$ A (rms), find $\mathbf{V}_{sa}$, $\mathbf{V}_{sb}$, and $\mathbf{V}_{sc}$.

12.57 A balanced Y-Δ system has $\measuredangle\,\mathbf{V}_{an} = 20^\circ$ and $\mathbf{I}_{aA} = 4\underline{/-10^\circ}$ A (rms). If the load absorbs 400 W per phase with a lagging power factor of 0.875 and the distribution line absorbs 20 W per phase, find (a) $\mathbf{V}_{an}$, (b) $\mathbf{Z}_{\text{line}}$, (c) $\mathbf{Z}_\Delta$, and (d) $\mathbf{V}_{AB}$.

12.58 A balanced, positive-phase-sequence Y-connected source with $\mathbf{V}_{an} = 120\underline{/0^\circ}$ V (rms) and negligible winding impedance supplies power to an unbalanced Δ-connected load via a distribution line having a per-phase impedance of $1 + j2\ \Omega$. Find the line currents if $\mathbf{Z}_{AB} = 30 + j40\ \Omega$, $\mathbf{Z}_{BC} = 40 - j30\ \Omega$, and $\mathbf{Z}_{CA} = 50\ \Omega$. (*Hint:* Transform the load into its Y-equivalent.)

12.5 Power in Three-Phase Systems

12.59 Show that in a balanced 3-ϕ system the total instantaneous power absorbed by a load is invariant with time and is given by $p = 3V_\phi I_\phi \cos\phi_\phi$, where V_ϕ and I_ϕ are the rms values of the phase voltage and current at the load, and ϕ_ϕ is the phase difference between the two.

12.60 A balanced Y-Y system has $\mathbf{V}_{an} = 120\underline{/45^\circ}$ V (rms) and $\mathbf{Z}_Y = 50\underline{/60^\circ}\ \Omega$. If the power-distribution line has a per-phase impedance of $2\underline{/50^\circ}\ \Omega$, find (a) the total complex power delivered to the load and (b) the percentage of the average power at the sending end of the line that is delivered to the load.

12.61   A balanced Y-Δ system has $V_{an} = 120\underline{/0°}$ V (rms) and $Z_\Delta = 40 + j30$ Ω. If the power-distribution line has a per-phase impedance of $0.5 + j1$ Ω, find (a) the total complex power delivered to the load and (b) the percentage of the average power at the sending end of the line that is delivered to the load.

12.62   A balanced load draws a complex power of $1200 + j900$ kVA at a line voltage of 4 kV (rms), 60 Hz. (a) Assuming a Δ-connected load, find a simple parallel equivalent for its phase impedance. (b) Assuming a Y-connected load, find a simple series equivalent for its phase impedance.

12.63   (a) What Δ-connected capacitor bank must be connected in parallel with the load of Problem 12.62 to achieve $pf = 1$? Assume the line voltage remains unchanged. (b) Repeat, but for a Y-connected bank.

12.64   A balanced 3-ϕ load draws a total of 20 kW from a balanced 3-ϕ source operating at 60 Hz. If the line voltage and current at the load are, respectively, 220 V (rms) and 70 A (rms), find (a) the complex power drawn by the load, (b) the power factor of the load, and (c) a Δ-connected capacitor bank that will raise the power factor of the load to 0.95, lagging.

12.65   A balanced load absorbs 360 kW at $pf = 0.8$, lagging. If the load is fed by a balanced source via a power-distribution line with a per-phase impedance of $0.12 + j0.04$ Ω and the line voltage at the load is 660 V (rms), find the power factor at the source.

12.66   A balanced source supplies 30 kVA at $pf = 0.8$, lagging, to a balanced load. If the line voltage at the source is 240 V (rms) and the distribution line has a per-phase impedance of $0.3 + j0.4$ Ω, find (a) the rms value of the line current, (b) the rms value of the line voltage at the load, and (c) the complex power at the load.

12.67   A balanced source supplies power to three balanced loads via a power-distribution line having a per-phase impedance of $0.05 + j0.1$ Ω. Load 1 absorbs 100 kVA at $pf = 0.85$, lagging; Load 2 absorbs 25 kVAR at $pf = 0.75$, leading; Load 3 dissipates 20 kW at $pf = 1$. If the line voltage at the load is 240 V (rms), find the line voltage and pf at the source.

12.68   A balanced source supplies power to three balanced loads via a power-distribution line having a per-phase impedance of $1 + j2$ Ω. Load 1 absorbs 100 kVA at $pf = 0.95$, leading; Load 2 absorbs 125 kW at $pf = 0.8$, lagging; Load 3 draws $150 + j50$ kVA. If the line voltage at the load is 3 kV

(rms), find the power factor at the source and the load, as well as the efficiency of the system.

12.69   A balanced Y-connected load is powered by a balanced source and draws a total of 120 kW at $pf = 0.8$, lagging. The line voltage at the load is 208 V (rms), 60 Hz. (a) If three 5-mF capacitances are connected in parallel with the load in a Y configuration, find the power factor of the combined load. (b) Repeat, but with the capacitances connected in a Δ configuration. Compare and comment.

12.70   A Δ-connected load with $Z_{AB} = 4 + j3$ Ω, $Z_{BC} = 12 - j8$ Ω, and $Z_{CA} = 10\underline{/45°}$ Ω is subjected to a balanced set of line voltages of 720 V (rms) each. Find the complex power of each phase as well as the complex power of the entire load.

12.71   Predict the wattmeter readings of Figure 12.26 if $V_{ab} = 240\underline{/-30°}$ V (rms), $V_{cb} = 240\underline{/-90°}$ V (rms), and $Z_\phi = 80 + j40$ Ω.

12.72   Let the wattmeter readings of Figure 12.26 be $P_{AB} = 1000$ W and $P_{CB} = -500$ W. If the line voltage is 208 V (rms), find (a) the total average power, (b) the power factor, and (c) the phase impedance of the load. Assume the system is balanced.

12.73   In the two-wattmeter arrangement of Figure 12.26 let the line voltage be 208 V (rms). Find the phase impedance of the load if the wattmeter readings are (a) $P_{AB} = P_{CB} = 1$ kW, (b) $P_{AB} = -P_{CB} = 1$ kW, (c) $P_{AB} = 0$ W, $P_{CB} = 1$ kW, (d) $P_{AB} = 1$ kW, $P_{CB} = 0$, and (e) $P_{AB} = 2P_{CB} = 1$ kW. Assume the system is balanced.

12.74   Assuming a positive-phase-sequence, balanced set of line voltages in Figure P12.74, show that the wattmeter reading P is proportional to the per-phase reactive power Q_ϕ associated with the load: $P = \sqrt{3}Q_\phi$.

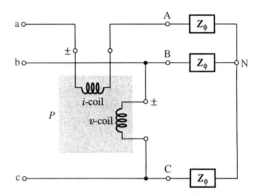

Figure P12.74

12.75 In the circuit of Figure P12.74 let $Z_\phi = 24 + j32\ \Omega$ and let the line voltages form a positive-phase-sequence, balanced set with a magnitude of 720 V (rms) and a frequency of 60 Hz. (a) Predict the wattmeter reading. (b) Specify a balanced Δ-connected load that, if connected in parallel with the existing Y-connected load, will drive the wattmeter reading to zero.

12.76 In Figure 12.26 the line voltages form a positive-phase-sequence, balanced set with a magnitude of 208 V (rms); however, the load is unbalanced and $Z_{AB} = 20 + j15\ \Omega$, $Z_{BC} = 30 - j40\ \Omega$, and $Z_{CA} = 25\underline{/-45^\circ}\ \Omega$. (a) Predict the

wattmeter readings. (b) Verify that the sum of the readings still equals the total power delivered to the load.

▼ 12.6 SPICE Analysis of 3-ϕ Systems

12.77 Use SPICE to predict the i-coil current and the v-coil voltage in the circuit of Figure P12.35. Hence, use the SPICE results to calculate the wattmeter reading.

12.78 Solve Problem 12.48 via SPICE.

12.79 Solve Problems 12.51 and 12.52 via SPICE.

12.80 Check Problem 12.54 via SPICE.

12.81 Solve Problem 12.76 via SPICE.

AC RESONANCE

I n Chapter 9 we found that *RLC* circuits possess the distinctive ability to provide natural responses of the oscillatory type. This ability stems from the flow of energy back and forth between the capacitance and the inductance. It is not surprising that *RLC* circuits also exhibit a distinctive ac behavior. This stems from the fact that inductance and capacitance have *dual* ac characteristics: The inductive reactance $X_L = \omega L$ is *positive* and is *directly* proportional to ω, whereas the capacitive reactance $X_C = -1/\omega C$ is *negative* and *inversely* proportional to ω. Depending on frequency, X_L or X_C may dominate, or the two reactances may *cancel* each other out, making a *series LC* combination act as a *short* circuit, and a *parallel LC* combination as an *open*. This cancellation provides the basis for a form of resonance known as **unity power-factor resonance** to reflect the fact that if the series or parallel *LC* combination is embedded in a resistive circuit, at resonance all voltages and currents in the circuit will be *in phase* with the applied source, thus yielding a power factor of unity.

Another distinguishing feature of resonant *RLC* circuits is the ability to provide a form of signal magnification known as **resonance signal rise,** a phenomenon not found in other passive ac circuits, where all signals in the circuit are of lesser magnitude than the applied signal.

In this chapter we first investigate the *series* and *parallel RLC* circuits, and we characterize their behavior at resonance in terms of the *resonance frequency* ω_0 and the *quality factor Q*. We also investigate the effect of practical coils upon the *Q* of a circuit.

Next, we examine a resonant circuit consisting of resistances and capacitances, but no inductances, and find that it always has $Q < 0.5$. However, with the inclusion of an op amp to provide a controlled amount of *positive feedback,* it is possible to change the *Q* of the circuit to any value, including $Q = \infty$ or even $Q < 0$. With $Q = \infty$ the circuit yields a *sustained oscillation,* and with $Q < 0$ it yields a *diverging oscillation.*

We conclude by discussing magnitude and frequency scaling.

13.1 SERIES RESONANCE

In this section we investigate the phenomenon of ac resonance for the case of
a capacitance and inductance connected in *series*. Once we reduce the remainder of the circuit to its Thévenin equivalent, we end up with the *series RLC*
configuration of Figure 13.1(a). The impedance seen by the source is

$$Z = R + j\left(\omega L - \frac{1}{\omega C}\right) = R + jX \qquad (13.1)$$

$$X = \omega L - \frac{1}{\omega C} \qquad (13.2)$$

where X is the *net reactance* of the circuit. Depending on the frequency ω of
the applied source, we have three possibilities:

(1) $\omega L < 1/\omega C$, or $X < 0$, indicating *capacitive* behavior.

(2) $\omega L > 1/\omega C$, or $X > 0$, indicating *inductive* behavior.

(3) $\omega L = 1/\omega C$, or $X = 0$. This is the borderline between the two preceding cases. Since we now have $Z = R + j0 = R$, the *RLC* circuit
behaves *resistively,* with I and V *in phase* with each other. This behavior, referred to as **unity power-factor resonance,** occurs at the special
frequency ω_0 that makes $\omega_0 L = 1/\omega_0 C$, or

$$\boxed{\omega_0 = \frac{1}{\sqrt{LC}}} \qquad (13.3)$$

For obvious reasons, ω_0 is called the **resonance frequency.** Even
though the individual impedances Z_L and Z_C are not zero, at resonance they *cancel* each other out, making the *LC series combination*
act as a *short circuit.* Figure 13.2 depicts the signals in the circuit at
resonance.

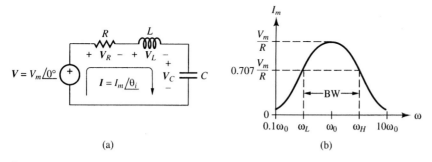

Figure 13.1 Series *RLC* circuit and frequency response of its current.

Frequency Response

We find it instructive to reexamine the frequency response of the series *RLC*
circuit using phasor analysis. To simplify the formalism, we again use the

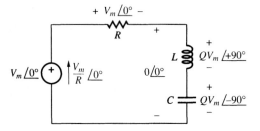

Figure 13.2 Series *RLC* circuit at resonance.

quality factor of the *series RLC circuit,*

$$Q = \frac{1}{R\sqrt{C/L}} \tag{13.4}$$

If this is combined with Equation (13.3), it is straightforward to verify that the reactances can be expressed as

$$\omega L = RQ\frac{\omega}{\omega_0} \tag{13.5a}$$

$$\frac{1}{\omega C} = RQ\frac{\omega_0}{\omega} \tag{13.5b}$$

Substituting into Equation (13.1) yields the more insightful form

$$\mathbf{Z} = R\left[1 + jQ\left(\frac{\omega}{\omega_0} - \frac{\omega_0}{\omega}\right)\right] \tag{13.6}$$

The current response is readily found as $\mathbf{I} = \mathbf{V}/\mathbf{Z}$, or

$$\mathbf{I} = \frac{\mathbf{V}}{R\left[1 + jQ\left(\dfrac{\omega}{\omega_0} - \dfrac{\omega_0}{\omega}\right)\right]} \tag{13.7}$$

Applying Phasor Rule 6 of Section 11.1 gives the amplitude and phase relationships between current and voltage as

$$I_m(\omega) = \frac{V_m}{R}\frac{1}{\sqrt{1 + Q^2\left(\dfrac{\omega}{\omega_0} - \dfrac{\omega_0}{\omega}\right)^2}} \tag{13.8a}$$

$$\theta_i(\omega) = -\tan^{-1}Q\left(\frac{\omega}{\omega_0} - \frac{\omega_0}{\omega}\right) \tag{13.8b}$$

Clearly, the derivations via phasors are much quicker than the time-domain derivations of Section 10.4.

As we know, the plot of I_m versus ω for a given V_m has the familiar bell-shaped profile of the **band-pass function.** Current approaches zero at both ends of the frequency spectrum because of the *open-circuit* action by L at low frequencies and by C at high frequencies. At the borderline frequency $\omega = \omega_0$ the reactances cancel each other out, minimizing $|\mathbf{Z}|$ and maximizing I_m, so $I_{m(\max)} = V_m/R$. At resonance the *average power P* dissipated in the resistance is also maximized, and

$$P_{\max} = \frac{1}{2}\frac{V_m^2}{R} = \frac{V_{\text{rms}}^2}{R} \qquad \textbf{(13.9)}$$

The frequencies at which P is down to half its maximum value are the *half-power frequencies* ω_L and ω_H. Recall from Section 10.4 that the *half-power bandwidth* is BW $= \omega_H - \omega_L$, and that

$$\boxed{\text{BW} = \frac{\omega_0}{Q}} \qquad \textbf{(13.10)}$$

If Equations (13.3) and (13.4) are combined, it is readily seen that

$$\boxed{\text{BW} = \frac{1}{L/R}} \qquad \textbf{(13.11)}$$

Equations (13.3) and (13.11) indicate that *in a series RLC circuit ω_0 is set by L and C, and* BW *is set by R and L.*

Exercise 13.1   A series *RLC* circuit has $R = 50\ \Omega$, $\omega_0 = 10^5$ rad/s, and BW $= 10^3$ rad/s. Find L and C.

ANSWER   50 mH, 2 nF.

▶ **Example 13.1**

A load consisting of a 20-Ω resistance in series with a 0.25-H inductance is to be driven by the utility power.

    (a) Specify a suitable capacitance that, when interposed between the source and the load, will maximize power transfer.

    (b) Compare with the case in which the load is connected to the utility power *directly,* without the coupling capacitance.

Solution

    (a) We want the circuit to resonate at $\omega_0 = 2\pi 60 = 377$ rad/s. Use a series capacitance $C = 1/(\omega_0^2 L) = 1/(377^2 \times 0.25) = 28.14\ \mu$F. Then, $P = P_{\max} = V_{\text{rms}}^2/R = 120^2/20 = 720$ W.

(b) Without the coupling capacitance, the source sees $Z_L = R + j\omega L = 20 + j377 \times 0.25 = 20 + j94.25$ Ω, so $I_{rms} = V_{rms}/|Z_L| = 120/\sqrt{20^2 + 94.25^2} = 1.246$ A. Hence, $P = RI_{rms}^2 = 20 \times 1.246^2 = 31.03$ W. Clearly, we now have $P \ll P_{max}$. ◄

Phasor Diagrams

We can gain additional insight into *RLC* behavior by constructing the phasor diagrams for $\omega < \omega_0$, $\omega = \omega_0$, and $\omega > \omega_0$. It is convenient to use the current phasor $I = I_m\underline{/\theta_i}$ as the reference phasor because current is common to all three elements. The phasors of the individual element voltages are $V_R = RI$, $V_L = j\omega LI$, and $V_C = (1/j\omega C)I$, or

$$V_R = RI_m\underline{/\theta_i} \tag{13.12a}$$

$$V_L = \omega L I_m\underline{/\theta_i + 90°} \tag{13.12b}$$

$$V_C = \frac{1}{\omega C}I_m\underline{/\theta_i - 90°} \tag{13.12c}$$

The phasor diagrams are shown in Figure 13.3. *Below* resonance ($\omega < \omega_0$) we have $|V_C| > |V_L|$ and $\theta_i > 0$, confirming *capacitive* behavior. *At* resonance ($\omega = \omega_0$) we have $V_C = -V_L$, so $V_R = V$ and $\theta_i = 0$. The resonant circuit behaves as if it consisted of just *R*. *Above* resonance ($\omega > \omega_0$) we have $|V_L| > |V_C|$ and $\theta_i < 0$, confirming *inductive* behavior.

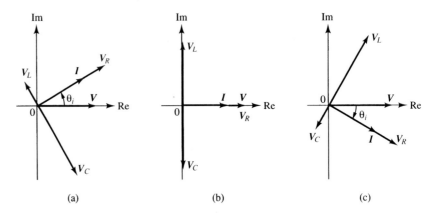

Figure 13.3 Phasor diagrams for the series *RLC* circuit: (a) $\omega < \omega_0$, (b) $\omega = \omega_0$, and (c) $\omega > \omega_0$.

► Example 13.2

In the circuit of Figure 13.1 let $V = 10\underline{/0°}$ V, $R = 400$ Ω, $L = 10$ mH, and $C = 10$ nF. Calculate the phasors for (a) $\omega = \omega_0/2$, (b) $\omega = \omega_0$, and (c) $\omega = 2\omega_0$.

Solution

We have $\omega_0 = 1/\sqrt{LC} = 1/\sqrt{10^{-2} \times 10^{-8}} = 10^5$ rad/s, and $Q = (1/R)\sqrt{L/C} = (1/400)\sqrt{10^{-2}/10^{-8}} = 2.5$.

(a) For $\omega = \omega_0/2 = 10^5/2$ rad/s we have $Q(\omega/\omega_0 - \omega_0/\omega) = 2.5 \times (0.5 - 2) = -3.75$. By Equations (13.8), $I_m = (10/400)/\sqrt{1 + 3.75^2}$ and $\theta_i = -\tan^{-1}(-3.75)$, or

$$I = 6.442\underline{/75.07°} \text{ mA}$$

Substituting into Equations (13.12) gives $V_R = (400 \times 6.442 \times 10^{-3})\underline{/75.07°}$, $V_L = [(10^5/2)10^{-2} \times 6.442 \times 10^{-3}]\underline{/75.07° + 90°}$, and $V_C = (1/[(10^5/2)10^{-8}])(6.442 \times 10^{-3})\underline{/75.07° - 90°}$, or

$$V_R = 2.577\underline{/75.07°} \text{ V}$$

$$V_L = 3.221\underline{/165.07°} \text{ V}$$

$$V_C = 12.88\underline{/-14.93°} \text{ V}$$

(b) For $\omega = \omega_0 = 10^5$ rad/s we obtain $I_m = V_m/R = 10/400$ and $\theta_i = -\tan^{-1} 0$, or

$$I = 25\underline{/0°} \text{ mA}$$

Performing analogous calculations yields

$$V_R = 10\underline{/0°} \text{ V}$$

$$V_L = 25\underline{/90°} \text{ V}$$

$$V_C = 25\underline{/-90°} \text{ V}$$

(c) For $\omega = 2\omega_0 = 2 \times 10^5$ rad/s, we likewise obtain

$$I = 6.442\underline{/-75.07°} \text{ mA}$$

$$V_R = 2.577\underline{/-75.07°} \text{ V}$$

$$V_L = 12.88\underline{/14.93°} \text{ V}$$

$$V_C = 3.221\underline{/-165.07°} \text{ V} \quad \blacktriangleleft$$

Resonance Voltage Rise

The phasor diagrams of Figure 13.3 and the results of Example 13.2 reveal an intriguing phenomenon, that is, depending on the operating frequency ω, the peak voltage amplitudes across one, the other, or both of the reactive elements may *exceed* the peak amplitude of the applied signal. We are particularly interested in the *situation at resonance,* where

$$\boxed{I = \frac{V_m}{R}\underline{/0°}} \qquad \textbf{(13.13)}$$

Letting $\omega = \omega_0$ in Equations (13.12) and using the fact that Equation (13.5) predicts $\omega_0 L = QR$ and $1/\omega_0 C = QR$, the phasors of the individual element voltages are, *at resonance,*

$$V_R = V_m \underline{/0°} \tag{13.14a}$$
$$V_L = QV_m \underline{/90°} \tag{13.14b}$$
$$V_C = QV_m \underline{/-90°} \tag{13.14c}$$

Clearly, if $Q > 1$, the peak voltages across the reactive elements will be *greater* than that of the applied source, a phenomenon known as **resonance voltage rise.** The amount of magnification coincides with the quality factor of the circuit. For instance, in the preceding example we have $Q = 2.5$, thus confirming the values $|V_L| = |V_C| = 2.5|V| = 2.5 \times 10 = 25$ V.

It is interesting to note that the smaller the resistance R in a series *RLC* circuit, the higher the value of Q and, hence, the greater the amount of resonance voltage rise. In the limit $R \to 0$ we would have, by Equation (13.4), $Q \to \infty$, and this would result in an infinite voltage rise.

Exercise 13.2   Find the peak voltage amplitudes across the coupling capacitor and the load of Example 13.1.

ANSWER   799.7 V, 817.5 V.

Exercise 13.3

(a) Using a 100-μH inductance, specify suitable values for R and C to implement a series *RLC* circuit that resonates at 1 MHz with a quality factor of 100.

(b) If the circuit is driven with a source having a peak amplitude of 1 V, find the power it dissipates at resonance, as well as the amplitudes of the voltages across its reactive elements.

ANSWER   (a) 6.283 Ω, 253.3 pF; (b) 79.58 mW, 100 V.

Resonance phenomena arise frequently in everyday life. A common example is offered by a child bouncing on a bed. The child exploits the ability to convert the potential energy of the bed springs into kinetic energy of his or her body, and vice versa, to achieve amazingly high bounces with little effort. The bouncing frequency at which effort is minimum is found empirically by the child: it is the resonance frequency of the mechanical system made up of the child's body and the bed springs. Any attempt to bounce at a higher or a lower frequency than the resonance frequency would require much more effort and would result in less spectacular bounces!

Energy at Resonance

The instantaneous energy stored in the inductance is $w_L(t) = (1/2)Li_L^2(t)$, and that stored in the capacitance is $w_C(t) = (1/2)Cv_C^2(t)$. The sum $w_L(t) + w_C(t)$ is called the *total stored energy.* This energy is *maximum* at resonance. We wish to find its value, aptly called the **maximum stored energy.**

At resonance, Equation (13.13) predicts $i_L(t) = (V_m/R)\cos\omega_0 t$. Moreover, Equation (13.5a) allows us to write $L = QR/\omega_0$, so

$$w_L(t) = \frac{Q}{\omega_0}\frac{V_{\text{rms}}^2}{R}\cos^2\omega_0 t \qquad \textbf{(13.15a)}$$

Still at resonance we have, by Equation (13.14c), $v_C(t) = QV_m\cos(\omega_0 t - 90°) = QV_m\sin\omega_0 t$. Moreover, Equation (13.5b) allows us to write $C = 1/\omega_0 QR$, so

$$w_C(t) = \frac{Q}{\omega_0}\frac{V_{\text{rms}}^2}{R}\sin^2\omega_0 t \qquad \textbf{(13.15b)}$$

The *maximum stored energy* is $w_L(t) + w_C(t) = (Q/\omega_0)(V_{\text{rms}}^2/R)(\cos^2\omega_0 t + \sin^2\omega_0 t)$, or

$$w_L(t) + w_C(t) = \frac{Q}{\omega_0}\frac{V_{\text{rms}}^2}{R} \qquad \textbf{(13.16)}$$

Clearly, this energy is *time invariant.* Physically, at resonance there is no energy exchange between the source and the LC pair. The capacitance and inductance simply exchange their energies *internally.* The only energy that the source needs to supply is that dissipated by the resistance.

We now wish to find an energetic interpretation for the quality factor Q. To this end, let us consider the **energy dissipated per cycle** at resonance, which we denote $w_{R(\text{cycle})}$ and find by dividing the energy dissipated in one second by the number of cycles contained in one second. The former is the maximum average power P_{max}, and the latter is the *resonance cyclical frequency* $f_0 = \omega_0/2\pi$. At resonance we thus have $w_{R(\text{cycle})} = P_{\text{max}}/f_0$, or

$$w_{R(\text{cycle})} = \frac{2\pi}{\omega_0}\frac{V_{\text{rms}}^2}{R} \qquad \textbf{(13.17)}$$

Dividing Equation (13.16) by Equation (13.17) pairwise we obtain $(w_L + w_C)/w_{R(\text{cycle})} = Q/2\pi$, which allows us to write

$$Q = 2\pi\,\frac{\text{maximum stored energy}}{\text{energy dissipated per cycle}} \qquad \textbf{(13.18)}$$

We thus have yet another interpretation for Q. Since it is based on energy considerations alone, this relation is applicable not only to electrical circuits, but also to nonelectrical systems such as acoustical and mechanical systems. It is thus generally regarded as the basic definition of the quality factor.

In summary, we have three different ways of defining Q:

(1) The definition of Equation (13.18), based on *energy* considerations.

(2) The definition of Equation (13.10), $Q = \omega_0/\text{BW}$, based on *frequency response* considerations.

(3) The definition of Equation (13.4), based on *circuit elements* and *topology*. As we shall see in the next section, changing the topology from *series* to *parallel* changes the expression for Q in terms of the circuit components. However, the meanings of Q in terms of energy or frequency response remain the same.

▶ **Example 13.3**

Referring to the circuit of Example 13.2, find the maximum stored energy, the energy dissipated per cycle, and the resonance frequency and half-power bandwidth, both in Hz.

Solution

Recall that $\omega_0 = 10^5$ rad/s and $Q = 2.5$. Then, $w_L + w_C = (Q/\omega_0)(V_{\text{rms}}^2/R) = (2.5/10^5)(10^2/400) = 6.25 \ \mu\text{J}; \ w_{R(\text{cycle})} = 2\pi(w_L + w_C)/Q = 2\pi \times 6.25/2.5 = 15.7 \ \mu\text{J}; \ f_0 = \omega_0/2\pi = 10^5/2\pi = 15.9$ kHz; BW $= f_0/Q = 15.9/2.5 = 6.366$ kHz.   ◀

Exercise 13.4   Find the maximum stored energy and the energy dissipated per cycle in the circuit of Exercise 13.3.

ANSWER   1.267 μJ, 79.58 nJ.

13.2 Parallel Resonance

We now turn our attention to the case of a capacitance and inductance connected in *parallel*. Once the remainder of the circuit is reduced to its Norton equivalent, we end up with the *parallel RLC* configuration of Figure 13.4(a). The admittance seen by the source is the sum of the three admittances, $Y = Y_R + Y_C + Y_L$, or

$$Y = G + j\left(\omega C - \frac{1}{\omega L}\right) \tag{13.19}$$

where $G = 1/R$. This expression is similar to that of Equation (13.1), provided Z and R are replaced with their reciprocals Y and G, and L and C are interchanged with each other. Of course, this could have been anticipated using *duality* considerations. As in the *series RLC* case, we expect a peculiar behavior when the susceptances ωC and $-1/\omega L$ cancel each other out.

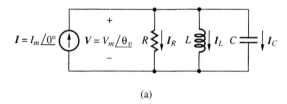

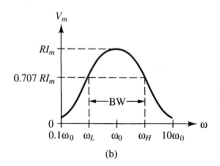

(b)

Figure 13.4 Parallel *RLC* circuit and frequency response of its voltage.

Retracing similar steps to the *series RLC* case, we can readily express the admittance of the *parallel RLC* circuit as

$$Y = G\left[1 + jQ\left(\frac{\omega}{\omega_0} - \frac{\omega_0}{\omega}\right)\right]$$ (13.20)

where

$$\omega_0 = \frac{1}{\sqrt{LC}}$$ (13.21)

is the **resonance frequency,** and

$$Q = R\sqrt{C/L}$$ (13.22)

is the **quality factor** of the **parallel *RLC* circuit.** Note that the expression for ω_0 is the same as for the *series RLC* circuit, but the expressions for Q are *reciprocal* of each other, another reflection of the duality principle.

We now wish to find the voltage response $V = V_m\underline{/\theta_v}$ to the applied current $I = I_m\underline{/0°}$. By Ohm's Law, $V = ZI = (1/Y)I$. Using Equation (13.20) with

$G = 1/R$, and applying Phasor Rule 6 of Section 11.1, we get

$$V_m(\omega) = R I_m \frac{1}{\sqrt{1 + Q^2 \left(\dfrac{\omega}{\omega_0} - \dfrac{\omega_0}{\omega} \right)^2}} \tag{13.23a}$$

$$\theta_v(\omega) = - \tan^{-1} Q \left(\frac{\omega}{\omega_0} - \frac{\omega_0}{\omega} \right) \tag{13.23b}$$

The plot of V_m versus ω is shown in Figure 13.4(b), where we see that the peak value is $V_{m(\max)} = R I_m$. Physically, voltage approaches zero at both ends of the frequency range because of the *short-circuit action* by L at low frequencies and by C at high frequencies. In fact, L dominates for $\omega < \omega_0$, making the parallel *RLC* circuit *inductive,* and C dominates for $\omega > \omega_0$, making it *capacitive.*

At the borderline frequency $\omega = \omega_0$, the susceptances ωC and $-1/\omega L$ cancel each other out in Equation (13.19), making **Y** *purely conductive,* $\mathbf{Y} = G + j0$. Even though the individual admittances $\mathbf{Y}_L$ and $\mathbf{Y}_C$ are nonzero, their sum vanishes at resonance, making the *parallel combination* of L and C effectively behave as an *open circuit.*

It is readily seen that since V_m is maximized at resonance, so is the *average power* dissipated in the resistance,

$$P_{\max} = \frac{1}{2} R I_m^2 = R I_{\mathrm{rms}}^2 \tag{13.24}$$

The frequencies at which P is down to half its maximum value are the *half-power frequencies* ω_L and ω_H, and the *half-power bandwidth,* $\mathrm{BW} = \omega_H - \omega_L$, is such that

$$Q = \frac{\omega_0}{\mathrm{BW}} \tag{13.25}$$

Combining Equations (13.21) and (13.22), we readily find that

$$\mathrm{BW} = \frac{1}{RC} \tag{13.26}$$

Equations (13.21) and (13.26) indicate that *in a parallel RLC circuit ω_0 is set by L and C, and* BW *is set by R and C.*

Exercise 13.5   A parallel *RLC* circuit has $R = 20$ kΩ, BW = 10^4 rad/s, and $Q = 100$. Find L, C, and ω_0.

ANSWER   200 μH, 5 nF, 10^6 rad/s.

Following the line of reasoning of the previous section, it is straightforward to show that in a *parallel RLC* circuit we have

$$I_R = \frac{V_m}{R} \underline{/\theta_v} \qquad\qquad \text{(13.27a)}$$

$$I_C = \omega C V_m \underline{/\theta_v + 90°} \qquad\qquad \text{(13.27b)}$$

$$I_L = \frac{1}{\omega L} V_m \underline{/\theta_v - 90°} \qquad\qquad \text{(13.27c)}$$

Moreover, at *resonance,* the following identities hold:

$$V = R I_m \underline{/0°} \qquad\qquad \text{(13.28)}$$

$$I_R = I_m \underline{/0°} \qquad\qquad \text{(13.29a)}$$

$$I_C = Q I_m \underline{/+90°} \qquad\qquad \text{(13.29b)}$$

$$I_L = Q I_m \underline{/-90°} \qquad\qquad \text{(13.29c)}$$

Exercise 13.6   Derive Equations (13.27) through (13.29).

Exercise 13.7   In the circuit of Figure 13.4(a) let $I_m = 1$ mA, $L = 10$ mH, and $C = 10$ nF. Find the value of R that yields $Q = 1$. Hence, sketch and label the phasor diagrams for the cases $\omega = \omega_0/2$, $\omega = \omega_0$, and $\omega = 2\omega_0$. Be neat and precise.

ANSWER   1 kΩ.

We again note that if $Q > 1$, the peak amplitudes of the currents through the reactive elements exceed the peak amplitude of the applied source, a phenomenon known as **resonant current rise.** The amount of magnification is Q. Comparing the expressions for the quality factors, we note that in a *series RLC* circuit the *smaller* the resistance the *greater* the resonance voltage rise; in a *parallel RLC* circuit, the *larger* the resistance, the *greater* the resonance *current* rise. Can you justify this physically?

The **maximum stored energy** at resonance, $w_L(t) + w_C(t)$, and the **power dissipated per cycle** at resonance, $w_{R(\text{cycle})}$, can be obtained from the *series RLC* expressions of Equations (13.16) and (13.17) using the duality principle. Thus, replacing V_{rms} with I_{rms} and $1/R$ with R in these equations yields

$$w_L(t) + w_C(t) = \frac{Q}{\omega_0} R I_{rms}^2 \qquad \text{(13.30)}$$

$$w_{R(cycle)} = \frac{2\pi}{\omega_0} R I_{rms}^2 \qquad \text{(13.31)}$$

As in the series case, the maximum stored energy is *time invariant,* indicating that at resonance there is no energy exchange between the source and the *LC* pair; the capacitance and inductance simply exchange their energies internally. The only energy that the source needs to supply is that dissipated by the resistance. As in the *series RLC* case, we have

$$Q = 2\pi \frac{\text{maximum stored energy}}{\text{energy dissipated per cycle}} \qquad \text{(13.32)}$$

As mentioned in the previous section, the expressions for Q based on *energy* or *frequency response* considerations hold regardless of the particular circuit or system. However, the expressions for Q in terms of the *circuit elements* depend on topology, as reflected by the fact that the expressions of Equations (12.4) and (11.22) are *reciprocals* of each other.

Exercise 13.8 Referring to the *RLC* circuit of Exercise 13.7, find the maximum stored energy and the energy dissipated per cycle, both in J.

ANSWER 5 nJ, 31.42 nJ.

Practical Resonant Circuits

The parallel *RLC* configuration finds application in a class of circuits known as *tuned circuits,* such as *tuned oscillators* and *tuned amplifiers* in radio communications.

Of the three components making up an *RLC* circuit, the inductor is the least ideal because of core losses and winding resistance. This departure from ideality can be modeled with a **stray resistance** R_s in series with the inductance L. A practical tuned circuit, also called a *tank circuit,* is thus more realistically represented as in Figure 13.5. We wish to investigate the impact of R_s upon the resonance frequency and the half-power bandwidth.

Regarding the R_s-L pair as a single impedance $R_s + j\omega L$, the admittance seen by the source is the sum of three admittances,

$$Y = \frac{1}{R} + j\omega C + \frac{1}{R_s + j\omega L} \qquad \text{(13.33)}$$

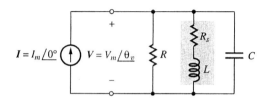

Figure 13.5 A practical tank circuit.

Rationalizing the last term as $(R_s - j\omega L)/(R_s^2 + \omega^2 L^2)$ and regrouping, we obtain

$$Y = \frac{1}{R} + \frac{R_s}{R_s^2 + (\omega L)^2} + j\omega\left(C - \frac{L}{R_s^2 + (\omega L)^2}\right)\qquad\textbf{(13.34)}$$

We are interested in the frequency $\omega_{\text{res}} \neq 0$ that makes the circuit *resonant,* that is, the frequency that makes **Y** *purely real.* This frequency must be such that

$$C - \frac{L}{R_s^2 + (\omega_{\text{res}}L)^2} = 0$$

or

$$\omega_{\text{res}} = \sqrt{\frac{1}{LC} - \frac{R_s^2}{L^2}}\qquad\textbf{(13.35)}$$

In the limit $R_s \to 0$, Equation (13.35) predicts $\omega_{\text{res}} \to 1/\sqrt{LC}$, as it should be. However, with $R_s \neq 0$ we have $\omega_{\text{res}} < 1/\sqrt{LC}$, indicating that the effect of R_s is a **downshift** in the resonance frequency. Clearly, the smaller R_s is as compared to $|\mathbf{Z}_L|$ at this frequency, the closer ω_{res} is to $1/\sqrt{LC}$.

An appropriate figure of merit for a practical inductor is the **quality factor of the coil,** defined as $Q_{\text{coil}} \triangleq |\mathbf{Z}_L(\omega_{\text{res}})|/R_s$, where ω_{res} is the frequency at which the coil is designed to resonate. Substituting $|\mathbf{Z}_L(\omega_{\text{res}})| = \omega_{\text{res}}L$, we obtain

$$Q_{\text{coil}} = \frac{\omega_{\text{res}}L}{R_s}\qquad\textbf{(13.36)}$$

For ω_{res} sufficiently close to $1/\sqrt{LC}$ we can approximate $R_s = \omega_{\text{res}}L/Q_{\text{coil}} \simeq (1/\sqrt{LC})(L/Q_{\text{coil}})$. Substituting this expression for R_s into Equation (13.35), we can express ω_{res} in the neater form

$$\omega_{\text{res}} = \frac{1}{\sqrt{LC}}\sqrt{1 - \frac{1}{Q_{\text{coil}}^2}}\qquad\textbf{(13.37)}$$

Clearly, the higher the value of Q_{coil}, the closer ω_{res} is to $1/\sqrt{LC}$.

▶ Example 13.4

In the circuit of Figure 13.5 let $L = 10$ mH and $C = 10$ nF. Find Q_{coil} and ω_{res} if (a) $R_s = 0$ and (b) $R_s = 100$ Ω.

Solution

(a) Applying Equations (13.36) and (13.37) yields $Q_{coil} = \infty$ and $\omega_{res} = 1/\sqrt{LC} = 1/\sqrt{10^{-2} \times 10^{-8}} = 10^5$ rad/s.

(b) $Q_{coil} = \omega_{res}L/R_s \simeq 10^5 \times 10^{-2}/100 = 10$; $\omega_{res} = (1/\sqrt{LC}) \times \sqrt{1 - 1/Q_{coil}^2} = 10^5\sqrt{1 - 1/10^2} = 99,499$ rad/s, indicating a 0.5% frequency downshift.   ◀

Exercise 13.9   In the circuit of Example 13.4 what is the maximum allowable value of R_s if the actual resonance frequency is to depart from 10^5 rad/s by no more than 1 percent?

ANSWER   139.7 Ω.

Figure 13.6 Series to parallel conversion of an *R-L* circuit.

To investigate the effect of R_s upon the half-power bandwidth, it is convenient to first perform the series-to-parallel conversion depicted in Figure 13.6. We claim that it is possible to select R_p and L_p so that, at a given frequency ω, the admittance of the parallel structure *equals* that of the series structure,

$$\frac{1}{R_p} + \frac{1}{j\omega L_p} = \frac{1}{R_s + j\omega L}$$

Rationalizing both sides and equating the real and imaginary parts pairwise yields the necessary conditions to make the two structures equivalent,

$$R_p = \frac{R_s^2 + (\omega L)^2}{R_s}$$

$$L_p = \frac{R_s^2 + (\omega L)^2}{\omega^2 L}$$

It is readily seen that for $\omega = 1/\sqrt{LC}$ these expressions simplify to

$$R_p = R_s(1 + Q_{coil}^2) \tag{13.38a}$$

$$L_p = L\left(1 + \frac{1}{Q_{coil}^2}\right) \tag{13.38b}$$

It is important to observe that these expressions hold only for $\omega = 1/\sqrt{LC}$. However, if we restrict ω to values sufficiently close to $1/\sqrt{LC}$, then R_p and L_p will remain essentially constant, indicating that we can refer to the all-parallel equivalent circuit of Figure 13.7 to estimate the resonance frequency and the half-power bandwidth of the original tank circuit of Figure 13.5. Thus, reusing Equation (13.21) with L_p in place of L, we obtain

$$\omega_{\text{res}} = \frac{1}{\sqrt{LC}} \frac{1}{\sqrt{1 + 1/Q_{\text{coil}}^2}} \qquad \textbf{(13.39)}$$

and reusing Equation (13.26) with $R \parallel R_p$ in place of R, we obtain

$$\text{BW} = \frac{1}{\{R \parallel [R_s(1 + Q_{\text{coil}}^2)]\}C} \qquad \textbf{(13.40)}$$

Clearly, in addition to downshifting the center frequency of the band-pass function, R_s also *broadens* its half-power bandwidth.

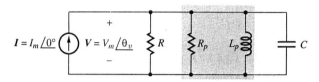

Figure 13.7 Approximate equivalent of the tank circuit of Figure 13.5.

▶ **Example 13.5**

A parallel *RLC* circuit has $R = 100$ kΩ, $L = 10$ mH, and $C = 10$ nF. Find ω_{res}, BW, and the quality factor $Q_{\text{circuit}} = \omega_{\text{res}}/\text{BW}$ for (a) $Q_{\text{coil}} = \infty$ and (b) $Q_{\text{coil}} = 50$.

Solution

(a) With $Q_{\text{coil}} = \infty$, we have

$$\omega_{\text{res}} = \frac{1}{\sqrt{LC}} = \frac{1}{\sqrt{10^{-2} \times 10^{-8}}} = 100 \text{ krad/s}$$

$$\text{BW} = \frac{1}{RC} = \frac{1}{10^5 \times 10^{-8}} = 1 \text{ krad/s}$$

$$Q_{\text{circuit}} = \frac{100}{1} = 100$$

(b) With $Q_{\text{coil}} = 50$, we have, by Equation (13.36), $R_s = \omega_{\text{res}}L/Q_{\text{coil}} \simeq 10^5 \times 10^{-2}/50 = 20 \ \Omega$. Then, Equations (13.39) and (13.40) yield

$$\omega_{\text{res}} = \frac{10^5}{\sqrt{1 + 1/50^2}} = 99.980 \text{ krad/s}$$

$$\text{BW} = \frac{1}{\{10^5 \parallel [20(1 + 50^2)]\}10^{-8}} = 2.999 \text{ krad/s}$$

$$Q_{\text{circuit}} = \frac{99,980}{2999} = 33.33$$

Remark   The finite value of Q_{coil} has also decreased the quality factor of the overall circuit from 100 to 33.33.   ◀

Exercise 13.10   Find ω_L and ω_H both in part (a) and in part (b) of Example 13.5.

ANSWER   (a) 99.501 krad/s, 100.501 krad/s; (b) 98.492 krad/s, 101.491 krad/s.

13.3 RESONANT OP AMP CIRCUITS

In Section 9.4 it was shown that adding an op amp to a second-order inductorless circuit allows us to achieve sustained or even diverging oscillations, effects that are not possible with purely passive circuits, let alone with passive circuits containing no inductances. In this section we show how the op amp can be exploited to bolster the quality factor Q of a circuit. We first examine a passive inductorless band-pass circuit and show that it cannot provide resonance signal rise. Then we add an op amp to provide a controlled amount of *positive feedback* and show that the Q of the circuit can be changed to any value we wish.

A Passive Inductorless Band-Pass Circuit

The circuit of Figure 13.8(a) consists of a low-pass R-C stage followed by a high-pass C-R stage, and it contains no inductances. We expect an insignificant output response at low frequencies because of the open-circuit action by C_2, and at high frequencies because of the short-circuit action by C_1. However, there is an intermediate frequency band over which C_1 no longer acts as a short and C_2 not yet as an open, indicating that the response will no longer be insignificant. We thus expect the bell-shaped frequency response typical of *band-pass* circuits.

To substantiate quantitatively, let us derive a phasor relationship between V_o and V_i. Applying the node method gives

$$\frac{V_i - V_1}{R_1} = \frac{V_1}{1/j\omega C_1} + \frac{V_1 - V_o}{1/j\omega C_2}$$

$$\frac{V_1 - V_o}{1/j\omega C_2} = \frac{V_o}{R_2}$$

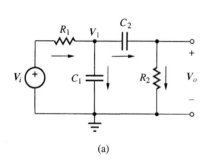 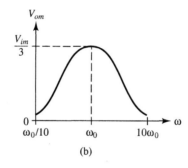

(a)   (b)

Figure 13.8   A passive inductorless band-pass circuit and its frequency response for $R_1 = R_2$ and $C_1 = C_2$.

Eliminating V_1 and rearranging yields, after simplification,

$$V_o = \frac{j\omega R_2 C_2}{1 + j\omega(R_1 C_1 + R_2 C_2 + R_1 C_2) - \omega^2 R_1 R_2 C_1 C_2} V_i$$

It is now a straightforward if tedious matter to verify that this expression can be manipulated into the standard band-pass form

$$V_o = \frac{A}{1 + jQ\left(\dfrac{\omega}{\omega_0} - \dfrac{\omega_0}{\omega}\right)} V_i \qquad \textbf{(13.41)}$$

where

$$A = \frac{1}{1 + (R_1/R_2)(1 + C_1/C_2)} \qquad \textbf{(13.42)}$$

$$\omega_0 = \frac{1}{\sqrt{R_1 R_2 C_1 C_2}} \qquad \textbf{(13.43)}$$

$$Q = \frac{\sqrt{R_1 R_2 C_1 C_2}}{R_1 C_1 + R_2 C_2 + R_1 C_2} \qquad \textbf{(13.44)}$$

The functional form of Equation (13.41) confirms the *band-pass* nature of our circuit. The bell-shaped plot of V_{om} versus ω reaches its maximum for $\omega = \omega_0$, where $V_{om(\text{max})} = A V_{im}$.

It is readily seen from Equation (13.42) that $A \leq 1$. Consequently, no matter how we specify the component values, we always have

$$V_{om(\text{max})} \leq V_{im} \qquad \textbf{(13.45)}$$

indicating that our circuit is *incapable of resonance signal rise*. Moreover, comparing Equation (13.44) with Equation (9.56), and using the condition of Equation (9.59), one can readily verify that

$$Q \leq 0.5 \qquad \textbf{(13.46)}$$

It can be proved that this condition holds for any second-order passive inductorless circuit, not just the one considered here.

▶ Example 13.6

Find expressions for A, ω_0, and Q for the case of equal-valued components in the circuit of Figure 13.8(a). Hence, show the plot of V_{om} versus ω.

Solution

Letting $R_1 = R_2 = R$ and $C_1 = C_2 = C$ in Equations (13.42) through (13.44) yields

$$A = \frac{1}{3} \qquad \omega_0 = \frac{1}{RC} \qquad Q = \frac{1}{3}$$

The frequency plot is shown in Figure 13.8(b). ◀

Exercise 13.11   Consider the circuit obtained from that of Figure 13.8(a) by interchanging R_1 with C_1, and R_2 with C_2. Show that V_o is still related to V_i by Equation (13.41). Hence, obtain expressions for A, ω_0, and Q, and find upper limits for A and Q.

ANSWER $A = 1/[1 + (C_2/C_1)(1 + R_2/R_1)]$, $\omega_0 = 1/\sqrt{R_1 R_2 C_1 C_2}$, $Q = \sqrt{R_1 R_2 C_1 C_2}/(R_1 C_1 + R_2 C_2 + R_1 C_2)$, $A \leq 1$, $Q \leq 0.5$.

An Active Inductorless Band-Pass Circuit

The circuit of Figure 13.9, consisting of two equal-valued R-C and C-R stages, is similar to that of Figure 13.8(a), except that V_o is now magnified with a noninverting amplifier having gain

$$K = 1 + \frac{R_2}{R_1} \tag{13.47}$$

and then reinjected via an additional resistance into the circuit itself. Since it is made up of R-C stages and an amplifier with gain K, the circuit is aptly

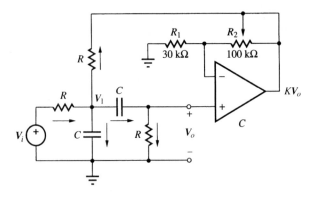

Figure 13.9 An active inductorless band-pass circuit.

referred to as a **KRC circuit.** We shall now see that thanks to this *positive feedback* action by the op amp, it is possible to bolster both A and Q.

Applying the node method again yields

$$\frac{V_i - V_1}{R} = \frac{V_1}{1/j\omega C} + \frac{V_1 - V_o}{1/j\omega C} + \frac{V_1 - KV_o}{R}$$

$$\frac{V_1 - V_o}{1/j\omega C} = \frac{V_o}{R}$$

Eliminating V_1 and rearranging yields, after simplification,

$$V_o = \frac{j\omega RC}{2 - (\omega RC)^2 + (4 - K)j\omega RC} V_i$$

Putting this, as usual, into the band-pass form of Equation (13.41) yields

$$A = \frac{1}{4 - K} \tag{13.48}$$

$$\omega_0 = \frac{\sqrt{2}}{RC} \tag{13.49}$$

$$Q = \frac{\sqrt{2}}{4 - K} \tag{13.50}$$

We observe that ω_0 is still set by R and C; however, both A and Q are now controlled by the amplifier gain. We can thus vary K to achieve a wide range of values for A and Q.

In the limit $K \to 4$ we obtain $A = \infty$ and $Q = \infty$. With an infinite resonance signal rise, the circuit would be capable of sustaining a nonzero output even with a zero input, indicating oscillator behavior. This is confirmed by the fact that since $\zeta = 1/2Q$, with $Q \to \infty$ we obtain $\zeta \to 0$, or undamped oscillation.

For $K > 4$, Q and ζ become *negative*. As we know, this situation leads to a diverging oscillation, which in a practical circuit is eventually limited by unavoidable amplifier nonlinearities.

▶ **Example 13.7**

In the circuit of Figure 13.9 specify suitable component values to make it resonate at 1 kHz with $Q = 10$. What is the amount of signal rise at resonance?

Solution

Arbitrarily choose $C = 10$ nF. Then, Equation (13.49) yields $R = \sqrt{2}/\omega_0 C = \sqrt{2}/(2\pi 10^3 \times 10^{-8}) = 22.51$ kΩ.

By Equation (13.50), $K = 4 - \sqrt{2}/Q = 4 - \sqrt{2}/10 = 3.859$. By Equation (13.47), the potentiometer must be set so that $R_2 = R_1(K - 1) = 30 \times (3.859 - 1) = 85.76$ kΩ.

The resonance signal rise is $A = 1/(4 - 3.858) = 7.071$. ◀

Exercise 13.12   In the circuit of Figure 13.9 specify suitable component values for a root pair at $s_{1,2} = -200 \pm j1990$ complex Np/s.

ANSWER $C = 10$ nF, $R = 70.71$ kΩ, and $R_2 = 81.51$ kΩ.

13.4 SCALING

A task engineers face daily is the *adaptation* of an existing circuit to a new set of specifications. Considering the amount of time and effort that may have gone into the original design, it is desirable to effect the adaptation with a minimum of computational effort. In the case of frequency-selective circuits such as tuned amplifiers, filters, and oscillators, this adaptation can often be effected via the simple technique of *scaling,* whereby the element values of the original circuit are multiplied or divided by suitable factors, or *scaled,* to obtain those of the new circuit. There are two types of scaling, *magnitude* scaling and *frequency* scaling.

Magnitude Scaling

In magnitude scaling the values of all impedances at a given frequency are multiplied by a common scaling factor k_m. Denoting the original component values as *old,* and those after scaling as *new,* we have $R_\text{new} = k_m R_\text{old}$, $j\omega L_\text{new} = k_m(j\omega L_\text{old})$, and $1/j\omega C_\text{new} = k_m/j\omega C_\text{old}$, or

$$R_\text{new} = k_m R_\text{old} \tag{13.51a}$$

$$L_\text{new} = k_m L_\text{old} \tag{13.51b}$$

$$C_\text{new} = C_\text{old}/k_m \tag{13.51c}$$

Magnitude scaling, also referred to as *impedance scaling,* leaves the frequency response of a circuit unchanged. It only affects signal amplitudes and power dissipation.

▶ **Example 13.8**

In Exercise 13.3 you found that a series *RLC* circuit with $R = 6.283$ Ω, $L = 100$ μH, and $C = 253.3$ pF resonates at 1 MHz with a quality factor of 100. Moreover, when powered with a 1-V peak amplitude source, it dissipates 79.58 mW at resonance.

You are now ready to try it out in the laboratory, but you discover that you have run out of 100-μH inductors, the only available value being 250 μH. Scale the component values to achieve the same frequency response but with the 250-μH inductor. What is the power dissipation of the new circuit at resonance?

Solution

$k_m = L_{new}/L_{old} = 250/100 = 2.5$, $R_{new} = k_m R_{old} = 2.5 \times 6.283 = 15.71\ \Omega$, $C_{new} = C_{old}/k_m = 253.3/2.5 = 101.3$ pF, $P_{(max)} = (1/2) \times 1^2/15.71 = 31.83$ mW.

Remark Magnitude scaling leaves ω_0, Q, and BW unaffected. However, it does affect power because of resistance scaling. ◀

Exercise 13.13 Scale the components of Exercise 13.3 so that power dissipation at resonance is 100 mW.

ANSWER $R = 5\ \Omega$, $L = 79.58\ \mu$H, $C = 318.3$ pF.

Frequency Scaling

In frequency scaling each element is changed so that its impedance at the new frequency is the same as at the original frequency. Letting

$$\omega_{new} = k_f \omega_{old} \tag{13.52}$$

we thus have $R_{new} = R_{old}$, $j\omega_{new}L_{new} = j\omega_{old}L_{old}$, and $1/j\omega_{new}C_{new} = 1/j\omega_{old}C_{old}$, or

$$R_{new} = R_{old} \tag{13.53a}$$

$$L_{new} = L_{old}/k_f \tag{13.53b}$$

$$C_{new} = C_{old}/k_f \tag{13.53c}$$

Frequency scaling shifts the frequency response of a circuit up or down the frequency spectrum while leaving signal amplitudes and power dissipation unchanged.

▶ Example 13.9

Scale the circuit of Exercise 13.3 so that it resonates at 100 kHz.

Solution

$k_f = \omega_{new}/\omega_{old} = f_{new}/f_{old} = 10^5/10^6 = 0.1$. Then $R_{new} = R_{old} = 6.283\ \Omega$, $L_{new} = L_{old}/0.1 = 1$ mH, and $C_{new} = C_{old}/0.1 = 2.533$ nF.

Remark Frequency scaling leaves Q and power dissipation unchanged. However, it does affect ω_0 and BW. ◀

FAMOUS ENGINEERS: NIKOLA TESLA

An Eccentric Inventor

Steve Martin has nothing on Nikola Tesla, the original wild and crazy guy. After all, Steve Martin never passed a million volts through his body to prove that alternating current is as "safe" as direct current. Steve Martin never held a vacuum tube in each hand, threw a switch, and ended up illuminated by two flaming swords of blue light just to prove that electrical power can travel without wires.

Nikola Tesla, inventor of the ac induction motor, is shown here in his laboratory calmly generating artificial lightning. (The Bettmann Archive)

Tesla's was one of the world's greatest scientific minds, but his inability to be practical about his discoveries (read: patent them and make money) and the eccentricities of his later years caused him to lose the respect of fellow scientists until after his death.

He was born on July 9, 1856, in Smiljan, Croatia, and educated in that country and in Austria. As a second-year student of electrical engineering in 1876, Tesla saw a demonstration of a Gramme machine, a dynamo/motor combination. Because the machine ran on direct current, sparks flew at the commutator as the alternating current generated by the coil was converted. Tesla argued that the conversion was unnecessary: just feed the alternating current from the dynamo right to the motor. "Impossible," said his professor. "You'll do great things, Tesla, but you'll never succeed in making a motor without a commutator." Well, of course Tesla did just that, his ac induction motor being one of the great inventions of the electrical age. His epochal contribution was summed up by the engineer B. A. Behrend, who presented the 1916 AIEE Edison Medal to Tesla: "Were we to seize and eliminate from our industrial world the results of Mr. Tesla's work, the wheels of industry would cease to turn, our electric cars and trains would stop, our towns would be dark."

After working for the Paris office of Consolidated Edison, Tesla moved to New York City in 1884 with a letter of introduction from his French supervisor to Edison. "I know two great men," the letter said, "and you are one of them; the other is this young man." Edison gave Tesla a job, but the two geniuses couldn't get along, and Tesla quit after only a year. Because his reputation was now established, he had no trouble getting investors for a research laboratory. The deal fell apart, though, and in 1886 and 1887 he had to work digging ditches. Lucky for both Tesla and the world, his foreman on the road crew believed Tesla's stories about his great inventions and introduced him to some investors who formed the Tesla Electric Company and gave the erstwhile ditchdigger a laboratory in lower Manhattan, very near that of his archrival Edison.

Those were the years, the late 1880s, of the ac versus dc fight, with Edison dc and George Westinghouse ac. Tesla aligned with Westinghouse, of course, since the Tesla motor ran on ac. On May 16, 1888, Tesla presented his ac ideas before the AIEE. So impressed was Westinghouse by this work that he offered Tesla $1,000,000 cash plus $1 per horsepower royalty for his ac patents. The two men signed a contract, but the Westinghouse board of directors

> *"Were we to seize and eliminate from our industrial world the results of Mr. Tesla's work, the wheels of industry would cease to turn, our electric cars and trains would stop, and our towns would be dark."*

later insisted the royalty part of the deal be voided. Instead of suing, Tesla said the 1880s equivalent of "No problem, George." All Tesla wanted was his ac system in use. If Westinghouse could do that only by a change in the original contract, so be it. Tesla would give up future millions in order to get his system installed.

In the 1890s Tesla was a magnificent figure in the New York social world. Six-foot-two, impeccably dressed, and unmarried, he had every mother of every marriage-age daughter scheming and planning to snare him. He remained a loner, though, the new century finding him devoting more and more time to "wild" schemes for wireless communication, cosmic

radiation, and robot ships for the Navy. His lack of what are today called interpersonal skills brought him a lot of detractors, and in the 1920s and 1930s he sank into eccentricity and obscurity. Destitute at times, he moved from hotel to hotel as the bills piled up until they were paid by such wealthy friends as John Jacob Astor and J. P. Morgan. In 1936 the Yugoslavian government established the Tesla Institute in Belgrade and paid this great man an annual stipend of $7200 for the remainder of his life.

A lifelong germ phobia grew worse as the years went by. Regardless of his finances, he would use a towel, a handkerchief, a pair of gloves only once before throwing them away. He refused to shake hands with anyone, and a familiar sight on the streets of lower Manhattan became this tall figure strolling with hands clasped tight behind his back lest he meet some acquaintance who might dare to shake hands. In the 1930s he could be seen by anyone walking along Fifth Avenue after midnight as he fed the pigeons at the Public Library and at St. Patrick's Cathedral.

He died in his sleep on January 7, 1943, and was discovered by the chambermaid of the hotel the following morning. Today his papers reside in the Nikola Tesla Museum in Belgrade (if it is still standing), and engineers continue to pore over his

> *"You'll do great things, Tesla, but you'll never succeed in making a motor without a commutator."*

notebooks to study his amazing insights. In 1956, to celebrate the centennial of his birth, the "tesla" was made the standard unit of magnetic flux density.

Magnitude and Frequency Scaling

A circuit can simultaneously be scaled in magnitude and frequency via

$$R_{\text{new}} = k_m R_{\text{old}} \tag{13.54a}$$

$$L_{\text{new}} = (k_m/k_f) L_{\text{old}} \tag{13.54b}$$

$$C_{\text{new}} = C_{\text{old}}/(k_m k_f) \tag{13.54c}$$

Simultaneous scaling is widely used in the design of frequency-selective networks, or filters, starting from *normalized prototypes*. These prototypes, having characteristic frequencies of 1 rad/s and resistance levels of 1 Ω, are tabulated in handbooks of proven filter designs. Application engineers then adapt these universal prototypes to their particular needs via scaling.

▶ Example 13.10

The circuit of Figure 13.10, top, is known as a *fifth-order Butterworth low-pass filter*. The component values, specified for a cutoff frequency of 1 rad/s, are available in filter handbooks. Scale the component values for a cutoff frequency of 20 kHz using 1-kΩ resistances.

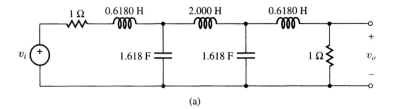

(a)

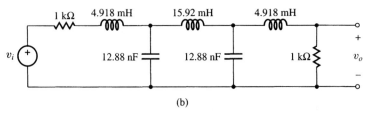

(b)

Figure 13.10 (a) Normalized filter prototype, and (b) scaled version of Example 13.10.

Solution

From Equation (13.54a), $k_m = R_{\text{new}}/R_{\text{old}} = 10^3/1 = 10^3$. From Equation (13.52), $k_f = \omega_{\text{new}}/\omega_{\text{old}} = 2\pi \times 20 \times 10^3/1 = 125.7 \times 10^3$. From Equation (13.54b), $L_{\text{new}}/L_{\text{old}} = 10^3/(125.7 \times 10^3) = 7.958 \times 10^{-3}$. The inductances thus scale to the values $7.958 \times 10^{-3} \times 0.6180 = 4.918$ mH and $7.958 \times 10^{-3} \times 2 = 15.92$ mH. The capacitances scale as

$C_{new} = C_{old}/(10^3 \times 125.7 \times 10^3) = 1.618/(125.7 \times 10^6) = 12.88$ nF. The new circuit is shown in Figure 13.10, bottom. ◄

Exercise 13.14   Scale the circuit of Figure 13.10, top, for a cutoff frequency of 100 kHz using 10-nF capacitances.

ANSWER   257.5 Ω, 253.3 μH and 819.7 μH, 10 nF.

Exercise 13.15   Scale the circuit of Exercise 13.3 so that it resonates at 500 kHz using a 500-μH inductance.

ANSWER   15.71 Ω, 500 μH, 202.6 pF.

▼ SUMMARY

- A one-port containing capacitances and inductances is said to be in *unity power-factor resonance* when its terminal voltage and current are in phase with each other or, equivalently, when its impedance becomes *purely resistive*.

- A *series RLC* circuit is in resonance when its capacitive and inductive *reactances cancel* each other out. Though the individual voltages across L and C are different from zero, the LC combination as a *pair* acts as a *short circuit*.

- A *parallel RLC* circuit is in resonance when its capacitive and inductive *susceptances cancel* each other out. Though the individual currents through L and C are not zero, the LC combination as a *pair* acts as an *open* circuit.

- The resonance frequency of both the *series* and the *parallel RLC* circuits is $\omega_0 = 1/\sqrt{LC}$. The *quality factor* of the series RLC circuit is $Q = 1/(R\sqrt{C/L})$, and that of the *parallel RLC* circuit is $Q = R\sqrt{C/L}$, the *reciprocal* of the former.

- For a series or parallel RLC circuit driven at its resonance frequency, the following hold: (a) current and voltage are *in phase*; (b) the average absorbed power is *maximum*; (c) the *maximum stored energy is time invariant*; (d) for $Q > 1$, we witness the phenomenon of *resonant signal rise*.

- The quality factor of a system can be defined (a) in terms of the frequency response as $Q = \omega_0/\text{BW}$, where ω_0 is the *resonance frequency* and BW is the *half-power bandwidth*; or (b) in terms of energy as $Q = 2\pi \times$ (*maximum stored energy*)/(*energy dissipated per cycle*).

- A figure of merit for a coil having inductance L and stray series resistance R_s is the *quality factor* of the *coil*, $Q_{coil} = \omega_{res}L/R_s$, where ω_{res} is the frequency at which the coil is designed to resonate.

- When a practical coil is used in a *parallel RLC* circuit, the actual resonance frequency is $\omega_{res} = \omega_0\sqrt{1 - 1/Q_{coil}^2}$, where $\omega_0 = 1/\sqrt{LC}$.

- A passive resonant circuit consisting of resistances and capacitances but no inductances always has $Q \leq 0.5$. However, with the inclusion of an op amp to provide a controlled amount of *positive feedback,* it is possible to change the Q of the circuit to any value, including $Q = \infty$ or even $Q < 0$.

- A circuit with $Q = \infty$ yields a *sustained oscillation,* and a circuit with $Q < 0$ yields a *diverging oscillation.*

- *Scaling* allows us to adapt an existing frequency-selective circuit to a new set of specifications by multiplying or dividing its component values by a suitable factor.

- *Magnitude* or *impedance* scaling changes the power dissipation in a circuit while leaving its frequency response unaltered. *Frequency* scaling shifts the circuit's frequency response up or down the frequency spectrum while leaving power dissipation unaltered.

▼ PROBLEMS

13.1 Series Resonance

13.1   In the series *RLC* circuit of Figure 13.1(a) let $R = 10\ \Omega$, $L = 1$ mH, and $C = 1$ nF. If $V_m = 10$ V and the source frequency is adjusted until the average power dissipated by R is maximized, find this power and frequency as well as the peak voltage across L and C.

13.2   A series *RLC* circuit is subjected to the voltage $v = 5 \cos 10^5 t$ V. Specify R, L, and C so that the capacitance current and voltage are $i_C = 10 \cos 10^5 t$ mA and $v_C = 1 \cos (10^5 t - 90°)$ kV.

13.3   (a) If a series *RLC* circuit has $R = 5\ \Omega$, find L and C so that the circuit resonates at 5 krad/s, with $Q = 50$. (b) Find ω_L, ω_H, and BW. (c) What are the values of $\mathbf{Z}$ for $\omega = \omega_L$ and $\omega = \omega_H$?

13.4   (a) Design a series *RLC* circuit meeting the following specifications: resonance frequency of 100 kHz, bandwidth of 2 kHz, and minimum impedance of 100 Ω. (b) Find the frequencies, in Hz, at which $|\mathbf{Z}| = 500\ \Omega$. (c) Find the frequencies at which $\measuredangle\mathbf{Z} = \pm 60°$.

13.5   A series *RLC* circuit with $C = 10$ nF is to resonate at $\omega_0 = 1$ Mrad/s with $\omega_L = 0.9512$ Mrad/s. Find ω_H, Q, BW, R, and L.

13.6   (a) Design a series *RLC* circuit meeting the following specifications: resonance frequency of 3 kHz, $|\mathbf{Z}|_{\min} = 500\ \Omega$, $\measuredangle\mathbf{Z} = 60°$ at 4 kHz. (b) Find the frequencies at which $|\mathbf{Z}| = 1$ kΩ.

13.7   (a) Design a series *RLC* circuit meeting the following specifications: resonance frequency of 100 kHz, $|\mathbf{Z}|_{\min} = 400\ \Omega$, and $|\mathbf{Z}| = 1$ kΩ at 200 kHz. (b) Find the frequencies at which $\measuredangle\mathbf{Z} = \pm 60°$.

13.8   In the series *RLC* circuit of Figure 13.1(a) let $R = 100\ \Omega$, $L = 1$ mH, and $C = 1$ nF. If $V_m = 100$ V, find the total stored energy and the energy dissipated per cycle if the

source frequency is adjusted for (a) $\omega = \omega_0$, (b) $\omega = \omega_L$, and (c) $\omega = \omega_H$.

13.9   Design a series *RLC* circuit that resonates at 100 kHz and, in response to the applied voltage $v = 8 \cos 2\pi 10^5 t$ V, stores a maximum energy of 10 μJ, and dissipates 2 μJ per cycle.

13.10   A practical source with $v_{oc} = 100 \cos \omega t$ V and $R_o = 20\ \Omega$ is connected in series with a resistance R, a capacitance C, and a coil having inductance $L = 100\ \mu$H and series resistance $R_s = 10\ \Omega$. (a) Sketch the circuit, and find R and C to make it resonate at $f_0 = 1$ MHz with BW $= 100$ kHz. (b) Find the voltage across the terminals of the source and those of the coil at resonance. (c) Find BW if $R = 0$.

13.11   Show that provided $R_2 > \sqrt{L/C}$, the impedance $\mathbf{Z}(\omega)$ seen by the source of Figure P13.11 becomes purely resistive for $\omega = \omega_{res} = (1/\sqrt{LC})\sqrt{1 - L/(R_2^2 C)}$, and that $\mathbf{Z}(\omega_{res}) = R_1 + L/(R_2 C)$.

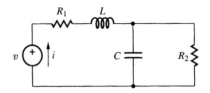

Figure P13.11

13.12   In the circuit of Figure P13.11 let $v = 100 \cos \omega t$ V, $R_1 = 4$ kΩ, $L = 8$ mH, $C = 80$ pF, and $R_2 = 50/3$ kΩ. (a) Using the results of Problem 13.11, find $i(t)$ for $\omega \ll \omega_{res}$, $\omega \gg \omega_{res}$, and $\omega = \omega_{res}$, where ω_{res} is the frequency at which i is in phase with v. (b) Find the peak voltages across L and across C for $\omega = \omega_{res}$.

13.13   (a) Show that provided $Q > 0.5$ in the series *RLC* circuit of Figure 13.1(a), the peak voltage $|\mathbf{V}_C|$ across C is maximized for $\omega = \omega_m = \omega_0\sqrt{1 - 1/(2Q^2)}$, where $\omega_0 = 1/\sqrt{LC}$. Moreover, $|\mathbf{V}_C|_{max} = QV_m/\sqrt{1 - 1/(2Q)^2}$. (b) If $V_m = 10$ V, $R = 800$ Ω, $L = 100$ μH, and $C = 100$ pF find the peak voltage across each element as well as the power factor seen by the source both for $\omega = \omega_0$ and $\omega = \omega_m$.

13.2 Parallel Resonance

13.14 In the parallel *RLC* circuit of Figure 13.4(a) let $R = 5$ kΩ, $L = 10$ mH, and $C = 1$ μF. (a) Find ω_0, ω_L, and ω_H. (b) If at resonance $V_m = 10$ V, find the peak current through L and C.

13.15 (a) Design a parallel *RLC* circuit with a resonance frequency of 10 MHz, a bandwidth of 100 kHz, and a maximum impedance of 100 kΩ. (b) Find the frequency range over which $V_m \geq 0.9V_{m(max)}$. What is the corresponding range for θ_v?

13.16 (a) Design a parallel *RLC* circuit meeting the following specifications: $f_0 = 100$ kHz, $\mathbf{Z}(jf_0) = 10$ kΩ, and $|\mathbf{Z}(j0.95f_0)| = 8$ kΩ where f_0 is the resonance frequency. (b) Find Q and BW.

13.17 In the parallel *RLC* circuit of Figure 13.4(a) let $I_m = 10$ mA and $R = 100$ Ω. (a) Specify L and C to make the circuit resonate at 10 krad/s with a peak capacitance current of 0.5 A. (b) Find the maximum stored energy as well as the energy dissipated per cycle.

13.18 The parallel *RLC* circuit of Figure 13.4 resonates at $f_0 = 100$ kHz with $Q = 6.25$. Moreover, at resonance, $V_m = 150$ V and the maximum stored energy is 1 mJ. Find R, L, and C. Find the energy dissipated per cycle as well as the average power dissipation for $f = f_0$.

13.19 In the circuit of Figure P13.19 find (a) ω_0, Q, ω_L, and ω_H. (b) If the input amplitude is adjusted for a 1 V output amplitude at resonance, what is the current amplitude through each element?

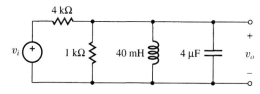

Figure P13.19

13.20 A practical source with $i_{sc} = 100 \cos \omega t$ μA and $R_o = 300$ kΩ is connected in parallel with an inductance $L = 8$ mH, and an unknown resistance R and capacitance C. (a) Find R and C to make the circuit resonate at 250 krad/s with $Q = 50$. (b) Find the bandwidth BW and the maximum peak amplitude of the voltage across the terminals of the source. (c) Repeat part (b), but with $R_o = \infty$.

13.21 A coil with inductance $L = 10$ mH and series resistance $R_s = 50$ Ω is in parallel with a capacitance $C = 1$ μF.

Find the unity power-factor resonance frequency of the parallel structure, as well as its impedance at that frequency.

13.22 In the tank circuit of Figure 13.5 let $I_m = 5$ μA, $R = 2$ kΩ, $L = 5$ mH, $R_s = 50$ Ω, and $C = 50$ nF. Find Q_{coil}, $Q_{circuit}$, ω_{res}, and $V_m(\omega_{res})$.

13.23 The tank circuit of Figure 13.5 utilizes a 500-μH coil with $Q_{coil} = 50$ at 1 Mrad/s. Specify R and C to achieve $\omega_{res} = 1$ Mrad/s with $Q_{circuit} = 40$.

13.24 In the circuit of Figure P13.24 ω is adjusted for $pf = 1$ at the source. Find Q_{coil}, $Q_{circuit}$, and $v(t)$.

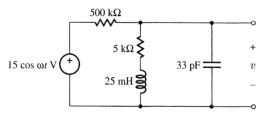

Figure P13.24

13.25 (a) Show that if $R_L = R_C = \sqrt{L/C}$, then the power factor seen by the source in the circuit of Figure P13.25 is unity for all ω. (b) If $L = 12$ mH, $C = 3$ nF, $R = 3$ kΩ, $R_L = R_C = \sqrt{L/C}$, and $i(t) = 5 \cos \omega t$ mA, find the impedance $\mathbf{Z}$ seen by the source, as well as $v(t)$.

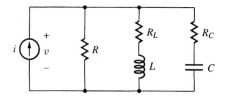

Figure P13.25

13.3 Resonant Op Amp Circuits

13.26 Obtain a relationship between the output and input phasors for the circuit of Figure P13.26. Then, put it in the form of Equation (13.41) to find ω_0, Q, and A.

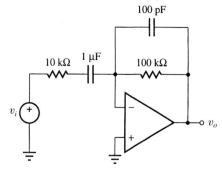

Figure P13.26

13.27   Repeat Problem 13.26 for the circuit of Figure P13.27.

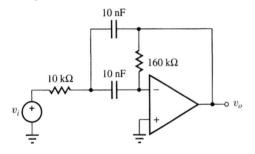

Figure P13.27

13.28   Repeat Problem 13.26 for the circuit of Figure P13.28.

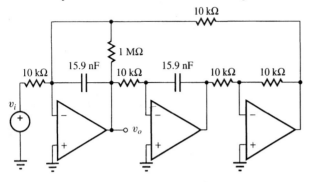

Figure P13.28

13.4 Scaling

13.29   Show that if a circuit is scaled both in magnitude and frequency, $Q_{new} = Q_{old}$ but $BW_{new} = k_f BW_{old}$.

13.30   (a) Specify the elements of a parallel-resonant *RLC* circuit for $\omega_0 = 1$ rad/s and $Q = 10$. (b) Then, scale it so that $\omega_0 = 1$ Mrad/s and $\mathbf{Z}(\omega_0) = 1$ kΩ.

13.31   Scale the circuit of Figure P13.19 so that the sum of all resistances in the circuit is 100 kΩ and ω_0 changes to 10 krad/s.

13.32   Scale the circuit of Figure P13.24 so that the unity power-factor resonance frequency is changed to 10 krad/s and the current drawn from the source at resonance has a peak amplitude of 0.1 mA.

13.33   Scale the op amp resonator of Figure P13.26 so that the sum of all resistances in the circuit is 330 kΩ and the resonance frequency becomes 1 Hz.

13.34   (a) Scale the op amp resonator of Figure P13.27 so that the feedback resistance is changed to 120 kΩ without altering ω_0. (b) Scale the same circuit so that the capacitances change to 1 nF and ω_0 is doubled.

NETWORK FUNCTIONS

S ince introducing energy-storage elements we have focused upon two particular but practically interesting responses, namely, the *transient response* and the *ac response.* The transient response, which is the response to a *dc* forcing signal of the type $x(t) = X_m$, was investigated using *time-domain* techniques. The ac response, which is the response to an *ac* forcing signal of the type $x(t) = X_m \cos(\omega t + \theta)$, was investigated using *frequency-domain* techniques (time-domain techniques are possible but generally much more laborious).

In the present chapter we expand our scope by investigating the response to a more general class of forcing signal, namely, **complex exponential signals** of the type $x(t) = Xe^{st}$, where X and s are complex variables. Though these signals cannot, in general, be duplicated in the laboratory because of these complex variables, they nonetheless constitute a most powerful analytical tool for the following reasons:

(1) The study of the response to complex exponentials evidences important information that is *characteristic* of the circuit, regardless of the applied signals or the internal stored energy. This information, expressed in the form of parameters known as the *natural, characteristic,* or *critical frequencies* of the circuit, or just the *roots,* is contained in a function arising in the study of complex exponential responses called the **network function.** Since design specifications are often given in terms of this function, this reason alone would justify our study of complex exponential responses!

(2) Real-world signals can be represented as *linear combinations of complex exponential functions,* indicating that once we know the responses to the individual exponential components, we can apply the superposition principle to find the response to the signal as a whole. This forms the basis of **transform theory,** the best-known examples of which are *Laplace's* and *Fourier's transforms,* to be covered in subsequent chapters.

(3) The derivation of the *network function* requires simple **algebraic techniques** of the type encountered in ac circuit analysis. Compared to the time-domain techniques of Chapters 7 through 9, which required formulating differential equations and then obtaining the critical frequencies as the roots of the corresponding characteristic equations, the network function method is much more efficient and less prone to error, especially with circuits of greater complexity.

As we proceed through the chapter we find that the network function concept provides a *unified approach* to the study of the *natural, forced, transient, steady-state,* and *complete* responses as well as the *frequency response.* Our study of natural responses has revealed a strong correlation between the *location of the roots* of the characteristic equation in the *s* plane and the *functional form* of the natural response. We suspect that root location will also have a profound impact on the *frequency response* of the circuit. In fact, we shall exploit the network function both to *locate* the critical frequencies in the *s* plane and to *visualize* the frequency response of the circuit via *frequency plots* known as *Bode plots.*

We conclude the chapter by demonstrating the use of PSpice to plot the various response types, both for the case in which the circuit is given in schematic form and the case in which only its network function is known.

14.1 COMPLEX FREQUENCY

In this section we investigate *complex exponential signals,* and in so doing we introduce the important concepts of *complex frequency s, generalized impedance* $\mathbf{Z}(s)$, and *s-domain analysis.* Our primary reason for studying circuit response to signals of this sort is not the response per se, but an important by-product known as the *network function.* This function, which is investigated in the following sections, constitutes a cornerstone concept in circuit theory because it contains all essential information about a circuit.

Complex Exponential Signals

A **complex exponential signal** is a signal of the type

$$x(t) = Xe^{st} \tag{14.1}$$

where *X* and *s* are *time-independent complex parameters* expressed, respectively, in polar and rectangular coordinates as

$$X = X_m e^{j\theta} \tag{14.2}$$
$$s = \sigma + j\omega \tag{14.3}$$

In these expressions,

X_m is the **magnitude** of $x(t)$, expressed either in V or in A

θ is the **phase angle,** either in *radians* or in *degrees*

σ, the real part of s, is called the **neper frequency** and is expressed in *nepers/s* (Np/s)

ω, the imaginary part of s, is called the **radian frequency** and is expressed in *radians/s* (rad/s)

We observe that $X = X_m\underline{/\theta}$ is a **phasor.** Moreover, since s is a complex quantity with ω as its imaginary part, it is called the **complex frequency.** Its units are *complex nepers per second* (complex Np/s). Substituting the expression for X and s, Equation (14.1) becomes $x(t) = (X_m e^{j\theta})e^{(\sigma+j\omega)t}$, or

$$x(t) = (X_m e^{\sigma t})e^{j(\omega t + \theta)}$$

Using Euler's identity, which is discussed in Appendix 3, we have the following alternate form for a complex exponential signal

$$\boxed{x(t) = X_m e^{\sigma t}[\cos(\omega t + \theta) + j\sin(\omega t + \theta)]} \qquad \textbf{(14.4)}$$

Because of its complex nature, $x(t)$ as given in Equations (14.1) or (14.4) cannot, in general, be duplicated in the laboratory, where signals are *real*. However, mathematicians have shown that real-world signals can be expressed as linear combinations of complex exponential signals. Thus, once we know how a circuit responds to individual complex exponential signals, we can apply the *superposition principle* to find its response to any linear combination of such signals and, hence, obtain the response to the real-world signal of interest.

To see an example, consider the following simple linear combination of x and its complex conjugate x^*, a combination that we shall denote as x,

$$x = \frac{1}{2}(x + x^*) = \text{Re}[x] \qquad \textbf{(14.5)}$$

By Equation (14.4), this combination represents the *real* signal

$$x(t) = X_m e^{\sigma t}\cos(\omega t + \theta) \qquad \textbf{(14.6)}$$

indicating that once we know how to predict the response to $x(t)$ and $x^*(t)$, we can also predict the response to $x(t)$. As we know from Chapter 9, this signal is a *damped sinusoid* if $\sigma < 0$, a *growing sinusoid* if $\sigma > 0$, and a *steady sinusoid* if $\sigma = 0$. The steady sinusoid

$$x(t) = X_m \cos(\omega t + \theta) \qquad \textbf{(14.7)}$$

is the *ac signal* that we have used as the basis for our study of the *ac response*. Conversely, with $\omega = \theta = 0$ in Equation (14.6), $x(t)$ represents a *decaying exponential* if $\sigma < 0$, a *growing exponential* if $\sigma > 0$, and a *steady function* if $\sigma = 0$. The steady function

$$x(t) = X_m \qquad \textbf{(14.8)}$$

is the *dc signal* at the basis of our *transient response* studies.

The signals of Equations (14.6) through (14.8) are *real* signals of the type that we encounter in the laboratory either as *applied signals* or as *natural responses* of circuits with energy-storage elements. It is apparent that complex exponential signals are *generalizations* of dc and ac signals as well as natural responses. It is fair to expect that the study of complex exponential signals will provide us with a *unified approach* to the natural, forced, transient, steady-state, and complete responses. We shall also find that it provides the framework for subsequent studies of transform techniques such as Fourier's and Laplace's transforms.

An Illustrative Example

To develop a quick feel for how to handle complex exponential signals, refer to the simple but instructive circuit example of Figure 14.1. We wish to find its response to the complex exponential signal

$$v_i(t) = V_i e^{st} \tag{14.9}$$

Assuming the response is also a complex exponential,

$$v_o(t) = V_o e^{st} \tag{14.10}$$

we seek a relationship between $v_o(t)$ and $v_i(t)$. Applying KCL at the rightmost upper node, along with Ohm's Law and the capacitance law yields $C\,d(v_i - v_o)/dt + (v_i - v_o)/R_1 = v_o/R_2$. Rearranging, we have

$$R_1 R_2 C \frac{dv_o}{dt} + (R_1 + R_2)v_o = R_1 R_2 C \frac{dv_i}{dt} + R_2 v_i$$

Substituting Equations (14.9) and (14.10) and using $d(e^{st})/dt = se^{st}$, we get

$$R_1 R_2 C s V_o e^{st} + (R_1 + R_2)V_o e^{st} = R_1 R_2 C s V_i e^{st} + R_2 V_i e^{st}$$

Collecting terms and using again Equations (14.9) and (14.10) gives

$$\boxed{v_o(t) = \frac{R_1 R_2 C s + R_2}{R_1 R_2 C s + (R_1 + R_2)} v_i(t)} \tag{14.11}$$

Once the component values are known, we can readily find $v_o(t)$ for a given $v_i(t)$ by calculating the right-hand side at the value of s appearing in the expression $v_i(t) = V_i e^{st}$.

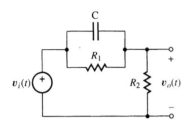

Figure 14.1 An illustrative example.

▶ **Example 14.1**

In the circuit of Figure 14.1 let $R_1 = 1\ \Omega$, $R_2 = 2\ \Omega$, and $C = 1$ F. Find the response to the following inputs:

(a) $v_i(t) = 12$ V

(b) $v_i(t) = 12e^{-3t}$ V

(c) $v_i(t) = 12e^{j2t}$ V

(d) $v_i(t) = 12e^{(-3+j2)t}$ V

Solution

Substituting the given component values into Equation (14.11) yields

$$v_o(t) = \frac{2s+2}{2s+3} v_i(t) \tag{14.12}$$

(a) To find the value of s at which to compute Equation (14.12), we must put $v_i(t)$ in the standard form of Equation (14.9). Writing $v_i(t) = (V_m \underline{/\theta}) e^{st} = 12$ V $= (12\underline{/0^\circ}) e^{0t}$ V reveals that $\theta = 0$ and $s = 0$. Substituting into Equation (14.12), we get

$$v_o(t) = \frac{2 \times 0 + 2}{2 \times 0 + 3} 12 = 8 \text{ V}$$

(b) It is easily seen that now $s = -3 + j0$ complex Np/s. Thus,

$$v_o(t) = \frac{2(-3)+2}{2(-3)+3} 12e^{-3t} = 16e^{-3t} \text{ V}$$

(c) Now $s = 0 + j2$ complex Np/s. Hence,

$$v_o(t) = \frac{2(j2)+2}{2(j2)+3} 12e^{j2t} = \frac{2+j4}{3+j4} 12e^{j2t} = \frac{24}{\sqrt{5}} \underline{/10.30^\circ} e^{j2t} \text{ V}$$

(d) Now $s = -3 + j2$ complex Np/s. Hence,

$$v_o(t) = \frac{2(-3+j2)+2}{2(-3+j2)+3} 12e^{(-3+j2)t} = 9.6\sqrt{2} \underline{/8.13^\circ} e^{(-3+j2)t} \text{ V} \blacktriangleleft$$

Exercise 14.1   Sketch the circuit obtained from that of Figure 14.1 by replacing C with an inductance L.

(a) Assuming $R_1 = 1$ Ω, $R_2 = 2$ Ω, and $L = 1$ H, obtain a relationship between $v_o(t)$ and $v_i(t)$.

(b) Find the response to $v_i(t) = 9\underline{/30^\circ} e^{(-2+j3)t}$ V.

ANSWER   (a) $v_o(t) = [(2s + 2)/(3s + 2)]v_i(t)$; (b) $v_o(t) = 5.779\underline{/24.47^\circ} e^{(-2+j3)t}$ V.

▶ Example 14.2

Find the response of the circuit of Example 14.1 to the damped sinusoid

$$v_i(t) = 10e^{-0.5t} \cos(1.5t + 30^\circ) \text{ V}$$

Solution

By Equations (14.5) and (14.6), we can write

$$v_i(t) = \text{Re}[\mathbf{v}_i(t)] \tag{14.13}$$

where

$$v_i(t) = 10\underline{/30°}\, e^{(-0.5+j1.5)t} \text{ V}$$

The *complex response* $v_o(t)$ to the *complex signal* $v_i(t)$ is

$$v_o(t) = \frac{2(-0.5 + j1.5) + 2}{2(-0.5 + j1.5) + 3}\, 10\underline{/30°}\, e^{(-0.5+j1.5)t}$$

$$= 8.771\underline{/45.26°}\, e^{(-0.5+j1.5)t} \text{ V}$$

Next, we obtain the *real response* $v_o(t)$ to the *real signal* $v_i(t)$ as

$$v_o(t) = \text{Re}[v_o(t)] \tag{14.14}$$

or

$$v_o(t) = 8.771e^{-0.5t}\cos(1.5t + 45.26°) \text{ V}$$

Clearly, the response is a *damped sinusoid* with the same damping factor σ and frequency ω as the applied signal, differing only in amplitude and phase angle. ◄

Exercise 14.2   Find the response of the circuit of Exercise 14.1 to the real signal $v_i(t) = 3e^{-t}\sin(2t + 60°)$ V.

ANSWER $v_o(t) = 1.973e^{-t}\cos(2t - 39.46°)$ V.

In our illustrative example we have obtained the relationship between the response $v_o(t)$ and the applied signal $v_i(t)$ via a *differential equation*. This tends to be a lengthy process, especially with circuits of greater complexity. We need a speedier procedure. As in the case of ac analysis, this procedure is based on *phasor algebra*. However, before presenting it, we must find the impedances of circuit elements and combinations thereof when subjected to complex exponential signals.

Generalized Impedance and Admittance

Let voltages and currents in a circuit have the complex exponential forms

$$v(t) = Ve^{st} \tag{14.15a}$$

$$i(t) = Ie^{st} \tag{14.15b}$$

Then, for *resistance* we have $v(t) = Ri(t)$, or $Ve^{st} = RIe^{st}$. Dropping the common term e^{st} yields

$$V = RI \tag{14.16a}$$

For *inductance* we have $v(t) = L\,di(t)/dt$, or $Ve^{st} = L\,d(Ie^{st})/dt = sLIe^{st}$. Dropping the common term e^{st} gives

$$V = (sL)I \tag{14.16b}$$

For *capacitance* we have $i(t) = C \, dv(t)/dt$, or $\boldsymbol{I}e^{st} = C \, d(\boldsymbol{V}e^{st})/dt = sC\boldsymbol{V}e^{st}$. Dropping the common term e^{st} and rearranging, we get

$$V = \frac{1}{sC}I \qquad \text{(14.16c)}$$

All three equations have the form of the **generalized Ohm's Law,**

$$\boxed{V = Z(s)I} \qquad \text{(14.17)}$$

where $Z(s)$ is called the **generalized impedance.** As usual, impedance is generally obtained as

$$\boxed{Z(s) \triangleq \frac{V}{I}} \qquad \text{(14.18)}$$

Its reciprocal is the **generalized admittance,** obtained as

$$\boxed{Y(s) = \frac{1}{Z} = \frac{I}{V}} \qquad \text{(14.19)}$$

Clearly, the simplest generalized impedances and admittances are those of the basic circuit elements,

$$
\begin{array}{ll}
Z_R = R & \text{(14.20a)} \\[6pt]
Z_L = sL & \text{(14.20b)} \\[6pt]
Z_C = \dfrac{1}{sC} & \text{(14.20c)} \\[10pt]
Y_G = \dfrac{1}{R} & \text{(14.21a)} \\[10pt]
Y_L = \dfrac{1}{sL} & \text{(14.21b)} \\[10pt]
Y_C = sC & \text{(14.21c)}
\end{array}
$$

Note the similarity with the impedances and admittances of ac circuits, except that we are now using s in place of $j\omega$.

s-Domain Circuit Analysis

The formal similarity of $Z(s)$ to $Z(j\omega)$ suggests that we can apply the techniques of phasor analysis developed for ac circuits to circuits subjected to *complex exponential signals.* Since it avoids the formulation of differential equations in favor of much simpler algebraic manipulations, this approach is quicker and less prone to error. To this end, we first redraw the circuit, but with signals replaced by their phasors and circuit elements replaced by their generalized impedances or admittances of Equations (14.20) and (14.21). Then, we apply standard circuit analysis techniques such as the *generalized Ohm's Law, series* and *parallel impedance combinations, voltage* and *current dividers,* the *proportionality analysis procedure,* the *node* and *loop methods,* Thévenin and *Norton reductions,* the *op amp rule,* and so forth, to obtain a relationship between the phasor of the response and that of the applied signal. It is fair to say that, aside from the replacement of $j\omega$ with s, this procedure is the same as that used in ac analysis.

▶ Example 14.3

Find the input-output relationship of the circuit of Figure 14.1, but via s-domain analysis.

Solution

First, redraw the circuit in s-domain form as in Figure 14.2. Then, use the voltage divider formula to obtain

$$V_o = \frac{Z_{R_2}}{Z_{R_2} + Z_p} V_i$$

$$Z_p = Z_{R_1} \parallel Z_C = \frac{R_1 \times 1/sC}{R_1 + 1/sC} = \frac{R_1}{R_1 Cs + 1}$$

Eliminating Z_p and simplifying gives

$$V_o = \frac{R_1 R_2 Cs + R_2}{R_1 R_2 Cs + (R_1 + R_2)} V_i \qquad (14.22)$$

Multiplying both sides by e^{st} again yields Equation (14.11), but without the tedious task of going through differential equations. ◀

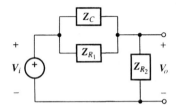

Figure 14.2 *s*-Domain representation of the circuit of Figure 14.1.

Exercise 14.3   Using s-domain analysis, find the input-output relationship of the circuit of Figure 14.1, but with the capacitance C replaced by an inductance L.

ANSWER $V_o = [(R_2 Ls + R_1 R_2)/(R_1 Ls + R_2 Ls + R_1 R_2)]V_i$.

▶ Example 14.4

Derive an expression for the generalized impedance $Z(s)$ of the one-port of Figure 14.3(a). Hence, find $Z(s)$ for $C = 2$ F, $L = 3$ H, $R = 4$ Ω, and $k_r = 5$ Ω.

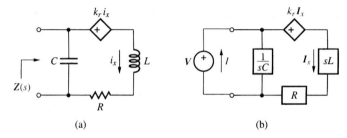

Figure 14.3 Finding the generalized impedance of a one-port.

Solution

First, redraw the circuit in phasor form as in Figure 14.3(b). Next, apply a test voltage V and find the resulting current I. Then, obtain the generalized impedance as $Z(s) = V/I$.

Applying KCL at the supernode surrounding the dependent source and using the generalized Ohm's Law, we obtain

$$I = \frac{V}{1/sC} + I_x = sCV + I_x$$

We need an additional equation to eliminate I_x. This is provided by KVL around the outer loop,

$$V = k_r I_x + sLI_x + RI_x = (k_r + R + sL)I_x$$

Eliminating I_x, rearranging, and taking the ratio V/I yields

$$Z(s) = \frac{V}{I} = \frac{Ls + k_r + R}{LCs^2 + (k_r + R)Cs + 1}$$

Finally, substituting the given component values gives

$$Z(s) = \frac{3s + 9}{6s^2 + 18s + 1}$$

◀

Exercise 14.4   Repeat Example 14.4, but with the CCVS changed to a VCVS of value $k_v v_x$, where $k_v = 8$ V/V, and v_x is the voltage across the inductance, positive side at the top.

ANSWER $Z(s) = (27s + 4)/(54s^2 + 8s + 1)$.

▶**Example 14.5**

If we apply the current

$$i(t) = 6e^{-5t}\cos(3t + 45°) \text{ A}$$

to the one-port of Example 14.4, what is the voltage $v(t)$ appearing across its terminals?

Solution

We have $i(t) = \text{Re}[\boldsymbol{i}(t)] = \text{Re}[\boldsymbol{I}e^{st}]$, with $\boldsymbol{I} = 6\underline{/45°}$ A, and $s = -5 + j3$ complex Np/s. Applying Ohm's Law and using the expression for $\boldsymbol{Z}(s)$ derived in Example 14.4,

$$\boldsymbol{V} = \boldsymbol{Z}(s)\boldsymbol{I}$$

$$= \frac{3(-5 + j3) + 9}{6(-5 + j3)^2 + 18(-5 + j3) + 1} 6\underline{/45°} = 0.5143\underline{/-104.49°} \text{ V}$$

Finally, letting $v(t) = \text{Re}[\boldsymbol{v}(t)] = \text{Re}[\boldsymbol{V}e^{st}]$ yields

$$v(t) = 0.5143e^{-5t}\cos(3t - 104.49°) \text{ V}$$

◀

Exercise 14.5 If we apply the voltage $v(t) = 3e^{-2t}\cos(4t + 45°)$ V to the one-port of Exercise 14.4, what is the current $i(t)$ drawn by the one-port?

ANSWER $i(t) = 26.82e^{-2t}\cos(4t + 161.61°)$ A

We will encounter many more examples of *s*-domain analysis as we proceed.

14.2 NETWORK FUNCTIONS

In spite of its simplicity, the illustrative example of the previous section has evidenced important features that hold for circuits with energy-storage elements in general, regardless of their complexity. We now wish to generalize our observations.

Given a network made up of *resistances, capacitances, inductances,* and possibly *dependent sources,* the applied signal $\boldsymbol{x}(t)$ and the response $\boldsymbol{y}(t)$ are in general related by a **linear differential equation** of the type

$$b_n\frac{d^n y}{dt^n} + b_{n-1}\frac{d^{n-1} y}{dt^{n-1}} + \cdots + b_1\frac{dy}{dt} + b_0 y = a_m\frac{d^m x}{dt^m} + a_{m-1}\frac{d^{m-1} x}{dt^{m-1}} + \cdots$$

$$+ a_1\frac{dx}{dt} + a_0 x \qquad \textbf{(14.23)}$$

where a_0 through a_m and b_0 through b_n are suitable coefficients whose expressions depend on the elements comprising the network and are thus *real* and *time-independent.*

With an applied signal of the complex exponential type

$$x(t) = Xe^{st} \qquad \text{(14.24)}$$

the response is also of the complex exponential type

$$y(t) = Ye^{st} \qquad \text{(14.25)}$$

and the value of s in the expression for $y(t)$ is the *same* as that given in the expression for $x(t)$. Substituting x and y into Equation (14.23) and exploiting the important exponential property

$$\frac{d^k(e^{st})}{dt^k} = s^k(e^{st}) \qquad \text{(14.26)}$$

we obtain

$$(b_n s^n + b_{n-1} s^{n-1} + \cdots + b_1 s + b_0)Ye^{st} = (a_m s^m + a_{m-1} s^{m-1} + \cdots + a_1 s + a_0)Xe^{st}$$

Eliminating the common term e^{st}, this can be abbreviated as

$$D(s)Y = N(s)X \qquad \text{(14.27)}$$

where $D(s)$ and $N(s)$ are **polynomials** in s with **real coefficients,**

$$D(s) = b_n s^n + b_{n-1} s^{n-1} + \cdots + b_1 s + b_0 \qquad \text{(14.28a)}$$

$$N(s) = a_m s^m + a_{m-1} s^{m-1} + \cdots + a_1 s + a_0 \qquad \text{(14.28b)}$$

and n and m are their **degrees.** These polynomials are, respectively, the left and right members of Equation (14.23), but with derivatives replaced by powers of s. Solving Equation (14.27) for the ratio

$$H(s) \triangleq \frac{Y}{X} \qquad \text{(14.29a)}$$

generally referred to as the **network function,** we obtain

$$H(s) = \frac{N(s)}{D(s)} = \frac{a_m s^m + a_{m-1} s^{m-1} + \cdots + a_1 s + a_0}{b_n s^n + b_{n-1} s^{n-1} + \cdots + b_1 s + b_0} \qquad \text{(14.29b)}$$

Since this function is in the form of a ratio of two polynomials with integer powers of s, it is a **rational function** of s.

Depending on the nature of x and y as well as the port at which we apply x and the port at which we observe y, the same circuit will generally yield a variety of different network functions. These functions take on different names, as follows:

(1) If the y-port is different from the x-port, then $H(s)$ is referred to as a **transfer function.** If x and y have the same physical dimensions, then $H(s)$ is referred to as a *gain function* and is expressed in V/V or in A/A. Otherwise, it has the dimensions of an impedance or an admittance, depending on whether x and y are, respectively, a current and a voltage, or a voltage and a current. Sometimes the expressions *transimpedance* and *transadmittance* are used for these types of transfer functions.

(2) If the y-port is the same as the x-port, then $H(s)$ is called a **driving-point function.** Specifically, it is a *driving-point impedance* if x is the current and y the voltage of the port, a *driving-point admittance* if x is the voltage and y the current.

Once $H(s)$ is known, the phasor Y of the response is readily found from the phasor X of the applied signal as

$$Y = H(s)X \tag{14.30}$$

Multiplying both sides by e^{st} yields

$$\boxed{y(t) = H(s)x(t)} \tag{14.31}$$

where $H(s)$ is calculated at the particular value of s appearing in the exponent of the applied signal $x(t) = Xe^{st}$.

Zeros and Poles

The special values of s that satisfy the equation $N(s) = 0$ are called the *roots* or *zeros* of $N(s)$ and are denoted as $z_1, z_2, \ldots, z_m$. Likewise, the special values of s that make $D(s) = 0$ are called the *roots* or *zeros* of $D(s)$ and are denoted as $p_1, p_2, \ldots, p_n$. Factoring out $N(s)$ and $D(s)$ in terms of their respective roots in Equation (14.29), we can express the network function in the form

$$\boxed{\begin{aligned} H(s) &= K \frac{(s - z_1)(s - z_2) \cdots (s - z_m)}{(s - p_1)(s - p_2) \cdots (s - p_n)} \tag{14.32a} \\ K &= \frac{a_m}{b_n} \tag{14.32b} \end{aligned}}$$

where K is a *scaling factor*. Since a_m and b_n are real, so is K.

The roots $z_1, z_2, \ldots, z_m$ are called the **zeros** of $H(s)$ because $H(s)$ becomes zero at these values of s. By contrast, the roots $p_1, p_2, \ldots, p_n$, which are the zeros of $D(s)$, are called the **poles** of $H(s)$ because $H(s)$ becomes infinite at these values of s. The zeros and the poles are collectively referred to as the **roots** or the **critical frequencies.** They are special complex-frequency values that depend solely on the circuit, regardless of the applied signals or the energy conditions of its reactive elements. Equation (14.32a) indicates that once the zeros and the poles are known, $H(s)$ is also known, aside from the scale factor K. In general the critical frequencies are complex quantities, though in many cases they may be real. Since the coefficients of $N(s)$ and $D(s)$ are real, it follows from polynomial theory that whenever the roots are complex, they always appear as **conjugate pairs.** We observe that a conjugate pair at

$$s = \alpha \pm j\omega \qquad \text{(14.33a)}$$

yields the polynomial term $P(s) = [s - (\alpha + j\omega)][s - (\alpha - j\omega)] = [(s - \alpha) - j\omega][(s - \alpha) + j\omega] = (s - \alpha)^2 + \omega^2$, or

$$P(s) = s^2 - 2\alpha s + \alpha^2 + \omega^2 \qquad \text{(14.33b)}$$

Conversely, given a polynomial term of the form

$$P(s) = s^2 + 2\alpha s + \omega_0^2 \qquad \text{(14.34a)}$$

the roots are

$$s = -\alpha \pm \sqrt{\alpha^2 - \omega_0^2} \qquad \text{(14.34b)}$$

These roots are *complex conjugate* if $\alpha^2 < \omega_0^2$, *real* if $\alpha^2 \geq \omega_0^2$, and *coincident* if $\alpha^2 = \omega_0^2$.

A polynomial term of the type

$$P(s) = (s + \alpha)^r \qquad \text{(14.35a)}$$

corresponds to a **repeated** root at

$$s = -\alpha \qquad \text{(14.35b)}$$

and r is called the **multiplicity** of that root.

The critical frequencies of $H(s)$ are conveniently visualized as points in the complex or s plane, with their real parts plotted against the horizontal axis and their imaginary parts against the vertical axis. In these plots, called **pole-zero plots,** a *zero* is denoted as o, and *a pole* as x. As we proceed, we shall see that essential circuit properties can be gleaned by just looking at the pole-zero pattern!

If $n > m$, Equation (14.29b) yields $\lim_{s \to \infty} H(s) = 0$, indicating that $H(s)$ has also $(n - m)$ *zeros* at infinity. Conversely, if $n < m$, we have $\lim_{s \to \infty} H(s) = \infty$, indicating that $H(s)$ has also $(m - n)$ *poles* at infinity. Clearly, in our pole-zero plots we are able to show only finite zeros and poles.

▶ Example 14.6

Show the pole-zero plot of the function

$$H(s) = 10 \frac{(s + 5)^2(s^2 + 4s + 13)}{s^2(s + 3)(s^2 + 2s + 2)} \qquad \textbf{(14.36)}$$

Solution

Using Equations (14.34) and (14.35), we find that $H(s)$ has a double zero at -5, a conjugate zero pair at $-2 \pm j3$, a double pole at the origin, a pole at -3, and a conjugate pole pair at $-1 \pm j1$. Moreover, it has a zero at infinity. Its pole-zero plot is shown in Figure 14.4.

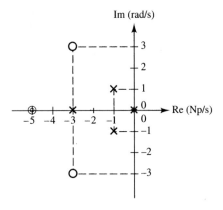

Figure 14.4 Pole-zero plot for the function of Equation (14.36).

Exercise 14.6   Find the values of the scale factor, zero, and poles of the impedance $Z(s)$ of Example 14.4. Hence, show its pole-zero plot.

ANSWER $K = 0.5$, $z_1 = -3$, $p_{1,2} = -1.5 \pm 2.5/\sqrt{3}$.

▶ Example 14.7

A function $H(s)$ has the following zero and pole values: $z_1 = -2$ with multiplicity 2; $z_{2,3} = -1 \pm j2$; $p_1 = 0$; $p_2 = -3$ with multiplicity 2;

$p_{3,4} = -2 \pm j1$; $p_{5,6} = 0 \pm j3$. Assuming a scale factor $K = 100$, calculate $H(s)$ at $s = -5 + j1$.

Solution

Using Equations (14.33) and (14.35), we find

$$H(s) = 100\,\frac{(s+2)^2(s^2+2s+5)}{s(s+3)^2(s^2+4s+5)(s^2+9)}$$

Substituting $s = -5 + j1$ yields $H(-5+j1) = 2.168\underline{/-124.72°}$. ◀

Exercise 14.7   A network function $H(s)$ has the following zero and pole values: $z_1 = -1$ with multiplicity 2; $p_1 = 0$ with multiplicity 3; $p_{2,3} = -2 \pm j4$. If it is found that $|H(j3)| = 0.1$, what is the scale factor K of this function?

ANSWER   4.395.

Physical Interpretation of Zeros and Poles

Each point of the s plane represents the complex frequency of a potential forcing signal. The phasor Y of the response is obtained by multiplying the phasor X of the forcing signal by the value of H at *that specific point,* or $Y = H(s)X$. This, in turn, splits into the separate magnitude and phase relationships,

$$|Y| = |H(s)| \times |X| \tag{14.37a}$$
$$\angle Y = \angle H(s) + \angle X \tag{14.37b}$$

Both $|H(s)|$ and $\angle H(s)$ can be plotted point by point versus s to obtain **magnitude** and **phase plots.** Since s is a two-dimensional parameter, we can plot $|H(s)|$ and $\angle H(s)$ as vertical distances from the s plane to obtain a pair of *surfaces.* We are especially interested in the *magnitude surface* because it facilitates understanding the significance of the critical frequencies.

Figure 14.5 shows the *magnitude plot* for a network function having a zero at the origin and a pole pair at $-3 \pm j4$. In general, a magnitude plot appears as *a tent pitched on the s plane.* The height of the tent becomes infinite for particular values of s aptly called *poles* and touches the s plane for values of s called *zeros.* These particular points are the *critical frequencies* of the circuit. We now wish to investigate their physical significance.

To find the meaning of a *zero* at $s = z_k$, suppose we apply a signal $x(t) = Xe^{z_k t}$. Then, by Equation (14.31) the response is $y(t) = H(z_k)x(t) = 0x(t) = 0$. We thus conclude that subjecting a circuit to a signal with complex frequency s equal to one of the zeros of its network function will result in a zero response, regardless of the amplitude and phase angle of the applied signal.

To find the meaning of a pole at $s = p_k$, suppose we apply a signal $x(t) = Xe^{st}$ with s *sufficiently close* to p_k to make $|H(s)|$ *fairly large.* This means that

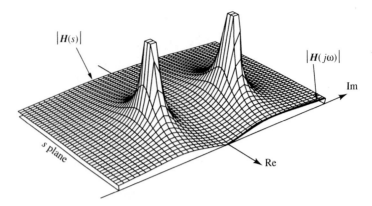

Figure 14.5   Magnitude plot of $H(s) = s/(s^2+6s+25)$. The intersection of $|H(s)|$ with the vertical plane passing through the imaginary axis yields the frequency response $|H(j\omega)|$.

in order to sustain a response $y(t)$ of a given amplitude $|Y|$, an applied signal $x(t)$ of *vanishingly small amplitude* $|X|$ will suffice. In fact, the closer s to p_k, the smaller $|X|$ for a given $|Y|$. In the limit $s \rightarrow p_k$ the circuit will supply a *nonzero response* even with *no applied signal!* As we know, this is a *source-free* or a *natural response component,* indicating that a pole p_k contributes a natural term of the type $y_k(t) = Y_k e^{p_k t}$. This contribution is made possible by the ability of the reactive elements in the circuit to release previously stored energy.

Procedure for Finding Network Functions

A network function is found using *s-domain phasor techniques.* These techniques are formally identical to the ω-domain techniques of ac circuits, except that we use s instead of $j\omega$. When deriving network functions, you should develop the habit of *always checking* your results. Checking is just as important as the derivation itself! Two simple checking techniques are *dimensional verification* and *asymptotic verification.*

Dimensional verification is based on the fact that all terms appearing in the same polynomial, be it $N(s)$ or $D(s)$, must have the *same physical dimensions.* Moreover, $H(s)$ must have the same dimensions as the ratio Y/X. Any dimensional inconsistency will indicate that an error was committed somewhere and that we need to go over our derivations more carefully. Denoting the physical dimensions of a quantity z as $[z]$, we can facilitate dimensional verification by observing that

$$[R] = [sL] = [1/sC] = \Omega \tag{14.38a}$$

$$[sRC] = [sL/R] = [s^2LC] = \text{dimensionless} \tag{14.38b}$$

Asymptotic verification is based on the use of *physical insight* to check the values of $H(s)$ in the limits $s \rightarrow 0$ and $s \rightarrow \infty$. Any inconsistency will again indicate that an error was committed somewhere. Physical insight is facilitated by the following observations stemming from Equation (14.20),

$$\lim_{s \to 0} \mathbf{Z}_L = 0 \tag{14.39a}$$

$$\lim_{s \to 0} \mathbf{Z}_C = \infty \tag{14.39b}$$

$$\lim_{s \to \infty} \mathbf{Z}_L = \infty \tag{14.40a}$$

$$\lim_{s \to \infty} \mathbf{Z}_C = 0 \tag{14.40b}$$

In words, in the low-frequency or dc limit an inductance acts as a short circuit and a capacitance as an open; in the high-frequency limit an inductance acts as an open and a capacitance as a short.

Following is the procedure for deriving and checking $H(s)$:

(1) Draw the circuit in phasor form, with each inductance L replaced by its generalized impedance sL and each capacitance C replaced by its generalized impedance $1/sC$.

(2) Using standard phasor analysis techniques (Ohm's Law, KCL and KVL, series/parallel impedance combinations, voltage and current dividers, the node and loop methods, Thévenin/Norton conversions, the op amp rule, and so forth), obtain the network function as the ratio $H(s) = Y/X$.

(3) Check that all numerator terms have the same dimensions, that all denominator terms have the same dimensions, and that the ratio of the two sets has the same dimensions as the ratio Y/X.

(4) Calculate $H(0) = \lim\limits_{s \to 0} H(s)$. Verify this value physically by analyzing the circuit with all capacitances open-circuited and all inductances short-circuited.

(5) Calculate $H(\infty) = \lim\limits_{s \to \infty} H(s)$. Verify this value physically by analyzing the circuit with all capacitances short-circuited and all inductances open-circuited.

Though these checks do not necessarily guarantee the correctness of your results, they nevertheless prove quite effective in flagging errors. Learn to apply them as a matter of course!

▶ **Example 14.8**

Find the network function for the circuit of Figure 14.6.

Solution

By inspection,

$$V_o = [(R_1 + sL) \parallel (1/sC + R_2)]I_i = \frac{(R_1 + sL)(1/sC + R_2)}{R_1 + sL + 1/sC + R_2}I_i$$

Multiplying out, simplifying, and taking the ratio V_o/I_i yields

$$H(s) = \frac{V_o}{I_i} = \frac{R_2 LCs^2 + (R_1 R_2 C + L)s + R_1}{LCs^2 + (R_1 + R_2)Cs + 1} \qquad (14.41)$$

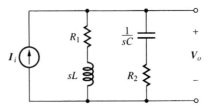

Figure 14.6 Circuit of Example 14.8.

Dimensional check. Each numerator term has the dimensions of resistance, each denominator term is dimensionless, and the ratio of the two sets is in Ω, the same dimensions as the ratio V_o/I_i. Our function thus passes the dimensional test. Suppose, however, we had erroneously obtained R_1 instead of 1 as the last denominator term. This would have caused a dimensional conflict, indicating that an error was committed somewhere.

Asymptotic check. Letting $s \rightarrow 0$ in Equation (14.41) yields

$$H(0) = R_1$$

To verify physically, we refer to the equivalent circuit of Figure 14.7(a), where it is readily seen that $V_o/I_i = R_1$, by Ohm's Law. Likewise, letting $s \rightarrow \infty$ in Equation (14.41) makes the terms in s^2 dominate, so that $H(\infty) = R_2 LCs^2 / LCs^2$, or

$$H(\infty) = R_2$$

a result that one can easily confirm physically by referring to the equivalent circuit of Figure 14.7(b). Consequently, our network function also passes the asymptotic test, giving us good reason to believe it is correct. Suppose, however, we had erroneously obtained R_2 instead of R_1 as the last numerator term in Equation (14.41). Then, this equation would have predicted $H(0) = R_2$, which contradicts the physical evidence provided by Figure 14.7(a) and, hence, indicates the presence of an error.

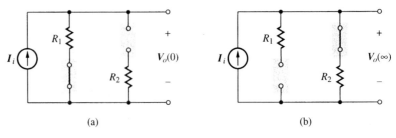

(a)   (b)

Figure 14.7 Circuit of Figure 14.6 in the limits $s \rightarrow 0$ and $s \rightarrow \infty$.

Exercise 14.8   Find the network function for the circuit of Figure 14.6 for the case in which the response is the current I_o through R_2, assumed to flow downward. Check your result!

ANSWER $(LCs^2 + R_1Cs)/[LCs^2 + (R_1 + R_2)Cs + 1]$.

▶ **Example 14.9**

Find the network function for the circuit of Figure 14.8. Check your result.

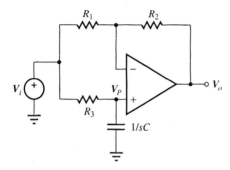

Figure 14.8   Circuit of Example 14.9.

Solution

Using the superposition principle, along with the inverting and noninverting amplifier formulas, we have

$$V_o = -\frac{R_2}{R_1}V_i + \left(1 + \frac{R_2}{R_1}\right)V_p \qquad \textbf{(14.42)}$$

But, by the voltage divider formula,

$$V_p = \frac{1/sC}{R_3 + 1/sC}V_i = \frac{1}{R_3Cs + 1}V_i$$

Eliminating V_p and collecting gives

$$H(s) = \frac{V_o}{V_i} = \frac{-R_2R_3Cs + R_1}{R_1R_3Cs + R_1} \qquad \textbf{(14.43)}$$

   Dimensional check. Both the numerator and denominator terms have the dimensions of resistance, and their ratio is dimensionless like the ratio V_o/V_i.

   Asymptotic check. Letting $s \to 0$ in Equation (14.43) yields

$$H(0) = 1 \text{ V/V}$$

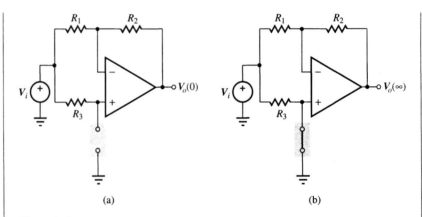

Figure 14.9 Circuit of Figure 14.8 in the limits $s \to 0$ and $s \to \infty$.

To confirm this result physically, refer to Figure 14.9(a), where it is readily seen that $V_p = V_i$. Substituting into Equation (14.42) yields $V_o = V_i$, or $H(0) = 1$ V/V. Likewise, letting $s \to \infty$ in Equation (14.43) yields

$$H(\infty) = -\frac{R_2}{R_1}$$

This is confirmed by Figure 14.9(b), where the op amp now acts as an inverting amplifier with gain $-R_2/R_1$. ◀

Exercise 14.9   Repeat Example 14.9, but with R_3 and C interchanged with each other. Check your result!

ANSWER $(R_1 R_3 C s - R_2)/(R_1 R_3 C s + R_1)$.

14.3 THE NATURAL RESPONSE USING $H(s)$

In this section we illustrate the use of the network function $H(s)$ to derive the *natural response* of a circuit. This is the response provided by the circuit with no applied signals,

$$x(t) = 0 \tag{14.44}$$

Also called the *source-free response,* it stems from the ability of the reactive elements in the circuit to store energy. Substituting Equation (14.44) into Equation (14.23) yields

$$b_n \frac{d^n y}{dt^n} + b_{n-1} \frac{d^{n-1} y}{dt^{n-1}} + \cdots + b_1 \frac{dy}{dt} + b_0 y = 0$$

For a circuit to yield a response $y(t) \neq 0$, the characteristic equation associated with this differential equation must vanish,

$$D(s) = 0 \tag{14.45}$$

But, the values of s satisfying this equation are precisely the *poles* p_1 through p_n of $H(s)$, indicating that the network function contains all the information needed to predict the functional form of the natural response. This is the real function

$$y(t) = A_1 e^{p_1 t} + A_2 e^{p_2 t} + \cdots + A_n e^{p_n t} \qquad (14.46)$$

where p_1 through p_n are the *poles* of $H(s)$, and A_1 through A_n are suitable *time-independent coefficients* reflecting the initial conditions in the circuit. We identify the following important cases:

(1) *Real poles.* If a pole p_k is *real,*

$$p_k = \alpha_k + j0 \qquad (14.47)$$

its contribution to the natural response is

$$y_k(t) = A_k e^{\alpha_k t} \qquad (14.48)$$

As we know, this is an *exponential decay* if the pole is *negative,* a *diverging exponential* if the pole is *positive,* and a *constant function* if the pole lies right at the origin of the s plane. In each case A_k is also *real* because so is y_k.

(2) *Complex pole pairs.* We know that if poles are *complex,* they appear in *conjugate pairs* because the coefficients of $D(s)$ are all real. In order for the contribution y_k to the natural response by a conjugate pair to be real, the coefficients of the corresponding exponential terms must also be complex conjugate. This means that given the pole

$$p_k = \alpha_k + j\omega_k \qquad (14.49)$$

yielding the complex exponential term $A_k e^{p_k t}$ with $A_k = A_k \underline{/\theta_k}$, the conjugate pole $p_k^* = \alpha_k - j\omega_k$ must yield the term $A_k^* e^{p_k^* t}$ with $A_k^* = A_k \underline{/-\theta_k}$. When combined together as a pair, their contribution to the natural response is

$$y_k(t) = A_k e^{p_k t} + A_k^* e^{p_k^* t} = A_k e^{p_k t} + \left(A_k e^{p_k t} \right)^*$$
$$= 2 \operatorname{Re}[A_k e^{p_k t}] = 2 \operatorname{Re}[A_k e^{(\alpha_k + j\omega_k)t}]$$

or

$$y_k(t) = 2 A_k e^{\alpha_k t} \cos(\omega_k t + \theta_k) \qquad (14.50)$$

As we know, this is a *damped sinusoid* if α_k, the real part of the pole pair, is *negative,* a *diverging sinusoid* if α_k is *positive,* and a *sustained* *sinusoid if $\alpha_k = 0$. In all cases the frequency of oscillation is ω_k, the imaginary part of p_k.*

(3) *Repeated poles.* If a pole p_k is a repeated root with multiplicity r_k, the r_k exponentials $e^{p_k t}$ are factored out as a common term, and the contribution by this multiple pole to the natural response takes on the form

$$y_k(t) = [A_{k,0} + A_{k,1}t + \cdots + A_{k,r_k-1}t^{r_k-1}]e^{p_k t} \qquad \text{(14.51)}$$

where p_k is the multiple pole and r_k is its multiplicity.

Figure 14.10 provides a visual correspondence between the location of a pole in the s plane and the functional form of its contribution to the natural response. We observe the following:

(1) Poles in the *left* half-plane yield *decaying* components; poles in the *right* half-plane yield *diverging* components.

(2) *Real* poles yield *nonoscillatory* components; *complex* pole pairs yield *oscillatory* components.

(3) A pole at the *origin* yields a *dc* component; a pole pair *right on the imaginary axis* yields an *ac* component.

In Chapter 8 we have seen that the pole of a **first-order** circuit is always *real*. If the circuit contains *no active elements* such as amplifiers, the pole lies on the *negative* real axis, and the response is a decaying exponential. The inclusion of active elements makes it possible to move the pole onto the *positive* real axis, where the response diverges. A pole right at the *origin* corresponds to a *constant* or *dc* response. In this case the circuit is called an *integrator,* and

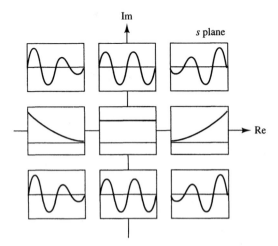

Figure 14.10 The location of poles in the s plane determines the functional form of the natural response.

the function it provides in the absence of any forcing signal is referred to as the *memory function*.

In Chapter 9 we have seen that the poles of a **second-order** circuit may be distinct, coincident, or complex conjugate, depending on the *damping condition*. In the absence of any active elements the poles lie in the *left half* of the *s* plane. However, with the inclusion of active elements the poles can be moved into the *right half*. If the poles are positioned right on the *imaginary axis*, the circuit is an *oscillator*.

▶ **Example 14.10**

If a circuit has the network function

$$H(s) = \frac{8(s^2 + 2s + 1)}{s(s + 4)^3(s + 5)(s^2 + 4s + 13)}$$

what is the functional form of its natural response?

Solution

Since $D(s)$ has degree 7, we have *seven* poles: $p_1 = 0$, $p_2 = -4$ with $r_2 = 3$, $p_3 = -5$, and the complex-conjugate pair $p_{4,5} = -2 \pm j3$. The form of the natural response is

$$y(t) = A_1 + (A_{2,0} + A_{2,1}t + A_{2,2}t^2)e^{-4t} + A_3e^{-5t}$$
$$+ 2A_4e^{-2t}\cos(3t + \theta_4)$$

All *seven* constants A_1, $A_{2,0}$, $A_{2,1}$, $A_{2,2}$, A_3, A_4, and θ_4 are real and depend on the initial conditions in the circuit.   ◀

Exercise 14.10   The natural response of a certain circuit is

$$y(t) = 9 + 7t - 5(1 + 0.5t)e^{-3t} + 8e^{-t} + 6e^{-2t}\cos(4t + 45°)$$

Find the denominator of its network function.

ANSWER $D(s) = s^2(s + 3)^2(s + 1)(s^2 + 4s + 20)$.

Critical Frequencies of Source-Free Circuits

Since the poles of the network function determine the functional form of the source-free response, the problem arises of how to derive such a function for a circuit that has no sources. We shall consider the case of a passive one-port. If its *driving-point impedance* is $\mathbf{Z}(s)$, we can express its terminal characteristic as

$$v(t) = \mathbf{Z}(s)i(t)$$

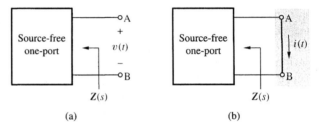

Figure 14.11 The *poles* of $\mathbf{Z}(s)$ determine the functional form of the open-circuit voltage $v(t)$, and the *zeros* of $\mathbf{Z}(s)$ that of the short-circuit current $i(t)$.

where $v(t)$ and $i(t)$ are, respectively, the complex exponential voltage and current at the terminals of the port. When $\mathbf{Z}(s)$ becomes very large, a voltage can exist with only a small applied current. In the limit $\mathbf{Z}(s) \rightarrow \infty$, such a voltage can exist even with *zero applied current*. As depicted in Figure 14.11(a), this is the *natural open-circuit voltage* response $v(t)$ of the port.

Expressing the terminal characteristic in the alternate form

$$i(t) = \mathbf{Y}(s)v(t)$$

where $\mathbf{Y}(s) = 1/\mathbf{Z}(s)$ is the *driving-point admittance* of the port, we observe that in the limit $\mathbf{Y}(s) \rightarrow \infty$ the circuit will provide a nonzero current response even with zero applied voltage. This is the *natural short-circuit current* response $i(t)$ of the one-port; see Figure 14.11(b).

Observing that when $\mathbf{Y}(s) \rightarrow \infty$ we have $\mathbf{Z}(s) \rightarrow 0$, we summarize our findings by saying that in a *source-free, passive one-port* having driving-point impedance $\mathbf{Z}(s)$,

   (1) **The *poles* of $\mathbf{Z}(s)$ determine the functional form of the *open-circuit voltage* response.**

   (2) **The *zeros* of $\mathbf{Z}(s)$ determine the functional form of the *short-circuit current* response.**

The following examples illustrate how to find the response of a source-free circuit.

▶ **Example 14.11**

In the source-free circuit of Figure 14.12 let $R_1 = 2$ kΩ, $C = 5$ nF, $L = 1$ mH, and $R_2 = 0.5$ kΩ. If the initial conditions are $v_C(0) = 10$ V and $i_L(0) = 0$, find $v(t \geq 0)$.

Solution

Since the open-circuit response is determined by the poles of the driving-point impedance of the port, our first task is to find such an impedance. By inspection,

$$\mathbf{Z}(s) = R_2 \parallel [sL + (R_1 \parallel 1/sC)]$$

Expanding and collecting, we get

$$Z(s) = \frac{R_2(R_1LCs^2 + Ls + R_1)}{R_1LCs^2 + (R_1R_2C + L)s + (R_1 + R_2)} \quad \textbf{(14.52)}$$

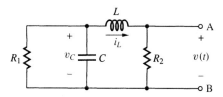

Figure 14.12 The open-circuit response of a source-free one-port.

Before proceeding, let us pause for dimensional and asymptotic checks. Each numerator term is in Ω^2 and each denominator term in Ω, indicating that their ratio is in Ω, as expected of an impedance. Moreover, physical inspection of the circuit reveals that $Z(0) = R_1 \parallel R_2$ and $Z(\infty) = R_2$. It is readily seen that Equation (14.52) agrees with these limiting values, giving us good reason to believe it is correct.

Next, substitute the given component values to obtain

$$Z(s) = 500\,\frac{s^2 + 10^5 s + 20 \times 10^{10}}{s^2 + 6 \times 10^5 s + 25 \times 10^{10}}\ \Omega \quad \textbf{(14.53)}$$

This function has a conjugate pole-pair $p_{1,2} = (-3 \pm j4)10^5$ complex Np/s. By Equation (14.50), the response is of the type

$$v(t) = 2Ae^{-3 \times 10^5 t} \cos\,(4 \times 10^5 t + \theta) \quad \textbf{(14.54)}$$

We now wish to find A and θ on the basis of the initial conditions.

Referring to Figure 14.12, we observe that $v = R_2 i_L$, so the first condition is

$$v(0) = R_2 i_L(0) \quad \textbf{(14.55)}$$

In our case, $v(0) = R_2 0 = 0$. But Equation (14.54) yields $v(0) = 2Ae^0 \cos\,(0 + \theta) = 2A\cos\theta$, so the first condition yields $2A\cos\theta = 0$. Since we are seeking a response with $A \neq 0$, we must have $\cos\theta = 0$, or $\theta = 90°$, so Equation (14.54) becomes

$$v(t) = 2Ae^{-3 \times 10^5 t} \sin\,(4 \times 10^5 t) \quad \textbf{(14.56)}$$

Next, we use the inductance law to write $dv/dt = R_2\,di_L/dt = R_2(v_C - v)/L$. The second condition is thus

$$\frac{dv(0)}{dt} = \frac{R_2}{L}[v_C(0) - v(0)] \quad \textbf{(14.57)}$$

In our case, $dv(0)/dt = (500/10^{-3})(10 - 0) = 5 \times 10^6$ V/s. But Equation (14.56) yields $dv(0)/dt = 2A(-3 \times 10^5 e^0 \sin 0 + 4 \times 10^5 e^0 \cos 0) = (8 \times 10^5)A$, so the second condition yields $(8 \times 10^5)A = 5 \times 10^6$, or

$A = 6.25$ V. Finally, substituting into Equation (14.56) yields

$$v(t) = 12.5e^{-3\times10^5 t}\sin(4\times10^5 t)\ \text{V}$$

This function is plotted using PSpice in Figure 14.38.

Exercise 14.11 Using the data of Example 14.11, find the response $v_C(t)$ across the capacitance.

ANSWER $v_C(t) = 11.18e^{-3\times10^5 t}\cos(4\times10^5 t - 26.57°)$ V.

▶ **Example 14.12**

Repeat Example 14.11, but with $R_2 = 1$ kΩ.

Solution

Substituting the given component values into Equation (14.52), we find that the poles of $\mathbf{Z}(s)$ are now *real* and *distinct*, $p_1 = -5\times10^5$ Np/s and $p_2 = -6\times10^5$ Np/s. The response is thus

$$v(t) = A_1e^{-5\times10^5 t} + A_2e^{-6\times10^5 t}$$

Imposing the initial conditions of Equations (14.55) and (14.57),

$$A_1 + A_2 = 0$$

$$-5\times10^5 A_1 - 6\times10^5 A_2 = 10^7\ \text{V/s}$$

that is, $A_1 = -A_2 = 100$ V. The natural response is thus

$$v(t) = 100(e^{-5\times10^5 t} - e^{-6\times10^5 t})\ \text{V}$$

Exercise 14.12 In the source-free circuit of Figure 14.12 let again $R_1 = 2$ kΩ, $C = 5$ nF, $L = 1$ mH, $v_C(0) = 10$ V, and $i_L(0) = 0$. For what value of R_2 are the poles *coincident?* What is now the response $v(t)$?

ANSWER $R_2 = 994.4$ Ω; $v(t) = 9.944\times10^6 te^{-5.472\times10^5 t}$ V.

▶ **Example 14.13**

If the one-port of Example 14.11 is short-circuited as in Figure 14.13, find the response $i(t)$.

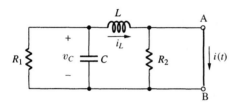

Figure 14.13 The short-circuit response of a source-free one-port.

Solution

The short-circuit response is determined by the *zeros* of the driving-point impedance $Z(s)$. Referring to Equation (14.53), we find that $Z(s)$ has the conjugate zero-pair $z_{1,2} = (-0.5 \pm j\sqrt{19.75})10^5$ complex Np/s. By Equation (14.50), the response is

$$i(t) = 2Ae^{-0.5 \times 10^5 t} \cos(10^5 \sqrt{19.75}\,t + \theta) \qquad \textbf{(14.58)}$$

To find A and θ, we first observe that

$$i(0) = i_L(0)$$

In our case $i(0) = 0$, so Equation (14.58) becomes

$$i(t) = 2Ae^{-0.5 \times 10^5 t} \sin(10^5 \sqrt{19.75}\,t) \qquad \textbf{(14.59)}$$

Next, we use the inductance law to write $di/dt = di_L/dt = v_L/L = v_C/L$ so that

$$\frac{di(0)}{dt} = \frac{v_C(0)}{L}$$

In our case, $di(0)/dt = 10/10^{-3} = 10^4$ A/s. But Equation (14.59) yields $di(0)/dt = 2A \times 10^5 \sqrt{19.75}$. Letting $2A \times 10^5 \sqrt{19.75} = 10^4$ yields $A = 11.25$ mA. Substituting into Equation (14.59), we finally obtain

$$i(t) = 22.50 e^{-0.5 \times 10^5 t} \sin(10^5 \sqrt{19.75}\,t) \text{ mA}$$

Remark   Short-circuiting the one-port has eliminated R_2 from the circuit, changing both the damping factor and the angular frequency.   ◀

Exercise 14.13   Sketch the one-port of Figure 14.13, but with L and C interchanged with each other. Assuming again $R_1 = 2$ kΩ, $L = 1$ mH, $C = 5$ nF, $R_2 = 0.5$ kΩ, $i_L(0) = 0$, and $v_C(0) = 10$ V, positive at the left, obtain an expression for the short-circuit response $i(t)$.

ANSWER $22.50 e^{-0.5 \times 10^5 t} \cos(10^5 \sqrt{19.75}\,t + 102.84°)$ mA.

CONCLUDING REMARKS

Looking back at the examples and exercises, we observe that finding the initial conditions for the response in terms of the initial conditions in the energy-storage elements can be a laborious process, especially if the circuit contains many such elements. We are interested in ways of expediting this process. In Chapter 16 we shall investigate a powerful analytical method known as the *Laplace transform method,* which takes the initial element conditions into account automatically.

14.4 THE COMPLETE RESPONSE USING $H(s)$

The network function $H(s)$ can be used to predict not only the natural response but also the *transient* and the *steady-state* responses and, hence, the *complete* response. As such, $H(s)$ provides a *unified approach,* but using simple algebra instead of differential equations. To investigate the various responses of a circuit, we subject it to the complex exponential signal

$$x(t) = Xe^{st}$$

and obtain the response by taking the product

$$y(t) = H(s)x(t) \qquad \text{(14.60)}$$

where $H(s)$ is calculated at the value of s supplied by the applied signal. Let us examine the most important response types.

The DC Steady-State Response

This is the response to a *dc signal* of the type

$$x = X_m$$

after *all transients have died out.* As we know, the complex exponential form of this signal is

$$x(t) = (X_m \underline{/0°})e^{0t}$$

so the complex exponential form of the response is

$$y(t) = H(0)(X_m \underline{/0°})e^{0t}$$

But this is the familiar *dc steady-state response,*

$$y_{ss} = H(0)X_m \qquad \text{(14.61)}$$

In words, to find the steady-state response to a dc signal of amplitude X_m, we calculate $H(s)$ at $s = 0$, and then multiply X_m by $H(0)$ to obtain y_{ss}. This forms the basis of the following rule:

DC Rule: In *dc analysis* $H(s)$ is calculated *at the origin of* the s plane.

Letting $s = 0$ makes $Z_L = sL = 0$ and $Z_C = 1/sC = \infty$, confirming that the dc steady-state response is the response with all *inductances* replaced by *short circuits* and all *capacitances* by *opens.*

▶**Example 14.14**

Find the steady-state response of the circuit of Figure 14.14 to

$$v_I = 5 \text{ V}$$

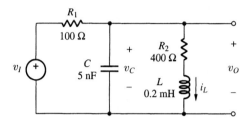

Figure 14.14 Circuit of Examples 14.14 through 14.16.

Solution

Denoting the input and output phasors as V_i and V_o, and applying the voltage divider formula, gives

$$V_o = \frac{(1/sC) \parallel (R_2 + sL)}{R_1 + [(1/sC) \parallel (R_2 + sL)]} V_i$$

Collecting terms and substituting the given component values, we get

$$H(s) = \frac{V_o}{V_i} = \frac{2 \times 10^{-4}s + 400}{10^{-10}s^2 + 4 \times 10^{-4}s + 500} \text{ V/V} \qquad \textbf{(14.62)}$$

The response is $v_O = H(0)v_I = \frac{400}{500}5$, or

$$v_O = 4 \text{ V}$$

◀

Exercise 14.14   Repeat Example 14.14, but with C and L interchanged with each other.

ANSWER $H(s) = 2 \times 10^{-4}s(2 \times 10^{-6}s + 1)/(5 \times 10^{-10}s^2 + 4 \times 10^{-4}s + 100)$ V/V; $v_O = 0$ V.

The AC Steady-State Response

This is the response to an *ac signal* of the type

$$x(t) = X_m \cos(\omega t + \theta_x)$$

after *all transients have died out.* As we know, this signal can be expressed as $x(t) = \text{Re}[\boldsymbol{x}(t)]$, where

$$\boldsymbol{x}(t) = (X_m \underline{/\theta_x}) e^{j\omega t}$$

The complex exponential response is

$$y(t) = H(j\omega)(X_m\underline{/\theta_x})e^{j\omega t}$$

and the *real* response is $y(t) = \text{Re}[y(t)]$. But this is the familiar *ac steady-state response*,

$$y_{ss} = |H(j\omega)|X_m \cos[\omega t + \theta_x + \measuredangle H(j\omega)] \qquad \text{(14.63)}$$

In words, to find the steady-state response to an ac signal of amplitude X_m, phase angle θ_x, and angular frequency ω, we calculate $H(s)$ at $s = j\omega$. We then multiply X_m by $|H(j\omega)|$ to obtain the amplitude of the response, and we augment θ_x by $\measuredangle H(j\omega)$ to obtain the phase angle. This forms the basis of the following rule:

AC Rule: In *ac analysis* $H(s)$ **is calculated on the** $j\omega$ **axis of the** s **plane.**

This is just what we have been doing all along in phasor analysis!

▶ **Example 14.15**

Find the steady-state response of the circuit of Figure 14.14 to

$$v_i = 10\cos(5 \times 10^6 t + 60°) \text{ V}$$

Solution

Calculating Equation (14.62) at $s = j\omega = j5 \times 10^6$ yields

$$H(j5 \times 10^6) = \frac{2 \times 10^{-4}(j5 \times 10^6) + 400}{10^{-10}(j5 \times 10^6)^2 + 4 \times 10^{-4}(j5 \times 10^6) + 500}$$

$$= \frac{400 + j1000}{-2000 + j2000} = 0.3808\underline{/-66.80°} \text{ V/V}$$

Applying Equation (14.63) yields $v_o = 0.3808 \times 10\cos(5 \times 10^6 t + 60° - 66.80°)$ V, or

$$v_o = 3.808\cos(5 \times 10^6 t - 6.80°) \text{ V} \qquad ◀$$

Exercise 14.15 Repeat Example 14.15, but with C and L interchanged with each other.

ANSWER $v_o = 8.001\cos(5 \times 10^6 t + 63.45°)$ V.

The Complete Response

Steady-state analysis focuses on the response after all transients have died out. In order for a circuit to reach this state, we implicitly assume that the forcing signal was turned on at a remote instant of the past. Mathematically, we say that the source was turned on at $t = -\infty$. If we are also interested in the *transient situation* following source turn-on and lasting until the circuit reaches its steady state, then we must look at the *complete response*. This too can be found via **H**(*s*).

As we know, the complete response consists of (a) a *transient* component taking on the form of the *natural response,* or Equation (14.46), and (b) a *steady-state* component taking on the form of the *applied signal,* or Equation (14.60). If the circuit is simultaneously subjected to more than one forcing signal, we first find the *individual* steady-state responses; then, we apply the *superposition principle* to add them up and obtain the *overall* steady-state response. Assuming ℓ forcing signals $x_1(t) = X_1 e^{s_1 t}$, $x_2(t) = X_2 e^{s_2 t}$, ..., $x_\ell(t) = X_\ell e^{s_\ell t}$, and a network function **H**(*s*) with *n* poles $p_1, p_2, \ldots, p_n$, the *complete response* is

$$
\begin{aligned}
y(t) = {} & B_1 e^{p_1 t} + B_2 e^{p_2 t} + \cdots + B_n e^{p_n t} + H(s_1) X_1 e^{s_1 t} \\
& + H(s_2) X_2 e^{s_2 t} + \cdots + H(s_\ell) X_\ell e^{s_\ell t}
\end{aligned}
\tag{14.64}
$$

where B_1 through B_n are suitable *real coefficients,* not necessarily the same as those of the *natural response,* to be determined on the basis of the initial conditions in the circuit. If a pole p_k has multiplicity r_k, the corresponding transient term takes on the form $(B_{k,0} + B_{k,1} t + \cdots + B_{k,r_k-1} t^{r_k-1}) e^{p_k t}$. For a complex-conjugate pole pair $\alpha_k \pm j\omega_k$, the transient term takes on the form $2 B_k e^{\alpha_k t} \cos(\omega_k t + \theta_k)$, where B_k is *real.*

It is worth noting that in Equation (14.64) we have two sets of frequencies that should not be confused with each other:

(1) p_1 through p_n are the **critical frequencies** of the circuit, pole frequencies in this case. Their values depend on circuit components and topology, regardless of the applied signals.

(2) s_1 through s_ℓ are the frequencies of the **applied signals.** If the applied signals are *ac signals* with angular frequencies $\omega_1, \omega_2, \ldots, \omega_\ell$, then $s_1 = j\omega_1$, $s_2 = j\omega_2$, ..., $s_\ell = j\omega_\ell$. Moreover, the complete response is obtained by taking the *real parts* of the complex exponential terms in the right-hand side of Equation (14.64).

▶ Example 14.16

Let the circuit of Figure 14.14 be subjected to the dc and ac signals of Examples 14.14 and 14.15 *simultaneously,*

$$
v_I(t) = 5 + 10 \cos(5 \times 10^6 t + 60°) \text{ V}
$$

Assuming zero initial stored energy in both reactive elements, find the complete response $v_O(t)$.

Solution

Equation (14.62) reveals that $H(s)$ has the conjugate pole pair $p_{1,2} = (-2 \pm j1)10^6$ complex Np/s. Taking the real part of Equation (14.64) and using the results of Examples 14.14 and 14.15, we can write

$$v_O(t) = 2Be^{-2 \times 10^6 t} \cos(10^6 t + \theta) + 4 + 3.808 \cos(5 \times 10^6 t - 6.80°) \text{ V}$$

To find the constants B and θ we must derive the initial conditions. Referring back to the circuit, it is readily seen that

$$v_O(0) = v_C(0)$$

$$\frac{dv_O(0)}{dt} = \frac{1}{C}\left[\frac{v_I(0) - v_C(0)}{R_1} - i_L(0)\right]$$

In the present case we have $v_O(0) = 0$ and $dv_O(0)/dt = (5 \times 10^{-9})^{-1}(5 + 10 \cos 60°)/100 = 2 \times 10^7$ V/s. Letting, respectively,

$$v(0) = 2Be^0 \cos\theta + 4 + 3.808 \cos(-6.80°) = 0 \text{ V}$$

$$\frac{dv_O(0)}{dt} = 2B(-2 \times 10^6 e^0 \cos\theta - 10^6 e^0 \sin\theta)$$
$$- 3.808 \times 5 \times 10^6 \sin(-6.80°) = 2 \times 10^7 \text{ V/s}$$

and solving for B and θ gives the complete response as

$$v_O(t) = -8.082 e^{-2 \times 10^6 t} \cos(10^6 t + 15.67°) + 4$$
$$+ 3.808 \cos(5 \times 10^6 t - 6.80°) \text{ V}$$

Remark   It is interesting to observe that 1×10^6 rad/s is the *natural frequency* of the circuit, whereas 5×10^6 rad/s is the frequency of the *applied signal*.   ◀

Exercise 14.16   Repeat Example 14.16, but with R_2 changed to 900 Ω. What is now the damping condition?

ANSWER $v_O(t) = 1.863 e^{-4 \times 10^6 t} - 10.10 e^{-2.5 \times 10^6 t} + 4.5 + 3.759 \times \cos(5 \times 10^6 t - 6.76°)$ V, overdamped.

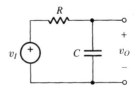

Figure 14.15 Circuit of Example 14.17.

▶ Example 14.17

In Figure 14.15 let $R = 10$ kΩ, $C = 50$ nF, and $v_I(t) = 10 \sin 2\pi 10^3 t$ V. Assuming *zero initial stored energy* in the capacitance, find $v_O(t)$. Hence, sketch and label v_I and v_O versus time for $t \geq 0$.

Solution

Using the voltage divider formula

$$H(s) = \frac{1/sC}{R + 1/sC} = \frac{1}{RCs + 1} = \frac{1}{5 \times 10^{-4}s + 1}$$

This function has the pole $p_1 = -2000$ Np/s. Denoting the applied frequency as $\omega_i = 2\pi 10^3$ rad/s, and the applied phasor as $V_i = V_{im}\underline{/\theta_i} = 10\underline{/-90°}$ V, the complete response takes on the form

$$v_O(t) = B_1 e^{p_1 t} + \text{Re}[H(j\omega_i)V_i e^{j\omega_i t}]$$

But, $H(j\omega_i) = 1/(5 \times 10^{-4} \times j2\pi 10^3 + 1) = 1/(1 + j\pi)$. Thus,

$$H(j\omega_i)V_i e^{j\omega_i t} = \frac{1}{1 + j\pi} 10\underline{/-90°} e^{j\omega_i t} = 3.033\underline{/-162.34°} e^{j\omega_i t} \text{ V}$$

Taking its real part and substituting gives

$$v_O(t) = B_1 e^{-2000t} + 3.033 \cos(2\pi 10^3 t - 162.34°) \text{ V}$$

To find B_1, we observe that since C is initially discharged, we have $v_O(0) = 0$. But $v_O(0) = B_1 e^0 + 3.033 \cos(0 - 162.34°)$, or $B_1 + 3.033 \cos(-162.34°) = 0$, which yields $B_1 = 2.890$ V. Finally,

$$v_0(t) = 2.890 e^{-2000t} + 3.033 \cos(2\pi 10^3 t - 162.34°) \text{ V}$$

The input and output waveforms for $t \geq 0$ are shown in Figure 14.16. For convenience, the complete response is repeated in Figure 14.17, along with its transient and steady-state components.

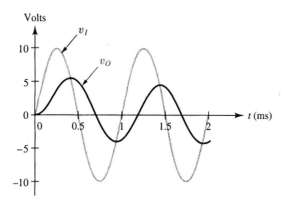

Figure 14.16 Waveforms for the *R-C* circuit of Example 14.17.

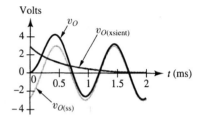

Figure 14.17 The complete response v_O of the *R-C* circuit of Example 14.17, along with its transient and steady-state components.

Remark   It is interesting to note the presence of the transient component even though the capacitance has zero initial stored energy. This component is needed to *satisfy the initial condition*. Had the initial capacitance voltage been *exactly* -2.890 V, then the transient component would have been absent as the circuit would already have been in its ac steady state at $t = 0$. ◀

Exercise 14.17   Repeat Example 14.17, but with R and C interchanged with each other.

ANSWER $v_O(t) = -2.890e^{-2000t} + 9.529 \cos(2\pi 10^3 t - 72.34°)$ V.

The Complete Response of *RC* and *RL* Circuits

The *RC* and *RL* circuits occur so often, especially as subcircuits, that their responses merit special attention. The applied signal, henceforth called the *input* v_I, will generally have both a *dc* component V_I and an *ac* component v_i,

$$v_I = V_I + v_i$$

where $v_i = V_{im} \cos(\omega t + \theta_i)$. The complete response, henceforth called the *output* v_O, takes on the form

$$v_O = Be^{-t/\tau} + V_O + v_o \qquad \textbf{(14.65)}$$

where V_O is the *dc* component and $v_o = V_{om} \cos(\omega t + \theta_o)$ is the *ac* component. Denoting the pole of $H(s)$ as p, the time constant is $\tau = -1/p$. Moreover, $V_O = H(0)V_I$, and $V_{om}\underline{/\theta_o} = H(j\omega)(V_{im}\underline{/\theta_i})$.

▶ Example 14.18

In the circuit of Figure 14.18 let $R = 15.9$ kΩ, $C = 0.1$ μF, and $v_S = 2 + 1 \sin 2\pi 10^3 t$ V, so that the input is

$$v_I(t \le 0^-) = 0$$

$$v_I(t \ge 0^+) = 2 + 1 \cos(2\pi 10^3 t - 90°)\ \text{V}$$

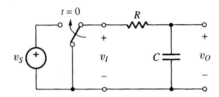

Figure 14.18 Turning on the input to an
R-C circuit.

Assuming the switch has been in the down position long enough to ensure zero initial stored energy, find v_O and plot both v_I and v_O versus t.

Solution

The transfer function is

$$H(s) = \frac{1}{RCs + 1} = \frac{1}{1.59 \times 10^{-3}s + 1}$$

and it has a pole $p = -1/(1.59 \times 10^{-3})$ Np/s. Thus, $\tau = -1/p = 1.59$ ms. Moreover, since $H(0) = 1$ and $H(j2\pi 10^3) \simeq 0.1\underline{/-84.3°}$, it follows that $V_O = 1 \times 2 = 2$ V, $V_{om} = 0.1 \times 1 = 0.1$ V, and $\theta_o = -84.3 - 90 = -174.3°$. Substituting into Equation (14.65) and using the initial condition $v_O(0^+) = v_C(0^+) = v_C(0^-) = 0$, we finally obtain

$$v_O(t \leq 0^-) = 0$$

$$v_O(t \geq 0^+) = -1.9e^{-t/(1.59 \text{ ms})} + 2 + 0.1\cos(2\pi 10^3 t - 174.3°) \text{ V}$$

The input and output waveforms are shown in Figure 14.19.

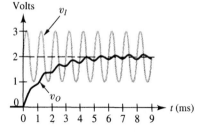

Figure 14.19 Waveforms for the *R-C* circuit of Example 14.19.

Exercise 14.18   Repeat Example 14.18, but with the *R-C* pair replaced by an *L-R* pair having $L = 10$ mH and $R = 20$ Ω.

ANSWER $v_O(t \leq 0^-) = 0$, $v_O(t \geq 0^+) = -1.711e^{-2000t} + 2 + 0.303\cos(2\pi 10^3 t - 162.3°)$ V.

▶ Example 14.19

Repeat Example 14.18, but with *R* and *C* interchanged with each other.

Solution

The circuit is shown in Figure 14.20. We have

$$H(s) = \frac{R}{R + 1/sC} = \frac{RCs}{RCs + 1} = \frac{1.59 \times 10^{-3}s}{1.59 \times 10^{-3}s + 1}$$

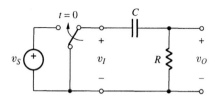

Figure 14.20 Turning on the input to a *C-R* circuit.

so $\tau = -1/p = 1.59$ ms. Moreover, since $H(0) = 0$ and $H(j2\pi 10^3) \simeq 0.995\underline{/5.7°}$, it follows that $V_O = 0 \times 2 = 0$ V, $V_{om} = 0.995 \times 1 = 0.995$ V, and $\theta_o = 5.7 - 90 = -84.3°$. Substituting into Equation (14.65) and using the initial condition $v_O(0^+) = v_I(0^+) - v_C(0^+) = v_I(0^+) - v_C(0^-) = 2 - 0 = 2$ V, we finally obtain

$$v_O(t \le 0^-) = 0$$

$$v_O(t \ge 0^+) = 1.9e^{-t/(1.59\,\text{ms})} + 0.995 \cos(2\pi 10^3 t - 84.3°) \text{ V}$$

The input and output waveforms are shown in Figure 14.21.

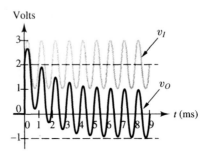

Figure 14.21 Waveforms for the C-R circuit of Example 14.20.

Exercise 14.19   Repeat Example 14.18, but with the RC pair replaced by an RL pair having $R = 20\ \Omega$ and $L = 10$ mH.

ANSWER $v_O(t \le 0^-) = 0$, $v_O(t \ge 0^+) = 1.711e^{-2000t} + 0.9529 \cos(2\pi 10^3 t - 72.34°)$ V.

• DC Passing and AC Blocking

Looking back at the response of Figure 14.19 we observe that once the transient condition has died out, the input dc component appears unchanged at the output. We thus say that once in steady state, the R-C circuit *passes the dc component in full,*

$$V_O = V_I \tag{14.66a}$$

Physically, we justify this behavior by noting that with respect to the dc component the capacitance acts as an *open circuit,* thus allowing R to fully transmit this component to the output.

By contrast, the ac component is attenuated by an amount that depends on its frequency ω as well as the time constant RC of the circuit. There are applications in which it is desired to *block out the ac component,*

$$v_o \simeq 0 \tag{14.66b}$$

Regarding the R-C circuit as an ac voltage divider, we observe that in order to provide a substantial attenuation at a given frequency ω, C *must behave as a short circuit compared to* R, that is, $|Z_C| \ll R$. Rephrasing as $1/\omega C \ll R$, this yields

$$\omega RC \gg 1 \qquad (14.67)$$

Summarizing, the R-C circuit will pass the dc component but block out *any ac signal* whose frequency is high enough to satisfy Equation (14.67). The discriminatory action by the R-C circuit upon its input components is depicted in Figure 14.22, where v_I is modeled with two separate sources in series. In steady state the output is thus $v_{O(ss)} = V_O + v_o \simeq V_I + 0$, or

$$v_{O(ss)} \simeq V_I \qquad (14.68)$$

A common application of the R-C circuit is in power supply distribution systems, where it is desired to block out any unwanted ac signals that may be present on the power supply rails. When used in this function, a capacitor is referred to as a **bypass capacitor.** In a well-designed printed-circuit board you will find a bypass capacitor next to each integrated circuit package.

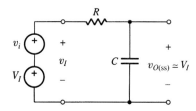

Figure 14.22 In steady state the R-C circuit *passes the dc component* of the input but *blocks out any ac component with* $\omega \gg 1/RC$.

▶ **Example 14.20**

A certain dc power supply yields 15 V with a 250-kHz, 0.5-V peak-to-peak spurious ac component called *ripple*. We wish to clean it up with an R-C network using $R = 22\ \Omega$.

(a) What bypass capacitance is required to ensure an output ripple of less than 2 mV peak-to-peak?

(b) After source turn-on, how long will it take for v_O to come within 1 mV of its steady-state dc value?

Solution

(a) Since $V_{om} = V_{im}/\sqrt{1 + (\omega RC)^2}$, it follows that $C = \sqrt{(V_{im}/V_{om})^2 - 1}/\omega R = \sqrt{(0.5/0.002)^2 - 1}/(2\pi \times 250 \times 10^3 \times 22) = 7.23\ \mu F$ (use 10 μF).

(b) We have $RC = 22 \times 10^{-5} = 0.22$ ms. The amount of time t it takes for v_O to rise to (15 V $-$ 1 mV) $= 14.999$ V can be estimated by imposing $14.999/15 = 1 - \exp[-t/(0.22\ \text{ms})]$. Solving for t yields $t = 0.22 \ln(15 \times 10^3) = 2.1$ ms. ◀

• **DC Blocking and AC Passing**

Figure 14.21 indicates that the C-R circuit provides the opposite function of the R-C circuit. Namely, once the transient condition has died out, it *blocks out the dc component* to yield

$$V_O = 0 \qquad (14.69a)$$

but it *passes any ac component* whose frequency is high enough to make C act as an ac short in comparison with R,

$$v_o \simeq v_i \qquad (14.69b)$$

The condition for full ac passing is again

$$\omega RC \gg 1 \qquad (14.70)$$

In steady state we thus have $v_{O(ss)} = V_O + v_o = 0 + v_o \simeq v_i$, or

$$v_{O(ss)} \simeq v_i \qquad (14.71)$$

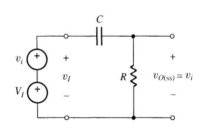

Figure 14.23 In steady state the *C-R* circuit *blocks out the dc component* of the input but *passes any ac component with $\omega \gg 1/CR$.*

The discriminatory action by the *C-R* circuit is depicted in Figure 14.23. A common application of the *C-R* circuit is **ac coupling** in audio amplifiers, where it is desired to pass all ac signals within the 20-Hz to 20-kHz audio range while blocking out any dc component that would only cause undue power dissipation.

▶ **Example 14.21**

The output of an audio amplifier is coupled to an 8-Ω loudspeaker via a capacitance C to block out any dc voltage that might be present at the amplifier output, as, for instance, in the event of amplifier malfunction. Specify C to ensure that the ac amplitude V_{om} across the loudspeaker never drops below $1/\sqrt{2} = 70.7\%$ of the amplitude V_{im} supplied by the amplifier.

Solution

For *C-R* circuits we have $V_{om} = V_{im}\omega RC/\sqrt{1+(\omega RC)^2}$. Imposing $V_{om} \geq V_{im}/\sqrt{2}$ requires $\omega RC/\sqrt{1+(\omega RC)^2} \geq 1/\sqrt{2}$, or $\omega RC \geq 1$. This condition must be met over the entire audio range. Imposing it at $\omega = \omega_{min} = 2\pi 20$ rad/s will automatically ensure its validity at all other audio frequencies. Thus, $C \geq 1/R\omega_{min} = 1/(8 \times 2\pi 20) = 995\ \mu\text{F}$. ◀

14.5 The Frequency Response Using $H(s)$

The manner in which the ac response varies as a function of frequency is called the **frequency response**. This response is of paramount importance in a variety of disciplines such as communications, signal processing, and control. The frequency response can be predicted *mathematically* using the network function $H(s)$. To this end, we first derive $H(s)$ using s-domain techniques; then we calculate it on the $j\omega$ axis by letting

$$\boxed{s \to j\omega} \qquad (14.72)$$

The frequency response can be visualized *graphically* by plotting the magnitude $|H(j\omega)|$ and phase $\sphericalangle H(j\omega)$ versus ω. It can also be observed *experimentally* by subjecting the circuit to an ac signal of fixed amplitude X_m and phase angle θ_x but variable frequency ω. The amplitude Y_m and phase angle θ_y of the response are measured with the oscilloscope for different values of ω, and the magnitude $|H(j\omega)| = Y_m/X_m$ and phase $\sphericalangle H(j\omega) = \theta_y - \theta_x$ are then plotted point by point versus ω.

The effect of letting $s \rightarrow j\omega$ is illustrated in Figure 14.5 for the magnitude case. As we know, $|H(s)|$ can be calculated anywhere in the s plane. However, when we seek the ac response, we confine ourselves to the $j\omega$ axis, indicating that the curve representing $|H(j\omega)|$ can be obtained as the *intersection* of the surface representing $|H(s)|$ with the vertical plane passing through the imaginary axis. Similar considerations hold for the phase curve.

Recall that the ac response is the response to an ac signal *after all transients have died out,* and that in order for the transients to die out, *all poles must lie in the left half* of the s plane. Section 14.3 revealed that the location of the critical frequencies has a profound impact on the natural response. We have good reasons to expect an impact upon the frequency response, too. For instance, Figure 14.5 indicates that the profile of the $|H(j\omega)|$ curve is strongly affected by the proximity of the poles to the imaginary axis. In the remainder of the chapter we illustrate the use of the network function $H(s)$ not only to *locate* the critical frequencies in the s plane but also to *visualize* the frequency response via suitable plots known as *Bode plots.*

For simplicity we restrict ourselves to network functions of the *gain* type. If a network function has the dimensions of an impedance, we can always express it as $RH(s)$, where R is a suitable *scaling resistance* and $H(s)$ is a dimensionless *gain function.* Likewise, a network function of the admittance type can be expressed as $GH(s)$, where G is a suitable *scaling conductance.*

Semilogarithmic Scales

The frequency ranges encountered in the study of frequency responses are often so wide that in order to visualize the response with an adequate degree of clarity over the entire range it is convenient to use **semilogarithmic scales.** Specifically, the magnitude $|H(j\omega)|$ and phase $\sphericalangle H(j\omega)$ are plotted on **linear scales** calibrated, respectively, in *decibels* and *degrees*. By contrast, ω is plotted on a **logarithmic scale** calibrated in *frequency decades.* Semilogarithmic plots are called **Bode plots** for the American engineer Hendrik W. Bode (pronounced *bodee*) (1905–1982) who pioneered them.

To sketch your plots you can purchase semilog graph paper in the bookstore. However, preparing your own semilog scales on plain engineering paper will make you better appreciate the peculiarities of these scales. To this end, you first make *equally spaced marks* on the vertical axis to represent *equal decibel intervals* such as $+40, +20, 0, -20, -40$ dB, or *equal degree intervals* such as $+180°, +90°, 0°, -90°, -180°$. Next, you make *equally spaced marks* on the horizontal axis to represent **decade** frequency intervals such as $1, 10, 10^2, 10^3, 10^4, \ldots$, rad/s. This is shown in Figure 14.24.

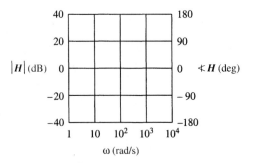

Figure 14.24 Semilog scales.

A decade interval is also called a **cycle.** Given a specific frequency value within the cycle $10^n < \omega < 10^{n+1}$ rad/s, its location ℓ within the cycle is

$$\ell = \log_{10} \frac{\omega}{10^n} \tag{14.73}$$

For instance, the location of $\omega = 320$ rad/s, which lies within the cycle $10^2 < \omega < 10^3$ rad/s, is $\ell = \log_{10}(320/10^2) \simeq 0.5$, or halfway between the 10 and 10^2 rad/s marks. Likewise, $\omega = 2000$ rad/s lies at $\ell = \log_{10}(2000/10^3) \simeq 0.3$, or one-third of the way between the 10^3 and 10^4 rad/s marks. Figure 14.25 shows the frequencies within the cycle $1 < \omega < 10$ rad/s.

Figure 14.25 Frequencies within the cycle $1 \le \omega \le 10$.

At times it is convenient to use the **normalized frequency** ω/ω_0, where ω_0 is the characteristic frequency of the circuit. The units of normalized frequency are *decades*. To visualize the frequency response at both sides of ω_0 we choose the marks for normalized frequency as . . . , $10^{-3}, 10^{-2}, 10^{-1}, 1, 10, 10^2, 10^3, \ldots$. It is apparent that the logarithmic nature of the frequency scale *compresses* high frequencies and *expands* low frequencies, a property that allows us to visualize the response at both extremes with a comparable level of detail.

Decibels

The decibel (or dB) value of a gain function H is defined as

$$|H|_{\text{dB}} \triangleq 20 \log_{10} |H| \tag{14.74}$$

TABLE 14.1 Common Gains and Their dB Values

$\|H\|$	$\|H\|_{dB}$	$\|H\|$	$\|H\|_{dB}$
1	0	1	0
$\sqrt{2}$	3	$1/\sqrt{2}$	-3
2	6	$1/2$	-6
$\sqrt{10}$	10	$1/\sqrt{10}$	-10
10	20	0.1	-20
100	40	0.01	-40
1000	60	0.001	-60
10^n	$20n$	10^{-n}	$-20n$

Conversion from dB values back to ordinary values is done via

$$|H| = 10^{|H|_{dB}/20} \tag{14.75}$$

For instance, the dB value of a gain of -400 is $20 \log 400 = 52.0$ dB. Conversely, 46 dB correspond to a gain of $10^{46/20} \simeq 200$. Table 14.1 summarizes some of the most commonly encountered gains and their dB values, which you are encouraged to commit to memory. Note that (a) *unity* gain corresponds to *zero* dB, (b) gain magnitudes *greater* than unity correspond to *positive* dBs, and (c) gain magnitudes *less* than unity correspond to *negative* dBs.

Using well-known properties of complex variables, you can readily verify that given two transfer functions H_1 and H_2,

$$|H_1 \times H_2|_{dB} = |H_1|_{dB} + |H_2|_{dB} \tag{14.76a}$$

$$\angle(H_1 \times H_2) = \angle H_1 + \angle H_2 \tag{14.76b}$$

$$|H_1/H_2|_{dB} = |H_1|_{dB} - |H_2|_{dB} \tag{14.77a}$$

$$\angle(H_1/H_2) = \angle H_1 - \angle H_2 \tag{14.77b}$$

$$|1/H|_{dB} = -|H|_{dB} \tag{14.78a}$$

$$\angle(1/H) = -\angle H \tag{14.78b}$$

We shall find these properties particularly useful in the manipulation of Bode plots, in Section 14.7.

An Illustrative Example

Let us illustrate these concepts with an example.

▶ **Example 14.22**

Sketch the pole-zero plot and the Bode plots for the driving-point impedance of the one-port of Figure 14.26.

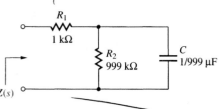

Figure 14.26 Circuit of Example 14.22.

Solution

By inspection, $Z(s) = R_1 + [R_2 \parallel (1/sC)]$. Expanding, it is readily seen that $Z(s)$ can be expressed as

$$Z(s) = (R_1 + R_2)H(s)$$

$$H(s) = \frac{(R_1 \parallel R_2)Cs + 1}{R_2Cs + 1}$$

where $H(s)$ is the frequency-dependent part of $Z(s)$. Substituting the given component values,

$$H(s) = \frac{s/10^6 + 1}{s/10^3 + 1} \tag{14.79}$$

$H(s)$ has a *zero* at $s = z_1 = -10^6$ Np/s, and a *pole* at $s = p_1 = -10^3$ Np/s. The pole-zero plot is shown at the top of Figure 14.27.

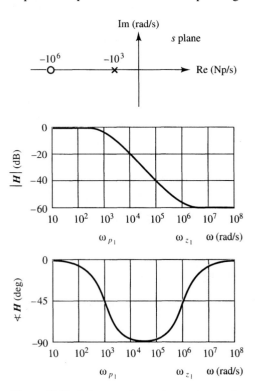

Figure 14.27 Pole-zero plot and Bode plots of the gain function of Equation (14.79).

To find the frequency response we let $s \rightarrow j\omega$ in Equation (14.79). After rearranging we obtain

$$H(j\omega) = \frac{1 + j(\omega/10^6)}{1 + j(\omega/10^3)}$$

Letting $|H| = \sqrt{[1 + (\omega/10^6)^2]/[1 + (\omega/10^3)^2]}$ in Equation (14.74) yields

$$|H|_{dB} = 10 \log_{10} \left[1 + \left(\frac{\omega}{10^6} \right)^2 \right] - 10 \log_{10} \left[1 + \left(\frac{\omega}{10^3} \right)^2 \right]$$

Moreover,

$$\angle H = \tan^{-1} \left(\frac{\omega}{10^6} \right) - \tan^{-1} \left(\frac{\omega}{10^3} \right)$$

Using a pocket calculator we evaluate these expressions for different values of ω and plot them point by point on semilog paper. The plots are shown in Figure 14.27, center and bottom.   ◀

Exercise 14.20   Repeat Example 14.22 if C is replaced with an inductance $L = 9.99$ mH.

ANSWER $Z(s) = R_1 H(s)$, $H(s) = [sL/(R_1 \parallel R_2) + 1]/[sL/R_2 + 1] = (s/10^5 + 1)/(s/10^8 + 1)$.

As we sweep ω from 0 to ∞, we note that both the *slope* of the magnitude curve and the phase curve *change* when ω reaches the critical frequencies $\omega = \omega_{p_1} = 10^3$ rad/s, and $\omega = \omega_{z_1} = 10^6$ rad/s. Namely, the *pole frequency* ω_{p_1} brings about an overall slope change of -20 dB/dec, and a phase change of $-90°$; the *zero frequency* ω_{z_1} brings about an overall slope change of $+20$ dB/dec, and a phase change of $+90°$. We summarize by stating that **a critical frequency brings about an overall *slope change* of ± 20 dB/dec and an overall *phase change* of $\pm 90°$, where the *plus* sign holds if the frequency corresponds to a *zero*, the *minus* sign if to a *pole*.**

In Figure 14.27 the slope of the magnitude curve changes from 0 dB/dec to -20 dB/dec at $\omega = \omega_{p_1}$, and from -20 dB/dec back to 0 dB/dec at $\omega = \omega_{z_1}$. For obvious reasons, the critical frequencies ω_{z_1} and ω_{p_1} are also referred to as **break frequencies.**

We also observe that the phase lag due to the pole starts at $\omega \simeq 0.1\omega_{p_1}$, reaches $-45°$ at $\omega = \omega_{p_1}$, and approaches $-90°$ at $\omega \simeq 10\omega_{p_1}$. Likewise, the phase lead due to the zero starts at $\omega \simeq 0.1\omega_{z_1}$, reaches $+45°$ at $\omega = \omega_{z_1}$, and approaches $+90°$ at $\omega \simeq 10\omega_{z_1}$.

We thus see a correlation between the *location* of a root in the s plane, where $H(s)$ goes either to *zero* or to *infinity,* and the corresponding *break frequency* on the $j\omega$ axis, where both slope and phase angle undergo either a *positive* or a *negative change.* It is interesting to observe that when dealing with *pole-zero plots,* we regard the critical frequencies p_1 and z_1 as *negative roots* and we express them in Np/s. By contrast, when dealing with *frequency plots,* we regard the corresponding break frequencies ω_{p_1} and ω_{z_1} as *positive frequencies* and we express them in rad/s.

14.6 NETWORK FUNCTION BUILDING BLOCKS

Even though the frequency response can be calculated and plotted *exactly,* it is found that *piecewise linear curve approximations* are generally sufficient to convey the essential features of a response. The advantage of piecewise plots, also called *idealized* or *linearized Bode plots,* is that they can be drawn quickly and without the tedious computations required of exact plots. However, before dealing with a general method for drawing piecewise plots, we must learn to plot the basic building blocks of network functions.

Frequency-Invariant Functions

If a gain function is independent of frequency,

$$H(s) = K \qquad \text{(14.80)}$$

then its magnitude plot is a *horizontal line* positioned at 0 dB if $|K| = 1$, *above* 0 dB if $|K| > 1$, and *below* 0 dB if $|K| < 1$. Moreover, the phase plot is a *horizontal line* positioned at 0° if K is *positive,* or at $\pm 180°$ if K is *negative.* These plots are shown in Figure 14.28 (here we have chosen $-180°$ for negative K).

Examples of circuits with constant gain functions are the ideal *noninverting amplifier,* for which $H = 1 + R_2/R_1 > 1$; the ideal *voltage follower,* for which $H = 1$; and the *voltage divider,* for which $H = R_2/(R_1 + R_2) < 1$. All these functions have $\angle H = 0$. By contrast, the ideal *inverting amplifier* has $H = -R_2/R_1$, indicating a phase of $\pm 180°$. Moreover, depending on the resistances used, we may have $|H| > 1$, $|H| = 1$, or $|H| < 1$. Just keep in mind that *amplification* implies *positive* dBs, and *attenuation negative* dBs.

Functions with a *Root at the Origin*

All network functions with just a root *at the origin* can be expressed in the standard form

$$H(s) = (s/\omega_0)^{\pm 1} \qquad \text{(14.81)}$$

where the *plus* sign holds if the root is a *zero,* and the *minus* sign if the root is a *pole.* As we know, the function $H = s/\omega_0$ is the *s*-domain counterpart of time **differentiation,** and $H = (s/\omega_0)^{-1}$ is the *s*-domain counterpart of time **integration.** An example of a function with a *zero* at the origin is the *inductive impedance* $Z(s) = sL$. This can be expressed as $Z(s) = R(sL/R) = R \times (s/\omega_0)$, where R is a scaling resistance and $\omega_0 = R/L$ is the normalizing frequency. An example of a function with a pole at the origin is the *capacitive impedance*

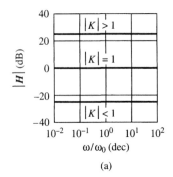

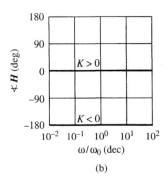

Figure 14.28 Bode plots of *frequency-invariant* functions.

$Z(s) = 1/sC = R/(sRC) = R \times (s/\omega_0)^{-1}$, where R is a scaling resistance and $\omega_0 = 1/RC$ is the normalizing frequency.

Letting $s \to j\omega$ in Equation (14.81) yields

$$H(j\omega) = (j\omega/\omega_0)^{\pm 1} \qquad \text{(14.82)}$$

so that

$$|H|_{dB} = \pm 20 \log_{10}(\omega/\omega_0) \qquad \text{(14.83a)}$$
$$\measuredangle H = \pm 90° \qquad \text{(14.83b)}$$

Since in a semilog magnitude plot the ordinate is $y = |H|_{dB}$ and the abscissa is $x = \log_{10}(\omega/\omega_0)$, Equation (14.83a) is of the type

$$y = \pm 20x$$

This represents a *straight line* with *slope* +20 dB/dec if the root is a *zero*, and −20 dB/dec if the root is a *pole*. The Bode plots of Equation (14.83) are shown in Figure 14.29. Since the magnitude curves go through the 0 dB or unity-gain axis at $\omega/\omega_0 = 1$, ω_0 is aptly called the **unity-gain frequency** of the integrator or differentiator circuit.

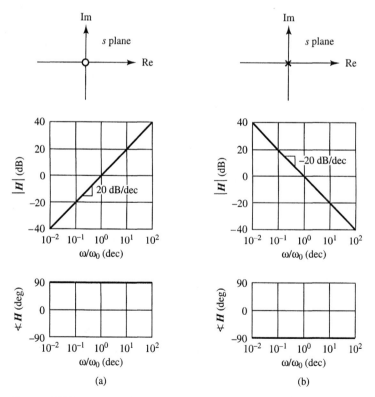

Figure 14.29 Root location and Bode plots of (a) the *differentiator* function $H = s/\omega_0$, and (b) the *integrator* function $H = 1/(s/\omega_0)$.

In audio applications, frequency intervals are expressed in **octaves** rather than decades, owing to the special significance that 2-to-1 frequency ratios have for the human ear. A slope of ± 20 dB/dec is thus equivalent to a slope of $\pm 20 \log_{10} 2 \simeq \pm 6$ dB/oct.

Functions with a *Real Negative Root*

All network functions with just a *real negative* root can be expressed as $KH(s)$, where K is a suitable scaling factor and

$$H(s) = (s/\omega_0 + 1)^{\pm 1} \qquad \textbf{(14.84)}$$

where the *plus* sign holds if the root is a *zero,* and the *minus* sign if the root is a *pole.* In either case the root is located at

$$s = -\omega_0 \qquad \textbf{(14.85)}$$

An example of a function with a real negative *zero* is the impedance of a resistance and inductance connected in series, $Z(s) = R + sL = R(1 + sL/R) = R \times (s/\omega_0 + 1)$, $\omega_0 = R/L$. An example of a function with a real negative *pole* is the impedance of a resistance and capacitance connected in parallel, $Z(s) = R \parallel (1/sC) = R \times (s/\omega_0 + 1)^{-1}$, $\omega_0 = 1/RC$.

Letting $s \to j\omega$ in Equation (14.84) and rearranging yields

$$H(j\omega) = [1 + j(\omega/\omega_0)]^{\pm 1} \qquad \textbf{(14.86)}$$

so that

$$|H|_{\mathrm{dB}} = \pm 10 \log_{10} [1 + (\omega/\omega_0)^2] \qquad \textbf{(14.87a)}$$

$$\sphericalangle H = \pm \tan^{-1} (\omega/\omega_0) \qquad \textbf{(14.87b)}$$

Plotting these functions versus ω/ω_0 yields the *exact* plots shown in black in Figure 14.30. Since for $\omega/\omega_0 = 1$ we have

$$|H| = \pm 3 \text{ dB} \qquad \textbf{(14.88a)}$$

$$\sphericalangle H = \pm 45° \qquad \textbf{(14.88b)}$$

ω_0 is also called the **3 dB frequency** or the **45° frequency** if the root is a *zero,* and the **−3 dB frequency** or the **−45° frequency** if the root is a *pole.*

Figure 14.30 suggests that we can approximate the actual *magnitude* curves with the *low-frequency asymptote*

$$|H|_{\mathrm{dB}} = 0 \qquad \textbf{(14.89a)}$$

for $\omega/\omega_0 \leq 1$, and with the *high-frequency asymptotes*

$$|H|_{\mathrm{dB}} = \pm 20 \log_{10} (\omega/\omega_0) \qquad \textbf{(14.89b)}$$

for $\omega/\omega_0 \geq 1$. In words, at low frequencies a function with a negative real root approaches the *unity gain* function; at high frequencies it approaches the

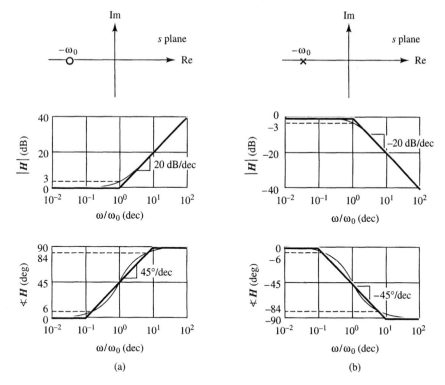

Figure 14.30 Root location and Bode plots of (a) $H = s/\omega_0 + 1$, and (b) $H = (s/\omega_0 + 1)^{-1}$. Thin curves are *exact* plots, and thick curves are their *piecewise linear approximations.*

differentiation function if the root is a *zero* or the *integration* function if the root is a *pole.*

Using Equation (14.87a) it is readily seen that the maximum deviation occurs at $\omega/\omega_0 = 1$, where it is 3 dB. For $\omega/\omega_0 = 2$ and $\omega/\omega_0 = 0.5$ the deviation is about 1 dB. For $\omega/\omega_0 \leq 0.1$ and $\omega/\omega_0 \geq 10$ the deviation can practically be ignored. Since the low-frequency and high-frequency asymptotes (shown in color) meet at $\omega/\omega_0 = 1$, ω_0 is also called the **corner frequency** or the **break frequency.**

Turning next to the *phase* curves, we can approximate them with the *low-frequency asymptote*

$$\sphericalangle H = 0° \tag{14.90a}$$

for $\omega/\omega_0 \leq 0.1$ and with the *high-frequency asymptotes*

$$\sphericalangle H = \pm 90° \tag{14.90b}$$

for $\omega/\omega_0 \geq 10$. Over the range $0.1 \leq (\omega/\omega_0) \leq 10$ we can approximate them with the *straight lines* going through 0° at $\omega/\omega_0 = 0.1$, and through $\pm 90°$ at $\omega/\omega_0 = 10$. The slopes of these lines are $\pm 45°$/dec. Using Equation (14.87b) it is readily seen that the maximum deviation occurs at $\omega/\omega_0 = 0.1$ and $\omega/\omega_0 = 10$, where it is about 6°.

Functions with a *Complex Root Pair*

As we know, complex roots always occur in *conjugate pairs*. All network functions with just a *complex root pair* can be expressed as $KH(s)$, where K is a suitable scaling factor and

$$H(s) = [(s/\omega_0)^2 + 2\zeta(s/\omega_0) + 1]^{\pm 1} \qquad \textbf{(14.91)}$$

where the *plus* sign holds if the roots are *zeros,* and the *minus* sign if they are *poles.* As we know from Chapter 9, ζ is the **damping ratio,** and ω_0 is the **undamped natural frequency.** For $\zeta > 1$ the roots are *real* and *distinct,* and for $\zeta = 0$ they are *real* and *coincident.* In either case the quadratic term of Equation (14.91) can be factored into the product of two terms of the type of Equation (14.84). We are interested in the case $\zeta < 1$, for then the roots are *complex conjugate,*

$$s/\omega_0 = -\zeta \pm j\sqrt{1 - \zeta^2} \qquad \textbf{(14.92)}$$

Letting $s \to j\omega$ in Equation (14.91) and rearranging yields

$$H(j\omega) = [1 - (\omega/\omega_0)^2 + 2\zeta j(\omega/\omega_0)]^{\pm 1} \qquad \textbf{(14.93)}$$

so that

$$|H|_{\mathrm{dB}} = \pm 10\log_{10}\left([1 - (\omega/\omega_0)^2]^2 + [2\zeta(\omega/\omega_0)]^2\right) \qquad \textbf{(14.94a)}$$

$$\sphericalangle H = \pm\tan^{-1}\left(\frac{2\zeta(\omega/\omega_0)}{1 - (\omega/\omega_0)^2}\right) \qquad \textbf{(14.94b)}$$

These functions are easily plotted versus ω/ω_0 to obtain the exact Bode plots. Shown in Figure 14.31 are such plots for the more common case of a *pole* pair; however, reflecting the magnitude and phase curves about, respectively, the 0 dB and 0° axes yields the plots for the case of a *zero* pair.

Depending on the value of ζ, the plots now consist of *families* of curves. We see that as long as $\zeta \geq 1/\sqrt{2}$, the magnitude curves lie *below* the 0 dB axis. The curve corresponding to

$$\zeta = \frac{1}{\sqrt{2}} = 0.707 \qquad \textbf{(14.95)}$$

is said to be **maximally flat.** For $\zeta < 1/\sqrt{2}$ there is a frequency band over which the magnitude is *greater* than 0 dB, that is, *greater* than unity. This phenomenon, referred to as **peaking,** is the familiar *resonant signal rise* investigated in Chapter 13. The smaller ζ, the more pronounced the amount of peaking. In the limit $\zeta \to 0$ we obtain $|H| \to \infty$, indicating that the circuit is now capable of providing an ac response even in the absence of any applied signal. As we know, this is a *sustained* oscillation. The poles now lie right on the imaginary axis, $s/\omega_0 = \pm j$.

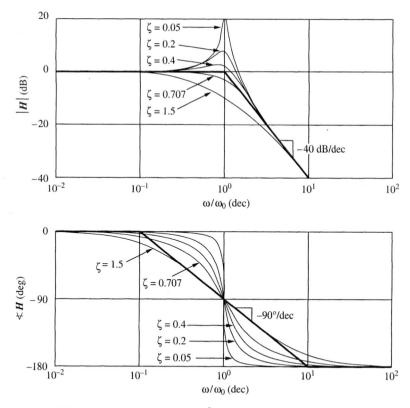

Figure 14.31   Bode plots of $H = [(s/\omega_0)^2 + 2\zeta(s/\omega_0) + 1]^{-1}$ for different values of ζ.

For values of ζ near $1/\sqrt{2}$ we can approximate the actual curves with straight lines. The *magnitude* plot is approximated with the *low-frequency asymptote*

$$|H|_{dB} = 0 \qquad\qquad \textbf{(14.96a)}$$

for $\omega/\omega_0 \leq 1$ and with the *high-frequency asymptotes*

$$|\mathbf{H}|_{dB} = \pm 40 \log_{10}(\omega/\omega_0) \qquad\qquad \textbf{(14.96b)}$$

for $\omega/\omega_0 \geq 1$, where the *plus* sign holds if the roots are *zeros,* and the *minus* sign if the roots are *poles.* Since the asymptotes meet at $\omega/\omega_0 = 1$, ω_0 is again called the **corner frequency.**

The *phase* plot is approximated with the *low-frequency asymptote*

$$\measuredangle H = 0° \qquad\qquad \textbf{(14.97a)}$$

for $\omega/\omega_0 \leq 0.1$ and with the *high-frequency asymptotes*

$$\measuredangle H = \pm 180° \qquad\qquad \textbf{(14.97b)}$$

for $\omega/\omega_0 \geq 10$. Over the range $0.1 \leq \omega/\omega_0 \leq 10$ we approximate with the *straight lines* going through $0°$ at $\omega/\omega_0 = 0.1$, and through $\pm 180°$ at $\omega/\omega_0 = 10$. We observe that since we now have a root *pair,* the magnitude and phase slopes (± 40 dB/dec, and $\pm 90°$/dec) are *twice* as large as those of a single root (± 20 dB/dec, and $\pm 45°$/dec).

These approximations are satisfactory for ζ near $1/\sqrt{2}$, but for smaller ζ values we may wish to apply suitable *corrections* to reflect the presence of peaking. It is customary to effect these corrections at the following significant points (refer to Figure 14.32):

(1) At the *corner frequency*, or $\omega/\omega_0 = 1$, Equation (14.94a) predicts

$$|H|_{dB} = \pm 20 \log_{10} 2\zeta \tag{14.98}$$

(2) The frequency at which $|H|_{dB}$ peaks out is found by differentiating Equation (14.94a) with respect to ω/ω_0 and then equating the result to zero. This frequency is readily found to be

$$\omega/\omega_0 = \sqrt{1 - 2\zeta^2} \tag{14.99a}$$

At this frequency Equation (14.94a) predicts

$$|H|_{dB} = \pm 10 \log_{10} [4\zeta^2(1 - \zeta^2)] \tag{14.99b}$$

(3) An *octave* below the corner frequency, or $\omega/\omega_0 = (1/2)$, Equation (14.94a) predicts

$$|H|_{dB} = \pm 10 \log_{10} (\zeta^2 + 0.75^2) \tag{14.100}$$

(4) The frequency at which the magnitude curve *crosses* the 0 dB axis is found, again using Equation (14.94a), and it is

$$\omega/\omega_0 = \sqrt{2(1 - 2\zeta^2)} \tag{14.101}$$

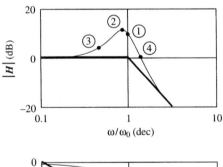

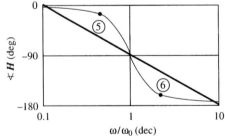

Figure 14.32 Significant points on the frequency plots of a complex pole pair.

(5) An octave *below* the corner frequency, or $\omega/\omega_0 = (1/2)$, Equation (14.94b) predicts

$$\measuredangle H = \pm \tan^{-1}(\zeta/0.75) \tag{14.102a}$$

(6) An octave *above* the corner frequency, where $\omega/\omega_0 = 2$, Equation (14.94b) predicts

$$\measuredangle H = \pm[180° - \tan^{-1}(\zeta/0.75)] \tag{14.102b}$$

▶ **Example 14.23**

Given the transfer function

$$H(s) = \frac{100}{s^2 + 5s + 100}$$

find ω_0, ζ, and the scaling factor K. Hence, compute $|H|_{\text{dB}}$ and $\measuredangle H$ at the significant points discussed earlier.

Solution

Factoring out the term 100 yields

$$H(s) = \frac{100}{100} \frac{1}{s^2/100 + 5s/100 + 1}$$

Clearly, $K = 100/100$, or

$$K = 1$$

Comparison with the standard form of Equation (14.91) yields $(s/\omega_0)^2 = s^2/100$, or

$$\omega_0 = 10 \text{ rad/s}$$

and $2\zeta(s/\omega_0) = 5s/100$, or

$$\zeta = 0.25$$

Using Equations (14.98) through (14.102), we have

(1) For $\omega = 10$ rad/s, $|H|_{\text{dB}} = -20\log_{10}(2 \times 0.25) = 6.0$ dB

(2) Magnitude peaks for $\omega = 10\sqrt{1 - 2 \times 0.25^2} = 9.35$ rad/s, and the peak value is $|H|_{\text{dB}} = -10\log_{10}[4 \times 0.25^2 \times (1 - 0.25^2)] = 6.3$ dB

(3) For $\omega = 5$ rad/s, $|H|_{\text{dB}} = -10\log_{10}(0.25^2 + 0.75^2) = 2.0$ dB

(4) The magnitude curve crosses the 0 dB axis for $\omega = 10 \times \sqrt{2(1 - 2 \times 0.25^2)} = 13.2$ rad/s

(5) For $\omega = 5$ rad/s, $\measuredangle H = -\tan^{-1}(0.25/0.75)] = -18.4°$

(6) For $\omega = 20$ rad/s, $\measuredangle H = -[180 - \tan^{-1}(0.25/0.75)] = -161.6°$ ◀

Exercise 14.21 $H(s)$ has a pole pair at $p_{1,2} = -10 \pm j50$ complex Np/s, and a scaling factor of 10. (a) Find ω_0 and ζ. (b) Compute $|H|$ and $\measuredangle H$ at the six significant points.

ANSWER   (a) 50.99 rad/s, 0.1961; (b) 28.13 dB, 28.30 dB, 22.21 dB, $|H|_{dB} = 0$ for $\omega = 168.4$ rad/s, $-14.65°$, $-165.35°$.

Multiple Roots

If a root or a complex root pair has multiplicity r, then the corresponding term has the form H^r. We thus have

$$|H^r|_{dB} = r \times |H|_{dB} \tag{14.103a}$$

$$\measuredangle(H^r) = r \times \measuredangle H \tag{14.103b}$$

indicating that the Bode plots of H^r can be obtained from those of H by multiplying the latter point by point by r. In particular, a slope of ± 20 dB/dec (or $\pm 45°$/dec) becomes a slope of $\pm 20r$ dB/dec (or $\pm r45°$/dec). Likewise, a slope of ± 40 dB/dec (or $\pm 90°$/dec) becomes a slope of $\pm 40r$ dB/dec (or $\pm r90°$/dec).

14.7 PIECEWISE-LINEAR BODE PLOTS

Having learned to plot the basic functional building blocks, we are now ready to sketch the piecewise-linear Bode plots of arbitrary gain functions. These plots are most useful when the roots are *well separated* from each other, say, by at least a decade. Even when this is not the case, these plots still provide a good starting point for a basic understanding of a frequency response.

In general, a gain function can be expressed as

$$H(s) = K\frac{N_1(s)N_2(s)\cdots N_k(s)}{D_1(s)D_2(s)\cdots D_\ell(s)} \tag{14.104}$$

where K is a suitable *scaling factor,* and the *numerator* terms $N_1(s)$ through $N_k(s)$ and the *denominator* terms $D_1(s)$ through $D_\ell(s)$ have one of the following standard forms:

$$s/\omega_0 \tag{14.105a}$$

$$s/\omega_0 + 1 \tag{14.105b}$$

$$(s/\omega_0)^2 + 2\zeta(s/\omega_0) + 1 \qquad \zeta < 1 \tag{14.105c}$$

The last form is used only if $\zeta < 1$. If $\zeta \geq 1$, the roots are real and the quadratic term can be factored into the *product* of two terms of the type of Equation (14.105b). By Equations (14.76) and (14.77),

$$|H|_{dB} = |K|_{dB} + |N_1|_{dB} + |N_2|_{dB} + \cdots + |N_k|_{dB} \qquad \textbf{(14.106a)}$$
$$\qquad - |D_1|_{dB} - |D_2|_{dB} - \cdots - |D_\ell|_{dB}$$
$$\angle H = \angle K + \angle N_1 + \angle N_2 + \cdots + \angle N_k$$
$$\qquad - \angle(D_1) - \angle(D_2) - \cdots - \angle(D_\ell) \qquad \textbf{(14.106b)}$$

indicating that the Bode plots of **H** can be obtained by first sketching the Bode plots of its individual terms and then adding them up geometrically. The procedure is summarized as follows:

(1) Given a gain function **H**(s), factor it into the *standard form* of Equation (14.104), with each numerator and denominator term having one of the forms of Equation (14.105).

(2) Separately sketch the linearized plots of the numerator terms as well as those of the *reciprocals* of the denominator terms. If a term has multiplicity r, then both the magnitude and phase slopes must be multiplied by r.

(3) Add up the individual magnitude plots and the individual phase plots to obtain the composite plots of **H**.

We observe that the effect of the scaling factor K upon magnitude is to shift the entire curve up or down, depending on whether $|K| > 1$ or $|K| < 1$. Moreover, its effect on phase is null if $K > 0$, or to shift it by $\pm 180°$ if $K < 0$.

▶ **Example 14.24**

Sketch and label the linearized Bode plots of the function

$$H(s) = \frac{\sqrt{10^5}(s + 10^2)}{(s + 10)(s + 10^3)}$$

Solution

First put each term in one of the standard forms of Equation (14.105),

$$H(s) = \frac{\sqrt{10^5} \times 10^2(s/10^2 + 1)}{10(s/10 + 1)10^3(s/10^3 + 1)} = \sqrt{10} \frac{s/10^2 + 1}{(s/10 + 1)(s/10^3 + 1)}$$

Clearly, we have a function of the type

$$H(s) = K \frac{N_1(s)}{D_1(s)D_2(s)}$$

with $K = \sqrt{10}$, a zero $z_1 = -10^2$ Np/s, and two poles $p_1 = -10$ Np/s and $p_2 = -10^3$ Np/s. We denote the corresponding break frequencies as ω_{z_1}, ω_{p_1}, and ω_{p_2}, and we mark them at the bottom of the diagrams for easy identification. We are now ready to sketch the plots of the individual terms, which we show as dotted lines.

Turning to magnitude first, we observe that $|K|_{dB}$ is a horizontal line positioned at $|\sqrt{10}|_{dB} = 10$ dB. The remaining terms share a common low-frequency asymptote at 0 dB. However, $|N_1|_{dB}$ starts to *rise* at ω_{z_1}, while $-|D_1|_{dB}$ and $-|D_2|_{dB}$ start to *drop* at ω_{p_1} and ω_{p_2}, respectively. The individual plots are shown in Figure 14.33, top. Starting at the left and proceeding toward the right, we add them up point by point to obtain the composite curve for $|H|_{dB}$, shown as a solid line.

Following a similar procedure for phase, we first plot the phases of the individual terms and then add them up, proceeding from left to right to obtain the composite curve for $\angle H$. This is shown in Figure 14.33, bottom.

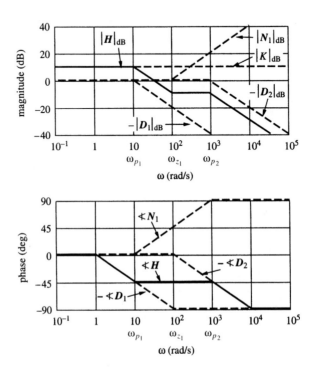

Figure 14.33 Bode plots for Example 14.24.

Exercise 14.22   Sketch and label the linearized Bode plots of

$$H(s) = \frac{10^5(s + 10)}{(s + 10^2)(s + 10^3)}$$

▶ **Example 14.25**

Sketch and label the linearized Bode plots of

$$H(s) = \frac{10^6(s + 10^3)^2}{s(s^2 + 1.1 \times 10^5 s + 10^9)}$$

Solution

It is readily seen that the quadratic term in the denominator has *real* roots at -10^4 and -10^5 Np/s, indicating that it can be factored as $(s + 10^4)(s + 10^5)$. In order to put each term in the standard form, we write

$$H(s) = \frac{10^6(s + 10^3)^2}{s(s + 10^4)(s + 10^5)} = \frac{10^6 \times (10^3)^2(s/10^3 + 1)^2}{s \times 10^4(s/10^4 + 1)10^5(s/10^5 + 1)}$$

$$= \frac{(s/10^3 + 1)^2}{(s/10^3)(s/10^4 + 1)(s/10^5 + 1)} = \frac{N_1^2(s)}{D_1(s)D_2(s)D_3(s)}$$

$H(s)$ has a zero $z_1 = -10^3$ Np/s with multiplicity 2, a pole p_1 at the origin with unity-gain frequency 10^3 Np/s, and the aforementioned pole pair $p_2 = -10^4$ Np/s and $p_3 = -10^5$ Np/s. We denote the break frequencies as ω_{z_1}, ω_{p_2}, and ω_{p_3}, and the unity-gain frequency as ω_{p_1}, and we mark them at the bottom of the diagrams.

Next, we sketch the plots of the individual terms as dotted lines. Then, we add them up proceeding from left to right. The resulting composite plots are shown as solid lines in Figure 14.34. Note that because its root has multiplicity 2, the magnitude and phase slopes of N_1^2 are, respectively, $2 \times 20 = 40$ dB/dec, and $2 \times 45 = 90°$/dec.

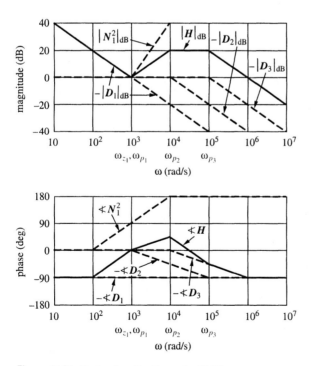

Figure 14.34 Bode plots for Example 14.25.

Exercise 14.23   Sketch and label the linearized Bode plots of

$$H(s) = \frac{10(s^2 + \sqrt{1210}s + 100)}{s(s + 10^4)}$$

▶ **Example 14.26**

Sketch and label the linearized Bode plots of

$$H(s) = \frac{18,600s^2}{(s^2 + 12s + 100)(6s + 1860)}$$

Solution

The quadratic term has $\zeta < 1$. We thus write

$$H(s) = \frac{18,600s^2}{100[s^2/100 + (12/100)s + 1]1860(6s/1860 + 1)}$$

$$= \frac{(s/3.16)^2}{[(s/10)^2 + 2 \times 0.6(s/10) + 1][s/316 + 1]} = \frac{N_1^2(s)}{D_1(s)D_2(s)}$$

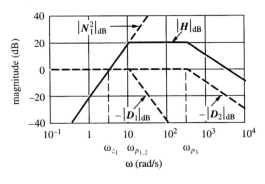

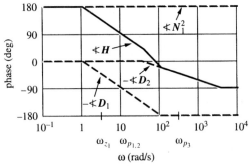

Figure 14.35 Bode plots for Example 14.26.

indicating that we have a zero at the origin with multiplicity 2, a complex pole pair, and a simple negative pole. The unity-gain frequency of the zero is $\omega_{z_1} = 3.16$ rad/s, the break frequency of the pole pair is $\omega_{p_{1,2}} = 10$ rad/s, and that of the remaining pole is $\omega_{p_3} = 316$ rad/s. These frequencies are marked at the bottom of the diagrams. Proceeding in the usual way with the plots of the individual terms first and then adding them up from left to right, we obtain the composite plots of Figure 14.35. ◀

Exercise 14.24   Sketch and label the linearized Bode plots of

$$H(s) = \frac{10^4(s^2 + 6s + 25)}{s^2(s^2 + 160s + 10^4)}$$

Plotting Magnitude Directly

Often only the magnitude plot is of interest. As you gain experience, you will be able to plot magnitude directly, without having to plot the magnitudes of the individual terms, as follows:

(1) Mark all break frequencies at the bottom of the diagram, as well as the unity-gain frequency of any root at the origin.

(2) If there are *no roots at the origin,* start at the point $|K|_{dB}$ on the axis of the ordinates and draw a horizontal line from left to right until you reach the first break frequency. At this point, *change* your slope by $+20r$ dB/dec if the corresponding root is a *zero* or by $-20r$ dB/dec if it is a *pole, r* being the multiplicity of the root. Continue toward the right with this new slope until you reach the next break frequency, where you again *change* your slope according to the type and multiplicity of the corresponding root. Keep going until all break frequencies have been exhausted.

(3) If there is *a root at the origin,* then, depending on whether it is a *zero* or a *pole,* start near the *lower left* or the *upper left* corner of the diagram and aim toward the unity-gain frequency point on the 0-dB axis with a slope of $+20r$ dB/dec or $-20r$ dB/dec, where r is the multiplicity of this root. Continue along this line until you reach the first break frequency, where you *change* your slope according to the nature and multiplicity of the corresponding root as in step 2. Keep going until all break frequencies have been exhausted.

Look back at the previous magnitude plots and convince yourself that they could have been sketched using this quicker procedure!

▶ Example 14.27

A gain function has a double zero $z_{1,2} = -10$ Np/s, two simple poles $p_1 = -10^2$ Np/s and $p_2 = -10^3$ Np/s, a complex pole pair $p_{3,4} = (-8 \pm j6)10^3$ complex Np/s, and one additional zero $z_3 = -1.75 \times 10^5$ Np/s. Assuming a scaling factor $K = 0.1$, sketch the magnitude plot.

Solution

First, mark all break frequencies. As shown at the bottom of Figure 14.36, these are $\omega_{z_{1,2}} = 10$ rad/s, $\omega_{p_1} = 10^2$ rad/s, $\omega_{p_2} = 10^3$ rad/s, $\omega_{p_{3,4}} = \sqrt{8^2 + 6^2} \times 10^3 = 10^4$ rad/s, and $\omega_{z_3} = 1.75 \times 10^5$ rad/s, where we have found $\omega_{p_{3,4}}$ via Equation (14.34b).

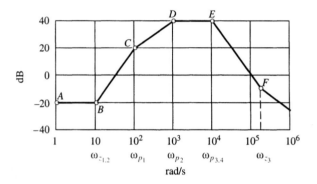

Figure 14.36 Plotting magnitude directly.

Next, starting at $|0.1|_{dB} = -20$ dB (point A), we proceed with 0 slope toward the right until the first break frequency (point B). Since this corresponds to a double zero, we change our slope from 0 to $0 + 20 \times 2 = 40$ dB/dec, and we rise with this new slope until the next break frequency (point C). Since this corresponds to a simple pole, we change our slope from $+40$ dB/dec to $+40 - 20 \times 1 = +20$ dB/dec, and proceed until point D. Since this corresponds to another simple pole, we change our slope from $+20$ dB/dec to $+20 - 20 \times 1 = 0$ dB/dec, and proceed toward point E. Due to the complex pole pair, slope now changes from 0 to $0 - 20 \times 2 = -40$ dB/dec. Finally, at point F, corresponding to a simple zero, slope changes once again from -40 to $-40 + 20 \times 1 = -20$ dB/dec. Henceforth, magnitude rolls off with ω at the rate of -20 dB/dec.

Exercise 14.25   Sketch the magnitude plot of a gain function having a simple zero at the origin with unity-gain frequency $\omega_{z_1} = 10$ rad/s, a simple zero $z_2 = -10^3$ Np/s, two simple poles $p_1 = -10^2$ Np/s and $p_2 = -\sqrt{10^5}$ Np/s, and a complex pole pair $p_{3,4} = (-8 \pm j6)10^5$ complex Np/s.

14.8 CIRCUIT RESPONSES USING SPICE

SPICE provides a powerful tool for visualizing the various responses of a circuit. Depending on the case, either the `.TRAN` or the `.AC` statement is used. In this section we direct our attention to the *source-free, complete,* and *frequency responses.*

▶ Example 14.28

Use PSpice to display the source-free voltage response at each node in the circuit of Example 14.11.

Solution

Refer to Figure 14.37. The input file is

```
SOURCE-FREE RESPONSE
R1 0 1 2K
C 1 0 5N IC=10
L 1 2 1M IC=0
R2 2 0 0.5K
.TRAN 0.5U 30U UIC
.PROBE
.END
```

After directing the Probe post-processor to display $V(1)$ and $V(2)$ we obtain the traces of Figure 14.38.

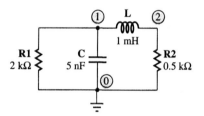

Figure 14.37 Circuit of Example 14.28.

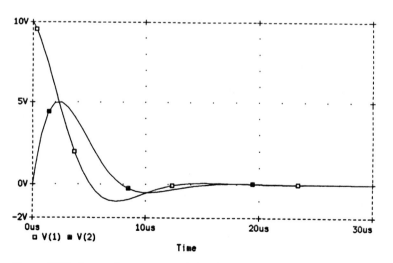

Figure 14.38 Source-free response of the circuit of Figure 14.37.

Exercise 14.26   Use SPICE to display the source-free response of the circuit of Example 14.13.

The Sinusoidal Function

When studying the *complete response* to an applied signal of the *ac* type, we wish to visualize such a response as a function of *time* rather than frequency. To this end, SPICE allows for any independent source to be a damped sinusoidal function of time. In PSpice a damped sinusoidal source is defined as

$$x(t) = X_O + X_A e^{-\alpha(t - t_D)} \sin\left[2\pi f(t - t_D) + \theta\right] \tag{14.107}$$

where X_O is the dc value or *offset*, X_A is the *peak amplitude*, α is the *damping coefficient*, t is *time*, t_D is a *time delay*, f is *frequency*, and θ is the *phase angle*. Setting $\alpha = 0$ yields an undamped sinusoid, that is, a sinusoid with constant peak amplitude.

The general forms of the PSpice statements for a sinusoidal voltage or a sinusoidal current source are

```
VXXX  N+  N-  SIN(VO  VA  FREQ  TD  ALPHA  THETA)
```
(14.108a)

```
IXXX  N+  N-  SIN(IO  IA  FREQ  TD  ALPHA  THETA)
```
(14.108b)

where `VO` and `IO` are the offset values, in V or A; `VA` and `IA` are the peak amplitudes, in V or A; `FREQ` is the frequency, in Hz; `TD` is the time delay, in s; `ALPHA` is the damping coefficient, in Np/s; and `THETA` is the phase angle, in degrees.

To direct SPICE to calculate the response as a function of time, the `.TRAN` statement must be used, and the resulting response is the *complete response*.

▶ Example 14.29

Use PSpice to verify the response of the ac-blocking circuit of Example 14.18.

Solution

Referring to Figure 14.39, we observe that $\tau = RC = 15.9 \times 10^3 \times 10^{-7} \simeq 1.6$ ms. Use a window of 9 ms to give the circuit sufficient time to reach its steady state. The input file is

```
AC BLOCKING CIRCUIT
VI 1 0 SIN(2 1 1K 0 0 0)
R 1 2 15.9K
C 2 0 0.1U IC=0
```

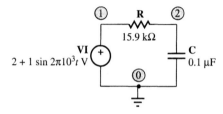

Figure 14.39 Circuit of Example 14.39.

```
.TRAN 0.05M 9M UIC
.PROBE
.END
```

After directing the Probe post-processor to display $V(1),V(2),0$, and 2, we obtain the traces of Figure 14.40.

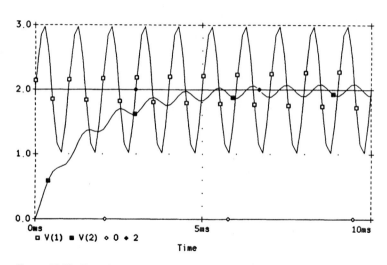

Figure 14.40 Complete response of the circuit of Figure 14.39.

Exercise 14.27   Use PSpice to verify the response of the dc blocking circuit of Example 14.19.

Bode Plots Using PSpice

A useful feature of PSpice, not available in SPICE, is the ability to specify the *value* of a VCVS or a VCCS as a *transfer function*. We can then use the .AC statement, along with the Probe post-processor, to generate Bode plots automatically. The general statements for these types of controlled sources are

```
EXXX  N+  N-  LAPLACE  {VCONTROL} = {H(S)}
```
$\qquad$ **(14.109a)**

```
GXXX  N+  N-  LAPLACE  {VCONTROL} = {H(S)}
```
$\qquad$ **(14.109b)**

These statements contain: the name of the source, which must begin with E if it is a VCVS, or with G if it is a VCCS; the nodes N+ and N- to which the source is connected; the keyword LAPLACE; the controlling voltage VCONTROL, which must appear between braces; the = sign; and the expression for the transfer function H(S), also between braces.

▶ Example 14.30

Use PSpice to generate the Bode plots of the network function having

$$z_1 = 0, \text{ with unity-gain frequency } \omega_{z_1} = 2 \text{ rad/s}$$
$$z_2 = -100 \text{ Np/s}$$
$$p_{1,2} = -6 \pm j8 \text{ complex Np/s}$$

Solution

The term contributing z_1 is $N_1(s) = s/2$, that contributing z_2 is $N_2(s) = s/100 + 1$, and that contributing $p_{1,2}$ is $D_1(s) = (s/10)^2 + 1.2(s/10) + 1 = (s/10)(s/10 + 1.2) + 1$. Thus,

$$H(s) = \frac{N_1(s)N_2(s)}{D_1(s)} = \frac{(s/2)(s/100 + 1)}{(s/10)(s/10 + 1.2) + 1}$$

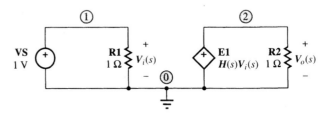

Figure 14.41 Circuit of Example 14.30.

To realize this function we use the circuit of Figure 14.41, where dummy resistances R1 and R2 serve the purpose of satisfying the SPICE requirement that every node have at least two element connections. The input file is

```
BODE PLOTS OF H(S)
VS 1 0 AC 1
R1 1 0 1
E1 2 0 LAPLACE {V(1,0)} =
```

```
+{(S/2)*(S/100+1)/[(S/10)*(S/10+1.2)+1]}
R2 2 0 1
.AC DEC 10 0.032 312
.PROBE
.END
```

After directing the Probe post-processor to plot `VDB(2)` and `VP(2)` we obtain the traces of Figure 14.42.

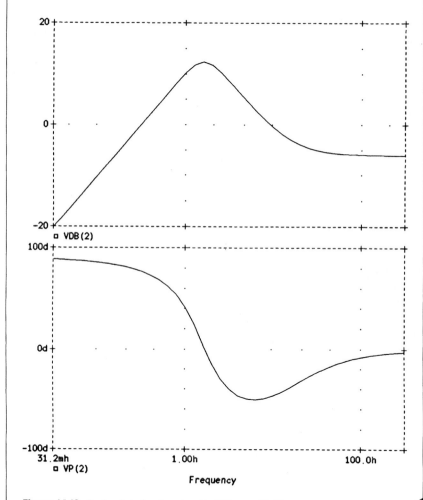

Figure 14.42 Bode plots for the circuit of Figure 14.41.

Exercise 14.28   Use PSpice to verify the plots of Example 14.24.

A CONVERSATION WITH
RICHARD WHITE

DIGITAL EQUIPMENT CORPORATION

Richard White, a customer consultant for Digital Equipment Corporation, also helps minority students pursue engineering careers. (Courtesy of Texas Instruments)

Richard White, an electrical engineer who graduated from the University of Massachusetts at Lowell in 1985, is a customer consultant with Digital Equipment in Landover, Maryland.

Let's start at the beginning. Can you remember when you first became interested in electricity?

I can't pinpoint it exactly, of course, but I do know that I was very young, still in grade school. What fascinated me was the idea that nonliving things would respond to some action of mine—throw a switch and electricity starts and stops, press a button and the doorbell makes a sound, responses like that. My parents say I was always taking things apart as a child, always trying to figure out what makes them work.

At first, I was equally interested in how everything worked—electrical and mechanical—but my father was an electronic technician for Raytheon, and that gave me a little more love for the electrical side of things. He'd let me use his tube tester sometimes, and no electrical appliance in the house was safe on those days. I'd pull them all apart just to get the tubes out and stick them in the tester.

Then when I was in high school and college, I fixed televisions, stereos, and other electronic equipment to make extra money, and that cinched it as far as what kind of engineer I would be.

So you knew what your college major would be right from the start?

Well, let's hedge a bit and say I mostly knew. I got bit pretty bad by the computer bug the first time I got to use one. (Used the computer, that is. I never used a computer bug!) My freshman year

I spent hours and hours in the computer lab, every spare minute, really. I have to admit I was tempted every now and then to switch to computer science, but looking back I can say I'm glad I didn't. I still love the science part of knowing as much as ever, but I feel more comfortable in the engineering part. I think the engineers of this world are the ones who make things better. The scientists figure out the theory, and then we engineers get to think up the applications.

What impressed you most about electronics when you were in college?

That's easy: transistors. The size thing really blew me away; so much could be accomplished with such a little thing. And the more I learned about them, the more amazed I got. For lots of things in life, you have that gee-whiz feeling when you first meet them, but

> ▼
>
> *"What fascinated me was the idea that nonliving things would respond to some action of mine ..."*
>
> ▲

then as you begin to understand how they work, you lose that awe and, sadly, some of the joy. It wasn't like that with me and transistors, though. Sometimes even now, after seven or eight years of working with them every day, I still get that feeling of amazement.

Once you graduated, you went directly to Digital?

Oh, no. My first job was at Raytheon, following in my father's footsteps, you might say. I was there for two years and worked on the cone of the Patriot missile. The cone is the missile's brain, so to speak, and I worked on all the equipment used to test it. I was as upset as the next person about the

Persian Gulf War, but I have to tell you, watching those guided missiles on TV hit their targets every time and knowing that I could have actually touched that very missile a few years back did give me a thrill.

My connection with Digital began while I was still in college because I won a Digital scholarship that was a great help in paying the bills. The summer I graduated I worked in Boston, running a computer camp for inner-city kids that was funded by Digital. I took that job because I wanted to be sure and give something back to the community. Digital helped me through school, and now I was able to turn around and pass on the favor. Besides, working with those kids was great. We didn't just show them how to turn on the machine and play Nintendo or use some word-processing software already in the machine to write about how they spent their summer vacations. No, we tore machines apart so that everyone learned something about what goes on inside the computer.

Then in April of 1987, I went to work full-time for Digital, first in North Andover, Massachusetts, in field service, and then in Hudson in the computer-integrated-manufacturing group, where they make the alpha chip, which is the fastest microprocessor around. I'm still with Digital, but now out of the Landover office in Maryland. I'm a consultant to companies that buy Digital equipment and need help either setting things up or designing ways to apply to task B the equipment they originally bought for task A. Two of my clients these days are Texas Instruments and Black & Decker, and I deliver solutions to any problems they have that involve Digital equipment.

Are you involved in any projects today that reflect the "return something to the community" philosophy you mentioned?

I'm glad you asked that question, and my answer is very loud "Yes!" I'm doing a lot of work with a computer-networking project called NSBENET sponsored by the National Society of Black Engineers. I wrote the software and designed some hardware for

(Continued)

(*Continued*)

the project, and we've just connected to Internet, a network that everyone uses—all the universities in the country, NASA, and just about everybody else. The point of NSBENET is to give minority students who are interested in engineering some help in finding out what's out theré—how the profession works, what scholarships are available, things like that.

What advice do you have for students just starting out in their EE studies?

Based on my own experience, I guess I have to say my advice is, if you need help, ask for it. Don't ever think you have to struggle through all the problems all by yourself—schoolwork problems, financial problems that worry you so much you can't study, whatever. I made my dreams of being an engineer come true by asking for help when I needed it. I went to a community organization in Boston called Freedom House and asked for help in getting into a college engineering program. They're the ones who steered me to summer school so I could get stronger in math and helped

me get the scholarships I needed. Lots of people know lots of things you might not know, so listen to people around you and ask for help when you need it. If you were drowning at the beach, you wouldn't say,

> *"...listen to people around you and ask for help when you need it. If you were drowning...you wouldn't say, 'I have to get out of this mess all by myself,' would you?"*

"I have to get out of this mess all by myself," would you? No, you'd yell for help from the lifeguard. The same idea holds in the rest of life, too. People are glad to help. Lots of other organizations like Freedom House are out there. Find out where they are if you ever need them and then just speak up and ask.

▼ Summary

- Real-world signals can be represented as suitable combinations of complex exponential signals of the type $x(t) = Xe^{st}$. In this expression, $X = X_m e^{j\theta}$ is the *phasor* associated with the signal and $s = \sigma + j\omega$ is its *complex frequency*.

- The function $\text{Re}[x(t)]$ represents a *decaying sinusoid* if $\sigma < 0$, a *growing sinusoid* if $\sigma > 0$, and a *steady sinusoid* if $\sigma = 0$. If $\omega = 0$, these functions reduce to a *decaying exponential* if $\sigma < 0$, a *growing exponential* if $\sigma > 0$, and a *dc signal* if $\sigma = 0$.

- In response to a current $i(t) = Ie^{st}$, a one-port consisting of resistances, capacitances, inductances, and dependent sources responds with a voltage $v(t) = Ve^{st}$. The ratio $Z(s) = v(t)/i(t) = V/I$ is called the *impedance* of the one-port. The impedances of the basic elements are R, sL, and $1/sC$.

- Using the concept of impedance, along with the generalized Ohm's Law and Kirchhoff's laws, we can analyze a circuit using the same techniques as in ac analysis, except that we are now using s instead of $j\omega$. This is referred to as *s-domain analysis*.

- A most important by-product of s-domain analysis is the *network function* $\boldsymbol{H}(s) = \boldsymbol{Y}/\boldsymbol{X}$, where $\boldsymbol{X}$ is the phasor of the applied signal, and $\boldsymbol{Y}$ is that of the response. This function contains essential information that is *characteristic* of the circuit, regardless of the applied signals or the energy stored in its reactive elements.

- $\boldsymbol{H}(s)$ is a *rational function* of s, or $\boldsymbol{H}(s) = N(s)/D(s)$, where $N(s)$ and $D(s)$ are suitable polynomials in s having *real* coefficients. The zeros of $N(s)$ are the *zeros* of $\boldsymbol{H}(s)$, and the zeros of $D(s)$ are the *poles* of $\boldsymbol{H}(s)$. Collectively, they are referred to as the *roots* or the *critical frequencies*.

- The roots of $\boldsymbol{H}(s)$ may be *real* or *complex*. When complex, they always come in *conjugate pairs*. Pole-zero plots allow us to visualize the roots as points of a complex plane called the s plane.

- $\boldsymbol{H}(s)$ provides a *unified approach* to the study of the *natural, forced, transient, steady-state,* and *complete responses,* as well as the *frequency response.*

- Given $\boldsymbol{H}(s)$, the *natural* or *source-free* response of a circuit takes on the form

$$y_{\text{natural}} = A_1 e^{p_1 t} + \cdots + A_n e^{p_n t}$$

where p_1 through p_n are the *poles* of $\boldsymbol{H}(s)$, and A_1 through A_n are suitable coefficients reflecting initial circuit conditions.

- Poles in the *left* half-plane yield *decaying* response components, and poles in the *right* half-plane yield *diverging* components; *real* poles yield *nonoscillatory* components, and *complex* pole pairs yield *oscillatory* components; a pole at the *origin* yields a *dc* component, and a conjugate pole pair *right on the imaginary axis* yields an ac component.

- Given $\boldsymbol{H}(s)$, the *dc steady-state* response is found by letting $s \to 0$, that is, by calculating $\boldsymbol{H}(s)$ at the *origin* of the s plane; the *ac steady-state* response is found by letting $s \to j\omega$, that is, by calculating $\boldsymbol{H}(s)$ on the imaginary axis.

- Given $\boldsymbol{H}(s)$, the *complete* response to ℓ simultaneous forcing functions $x_1(t) = X_1 e^{s_1 t}$ through $x_\ell(t) = X_\ell e^{s_\ell t}$ takes on the form

$$y_{\text{complete}} = B_1 e^{p_1 t} + \cdots + B_n e^{p_n t} + \boldsymbol{H}(s_1) X_1 e^{s_1 t} + \cdots + \boldsymbol{H}(s_\ell) X_\ell e^{s_\ell t}$$

where p_1 through p_n are the poles of $\boldsymbol{H}(s)$, and B_1 through B_n are initial-condition coefficients.

- An *R-C* circuit passes the dc component of the input while blocking out any ac component with frequency $\omega \gg 1/RC$. A *C-R* circuit blocks out the dc component of the input while passing any ac component with $\omega \gg 1/RC$.

- The manner in which the ac response varies with frequency is called the *frequency response.* Given $\boldsymbol{H}(s)$, this response is found by letting $s \to j\omega$ in the expression for $\boldsymbol{H}(s)$ and then plotting $|\boldsymbol{H}(j\omega)|$ and $\sphericalangle \boldsymbol{H}(j\omega)$ versus ω.

- If $|\boldsymbol{H}(j\omega)|$ is expressed in decibels, $\sphericalangle \boldsymbol{H}(j\omega)$ in degrees, and ω in frequency decades, the corresponding plots are called *Bode plots.* These plots are sketched using *semilog* paper.

- A network function $H(s)$ can always be factored into the product or ratio of basic terms of the type K, s/ω_0, s/ω_0+1, and $(s/\omega_0)^2+2\zeta(s/\omega_0)+1$, $\zeta < 1$. Once the magnitude and phase plots of the individual terms are known, we add them up to obtain the composite plots of $H(s)$.

- PSpice provides a powerful tool for investigating the *natural, forced, transient, steady-state,* and *complete* response of a circuit, as well for plotting its *frequency response.*

- Given a network function $H(s)$, it is possible to display Bode plots automatically using PSpice.

▼ PROBLEMS

14.1 Complex Frequency

14.1   Find V and s if (a) $v(t) = 5$ V and (b) $v(t) = 2e^{-t/10}$ V. Find I and s if (c) $i(t) = 5e^{-3t}\cos(4t+30°)$ A and (d) $i(t) = 2\sin(10^5 t - 30°)$ A.

14.2   Find $i(t)$ if (a) $I = 10/\underline{180°}$ mA and $s = 0$, and (b) $I = 5/\underline{0°}$ mA and $s = -10^6$ Np/s. Find $v(t)$ if (c) $V = 15/\underline{60°}$ V and $s = j1$ Mrad/s, and (d) $V = 2/\underline{-45°}$ V and $s = 5/\underline{180°} - \tan^{-1}(4/3)$ complex Np/s.

14.3   A source $v_S(t)$, a resistance $R = 5$ Ω, an inductance $L = 3$ H, and a capacitance $C = 0.2$ F are connected in series. Sketch the circuit; hence, find the current $i(t)$ supplied by the source if $v_S(t) = $ (a) e^{-5t} V, (b) $2\sin 5t$ V, and (c) $10e^{-3t}\cos(4t+45°)$ V.

14.4   A source $i_S(t)$, a resistance $R = 10$ Ω, an inductance $L = 2$ H, and a capacitance $C = 1/20$ F are connected in parallel. Sketch the circuit; hence, (a) find the voltage $v(t)$ across the source if $i_S(t) = e^{-5t}\cos(10t - 30°)$ A (↑); (b) find $i_S(t)$ if the current through R is $i(t) = 5e^{-3t}$ A (↓); (c) find the current through C if the current through L is $i(t) = 10\cos(5t+45°)$ A (↓).

14.5   Use s-domain analysis to find $v(t)$ in the circuit of Figure P14.5.

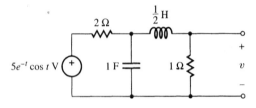

Figure P14.5

14.6   Using s-domain analysis, find $i(t)$ so that $v(t) = 5e^{-10t} \times \cos 5t$ V in the circuit of Figure P14.6.

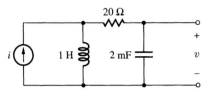

Figure P14.6

14.7   (a) Using the proportionality analysis procedure, find a relationship between $v_o(t)$ and $v_i(t)$ in the circuit of Figure P14.7. (b) Find $v_o(t)$ if $v_i(t) = 2e^{-t}\cos 2t$ V.

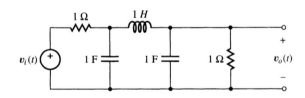

Figure P14.7

14.8   Using s-domain loop analysis, find the currents out of the positive terminals of the sources in Figure P14.8.

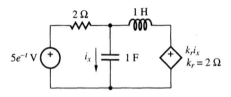

Figure P14.8

14.9   Using s-domain nodal analysis, find the voltages across the resistances in Figure P14.9.

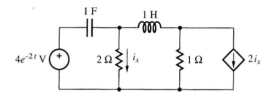

Figure P14.9

14.10 Find the generalized impedance $Z(s)$ seen by the source in Figure P14.9; hence, find the current supplied by the source.

14.11 Find the s-domain Thévenin equivalent of the circuit of Figure P14.11.

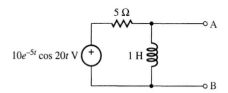

Figure P14.11

14.12 (a) Find the s-domain Norton equivalent of the circuit of Figure P14.11. (b) Repeat, but with the resistance and inductance interchanged with each other.

14.2 Network Functions

14.13 (a) Sketch the pole-zero plot of the function

$$H(s) = \frac{y(t)}{x(t)} = \frac{20s + 5}{s^3 + 3s^2 + 7s + 5}$$

(b) Write the differential equation relating $y(t)$ to $x(t)$.

14.14 (a) Sketch the pole-zero plot for the impedance

$$Z(s) = R\frac{s^2 + 4s + 13}{(s + 1)^2(s^2 + 2s + 2)}$$

(b) Find R so that $|Z| = 1$ kΩ at $s = -1 + j2$ complex Np/s. What is $\angle Z$ at this frequency?

14.15 Obtain an expression of the type $H(s) = N(s)/D(s)$ if (a) $H(s)$ has a pole, $p_1 = -5$, and $H(0) = 10$; (b) $H(s)$ has a zero, $z_1 = -10^3$, and $H(0) = 1$; (c) $H(s)$ has the roots $z_1 = 0$, $z_2 = -3$, $p_1 = p_2 = -1$, $p_3 = -7$, and $H(2) = 1/9$; (d) $H(s)$ has the roots $z_{1,2} = \pm j3$, $p_1 = 0$, $p_{2,3} = \pm j1$, $p_{4,5} = -3 \pm j4$, and $H(-1) = 1$.

14.16 Find $Z(s)$ for the circuit of Figure P14.16; hence, sketch the pole-zero plot. Check dimensionally and asymptotically.

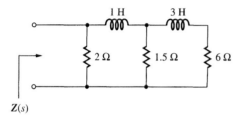

$Z(s)$

Figure P14.16

14.17 Find $H(s) = V_o/I_i$ for the circuit of Figure P14.17; hence, find the critical frequencies. Check dimensionally and asymptotically.

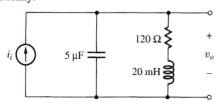

Figure P14.17

14.18 Find $H(s) = V_o/V_i$ for the circuit of Figure P14.18; hence, sketch the pole-zero plot. Don't forget to check.

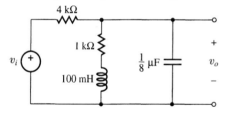

Figure P14.18

14.19 Find $H(s) = I_o/V_i$ for the circuit of Figure P14.19; hence, find the critical frequencies. Don't forget to check.

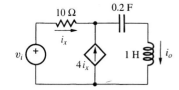

Figure P14.19

14.20 Find $H(s) = V_o/V_i$ for the circuit of Figure P14.20; hence, sketch the pole-zero plot. Don't forget to check.

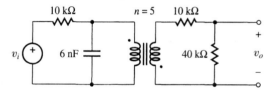

Figure P14.20

14.21   Find $H(s) = V_o/V_i$ for the circuit of Figure P14.21; hence, find the critical frequencies. Don't forget to check.

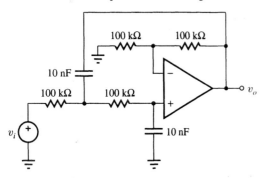

Figure P14.21

14.22   Repeat Problem 14.21, but for the op amp circuit of Figure P13.26.

14.23   Repeat Problem 14.21, but for the op amp circuit of Figure P13.27.

14.24   Find the transfer function of the op amp circuit of Figure P13.28. Show the pole-zero plot.

14.3 The Natural Response Using $H(s)$

14.25   Repeat Example 14.11, but with $v_C(0) = 0$ and $i_L(0) = $ 6 mA.

14.26   (a) Find $v(t)$ in the circuit of Figure P14.26(a) if $i_L(0) = 10$ mA. (b) Find $i(t)$ in the circuit of Figure P14.26(b) if $v_C(0) = 10$ V.

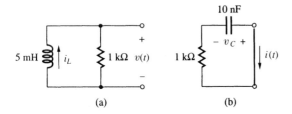

Figure P14.26

14.27   Assuming the circuit of Figure P14.27 is in steady state prior to switch activation, find $v(t)$.

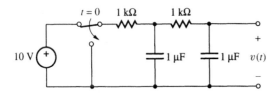

Figure P14.27

14.28   Find $i(t)$ in the circuit of Figure P14.28 if $v_C(0) = 12$ V and $i_L(0) = 3$ A.

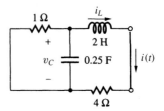

Figure P14.28

14.29   Repeat Problem 14.28, but with the 4-Ω resistance changed to 2 Ω.

14.30   Find $v(t)$ in the circuit of Figure P14.30 if $v_C(0) = 10$ V and both inductances have zero initial stored energy.

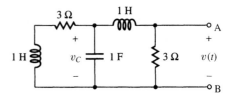

Figure P14.30

14.31   Find the open-circuit voltage and the short-circuit current supplied by the one-port of Problem 14.30 if the leftmost inductance is replaced by a wire.

14.32   Assuming the circuit of Figure P14.32 is in steady state prior to switch activation, find $v(t)$ if $R_1 = R_2 = 100$ kΩ, $C_1 = 4$ nF, and $C_2 = 1$ nF.

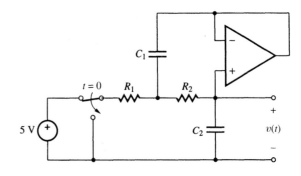

Figure P14.32

14.33   Repeat Problem 14.32, but with (a) $R_1 = R_2 = 100$ kΩ, $C_1 = 1$ nF, and $C_2 = 4$ nF, and (b) $R_1 = R_2 = 100$ kΩ, and $C_1 = C_2 = 1$ nF.

14.4 The Complete Response Using $H(s)$

14.34 Assuming zero initial stored energy in the capacitance of Figure P14.34, find $v_O(t)$ if $v_I = 12 \sin 2\pi 10^3 t$ V.

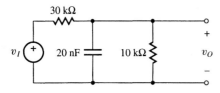

Figure P14.34

14.35 Repeat Problem 14.34, but with the capacitance in parallel with the 30-kΩ resistance.

14.36 Assuming zero initial stored energy in the capacitance of Figure P14.36, find $v_O(t)$ if $v_I = 1 \sin 2\pi 10^3 t$ V.

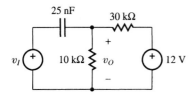

Figure P14.36

14.37 Assuming zero initial stored energy in each reactive element of Figure P14.37, find $v_O(t)$ if $v_I = 10 + 20 \sin 5t$ V.

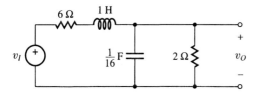

Figure P14.37

14.38 Repeat Problem 14.37, but with the capacitance and inductance interchanged with each other.

14.39 Assuming zero initial stored energy in the capacitances of Figure P14.39, find $v_O(t)$ if $v_I = 5(1 + \sin 10^4 t)$ V.

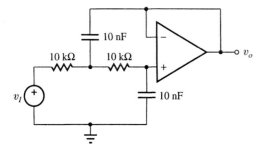

Figure P14.39

14.40 Repeat Problem 14.39, but with the rightmost resistance and capacitance interchanged with each other. Comment.

14.5 The Frequency Response Using $H(s)$

14.41 Exploiting the properties $|H_1 \times H_2|_{dB} = |H_1|_{dB} + |H_2|_{dB}$ and $|H_1/H_2|_{dB} = |H_1|_{dB} - |H_2|_{dB}$, and using the entries of Table 14.1, but *no calculator*, find $|H|_{dB}$ if (a) $H = 400$, (b) $H = -200\sqrt{2}$, and (c) $H = 1/\sqrt{500}$; find $|H|$ if (d) $|H|_{dB} = 19$ dB, (e) $|H|_{dB} = -34$ dB, and (f) $|H|_{dB} = -87$ dB.

14.42 Find $|H|_{dB}$ at $\omega = 10$ rad/s if (a) $H(s) = \sqrt{3}/(s+20)$ and (b) $H(s) = s/(s^2+5s+20)$. Find $|H|_{dB}$ in the limits $\omega \to 0$, $\omega \to \infty$, and $\omega = 10$ rad/s if $H(s) = (s^2+1)/(s^2+2s+100)$.

14.43 (a) Find the impedance $Z(s)$ of a series RLC circuit having $R = 1$ kΩ, $L = 10$ mH, and $C = 10$ nF; hence, sketch its magnitude Bode plot over the range $10^3 \le \omega \le 10^7$ rad/s. (b) Repeat, but for the impedance of a parallel RLC circuit having the same element values.

14.44 Estimate $H(s)$ if its magnitude plot is that of Figure P14.44.

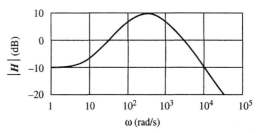

Figure P14.44

14.6 Network Function Building Blocks

14.45 Find the transfer function of the circuit of Figure P14.45; hence, sketch and label its pole-zero plot and its Bode plots.

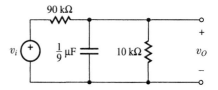

Figure P14.45

14.46 Repeat Problem 14.45, but with the capacitance replaced by a 0.9-mH inductance.

14.47 (a) Sketch and label the pole-zero plot and the Bode plots of the circuit of Figure P14.47. (b) Repeat, but with the 10-kΩ resistance replaced by a short circuit.

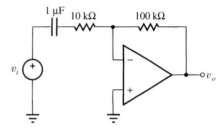

Figure P14.47

14.48 (a) Sketch and label the pole-zero plot and the Bode plots of the circuit of Figure P14.48. (b) Repeat, but with a 100-kΩ resistance in parallel with the capacitance.

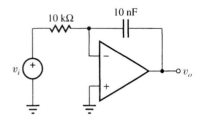

Figure P14.48

14.49 Sketch and label the pole-zero plot and the Bode plots of the op amp circuit of Figure P11.64.

14.50 Repeat Problem 14.49 for the op amp circuit of Figure P11.65.

14.51 Given the transfer function

$$H(s) = \frac{2500}{s^2 + 20s + 2500}$$

find (a) the frequencies at which $|H| = 1$, (b) the frequency at which $|H|$ is maximum, and (c) $|H|_{max}$.

14.52 (a) Sketch and label the Bode plots for the op amp circuit of Figure P14.21. (b) Calculate the plots at the six significant points. (c) At what frequency is $|H|_{dB} = 0$?

14.7 Piecewise-Linear Bode Plots

14.53 Sketch and label the linearized Bode plots of the function

$$H(s) = \frac{10(s + 10)}{s + 100}$$

14.54 Sketch and label the linearized Bode plots of the function

$$H(s) = \frac{10^6(s + 10)^2}{s^2(s + 10^4)}$$

14.55 (a) Sketch and label the linearized Bode plots of

$$H(s) = \frac{10^3(s + 1)}{(s + 100)(s + 10)}$$

(b) Calculate $|H(j\omega)|_{dB}$ and $\angle H(j\omega)$ at $\omega = \sqrt{10^3}$ rad/s.
(c) Compare with the values predicted by the linearized plots.

14.56 Sketch and label the linearized magnitude Bode plot of

$$H(s) = \frac{25 \times 10^7(s^2 + 33s + 90)}{s(s + 300)(s^2 + 4 \times 10^3 s + 25 \times 10^6)}$$

14.57 Sketch and label the linearized magnitude Bode plot of

$$H(s) = \frac{10s(s + 300)}{(s + 1)(s^2 + 6s + 400)}$$

14.58 Sketch and label the linearized Bode plots of V_o/I_i in the circuit of Figure 14.6 if $R_1 = 1$ kΩ, $R_2 = 1$ Ω, $L = 1$ mH, and $C = 1$ μF. Comment.

14.59 Sketch and label the linearized Bode plots of the transfer functions of the circuits of Figure P14.59(a) and (b).

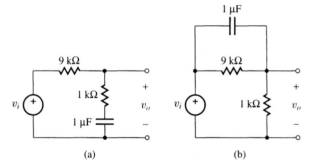

(a)   (b)

Figure P14.59

14.60 (a) Sketch and label the Bode plots of the transfer function of the op amp circuit of Figure 14.8 if $R_1 = R_2 = R_3 = 100$ kΩ, and $C = 10$ nF. (b) Repeat (a), but with R_1 changed to 10 kΩ. (c) Repeat (a), but with R_2 changed to 10 kΩ.

14.61 The plot of Figure P14.61, known as the RIAA playback equalization curve, represents the transfer function of a phono preamplifier. (a) Find $H(s)$. (b) Calculate $|H|_{dB}$ at the three break frequencies.

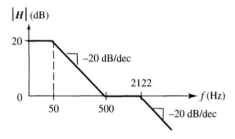

Figure P14.61

14.62 (a) Find $H(s)$ if its magnitude Bode plot is as shown in Figure P14.62. (b) Calculate $|H|_{dB}$ and $\angle H$ at $\omega = 10^2\sqrt{2}$ rad/s.

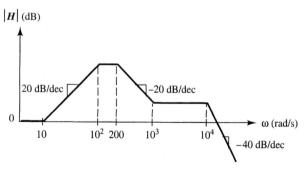

Figure P14.62

14.63 Using the direct technique, sketch the Bode plot of $|H|_{dB}$ if (a) $K = 1$, $\omega_{z_1} = 1$ rad/s, $\omega_{p_1} = 10$ rad/s, and $\omega_{p_2} = 100$ rad/s; (b) $K = 10$, $\omega_{p_1} = 1$ rad/s, $\omega_{z_1} = 10$ rad/s, and $\omega_{p_2} = 100$ rad/s; and (c) $K = -10$, $\omega_{p_1} = 1$ rad/s, $\omega_{p_2} = 10$ rad/s, and $\omega_{z_1} = 100$ rad/s.

14.64 (a) Using the direct technique, sketch and label the magnitude plot of

$$H(s) = 10^5 \frac{(s + 1)(s^2 + 2s + 400)}{s(s + 100)^2(s + 400)}$$

(b) Show the correction at the break point corresponding to the complex zero pair.

14.8 Circuit Responses Using SPICE

14.65 Use the Probe post-processor of PSpice to display $i(t)$ in the source-free RLC circuit of Problem 14.28.

14.66 Use the Probe post-processor of PSpice to display $v(t)$ in the op amp circuit of Problem 14.32.

14.67 Use the Probe post-processor of PSpice to display the complete response $v_O(t)$ of the circuit of Problem 14.36.

14.68 Use the Probe post-processor of PSpice to display the complete response $v_O(t)$ of the circuit of Problem 14.37.

14.69 Use the Probe post-processor of PSpice to verify the Bode plots of Example 14.26.

14.70 Use the Probe post-processor of PSpice to display the amplitude Bode plots of the gain function of Example 14.27.

CHAPTER 15

Two-Port Networks and Magnetically Coupled Coils

T he concept of a two-port was informally introduced in Chapter 5 to provide a common framework for the study of transformers and amplifiers. We now reexamine this concept from a more general and systematic viewpoint, using the network function concept developed in the previous chapter. We begin by introducing the two-port as a generalization of the one-port and find that there are *six* different ways of expressing its terminal characteristics, namely, via what are known as the *z, y, a, b, h,* and *g* parameter sets. We then illustrate the derivation of the four most common parameter sets, and we study two-port *models, termination,* and *interconnections.*

In the second part of the chapter we study an important two-port example, the *magnetically coupled coil pair.* Just as the two-port is a generalization of the one-port, the magnetically coupled pair is a generalization of the inductance, and a new concept emerges from this generalization, that of *mutual inductance.*

We conclude by demonstrating the use of SPICE to find two-port parameters as well as to investigate the transient and ac responses of magnetically coupled coils and transformers.

15.1 Two-Port Parameters

Before undertaking the study of two-ports, it is worth reviewing the concept of a one-port, as the former can be regarded as a generalization of the latter. Recall

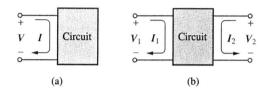

Figure 15.1 (a) One-port and (b) two-port.

that a one-port, depicted in Figure 15.1(a), is a circuit with a pair of terminals, or *port,* through which signals and energy may enter or leave the circuit. Its distinguishing feature is a *unique relationship* between the voltage V and the current I at these terminals. If the circuit contains *no independent sources,* this relationship is simply the generalized Ohm's Law. If the applied signal, also called the *independent variable,* is a current, then the response, also called the *dependent variable,* is a voltage such that

$$V = Z(s)I$$

where $Z(s)$ is the *driving-point impedance* of the port and s is the *complex frequency.* Conversely, if the independent variable is a voltage, then the dependent variable is a current such that

$$I = Y(s)V$$

where $Y(s)$ is the *driving-point admittance* of the port. $Z(s)$ and $Y(s)$ are not independent of each other but are *reciprocals* of each other. Once one is known, the other is found as

$$Y(s) = 1/Z(s)$$

$$Z(s) = 1/Y(s)$$

Let us now turn to the two-port depicted in Figure 15.1(b). To distinguish one port from the other, we use the subscript 1 to identify the variables of the left port, also called the *input* port, and the subscript 2 to identify those of the right port, or *output* port. For reasons of symmetry it is convenient to assume the *passive sign convention* for both ports. This convention, universally accepted in the literature, makes it easier to generalize the analysis of complex networks in terms of simpler two-port building blocks. To simplify such an analysis further, we assume that **the current entering the upper terminal exits the lower terminal of the same port.** It is important to realize that for this assumption to hold, *no external connections may be made between terminals of different ports,* though external connections between the terminals of the same port are perfectly acceptable.

With these considerations in mind, we observe that of the four terminal variables V_1, I_1, V_2, and I_2, only two are *independent;* the remaining two are *dependent* in a manner established by the internal circuitry. Consequently, a two-port network can be described with just a *pair* of simultaneous equations. There are *six* possible ways of selecting two independent variables out of a group of four, indicating six possible pairs of equations. If the two-port contains *no independent sources,* which we shall assume to be always the case, the six pairs of equations are

$$V_1 = z_{11}I_1 + z_{12}I_2 \qquad \text{(15.1a)}$$
$$V_2 = z_{21}I_1 + z_{22}I_2 \qquad \text{(15.1b)}$$
$$I_1 = y_{11}V_1 + y_{12}V_2 \qquad \text{(15.2a)}$$
$$I_2 = y_{21}V_1 + y_{22}V_2 \qquad \text{(15.2b)}$$
$$V_1 = a_{11}V_2 - a_{12}I_2 \qquad \text{(15.3a)}$$
$$I_1 = a_{21}V_2 - a_{22}I_2 \qquad \text{(15.3b)}$$
$$V_2 = b_{11}V_1 - b_{12}I_1 \qquad \text{(15.4a)}$$
$$I_2 = b_{21}V_1 - b_{22}I_1 \qquad \text{(15.4b)}$$
$$V_1 = h_{11}I_1 + h_{12}V_2 \qquad \text{(15.5a)}$$
$$I_2 = h_{21}I_1 + h_{22}V_2 \qquad \text{(15.5b)}$$
$$I_1 = g_{11}V_1 + g_{12}I_2 \qquad \text{(15.6a)}$$
$$V_2 = g_{21}V_1 + g_{22}I_2 \qquad \text{(15.6b)}$$

In the following we refer to the coefficients appearing in these equations as the *z*, *y*, *a*, *b*, *h*, and *g parameters,* respectively. Though in general these parameters are functions of the complex frequency *s*, we omit indicating this dependence explicitly in order to streamline our formulas. It is often convenient to visualize a parameter set in *matrix* form. For example, we often represent the *z* parameters as

$$[z] = \begin{bmatrix} z_{11} & z_{12} \\ z_{21} & z_{22} \end{bmatrix}$$

A similar representation is used for any of the other sets.

Which pair of equations best describes a two-port depends on the application. The *z* parameters of Equation (15.1), relating voltages to currents, have the dimensions of impedances and are thus referred to as the *impedance parameters.* Likewise, the *y* parameters of Equation (15.2), relating currents to voltages, have the dimensions of admittances and are thus called the *admittance parameters.* Collectively, the *z* and *y* parameters are referred to as the **immittance parameters.**

The *a* and *b* parameters of Equations (15.3) and (15.4), relating the current and voltage of one port to the current and voltage of the other port, are collectively referred to as the **transmission parameters.** They give a measure of how a circuit transmits a voltage and a current from a source to a load. These equations contain negative signs to reflect the fact that the load current usually flows *out* of the positive terminal of the corresponding port and its direction is thus *opposite* to that chosen in Figure 15.1(b).

Finally, the *h* and *g* parameters relate cross-variables, that is, they relate the voltage of one port and current of the other port to the current of the former and the voltage of the latter. For this reason they are collectively referred to as the **hybrid parameters.** A common application of these parameters is in the modeling of *bipolar junction transistors* (BJTs).

We shall investigate only the most widely used parameter sets.

The T and π Networks

To develop a feel for the preceding concepts, consider the widely used two-port topologies of Figure 15.2. These networks belong to an important subclass of two-ports known as **three-terminal networks** because both ports share the same reference terminal. In response to two externally applied currents I_1 and I_2, the T network yields

$$V_1 = Z_A I_1 + Z_C(I_1 + I_2) = (Z_A + Z_C)I_1 + Z_C I_2$$

$$V_2 = Z_C(I_1 + I_2) + Z_B I_2 = Z_C I_1 + (Z_B + Z_C)I_2$$

Comparing with Equation (15.1), it is readily seen that the z parameters of the T network are

$$[z] = \begin{bmatrix} Z_A + Z_C & Z_C \\ Z_C & Z_B + Z_C \end{bmatrix} \qquad \textbf{(15.7)}$$

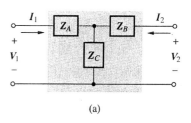

(a)

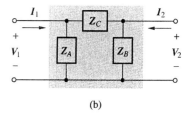

(b)

Figure 15.2 (a) T network and (b) π network.

▶ **Example 15.1**

Let the T network of Figure 15.2(a) consist of an inductance $Z_A = 2s$ Ω, a resistance $Z_B = 4$ Ω, and a capacitance $Z_C = 1/3s$ Ω. Find the z parameters.

Solution

By Equation (15.7), $z_{11} = 2s + 1/3s = (6s^2 + 1)/3s$ Ω, $z_{12} = z_{21} = 1/3s$ Ω, and $z_{22} = 4 + 1/3s = (12s + 1)/3s$ Ω. ◀

Turning next to the π network, we observe that in response to two externally applied voltages V_1 and V_2, it yields

$$I_1 = V_1/Z_A + (V_1 - V_2)/Z_C = (1/Z_A + 1/Z_C)V_1 - (1/Z_C)V_2$$

$$I_2 = (V_2 - V_1)/Z_C + V_2/Z_B = -(1/Z_C)V_1 + (1/Z_B + 1/Z_C)V_2$$

Comparing with Equation (15.2), it is readily seen that the y parameters of the π network are

$$[y] = \begin{bmatrix} 1/Z_A + 1/Z_C & -1/Z_C \\ -1/Z_C & 1/Z_B + 1/Z_C \end{bmatrix} \qquad \textbf{(15.8)}$$

Exercise 15.1   Let the π network of Figure 15.2(b) be an ac network with $Z_A = Z_B = j10$ Ω and $Z_C = 20$ Ω. Find its y parameters.

ANSWER $y_{11} = y_{22} = 0.05 - j0.1$ Ω^{-1}, $y_{12} = y_{21} = -0.05$ Ω^{-1}.

Parameter Conversion

The parameter sets appearing in Equations (15.1) through (15.6) are not independent. In fact, once one set is known, all other sets can be found from this known set via algebraic manipulations. As an example, consider Equation (15.1). Solving for I_1 and I_2 via Cramer's rule we obtain

$$I_1 = \frac{\begin{vmatrix} V_1 & z_{12} \\ V_2 & z_{22} \end{vmatrix}}{\begin{vmatrix} z_{11} & z_{12} \\ z_{21} & z_{22} \end{vmatrix}} = \frac{V_1 z_{22} - V_2 z_{12}}{z_{11} z_{22} - z_{21} z_{12}} = \frac{z_{22}}{\Delta z} V_1 - \frac{z_{12}}{\Delta z} V_2$$

and

$$I_2 = \frac{\begin{vmatrix} z_{11} & V_1 \\ z_{21} & V_2 \end{vmatrix}}{\begin{vmatrix} z_{11} & z_{12} \\ z_{21} & z_{22} \end{vmatrix}} = \frac{z_{11} V_2 - z_{21} V_1}{z_{11} z_{22} - z_{21} z_{12}} = -\frac{z_{21}}{\Delta z} V_1 + \frac{z_{11}}{\Delta z} V_2$$

where we have set

$$\Delta z = z_{11} z_{22} - z_{21} z_{12}$$

Comparing with Equation (15.2), we find that the y parameters can be obtained from the z parameters as

$$y_{11} = \frac{z_{22}}{\Delta z} \qquad y_{12} = -\frac{z_{12}}{\Delta z}$$

$$y_{21} = -\frac{z_{21}}{\Delta z} \qquad y_{22} = \frac{z_{11}}{\Delta z}$$

Applying similar techniques, we can express any parameter set in terms of any other set. The rules are summarized in Table 15.1.

▶ Example 15.2

In the T network of Figure 15.2(a) let $Z_A = 10 \ \Omega$, $Z_B = 20 \ \Omega$, and $Z_C = 30 \ \Omega$. Find the network's y parameters.

Solution

First use Equation (15.7) to find the z parameters, then use Table 15.1 to convert to the y parameters. Thus,

$$z_{11} = 10 + 30 = 40 \ \Omega$$

$$z_{12} = z_{21} = 30 \ \Omega$$

$$z_{22} = 20 + 30 = 50 \ \Omega$$

Next, $\Delta z = z_{11} z_{22} - z_{21} z_{12} = 40 \times 50 - 30 \times 30 = 1100 \ \Omega^2$. Hence,

$$y_{11} = \frac{z_{22}}{\Delta z} = \frac{50}{1100} = \frac{1}{22} \ \Omega^{-1}$$

$$y_{12} = y_{21} = \frac{-z_{12}}{\Delta z} = -\frac{30}{1100} = -\frac{3}{110} \; \Omega^{-1}$$

$$y_{22} = \frac{z_{11}}{\Delta z} = \frac{40}{1100} = \frac{2}{55} \; \Omega^{-1}$$

◄

Exercise 15.2   In the π network of Figure 15.2(b) let $Z_A = 2$ kΩ, $Z_B = 4$ kΩ, and $Z_C = 5$ kΩ. Find the network's z parameters.

ANSWER $z_{11} = \dfrac{18}{11}$ kΩ, $z_{12} = z_{21} = \dfrac{8}{11}$ kΩ, and $z_{22} = \dfrac{28}{11}$ kΩ.

TABLE 15.1 Parameter Conversion Table

z_{11}	z_{12}	$\dfrac{y_{22}}{\Delta y}$	$-\dfrac{y_{12}}{\Delta y}$	$\dfrac{a_{11}}{a_{21}}$	$\dfrac{\Delta a}{a_{21}}$	$\dfrac{b_{22}}{b_{21}}$	$\dfrac{1}{b_{21}}$	$\dfrac{\Delta h}{h_{22}}$	$\dfrac{h_{12}}{h_{22}}$	$\dfrac{1}{g_{11}}$	$-\dfrac{g_{12}}{g_{11}}$
z_{21}	z_{22}	$-\dfrac{y_{21}}{\Delta y}$	$\dfrac{y_{11}}{\Delta y}$	$\dfrac{1}{a_{21}}$	$\dfrac{a_{22}}{a_{21}}$	$\dfrac{\Delta b}{b_{21}}$	$\dfrac{b_{11}}{b_{21}}$	$-\dfrac{h_{21}}{h_{22}}$	$\dfrac{1}{h_{22}}$	$\dfrac{g_{21}}{g_{11}}$	$\dfrac{\Delta g}{g_{11}}$
$\dfrac{z_{22}}{\Delta z}$	$-\dfrac{z_{12}}{\Delta z}$	y_{11}	y_{12}	$\dfrac{a_{22}}{a_{12}}$	$-\dfrac{\Delta a}{a_{12}}$	$\dfrac{b_{11}}{b_{12}}$	$-\dfrac{1}{b_{12}}$	$\dfrac{1}{h_{11}}$	$-\dfrac{h_{12}}{h_{11}}$	$\dfrac{\Delta g}{g_{22}}$	$\dfrac{g_{12}}{g_{22}}$
$-\dfrac{z_{21}}{\Delta z}$	$\dfrac{z_{11}}{\Delta z}$	y_{21}	y_{22}	$-\dfrac{1}{a_{12}}$	$\dfrac{a_{11}}{a_{12}}$	$-\dfrac{\Delta b}{b_{12}}$	$\dfrac{b_{22}}{b_{12}}$	$\dfrac{h_{21}}{h_{11}}$	$\dfrac{\Delta h}{h_{11}}$	$-\dfrac{g_{21}}{g_{22}}$	$\dfrac{1}{g_{22}}$
$\dfrac{z_{11}}{z_{21}}$	$\dfrac{\Delta z}{z_{21}}$	$-\dfrac{y_{22}}{y_{21}}$	$-\dfrac{1}{y_{21}}$	a_{11}	a_{12}	$\dfrac{b_{22}}{\Delta b}$	$\dfrac{b_{12}}{\Delta b}$	$-\dfrac{\Delta h}{h_{21}}$	$-\dfrac{h_{11}}{h_{21}}$	$\dfrac{1}{g_{21}}$	$\dfrac{g_{22}}{g_{21}}$
$\dfrac{1}{z_{21}}$	$\dfrac{z_{22}}{z_{21}}$	$-\dfrac{\Delta y}{y_{21}}$	$-\dfrac{y_{11}}{y_{21}}$	a_{21}	a_{22}	$\dfrac{b_{21}}{\Delta b}$	$\dfrac{b_{11}}{\Delta b}$	$-\dfrac{h_{22}}{h_{21}}$	$-\dfrac{1}{h_{21}}$	$\dfrac{g_{11}}{g_{21}}$	$\dfrac{\Delta g}{g_{21}}$
$\dfrac{z_{22}}{z_{12}}$	$\dfrac{\Delta z}{z_{12}}$	$-\dfrac{y_{11}}{y_{12}}$	$-\dfrac{1}{y_{12}}$	$\dfrac{a_{22}}{\Delta a}$	$\dfrac{a_{12}}{\Delta a}$	b_{11}	b_{12}	$\dfrac{1}{h_{12}}$	$\dfrac{h_{11}}{h_{12}}$	$-\dfrac{\Delta g}{g_{12}}$	$-\dfrac{g_{22}}{g_{12}}$
$\dfrac{1}{z_{12}}$	$\dfrac{z_{11}}{z_{12}}$	$-\dfrac{\Delta y}{y_{12}}$	$-\dfrac{y_{22}}{y_{12}}$	$\dfrac{a_{21}}{\Delta a}$	$\dfrac{a_{11}}{\Delta a}$	b_{21}	b_{22}	$\dfrac{h_{22}}{h_{12}}$	$\dfrac{\Delta h}{h_{12}}$	$-\dfrac{g_{11}}{g_{12}}$	$-\dfrac{1}{g_{12}}$
$\dfrac{\Delta z}{z_{22}}$	$\dfrac{z_{12}}{z_{22}}$	$\dfrac{1}{y_{11}}$	$-\dfrac{y_{12}}{y_{11}}$	$\dfrac{a_{12}}{a_{22}}$	$\dfrac{\Delta a}{a_{22}}$	$\dfrac{b_{12}}{b_{11}}$	$\dfrac{1}{b_{11}}$	h_{11}	h_{12}	$\dfrac{g_{22}}{\Delta g}$	$-\dfrac{g_{12}}{\Delta g}$
$-\dfrac{z_{21}}{z_{22}}$	$\dfrac{1}{z_{22}}$	$\dfrac{y_{21}}{y_{11}}$	$\dfrac{\Delta y}{y_{11}}$	$-\dfrac{1}{a_{22}}$	$\dfrac{a_{21}}{a_{22}}$	$-\dfrac{\Delta b}{b_{11}}$	$\dfrac{b_{21}}{b_{11}}$	h_{21}	h_{22}	$-\dfrac{g_{21}}{\Delta g}$	$\dfrac{g_{11}}{\Delta g}$
$\dfrac{1}{z_{11}}$	$\dfrac{z_{12}}{z_{11}}$	$\dfrac{\Delta y}{y_{22}}$	$\dfrac{y_{12}}{y_{22}}$	$\dfrac{a_{21}}{a_{11}}$	$-\dfrac{\Delta a}{a_{11}}$	$\dfrac{b_{21}}{b_{22}}$	$-\dfrac{1}{b_{22}}$	$\dfrac{h_{22}}{\Delta h}$	$-\dfrac{h_{12}}{\Delta h}$	g_{11}	g_{12}
$\dfrac{z_{21}}{z_{11}}$	$\dfrac{\Delta z}{z_{11}}$	$\dfrac{y_{21}}{y_{22}}$	$\dfrac{1}{y_{22}}$	$\dfrac{1}{a_{11}}$	$\dfrac{a_{12}}{a_{11}}$	$\dfrac{\Delta b}{b_{22}}$	$\dfrac{b_{12}}{b_{22}}$	$-\dfrac{h_{21}}{\Delta h}$	$\dfrac{h_{11}}{\Delta h}$	g_{21}	g_{22}

Δz	$=$	$z_{11}z_{22} - z_{21}z_{12}$	Δy	$=$	$y_{11}y_{22} - y_{21}y_{12}$
Δb	$=$	$b_{11}b_{22} - b_{21}b_{12}$	Δh	$=$	$h_{11}h_{22} - h_{21}h_{12}$

Δa	$=$	$a_{11}a_{22} - a_{21}a_{12}$
Δg	$=$	$g_{11}g_{22} - g_{21}g_{12}$

Reciprocal Two-Ports

A two-port is said to be **reciprocal** if interchanging an *ideal voltage source V* at one port with an *ideal ammeter* at the other port produces the same ammeter reading *I*. This is illustrated in Figure 15.3. Similarly, a two-port is reciprocal if interchanging an *ideal current source I* at one port with an *ideal voltmeter* at the other port produces the same voltmeter reading *V*.

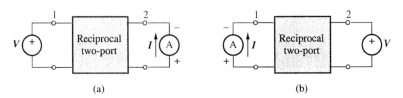

(a)   (b)

Figure 15.3 In a reciprocal two-port, interchanging a voltage source *V* at one port with an ideal ammeter A at the other port yields the same reading *I*.

We now wish to find the parameter conditions that make a two-port reciprocal. When connected as in Figure 15.3(a), a reciprocal two-port yields, by Equation (15.2b), $I = y_{21}V$; when connected as in Figure 15.3(b), the same two-port yields, by Equation (15.2a), $I = y_{12}V$. This is possible only if the condition $y_{12} = y_{21}$ is met. Using the entries in the top two rows of Table 15.1, it is readily seen that the following hold for a reciprocal network:

$$z_{12} = z_{21} \tag{15.9a}$$

$$y_{12} = y_{21} \tag{15.9b}$$

$$\Delta a = 1 \tag{15.9c}$$

$$\Delta b = 1 \tag{15.9d}$$

$$h_{12} = -h_{21} \tag{15.9e}$$

$$g_{12} = g_{21} \tag{15.9f}$$

Looking back at Equations (15.7) and (15.8), we note that both the *T* network and the *π* network are reciprocal. This is not a mere coincidence; it stems from the fact that the elements making up these networks are *bilateral,* that is, they can be placed in either direction in the circuit to yield the same behavior. Were the circuit to contain a nonbilateral element such as a dependent source, this property would no longer hold. We shall see examples later.

It is apparent that in order to characterize a reciprocal network we need only *three* parameters instead of four. Moreover, any reciprocal network can be modeled either with a *T* or a *π* network.

▶ Example 15.3

Specify the elements of a *T* network so that

$$[z] = \begin{bmatrix} 20 & 5 \\ 5 & 30 \end{bmatrix} \text{ k}\Omega$$

Solution

Comparing with Equation (15.7), we need

$$\mathbf{Z}_A + \mathbf{Z}_C = 20 \text{ k}\Omega$$

$$\mathbf{Z}_C = 5 \text{ k}\Omega$$

$$\mathbf{Z}_B + \mathbf{Z}_C = 30 \text{ k}\Omega$$

Solving, we obtain $\mathbf{Z}_A = 15$ kΩ, $\mathbf{Z}_B = 25$ kΩ, and $\mathbf{Z}_C = 5$ kΩ. ◀

A *reciprocal* network with $y_{11} = y_{22}$ is said to be **symmetrical** because its ports can be interchanged without altering its terminal currents and voltages. Using the entries in the top two rows of Table 15.1, it is readily seen that for a symmetrical network the following relationships hold:

$$z_{11} = z_{22} \qquad \text{(15.10a)}$$

$$y_{11} = y_{22} \qquad \text{(15.10b)}$$

$$a_{11} = a_{22} \qquad \text{(15.10c)}$$

$$b_{11} = b_{22} \qquad \text{(15.10d)}$$

$$\Delta h = 1 \qquad \text{(15.10e)}$$

$$\Delta g = 1 \qquad \text{(15.10f)}$$

Clearly, to characterize a symmetrical network we need only *two* parameters instead of four.

▶ Example 15.4

The following measurements are performed on a symmetrical, passive resistive network: (a) with port 2 open, $V_1 = 8$ V and $I_1 = 2$ mA; (b) with port 2 shorted, $V_1 = 7$ V and $I_2 = -3$ mA. Find the z parameters of this network.

Solution

With port 2 open, $I_2 = 0$. Then Equation (15.1a) yields

$$8 = z_{11} \times 2 + z_{12} \times 0$$

or $z_{11} = 8/2 = 4$ kΩ. With port 2 shorted, $V_2 = 0$. Considering that for a symmetrical network we have $z_{11} = z_{22}$ and $z_{12} = z_{21}$, Equations (15.1) can be written as

$$7 = 4I_1 + z_{12}(-3)$$

$$0 = z_{12}I_1 + 4(-3)$$

Eliminating I_1, solving for z_{12}, and retaining the positive solution, which is the only physically acceptable one, we obtain $z_{12} = 3$ kΩ. In summary, $z_{11} = z_{22} = 4$ kΩ and $z_{12} = z_{21} = 3$ kΩ. ◀

> **Exercise 15.3**   Specify the elements of a π network so that
>
> $$[y] = \begin{bmatrix} 0.5 & -0.1 \\ -0.1 & 0.5 \end{bmatrix} \; \Omega^{-1}$$
>
> **ANSWER** $Z_A = Z_B = 2.5 \; \Omega, \; Z_C = 10 \; \Omega.$

Nonexistence of Parameters

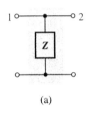

(a)

(b)

Figure 15.4
Single-element
networks:
(a) *shunt* and
(b) *series*.

There are networks for which the z and y parameters do not both exist. The simplest examples are offered by the single-element networks of Figure 15.4. The *shunt* element may be regarded as a T network with $Z_A = Z_B = 0$ and $Z_C = Z$. Hence, Equation (15.7) yields

$$[z_{\text{shunt}}] = \begin{bmatrix} Z & Z \\ Z & Z \end{bmatrix} \tag{15.11}$$

Since the determinant Δz is zero, the y parameters as determined from the z parameters via Table 15.1 do not exist.

By similar reasoning, the *series* network may be regarded as a π network with $Z_A = Z_B = \infty$, and $Z_C = Z$. Hence, Equation (15.8) yields

$$[y_{\text{series}}] = \begin{bmatrix} 1/Z & -1/Z \\ -1/Z & 1/Z \end{bmatrix} \tag{15.12}$$

Since the determinant Δy is zero, the z parameters as determined from the y parameters via Table 15.1 do not exist.

Even though the z parameter set does not exist for the series element, other sets exist besides the y set. For example, using Table 15.1, we find $h_{11} = Z$, $h_{12} = 1$, $h_{21} = -1$, and $h_{22} = 0$. One can likewise verify that even though the y parameters do not exist for the shunt element, other parameters, such as the a parameters, do.

15.2 THE z, y, a, AND h PARAMETERS

Any one of the six parameter sets may be found directly, either by calculation or by measurement. Though we focus upon the z, y, a, and h sets because they are the most widely used ones, the techniques of this section can readily be applied to the remaining sets.

The z Parameters

By Equation (15.1), the z parameters of a linear two-port are found as

$$z_{11} = \frac{V_1}{I_1}\bigg|_{I_2=0} \qquad z_{21} = \frac{V_2}{I_1}\bigg|_{I_2=0} \tag{15.13a}$$

$$z_{12} = \frac{V_1}{I_2}\bigg|_{I_1=0} \qquad z_{22} = \frac{V_2}{I_2}\bigg|_{I_1=0} \tag{15.13b}$$

We observe that z_{11} and z_{22} are *driving-point* impedances: z_{11} is the impedance of port 1 with port 2 open, and z_{22} is that of port 2 with port 1 open. By contrast, z_{12} and z_{21} are *transfer* impedances: z_{12} relates the open-circuit voltage response of port 1 to the current applied at port 2, and z_{21} relates the open-circuit voltage response of port 2 to the current applied at port 1. It is precisely the *transfer* impedance that makes two-ports more powerful circuits than one-ports. For obvious reasons, the z parameters are also called the **open-circuit impedance parameters.**

▶ Example 15.5

Find [z] for the circuit of Figure 15.5(a).

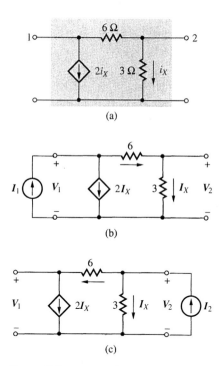

(a)

(b)

(c)

Figure 15.5 (a) Two-port of Example 15.5, and test circuits to find (b) z_{11} and z_{21}, and (c) z_{12} and z_{22}.

Solution

To find z_{11} and z_{21} we must, by Equation (15.13a), apply a *test current* I_1 to port 1 and leave port 2 *open-circuited* to ensure $I_2 = 0$. The situation is depicted in Figure 15.5(b). Observing that $I_X = V_2/3$, we apply the node method and write

$$I_1 = 2I_X + \frac{V_1 - V_2}{6} = 2\frac{V_2}{3} + \frac{V_1 - V_2}{6}$$

$$\frac{V_1 - V_2}{6} = I_X = \frac{V_2}{3}$$

Solving for V_1 and V_2 in terms of I_1 yields $V_1 = 3I_1$ and $V_2 = 1I_1$, so $z_{11} = 3 \ \Omega$, and $z_{21} = 1 \ \Omega$.

To find z_{12} and z_{22} we must, by Equation (15.13b), apply a *test current* I_2 to port 2 and leave port 1 *open-circuited* to ensure $I_1 = 0$. This is depicted in Figure 15.5(c). Again applying the node method, along with $I_X = V_2/3$, we have

$$I_2 = I_X + \frac{V_2 - V_1}{6} = \frac{V_2}{3} + \frac{V_2 - V_1}{6}$$

$$\frac{V_2 - V_1}{6} = 2I_X = 2\frac{V_2}{3}$$

Solving gives $V_1 = -3I_2$ and $V_2 = 1I_2$, so $z_{12} = -3 \ \Omega$ and $z_{22} = 1 \ \Omega$. In summary,

$$[z] = \begin{bmatrix} 3 & -3 \\ 1 & 1 \end{bmatrix} \ \Omega$$

Remark   Since $z_{12} \neq z_{21}$, our network is not reciprocal, even though it has a π topology. This stems from the presence of the dependent source, which is not a bilateral element.   ◀

Exercise 15.4   Repeat Example 15.5, but with a 4-Ω resistance in parallel with the dependent source.

ANSWER $z_{11} = -z_{12} = 12/7 \ \Omega$, $z_{21} = 4/7 \ \Omega$, and $z_{22} = 10/7 \ \Omega$.

Models of z Parameters

Having illustrated how to find the z parameters once a network is given, we now turn to the inverse task of finding a network that realizes a given set of z parameters. Since this is a synthesis problem, it does not admit a unique solution. One possible solution is shown in Figure 15.6. You are encouraged to verify that it does indeed satisfy Equation (15.1). Note that if $z_{12} = z_{21}$ the network is reciprocal and it reduces to the familiar T network.

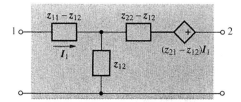

Figure 15.6 Model of z parameters.

▶ Example 15.6

Find a two-port with

$$[z] = \begin{bmatrix} 6 + 1/2s & 1 + 1/2s \\ 4 + 1/2s & 4s + 1 + 1/2s \end{bmatrix} \ \Omega$$

Solution

Since $z_{12} \neq z_{21}$, this is a nonreciprocal network. The element values of Figure 15.6 are $z_{11} - z_{12} = 6 + 1/2s - (1 + 1/2s) = 5\ \Omega$ (a resistance), $z_{22} - z_{12} = 4s + 1 + 1/2s - (1 + 1/2s) = 4s\ \Omega$ (an inductance), $z_{12} = 1 + 1/2s\ \Omega$ (a resistance in series with a capacitance), and $(z_{21} - z_{12})I_1 = [4 + 1/2s - (1 + 1/2s)]I_1 = (3\ \Omega)I_1$ (a CCVS). The circuit is shown in Figure 15.7.

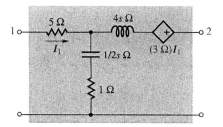

Figure 15.7 Circuit of Example 15.6.

Exercise 15.5 Find an ac network with

$$[z] = \begin{bmatrix} 3 - j4 & 3 - j2 \\ 8 - j2 & 3 - j4 \end{bmatrix} \ \text{k}\Omega$$

ANSWER $-j2\ \text{k}\Omega$, $-j2\ \text{k}\Omega$, $3 - j2\ \text{k}\Omega$, $(5\ \text{k}\Omega)I_1$ (a CCVS).

The *y* Parameters

By Equation (15.2), the *y* parameters can be found as

$$
y_{11} = \left.\frac{I_1}{V_1}\right|_{V_2=0} \qquad y_{21} = \left.\frac{I_2}{V_1}\right|_{V_2=0} \tag{15.14a}
$$

$$
y_{12} = \left.\frac{I_1}{V_2}\right|_{V_1=0} \qquad y_{22} = \left.\frac{I_2}{V_2}\right|_{V_1=0} \tag{15.14b}
$$

We observe that y_{11} and y_{22} are *driving-point* admittances, whereas y_{12} and y_{21} are *transfer* admittances. For obvious reasons, the *y* parameters are also called the **short-circuit admittance parameters.**

▶ **Example 15.7**

Find the *y* parameters for the circuit of Figure 15.8(a).

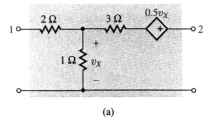

(a)

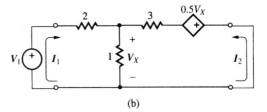

(b)

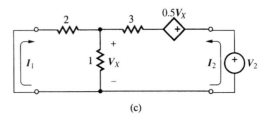

(c)

Figure 15.8 (a) Two-port of Example 15.7, and test circuits to find (b) y_{11} and y_{21}, and (c) y_{12} and y_{22}.

Solution

To find y_{11} and y_{21} we apply a *test voltage* V_1 to port 1 and *short-circuit* port 2 to ensure $V_2 = 0$. This is shown in Figure 15.8(b). Applying the loop method, along with Ohm's Law, yields

$$V_1 = 2I_1 + 1(I_1 + I_2)$$

$$0 = 0.5V_X + 3I_2 + 1(I_1 + I_2)$$

$$V_X = 1(I_1 + I_2)$$

Eliminating V_X and solving gives $I_1 = (3/8)V_1$ and $I_2 = (-1/8)V_1$. Thus, $y_{11} = 3/8 \ \Omega^{-1}$ and $y_{21} = -1/8 \ \Omega^{-1}$.

To find y_{12} and y_{22} we apply a *test voltage* V_2 to port 2 and *short-circuit* port 1 to ensure $V_1 = 0$. Referring to Figure 15.8(c), we have

$$V_2 = 0.5V_X + 3I_2 + 1(I_1 + I_2)$$

$$0 = 2I_1 + 1(I_1 + I_2)$$

$$V_X = 1(I_1 + I_2)$$

Eliminating V_X yields $I_1 = (-1/12)V_2$, and $I_2 = (1/4)V_2$. Thus, $y_{12} = -1/12 \ \Omega^{-1}$ and $y_{22} = 1/4 \ \Omega^{-1}$. In summary, we have

$$[y] = \begin{bmatrix} 3/8 & -1/12 \\ -1/8 & 1/4 \end{bmatrix} \ \Omega^{-1}$$

◀

Exercise 15.6   Repeat Example 15.7, but with the 1-Ω resistance removed from the circuit, everything else being the same.

ANSWER $y_{11} = 1/4 \ \Omega^{-1}$, $y_{12} = -1/6 \ \Omega^{-1}$, $y_{21} = -1/4 \ \Omega^{-1}$, and $y_{22} = 1/6 \ \Omega^{-1}$.

Exercise 15.7   Find the y parameters for the circuit of Figure 15.5(a), and check using Table 15.1.

ANSWER $y_{11} = -y_{21} = 1/6 \ \Omega^{-1}$, $y_{12} = y_{22} = 1/2 \ \Omega^{-1}$.

Models of *y* Parameters

Given a y parameter set, Figure 15.9 shows a possible network to realize it. Note that if $y_{12} = y_{21}$ the network is reciprocal and reduces to the familiar π network.

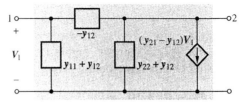

Figure 15.9 Model of y parameters.

Example 15.8

Find a network such that

$$[y] = \begin{bmatrix} 2s + 5 & -2 \\ -4 & 3s + 2 \end{bmatrix} \; \Omega^{-1}$$

Solution

Referring to Figure 15.9, we have $y_{11} + y_{12} = 2s + 3 \; \Omega^{-1}$ (a 2-F capacitance in parallel with a 1/3-Ω resistance), $-y_{12} = 2 \; \Omega^{-1}$ (a 1/2-Ω resistance), $y_{22} + y_{12} = 3s \; \Omega^{-1}$ (a 3-F capacitance), and $(y_{21} - y_{12})V_1 = (-2 \; \Omega^{-1})V_1$, indicating that the VCCS must point upward. The circuit is shown in Figure 15.10.

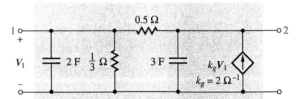

Figure 15.10 Circuit of Example 15.8.

Exercise 15.8 Find an ac network such that at $\omega = 10^4$ rad/s it has $y_{11} = y_{22} = (4 - j4)10^{-3}\Omega^{-1}$, $y_{12} = -2 \times 10^{-3}\Omega^{-1}$, and $y_{21} = -10^{-3}\Omega^{-1}$.

ANSWER $y_{11} + y_{12}$: 100 Ω in series with 20 mH; $-y_{12}$: 500 Ω; $y_{22} + y_{12}$: 100 Ω in series with 20 mH; VCCS: $(10^{-3} \; \Omega^{-1})V_1$.

The *a* Parameters

By Equation (15.3), the *a* parameters, also called the *ABCD* parameters, are found as

$$a_{11} = \frac{V_1}{V_2}\bigg|_{I_2=0} \qquad a_{21} = \frac{I_1}{V_2}\bigg|_{I_2=0} \qquad \text{(15.15a)}$$

$$a_{12} = -\frac{V_1}{I_2}\bigg|_{V_2=0} \qquad a_{22} = -\frac{I_1}{I_2}\bigg|_{V_2=0} \qquad \text{(15.15b)}$$

We observe that all parameters are functions of the *transfer* type. Namely, a_{11} and a_{22} are *gain* functions, a_{12} is a transfer *impedance,* and a_{21} is a transfer *admittance.*

▶ Example 15.9

Find the *a* parameters for the circuit of Figure 15.11(a).

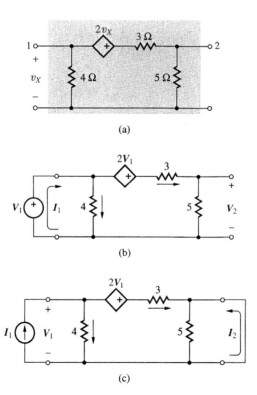

(a)

(b)

(c)

Figure 15.11 (a) Two-port of Example 15.9, and test circuits to find (b) a_{11} and a_{21}, and (c) a_{12} and a_{22}.

Solution

Following Equation (15.15a), we leave port 2 open and apply a test voltage V_1 to port 1, as shown in Figure 15.11(b). Applying KVL, the voltage divider formula, and KCL, we get

$$V_2 = \frac{5}{3+5}(V_1 + 2V_1) = \frac{15}{8}V_1$$

$$I_1 = \frac{V_1}{4} + \frac{(V_1 + 2V_1) - V_2}{3} = \frac{5}{8}V_1$$

Then, $a_{11} = V_1/V_2 = \frac{8}{15}$, and $a_{21} = I_1/V_2 = (I_1/V_1) \times (V_1/V_2) = \frac{5}{8} \times \frac{8}{15} = \frac{1}{3}\ \Omega^{-1}$.

Following Equation 15.15b, we short port 2 and apply a test current I_1 to port 1, as shown in Figure 15.11(c). Then,

$$I_2 = -\frac{V_1 + 2V_1}{3} = -V_1$$

$$I_1 = \frac{V_1}{4} + \frac{V_1 + 2V_1}{3} = \frac{5}{4}V_1$$

so $a_{12} = -(V_1/I_2) = 1\ \Omega$ and $a_{22} = -(I_1/I_2) = -(I_1/V_1) \times (V_1/I_2) = \left(-\frac{5}{4}\right)(-1) = \frac{5}{4}$. In summary, we have

$$[a] = \begin{bmatrix} \frac{8}{15} & 1\ \Omega \\ \frac{1}{3}\ \Omega^{-1} & \frac{5}{4} \end{bmatrix}$$ ◄

Exercise 15.9   Find the a parameters for the circuit of Figure 15.5(a), and check using Table 15.1.

ANSWER $a_{11} = 3, a_{12} = 6\ \Omega, a_{21} = 1\ \Omega^{-1}, a_{22} = 1$.

Exercise 15.10   Find the a parameters for the circuit of Figure 15.8(a), and check using Table 15.1.

ANSWER $a_{11} = 2, a_{12} = 8\ \Omega, a_{21} = 2/3\ \Omega^{-1}, a_{22} = 3$.

Exercise 15.11   Find the a parameters of (a) a T network and (b) a π network, each consisting of three 10-Ω resistances.

ANSWER   (a) $a_{11} = a_{22} = 2, a_{12} = 30\ \Omega, a_{21} = 1/10\ \Omega^{-1}$; (b) $a_{11} = a_{22} = 2, a_{12} = 10\ \Omega, a_{21} = 3/10\ \Omega^{-1}$.

▶ **Example 15.10**

Find the *a* parameters for the single-element networks of Figure 15.4.

Solution

(a) Applying a test voltage V_1 to port 1 of the *shunt* element with port 2 open-circuited yields $V_1 = V_2$ and $I_1 = (1/Z)V_2$, so $a_{11} = 1$ and $a_{21} = 1/Z$. Moreover, subjecting port 1 to a test current I_1 with port 2 short-circuited yields $V_1 = 0$ and $I_2 = -I_1$, so $a_{12} = 0$ and $a_{22} = 1$. Consequently,

$$[a_{\text{shunt}}] = \begin{bmatrix} 1 & 0 \\ 1/Z & 1 \end{bmatrix}$$ (15.16a)

(b) Repeating the procedure for the *series* element yields

$$[a_{\text{series}}] = \begin{bmatrix} 1 & Z \\ 0 & 1 \end{bmatrix}$$ (15.16b) ◀

Exercise 15.12   Show that for an *ideal transformer* we have

$$[a] = \begin{bmatrix} 1/n & 0 \\ 0 & n \end{bmatrix}$$

▶ **Example 15.11**

Two sets of measurements are performed on an unknown resistive two-port using a 10-V source connected to port 1, and a multimeter. If (a) with port 2 open-circuited we read $I_1 = 1$ A and $V_2 = 20$ V, and (b) with port 2 short-circuited we read $I_1 = 0.75$ A and $I_2 = -0.5$ A, what are the *z* parameters of this network?

Solution

The first set of measurements allows us to write

$$a_{11} = V_1/V_2|_{I_2=0} = 10/20 = 0.5$$

$$a_{21} = I_1/V_2|_{I_2=0} = 1/20 = 0.05 \ \Omega^{-1}$$

The second set yields

$$a_{12} = -V_1/I_2|_{V_2=0} = -10/(-0.5) = 20 \ \Omega$$

$$a_{22} = -I_1/I_2|_{V_2=0} = -0.75/(-0.5) = 1.5$$

Finally, using Table 15.1 to convert to the z parameters, we get

$$[z] = \begin{bmatrix} 10 & -5 \\ 20 & 30 \end{bmatrix} \ \Omega \qquad \blacktriangleleft$$

Exercise 15.13   Two sets of measurements are performed on an unknown network using an ac source $v(t) = 10 \sin 2\pi 10^3 t$ V; (a) with the source connected to port 1 and port 2 short-circuited, we find $i_1(t) = 5 \cos (2\pi 10^3 t + 45°)$ mA and $i_2(t) = 2 \cos (2\pi 10^3 t - 60°)$ mA; (b) with the source connected to port 2 and port 1 short-circuited, $i_1(t) = 10 \cos (2\pi 10^3 t - 45°)$ mA, and $i_2(t) = 5 \cos (2\pi 10^3 t + 30°)$ mA. Find a suitable parameter set.

ANSWER $y_{11} = 0.5 \times 10^{-3} \underline{/135°} \ \Omega^{-1}$, $y_{12} = 10^{-3} \underline{/45°} \ \Omega^{-1}$, $y_{21} = 0.2 \times 10^{-3} \underline{/30°} \ \Omega^{-1}$, $y_{22} = 0.5 \times 10^{-3} \underline{/120°} \ \Omega^{-1}$.

The h Parameters

By Equation (15.5), the h parameters are found as

$$h_{11} = \left. \frac{V_1}{I_1} \right|_{V_2=0} \qquad h_{21} = \left. \frac{I_2}{I_1} \right|_{V_2=0} \qquad \text{(15.17a)}$$

$$h_{12} = \left. \frac{V_1}{V_2} \right|_{I_1=0} \qquad h_{22} = \left. \frac{I_2}{V_2} \right|_{I_1=0} \qquad \text{(15.17b)}$$

The driving-point parameters h_{11} and h_{22} are called, respectively, the *input impedance,* in Ω, and the *output admittance,* in Ω^{-1}. The transfer parameters h_{12} and h_{21} are called, respectively, the *reverse voltage gain,* in V/V, and the *forward current gain,* in A/A. To emphasize this terminology, they are often denoted as

$$[h] = \begin{bmatrix} h_i & h_r \\ h_f & h_o \end{bmatrix} \qquad \text{(15.18)}$$

▶ **Example 15.12**

Find the h parameters for the circuit of Figure 15.12(a).

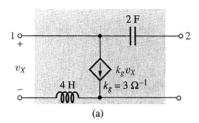

(a)

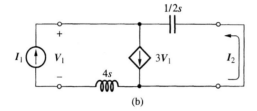

(b)

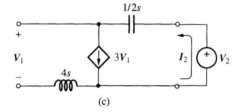

(c)

Figure 15.12 (a) Two-port of Example 15.12, and test circuits to find (b) h_{11} and h_{21}, and (c) h_{12} and h_{22}.

Solution

Equation (15.17a) leads to the circuit of Figure 15.12(b). Applying KCL at the upper leftmost node and KVL around the outer loop clockwise yields

$$I_1 + I_2 = 3V_1$$

$$V_1 + \frac{1}{2s}I_2 = 4sI_1$$

Eliminating I_2 yields $V_1 = [(8s^2 + 1)/(2s + 3)]I_1$. Eliminating V_1 yields $I_2 = [2s(12s - 1)/(2s + 3)]I_1$.

Equation (15.17b) leads to the circuit of Figure 15.12(c), where

$$I_2 = 3V_1$$

$$V_2 - V_1 = I_2/2s$$

so $V_1 = [2s/(2s + 3)]V_2$ and $I_2 = [6s/(2s + 3)]V_2$. Hence,

$$[h] = \begin{bmatrix} (8s^2 + 1)/(2s + 3) \ \Omega & 2s/(2s + 3) \\ 2s(12s - 1)/(2s + 3) & 6s/(2s + 3) \ \Omega^{-1} \end{bmatrix}$$

◀

Exercise 15.14   Find the h parameters for the circuit of Figure 15.5(a), and check using Table 15.1.

ANSWER $h_{11} = 6 \ \Omega$, $h_{12} = -3$, $h_{21} = -1$, $h_{22} = 1 \ \Omega^{-1}$.

Exercise 15.15   Find the h parameters for the circuit of Figure 15.8(a), and check using Table 15.1.

ANSWER $h_{11} = 8/3 \ \Omega$, $h_{12} = 2/9$, $h_{21} = -1/3$, $h_{22} = 2/9 \ \Omega^{-1}$.

Exercise 15.16   Find the h parameters for the circuit of Figure 15.11(a), and check using Table 15.1.

ANSWER $h_{11} = 0.8 \ \Omega$, $h_{12} = 4/15$, $h_{21} = -0.8$, $h_{22} = 4/15 \ \Omega^{-1}$.

Models of h Parameters

Given a set of h parameters, a circuit that realizes them is shown in Figure 15.13. This circuit is used to model the *small-signal behavior* of bipolar junction transistors (BJTs).

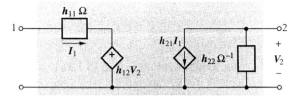

Figure 15.13 Model of h parameters.

Exercise 15.17   Figure 15.14 shows the small-signal model of a *common-emitter* (CE) *BJT amplifier*. Show that its h parameters are

$$[h_e] = \begin{bmatrix} h_{ie} & h_{re} \\ h_{fe} & h_{oe} \end{bmatrix} = \begin{bmatrix} r_\pi + (\beta + 1)(R_E \parallel r_o) & \dfrac{R_E}{R_E + r_o} \\ \beta - (\beta + 1)\dfrac{R_E}{R_E + r_o} & \dfrac{1}{R_E + r_o} \end{bmatrix}$$

where the subscript e is meant to signify common "emitter."

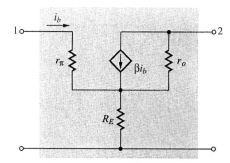

Figure 15.14 Small-signal model of the common-emitter BJT amplifier of Exercise 15.17.

15.3 TWO-PORT INTERCONNECTIONS

In the typical application shown in Figure 15.15, a two-port is driven by a source V_s with internal impedance Z_s, and it drives a load impedance Z_L. Z_s and Z_L are referred to as the *terminations*. The functions of interest are the *current gain* I_2/I_1, the *source-to-load voltage gain* V_2/V_s, and the impedances Z_1 and Z_2 seen looking into the input and output ports.

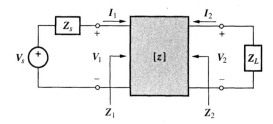

Figure 15.15 Terminated two-port.

If the two-port is described in terms of its z parameters, as shown, the equations governing the circuit are

$$V_1 = z_{11}I_1 + z_{12}I_2 \tag{15.19}$$

$$V_2 = z_{21}I_1 + z_{22}I_2 \tag{15.20}$$

$$V_1 = V_s - Z_s I_1 \tag{15.21}$$

$$V_2 = -Z_L I_2 \tag{15.22}$$

By Equations (15.20) and (15.22), $z_{21}I_1 + z_{22}I_2 = -Z_L I_2$, or

$$\boxed{\frac{I_2}{I_1} = \frac{-z_{21}}{z_{22} + Z_L}} \tag{15.23}$$

MAYBE THERE'S AN EV IN YOUR FUTURE

A —if not *the*—better mousetrap of the 21st century will be a practical electric automobile. The combined pressures of air pollution and oil shortages have made this elusive dream fashionable once again.

Like most other things under the sun, electric cars are not new. The first U.S. model is credited to William Morrison and was built in Des Moines in 1891. It used 24 storage batteries and could run for 13 hours at a top speed of 14 miles per hour. Electric taxicabs were introduced in New York City in 1897, and growth continued steadily: By 1912 there were 34,000 electric cars on U.S. roads and only 20,000 gasoline-powered ones.

Such parity was maintained only while top speeds remained in the 20-mph range, though. Once technology advanced enough to allow greater speeds, the internal combustion engine became king of the road, with the result that for years electrically powered vehicles were relegated to such small-range, low-power applications as golf carts and forklifts.

Today, as people all over the world become more and more concerned about air pollution and the price of gasoline, electric cars are once again singing their siren song. A 1976 near-term electric vehicle program sponsored by the U.S. Department of Energy made an ambitious start in designing prototypes but for the most part was an idea ahead of its time. Technology had some catching up to do. Electric motors needed to be made more efficient, electronic control devices needed to be made more reliable, and batteries needed to learn how to store more energy (for range) and power (for acceleration).

Today, some relatively powerful motors have been developed—BMW has a permanent-magnet dc motor rated at 45 horsepower and GM has an ac induction motor that delivers 30 hp on average and peaks at 57 hp—and integrated circuits have vastly improved the reliability of the electronic control devices all cars need. Various types of batteries hold promise—the old stand-bys lead-acid and nickel-cadmium as well as zinc-air, nickel-hydride, and lithium-polymer. Each type has some advantages and some drawbacks, so trade-offs have to be made. Lead-acids have sufficient power to give good acceleration but do not rate high in the range column, for instance, and zinc-airs store enough energy to give excellent range but have very little power, so acceleration is a problem.

Not able to wait patiently for the automotive companies to develop electric cars, many buffs opted for the "Just do it!" school of thought and have built their own. Marianne Walpert of Belmont, California, spent the early part of 1992 converting a 1981 Volkswagen Rabbit with a kit purchased from a Santa Cruz company called Electro Automotive. Working in her garage, Walpert took three weeks for the conversion and had her EV ready for the Electric 200, a 200-kilometer race held at Phoenix International

Raceway in April 1992. In relatively hilly Belmont, the VW gets 55 miles per charge when cruise speed is held to 45 or 50 mph. The car can do 60, but the 55-mile range figure drops a bit. The car is used every day, and an overnight charge in the garage on a 220-V line costs less than $1. Because Walpert's major interest is photovoltaic cells, she has mounted a solar panel on the VW's roof and fills about 10% of her energy needs that way. The battery pack is sixteen 6-volt lead-acid cells that cost about $700 and will have to be replaced every 4 to 5 years.

Bob Batson, a mechanical engineer from Maynard, Massachusetts, believes in EVs so much that he has formed his own company, Electric Vehicles of America, to sell conversion kits and do conversions

> "A—if not the—better mousetrap of the 21st century will be a practical electric automobile."

for customers. Batson owns two EVs these days: a Bradley GT-II Sports Car born as an EV and made by a company no longer in business, and an '87 Dodge pickup truck that he converted himself in about 70 hours. The Bradley is reserved for special occasions, but the pickup is his everyday workhorse. It has a range of about 60 miles when driven at 45 mph and charges in about 10 hours on a 220-volt line.

A conversion kit has three major components: the electric motor, a controller box requiring four wires to be connected, and a potentiometer requir-

ing five. You pull the gasoline engine, radiator, and exhaust system, leaving in place the 12-volt battery the car came with (to run headlights and so on) and the transmission. Connect the new motor, controller, and potentiometer, and you're in business. No oil changes, no tuneups, no catalytic converter to break and cause you to flunk inspection.

Electric cars are not totally innocent in the air pollution department, of course, because the electricity they need has to be generated at a power plant somewhere. If that plant uses hydro power, there's no problem, but most likely it will be either coal-fired or nuclear with all the attendant problems. It is much easier to control emissions at

> *The first electric car dealership in Hollywood opened in the summer of 1992...*

one central plant, however, than to control the emissions from millions of cars and trucks on the road. Today California has a law in place that says by 1998 any company that sells 5000 cars in the state must have 2 percent of those sales be ZEVs—zero-emission vehicles (zero!)—and the ratio increases gradually until topping out at 10 percent in 2003. The first electric car dealership in Hollywood opened in the summer of 1992, and several charging stations have been opened in various parts of Los Angeles by entrepreneurs determined to be in on the beginning of a new industry.

where the negative sign indicates that I_2 actually flows out of the port 2 positive terminal, as one might have expected. The impedance seen looking into port 1 is found as $Z_1 = V_1/I_1$. Solving Equation (15.23) for I_2, substituting into Equation (15.19), and solving for the ratio V_1/I_1 yields

$$Z_1 = z_{11} - \frac{z_{12}z_{21}}{z_{22} + Z_L} \tag{15.24}$$

Writing $V_s = (Z_s + Z_1)I_1$, we have $V_2/V_s = -Z_L I_2/[(Z_s + Z_1)I_1] = -[Z_L/(Z_s + Z_1)] \times (I_2/I_1)$. Using Equations (15.23) and (15.24),

$$\frac{V_2}{V_s} = \frac{z_{21}Z_L}{(z_{11} + Z_s)(z_{22} + Z_L) - z_{12}z_{21}} \tag{15.25}$$

Suppressing V_s, subjecting port 2 to a test voltage V_2, and taking the ratio V_2/I_2 we readily obtain

$$Z_2 = z_{22} - \frac{z_{12}z_{21}}{z_{11} + Z_s} \tag{15.26}$$

As we might have expected, both impedances depend on the z parameters. Moreover, Z_1 depends also on Z_L, and Z_2 also on Z_s.

▶ **Example 15.13**

In the circuit of Figure 15.15 let $V_s = 10$ V, $Z_s = 5$ Ω, $z_{11} = z_{22} = 20$ Ω, and $z_{12} = z_{21} = 10$ Ω.

(a) Find Z_L for maximum power transfer.
(b) Find the power dissipated by Z_L.
(c) Find the power delivered by V_s.

Solution

(a) Power transfer is maximized when $Z_L = Z_2^*$. By Equation (15.26), $Z_2 = 20 - 10^2/(20 + 5) = 16$ Ω. Hence, $Z_L = 16$ Ω.

(b) By Equation (15.25), $V_2/10 = (10 \times 16)/[(20+5)(20+16) - 10^2]$, or $V_2 = 2$ V. Then, $P_L = V_2^2/R_L = 2^2/16 = 0.25$ W.

(c) By Equation (15.24), $Z_1 = 20 - 10^2/(20 + 16) = 155/9$ Ω. Then, $P_s = V_s^2/(Z_s + Z_1) = 10^2/(5 + 155/9) = 4.5$ W. ◀

A two-port with $z_{11} = z_{22} = 0$ and $z_{21} = -z_{12} = r$ is called a **gyrator,** and r is called the *gyration resistance.* The distinguishing feature of the gyrator is that if its output port is terminated on an impedance Z_2 as in Figure 15.16, then the impedance seen looking into the input port is, by Equation (15.24),

$$Z_1 = \frac{r^2}{Z_2} \qquad\qquad \textbf{(15.27)}$$

that is, Z_1 is proportional to the *reciprocal* of Z_2. A common gyrator application is the synthesis of an inductance using a capacitance. Indeed, if $Z_2 = 1/sC$, then $Z_1 = r^2 sC = sL_{eq}$, where $L_{eq} = r^2 C$. A gyrator can be built using op amps (see Problem 15.27).

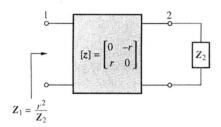

Figure 15.16 Gyrator.

Exercise 15.18   Show that if the two-port of Figure 15.15 is described in terms of its h parameters,

$$Z_1 = h_{11} - \frac{h_{12}h_{21}}{h_{22} + 1/Z_L} \qquad\qquad \textbf{(15.28)}$$

$$Z_2 = \frac{Z_s + h_{11}}{h_{22}Z_s + \Delta h} \qquad\qquad \textbf{(15.29)}$$

$$\frac{I_2}{I_1} = \frac{h_{21}}{1 + h_{22}Z_L} \qquad\qquad \textbf{(15.30)}$$

$$\frac{V_2}{V_s} = \frac{-h_{21}}{(h_{11} + Z_s)(h_{22} + 1/Z_L) - h_{12}h_{21}} \qquad\qquad \textbf{(15.31)}$$

Exercise 15.19   In the CE BJT amplifier of Figure 15.14 let $r_\pi = 3.9$ kΩ, $R_E = 1$ kΩ, $r_o = 100$ kΩ, and $\beta = 150$. Find the source-to-load gain if the circuit is driven by a source with $Z_s = 5$ kΩ and drives a load $Z_L = 10$ kΩ.

ANSWER -9.318 V/V.

A two-port with $h_{11} = h_{22} = 0$ and $h_{21} = h_{12} = k$ is called a **negative impedance converter** (NIC) because the impedance seen looking into one port is proportional to the *negative* of the terminating impedance of the other port.

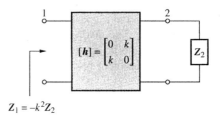

$Z_1 = -k^2 Z_2$

Figure 15.17   Negative impedance converter (NIC).

The NIC is depicted in Figure 15.17 and it can be built using op amps (see Problem 15.30).

Series-Connected Two-Ports

Two-ports may be interconnected to create more complex networks. The connection of Figure 15.18 is a *series* connection because the input ports of the two subnetworks carry the same current I_1 and the output ports carry the same current I_2. Using subscripts a and b to identify the two subnetworks, we have

$$V_1 = V_{a1} + V_{b1} = (z_{a11}I_1 + z_{a12}I_2) + (z_{b11}I_1 + z_{b12}I_2)$$

$$= (z_{a11} + z_{b11})I_1 + (z_{a12} + z_{b12})I_2 = z_{11}I_1 + z_{12}I_2$$

$$V_2 = V_{a2} + V_{b2} = (z_{a21}I_1 + z_{a22}I_2) + (z_{b21}I_1 + z_{b22}I_2)$$

$$= (z_{a21} + z_{b21})I_1 + (z_{a22} + z_{b22})I_2 = z_{21}I_1 + z_{22}I_2$$

where we have set

$$\boxed{z_{jk} = z_{ajk} + z_{bjk}} \tag{15.32}$$

$j = 1, 2$ and $k = 1, 2$. The *series* connection of two-port subnetworks is still a two-port, and its *z parameters* are the *sums* of the corresponding z parameters of the individual subnetworks. This can be exploited to facilitate the analysis of a complex two-port.

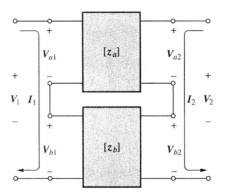

Figure 15.18   Series-connected two-ports.

▶ **Example 15.14**

Find the z parameters of the CE BJT amplifier of Figure 15.19(a).

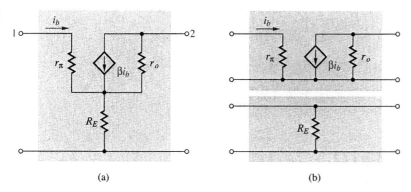

(a)   (b)

Figure 15.19 Decomposing a network into two series-connected subnetworks for Example 15.14.

Solution

We can regard the circuit as the series connection of the two simpler networks of Figure 15.19(b). First we find their individual z parameters, then we add them up as follows:

$$\begin{bmatrix} r_\pi & 0 \\ -\beta r_o & r_o \end{bmatrix} + \begin{bmatrix} R_E & R_E \\ R_E & R_E \end{bmatrix} = \begin{bmatrix} r_\pi + R_E & R_E \\ -\beta r_o + R_E & r_o + R_E \end{bmatrix}$$

◀

Parallel-Connected Two-Ports

The connection of Figure 15.20 is a *parallel* connection because the input ports of the two subnetworks experience the same voltage V_1 and the output ports the same voltage V_2. Proceeding as in the case of the series connection, but using the y parameters, it is readily seen that we now have

$$\boxed{y_{jk} = y_{ajk} + y_{bjk}} \qquad \text{(15.33)}$$

$j = 1, 2$ and $k = 1, 2$. The *parallel* connection of two-port subnetworks is still a two-port, and its *y parameters* are the *sums* of the corresponding y parameters of the individual subnetworks.

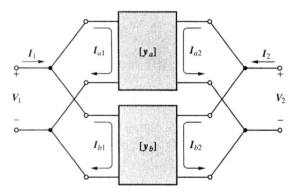

Figure 15.20 Parallel-connected two-ports.

▶ **Example 15.15**

Find the y parameters of the network of Figure 15.21(a).

Solution

Regarding the circuit as the parallel connection of the two subnetworks of Figure 15.21(b), we first find their individual y parameters, then we add them up as follows:

$$\begin{bmatrix} j\frac{1}{4} & -j\frac{1}{4} \\ -j\frac{1}{4} & j\frac{1}{4} \end{bmatrix} + \begin{bmatrix} \frac{4}{11} & -\frac{1}{11} \\ -\frac{1}{11} & \frac{3}{11} \end{bmatrix} = \begin{bmatrix} \frac{16+j11}{44} & -\frac{4+j11}{44} \\ -\frac{4+j11}{44} & \frac{12+j11}{44} \end{bmatrix} \ \Omega^{-1}$$

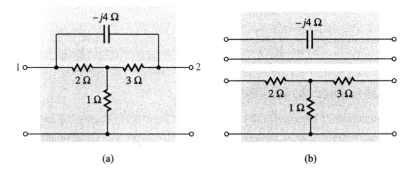

(a)   (b)

Figure 15.21 Decomposing a network into two parallel-connected subnetworks for Example 15.15.   ◀

Exercise 15.20   In the parallel network of Figure 15.20 let the a and b subnetworks be, respectively, a T and a π network, each consisting of three 10-kΩ resistances. Find the z parameters of the composite network. (*Hint:* First find the y parameters; then convert to the z parameters.)

ANSWER $z_{11} = z_{22} = 5$ kΩ, $z_{12} = z_{21} = 2.5$ kΩ.

Cascade-Connected Two-Ports

The connection of Figure 15.22 is called a *cascade* connection because the output port of one subnetwork is connected to the input port of the other subnetwork so that they share the same voltage V and current I. Using Equation (15.3), we have

$$V_1 = a_{a11}V + a_{a12}I = a_{a11}(a_{b11}V_2 - a_{b12}I_2) + a_{a12}(a_{b21}V_2 - a_{b22}I_2)$$

$$= (a_{a11}a_{b11} + a_{a12}a_{b21})V_2 - (a_{a11}a_{b12} + a_{a12}a_{b22})I_2 = a_{11}V_2 - a_{12}I_2$$

$$I_1 = a_{a21}V + a_{a22}I = a_{a21}(a_{b11}V_2 - a_{b12}I_2) + a_{a22}(a_{b21}V_2 - a_{b22}I_2)$$

$$= (a_{a21}a_{b11} + a_{a22}a_{b21})V_2 - (a_{a21}a_{b12} + a_{a22}a_{b22})I_2 = a_{21}V_2 - a_{22}I_2$$

where we have set

$$a_{11} = a_{a11}a_{b11} + a_{a12}a_{b21} \qquad \text{(15.34a)}$$
$$a_{12} = a_{a11}a_{b12} + a_{a12}a_{b22} \qquad \text{(15.34b)}$$
$$a_{21} = a_{a21}a_{b11} + a_{a22}a_{b21} \qquad \text{(15.34c)}$$
$$a_{22} = a_{a21}a_{b12} + a_{a22}a_{b22} \qquad \text{(15.34d)}$$

The *cascade* connection of two subnetworks is still a two-port, and its a parameters are found by *multiplying* the a parameters of the individual subnetworks according to Equation (15.34). We wish to point out that the *order* in which the subnetworks appear in the cascade is important. Interchanging one network with the other generally results in a different set of a parameters.

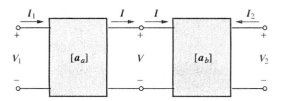

Figure 15.22   Cascade-connected two-ports.

▶ Example 15.16

Find the a parameter of the network of Figure 15.23.

Solution

The circuit consists of a T subnetwork followed by a π subnetwork. Following are their individual a matrices, along with the product a matrix

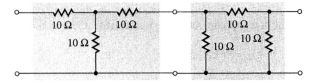

Figure 15.23 Cascade connection of Example 15.16.

obtained via Equation (15.34):

$$\begin{bmatrix} 2 & 30 \\ 0.1 & 2 \end{bmatrix} \times \begin{bmatrix} 2 & 10 \\ 0.3 & 2 \end{bmatrix} = \begin{bmatrix} 13 & 80\ \Omega \\ 0.8\ \Omega^{-1} & 5 \end{bmatrix}$$

Exercise 15.21   Repeat Example 15.16 if the *T* and *π* subnetworks are interchanged with each other.

ANSWER $a_{11} = 5$, $a_{12} = 80\ \Omega$, $a_{21} = 0.8\ \Omega^{-1}$, $a_{22} = 13$.

The properties of the cascade interconnection can be exploited in the analysis of *ladder networks*. Regarding a ladder as the cascade connection of *series* and *shunt* elements, we first find the matrices of its individual elements; then we multiply them out, one pair at a time, to obtain the *a* matrix of the entire ladder.

▶ **Example 15.17**

Find the *a* parameters of the ladder of Figure 15.24.

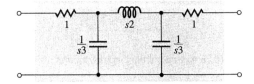

Figure 15.24 Ladder of Example 15.17.

Solution

By Equations (15.16), the *a* matrix of the ladder is

$$[a] = \begin{bmatrix} 1 & 1 \\ 0 & 1 \end{bmatrix} \times \begin{bmatrix} 1 & 0 \\ 3s & 1 \end{bmatrix} \times \begin{bmatrix} 1 & 2s \\ 0 & 1 \end{bmatrix} \times \begin{bmatrix} 1 & 0 \\ 3s & 1 \end{bmatrix} \times \begin{bmatrix} 1 & 1 \\ 0 & 1 \end{bmatrix}$$

Multiplying out the first matrix by the second and the third by the fourth yields

$$[a] = \begin{bmatrix} 1+3s & 1 \\ 3s & 1 \end{bmatrix} \times \begin{bmatrix} 1+6s^2 & 2s \\ 3s & 1 \end{bmatrix} \times \begin{bmatrix} 1 & 1 \\ 0 & 1 \end{bmatrix}$$

Multiplying out the first two matrices yields

$$[a] = \begin{bmatrix} 18s^3+6s^2+6s+1 & 6s^2+2s+1 \\ 18s^3+6s & 6s^2+1 \end{bmatrix} \times \begin{bmatrix} 1 & 1 \\ 0 & 1 \end{bmatrix}$$

Finally, multiplying out this last matrix pair, we get

$$[a] = \begin{bmatrix} 18s^3+6s^2+6s+1 & 18s^3+12s^2+8s+2 \\ 18s^3+6s & 18s^3+6s^2+6s+1 \end{bmatrix}$$

Exercise 15.22 Verify the results of Example 15.16, but using the ladder approach.

15.4 MAGNETICALLY COUPLED COILS

Just as a two-port is a generalization of a one-port, a pair of magnetically coupled coils is a generalization of the inductor. Recall that forcing a current i through a coil results in a magnetic flux ϕ in the core and, hence, a flux linkage $N\phi$ with the coil itself, where N is the number of turns. The direction of ϕ relative to i is found via the rule depicted in Figure 15.25:

Right-Hand Rule: If we curl the fingers of our right hand around the coil so that the fingers point in the direction of i, then ϕ is in the direction of the thumb.

The rate at which $N\phi$ varies with i is called the *self-inductance* of the coil, or $L = N\,d\phi/di$. By Faraday's law, any change in flux linkage induces across the terminals of the coil a voltage $v = N\,d\phi/dt$, as depicted in Figure 15.26(a). Eliminating $N\,d\phi$ yields

$$v = L\frac{di}{dt}$$

or, in s-domain form,

$$V = sLI$$

We also know that the inductor stores the energy

$$w(t) = \frac{1}{2}Li^2(t)$$

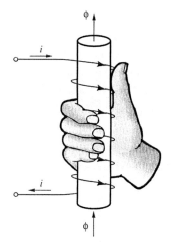

Figure 15.25 Illustrating the right-hand rule.

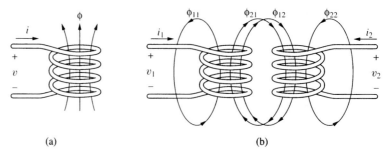

Figure 15.26 (a) Single coil and (b) magnetically coupled coil pair.

When two coils are placed in proximity to each other as in Figure 15.26(b), they share magnetic flux. Current i_1 produces flux inside coil 1 pointing upward. This flux can be expressed as $\phi_1 = \phi_{11} + \phi_{21}$, where ϕ_{11}, called the *leakage flux*, is the flux linking coil 1 but *not* coil 2, and ϕ_{21}, called the *mutual flux*, is the flux linking *both* coil 1 and coil 2. By Faraday's law, these flux components induce, respectively, a voltage component $v_{11} = N_1\, d\phi_{11}/dt = L_1\, di_1/dt$ in coil 1, and a voltage component $v_{21} = N_2\, d\phi_{21}/dt = M_{21}\, di_1/dt$ in coil 2, where N_1 and N_2 are the turns in the two coils. Likewise, the coil 2 current i_2 produces a flux $\phi_2 = \phi_{22} + \phi_{12}$ pointing downward inside this coil and consisting of the leakage flux ϕ_{22}, linking coil 2 but not coil 1, and the mutual flux ϕ_{12}, linking both coils. By Faraday's law, these fluxes induce, respectively, a voltage component $v_{22} = N_2\, d\phi_{22}/dt = L_2\, di_2/dt$ in coil 2 and a voltage component $v_{12} = N_1\, d\phi_{12}/dt = M_{12}\, di_2/dt$ in coil 1. Note that the mutual fluxes ϕ_{21} and ϕ_{12} have the same direction inside either coil. This congruence is signified by dot markings on the two coils, a subject to which we shall return soon. Letting $v_1 = v_{11} + v_{12}$ and $v_2 = v_{21} + v_{11}$ yields

$$v_1 = L_1 \frac{di_1}{dt} + M_{12} \frac{di_2}{dt} \tag{15.35a}$$

$$v_2 = M_{21} \frac{di_1}{dt} + L_2 \frac{di_2}{dt} \tag{15.35b}$$

or, in *s*-domain form,

$$V_1 = sL_1I_1 + sM_{12}I_2 \tag{15.36a}$$

$$V_2 = sM_{21}I_1 + sL_2I_2 \tag{15.36b}$$

The last set of equations indicates that a pair of coupled coils constitutes a two-port with $z_{11} = sL_1$, $z_{12} = sM_{12}$, $z_{21} = sM_{21}$, and $z_{22} = sL_2$. L_1 and L_2, each representing the ability of a coil to produce a voltage in response to a change in its own current, are **self-inductances.** By contrast, M_{12} and M_{21}, each representing the ability of one coil to produce a voltage in response to a changing current in the *other* coil, are called **mutual inductances.** All four parameters are expressed in *henrys* (H).

Mutual Inductance

Since coupled coils are based on the concept of magnetic inductance, which is *bilateral,* we have good reasons to expect that they form a *reciprocal* two-port, that is, $z_{12} = z_{21}$, or $M_{12} = M_{21}$. The following proof is based on an article by P. M. Lin,* which you may wish to look up in the library to appreciate how what might appear to be well-established concepts still continue to occupy the minds of scholars.

Suppose we excite the two coils with the currents $i_1 = \sin t$ and $i_2 = \cos t$. The instantaneous power $p(t)$ delivered to the coils is found with the help of Equation (15.35) as

$$p(t) = v_1 i_1 + v_2 i_2 = L_1 \sin t \cos t - M_{12}\sin^2 t + M_{21} \cos^2 t + L_2 \cos t \sin t$$

The *average power P* delivered to the coils is obtained by integrating $p(t)$ over one period and then dividing the result by the period itself, which in the present case is $T = 2\pi$. Using the fact that $\int_0^{2\pi} \sin t \cos t \, dt = 0$ and $\int_0^{2\pi} \sin^2 t = \int_0^{2\pi} \cos^2 t \, dt = \pi$, we obtain

$$P = \frac{1}{2}(M_{21} - M_{12})$$

Repeating the calculations with $i_1 = \cos t$ and $i_2 = \sin t$ yields

$$P = \frac{1}{2}(M_{12} - M_{21})$$

Since a network of coupled coils is a *passive* network, *P cannot be negative.* This implies that we cannot have $M_{12} > M_{21}$ or $M_{21} > M_{12}$. The only way to avoid both inequalities is for M_{12} and M_{21} to be *equal.*

Denoting the common value of the mutual inductance as $M_{12} = M_{21} = M$, Equations (15.35) and (15.36) take on the simpler forms

$$v_1 = L_1 \frac{di_1}{dt} + M \frac{di_2}{dt} \tag{15.37a}$$

$$v_2 = M \frac{di_1}{dt} + L_2\frac{di_2}{dt} \tag{15.37b}$$

$$V_1 = sL_1I_1 + sMI_2 \tag{15.38a}$$

$$V_2 = sMI_1 + sL_2I_2 \tag{15.38b}$$

Figure 15.27 shows the circuit symbol for coupled coils, along with a model with two dependent voltage sources to account for the effect of the mutual inductance.

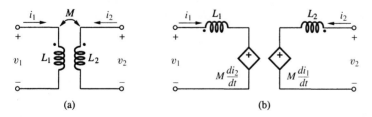

Figure 15.27 Circuit symbol and model for coupled coils.

*"Proof of $M_{12} = M_{21}$ Based on Energy Storage Is Incorrect," *IEEE Trans. on Circuits and Systems,* vol. 36, no. 9, Sept. 1989, pp. 1153–1158.

▶ **Example 15.18**

In the circuit of Figure 15.28 let $V_S = 10$ V, $R = 5$ Ω, $M = 2$ H, $L_1 = 4$ H, and $L_2 = 6$ H. Find v_2 if $i_1(0^-) = 0$.

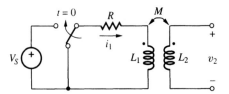

Figure 15.28 Circuit of Example 15.18.

Solution

Using Equation (15.37a) along with KVL gives

$$V_S = Ri_1 + L_1 \frac{di_1}{dt} + M \frac{di_2}{dt}$$

Since $i_2 = 0$ because coil 2 is open, this equation becomes

$$10 = 5i_1 + 4 \frac{di_1}{dt}$$

Dividing both sides by 5 to put it in the standard form of Equation (7.25), we obtain $0.8 \, di_1/dt + i_1 = 2$, indicating that $\tau = 0.8$ s. Applying Equation (7.41), we get

$$i_1(t \geq 0^+) = 2 \left(1 - e^{-t/0.8}\right) \text{ A}$$

To find v_2, use Equation (15.37b), which in the present case becomes

$$v_2 = M \frac{di_1}{dt}$$

Thus, $v_2 = 2(2e^{-t/0.8}/0.8)$, or

$$v_2(t \geq 0^+) = 5e^{-t/0.8} \text{ V}$$

◀

Exercise 15.23   In the circuit of Example 15.18 find i_1 if coil 2 is short-circuited.

ANSWER $i_1(t \geq 0^+) = 2 \left(1 - e^{-1.5t}\right)$ A.

The Dot Convention

The coil winding directions of Figure 15.26(b) were deliberately chosen so as to make the mutual voltage terms $M \, di_2/dt$ and $M \, di_1/dt$ come out *positive* for the terminal voltage and current assignments shown. In a circuit diagram, however, it is impractical to show the winding details, and a dot convention is used in order to keep track of the polarities of the mutual voltage terms when formulating circuit equations. The rule for using such a convention is

Dot Rule: When the reference current direction is *into the dotted terminal* of one coil, the reference polarity of the voltage term induced in the other coil is *positive* at its *dotted terminal.*

Conversely, when the current is *out* of the dotted terminal of one coil, the mutual voltage term is *negative* at the dotted terminal of the other. This rule is visualized with the aid of the circuit model of Figure 15.27(b).

▶ Example 15.19

In the circuit of Figure 15.29 find the steady-state ac voltage $v(t)$.

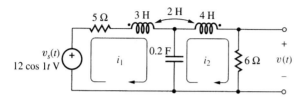

Figure 15.29   Circuit of Example 15.19.

Solution

Analyze the circuit in the s domain using the loop method. Then, find the ac response by letting $s \rightarrow j\omega = j1$ rad/s.

Going around the mesh labeled i_1 we encounter a voltage *rise* V_s, three voltage *drops* $5I_1$, $s3I_1$, and $(1/0.2s) \times (I_1 - I_2)$, and the mutual voltage term $s2I_2$. Is this term a rise or a drop? Since i_2 flows *out* of the dotted terminal of the 4-H coil, the mutual voltage term at the dotted terminal of the 3-H coil is *negative,* indicating that it appears as a *rise.* Equating the sum of the rises to the sum of the drops around the first mesh yields

$$V_s + s2I_2 = 5I_1 + s3I_1 + \frac{1}{s0.2}(I_1 - I_2)$$

or

$$(3s^2 + 5s + 5)I_1 - (2s^2 + 5)I_2 = sV_s$$

Going around the mesh labeled i_2 we encounter three voltage *drops* $(1/0.2s)(I_2 - I_1)$, $s4I_2$, and $6I_2$, and the mutual voltage term $s2I_1$. Since i_1 flows *into* the dotted terminal of the 3-H coil, the mutual term at the dotted terminal of the 4-H coil is *positive,* indicating that it appears as a *rise.* Thus,

$$s2I_1 = \frac{1}{s0.2}(I_2 - I_1) + s4I_2 + 6I_2$$

or

$$(2s^2 + 5)I_1 = (4s^2 + 6s + 5)I_2$$

Eliminating I_1, solving for I_2, and simplifying yields

$$I_2 = \frac{2s^2 + 5}{8s^3 + 38s^2 + 45s + 55} V_s$$

Before proceeding, perform asymptotic checks. Mathematically, $\lim_{s \to 0} I_2 = V_s/11$ and $\lim_{s \to \infty} I_2 = 0$. Physically, at low frequencies the coils act as shorts and the capacitor as an open, yielding $I_2 = V_s/(5 + 6) = V_s/11$. At high frequencies the coils act as opens, yielding $I_2 = 0$. Our result thus passes the asymptotic tests.

Letting $s \to j1$ and $V_s = 12/\underline{0°}$ V yields $I_2 = 36/(17 + j37) = 0.8841/\underline{-65.32°}$ A. Thus $V = 6I_2 = 5.305/\underline{-65.32°}$ V, so

$$v(t) = 5.305 \cos{(1t - 65.32°)} \text{ V}$$

Exercise 15.24   Repeat Example 15.19 with the terminals of the 4-H coil interchanged so that the dot marking appears at the left.

ANSWER $6.392 \cos{(1t - 77.55°)}$ V.

Dot Markings Assignment

A common task facing the user of coupled coils is to assign a proper set of dot markings once the winding arrangement is known. Following is a procedure to achieve this goal:

(1) Arbitrarily select one terminal of one coil, mark it with a dot, and assign a current *into* this terminal.

(2) Use the right-hand rule to find the direction of the mutual flux produced by the current of step 1 inside the second coil.

(3) Curl your right hand around the second coil so that the thumb points in the mutual flux direction found in step 2.

(4) Assign a current in the direction of the fingers, and mark with a dot the terminal of *entry* of this current.

▶ Example 15.20

Find a proper set of dot markings for the coils of Figure 15.30(a).

Solution

Pick terminal A as the terminal of current entry, and mark it with a dot. Next, use the right-hand rule to find that ϕ due to this current is directed

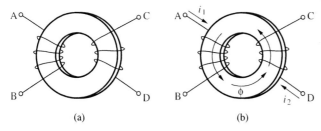

Figure 15.30 Coupled coils of Example 15.20.

downward inside this coil and, hence, *upward* inside the second coil. Curl your right hand around the second coil with the thumb *upward* to find that its current must *enter* via terminal *D*. Hence, place a dot on terminal *D*. In summary, *A* and *D* are dotted. ◀

Exercise 15.25   Find a proper set of dot markings for the coils of Figure 15.31.

ANSWER   (a) Either *A* and *D* or *B* and *C*; (b) either *A* and *D* or *B* and *C*.

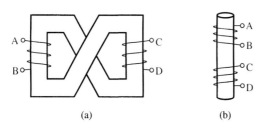

Figure 15.31 Coupled coils of Exercise 15.25.

Coupled coils are often encapsulated or not easily accessible for inspection, indicating that the dot markings must be determined *experimentally*. Figure 15.28 provides a suitable arrangement for this purpose. First, we mark the coil 1 terminal facing the resistance. Next, we observe the response v_2 with an oscilloscope or a voltmeter. If flipping the switch up yields a *positive* response, we mark the coil 2 terminal connected to the positive terminal of the instrument; otherwise we mark the other terminal of coil 2.

Energy in Coupled Coils

We now wish to find the instantaneous stored energy $w(t)$ in a pair of coupled coils. Assuming $i_1(0) = i_2(0) = 0$ so that the initial stored energy is zero, $w(t)$ is found by integrating the total power p delivered to the two coils,

$$w(t) = \int_0^t p(\xi)\,d\xi \qquad (15.39)$$

where ξ is a dummy variable of integration, and

$$p(\xi) = p_1(\xi) + p_2(\xi) = v_1(\xi)i_1(\xi) + v_2(\xi)i_2(\xi)$$

Using Equation (15.37), we can write

$$p(\xi) = \left(L_1\frac{di_1}{d\xi} + M\frac{di_2}{d\xi} \right) i_1 + \left(M\frac{di_1}{d\xi} + L_2\frac{di_2}{d\xi} \right) i_2$$

$$= L_1 i_1 \frac{di_1}{d\xi} + L_2 i_2 \frac{di_2}{d\xi} + M\frac{d}{d\xi}(i_1 i_2)$$

Substituting into Equation (15.39) yields

$$w(t) = L_1 \int_0^{i_1(t)} i_1(\xi)\,di_1(\xi) + L_2 \int_0^{i_2(t)} i_2(\xi)\,di_2(\xi) + M \int_0^{i_1(t)i_2(t)} d[i_1(\xi)i_2(\xi)]$$

or

$$\boxed{w(t) = \frac{1}{2}L_1 i_1^2(t) + \frac{1}{2}L_2 i_2^2(t) + M i_1(t) i_2(t)} \qquad \textbf{(15.40)}$$

The total stored energy in a pair of coupled coils is the sum of the energies stored in its self-inductances and that stored in its mutual inductance. Although the former are always positive, the latter may be positive or negative, depending on the relative polarities of i_1 and i_2, whose reference directions are assumed *into* the dotted terminals, as we know.

▶ **Example 15.21**

In the coupled coils of Example 15.19 find $w(t = 0.5\text{s})$.

Solution

We have seen that $I_2 = 0.8841\underline{/-65.32°}$ A. Moreover, from the second mesh equation we obtain

$$I_1 = \frac{4s^2 + 6s + 5}{2s^2 + 5}I_2 = \frac{1 + j6}{3}0.8841\underline{/-65.32°} = 1.793\underline{/15.21°}\ \text{A}$$

Thus, $i_1(0.5) = 1.793\cos(1 \times 0.5 + 15.21°) = 1.293$ A and $i_2(0.5) = 0.8841\cos(0.5 - 65.32°) = 0.7091$ A. Using Equation (15.40), $w(0.5) = 0.5[3 \times 1.293^2 + 4 \times 0.7091^2 + 2 \times 1.293 \times (-0.7091)] = 2.597$ J, where the negative sign stems from the reference direction of i_2, which was assumed *out* of the dotted terminal of the 4-H coil. ◀

Exercise 15.26   Find $w(0)$ in the circuit of Exercise 15.24.

ANSWER   1.6 J.

The Coefficient of Coupling

Adding and subtracting the term $\frac{1}{2}M^2 i_1^2(t)/L_2$ in the right-hand side of Equation (15.40) allows us to express the stored energy in a coupled pair as

$$w(t) = \frac{1}{2}\left[\left(L_1 - \frac{M^2}{L_2}\right)i_1^2(t) + L_2\left(i_2 + \frac{M}{L_2}i_1\right)^2\right]$$

To ensure $w(t) \geq 0$ for any i_1 and i_2, we must require that $L_1 - M^2/L_2 \geq 0$, or

$$M \leq \sqrt{L_1 L_2} \qquad \textbf{(15.41)}$$

This important inequality provides an upper limit on the value of the mutual inductance in terms of the self-inductances. The extent to which this inequality is met is expressed via the parameter

$$k \triangleq \frac{M}{\sqrt{L_1 L_2}} \qquad \textbf{(15.42)}$$

called the **coefficient of coupling** between the two inductors L_1 and L_2. This coefficient gives a measure of how much of the magnetic flux due to one coil links the other coil. In view of Equation (15.41), the range of values of k is

$$0 \leq k \leq 1 \qquad \textbf{(15.43)}$$

We identify three important cases.

(1) $k = 0$, or $M = 0$. In this case *none* of the flux from either coil links the other. The two coils act as a pair of independent inductances.

(2) $k = 1$. Now *all* the flux from either coil links the other, and $M = \sqrt{L_1 L_2}$, that is, the mutual inductance equals the *geometric mean* of the self-inductances. The coils are said to be *perfectly coupled,* or to form a **perfect transformer.** To approach perfect coupling, the coils are wound around a common core of highly permeable material such as iron or ferrite. We shall see later that the perfect transformer forms the basis of the *ideal* transformer introduced in Chapter 5.

(3) With $0 < k < 1$, only *some* of the flux from either coil links the other, and the coils are said to be *loosely coupled.* To achieve loose coupling, the coils are usually wound around a core of nonmagnetic material. Since these materials exhibit a *linear* relationship between flux and current, a pair of loosely coupled coils is also referred to as a **linear transformer.**

Linear and perfect transformers are investigated in the next section.

The stored energy in an inductance is $w = \frac{1}{2}Li^2$, indicating that for $w = 0$ we must have $i = 0$. We now wish to find the conditions for the stored energy of coupled coils to go to zero. Letting $w = 0$ in Equation (15.40) and solving the quadratic equation for i_1, we obtain, after simple manipulations,

$$i_1 = \sqrt{L_2/L_1}\left(-k \pm \sqrt{k^2 - 1}\right)i_2$$

Since both i_1 and i_2 are real, we see that for $w = 0$ we must have

$$i_1 = i_2 = 0 \tag{15.44a}$$

if $k < 1$, and

$$i_1 = -i_2\sqrt{L_2/L_1} \tag{15.44b}$$

if $k = 1$. For $w = 0$ in *loosely* coupled coils, both currents must be zero, just as in ordinary inductances. However, for $w = 0$ in *perfectly* coupled coils, the currents need not necessarily be zero. All that is required is that their polarities relative to the reference directions of Figure 15.27 be opposite and that their magnitudes be in the ratio $|i_1|/|i_2| = \sqrt{L_2/L_1}$. In the next section we shall see that the most striking consequence of Equation (15.44) is that currents in *perfectly* coupled coils *can change instantaneously,* a feature not possible in independent or in loosely coupled coils.

Coupled Coil Measurements

The parameters of a coupled pair are readily measured with an impedance bridge. L_1 is found by measuring the inductance of coil 1 with coil 2 open, L_2 by measuring the inductance of coil 2 with coil 1 open. M is found indirectly by measuring the inductance of either a series or a parallel combination of the two coils.

Exercise 15.27

(a) Show that if two coupled coils are connected in *series* as in Figure 15.32(a), then the net inductance is

$$L_s = L_1 + L_2 + 2M \tag{15.45}$$

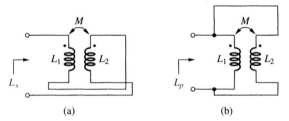

Figure 15.32 Series and parallel connections of coupled coils.

(b) Show that if two coupled coils are connected in *parallel* as in Figure 15.32(b), the net inductance is now

$$L_p = \frac{L_1 L_2 - M^2}{L_1 + L_2 - 2M} \tag{15.46}$$

(c) Show that if the connections of coil 2 are reversed, Equations (15.45) and (15.46) still hold, but with M replaced by $-M$.

▶ **Example 15.22**

The impedance bridge measurements of the coupled pair of Figure 15.32(a) yield $L_1 = 2.0$ mH, $L_2 = 3.0$ mH, and $L_s = 8.0$ mH. What is the value of k?

Solution

By Equation (15.45),

$$M = \frac{1}{2}(L_s - L_1 - L_2) \qquad \textbf{(15.47)}$$

or $M = \frac{1}{2}(8 - 2 - 3) = 1.5$ mH. Thus, $k = 1.5/\sqrt{2 \times 3} = 0.6124$. ◀

15.5 TRANSFORMERS

In this section we examine *linear* and *perfect* transformers in greater detail. We also identify the conditions under which two coupled coils form the *ideal* transformer introduced in Section 5.3.

The Linear Transformer

Figure 15.33 shows the s-domain representation of a typical linear transformer termination. The primary is driven by a source with series impedance $\mathbf{Z}_s$, and the secondary drives a load impedance $\mathbf{Z}_L$. Considering that the transformer is a two-port with $z_{11} = sL_1$, $z_{12} = z_{21} = sM$, and $z_{22} = sL_2$, we can directly apply Equations (15.23) through (15.26). The impedances seen looking into the primary and secondary are, respectively,

$$\mathbf{Z}_1 = sL_1 - \frac{s^2 M^2}{sL_2 + \mathbf{Z}_L} \qquad \textbf{(15.48)}$$

$$\mathbf{Z}_2 = sL_2 - \frac{s^2 M^2}{sL_1 + \mathbf{Z}_s} \qquad \textbf{(15.49)}$$

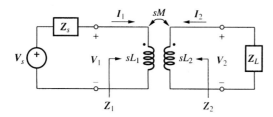

Figure 15.33 Terminated linear transformer.

Moreover, the current gain and the source-to-load voltage gain are

$$\frac{I_2}{I_1} = \frac{-sM}{sL_2 + Z_L} \tag{15.50}$$

$$\frac{V_2}{V_s} = \frac{sMZ_L}{s^2(L_1L_2 - M^2) + s(L_1Z_L + L_2Z_s) + Z_sZ_L} \tag{15.51}$$

▶ **Example 15.23**

In the circuit of Figure 15.33 let $L_1 = 3$ H, $L_2 = 2$ H, $M = 1$ H, $Z_s = 2\ \Omega$, and $Z_L = 1\ \Omega$. Find $v_2(t)$ if $v_S(t)$ is a 10-V source that is turned on at $t = 0$. Assume zero initial energy in the transformer.

Solution

Find the complete response $v_2(t)$ via the network function techniques of Section 14.4. Substituting the given component values into Equation (15.51) yields

$$V_2 = \frac{s}{5s^2 + 7s + 2} V_s$$

The network function has the poles $p_1 = -1$ Np/s and $p_2 = -0.4$ Np/s. Since $\lim_{s \to 0} V_2 = 0$, the dc steady-state value of $v_2(t)$ must be zero, indicating a response of the type

$$v_2(t \geq 0^+) = A_1 e^{-1t} + A_2 e^{-0.4t} \tag{15.52}$$

where A_1 and A_2 are to be found on the basis of the initial circuit conditions. The first condition stems from $w(0^+) = w(0^-) = 0$. Since $k = 1/\sqrt{3 \times 2} = 0.4082 < 1$, it follows from Equation (15.44a) that

$$i_1(0^+) = i_2(0^+) = 0 \tag{15.53}$$

Imposing $v_2(0^+) = -Z_L i_2(0^+) = 0$ yields $A_1 e^0 + A_2 e^0 = 0$, or

$$A_1 + A_2 = 0 \tag{15.54}$$

To obtain the second condition, apply the loop method at $t = 0^+$,

$$10 = 2i_1(0^+) + 3\frac{di_1(0^+)}{dt} + 1\frac{di_2(0^+)}{dt}$$

$$0 = 1i_2(0^+) + 1\frac{di_1(0^+)}{dt} + 2\frac{di_2(0^+)}{dt}$$

Using Equation (15.53) and eliminating $di_1(0^+)/dt$, we obtain

$$\frac{di_2(0^+)}{dt} = -2 \text{ A/s} \tag{15.55}$$

By Ohm's Law, $v_2(0^+) = -1i_2(0^+)$, so $dv_2(0^+)/dt = -1di_2(0^+)/dt = 2$ V/s. But, by Equation (15.52), $dv_2(0^+)/dt = -1A_1e^0 - 0.4A_2e^0 = -A_1 - 0.4A_2$. Hence,

$$-A_1 - 0.4A_2 = 2 \qquad (15.56)$$

Combining with Equation (15.54) yields $A_2 = -A_1 = 10/3$ V. Substituting into Equation (15.52) finally yields

$$v_2(t \geq 0^+) = \frac{10}{3}(e^{-0.4t} - e^{-1t}) \text{ V} \qquad (15.57) \blacktriangleleft$$

Exercise 15.28 Find $i_1(t)$ in the circuit of Example 15.23.

ANSWER $i_1(t \geq 0^+) = 5 - \frac{5}{3}e^{-0.4t} - \frac{10}{3}e^{-1t}$ A.

▶ Example 15.24

Repeat Example 15.23 with $M = \sqrt{L_1 L_2} = \sqrt{6}$ H.

Solution

Now we have $k = 1$, and Equation (15.51) yields

$$V_2 = \frac{\sqrt{6}s}{7s + 2}V_s$$

With the single pole $p_1 = -2/7$ Np/s, the response is of the type

$$v_2(t \geq 0^+) = Ae^{-2t/7} \qquad (15.58)$$

To find A, we observe that since $w(0^+) = w(0^-) = 0$ and $k = 1$, we must have, by Equation (15.44b),

$$i_1(0^+) = -\sqrt{2/3}i_2(0^+) \qquad (15.59)$$

Applying again the loop method at $t = 0^+$,

$$10 = 2i_1(0^+) + 3\frac{di_1(0^+)}{dt} + \sqrt{6}\frac{di_2(0^+)}{dt}$$

$$0 = 1i_2(0^+) + 2\frac{di_2(0^+)}{dt} + \sqrt{6}\frac{di_1(0^+)}{dt}$$

Multiplying the second equation by $\sqrt{3/2}$ throughout and subtracting pairwise from the first equation yields

$$10 = 2i_1(0^+) - \sqrt{3/2}i_2(0^+)$$

Combining with Equation (15.59) gives

$$i_2(0^+) = \frac{-10\sqrt{6}}{7} \text{ A} \qquad (15.60)$$

Hence, $v_2(0^+) = -1i_2(0^+) = 10\sqrt{6}/7 = 3.499$ V. But, by Equation (15.58), $v_2(0^+) = Ae^0 = A$. Thus,

$$v_2(t \geq 0^+) = 3.499e^{-2t/7} \text{ V} \qquad \textbf{(15.61)}$$

Remark   Note that at $t = 0$, i_2 changes abruptly from 0 to $-10\sqrt{6}/7$ A in order to ensure $w(0^+) = w(0^-)$.   ◀

Exercise 15.29   Find $w(t \geq 0^+)$ in Example 15.24.

ANSWER $w(t \geq 0^+) = 37.5(1 - 2e^{-2t/7} + e^{-4t/7})$ J.

▶ Example 15.25

In the circuit of Figure 15.33 let $L_1 = 2$ H, $L_2 = 1$ H, $M = 1$ H, and $Z_s = 2$ Ω, and let the source be an ac source with $V_s = 10\underline{/0°}$ V and $\omega = 1$ rad/s. Find (a) Z_L for maximum average power transfer and (b) the maximum average power.

Solution

(a) Power transfer is maximized when $Z_L = Z_2^*$. Substituting the given component values into Equation (15.49) and letting $s \to j1$ yields $Z_2 = 0.25 + j0.75$ Ω. We thus need a capacitive load

$$Z_L = 0.25 - j0.75 = 0.79\underline{/-71.57°} \text{ Ω}$$

(b) Substituting into Equation (15.51) gives $V_2 = 5.590\underline{/-26.57°}$ V. By Equation (12.12), $P = [(5.590/\sqrt{2})\cos(-71.57°)]^2/0.25 = 6.250$ W.   ◀

Reflected Impedance

We wish to take a closer look at the manner in which a linear transformer reflects the impedance of one port to the other. Letting $s \to j\omega$ in Equation (15.48) yields

$$Z_1 = j\omega L_1 + \frac{\omega^2 M^2}{j\omega L_2 + Z_L} \qquad \textbf{(15.62)}$$

The first component, $j\omega L_1$, is simply the impedance of coil 1. The second component, due to mutual coupling, is aptly called the **reflected impedance** by the secondary to the primary,

$$Z_{\text{reflected}} = \frac{\omega^2 M^2}{j\omega L_2 + Z_L} \qquad (15.63)$$

Expressing the load impedance as $Z_L = R_L + jX_L$ and rationalizing,

$$Z_{\text{reflected}} = \frac{\omega^2 M^2}{R_L^2 + (X_L + \omega L_2)^2}[R_L - j(X_L + \omega L_2)] \qquad (15.64)$$

The reflected impedance may be *capacitive, inductive,* or *resistive,* depending on whether the reactance $(X_L + \omega L_2)$ is, respectively, *positive, negative,* or *zero.*

Exercise 15.30 In the circuit of Figure 15.33 let $L_1 = 10$ mH, $L_2 = 16$ mH, and $M = 8$ mH, and let Z_L consist of a 100-Ω resistance in series with a 0.1-μF capacitance. (a) Find ω for $\sphericalangle Z_{\text{reflected}} = 0$. (b) Find $Z_{\text{reflected}}$ as well as Z_1 at this frequency.

ANSWER (a) 25 krad/s, (b) 400 Ω, $400 + j250$ Ω.

The Ideal Transformer

When two coils are *perfectly coupled* and their reactances are made to approach *infinity* compared to the external impedances, they are said to form an **ideal** transformer. These two conditions are achieved by winding the coils around a common core of highly permeable material. These materials have the ability to concentrate virtually all magnetic flux within the core, thus approaching the conditions of perfect coupling. Moreover, in the limit $\mu \rightarrow \infty$ the current needed to establish a given flux will approach zero, making the self-inductances of the coils approach infinity. As shown in Figure 15.34(a), the presence of the magnetic core is signified by means of two bars.

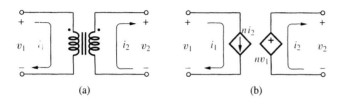

(a)　　　　　　　　(b)

Figure 15.34 Circuit symbol and model of the ideal transformer.

It is convenient to redefine the direction of i_2 so that it flows *out* of the dotted terminal of the secondary. Letting $M = \sqrt{L_1 L_2}$ in Equations (15.38), the perfect-transformer equations are

$$V_1 = sL_1I_1 - s\sqrt{L_1L_2}I_2 \qquad \text{(15.65a)}$$

$$V_2 = s\sqrt{L_1L_2}I_1 - sL_2I_2 \qquad \text{(15.65b)}$$

Multiplying both sides of Equation (15.65a) by $\sqrt{L_2}$ and both sides of Equation (15.65b) by $-\sqrt{L_1}$, and then adding them up pairwise yields $\sqrt{L_2}V_1 - \sqrt{L_1}V_2 = 0$, or

$$\frac{V_2}{V_1} = \sqrt{L_2/L_1}$$

Recall that the inductance of a coil is proportional to the number of its turns squared. Consequently, $L_2/L_1 = N_2^2/N_1^2 = n^2$, where N_1 and N_2 are the turns of the two coils, and

$$n = \frac{N_2}{N_1} \qquad \text{(15.66)}$$

is the *turns ratio*. Substituting yields the *perfect-transformer* relation

$$V_2 = nV_1 \qquad \text{(15.67)}$$

Solving Equation (15.65a) for I_1 and using $\sqrt{L_2/L_1} = n$ yields

$$I_1 = \frac{V_1}{sL_1} + nI_2$$

In the ideal limit $sL_1 \to \infty$ we obtain the *ideal-transformer* relation

$$I_1 = nI_2 \qquad \text{(15.68)}$$

As we know, Equations (15.67) and (15.68) are modeled by the equivalent of Figure 15.34(b).

Using this equivalent, we generalize the resistance transformation properties of Section 5.3 to impedances and state that if the secondary of an ideal transformer is terminated on an impedance Z_2, the equivalent impedance seen looking into the primary is

$$Z_1 = \frac{Z_2}{n^2} \qquad \text{(15.69)}$$

If Z_2 is a resistance R_2, an inductive impedance sL_2, or a capacitive impedance $1/sC_2$, it transforms, respectively, as $Z_1 = R_2/n^2$, $Z_1 = sL_2/n^2 = s(L_2/n^2)$, and $Z_1 = (1/sC_2)/n^2 = 1/s(n^2C_2)$. Clearly, when reflected to the input port, the resistance and the inductance are *divided* by n^2, but the capacitance is *multiplied* by n^2.

Comparing Equation (15.68) with Equation (15.44b), we observe that the stored energy in an ideal transformer is always zero. As we know, any energy entering the primary is entirely transferred to the load via the secondary. This is also confirmed by the fact that $V_1I_1 = V_2I_2$, by Equations (15.67) and (15.68).

15.6 SPICE ANALYSIS OF TWO-PORTS AND COUPLED COILS

SPICE provides an effective means for finding any one of the six parameter sets of a two-port, as well as for performing the transient and ac analysis of circuits containing coupled coils.

Two-Ports

Our study of two-ports revealed that the derivation of each parameter set requires *two* sets of measurements on the same network. Consequently, to avoid repeating the SPICE description of the circuit for each measurement, it is convenient to use the .SUBCKT facility introduced in Section 6.6.

Each set of measurements requires the application of either a *test voltage* or a *test current* to one port, and either an *open circuit* or a *short circuit* to the other port. To simplify the interpretation of the SPICE results, it is convenient to use test signals with *unity values*. With this artifice the value of each parameter will simply coincide with the value of the corresponding response or, in the case of transmission parameters, with its reciprocal. Moreover, to satisfy the SPICE requirements that each node have at least two connections, we use a *dummy* 0-V source to implement a *short circuit,* and a *dummy* 0-A source to implement an open circuit. Some examples will better illustrate.

► Example 15.26

Use SPICE to confirm the z parameters of Example 15.5.

Solution

The two-port subcircuit is shown in Figure 15.35(a), where we use the dummy source VD to monitor the controlling current i_X. The two test circuits are shown, respectively, in Figure 15.35(b) and (c), and the input file is

```
FINDING THE Z PARAMETERS
.SUBCKT TWO-PORT 1 2
F1 1 0 VD 2
R1 1 2 6
R2 2 3 3
VD 3 0 DC 0
.ENDS TWO-PORT
*Z11 = V(1), Z21 = V(2)
```

```
I1 0 1 DC 1
I2 0 2 DC 0
X1 1 2 TWO-PORT
*Z12 = V(3), Z22 = V(4)
I3 0 3 DC 0
I4 0 4 DC 1
X2 3 4 TWO-PORT
.END
```

After SPICE is run, the output file contains the following:

```
NODE VOLTAGE NODE VOLTAGE NODE VOLTAGE NODE VOLTAGE
(1)  3.0000  (2)  1.0000  (3)  -3.0000 (4)  1.0000
```

We interpret this as $z_{11} = V(1) = 3$, $z_{21} = V(2) = 1$, $z_{12} = V(3) = -3$, and $z_{22} = V(4) = 1$, all in Ω, in agreement with Example 15.5.

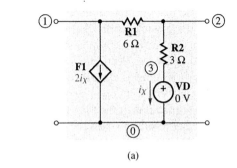

(a)

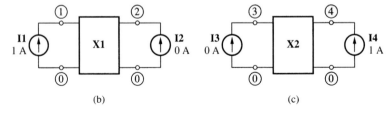

(b)   (c)

Figure 15.35 (a) Two-port subcircuit of Example 15.26, and (b), (c) test circuits to find its z parameters.

 ▶ **Example 15.27**

Use SPICE to confirm the y parameters of Example 15.7.

Solution

The two-port subcircuit is shown in Figure 15.36(a), and the two test circuits are shown, respectively, in Figure 15.36(b) and (c). The reason for specifying the polarities of the sources as shown is to conform to the SPICE convention which assumes current to flow into the positive terminal, through the source, and out of the negative terminal. The input file is

```
FINDING THE Y PARAMETERS
.SUBCKT TWO-PORT 1 2
R1 1 3 2
R2 3 0 1
R3 3 4 3
E1 2 4 3 0 0.5
.ENDS TWO-PORT
*Y11 = I(V1), Y21 = I(V2)
V1 0 1 DC -1
V2 0 2 DC 0
X1 1 2 TWO-PORT
*Y12 = I(V3), Y22 = I(V4)
V3 0 3 DC 0
V4 0 4 DC -1
X2 3 4 TWO-PORT
.END
```

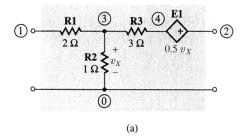

(a)

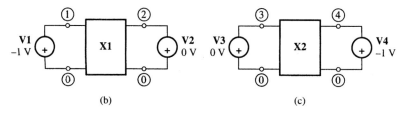

(b)   (c)

Figure 15.36 (a) Two-port subcircuit of Example 15.27, and (b), (c) test circuits to find its y parameters.

After SPICE is run, the output file contains the following:

```
VOLTAGE SOURCE CURRENTS:
NAME          CURRENT
 V1           3.750E-01
 V2          -1.250E-01
 V3          -8.333E-02
 V4           2.500E-01
```

We interpret this as $\mathbf{y}_{11} = \mathrm{I\,(V1)} = 0.375 = 3/8$, $\mathbf{y}_{21} = \mathrm{I\,(V2)} = -0.125 = -1/8$, $\mathbf{y}_{12} = \mathrm{I\,(V3)} = -0.08333 = -1/12$, and $\mathbf{y}_{22} = \mathrm{I\,(V4)} = 0.25 = 1/4$, all in Ω^{-1}. This confirms Example 15.7.   ◀

Exercise 15.31   Use SPICE to confirm the *a* parameters of Example 15.9.

Exercise 15.32   Use SPICE to confirm the *h* parameters of Exercise 15.16.

Magnetically Coupled Coils

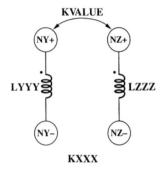

KVALUE

Figure 15.37 SPICE terminology for coupled coils.

The general form of the statement for a pair of magnetically coupled coils, depicted in Figure 15.37, is

```
KXXX   LYYY   LZZZ   KVALUE
```
(15.70)

where `KXXX` is the name of the coupled pair, which must begin with the letter K; `LYYY` and `LZZZ` are the names of its individual inductances; and `KVALUE` is the coefficient of coupling, which must lie in the range $0 \le \mathrm{KVALUE} \le 1$. The inductances `LYYY` and `LZZZ` also appear elsewhere in the SPICE program, in statements of the usual type

```
LYYY   NY+   NY-   VALUEY   IC=ICONDY
LZZZ   NZ+   NZ-   VALUEZ   IC=ICONDZ
```

The dot convention holds also in SPICE circuits, and the dotted nodes should always be the `NY+` and `NZ+` nodes.

Ⓢ ▶ **Example 15.28**

Use PSpice to verify the transient response of Example 15.23.

Solution

The circuit is shown in Figure 15.38, where we have $k = 1/\sqrt{L_1L_2} = 1/\sqrt{3 \times 2} = 0.40825$. The input file is

```
XSIENT RESPONSE OF A TERMINATED LINEAR XFORMER
VS 3 0 DC 10V
RS 3 1 2
*HERE ARE THE COUPLED COILS:
L1 1 0 3 IC=0
L2 2 0 2 IC=0
K12 L1 L2 0.40825
RL 2 0 1
```

```
.TRAN 0.2 6 UIC
.PROBE
.END
```

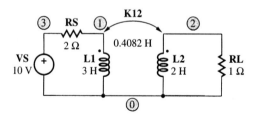

Figure 15.38 Circuit to investigate the transient response of a terminated linear transformer.

After directing the Probe post-processor to display `V(2)` we obtain the response of Figure 15.39, in agreement with Equation (15.57).

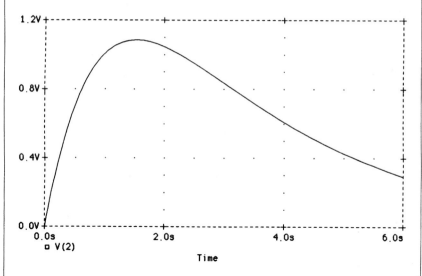

Figure 15.39 PSpice plot of the transient response of the circuit of Figure 15.38. ◀

Exercise 15.33   Use PSpice to verify the results of Example 15.24. Show the discontinuity of i_2.

▶**Example 15.29**

Use SPICE to verify the ac response of Example 15.19.

Solution

The circuit is shown in Figure 15.40, where we have $k = 2/\sqrt{3 \times 4} = 0.5774$. The input file is

```
AC RESPONSE OF A CIRCUIT WITH COUPLED COILS
VS 1 0 AC 12V
R1 1 2 5
*HERE ARE THE COUPLED COILS:
L1 2 3 3
L2 4 3 4
K12 L1 L2 0.5774
C 3 0 0.2
R2 4 0 6
.AC LIN 1 0.1592 0.1592
.PRINT AC VM(4) VP(4)
.END
```

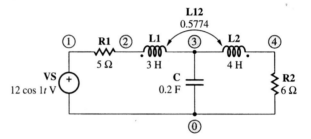

Figure 15.40 Circuit of Example 15.29.

After SPICE is run, the output file contains the following:

```
 FREQ          VM(4)          VP(4)
1.592E-01    5.305E+00     -6.532E+01
```

This confirms the results of Example 15.19. ◀

Exercise 15.34 Use SPICE to verify the ac response of Exercise 15.24.

▼ SUMMARY

- A linear two-port is a circuit designed to establish a prescribed set of linear relationships among its terminal voltages and currents. Of these four variables, only two can be chosen as *independent*. The other two are related to the former by a pair of *linear equations*.

- There are *six* different ways of choosing the independent variable pair. The parameters defining the corresponding equations are called z, y, a, b, h, and g parameters. The z and y parameters are the *immittance* parameters, a and b are the *transmission* parameters, and h and g are the *hybrid* parameters.

- Each parameter set is found via suitable *open-circuit* or *short-circuit* tests. Once a parameter set is known, it is generally possible to find any other set using appropriate conversion tables. However, not all parameter sets may exist for a given two-port.

- The simplest two-ports consist of just one element, which may be either a *series* or a *shunt* element. Other popular two-ports are the *T* network and the π network, each consisting of three elements. These networks are often used to model other two-ports.

- A two-port is *reciprocal* if interchanging a voltage source at one port with an ammeter at the other produces the *same* ammeter reading. A reciprocal two-port is *symmetrical* if interchanging its ports does not alter its terminal currents and voltages.

- When two-ports are connected in *series*, the *z* parameters of the composite two-port are the *sums* of the *z* parameters of the individual two-ports. When two-ports are connected in *parallel*, the *y* parameters of the composite two-port are the *sums* of the individual *y* parameters. When two-ports are connected in *cascade*, the *a* parameters of the composite two-port are combinations of *products* of the individual *a* parameters.

- When a two-port is terminated on a source and a load, it is of interest to know the *impedances* seen looking into the individual ports as well as the source-to-load *gain*. In general, these quantities are functions not only of the two-port parameters, but also of the terminating impedances.

- An important example of a two-port is the *coupled coil pair*. In addition to the *self-inductances* L_1 and L_2 of the individual coils, we also have the *mutual inductance M* stemming from the flux linkage in one coil due to the current in the other coil.

- The *dot convention* states that when the reference current direction is *into* the dotted terminal of one coil, the reference polarity of the induced voltage term at the other coil is *positive* at the dotted terminal.

- The energy stored in a coupled coil pair is the sum of the energies in its self-inductances as well as in its mutual inductance.

- The mutual and self-inductances are related as $M = k\sqrt{L_1L_2}$, where k is the *coefficient of coupling*, and $0 \le k \le 1$. If $k = 0$, there is *no coupling* between the coils; if $k = 1$ the coils form a *perfect* transformer; for intermediate values of k the coils form a *linear* transformer.

- An *ideal transformer* is a perfect transformer whose coils have sufficiently large self-inductances to make their reactances approach infinity in comparison with the external impedances.

- A distinguishing feature of transformers is their ability to effect *impedance transformations.*

- SPICE can be used to find two-port parameters as well as to display both the transient and frequency response of circuits containing coupled coils.

▼ PROBLEMS

15.1 Two-Port Parameters

15.1   Specify the element values (in Ω, F, and H) of a T network to realize the following z parameters at $\omega = 1$ Mrad/s:

$$[z] = \begin{bmatrix} 2 - j1 & 1 \\ 1 & 2 + j1 \end{bmatrix} k\Omega$$

15.2   Find π networks realizing the following y-parameter sets:

$$(a)\ [y] = \begin{bmatrix} 2s + 3 & -1 \\ -1 & 3s + 2 \end{bmatrix} \Omega^{-1}$$

$$(b)\ [y] = \begin{bmatrix} -j/15 & j/7.5 \\ j/7.5 & -j/15 \end{bmatrix} \Omega^{-1}$$

15.3   Given a T network with $Z_A = j5\ \Omega$, $Z_B = -j5\ \Omega$, and $Z_C = 1\ \Omega$, find its π equivalent.

15.4   Given a network with the z parameters of Problem 15.1, find its y, b, and g parameters.

15.5   (a) Predict the reading of the ideal voltmeter V in Figure P15.5. (b) Repeat, but with the voltmeter V and the 4-A current source interchanged with each other. What do you conclude?

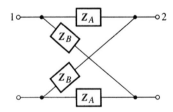

Figure P15.5

15.6   (a) Show that the network of Figure P15.6, known as a *lattice network,* is a symmetrical network with $z_{11} = z_{22} = \frac{1}{2}(Z_B + Z_A)$ and $z_{12} = z_{21} = \frac{1}{2}(Z_B - Z_A)$. (b) Show that $y_{11} = y_{22} = \frac{1}{2}(Y_B + Y_A)$, and $y_{12} = y_{21} = \frac{1}{2}(Y_B - Y_A)$, where $Y_A = 1/Z_A$, and $Y_B = 1/Z_B$.

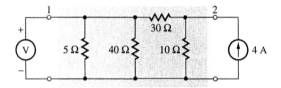

Figure P15.6

15.2 The z, y, a, and h Parameters

15.7   Find the z parameters of the network of Figure P15.7. Is it reciprocal? Symmetrical?

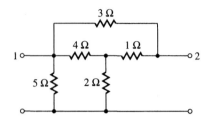

Figure P15.7

15.8   Find the z parameters of the network of Figure P15.8. Is it reciprocal? Why?

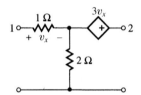

Figure P15.8

15.9   Find the z parameters of the network of Figure P15.9. Is it reciprocal? Why?

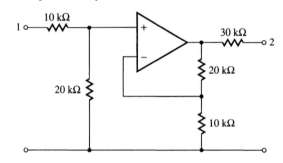

Figure P15.9

15.10   Find a T equivalent for the network of Figure P15.10.

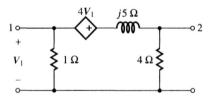

Figure P15.10

15.11   The following measurements are performed on a two-port: (a) with port 1 open and $V_2 = 10$ V, we find $V_1 = 5$ mV and $I_2 = 0.5$ mA; (b) with port 1 shorted and $V_2 = 10$ V, we find $I_1 = -2.5\ \mu$A and $I_2 = 0.75$ mA. Find the z parameters of this two-port.

15.12   Find the y parameters of the network of Figure P15.12.

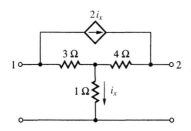

Figure P15.12

15.13 Find the *y* parameters of the network of Figure P15.13. Is it reciprocal? Why?

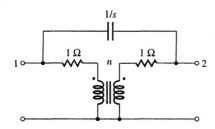

Figure P15.13

15.14 A 10-V source is connected to port 1 of a two-port, and the following measurements are taken: (a) with port 2 open, $V_2 = 20$ V and $I_1 = 6$ mA; (b) with port 2 shorted, $I_2 = -1$ mA and $I_1 = 2$ mA. Find the *y* parameters of this two-port.

15.15 Find the *a* parameters of the network of Figure P15.15.

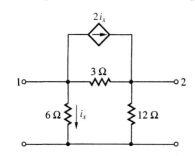

Figure P15.15

15.16 Find the *a* parameters of the network of Figure P15.16.

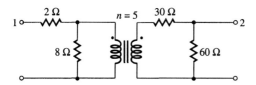

Figure P15.16

15.17 Find the *a* parameters of the network of Figure P15.12.

15.18 Find the *b* parameters of the network of Figure P15.15.

15.19 Find the *h* parameters for the network of Figure P15.19.

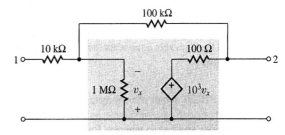

Figure P15.19

15.20 Find the *h* parameters for the network of Figure P15.20.

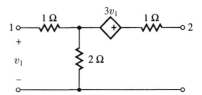

Figure P15.20

15.21 (a) Find the *h* parameters of the op amp circuit of Figure P15.21. (b) Under what condition do we have $\Delta h = 0$? What are the parameter values under this condition?

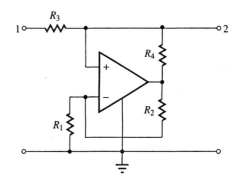

Figure P15.21

15.22 The following measurements are performed on a two-port: (a) with port 1 open and $V_2 = 10$ mV, $V_1 = 25$ μV and $I_2 = 5$ μA; (b) with port 2 shorted and $V_1 = 20$ mV, $I_2 = 1$ mA and $I_1 = 10$ μA. Find the *h* and *g* parameters of this two-port.

15.23 Find the *g* parameters for the network of Figure P15.20.

15.24 Find a network that realizes the following *g* parameters:

$$[g] = \begin{bmatrix} s + 1 \ \Omega^{-1} & 0 \\ 100 & 5 \ \Omega \end{bmatrix}$$

15.3 Two-Port Interconnections

15.25   A two-port network with $z_{11} = 20\ \Omega$, $z_{12} = 2\ \Omega$, $z_{21} = 40\ \Omega$, and $z_{22} = 10\ \Omega$ is driven by a 10-V dc source having an internal resistance of 5 Ω and is terminated on a 25-Ω load. Find the Thévenin equivalent seen (a) by the source and (b) by the load.

15.26   Figure P15.26 shows the z-parameter representation of a BJT amplifier configuration known as the *common-base* (CB) configuration. Find an expression for its source-to-load voltage gain, and compute it for $R_s = 50\ \Omega$, $r_e = 26\ \Omega$, $r_b = 200\ \Omega$, $r_o = 100\ k\Omega$, $\alpha = 0.99$, and $R_L = 10\ k\Omega$.

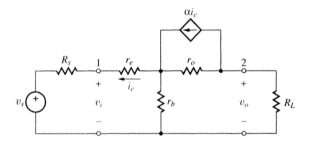

Figure P15.26

15.27   (a) Show that the circuit of Figure P15.27, known as Antoniu's gyrator, yields $\mathbf{Z}_1 = R^2/\mathbf{Z}_2$. (b) Assuming $R = 10\ k\Omega$ and $\mathbf{Z}_2 = 1/sC$, find C so that $\mathbf{Z}_1 = sL$, with $L = 1$ H.

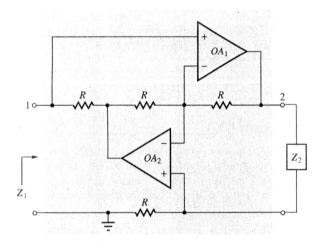

Figure P15.27

15.28   Find the y parameters for the gyrator circuit of Figure P15.28.

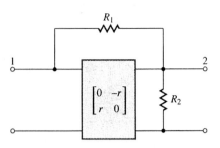

Figure P15.28

15.29   A 10-V dc source with an internal resistance of 20 Ω drives a two-port having $\mathbf{y}_{11} = 1/(80\ \Omega)$, $\mathbf{y}_{12} = -1/(2\ k\Omega)$, $\mathbf{y}_{21} = -1/(8\ \Omega)$, and $\mathbf{y}_{22} = 1/(50\ \Omega)$, and the two-port is terminated on a load R_L. (a) Find R_L for maximum power transfer to the load. (b) Find the maximum power delivered to the load.

15.30   Find the h parameters for the circuit of Figure P15.30, known as a *negative impedance converter* (NIC).

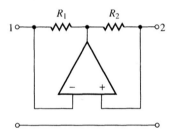

Figure P15.30

15.31   Find the y parameters of the network of Figure P15.31, which consists of two parallel-connected T subnetworks.

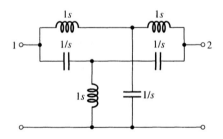

Figure P15.31

15.32   Find the z parameters of the network of Figure P15.32 as follows: first find the z parameters of its individual T subnetworks; next, convert to the y parameters and find the overall y parameters by exploiting the parallel-connection of the two subnetworks; finally, convert back to the z parameters.

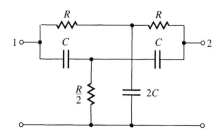

Figure P15.32

15.33 Find the *a* parameters of the ladder network of Figure P15.33.

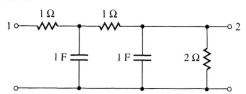

Figure P15.33

15.34 Find the transfer function $H(s) = V_2/V_1$ of the ladder network of Figure P15.34.

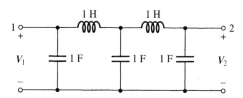

Figure P15.34

15.35 Two identical amplifiers, each with $h_{11} = 1$ kΩ, $h_{12} = 0.003$, $h_{21} = 50$, and $h_{22} = (10$ kΩ$)^{-1}$, are connected in cascade. If the first amplifier is driven by a voltage source having a 200-Ω series resistance and the second amplifier drives a 2-kΩ load, find the source-to-load voltage gain.

15.4 Magnetically Coupled Coils

15.36 Find the dot marking of coil L_2 in the circuit of Figure P15.36, given that whenever the switch is opened, the dc voltmeter V kicks upscale.

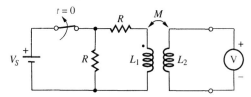

Figure P15.36

15.37 Obtain the transfer function $H(s) = V_o/V_i$ of the circuit of Figure P15.37. Hence, sketch its pole-zero plot.

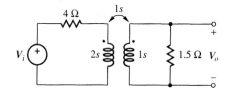

Figure P15.37

15.38 (a) Assuming that the circuit of Figure P15.38 is in steady state prior to switch activation, find $v(t)$. (b) Repeat, but with the resistance increased to 250/3 Ω.

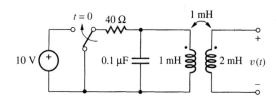

Figure P15.38

15.39 Suppose the terminals of the 2-mH coil of Figure P15.38 are shorted together. Assuming zero stored energy prior to switch activation, find the current out of the dotted terminal of the 2-mH coil. Comment.

15.40 (a) Find **V** in the circuit of Figure P15.40. (b) Repeat, but with the 3-Ω resistance replaced by a $-j5$-Ω capacitance.

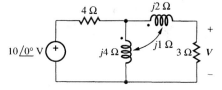

Figure P15.40

15.41 Apply *s*-domain loop analysis to the circuit of Figure P15.41 to find the currents through the resistances and the capacitance as functions of V_s for $\omega = 1$ rad/s.

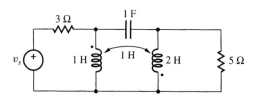

Figure P15.41

15.42 Find $i(t)$ in the circuit of Figure P15.42.

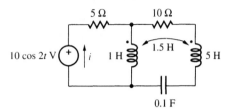

Figure P15.42

15.43 Find Z in the circuit of Figure P15.43.

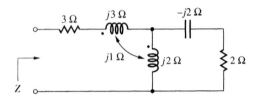

Figure P15.43

15.44 Figure P15.44 shows the equivalent circuit of a source driving a load via a coaxial cable. The cable conductors are modeled with a pair of resistances and mutually coupled inductances. Find the average power transferred to the load at (a) $f = 1$ kHz and (b) $f = 1$ MHz.

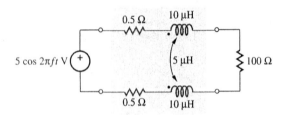

Figure P15.44

15.45 Find proper sets of dot markings for each of the coupled pairs of Figure P15.45.

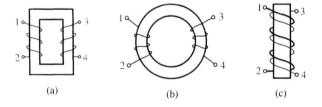

(a)   (b)   (c)

Figure P15.45

15.46 In the coupled pair of Figure 15.27(a) let $i_1(t) = 10\cos 2\pi 60t$ mA and $i_2(t) = 5\cos(2\pi 60t + 150°)$ mA. Given that $L_1 = 1$ H, $L_2 = 3$ H, and $M = 1$ H, find $v_1(t)$, $v_2(t)$, and $w(t = 1$ ms).

15.47 Find the energy stored in the coupled coils of Figure P15.47 at $t = 0$ if the load is (a) an open circuit, (b) a short circuit, and (c) a 5-Ω resistance.

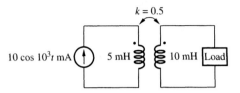

Figure P15.47

15.48 Assuming the circuit of Figure P15.48 is in steady state prior to switch activation, find (a) $w(0^+)$ and (b) the time it takes for the resistance to dissipate half this energy.

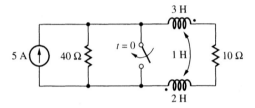

Figure P15.48

15.49 In the circuit of Figure P15.49 let $v_s = 100\sin 10^4 t$ V. Find the load that will absorb the maximum average power, as well as the power itself.

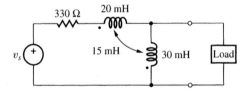

Figure P15.49

15.50 Find the average power dissipated in the 5-Ω resistance of Figure P15.50 if $|V_s| = 100$ V (rms).

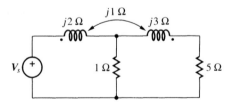

Figure P15.50

15.51 In the circuit of Figure P15.51 let $i_1 = 10\sin 100t$ A and $i_2 = 5\cos 100t$ A. (a) Find i and v. (b) Is average power flow in the center of the circuit toward the left or toward the right? What is this power?

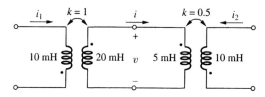

Figure P15.51

15.52 In the coupled pair of Figure P15.52, find (a) L_{AD} with B connected to C, (b) L_{AC} with B connected to D; (c) L_{AB} with C connected to D, (d) L_{CD} with A connected to B, (e) L_{AB} with A connected to C and B to D, (f) L_{AB} with A connected to D and B to C, (g) L_{AB} with no other connections, and (h) L_{CD} with no other connections.

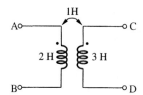

Figure P15.52

15.53 In the circuit of Figure P15.53 let $v_s = 10\cos\omega t$ V, $L_1 = 15$ mH, $L_2 = 20$ mH, $M = 10$ mH, and $C = 6$ μF. (a) Find the resonance frequency ω_0. (b) Find R for a 100-V peak voltage across the capacitance at ω_0. (c) What are the peak voltages across the coils at ω_0?

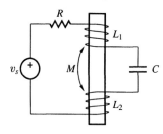

Figure P15.53

15.5 Transformers

15.54 (a) Fill in the blanks in the following table:

L_1	L_2	M	k
3 mH	4 mH	2mH	
40 μH	30 μH		0.5
	4 mH	2 mH	0.2
100 mH		125 mH	0.2

(b) What is the maximum mutual inductance M that an air-core transformer with $L_1 = 20$ μH and $L_2 = 40$ μH can have?

15.55 In the circuit of Figure 15.33 let $V_s = 120\underline{/0°}$ V (rms), $\omega = 100$ rad/s, $Z_s = 50$ Ω, and $Z_L = 10 - j10$ Ω. (a) If $L_1 = 0.5$ H, $L_2 = 0.2$ H, and $M = 0.2$ H, find Z_1, I_1, I_2, V_1 and V_2. (b) Repeat, but with the dot marking of the secondary at the bottom.

15.56 In the circuit of Figure P15.56 let $v_s = 5\sin 10t$ V, $R_1 = 5$ Ω, $L_1 = 1$ H, $k = 0.8$, $L_2 = 0.5$ H and $R_2 = 10$ Ω. Find the average power supplied by the source and the corresponding power factor.

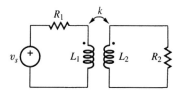

Figure P15.56

15.57 In the circuit of Figure P15.56 let $v_s = 100\sqrt{2} \times \cos 10^3 t$ V, $R_1 = 10$ Ω, $L_1 = 9$ mH, $L_2 = 4$ mH, and $R_2 = 20$ Ω. Find the average power dissipated in the 20-Ω resistance as a function of k, and plot it versus k for $0 \le k \le 1$.

15.58 Find V in the circuit of Figure P15.58.

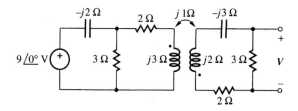

Figure P15.58

15.59 Find v in the circuit of Figure P15.38 if the 2-mH coil is terminated on a 100-Ω resistance and M is increased to $\sqrt{2}$ mH. Assume the circuit is in steady state prior to switch activation.

15.60 In the circuit of Figure P15.60, let $v_s = 100\cos 5000t$ V. If the load has been adjusted for maximum average power transfer, find the percentage of the power delivered by the source that is transferred to the load.

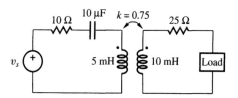

Figure P15.60

15.61   Find the value of k that will maximize the amplitude of the current out of the positive terminal of the source of Figure P15.61. What is this amplitude?

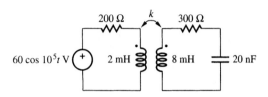

Figure P15.61

15.62   Find the value of k for which $\mathbf{Z}$ in the circuit of Figure P15.62 becomes purely resistive at $\omega = 5$ krad/s. What is $\mathbf{Z}$ at this frequency?

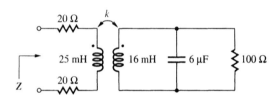

Figure P15.62

15.63   Find the frequencies at which $\mathbf{Z}$ in the circuit of Figure P15.63 is purely resistive.

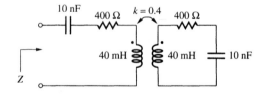

Figure P15.63

15.64   (a) Verify that the coupled coil pair of Figure 15.27(a) admits the T equivalent of Figure P15.64(a). To this end, express v_1 and v_2 in terms of i_1 and i_2, and show that they obey Equation (15.37). (b) Verify that the coupled coil pair of Figure 15.27(a) admits the phasor-domain π equivalent of Figure P15.64(b).

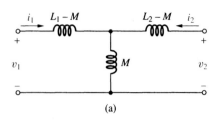

(a)

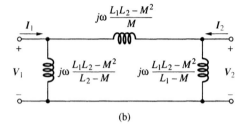

(b)

Figure P15.64

15.65   Shown in Figure P15.65 is the circuit model of a practical transformer. In this model R_1 and R_2 account for the resistances of the windings, L_1 and L_2 for the flux leakages from the coils, and R_c and L_m for core losses and the magnetizing current. If $R_1 = 5\ \Omega$, $R_2 = 1\ \Omega$, $X_{L_1} = 6\ \Omega$, $X_{L_2} = 2\ \Omega$, $R_c = 2\ \text{k}\Omega$, $X_{L_m} = 1\ \text{k}\Omega$, and $n = 2$, find $\mathbf{V}_s$ so that $\mathbf{Z}_L$ dissipates 1 kW at $220\underline{/0^\circ}$ V (rms) with $pf = 0.8$, lagging.

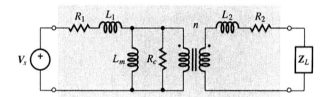

Figure P15.65

⑨ 15.6 SPICE Analysis of Two-Ports and Coupled Coils

15.66   Use SPICE to find the z parameters of the circuit of Figure P15.8.

15.67   Use SPICE to find the y parameters of the circuit of Figure P15.12.

15.68   Use SPICE to find the a parameters of the circuit of Figure P15.15.

15.69   Use SPICE to find the h parameters of the circuit of Figure P15.19.

15.70   Use the Probe post-processor of PSpice to display $v(t)$ in the coupled-coil circuit of Problem 15.38.

15.71   Assuming an operating frequency of 1 rad/s in the linear transformer circuit of Figure P15.58, use SPICE to find $\mathbf{V}$.

THE LAPLACE TRANSFORM

Frequency-domain analysis techniques, introduced in Chapter 11 to find the *steady-state ac response* and generalized in Chapter 14 to find, in addition to the *steady-state,* the *transient* and, hence, the *complete* response, allow us to avoid differential equations in favor of algebraic equations. However, the techniques of Chapter 14 succeed in predicting only the *functional form* of the response. The unknown coefficients appearing in the transient component must subsequently be found on the basis of the initial conditions in the circuit. This can be a tedious and time-consuming task.

In the present chapter we present a powerful analytical tool known as the *Laplace transform.* This not only retains the advantage of transforming differential equations into *algebraic equations,* it also takes the *initial circuit conditions* into account automatically. Since these conditions are an inherent part of the transform process, the Laplace method provides the expression for the complete response *explicitly.*

After introducing the Laplace transform, we investigate *functional* and *operational* transforms as well as the *inverse* Laplace transform. We then demonstrate the application of Laplace methods to the solution of *differential circuit equations,* both for the case in which the initial conditions are specified in terms of the response itself and the case in which they are specified in terms of the initial stored energies in the reactive elements of the circuit. Next, we illustrate the use of *convolution* to predict circuit responses in those situations in which the characteristics of the applied signal or of the circuit are available only in the time domain. We conclude by demonstrating the use of PSpice to plot the impulse and step responses using $H(s)$.

16.1 THE STEP AND IMPULSE FUNCTIONS

Before undertaking the study of Laplace transforms we need to introduce a pair of important functions as well as other mathematical concepts that will facilitate such a study.

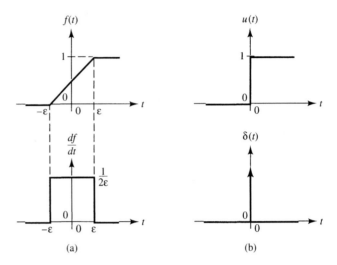

Figure 16.1 In the limit $\varepsilon \to 0$, the functions $f(t)$ and $df(t)/dt$ of (a) tend, respectively, to the functions $u(t)$ and $\delta(t)$ of (b).

Consider the function $f(t)$ and its derivative $df(t)/dt$ depicted, respectively, at the top and bottom of Figure 16.1(a). The function $f(t)$ is 0 for $t < \varepsilon$, increases uniformly from 0 to 1 for $-\varepsilon < t < \varepsilon$, and is 1 for $t > \varepsilon$. It is readily seen that

$$f(t) = \begin{cases} 0 & t < -\varepsilon \\ \dfrac{t + \varepsilon}{2\varepsilon} & -\varepsilon < t < \varepsilon \\ 1 & t > \varepsilon \end{cases} \tag{16.1}$$

The derivative of $f(t)$ is different from zero only over the interval $-\varepsilon < t < \varepsilon$ where, by Equation (16.1), $df(t)/dt = 1/2\varepsilon$. Thus,

$$\frac{df(t)}{dt} = \begin{cases} 0 & t < -\varepsilon \\ \dfrac{1}{2\varepsilon} & -\varepsilon < t < \varepsilon \\ 0 & t > \varepsilon \end{cases} \tag{16.2}$$

The profile of $df(t)/dt$ is a *pulse* centered around $t = 0$ and having width 2ε, magnitude $1/2\varepsilon$, and area $= 2\varepsilon \times 1/2\varepsilon$, or

$$\text{Area} = 1 \tag{16.3}$$

These considerations hold regardless of the size of ε. If we decrease ε, f will change from 0 to 1 more abruptly while the df/dt pulse will become narrower and higher; its area A, however, remains *unity* by Equation (16.3). In the limit $\varepsilon \to 0$, $f(t)$ approaches a function called the **unit step function**

$$u(t) = \lim_{\varepsilon \to 0} f(t) \tag{16.4}$$

and df/dt approaches a function called the **unit impulse function**

$$\delta(t) = \lim_{\varepsilon \to 0} \frac{df(t)}{dt} \tag{16.5}$$

These functions are shown in Figure 16.1(b). Clearly, we have

$$u(t) = \begin{cases} 0 & t < 0 \\ 1 & t > 0 \end{cases} \tag{16.6}$$

$$\delta(t) = \begin{cases} 0 & t < 0 \\ \infty & t = 0 \\ 0 & t > 0 \end{cases} \tag{16.7}$$

We use an arrow to signify that the impulse has infinite magnitude. It is important to keep in mind that the impulse, though infinitely large in magnitude, is also infinitely narrow in width in such a way as to ensure *unity area*. We express this mathematically as

$$\int_{-\infty}^{\infty} \delta(t)\, dt = 1 \tag{16.8}$$

Moreover, $\delta(t)$ is related to $u(t)$ as

$$\delta(t) = \frac{du(t)}{dt} \tag{16.9}$$

It is also possible to define the time derivatives of the impulse function. To this end, refer to the function $g(t)$ of Figure 16.2, whose area is $(2\varepsilon \times 1/\varepsilon)/2 = 1$, regardless of ε. We could easily have used $g(t)$ as an alternative to the function df/dt of Figure 16.1 to define the impulse function as

$$\delta(t) = \lim_{\varepsilon \to 0} g(t) \tag{16.10}$$

The derivative of $g(t)$, pictured at the bottom, consists of a pair of adjacent pulses of opposite polarities, called a **doublet.** It is apparent that just as $g(t)$ approaches $\delta(t)$, its derivative will approach that of $\delta(t)$,

$$\frac{d\delta}{dt} = \lim_{\varepsilon \to 0} \frac{dg}{dt} \tag{16.11}$$

The derivative of the impulse function is also called the **moment function.** Having defined $u(t)$ and $\delta(t)$ mathematically, we now wish to investigate how they relate to physical circuits.

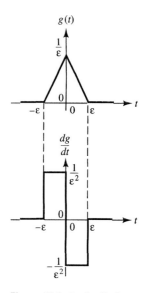

Figure 16.2 In the limit $\varepsilon \to 0$, the functions $g(t)$ and $dg(t)/dt$ tend, respectively, to the functions $\delta(t)$ and $d\delta(t)/dt$.

The Physical Significance of $u(t)$ and $\delta(t)$

The step function $u(t)$ provides a concise mathematical representation for a signal that changes *abruptly,* such as a voltage or a current change brought about by the closure of a switch. Figure 16.3(a) illustrates the effect of applying a unit voltage step to an *RC* circuit with the capacitance initially discharged, or in its *zero state.* As we know, the response is

$$v(t > 0) = 1 - e^{-t/RC} \text{ V}$$

We can thus state that the physical effect of applying $u(t)$ to a circuit is to induce its **forced response** to a unit dc forcing function.

We now wish to investigate the response of the same circuit to a voltage impulse $\delta(t)$, a response that we shall aptly call the **impulse response.** To this end, it is convenient to start out with a unity-area pulse of width 2ε s and magnitude $1/2\varepsilon$ V, as shown in Figure 16.4, and then let $\varepsilon \to 0$. During the pulse, that is, for $-\varepsilon < t < \varepsilon$, the capacitance voltage builds up according to

$$v(t) = \frac{1}{2\varepsilon} \left(1 - e^{-(t+\varepsilon)/RC}\right)$$

Hence, $V_m = v(\varepsilon) = \left(1 - e^{-2\varepsilon/RC}\right)/2\varepsilon$. In the limit $\varepsilon \to 0$ we can approximate $e^{-2\varepsilon/RC} \simeq 1 - 2\varepsilon/RC$, so $V_m \to (1 - 1 + 2\varepsilon/RC)/2\varepsilon$, or

$$V_m \to \frac{1}{RC}$$

Since in the limit $\varepsilon \to 0$ the pulse becomes the impulse function $\delta(t)$, we can state that the effect of applying $\delta(t)$ to the circuit is to *instantaneously charge the capacitance* from 0 V to $1/RC$ V. This abrupt change is made possible by

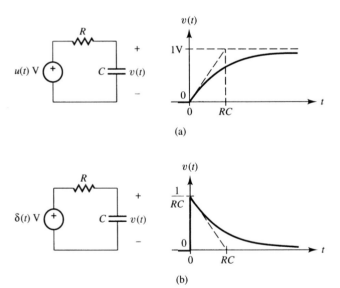

Figure 16.3 (a) Step response, and (b) impulse response of the *RC* circuit.

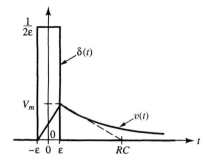

Figure 16.4 *RC* circuit response to a narrow pulse of unity area.

the infinite magnitude of the impulse, even though its width is infinitely narrow. After the impulse has occurred, the voltage will decay exponentially as

$$v(t > 0) = (1/RC)e^{-t/RC} \text{ V}$$

This is the familiar natural response depicted in Figure 16.3(b). In summary, the physical effect of applying $\delta(t)$ to a circuit is to *inject energy* and thus induce the **natural response.** It is worth pointing out that the impulse response can be obtained mathematically by differentiating the step response.

Practical Considerations

It is apparent that both $u(t)$ and $\delta(t)$ are mathematical conceptualizations of situations that in practice can only be approximated. Practical pulse generators have nonzero rise times. Moreover, an infinitely narrow yet infinitely large pulse is inconceivable in the laboratory. We can nonetheless approximate it with a pulse of sufficiently short duration to appear *instantaneous* in comparison with the time constant of the circuit, and of sufficient magnitude to inject enough energy into the circuit to make its response observable with an adequately sensitive instrument.

▶ **Example 16.1**

We wish to observe the impulse response of an *RC* circuit with $R = 10 \text{ k}\Omega$ and $C = 10 \text{ nF}$ using an ordinary pulse generator and an oscilloscope. How do we go about this?

Solution

Since $\tau = RC = 10^4 \times 10^{-8} = 100 \ \mu s$, select a pulse width $2\varepsilon \ll \tau$, say, $2\varepsilon = 1 \ \mu s$.

Assuming a pulse amplitude of 5 V, which is typical of most pulse generators, we have $V_m = v(1 \ \mu s) = 5[1 - e^{-(1 \ \mu s)/(100 \ \mu s)}] \simeq 50 \text{ mV}$. The

> response is thus $v(t) \simeq 50 \exp\left[-t/(100\ \mu s)\right]$ mV, which can comfortably be observed with an ordinary oscilloscope. ◀

Even though the step and impulse functions can, in practice, only be approximated, they nevertheless constitute powerful analytical tools, as we shall see shortly.

Scaling

Multiplying $\delta(t)$ by a positive scaling factor k yields the function $k\delta(t)$, which represents, by Equation (16.8), an impulse with

$$\text{Area} = k \qquad\qquad \textbf{(16.12a)}$$

We say that such an impulse has *strength k,* and we indicate this by means of the symbol k next to the arrow. Likewise, multiplying $u(t)$ by k yields the function $ku(t)$ such that $ku(t) = 0$ for $t < 0$, and $ku(t) = k$ for $t > 0$. Clearly, this is a step function of *amplitude k*. These functions are shown in Figure 16.5(a) and (b).

If k is negative, we have $ku(t) = 0$ for $t < 0$ and $ku(t) = -|k|$ for $t > 0$. Clearly, this is a *negative* step of magnitude $|k|$. Likewise, $k\delta(t)$ is a *negative* impulse with

$$\text{Area} = -|k| \qquad\qquad \textbf{(16.12b)}$$

These functions are shown in Figure 16.5(c) and (d).

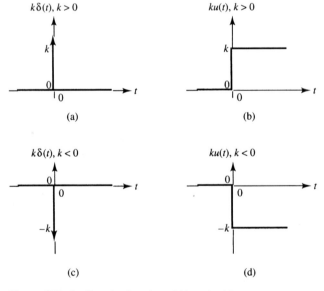

Figure 16.5 Scaling the functions $\delta(t)$ and $u(t)$.

Time-Shifting

So far we have considered impulses and step transitions that occur at $t = 0$; however, there will be instances when we will want to shift them right or left along the t axis by a given amount a, in seconds. As shown in Figure 16.6(a), the function $\delta(t - a)$ is infinite when its argument vanishes, that is, when $t - a = 0$, or $t = a$; it is zero for $t \neq a$. Likewise, the function $u(t - a)$ depicted in Figure 16.6(b) changes from 0 to 1 when $t - a = 0$, or $t = a$.

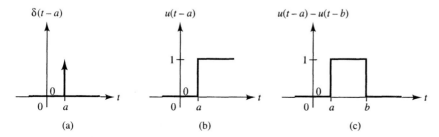

Figure 16.6 Delayed (a) impulse and (b) step functions. (c) Pulse representation in terms of step functions.

Time-shifted step functions can be used as mathematical building blocks for other discontinuous functions such as pulses and pulse trains. As an example, the pulse of Figure 16.6(c) can be obtained by creating a *positive* step at $t = a$ and a *negative* step at $t = b$, and then adding them up algebraically to form the pulse $p(t) = u(t - a) - u(t - b)$. Clearly, $p(t) = 1$ for $a < t < b$, and $p(t) = 0$ for $t < a$ and $t > b$.

The Sifting Property

An important property of the impulse function is the **sifting property,** also called the **sampling property,**

$$\int_{t_1}^{t_2} f(t)\delta(t - a)\, dt = f(a) \qquad \text{(16.13)}$$

$t_1 < a < t_2$, where we assume $f(t)$ to be continuous at $t = a$. To verify this property, note that since $\delta(t - a)$ is zero everywhere except at $t = a$, we can change the limits of the integral from t_1 to a^-, the instant just preceding a, and from t_2 to a^+, the instant just following a. Then,

$$\int_{a^-}^{a^+} f(t)\delta(t - a)\, dt = f(a) \int_{a^-}^{a^+} \delta(t - a)\, dt = f(a) \times 1 = f(a)$$

Clearly, the operation of Equation (16.13) sifts out everything except the value of the function right at $t = a$.

It is interesting to observe that Equation (16.13) could actually be used to *define* the impulse function $\delta(t - a)$ as well as its *n*th *derivative, $d^n \delta(t - a)/dt^n$,* as the function satisfying the property

$$\int_{t_1}^{t_2} f(t) \frac{d^n \delta(t - a)}{dt^n} \, dt = (-1)^n \frac{d^n f(a)}{dt^n}$$ **(16.14)**

$t_1 < a < t_2$, where $d^n f(a)/dt^n$ is the nth derivative of $f(t)$ computed at $t = a$.

16.2 THE LAPLACE TRANSFORM

In our study of *ac circuits* we have found that working with phasors instead of actual signals allowed us to avoid differential equations in favor of algebraic equations. In switching from an ac signal such as $v(t) = V_m \cos(\omega t + \theta)$ to its phasor $\mathbf{V}(\omega) = V_m \underline{/\theta}$ we in effect perform a *transformation* from a function of *time* to a function of *frequency*, $v(t) \rightarrow \mathbf{V}(\omega)$. Likewise, when we switch back from a phasor to the corresponding ac signal we perform the *inverse transformation*, $\mathbf{V}(\omega) \rightarrow v(t)$.

The Laplace transform, named for the French mathematician Pierre Simon, Marquis de Laplace (1749–1827), effects a transformation from functions of *time t* to functions of the *complex frequency s*, $f(t) \rightarrow \mathbf{F}(s)$, according to

$$\mathbf{F}(s) = \int_{-\infty}^{\infty} f(t) e^{-st} \, dt$$ **(16.15)**

where

$$\boxed{s = \sigma + j\omega}$$ **(16.16)**

is the complex frequency, in complex Np/s. Since the integration limits are $-\infty$ and $+\infty$, Equation (16.15) is referred to as the **double-sided** or **bilateral Laplace transform.**

Just as the phasor transformation applies only to ac signals, we must stipulate what kinds of signals we are going to subject to the Laplace transformation. Physically realizable circuits do not respond to a signal until it is actually applied and are thus referred to as *causal systems* to signify that the effect never precedes the cause. When observing the response to a particular forcing signal, it is convenient to take $t = 0$ as the instant at which the signal is applied. If the circuit is in its zero state, the response is zero for $t < 0$, and it manifests itself only for $t \geq 0$. Signals of this nature are referred to as **causal signals,** or **positive-time signals.** It is precisely this class of signals that shall be the subject of our Laplace transformations.

Two common examples of causal signals are the step and impulse responses encountered in Figure 16.3. For instance, the impulse response is $v(t) = 0$ for $t < 0$, and $v(t) = (1/RC)e^{-t/RC}$ for $t > 0$. Exploiting the availability of the function $u(t)$, we can represent the impulse response concisely as

$$v(t) = \frac{1}{RC} e^{-t/RC} u(t)$$

for any t. Note that the function $f(t) = (1/RC)e^{-t/RC}$ is not causal, whereas the function $v(t) = f(t)u(t)$ is causal. This difference is illustrated in Figure 16.7.

Since we assume $f(t) = 0$ for $t < 0$, we can change the lower limit of integration in Equation (16.15) from $-\infty$ to 0 and make the transform *single*

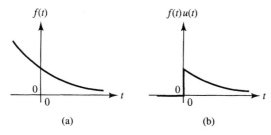

Figure 16.7 (a) Noncausal and (b) causal function.

sided. Actually, to make it possible for our list of Laplace-transformable functions to include also the impulse function $\delta(t)$ and higher-order singularities at the origin, we shall take the lower limit of integration as 0^-, the instant *just preceding* $t = 0$. In summary, we define the **single-sided** or **unilateral Laplace transform** of a causal function of time $f(t)$ as

$$\mathscr{L}\{f(t)\} = \int_{0^-}^{\infty} f(t)e^{-st}dt \qquad \textbf{(16.17)}$$

It is apparent that once we evaluate the integral of Equation (16.17) we are left with a function of s, which we denote as $F(s)$,

$$F(s) = \mathscr{L}\{f(t)\} \qquad \textbf{(16.18)}$$

Once again we stress that the Laplace transform of Equation (16.17) ignores $f(t)$ for $t < 0^-$. Whatever happened prior to $t = 0^-$ is accounted for by the initial conditions, which are memorized by the circuit in its energy-storage elements.

Just as $F(s)$ is said to be the Laplace transform of $f(t)$, $f(t)$ is said to be the *inverse* Laplace transform of $F(s)$, denoted as

$$f(t) = \mathscr{L}^{-1}\{F(s)\} \qquad \textbf{(16.19)}$$

Though the inverse transform may also be expressed as a suitable integral, it is more conveniently obtained via an algebraic technique known as the *partial fraction expansion*, to be presented in Section 16.4. The function $f(t)$ and its Laplace transform $F(s)$ are said to constitute a *transform pair*, denoted as

$$f(t) \leftrightarrow F(s) \qquad \textbf{(16.20)}$$

Mathematicians have shown that no two distinct functions have the same Laplace transform, and vice versa, a property referred to as the *uniqueness* of the Laplace transform and its inverse.

Transform Pairs

Using Equation (15.17), we can readily apply our mathematical skills to the calculation of a variety of transform pairs. We are especially curious about the transforms of the *impulse* function $\delta(t)$ and the *step* function $u(t)$. The former is

$$\mathcal{L}\{\delta(t)\} = \int_{0^-}^{\infty} \delta(t)e^{-st}\, dt = e^{-s0} = 1 \qquad \text{(16.21)}$$

The latter is

$$\mathcal{L}\{u(t)\} = \int_{0^-}^{\infty} u(t)e^{-st}\, dt = \int_{0^-}^{\infty} 1e^{-st}\, dt = -\frac{1}{s}e^{-st}\Big|_{0^-}^{\infty}$$

$$= -\frac{1}{s}e^{-(\sigma+j\omega)t}\Big|_{0^-}^{\infty} = \frac{1}{s} - \lim_{t\to\infty}\frac{1}{s}e^{-\sigma t}e^{-j\omega t}$$

If $\sigma > 0$, this simplifies as

$$\mathcal{L}\{u(t)\} = \frac{1}{s} \qquad \text{(16.22)}$$

Another function of great interest is the *decaying exponential* $e^{-at}u(t)$, which appears in the transient response of first-order circuits and overdamped second-order circuits. We have

$$\mathcal{L}\{e^{-at}\} = \int_{0^-}^{\infty} e^{-at}e^{-st}\, dt = \int_{0^-}^{\infty} e^{-(s+a)t}\, dt = -\frac{1}{s+a}e^{-(s+a)t}\Big|_{0^-}^{\infty}$$

$$= \frac{1}{s+a} - \frac{1}{s+a}\lim_{t\to\infty} e^{-(\sigma+a)t}e^{-j\omega t}$$

If $(\sigma + a) > 0$, or $\sigma > -a$, this simplifies to

$$\mathcal{L}\{e^{-at}\} = \frac{1}{s+a} \qquad \text{(16.23)}$$

Equally interesting is the *cosine* function $(\cos \omega t)u(t)$, which forms the basis of the ac response. Using Euler's identity,

$$\mathcal{L}\{\cos \omega t\} = \int_{0^-}^{\infty} \frac{1}{2}\left(e^{j\omega t} + e^{-j\omega t}\right)e^{-st}\, dt$$

Applying Equation (16.23) twice, first with $a = -j\omega$ and then with $a = j\omega$, we obtain

$$\mathcal{L}\{\cos \omega t\} = \frac{1}{2}\left(\frac{1}{s - j\omega} + \frac{1}{s + j\omega}\right) = \frac{s}{s^2 + \omega^2} \qquad \text{(16.24)}$$

Finally, the transform of the *damped sinusoid* $(e^{-at} \cos \omega t)u(t)$, which arises in second-order underdamped responses, is

$$\mathcal{L}\{e^{-at} \cos \omega t\} = \int_{0^-}^{\infty} e^{-at} \frac{1}{2} \left(e^{j\omega t} + e^{-j\omega t} \right) e^{-st} \, dt$$

$$= \int_{0^-}^{\infty} \frac{1}{2} \left(e^{-(a-j\omega)t} + e^{-(a+j\omega)t} \right) e^{-st} \, dt$$

Applying Equation (16.23) twice, first with a replaced by $(a - j\omega)$, and then with a replaced by $(a + j\omega)$, we obtain

$$\mathcal{L}\{e^{-at} \cos \omega t\} = \frac{1}{2} \left(\frac{1}{s + a - j\omega} + \frac{1}{s + a + j\omega} \right)$$

or

$$\mathcal{L}\{e^{-at} \cos \omega t\} = \frac{s + a}{(s + a)^2 + \omega^2} \tag{16.25}$$

Exercise 16.1   Find the Laplace transform of the *damped ramp* $f(t) = te^{-at}u(t)$, which arises in second-order critically damped responses.

ANSWER $1/(s + a)^2$.

Exercise 16.2   Find the Laplace transforms of the following functions: (a) $f(t) = (\cosh at)u(t)$; (b) $f(t) = [\cos (\omega t + \phi)]u(t)$.

ANSWER   (a) $s/(s^2 - a^2)$; (b) $(s \cos \phi - \omega \sin \phi)/(s^2 + \omega^2)$.

The transform pairs just discussed are listed in Table 16.1. Also included are two additional transforms of mathematical significance, namely, the functions $F(s) = 1/s^n$ and $F(s) = s^n$, where n is an integer and $n \geq 1$. Since s^n is listed as the transform of $d^n \delta(t)/dt^n$, we wish to take a closer look at this kind of function. Referring back to Figure 16.2, we observe that since $d\delta/dt = \lim_{\varepsilon \to 0} dg/dt$, the Laplace transform of the moment function can be found as

$$\mathcal{L}\left\{ \frac{d\delta}{dt} \right\} = \lim_{\varepsilon \to 0} \left[\int_{-\varepsilon}^{0^-} \frac{1}{\varepsilon^2} e^{-st} \, dt + \int_{0^+}^{\varepsilon} -\frac{1}{\varepsilon^2} e^{-st} \, dt \right] = \lim_{\varepsilon \to 0} \frac{e^{s\varepsilon} + e^{-s\varepsilon} - 2}{s\varepsilon^2}$$

Since this limit yields the indeterminate result 0/0, we apply l'Hôpital's rule twice to obtain $\mathcal{L}\{d\delta/dt\} = \lim_{\varepsilon \to 0} (se^{s\varepsilon} - se^{-s\varepsilon})/2s\varepsilon = \lim_{\varepsilon \to 0} (s^2 e^{s\varepsilon} + s^2 e^{s\varepsilon})/2s =$

TABLE 16.1 Some Laplace Transform Pairs

Name	$f(t),\ t > 0^-$	$F(s)$
Impulse	$\delta(t)$	1
Step	$u(t)$	$\dfrac{1}{s}$
Ramp	t	$\dfrac{1}{s^2}$
Exponential	e^{-at}	$\dfrac{1}{s+a}$
Sine	$\sin \omega t$	$\dfrac{\omega}{s^2 + \omega^2}$
Cosine	$\cos \omega t$	$\dfrac{s}{s^2 + \omega^2}$
Damped ramp	te^{-at}	$\dfrac{1}{(s+a)^2}$
Damped sine	$e^{-at} \sin \omega t$	$\dfrac{\omega}{(s+a)^2 + \omega^2}$
Damped cosine	$e^{-at} \cos \omega t$	$\dfrac{s+a}{(s+a)^2 + \omega^2}$
Inverse of $1/s^n$	$\dfrac{t^{n-1}}{(n-1)!}$	$\dfrac{1}{s^n}$
Inverse of s^n	$\dfrac{d^n \delta(t)}{dt^n}$	s^n

$2s^2/2s$, or

$$\mathscr{L}\left\{ \frac{d\delta}{dt} \right\} = s \tag{16.26}$$

We use the concept of derivatives of $\delta(t)$ in Section 16.4.

Convergence

Though we have seen various examples of functions and their Laplace transforms, there are functions for which the Laplace transform does not exist. Even when it exists, it does so only under suitable restrictions, as the examples of Equations (16.22) and (16.23) have already revealed. Existence problems stem from the fact that the upper limit in the integral of Equation (16.17) is infinity, thus raising the question as to whether or not the integral converges and, hence, $F(s)$ exists. For instance, if $f(t)$ is of *exponential order,* that is, if a *real number* σ exists such that

$$\lim_{t \to \infty} |f(t)| e^{-\sigma t} = 0 \tag{16.27}$$

then the Laplace transform exists. For each $f(t)$, Equation (16.27) places a limitation on the possible values of σ. The **abscissa of convergence** is defined as the value σ_c such that for

$$\sigma > \sigma_c \qquad \text{(16.28)}$$

Equation (16.27) holds. The range of values defined by Equation (16.28) is called the **region of convergence,** and the Laplace transform exists only for values of $s = \sigma + j\omega$ confined to this region. Regions of convergence appear as vertical strips in the s plane.

As an example, let us find σ_c for $f(t) = e^{at}$. We are seeking a range of values for σ such that $\lim_{t \to \infty} |e^{at}|e^{-\sigma t} = 0$, or $\lim_{t \to \infty} e^{(a-\sigma)t} = 0$. Clearly, any number $\sigma > a$ will meet this requirement, indicating that this function has $\sigma_c = a$. The region of convergence is the portion of the s plane to the right of the vertical line positioned at $\sigma = a$. If $a > 0$, this line lies to the right of the $j\omega$ axis; if $a < 0$, it lies to the left.

An example of a function for which the Laplace transform does not exist is $f(t) = e^{t^2}$. This is so because no σ can be found that will make $\lim_{t \to \infty} |e^{t^2}|e^{-\sigma t}$ approach zero. Fortunately, the functions of interest in engineering do admit Laplace transforms.

Exercise 16.3 Find the abscissa of convergence of (a) $\delta(t)$, (b) $te^{-at}u(t)$, (c) $(e^{-at}\cos bt)u(t)$, and (d) $(\cosh bt)u(t)$.

ANSWER (a) $\sigma_c = 0$; (b) $\sigma_c = -a$; (c) $\sigma_c = -a$; (d) $\sigma_c = |b|$.

16.3 OPERATIONAL TRANSFORMS

The Laplace transform and its inverse affect not only *functions* but also *operations* upon functions. In this section we investigate how *time operations* upon $f(t)$ translate into *complex-frequency operations* upon $F(s)$, and vice versa.

Linearity

Given two causal functions $f_1(t)$ and $f_2(t)$ having Laplace transforms $F_1(s)$ and $F_2(s)$, we wish to find the Laplace transform of their *linear combination* $af_1(t) + bf_2(t)$, where a and b are constants. By definition,

$$\mathcal{L}\{af_1(t) + bf_2(t)\} = \int_{0^-}^{\infty} [af_1(t) + bf_2(t)]e^{-st}\, dt$$

$$= a\int_{0^-}^{\infty} f_1(t)e^{-st}\, dt + b\int_{0^-}^{\infty} f_2(t)e^{-st}\, dt$$

or

$$\mathcal{L}\{af_1(t) + bf_2(t)\} = aF_1(s) + bF_2(s) \qquad \text{(16.29)}$$

In words, **the Laplace transform of a *linear combination* is the *linear combination* of the individual transforms.**

▶ **Example 16.2**

Find the Laplace transform of

$$f(t) = 2(1 - 5e^{-3t})u(t)$$

Solution

Using linearity, along with Table 16.1, we have

$$\mathscr{L}\{f(t)\} = 2\left(\mathscr{L}\{u(t)\} - 5\mathscr{L}\{e^{-3t}u(t)\}\right) = 2\left(\frac{1}{s} - \frac{5}{s+3}\right) = \frac{-8s + 6}{s(s+3)} \blacktriangleleft$$

Exercise 16.4   Find the Laplace transform of the function $f(t) = [5 + (t + \cos 3t)e^{-2t}]u(t)$.

ANSWER $5/s + 1/(s+2)^2 + (s+2)/(s^2 + 4s + 13)$.

▶ **Example 16.3**

Find the inverse Laplace transform of

$$F(s) = \frac{8}{s^3} + \frac{3}{s+2}$$

Solution

Using linearity, we can write

$$f(t) = \mathscr{L}^{-1}\{F(s)\} = 8\mathscr{L}^{-1}\left\{\frac{1}{s^3}\right\} + 3\mathscr{L}^{-1}\left\{\frac{1}{s+2}\right\}$$

Using Table 16.1, we find

$$f(t) = \left(4t^2 + 3e^{-2t}\right)u(t)$$

where we use $u(t)$ to make $f(t)$ causal.

Remark   This example suggests a method for obtaining inverse Laplace transforms. We first express $F(s)$ as a linear combination of terms of the type of Table 16.1. Next we look up the corresponding inverse transforms. Finally, we obtain $f(t)$ as a suitable combination of these inverses. We shall return to this method in Section 16.4.

Exercise 16.5 Find the inverse Laplace transform of $F(s) = 5 - 2/s + 6/(s^2 + 4)$.

ANSWER $5\delta(t) + (3 \sin 2t - 2)u(t)$.

Differentiation

Given a causal function $f(t)$ having the Laplace transform $F(s)$, we wish to find the transform of its *derivative* df/dt; that is, given the t-domain operation of *differentiation* of $f(t)$, we ask what is the corresponding s-domain operation on $F(s)$. Using the definition of Laplace transform and integrating by parts,

$$\mathscr{L}\left\{\frac{df}{dt}\right\} = \int_{0^-}^{\infty} \frac{df}{dt}e^{-st}\,dt = f(t)e^{-st}\Big|_{0^-}^{\infty} + s\int_{0^-}^{\infty} f(t)e^{-st}\,dt$$

or

$$\boxed{\mathscr{L}\left\{\frac{df}{dt}\right\} = sF(s) - f(0^-)} \tag{16.30}$$

where we have assumed $\lim_{t\to\infty} f(t)\,e^{-st} = 0$. In words, **differentiation in the t domain corresponds to *multiplication by* s in the s domain, followed by the subtraction of the initial value** $f(0^-)$. As in the case of phasors, the *differentiation* operation is changed into an *algebraic* operation. However, unlike the phasor, the Laplace transform accounts also for the *initial value* of the function.

Equation (16.30) can be applied to find the Laplace transform of the second derivative,

$$\mathscr{L}\left\{\frac{d^2 f}{dt^2}\right\} = s\mathscr{L}\left\{\frac{df}{dt}\right\} - \frac{df(0^-)}{dt} = s[sF(s) - f(0^-)] - \frac{df(0^-)}{dt}$$

$$= s^2 F(s) - sf(0^-) - \frac{df(0^-)}{dt} \tag{16.31}$$

Using similar reasoning, we can generalize to the nth derivative,

$$\mathscr{L}\left\{\frac{d^n f}{dt^n}\right\} = s^n F(s) - s^{n-1} f(0^-) - s^{n-2}\frac{df(0^-)}{dt} - \cdots - \frac{d^{n-1} f(0^-)}{dt^{n-1}}$$

To simplify the notation it is convenient to express derivatives as

$$f^{(n)} \triangleq \frac{d^n f}{dt^n}$$

after which we can write

$$\mathscr{L}\{f^{(n)}(t)\} = s^n F(s) - s^{n-1} f(0^-) - s^{n-2} f^{(1)}(0^-)$$
$$- \cdots - f^{(n-1)}(0^-)$$

(16.32)

In Section 16.5 we apply Equation (16.32) to the solution of differential equations via Laplace transforms. It is apparent that the initial conditions are automatically taken into consideration with this method.

▶ **Example 16.4**

Show that

$$\mathscr{L}\left\{\frac{d^n \delta(t)}{dt^n}\right\} = s^n$$

Solution

Exploiting the fact that $\delta(t)$ as well as its derivatives are zero for $t = 0^-$, Equation (16.32) allows us to write $\mathscr{L}\{d^n \delta(t)/dt^n\} = s^n \mathscr{L}\{\delta(t)\} = s^n 1 = s^n$. ◀

Exercise 16.6   Show that *differentiation* in the *s* domain corresponds to *multiplication by* $(-t)$ in the *t* domain,

$$\mathscr{L}^{-1}\left\{\frac{dF(s)}{ds}\right\} = -tf(t)$$

(16.33)

(*Hint:* Differentiate Equation (16.17) with respect to *s* throughout.)

Integration

Given a causal function $f(t)$ having the transform $F(s)$, we wish to find the transform of its *integral* $\int_{0^-}^t f(\xi) \, d\xi$, where ξ is a dummy variable of integration. Noting that

$$f(t) = \frac{d}{dt} \int_{0^-}^t f(\xi) \, d\xi$$

we can apply the differentiation rule of Equation (16.30) and write

$$F(s) = \mathscr{L}\{f(t)\} = \mathscr{L}\left\{\frac{d}{dt} \int_{0^-}^t f(\xi) \, d\xi\right\} = s\mathscr{L}\left\{\int_0^t f(\xi) \, d\xi\right\} - \int_{0^-}^{0^-} f(\xi) \, d\xi$$

Since the last term is zero, we divide by s throughout to obtain

$$\mathscr{L}\left\{\int_{0^-}^t f(\xi)\,d\xi\right\} = \frac{F(s)}{s} \qquad\qquad \textbf{(16.34)}$$

In words, ***integration* in the t domain corresponds to *division* by s in the s domain.** Note the similarity with the case of phasors.

Exercise 16.7   Given that $\mathscr{L}\{tu(t)\} = 1/s^2$, use Equation (16.34) to find $\mathscr{L}\{6t^2u(t)\}$.

ANSWER $12/s^3$.

Exercise 16.8   Show that ***integration* in the s domain corresponds to *division* by t in the t domain,**

$$\mathscr{L}^{-1}\left\{\int_s^\infty F(s)\,ds\right\} = \frac{f(t)}{t} \qquad\qquad \textbf{(16.35)}$$

(*Hint:* Integrate Equation (16.17) from s to ∞ throughout.)

Time Shifting

Given a causal function $f(t)$ having the transform $F(s)$, we wish to find the transform of its *time-shifted* version $f(t-a)u(t-a)$, where $a > 0$ represents the time shift, in seconds. By definition,

$$\mathscr{L}\{f(t-a)u(t-a)\} = \int_{0^-}^\infty f(t-a)u(t-a)e^{-st}\,dt = \int_a^\infty f(t-a)e^{-st}\,dt$$

where we have exploited the fact that $u(t-a) = 0$ for $t < a$. We now introduce the new variable of integration $\xi = t - a$, so that as t varies from a to ∞, ξ varies from $a - a = 0$ to $\infty - a = \infty$. Moreover, since $d\xi = dt$ and $t = a + \xi$, the preceding equation becomes

$$\mathscr{L}\{f(t-a)u(t-a)\} = \int_0^\infty f(\xi)e^{-s(a+\xi)}\,d\xi = e^{-sa}\int_0^\infty f(\xi)e^{-s\xi}\,d\xi$$

or

$$\mathscr{L}\{f(t-a)u(t-a)\} = e^{-as}F(s) \qquad\qquad \textbf{(16.36)}$$

In words, ***shifting* by $a > 0$ in the t domain corresponds to *multiplication* by e^{-as} in the s domain.**

▶**Example 16.5**

Find the Laplace transform of the following rectangular pulse

$$f(t) = 10[u(t-3) - u(t-5)]$$

Solution

Applying linearity and time shifting,

$$\mathcal{L}\{f(t)\} = 10\left(e^{-3s}\frac{1}{s} - e^{-5s}\frac{1}{s}\right) = \frac{10}{s}\left(e^{-3s} - e^{-5s}\right)$$ ◀

Frequency Shifting

Using once again the definition of Laplace transform, one can easily show that

$$\mathcal{L}^{-1}\{F(s+a)\} = e^{-at}f(t) \qquad \textbf{(16.37)}$$

that is, **shifting by $a > 0$ in the s domain corresponds to *multiplication* by e^{-at} in the t domain,** where a is the amount of shift, in complex Np/s.

Exercise 16.9   Using the results of Exercise 16.2(b), find $\mathcal{L}\{[e^{-at}\cos(\omega t + \phi)]u(t)\}$.

ANSWER $[(s+a)\cos\phi - \omega\sin\phi]/[(s+a)^2 + \omega^2]$.

Scaling

Given a causal function $f(t)$ having the Laplace transform $F(s)$, we wish to find the transform of $f(at)$, obtained by scaling time by the dimensionless constant $a > 0$. By definition,

$$\mathcal{L}\{f(at)\} = \int_{0^-}^{\infty} f(at)e^{-st}\, dt$$

Letting $\xi = at$, so that $t = \xi/a$ and $dt = d\xi/a$, we obtain

$$\mathcal{L}\{f(at)\} = \frac{1}{a}\int_{0^-}^{\infty} f(\xi)e^{-(s/a)\xi}\, d\xi$$

or

$$\mathcal{L}\{f(at)\} = \frac{1}{a}F\left(\frac{s}{a}\right) \qquad \textbf{(16.38)}$$

Likewise,

$$\mathscr{L}^{-1}\{F(as)\} = \frac{1}{a} f\left(\frac{t}{a}\right) \tag{16.39}$$

In words, *scaling* **by** $a > 0$ **in one domain corresponds to** *scaling* **by** $1/a$ **in the other domain, followed by** *multiplication* **by** $1/a$.

▶ Example 16.6

Given that $\mathscr{L}\{(\cos t)u(t)\} = s/(s^2 + 1)$, find the Laplace transform of a 1-MHz cosine wave.

Solution

$\mathscr{L}\{(\cos 2\pi 10^6 t)u(t)\} = (1/2\pi 10^6) \times (s/2\pi 10^6)/[(s/2\pi 10^6)^2 + 1] = s/[s^2 + (2\pi 10^6)^2] = s/(s^2 + 4\pi^2 10^{12})$. ◀

Convolution

Given two functions $F(s)$ and $G(s)$ having inverse Laplace transforms $f_1(t)u(t)$ and $f_2(t)u(t)$, we wish to find the inverse Laplace transform of their *product* $F(s)G(s)$. Products of Laplace transforms arise quite frequently in systems analysis, as when we express the output $Y(s)$ in terms of the input $X(s)$ via the network function $H(s)$ as $Y(s) = H(s)X(s)$. We have,

$$F(s)G(s) = \left(\int_{0^-}^{\infty} f(\xi)e^{-s\xi}\,d\xi\right)G(s) = \int_{0^-}^{\infty} f(\xi)\left[G(s)e^{-s\xi}\right]d\xi$$

Using the time-shifting property of Equation (16.36), we can write

$$F(s)G(s) = \int_{0^-}^{\infty} f(\xi)\mathscr{L}\{g(t-\xi)u(t-\xi)\}\,d\xi$$

$$= \int_{0^-}^{\infty} f(\xi)\left(\int_{0^-}^{\infty} g(t-\xi)u(t-\xi)e^{-st}\,dt\right)d\xi$$

Interchanging the order of integration and exploiting the fact that $u(t-\xi)$, viewed as a function of ξ, satisfies $u(t-\xi) = 0$ for $\xi > t$, we can lower the upper limit of the second integral from ∞ to t and write

$$F(s)G(s) = \int_{0^-}^{\infty} \left(\int_{0^-}^{t} f(\xi)g(t-\xi)\,d\xi\right)e^{-st}\,dt$$

If we now define the **convolution** of $f(t)$ and $g(t)$ as

$$f(t) * g(t) \triangleq \int_{0^-}^{t} f(\xi)g(t-\xi)\,d\xi \tag{16.40}$$

it is apparent that the inverse transform of the product $F(s)G(s)$ is the convolution of their individual inverse transforms. We thus have the following additional transform pair:

$$f(t) * g(t) \leftrightarrow F(s)G(s)$$

(16.41)

In words, *convolution* **in the** t **domain corresponds to** *multiplication* **in the** s **domain.**

Making the change of variable $\xi = t - x$ in Equation (16.41), we can readily verify that the operation of convolution is *commutative*

$$f(t) * g(t) = g(t) * f(t)$$

(16.42)

where

$$g(t) * f(t) \triangleq \int_{0^-}^{t} g(\xi) f(t - \xi) \, d\xi$$

(16.43)

Clearly, the order in which the functions $f(t)$ and $g(t)$ appear in the convolution is immaterial. We shall of course choose the order that renders the calculation of the integral easier. The application of convolution to circuit analysis is shown in Section 16.7.

▶ **Example 16.7**

Use convolution to find

$$\mathcal{L}^{-1} \left\{ \frac{1}{s(s + 2)} \right\}$$

Solution

The function to be inverse-transformed is the product of the function $F(s) = 1/s$, whose inverse transform is $f(t) = u(t)$, and the function $G(s) = 1/(s + 2)$, whose inverse transform is $g(t) = e^{-2t}u(t)$. Using convolution, we get

$$\mathcal{L}^{-1} \left\{ \frac{1}{s(s + 2)} \right\} = g(t) * f(t) = \int_{0^-}^{t} g(\xi) f(t - \xi) \, d\xi$$

$$= \int_{0^-}^{t} e^{-2\xi} u(\xi) u(t - \xi) \, d\xi = \int_{0^-}^{t} e^{-2\xi} \, d\xi$$

$$= 0.5(1 - e^{-2t})u(t)$$

◀

Exercise 16.10   Use convolution to find $\mathscr{L}^{-1}\{4/[s^2(s+2)]\}$.

ANSWER $(2t + e^{-2t} - 1)u(t)$.

Periodic Functions

As we know, a function $f(t)$ is periodic with *period T* if

$$f(t) = f(t \pm nT)$$

for $n = 1, 2, 3, \ldots,$ and for all t. The Laplace transform of such a function is obtained by repeated application of the time-shifting property of Equation (16.36). To this end, let us define the function

$$f_1(t) = f(t)[u(t) - u(t - T)]$$

which is nonzero only over the first period. Then, a causal periodic function $f(t)$ can be represented as the sum of an infinite number of such functions, delayed by integral multiples of T,

$$f(t) = f_1(t) + f_1(t - T) + f_1(t - 2T) + f_1(t - 3T) + \cdots$$

Letting $F(s) = \mathscr{L}\{f(t)\}$ and $F_1(s) = \mathscr{L}\{f_1(t)\}$, we can apply Equations (16.36) and (16.29) to write

$$F(s) = F_1(s) + e^{-sT}F_1(s) + e^{-2sT}F_1(s) + e^{-3sT}F_1(s) + \cdots$$

$$= (1 + e^{-sT} + e^{-2sT} + e^{-3sT} + \cdots)F_1(s)$$

Applying the binomial theorem to the expression within parentheses,

$$\boxed{F(s) = \frac{F_1(s)}{1 - e^{-sT}}} \qquad \textbf{(16.44)}$$

where

$$F_1(s) = \int_{0^-}^{\infty} f_1(t)e^{-st}\, dt = \int_{0^-}^{T} f_1(t)e^{-st}\, dt$$

In words, **making a function $f_1(t)$ periodic with period T is equivalent to dividing its Laplace transform $F_1(s)$ by $1 - e^{-sT}$.**

▶ Example 16.8

Find the Laplace transform of the causal pulse train of Figure 16.8(a).

Solution

First, consider the function $f_1(t)$ shown in Figure 16.8(b). We have

$$F_1(s) = \int_{0^-}^{\infty} Ae^{-st}\, dt = \int_{0^-}^{W} Ae^{-st}\, dt = \frac{A}{s}\left(1 - e^{-sW}\right)$$

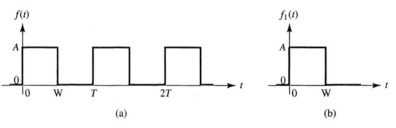

(a)   (b)

Figure 16.8 A periodic function $f(t)$, and the corresponding first-period function $f_1(t)$.

Then, applying Equation (16.44),

$$F(s) = \frac{A}{s} \frac{1 - e^{-sW}}{1 - e^{-sT}}$$

◀

Exercise 16.11   Find the Laplace transform of the causal sawtooth waveform of Figure 16.9.

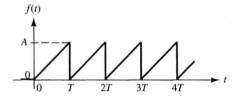

Figure 16.9 Sawtooth waveform of Exercise 16.11.

ANSWER $[A/(s^2 T)](e^{sT} - 1 - sT)/(e^{sT} - 1)$.

Exercise 16.12   Sketch and label the function $f(t) = A \times |\sin \omega t| u(t)$. Hence, find $\mathscr{L}\{f(t)\}$.

ANSWER $[\omega/(s^2 + \omega^2)] \coth (\pi s /2\omega)$.

Initial and Final Values

Given the pair $f(t) \leftrightarrow F(s)$, we are interested in ways of finding $f(0^+) = \lim_{t \to 0^+} f(t)$ and $f(\infty) = \lim_{t \to \infty} f(t)$ once $F(s)$ is known. These limiting values provide useful checks for our calculations as we transform from one domain to the other.

Consider first the transform of the derivative, which can be expanded as

$$\mathscr{L}\left\{\frac{df}{dt}\right\} = sF(s) - f(0^-)$$

$$= \int_{0^-}^{\infty} \frac{df}{dt} e^{-st}\, dt = \int_{0^-}^{0^+} \frac{df}{dt} e^{s0}\, dt + \int_{0^+}^{\infty} \frac{df}{dt} e^{-st}\, dt$$

$$= \int_{0^-}^{0^+} df + \int_{0^+}^{\infty} \frac{df}{dt} e^{-st}\, dt = f(0^+) - f(0^-) + \int_{0^+}^{\infty} \frac{df}{dt} e^{-st}\, dt$$

Simplifying, we have

$$sF(s) = f(0^+) + \int_{0^+}^{\infty} \frac{df}{dt} e^{-st}\, dt$$

If we now let $s \to \infty$ throughout, the last term will vanish, subject to the condition that df/dt be transformable. Moreover, $\lim_{s\to\infty} f(0^+) = f(0^+)$ because $f(0^+)$ is independent of s. We thus obtain

$$\boxed{f(0^+) = \lim_{s\to\infty} sF(s)}$$

(16.45)

a result referred to as the **initial-value theorem.** For it to hold, the limit must exist. If $F(s)$ is a *rational function* of s, this requires that it be of the *proper* type.

Considering again the transform of the derivative, but in the limit $s \to 0$, we can write

$$\lim_{s\to 0}[sF(s) - f(0^-)] = \lim_{s\to 0} \int_{0^-}^{\infty} \frac{df}{dt} e^{-st}\, dt = \int_{0^-}^{\infty} \frac{df}{dt} e^{0t}\, dt$$

$$= \int_{0^-}^{\infty} df = f(\infty) - f(0^-)$$

Exploiting the fact that $\lim_{s\to 0} f(0^-) = f(0^-)$ because $f(0^-)$ is independent of s, we can simplify and obtain

$$\boxed{f(\infty) = \lim_{s\to 0} sF(s)}$$

(16.46)

a result referred to as the **final-value theorem.** For it to hold, the limit must exist. If $F(s)$ is a *rational function* of s, this requires that all its poles be located in the *left half* of the s plane, with the possible exception of a single pole at the origin.

The findings of this section are summarized in Table 16.2.

TABLE 16.2 Some Operational Transforms

Name	Time Domain	Frequency Domain
Linearity	$af_1(t) + bf_2(t)$	$a\mathbf{F}_1(s) + b\mathbf{F}_2(s)$
t differentiation	$\dfrac{df(t)}{dt}$	$s\mathbf{F}(s) - f(0^-)$
t integration	$\int_{0^-}^{t} f(\xi)\,d\xi$	$\dfrac{\mathbf{F}(s)}{s}$
s differentiation	$-tf(t)$	$\dfrac{d\mathbf{F}(s)}{ds}$
s integration	$\dfrac{f(t)}{t}$	$\int_{s}^{\infty} \mathbf{F}(\xi)\,d\xi$
t shifting	$f(t - a)u(t - a)$	$e^{-as}\mathbf{F}(s)$
s shifting	$e^{-at}f(t)$	$\mathbf{F}(s + a)$
t scaling	$f(at)$	$\dfrac{1}{a}\mathbf{F}\left(\dfrac{s}{a}\right)$
s scaling	$\dfrac{1}{t}f\left(\dfrac{t}{a}\right)$	$\mathbf{F}(as)$
Convolution	$f(t) * g(t)$	$\mathbf{F}(s)\mathbf{G}(s)$
t periodicity	$f(t) = f(t + nT)$	$\dfrac{1}{1 - e^{-sT}}\mathbf{F}_1(s)$
Initial value	$f(0^+)$	$\lim\limits_{s \to \infty} s\mathbf{F}(s)$
Final value	$f(\infty)$	$\lim\limits_{s \to 0} s\mathbf{F}(s)$

▶ **Example 16.9**

Find the Laplace transform of

$$f(t) = (5 + 6te^{-3t} - 4e^{-t}\cos 2t)u(t)$$

Then use the initial-value and final-value theorems to check your results.

Solution

Using linearity, along with Table 16.1, we obtain

$$\mathbf{F}(s) = \frac{5}{s} + \frac{6}{(s + 3)^2} - 4\frac{s + 1}{(s + 1)^2 + 2^2}$$

$$s\mathbf{F}(s) = \frac{s^4 + 18s^3 + 82s^2 + 234s + 225}{s^4 + 8s^3 + 26s^2 + 48s + 45}$$

By inspection, $f(0^+) = 5 + 6 \times 0 - 4e^0\cos 0 = 1$, and $\lim\limits_{s \to \infty} s\mathbf{F}(s) = s^4/s^4 = 1$, in agreement with each other. Likewise, $f(\infty) = 5$ and $\lim\limits_{s \to 0} s\mathbf{F}(s) = 225/45 = 5$, also in agreement.

Exercise 16.13   Find $\mathscr{L}^{-1}\{2/s - 1/(s + 1)^3 - 3/(s + 4)\}$. Then use the initial-value and final-value theorems to check your results.

ANSWER $(2 - 0.5t^2 e^{-t} - 3e^{-4t})u(t)$.

16.4 THE INVERSE LAPLACE TRANSFORM

As evidenced by Tables 16.1 and 16.2, Laplace transforms may take on a variety of different functional forms. However, in the next sections we shall find that the functions arising when applying the Laplace transform to circuit analysis are *rational functions* of s, that is, functions that can be expressed as ratios of two polynomials $N(s)$ and $D(s)$,

$$F(s) = \frac{N(s)}{D(s)} = \frac{a_m s^m + a_{m-1} s^{m-1} + \cdots + a_1 s + a_0}{b_n s^n + b_{n-1} s^{n-1} + \cdots + b_1 s + b_0} \qquad \text{(16.47)}$$

Since this is the familiar form already investigated in Section 14.2, the terminology introduced there applies here as well. The coefficients a_i, $i = 1, 2, \ldots,$ m, and b_k, $k = 1, 2, \ldots, n$, are *real*. The roots of the equation $D(s) = 0$ are called the **poles** of $F(s)$. If $b_k \geq 0$, $k = 1, 2, \ldots, n$, the poles lie in the *left half* of the s plane or, at most, on the imaginary axis, but not in the right half. Moreover, $F(s)$ is said to be **proper** if $m < n$, and **improper** if $m \geq n$.

The goal of this section is to present techniques for finding $f(t) = \mathscr{L}^{-1}\{F(s)\}$ using the known transform pairs of Tables 16.1 and 16.2. These techniques require that $F(s)$ first be put in a suitable functional form. We begin with *proper* functions; then we proceed to examine *improper* functions. Both cases are in turn broken down into subcases, depending on whether the poles of $F(s)$ are real or complex, distinct or repeated.

Real and Distinct Poles

If the poles of $F(s)$ are *real* and *distinct*, its denominator polynomial can be factored out as

$$F(s) = \frac{N(s)}{(s - p_1)(s - p_2) \cdots (s - p_n)} \qquad \text{(16.48)}$$

where $p_1, p_2, \ldots, p_n$ are the poles. If $F(s)$ is *proper*, it is possible to express it in a form known as a **partial fraction expansion,**

$$F(s) = \frac{A_1}{s - p_1} + \frac{A_2}{s - p_2} + \cdots + \frac{A_n}{s - p_n} \qquad \text{(16.49)}$$

where A_1 through A_n are suitable coefficients, independent of s, known as the **residues** of $F(s)$. Once their values are known, we can exploit linearity, along with the property

$$\mathscr{L}^{-1}\left\{\frac{1}{s+a}\right\} = e^{-at}u(t)$$

to write

$$f(t) = (A_1e^{p_1t} + A_2e^{p_2t} + \cdots + A_ne^{p_nt})u(t) \qquad \textbf{(16.50)}$$

The method for finding the residues is straightforward and it works as follows. To find A_1, first multiply Equation (16.49) throughout by $(s - p_1)$ to eliminate the denominator of A_1,

$$(s - p_1)F(s) = A_1 + (s - p_1)\left[\frac{A_2}{(s - p_2)} + \cdots + \frac{A_n}{(s - p_n)}\right]$$

Then, let $s = p_1$ throughout to leave only A_1 at the right-hand side. Hence,

$$A_1 = (s - p_1)F(s)\Big|_{s=p_1}$$

The process is readily generalized to all residues as

$$A_k = (s - p_k)F(s)\Big|_{s=p_k} \qquad \textbf{(16.51)}$$

$k = 1, 2, \ldots, n$. In words, **the residue A_k of $F(s)$ at a simple real pole p_k is found by taking the product $(s - p_k)F(s)$ and computing it at $s = p_k$.**

Once all the residues have been obtained, it is important to check that they are correct. To this end, we substitute their values in Equation (16.49) and verify that both the original expression for $F(s)$ and its partial fraction expansion achieve the same values at some arbitrarily chosen values of s. To avoid the indeterminacy of infinity, the values of s must be *different* from the poles. If applicable, it is wise to use also the initial-value and final-value theorems as additional checks.

▶ **Example 16.10**

Find the inverse Laplace transform of

$$F(s) = \frac{2s^2 + s - 3}{s(s^2 + 4s + 3)}$$

Solution

This function has three distinct real poles, $p_1 = 0$, $p_2 = -1$, and $p_3 = -3$, all in Np/s. We can thus write

$$F(s) = \frac{2s^2 + s - 3}{s(s+1)(s+3)} = \frac{A_1}{s} + \frac{A_2}{s+1} + \frac{A_3}{s+3}$$

The residues are

$$A_1 = sF(s)\Big|_{s=0} = \frac{2s^2 + s - 3}{(s+1)(s+3)}\Big|_{s=0} = \frac{2 \times 0^2 + 0 - 3}{(0+1)(0+3)} = -1$$

$$A_2 = (s+1)F(s)\Big|_{s=-1} = \frac{2s^2 + s - 3}{s(s+3)}\Big|_{s=-1} = \frac{2 \times (-1)^2 - 1 - 3}{(-1)(-1+3)} = 1$$

$$A_3 = (s+3)F(s)\Big|_{s=-3} = \frac{2s^2 + s - 3}{s(s+1)}\Big|_{s=-3} = \frac{2 \times (-3)^2 - 3 - 3}{(-3)(-3+1)} = 2$$

Hence,

$$F(s) = \frac{2s^2 + s - 3}{s(s^2 + 4s + 3)} = \frac{-1}{s} + \frac{1}{s+1} + \frac{2}{s+3}$$

Before proceeding, check that the expansion is correct. For example, pick $s = 1$, which is not a pole. The original expression then yields $F(1) = (2 \times 1^2 + 1 - 3)/[1(1^2 + 4 \times 1 + 3)] = 0$, and the partial fraction expansion yields $F(1) = -1/1 + 1/(1+1) + 2/(1+3) = 0$. Applying a similar check at $s = 2$, which also is not a pole, we obtain in both cases $F(2) = 7/30$, indicating that we have good reasons to believe our expansion to be correct.

Finally, applying Equation (16.50), we have

$$f(t) = (-1 + 1e^{-t} + 2e^{-3t})u(t)$$

As a final check, we observe that $f(0^+) = 2$ and $\lim_{s \to \infty} sF(s) = 2$; $f(\infty) = -1$ and $\lim_{s \to 0} sF(s) = -1$. This further corroborates our results. ◀

Exercise 16.14   Find the inverse Laplace transform of $F(s) = (7 - 4s)/(5s^3 + 3s^2 + s/4)$. Check your results.

ANSWER $(28 - 37e^{-0.1t} + 9e^{-0.5t})u(t)$.

Complex-Conjugate Poles

If $F(s)$ has a *complex-conjugate pole pair*, $p_1 = \alpha + j\beta$ and $p_2 = \alpha - j\beta$, then its denominator $D(s)$ will contain the quadratic term $(s - p_1)(s - p_2) = (s - \alpha - j\beta)(s - \alpha + j\beta) = (s - \alpha)^2 + \beta^2$. For the sake of simplicity, we assume $F(s)$ has just this single pole pair and we later show how to generalize.

Thus, let

$$F(s) = \frac{N(s)}{(s-\alpha)^2 + \beta^2} = \frac{N(s)}{(s-\alpha-j\beta)(s-\alpha+j\beta)} = \frac{\boldsymbol{A_1}}{s-\alpha-j\beta} + \frac{\boldsymbol{A_2}}{s-\alpha+j\beta}$$

where we have expressed the residues in boldface in anticipation of their complex nature. Adapting Equation (16.51) to the present case, we have

$$\boldsymbol{A_1} = (s-\alpha-j\beta)\boldsymbol{F}(s)\Big|_{s=\alpha+j\beta} = \frac{N(\alpha+j\beta)}{2j\beta}$$

$$\boldsymbol{A_2} = (s-\alpha+j\beta)\boldsymbol{F}(s)\Big|_{s=\alpha-j\beta} = \frac{N(\alpha-j\beta)}{-2j\beta} = \boldsymbol{A_1^*}$$

Since $\boldsymbol{A_1}$ and $\boldsymbol{A_2}$ are *conjugate* of each other, we need to calculate only one of them. In the following we shall *always* calculate $\boldsymbol{A_1}$, the residue at the *upper pole* in the s plane,

$$p_1 = \alpha + j\beta \tag{16.52}$$

This residue is

$$\boxed{\boldsymbol{A_1} = (s-\alpha-j\beta)\boldsymbol{F}(s)\Big|_{s=\alpha+j\beta}} \tag{16.53}$$

Letting $\boldsymbol{A_1} = |\boldsymbol{A_1}|/\!\!\underline{\measuredangle\boldsymbol{A_1}}$ and $\boldsymbol{A_2} = |\boldsymbol{A_1}|/\!\!\underline{-\measuredangle\boldsymbol{A_1}}$, and using Equation (16.50), we can write

$$f(t) = \left(|\boldsymbol{A_1}|e^{j\measuredangle\boldsymbol{A_1}}e^{(\alpha+j\beta)t} + |\boldsymbol{A_1}|e^{-j\measuredangle\boldsymbol{A_1}}e^{(\alpha-j\beta)t}\right)u(t)$$

$$= |\boldsymbol{A_1}|e^{\alpha t}[e^{j(\beta t+\measuredangle\boldsymbol{A_1})} + e^{-j(\beta t+\measuredangle\boldsymbol{A_1})}]u(t)$$

Using the identity $e^{j\theta} + e^{-j\theta} = 2\cos\theta$, we can summarize our findings by writing

$$\boxed{\begin{aligned}\mathscr{L}^{-1}&\left\{\frac{\boldsymbol{A_1}}{s-\alpha-j\beta} + \frac{\boldsymbol{A_1^*}}{s-\alpha+j\beta}\right\}\\ &= 2|\boldsymbol{A_1}|e^{\alpha t}[\cos(\beta t + \measuredangle\boldsymbol{A_1})]u(t)\end{aligned}} \tag{16.54}$$

As we might have expected, the inverse transform of a function with a complex pole pair in the left half of the s plane, where $\alpha < 0$ is a *damped sinusoid*. The parameters of this sinusoid are determined by the *upper pole* $p_1 = \alpha + j\beta$ and by the corresponding residue $\boldsymbol{A_1}$ as follows:

$$\text{negative of the damping coefficient} = \alpha$$

$$\text{angular frequency} = \beta$$

$$\text{amplitude} = 2|\boldsymbol{A_1}|$$

$$\text{phase angle} = \measuredangle\boldsymbol{A_1}$$

These observations are readily generalized to the case in which $\boldsymbol{F}(s)$ has additional poles, besides the complex pair.

▶ **Example 16.11**

Find the inverse Laplace transform of

$$F(s) = \frac{40(s+1)}{(s+5)(s^2+6s+25)}$$

Solution

This function has one real pole, $p_1 = -5$ Np/s, and two complex-conjugate poles, $p_2 = -3 + j4$ and $p_3 = -3 - j4$, both in complex Np/s. The inverse transform thus consists of a decaying exponential due to p_1 and a damped sinusoid due to the complex pole pair. We have

$$F(s) = \frac{40(s+1)}{(s+5)(s+3-j4)(s+3+j4)}$$

$$= \frac{A_1}{s+5} + \frac{A_2}{s+3-j4} + \frac{A_2^*}{s+3+j4}$$

We are only interested in the first two residues,

$$A_1 = (s+5)F(s)\Big|_{s=-5} = \frac{40(s+1)}{s^2+6s+25}\Big|_{s=-5} = -8$$

$$A_2 = (s+3-j4)F(s)\Big|_{s=-3+j4} = \frac{40(s+1)}{(s+5)(s+3+j4)}\Big|_{s=-3+j4}$$

$$= \frac{40(-3+j4+1)}{(-3+j4+5)j8} = 5\frac{1-j2}{2-j1} = 5\underline{/-36.87°}$$

Hence,

$$f(t) = \left[-8e^{-5t} + 10e^{-3t}\cos(4t - 36.87°)\right]u(t)$$

which can be checked with the initial-value and final-value theorems.   ◀

Exercise 16.15   Find the inverse Laplace transform of $F(s) = 5s/[(s+1)(s^2+2s+2)]$. Check your results.

ANSWER $5e^{-t}[\sqrt{2}\cos(t - 45°) - 1]u(t)$.

Repeated Real Poles

Repeated poles may be either real or complex conjugate. Let us consider first the case of a *repeated real pole* at $s = p$ with multiplicity r. The denominator $D(s)$ thus contains the term $(s-p)^r$. For the sake of simplicity, assume $F(s)$ has just this single repeated pole, though we shall show later how to generalize.

Repeated poles require that the partial fraction expansion contain all powers of $(s - p)$ from r down to 1,

$$F(s) = \frac{N(s)}{(s - p)^r} = \frac{A_r}{(s - p)^r} + \frac{A_{r-1}}{(s - p)^{r-1}} + \cdots + \frac{A_1}{(s - p)} \qquad \textbf{(16.55)}$$

Once A_r through A_1 are known, we exploit the property

$$\mathscr{L}^{-1}\left\{ \frac{1}{(s + a)^k} \right\} = \frac{t^{k-1}e^{-at}}{(k - 1)!}u(t)$$

to write

$$f(t) = \left[\frac{A_r t^{r-1}}{(r - 1)!} + \cdots + A_3 t^2 + A_2 t + A_1 \right] e^{pt}u(t) \qquad \textbf{(16.56)}$$

We now wish to find A_r through A_1. To find A_r we multiply Equation (16.55) throughout by $(s - p)^r$ to get rid of the denominator of A_r. Then we let $s = p$ to get rid of all the right-hand side terms except A_r itself,

$$A_r = (s - p)^r F(s) \Big|_{s=p} \qquad \textbf{(16.57a)}$$

To find A_{r-1} we again multiply Equation (16.55) throughout by $(s - p)^r$; however, to get rid of the constant A_r, we now *differentiate* with respect to s throughout, and to get rid of all remaining terms except A_{r-1} we then let $s = p$,

$$A_{r-1} = \frac{d}{ds}(s - p)^r F(s) \Big|_{s=p}$$

The procedure is readily generalized as

$$A_i = \frac{1}{(r - i)!} \frac{d^{r-i}}{ds^{r-i}}(s - p)^r F(s) \Big|_{s=p} \qquad \textbf{(16.57b)}$$

$i = 1, 2, \ldots, r - 1$. The following example illustrates the application of these concepts for the case in which $F(s)$ contains other poles besides a repeated pole.

▶ **Example 16.12**

Find the inverse Laplace transform of

$$F(s) = \frac{8(s+2)}{(s+1)^3(s+3)}$$

Solution

This function has a repeated real pole, $p_1 = -1$ Np/s, with multiplicity 3, and a simple pole, $p_2 = -3$ Np/s. Thus,

$$F(s) = \frac{A_{1,3}}{(s+1)^3} + \frac{A_{1,2}}{(s+1)^2} + \frac{A_{1,1}}{s+1} + \frac{A_2}{s+3}$$

Applying Equations (16.57) and (16.51), we have

$$A_{1,3} = (s+1)^3 F(s)\Big|_{s=-1} = \frac{8(s+2)}{s+3}\Big|_{s=-1} = 4$$

$$A_{1,2} = \frac{d}{ds}\frac{8(s+2)}{s+3}\Big|_{s=-1} = \frac{8}{(s+3)^2}\Big|_{s=-1} = 2$$

$$A_{1,1} = \frac{1}{2}\frac{d}{ds}\frac{8}{(s+3)^2}\Big|_{s=-1} = \frac{1}{2}\frac{-16}{(s+3)^3}\Big|_{s=-1} = -1$$

$$A_2 = (s+3)F(s)\Big|_{s=-3} = \frac{8(s+2)}{(s+1)^3}\Big|_{s=-3} = 1$$

Finally, combining Equations (16.56) and (16.50), we get

$$f(t) = \left[(4t^2 + 2t - 1)e^{-t} + e^{-3t}\right]u(t)$$

which can be checked with the initial-value and final-value theorems.   ◀

Exercise 16.16   Find $\mathcal{L}^{-1}\{3(s+2)/[s^2(s+1)]\}$.

ANSWER $3(2t - 1 + e^{-t})u(t)$.

Repeated Complex Pole Pairs

Repeated complex pole pairs are treated in similar fashion, except that the calculations now involve complex algebra.

▶ **Example 16.13**

Find the inverse Laplace transform of

$$F(s) = \frac{2}{(s^2 + 4s + 5)^2}$$

Solution

This function has a repeated pole pair, $p_{1,2} = -2 \pm j1$ complex Np/s, with multiplicity 2. Thus,

$$F(s) = \frac{2}{(s+2-j1)^2(s+2+j1)^2}$$

$$= \frac{A_{1,2}}{(s+2-j1)^2} + \frac{A_{1,1}}{(s+2-j1)} + \frac{A_{1,2}^*}{(s+2+j1)^2} + \frac{A_{1,1}^*}{(s+2+j1)}$$

We are only interested in the first two residues,

$$A_{1,2} = (s+2-j1)^2 F(s)\Big|_{s=-2+j1} = \frac{2}{(s+2+j1)^2}\Big|_{s=-2+j1} = -0.5$$

$$A_{1,1} = \frac{d}{ds}\frac{2}{(s+2+j1)^2}\Big|_{s=-2+j1} = \frac{-2}{(s+2+j1)^3}\Big|_{s=-2+j1} = 0.5\underline{/-90°}$$

Hence, $f(t) = [-te^{-2t}\cos t + e^{-2t}\cos(t-90°)]u(t)$, or

$$f(t) = (\sin t - t\cos t)e^{-2t}u(t)$$ ◀

Exercise 16.17   Find $\mathscr{L}^{-1}\{(s+1)/[s(s^2+2s+2)^2]\}$.

ANSWER $0.25\left(1 + [\sqrt{2}t\cos(t+135°) - \cos t]e^{-t}\right)u(t)$.

Improper Rational Functions

If $m \geq n$ in Equation (16.47), $F(s)$ is *improper* and we must perform *long division* to first put it in the form

$$F(s) = \frac{N(s)}{D(s)} = Q(s) + \frac{R(s)}{D(s)} \tag{16.58}$$

where $Q(s)$ is the *quotient* and $R(s)$ is the *remainder* of the long division, and the degree of $R(s)$ is *less* than the degree of $D(s)$. We then perform the usual partial fraction expansion of the ratio $R(s)/D(s)$, which is *proper,* and we finally exploit linearity, along with the appropriate entries of Table 16.1, to find $f(t)$.

▶ Example 16.14

Find the inverse Laplace transform of

$$F(s) = \frac{2s^3 + 11s^2 + 14s - 5}{s^2 + 5s + 6}$$

Solution

Since $m = 3$ and $n = 2$, this rational function is improper and we must first perform long division as follows:

$$s^2 + 5s + 6 | 2s^3 + 11s^2 + 14s - 5 | 2s + 1 \longleftarrow Q(s)$$

$$\underline{2s^3 + 10s^2 + 12s}$$

$$s^2 + 2s - 5$$

$$\underline{s^2 + 5s + 6}$$

$$-3s - 11 \longleftarrow R(s)$$

Using the form of Equation (16.58) and expanding, we have

$$F(s) = 2s + 1 + \frac{-3s - 11}{s^2 + 5s + 6} = 2s + 1 + \frac{2}{s + 3} - \frac{5}{s + 2}$$

Finally, applying linearity and using Table 16.1 we obtain

$$f(t) = 2\frac{d\delta(t)}{dt} + \delta(t) + (2e^{-3t} - 5e^{-2t})u(t)$$ ◀

Exercise 16.18   Find $\mathscr{L}^{-1}\{(2s)^2/(s^2 + 3s + 2)\}$.

ANSWER $4[\delta(t) + (e^{-t} - 4e^{-2t})u(t)]$.

16.5 APPLICATION TO DIFFERENTIAL EQUATIONS

A most elegant application of Laplace transforms is the solution of *linear differential equations.* As we know, in a linear circuit the applied signal $x(t)$ and the response $y(t)$ are related by a differential equation of the type

$$b_n \frac{d^n y}{dt^n} + b_{n-1} \frac{d^{n-1} y}{dt^{n-1}} + \cdots + b_1 \frac{dy}{dt} + b_0 y = a_m \frac{d^m x}{dt^m} + a_{m-1} \frac{d^{m-1} x}{dt^{m-1}} + \cdots$$

$$+ a_1 \frac{dx}{dt} + a_0 x \qquad \text{(16.59)}$$

where a_0 through a_m and b_0 through b_n are *real, time-independent* coefficients. In Section 11.1 we solved differential equations using *phasors;* however, this method is applicable only to circuits in the *ac steady state.* In Chapter 14 we generalized the phasor method by introducing the concept of the *network function;* however, this function provides only the *functional form* of the solution. Finding the unknown coefficients appearing in this function requires additional calculations on the basis of the initial conditions. In the present section we solve differential equations using *Laplace transforms,* and we find that this approach takes the *initial conditions* into account automatically, thus providing the complete solution without requiring any additional calculations. Moreover, we find a new meaning for the network function itself, namely, that its inverse transform is the *impulse response* of the circuit.

The Forced and Natural Response Components

We assume the applied signal to be *causal* so that both $x(t)$ and its derivatives are zero for $t \le 0^-$. By contrast, the initial values of the response $y(t)$ and its derivatives are not necessarily zero because of initial stored energy in the capacitances and inductances. Taking the Laplace transforms of both sides of Equation (16.59) and using Equation (16.32), we obtain, after rearranging,

$$(b_n s^n + b_{n-1} s^{n-1} + \cdots + b_1 s + b_0) Y(s) - P_{\text{ic}}(s) = (a_m s^m + a_{m-1} s^{m-1} + \cdots$$
$$+ a_1 s + a_0) X(s)$$

where $Y(s) = \mathscr{L}\{y(t)\}$, $X(s) = \mathscr{L}\{x(t)\}$, and $P_{\text{ic}}(s)$ is a polynomial of degree $n - 1$ in s consisting of linear combinations of the *initial values* of the function, $y(0^-)$, and of its first $(n - 1)$ derivatives, $y^{(1)}(0^-)$, $y^{(2)}(0^-)$, ..., $y^{(n-1)}(0^-)$. Solving for $Y(s)$ yields

$$Y(s) = H(s)X(s) + Y_{\text{ic}}(s) \qquad (16.60)$$

where

$$H(s) = \frac{N(s)}{D(s)} = \frac{a_m s^m + a_{m-1} s^{m-1} + \cdots + a_1 s + a_0}{b_n s^n + b_{n-1} s^{n-1} + \cdots + b_1 s + b_0} \qquad (16.61)$$

is the familiar **network function** introduced in Section 14.2, and

$$Y_{\text{ic}}(s) = \frac{P_{\text{ic}}(s)}{D(s)} \qquad (16.62)$$

is a rational function of s that depends on the **initial conditions** in $y(t)$ and its derivatives, and that *vanishes* if these conditions are zero. To get a feel for its functional form, we show $Y_{\text{ic}}(s)$ explicitly for the cases $n = 1$, 2, and 3:

$$Y_{\text{ic}}(s) = \frac{b_1 y(0^-)}{b_1 s + b_0} \qquad (16.63a)$$

$$Y_{\text{ic}}(s) = \frac{(b_2 s + b_1) y(0^-) + b_2 y^{(1)}(0^-)}{b_2 s^2 + b_1 s + b_0} \qquad (16.63b)$$

$$Y_{\text{ic}}(s) = \frac{(b_3 s^2 + b_2 s + b_1) y(0^-) + (b_3 s + b_2) y^{(1)}(0^-) + b_3 y^{(2)}(0^-)}{b_3 s^3 + b_2 s^2 + b_1 s + b_0} \qquad (16.63c)$$

The complete response $y(t)$ to the applied signal $x(t)$ is found by taking the inverse Laplace transforms of both sides of Equation (16.60),

$$y(t) = \mathscr{L}^{-1}\{H(s)X(s)\} + \mathscr{L}^{-1}\{Y_{\text{ic}}(s)\} \qquad (16.64)$$

This response can be expressed as the sum of two components,

$$y(t) = y_{\text{forced}} + y_{\text{natural}} \qquad\qquad \textbf{(16.65)}$$

The first component,

$$y_{\text{forced}} = \mathscr{L}^{-1}\{H(s)X(s)\} \qquad\qquad \textbf{(16.66)}$$

depends on the applied signal $x(t)$ regardless of the initial conditions, and is thus the **forced** component. As we know, it is also called the **zero-state** component because this is the response that the circuit would yield with *zero initial energy* in its reactive elements. The second component,

$$y_{\text{natural}} = \mathscr{L}^{-1}\{Y_{\text{ic}}(s)\} \qquad\qquad \textbf{(16.67)}$$

depends on the initial conditions regardless of the applied signal $x(t)$, and is thus the **natural** component, also called the **zero-input** or the **source-free** component. As we know, the circuit produces this component using the energy initially stored in its reactive elements. It is precisely the ability of the Laplace method to incorporate the initial conditions explicitly into the solution that makes it superior to the network function techniques of Chapter 14!

▶ **Example 16.15**

Use the Laplace method to solve the equation

$$\frac{d^2 y(t)}{dt^2} + 3\frac{dy(t)}{dt} + 2y(t) = 2\frac{dx(t)}{dt} + x(t)$$

where

$$x(t) = 2e^{-3t}u(t)$$

subject to the initial conditions $y(0^-) = 1$, and $y^{(1)}(0^-) = 2$.

Solution

Taking the Laplace transforms of both sides yields $s^2 Y(s) - sy(0^-) - y^{(1)}(0^-) + 3sY(s) - 3y(0^-) + 2Y(s) = 2sX(s) + X(s)$. Solving for $Y(s)$, we get

$$Y(s) = \frac{2s+1}{s^2+3s+2}X(s) + \frac{(s+3)y(0^-) + y^{(1)}(0^-)}{s^2+3s+2} \qquad\qquad \textbf{(16.68)}$$

Substituting $X(s) = \mathscr{L}\{2e^{-3t}u(t)\} = 2/(s+3)$, along with $y(0^-) = 1$ and $y^{(1)}(0^-) = 2$, yields

$$Y(s) = \frac{2s+1}{s^2+3s+2} \times \frac{2}{s+3} + \frac{s+5}{s^2+3s+2} \qquad\qquad \textbf{(16.69)}$$

We now have two options, depending on whether we want to find the complete response or its individual components. Let us pursue both approaches and then check that our results satisfy Equation (16.65).

(a) To find the *complete* response, combine $Y(s)$ into a single expression and then perform a partial fraction expansion,

$$Y(s) = \frac{s^2 + 12s + 17}{(s^2 + 3s + 2)(s + 3)} = \frac{3}{s + 1} + \frac{3}{s + 2} - \frac{5}{s + 3}$$

The complete response is thus

$$y(0^-) = 1 \tag{16.70a}$$

$$y(t > 0) = 3e^{-t} + 3e^{-2t} - 5e^{-3t} \tag{16.70b}$$

As a check, apply the initial-value and final-value theorems to find that $y(0^+) = 1$ agrees with $\lim_{s \to \infty} sY(s) = s^3/s^3 = 1$ and $y(\infty) = 0$ agrees with $\lim_{s \to 0} sY(s) = 0/(2 \times 3) = 0$.

(b) To find the individual response components, expand the terms comprising $Y(s)$ *separately*,

$$\frac{2s + 1}{s^2 + 3s + 2} \times \frac{2}{s + 3} = -\frac{1}{s + 1} + \frac{6}{s + 2} - \frac{5}{s + 3}$$

$$\frac{s + 5}{s^2 + 3s + 2} = \frac{4}{s + 1} - \frac{3}{s + 2}$$

Then, the individual response components are, for $t > 0$,

$$y_{\text{forced}} = -e^{-t} + 6e^{-2t} - 5e^{-3t} \tag{16.71a}$$

$$y_{\text{natural}} = 4e^{-t} - 3e^{-2t} \tag{16.71b}$$

As a check, you can verify that the sum of y_{forced} and y_{natural} coincides with $y(t)$ of Equation (16.70), as it should. ◀

Exercise 16.19   Use the Laplace method to solve the equation $d^2y/dt + 4dy/dt + 3y = x(t)$, where $x(t) = 5e^{-t}u(t)$, $y(0^-) = 2$, and $y^{(1)}(0^-) = 2$. Don't forget to check your results!

ANSWER $y(t > 0) = (2.5t + 2.75)e^{-t} - 0.75e^{-3t}$.

▶ **Example 16.16**

(a) Repeat Example 16.15, but with the initial conditions changed to $y(0^-) = 3$ and $y^{(1)}(0^-) = 1$.

(b) Repeat Example 16.15, but with the applied signal changed to $x(t) = 2(\cos 3t)u(t)$.

Solution

(a) With a changed set of initial conditions, only $Y_{ic}(s)$ and hence $y_{natural}$ will change. Thus,

$$Y_{ic}(s) = \frac{(s+3)y(0^-) + y^{(1)}(0^-)}{s^2 + 3s + 2} = \frac{3s + 10}{s^2 + 3s + 2} = \frac{7}{s+1} - \frac{4}{s+2}$$

so that $y_{natural} = (7e^{-t} - 4e^{-2t})u(t)$. Letting $y(t > 0) = y_{forced} + y_{natural}$, with y_{forced} as given in Equation (16.71a), we obtain

$$y(t > 0) = 6e^{-t} + 2e^{-2t} - 5e^{-3t}$$

(b) With a changed applied signal, only y_{forced} will change. Letting $X(s) = 2s/(s^2 + 3^2)$, we obtain

$$H(s)X(s) = \frac{2s+1}{s^2 + 3s + 2} \times \frac{2s}{s^2 + 9} = \frac{0.2}{s+1} - \frac{12/13}{s+2}$$

$$+ \frac{\sqrt{37/130}\underline{/-47.34°}}{s - j3} + \frac{\sqrt{37/130}\underline{/47.34°}}{s + j3}$$

Thus, $y_{forced} = [0.2e^{-t} - (12/13)e^{-2t} + \sqrt{74/65}\cos(3t - 47.34°)] \times u(t)$. Letting $y(t > 0) = y_{forced} + y_{natural}$, with $y_{natural}$ as given in Equation (16.71b), we obtain

$$y(t > 0) = 4.2e^{-t} - (51/13)e^{-2t} + \sqrt{74/65}\cos(3t - 47.34°) \quad \blacktriangleleft$$

Exercise 16.20

(a) Repeat Exercise 16.19, but with the initial conditions changed to $y(0^-) = 2$ and $y^{(1)}(0^-) = -2$.

(b) Repeat Exercise 16.19, but with the applied signal changed to $x(t) = 2(\sin t)u(t)$.

ANSWER   (a) $y(t > 0) = (2.5t + 0.75)e^{-t} + 1.25e^{-3t}$; (b) $y(t > 0) = 4.5e^{-t} - 2.1e^{-3t} + (1/\sqrt{5})\cos(t - 153.43°)$.

▶ **Example 16.17**

Use Laplace transforms to solve the integrodifferential equation

$$\frac{dy(t)}{dt} + 4y(t) + 5\int_{0^-}^{t} y(\xi)\,d\xi = 3u(t)$$

subject to the initial condition $y(0^-) = 1$.

Solution

Transforming,

$$sY(s) - 1 + 4Y(s) + \frac{5}{s}Y(s) = \frac{3}{s}$$

or

$$Y(s) = \frac{s + 3}{s^2 + 4s + 5} = \frac{0.5 - j0.5}{s + 2 - j1} + \frac{0.5 + j0.5}{s + 2 + j1}$$

Thus,

$$y(t > 0) = \sqrt{2}e^{-2t}\cos(t - 45°)$$

◀

Exercise 16.21 Use Laplace transforms to solve the integrodifferential equation $dy(t)/dt + \int_{0^-}^{t} y(\xi)e^{-2(t-\xi)} d\xi = 2u(t)$, subject to the initial condition $y(0^-) = 1$.

ANSWER $y(t > 0) = 4 - (t + 3)e^{-t}$.

▶ **Example 16.18**

Use Laplace transforms to solve the simultaneous equations

$$\frac{dy_1(t)}{dt} + 2y_1(t) - y_2(t) = u(t)$$

$$y_1(t) - \frac{dy_2(t)}{dt} - 2y_2(t) = 0$$

subject to the initial conditions $y_1(0^-) = 2$, and $y_2(0^-) = 0$.

Solution

Taking the Laplace transforms of both equations,

$$sY_1(s) - 2 + 2Y_1 - Y_2 = 1/s$$

$$Y_1(s) - (sY_2 - 0) - 2Y_2(s) = 0$$

Solving by Cramer's rule yields

$$Y_1 = \frac{(s + 2)(2s + 1)}{s(s^2 + 4s + 3)} = \frac{2/3}{s} + \frac{1/2}{s + 1} + \frac{5/6}{s + 3}$$

$$Y_2 = \frac{2s + 1}{s(s^2 + 4s + 3)} = \frac{1/3}{s} + \frac{1/2}{s + 1} - \frac{5/6}{s + 3}$$

Taking the inverse Laplace transforms,

$$y_1(t > 0) = \frac{1}{6}(4 + 3e^{-t} + 5e^{-3t})$$

$$y_2(t > 0) = \frac{1}{6}(2 + 3e^{-t} - 5e^{-3t})$$

◀

Exercise 16.22   Use Laplace transforms to solve the simultaneous equations $dy_1(t)/dt + y_1(t) + dy_2(t)/dt + y_2(t) = e^{-2t}u(t)$, and $dy_1(t)/dt - y_1(t) - 2y_2(t) = 0$, subject to the initial conditions $y_1(0^-) = 0$ and $y_2(0^-) = 1$.

ANSWER $y_1(t > 0) = 2[e^{-2t} + (2t - 1)e^{-t}]$, and $y_2(t > 0) = -3e^{-2t} - 4(t - 1)e^{-t}$.

▶ **Example 16.19**

A circuit has the network function

$$H(s) = \frac{2s + 3}{s^2 + 2s + 5}$$

Find its response $y(t)$ to the applied signal $x(t) = 2tu(t)$, subject to the initial conditions $y(0^-) = 1$, and $y^{(1)}(0^-) = -1$.

Solution

To find $y(t)$ we must first put $Y(s)$ in the form of Equation (16.60). By inspection, $b_2 = 1$, and $b_1 = 2$. Hence, using Equation (16.63b),

$$Y(s) = \frac{2s + 3}{s^2 + 2s + 5}X(s) + \frac{(s + 2)y(0^-) + y^{(1)}(0^-)}{s^2 + 2s + 5}$$

Substituting $X(s) = \mathcal{L}\{2tu(t)\} = 2/s^2$, along with $y(0^-) = 1$ and $y^{(1)}(0^-) = -1$, we obtain, after collecting and expanding,

$$Y(s) = \frac{s^3 + s^2 + 4s + 6}{(s^2 + 2s + 5)s^2}$$

$$= \frac{6/5}{s^2} + \frac{8/25}{s} + \frac{(\sqrt{26}/10)\underline{/48.18°}}{s + 1 - j2} + \frac{(\sqrt{26}/10)\underline{/-48.18°}}{s + 1 + j2}$$

The complete response is thus

$$y(0^-) = 1$$

$$y(t > 0) = \frac{1}{5}[6t + 1.6 + \sqrt{26}e^{-t} \cos(2t + 48.18°)]$$

◀

Exercise 16.23 If a circuit has $H(s) = s/[(s + 2)(s + 3)]$, find its response $y(t)$ to the applied signal $x(t) = (e^{-2t} \sin 3t)u(t)$, subject to the initial conditions $y(0^-) = 0$, and $y^{(1)}(0^-) = 1$.

ANSWER $y(t) = \frac{1}{3}[e^{-2t} - 0.3e^{-3t} + \sqrt{13/10}e^{-2t} \cos(3t - 127.87°)]u(t)$.

The Network Function $H(s)$

The study of complex exponential signals in Chapter 14 has led us to formulate the important concept of *network function*, defined as the ratio of the response $Y(s)$ to the applied signal $X(s)$, that is, $H(s) \triangleq Y(s)/X(s)$. Can the network function be regarded also as the ratio of the *Laplace transforms* of the response and the applied signal? To find the answer, we rewrite Equation (16.60) as

$$H(s) = \frac{Y(s) - Y_{ic}(s)}{X(s)}$$

and observe that **if all initial conditions in a circuit are zero, then the ratio of the Laplace transforms of the response and the applied signal is simply the network function of the circuit,**

$$H(s) = \frac{Y(s)}{X(s)}\bigg|_{Y_{ic}(s)=0} \tag{16.72}$$

Since $H(s)$ is a function of the complex frequency s, it is interesting to investigate the significance of its *inverse transform*

$$h(t) = \mathcal{L}^{-1}\{H(s)\} \tag{16.73}$$

This task is accomplished by considering the special case in which the circuit is in its *zero state*, so that $Y_{ic}(s) = 0$, and is subjected to the *impulse* function

$$x(t) = \delta(t)$$

so that $X(s) = 1$. The ensuing response $y(t)$, aptly called the *impulse response*, is found as $y(t) = \mathcal{L}^{-1}\{Y(s)\} = \mathcal{L}^{-1}\{H(s)1 + 0\} = \mathcal{L}^{-1}\{H(s)\}$, or

$$y(t) = h(t)$$

Clearly, the inverse transform of the network function of a circuit is simply its **impulse response.** Once we know $H(s)$, this response may be found by taking its inverse transform. Conversely, if we know the impulse response $h(t)$, we can find the network function by taking its Laplace transform. This is particularly important when the network function must be found *experimentally.* We now have an important new interpretation for the network function!

▶ Example 16.20

The network function of a certain circuit has a pole pair $p_{1,2} = -3 \pm j4$ complex Np/s, and a zero $z_1 = -5$ Np/s. Moreover, $H(0) = 10$. Find the impulse response of this circuit.

Solution

By inspection,

$$H(s) = \frac{50(s+5)}{(s+3-j4)(s+3+j4)}$$

The residue at the upper pole is $A_1 = 12.5\sqrt{5}\underline{/-26.57°}$. Hence,

$$h(t) = 25\sqrt{5}e^{-3t}[\cos(4t - 26.57°)]u(t) \qquad ◀$$

▶ Example 16.21

If the impulse response of a circuit is $h(t) = 10e^{-2t}u(t)$, what is its response to a unit step?

Solution

$H(s) = \mathcal{L}\{h(t)\} = 10/(s+2)$. Then,

$$Y(s) = H(s)X(s) = \frac{10}{s+2}\frac{1}{s} = \frac{5}{s} - \frac{5}{s+2}$$

Hence,

$$y(t) = 5(1 - e^{-2t})u(t) \qquad ◀$$

Exercise 16.24   If the unit step response of a circuit is $y(t) = 10e^{-2t}$, what is its impulse response?

ANSWER $h(t) = 10\delta(t) - 20e^{-2t}u(t)$.

A CONVERSATION WITH
LOUIS S. HURESTON

PACIFIC BELL

Louis Hureston found that his electrical engineering degree was a great platform to becoming a telecommunications manager at Pacific Bell.

After earning his BS degree in electrical engineering from the Illinois Institute of Technology (1982) and his MS in electrical engineering from Georgia Institute of Technology (1983), Louis S. Hureston worked for Bell Labs in Chicago before accepting a management position with Pacific Bell in 1987. He completed the Executive Development Program at Northwestern University in 1989 and is currently director of the Information Technology and Engineering Consulting Group at Pacific Bell in San Ramon, California.

What is the main responsibility of your group, Mr. Hureston?

The short answer is "design software and propose hardware." The long answer covers a lot of ground with those five words because we offer a whole menu of services to the operational units of Pacific Bell. The company is divided into regional business units, each unit operating a segment of the California public telephone network. My organization offers re-engineering of business processes and development of knowledge-based systems for these business units. A typical request might be, "Design a computer system that improves the efficiency of our service representatives," or perhaps, "Give us an intelligent workstation that will help our repair department handle service calls more effectively."

Then my people have to come up with a program that gets the job done, analyze all the cost, performance, and scheduling trade-offs, and end up with the best package possible for the task.

How does your EE background enter into this work?

Well, because most of our assignments boil down to revamping telecommunications systems, training in EE helps in analyzing those systems. Also, the EE work I did for my master's had a heavy emphasis on the combination of computer systems and communication theory because I knew I'd be going into telecommunications from the time I had my first summer job with AT&T in Chicago in 1978.

When did you know you wanted to major in EE?

Up until I was a junior in high school, I always thought I would be a doctor. I did well in math and science and felt, the way a lot of other idealistic kids do, that I wanted to help people. So everyone I talked to told me I was a natural for being a doctor. During my senior year, though, when I started exploring the college catalogs more closely, I became intrigued by the engineering courses, and that's the route I ended up taking.

Any regrets?

Not at all. I think you get more career choices with an engineering base than with premed. For instance, if I had continued to want to go to med school, I would have stressed biomedical engineering as an undergraduate. As things turned out, I ended up in management, and engineering offered a great platform for making that jump. In general, an engineering degree is a prized asset in business careers. It's common for about 30 percent of students in the top MBA programs to have an undergraduate degree in engineering or one of the sciences, and a significant number of senior managers in Fortune 500 companies have a technical degree.

Once I had a few engineering courses under my belt in college, I said goodbye to med school plans. Now the service-to-people part of me gets satisfaction from knowing that engineers can use design to improve conditions in society across all disciplines.

Plus I'm active in the National Society of Black Engineers, an organization I joined when I was a junior in college. I speak at high schools and colleges and try to get African-American students enthusiastic about

> *"Enjoy the learning and...the applications will come....don't be concerned if you finish school with more questions than answers."*

engineering. One of the NSBE projects I have been involved in is the design and implementation of an electronic mail system that students and alumni members of NSBE can use as a communication tool. As an added benefit, projects like this are great for creating enthusiasm and awareness among high school and even elementary school students, making these kids aware of the opportunities in engineering.

Can you remember anything about your circuits course you'd like to mention?

It's the first time most science students see the abstract laws they learned in physics applied to the real world. In that respect, this course gave me a new feel for the meaning of the core science program my teachers had been talking about since grade school. Sort of "Aha, so this is what all that work was for, how to bring electricity into people's lives."

What advice would you offer to someone taking the circuits course today, in other words, someone just starting out on the EE road?

Enjoy the learning and accept that the applications will come. Take as many social science and humanities courses as you can to balance the math and

(Continued)

(*Continued*)

science courses. Doing so will show you the people side of the picture, which you will need as much as you need the computer side. If you think you might like to go into management, take some economics courses, too, to help you appreciate the financial dimension of any project.

Also, don't be concerned if you finish school with more questions than answers. I naively expected that grad school would answer all the EE questions I had accumulated in college. I ended up with so many new questions, it took me five years of work experience to get most of the answers. And note I said "most." I still have questions—and hope I always will, for that matter.

Let's talk for a minute about being a manager. How does it compare with being a working engineer?

My first six months as a manager at Pacific Bell were, shall we say, challenging, and in many respects I felt quite uncomfortable. But then I began to see how I could have a positive impact on other people's potential. Instead of working alone and solving one problem in the course of a week, say, or a month, I could get ten people to solve 25 problems. Once I realized that, I became quite comfortable with my job.

Being a manager is as demanding as being an engineer but in a different way. A manager's interpersonal skills and communications skills are needed every minute of the day. An engineer needs these skills, too, of course, but they come into play when she or he is writing a report, say, or working on a team project—activities that are only part of the engineer's job. For managers, communication is a greater part of the job. In other words, managers spend a significant portion of their time coaching and acting as mentors.

The questions I deal with these days are not "What series resistors will get the job done here?" but "How can I tap into the human potential of my staff?" or "How can I help this engineer develop his critical thinking skills?" or "What will motivate this engineer to handle her time more efficiently?"

What's the future look like for information technology?

Growth, growth, growth. The information industry worldwide is expected to have hundreds of billions of dollars of growth during the 1990s. It's almost

> *"It's almost indescribable how much growth potential there is in information technology, and...circuits are the hub...hardware design boils down to circuit design."*

indescribable how much growth potential there is in information technology, and when you think about it, circuits are the hub of the whole issue: The whole area of hardware design boils down to circuit design.

16.6 APPLICATION TO CIRCUIT ANALYSIS

The Laplace techniques of the previous section allow us to solve a differential equation in which the initial conditions are specified in terms of the *solution itself* and its first $n - 1$ *derivatives*. In circuit analysis, however, it is more sensible to have these conditions specified in terms of the initial states of its *reactive elements.* Moreover, it is desirable to formulate circuit equations directly in *algebraic* form, bypassing the derivations of any differential equations. We now develop a method to simultaneously achieve both goals. But first, let us investigate how Laplace transforms affect the constitutive laws of the basic circuit elements.

Circuit Element Models

The constitutive law for **resistance** is $v(t) = Ri(t)$. Taking the Laplace transform of both sides yields

$$V(s) = RI(s) \tag{16.74}$$

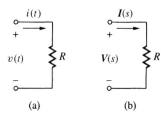

Figure 16.10 (a) *t*-domain and (b) *s*-domain representations for resistance.

where $V(s) = \mathcal{L}\{v(t)\}$ and $I(s) = \mathcal{L}\{i(t)\}$. Clearly, Ohm's Law holds also for the Laplace transforms of voltage and current. Note, however, that the physical units of $V(s)$ and $I(s)$ are, respectively, *volts × seconds* (V-s) and *amperes × seconds* (A-s); this follows from the Laplace transform definition of Equation (16.17). The dimensions of the ratio $V(s)/I(s)$ are *ohms.* As depicted in Figure 16.10, the *s*-domain model for resistance is again a resistance of R Ω, but carrying a current of $I(s)$ A-s and developing a voltage of $V(s)$ V-s.

The constitutive law for **inductance**, shown in Figure 16.11(a), is $v(t) = L\,di(t)/dt$. Taking the Laplace transform of both sides yields

$$V(s) = sLI(s) - Li_L(0^-) \tag{16.75}$$

where we have used the subscript L to emphasize that the initial condition accounted for by the Laplace transformation is the *inductance* current. Note that $Li_L(0^-)$ has the same dimensions as $V(s)$, namely, V-s. Equation (16.75) suggests the *s*-domain model of Figure 16.11(b), which consists of an impedance of sL Ω in series with a voltage source of $Li_L(0^-)$ V-s and with polarity *opposite* to $V(s)$.

Equation (16.75) can be turned around to yield

$$I(s) = \frac{V(s)}{sL} + \frac{i_L(0^-)}{s} \tag{16.76}$$

This suggests the alternative *s*-domain model of Figure 16.11(c), consisting of an impedance of sL Ω in parallel with a current source of $i_L(0^-)/s$ A-s and with the *same* direction as $I(s)$.

The constitutive law for **capacitance**, shown in Figure 16.12(a), is $i(t) = C\,dv(t)/dt$, and it transforms as $I(s) = sCV(s) - Cv_C(0^-)$, or

$$I(s) = sCV(s) - Cv_C(0^-) \tag{16.77}$$

where we have used the subscript C to emphasize that the initial condition is now the *capacitance* voltage. Since $Cv_C(0^-)$ has the same dimensions as $I(s)$,

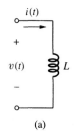

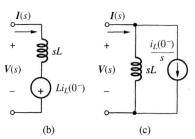

Figure 16.11 (a) *t*-domain and (b), (c) *s*-domain representations for inductance.

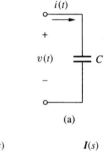

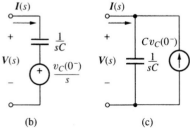

Figure 16.12 (a) *t*-domain and (b), (c) *s*-domain representations for capacitance.

or A-s, we can use the *s*-domain model of Figure 16.12(c), consisting of an impedance of $1/sC$ Ω in parallel with a current source of $Cv_C(0^-)$ A-s and with *opposite* direction as $I(s)$. Turning Equation (16.77) around as

$$V(s) = \frac{I(s)}{sC} + \frac{v_C(0^-)}{s}$$ (16.78)

suggests the alternative *s*-domain model of Figure 16.12(b), where the impedance of $1/sC$ Ω is now in series with a voltage source of $v_C(0^-)/s$ V-s and with the *same* polarity as $V(s)$.

Finally, the constitutive laws for **magnetically coupled coils** are $v_1(t) = L_1\, di_1(t)/dt + M\, di_2(t)/dt$, and $v_2(t) = M\, di_1(t)/dt + L_2\, di_2(t)/dt$, and they transform as

$$V_1(s) = sL_1I_1(s) + sMI_2(s) - [L_1i_1(0^-) + Mi_2(0^-)]$$ (16.79a)

$$V_2(s) = sMI_1(s) + sL_2I_2(s) - [Mi_1(0^-) + L_2i_2(0^-)]$$ (16.79b)

This suggests the *s*-domain model of Figure 16.13(b).

In dealing with *s*-domain circuit representations, careful attention must be paid to the polarities of the initial-condition sources relative to the polarities of $V(s)$ and $I(s)$. If the polarity of $i_L(0^-)$ or $v_C(0^-)$ is reversed, then the corresponding source must also be reversed.

We observe that if *all initial conditions are zero,* then all corresponding sources vanish and the *s*-domain element models reduce to those developed in Chapter 14, or $V(s) = RI(s)$, $V(s) = sLI(s)$, and $V(s) = I(s)/sC$. It is apparent that thanks to its ability to account for nonzero initial conditions, the Laplace technique is a generalization of the network function techniques of Chapter 14.

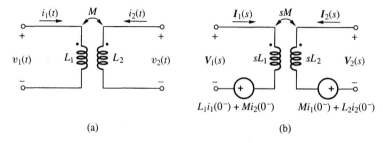

Figure 16.13 (a) t-domain and (b) s-domain representations for magnetically coupled coils.

Circuit Analysis Using Laplace Transforms

We are now ready to analyze a circuit and find its response in terms of the initial states of its reactive elements. Our analysis is based upon the s-domain element laws of Equations (16.74) through (16.79), along with KCL and KVL, which hold also in s-domain form,

$$\sum_n \boldsymbol{I}_{\text{in}}(s) = \sum_n \boldsymbol{I}_{\text{out}}(s) \qquad \textbf{(16.80a)}$$

$$\sum_\ell \boldsymbol{V}_{\text{rise}}(s) = \sum_\ell \boldsymbol{V}_{\text{drop}}(s) \qquad \textbf{(16.80b)}$$

Circuit analysis via Laplace methods involves the following steps:

(1) Use the t-domain form of the circuit to find $i_L(0^-)$ through each inductance, and $v_C(0^-)$ across each capacitance.

(2) Redraw the circuit in s-domain form, with each voltage and current source replaced by its Laplace transform, and with each element replaced by its s-domain equivalent. For *loop* analysis, the models using initial-condition *voltage* sources are generally preferable; for *node* analysis, the models with *current* sources are generally preferable.

(3) Analyze the transformed circuit as if it were purely resistive, except that the signals are Laplace transforms, in V-s or A-s, the elements are impedances, in Ω, and the circuit contains additional sources to account for the initial conditions in its reactive elements. These sources will be absent if the circuit is in its zero state. Depending on the case, the analysis may require the use of the series or parallel impedance formulas, the voltage or current divider formulas, the superposition principle, the node or loop methods, Thévenin or Norton reductions, the op amp rule, and so forth. The outcome is one or more *algebraic equations* relating the Laplace transform $\boldsymbol{Y}(s)$ of the response to those of the applied sources. The coefficients of these equations are functions of the complex impedances R, sL, and $1/sC$, as well as the initial conditions $i_L(0^-)$ and $v_C(0^-)$ of the reactive elements.

Once we know $Y(s)$, the time-domain response is obtained as

$$y(t) = \mathcal{L}^{-1}\{Y(s)\}$$

(16.81)

Of particular interest is the case of a *single* applied source $X(s)$, for then the response $Y(s)$ takes on the general form

$$Y(s) = H(s)X(s) + Y_{\text{natural}}(s)$$

(16.82)

where $H(s)$ is the *network function,* and $Y_{\text{natural}}(s)$ is a function arising from the *initial stored energies* in the reactive elements. The complete response is the sum of the *forced* and *natural* components, $y(t) = y_{\text{forced}} + y_{\text{natural}}$, where

$$y_{\text{forced}} = \mathcal{L}^{-1}\{H(s)X(s)\}$$

(16.83a)

$$y_{\text{natural}} = \mathcal{L}^{-1}\{Y_{\text{natural}}(s)\}$$

(16.83b)

Let us illustrate the Laplace method with actual examples.

▶ Example 16.22

Find $v(t)$ in the circuit of Figure 16.14(a).

Solution

We could analyze the circuit via the techniques of Chapter 8; however, we shall use the Laplace method to see how it automatically accounts for the initial conditions.

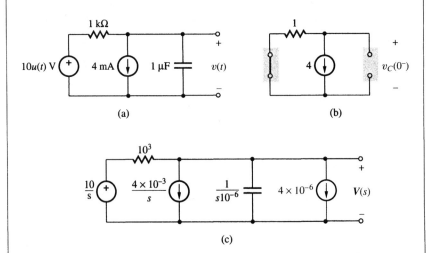

Figure 16.14 (a) Circuit of Example 16.22; (b) circuit to find the initial conditions; (c) s-domain representation.

Referring to Figure 16.14(b), where the units are [V], [A], and [Ω], we observe that

$$v_C(0^-) = -1 \times 4 = -4 \text{ V}$$

indicating that the *s*-domain model of the capacitance must include a parallel current source $Cv_C(0^-) = 10^{-6} \times 4$ A-s pointing *downward* because of the polarity of $v_C(0^-)$.

Next, redraw the circuit in *s*-domain form as in Figure 16.14(c), where the units are now [V-s], [A-s], and [Ω]. Note that the original sources are replaced by their Laplace transforms $10/s$ and $4 \times 10^{-3}/s$, the capacitance by its impedance $1/(s10^{-6})$, and the response by its transform $V(s)$. Applying KCL, we have

$$\frac{10/s - V(s)}{10^3} = \frac{4 \times 10^{-3}}{s} + \frac{V(s)}{1/(s10^{-6})} + 4 \times 10^{-6}$$

Collecting and performing a partial fraction expansion gives

$$V(s) = \frac{6 \times 10^3 - 4s}{s(s + 10^3)} = \frac{6}{s} - \frac{10}{s + 10^3}$$

Taking the inverse Laplace transform yields

$$v(t > 0) = 6 - 10e^{-10^3 t} \text{ V} \qquad \blacktriangleleft$$

Exercise 16.25   Repeat Example 16.22, but with the capacitance replaced by a 10-mH inductance.

ANSWER $v(t > 0) = 10e^{-10^5 t}$ V.

▶ **Example 16.23**

Find $v(t)$ in the circuit of Figure 16.15(a).

Solution

We could use the techniques of Chapter 14; however, we shall employ the Laplace method to obtain the complete response with the initial conditions automatically accounted for.

To find these conditions, refer to Figure 16.15(b), where the units are [V], [A], and [Ω]. By Ohm's Law and the current divider formula,

$$v_C(0^-) = (4 \parallel 1)2.5 = 2 \text{ V}$$

$$i_L(0^-) = \frac{4}{4 + 1} 2.5 = 2 \text{ A}$$

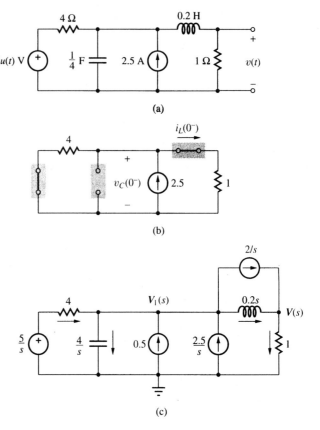

Figure 16.15 (a) Circuit of Example 16.23; (b) circuit to find the initial conditions; (c) *s*-domain representation.

Next, redraw the circuit in *s*-domain form as in Figure 16.15(c), where the units are [V-s], [A-s], and [Ω]. Anticipating the use of the node method, we model the initial conditions with the *current* sources $Cv_C(0^-) = \frac{1}{4} \times 2 = 0.5$ A-s (↑), and $i_L(0^-)/s = 2/s$ (→). We also select the reference node, label the essential nodes, and select reference directions for the branch currents.

Applying KCL at nodes $V_1(s)$ and $V(s)$ yields

$$\frac{5/s - V_1(s)}{4} + 0.5 + \frac{2.5}{s} = \frac{V_1(s)}{4/s} + \frac{2}{s} + \frac{V_1(s) - V(s)}{0.2s}$$

$$\frac{2}{s} + \frac{V_1(s) - V(s)}{0.2s} = \frac{V(s)}{1}$$

Eliminating $V_1(s)$, solving for $V(s)$, and performing a partial fraction expansion, we obtain

$$V(s) = \frac{2s^2 + 12s + 75}{s(s^2 + 6s + 25)} = \frac{3}{s} + \frac{0.625\underline{/143.13°}}{s + 3 - j4} + \frac{0.625\underline{/-143.13°}}{s + 3 + j4}$$

Taking the inverse Laplace transform,

$$v(t > 0) = 3 + 1.25e^{-3t}\cos(4t + 143.13°) \text{ V}$$

◄

Exercise 16.26   Find $v(t)$ in the circuit of Figure 16.15(a) if the capacitance and inductance are interchanged with each other.

ANSWER $v(t) = [1.25e^{-2.4t}\cos(3.2t + 36.87°)]u(t) \text{ V}.$

► **Example 16.24**

   (a) Find $v(t)$ in Figure 16.15(a), but with the 2.5-A source removed from the circuit.

   (b) Repeat part (a) if the applied source is changed to $(\sin 2t)u(t) \text{ V}$.

Solution

   (a) Without the 2.5-A source we have $v_C(0^-) = 0$ and $i_L(0^-) = 0$, so the s-domain circuit assumes the simpler form of Figure 16.16. Since now $V_{\text{natural}}(s) = 0$, we can solve our circuit by working directly with the network function. Applying the voltage divider formula twice, we get

$$H(s) = \frac{1}{0.2s + 1} \times \frac{(4/s) \parallel (0.2s + 1)}{4 + [(4/s) \parallel (0.2s + 1)]} = \frac{5}{s^2 + 6s + 25}$$

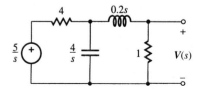

Figure 16.16 Circuit of Example 16.24.

   Hence,

$$V(s) = H(s)\frac{5}{s} = \frac{25}{s(s^2 + 6s + 25)}$$

$$= \frac{1}{s} + \frac{0.625\underline{/143.13°}}{s + 3 - j4} + \frac{0.625\underline{/-143.13°}}{s + 3 + j4}$$

   Taking the inverse Laplace transform,

$$v(t) = [1 + 1.25e^{-3t}\cos(4t + 143.13°)]u(t) \text{ V}$$

   (b) The Laplace transform of the applied signal is now $2/(s^2 + 2^2)$. Hence,

$$V(s) = H(s)\frac{2}{s^2 + 4} = \frac{10}{(s^2 + 6s + 25)(s^2 + 4)}$$

After performing the usual partial fraction expansion we find

$$v(t) = \frac{5}{6\sqrt{65}}[e^{-3t}\cos(4t+7.13°)+2\cos(2t-119.75°)]u(t) \text{ V} \blacktriangleleft$$

Exercise 16.27   Consider the circuit obtained from that of Figure 16.15(a) by removing the 2.5-A source.

(a) Find $v(t)$ if the 1-Ω resistance is changed to 2 Ω. What is the damping condition?

(b) Find the value to which the 1-Ω resistance must be changed in order to achieve critical damping. Hence, find $v(t)$.

ANSWER   (a) $v(t) = \frac{5}{3}(1 - 6e^{-5t} + 5e^{-6t})u(t)$ V, overdamped; (b) $0.2(1+4\sqrt{5})$ Ω, $v(t) = [1.660 - (9.086t + 1.660)e^{-5.472t}]u(t)$ V.

▶ Example 16.25

Assuming the switch in the circuit of Figure 16.17(a) has been open for a long time, find $i(t > 0)$.

Solution

Referring to Figure 16.17(b), where the units are [V], [A], and [Ω], we have, by KCL,

$$\frac{8 - v_X}{2} = \frac{v_X}{1} + \frac{v_X}{2}$$

or $v_X = 2$ V. The initial conditions are

$$v_C(0^-) = v_X = 2 \text{ V}$$

$$i_L(0^-) = \frac{8 - v_X}{2} = 3 \text{ A}$$

so $Cv_C(0^-) = \frac{1}{4} \times 2 = 0.5$ A-s, and $i_L(0^-)/s = 3/s$ A-s. This leads to the s-domain representation of Figure 16.17(c), where the units are [V-s], [A-s], and [Ω]. Because the switch is closed, we have $V_X(s) = 8/s$. Applying KCL and Ohm's Law,

$$\frac{3}{s} + \frac{8/s - V_1(s)}{0.5s} + 0.5 = \frac{V_1(s)}{1} + \frac{8/s}{2} + \frac{V_1(s)}{4/s}$$

$$I(s) = \frac{8/s - V_1(s)}{0.5s} + \frac{3}{s}$$

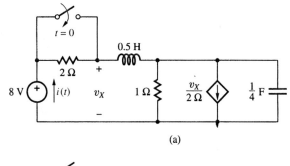

(a)

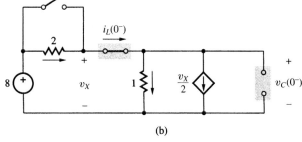

(b)

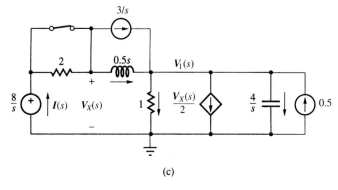

(c)

Figure 16.17 Circuits of Example 16.25.

Eliminating $V_1(s)$ and performing a partial fraction expansion,

$$I(s) = \frac{3s^2 + 24s + 96}{s(s^2 + 4s + 8)} = \frac{12}{s} + \frac{1.5\sqrt{10}\underline{/161.56^\circ}}{s + 2 - j2} + \frac{1.5\sqrt{10}\underline{/-161.56^\circ}}{s + 2 + j2}$$

Hence,

$$i(t > 0) = 3[4 + \sqrt{10}e^{-2t}\cos(2t + 161.57^\circ)] \text{ A}$$

Exercise 16.28   Repeat Example 16.25 if the switch, after having been closed for a long time, is opened at $t = 0$.

ANSWER $i(t > 0) = 3[1 + \sqrt{10}e^{-4t}\cos(4t + 18.43^\circ)]$ A.

▶Example 16.26

Find the response of the circuit of Figure 16.18(a) if $v_I(t)$ is a 1-V, 1.5-s pulse.

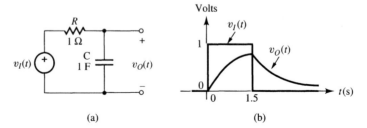

Figure 16.18 Circuit of Example 16.26 and its pulse response.

Solution

Since the initial conditions are zero, we have $Y_{\text{natural}}(s) = 0$, and we can thus write

$$V_O(s) = H(s)V_I(s) = \frac{1/sC}{R + 1/sC} \frac{1}{s}(1 - e^{-1.5s})$$

$$= \left(\frac{1}{s} - \frac{1}{s+1}\right)(1 - e^{-1.5s})$$

where we have used voltage division, along with Equation (16.36). The response is thus

$$v_O(t) = (1 - e^{-t})u(t) - (1 - e^{-(t-1.5)})u(t - 1.5) \text{ V}$$

and is shown, along with the applied pulse, in Figure 16.18(b).   ◀

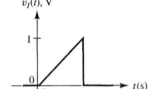

Figure 16.19 Signal of Exercise 16.30.

Exercise 16.29   Repeat Example 16.26, but with the capacitance replaced by an inductance $L = 1$ H.

ANSWER $v_O(t) = e^{-t}u(t) - e^{-(t-1.5)}u(t - 1.5)$ V.

Exercise 16.30   Repeat Example 16.26 if the applied signal is that of Figure 16.19.

ANSWER $v_O(t) = (t - 1 + e^{-t})u(t) - (t - 1)u(t - 1)$ V.

▶ **Example 16.27**

Assuming the switch in the circuit of Figure 16.20(a) has been open for a long time, find $v(t > 0)$.

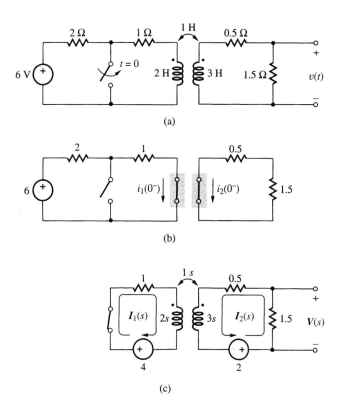

Figure 16.20 Circuits of Example 16.27.

Solution

Referring to Figure 16.20(b), where the units are [V], [A], and [Ω], we have

$$i_1(0^-) = 6/(2+1) = 2 \text{ A}$$

$$i_2(0^-) = 0$$

Consequently, the s-domain representation must include a source $-L_1 i_1(0^-) = -2 \times 2 = -4$ V-s in series with L_1, and a source $-M i_1(0^-) = -1 \times 2 = -2$ V-s in series with L_2. This is depicted in Figure 16.20(c), where the units are [V-s], [A-s], and [Ω]. Applying the loop method, we get

$$4 = 1I_1 + 2sI_1 + 1sI_2$$

$$2 = (1.5 + 0.5)I_2 + 3sI_2 + 1sI_1$$

Eliminating I_1 and performing a partial fraction expansion, we get

$$I_2 = \frac{2}{5s^2 + 7s + 2} = \frac{2/3}{s + 0.4} - \frac{2/3}{s + 1}$$

The current response is thus $i_2(t) = \frac{2}{3}(e^{-0.4t} - e^{-t})u(t)$ A. Letting $v(t) = -1.5i_2(t)$ yields

$$v(t) = (e^{-t} - e^{-0.4t})u(t) \text{ V}$$

◀

Exercise 16.31   Repeat Example 16.27 if the switch, after having been closed for a long time, is opened at $t = 0$.

ANSWER $v(t) = \frac{9}{7}(e^{-0.6t} - e^{-2t})u(t)$ V.

▶ **Example 16.28**

(a) Find the impulse response $h(t)$ of the circuit of Figure 16.21.

(b) What is the amount of energy injected by the impulse into the circuit?

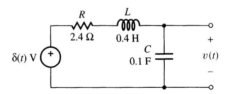

Figure 16.21   Circuit of Example 16.28.

Solution

(a) $v(t) = \mathscr{L}^{-1}\{H(s)X(s)\} = \mathscr{L}^{-1}\{H(s)1\} = h(t)$. By the voltage divider formula,

$$H(s) = \frac{1/sC}{R + sL + 1/sC} = \frac{1}{LCs^2 + RCs + 1}$$

Substituting the given component values and performing a partial fraction expansion, we have

$$H(s) = \frac{25}{s^2 + 6s + 25} = \frac{-j25/8}{s + 3 - j4} + \frac{j25/8}{s + 3 + j4}$$

Hence, $h(t) = \frac{25}{4}e^{-3t}\cos(4t - 90°)u(t)$, or

$$h(t) = (6.25e^{-3t}\sin 4t)u(t) \text{ V}$$

(b) The total initial energy is the sum of the capacitive and inductive energies:

$$w(0^+) = w_C(0^+) + w_L(0^+) = \frac{1}{2}Cv_C^2(0^+) + \frac{1}{2}Li_L^2(0^+)$$

By inspection, $v_C(0^+) = h(0^+) = 6.25e^0 \sin 0 = 0$, so $w_C(0^+) = 0$. By KCL and the capacitance law, $i_L(0^+) = i_C(0^+) = C\, dh(0^+)/dt = 6.25C(-3e^0 \sin 0 + 4e^0 \cos 0) = 25C = 2.5$ A. Then,

$$w(0^+) = w_L(0^+) = \frac{1}{2}0.4 \times 2.5^2 = 1.25 \text{ J}$$

Remark   Note that the energy injected by the impulse is initially stored entirely in L, and that the amount is independent of the value of R. Can you justify this physically?   ◀

Exercise 16.32   Find the impulse response of the circuit of Figure 16.21 if (a) $R = 4\ \Omega$ and (b) $R = \sqrt{20}\ \Omega$.

ANSWER   (a) $h(t) = 25te^{-5t}u(t)$ V; (b) $h(t) = 5(e^{-3.090t} - e^{-8.090t})u(t)$ V.

▶ Example 16.29

Find the response $i(t)$ of the circuit of Figure 16.22(a) if $v_C(0^-) = 5$ V and $i_L(0^-) = 2.5$ A.

Solution

Since this is a source-free circuit, we have $i(t) = i_{\text{natural}}$. Anticipating the use of the loop method, we redraw the circuit as in Figure 16.22(b), where the units are [V-s], [A-s], and [Ω]. Applying KVL around the loop yields

$$1 = \frac{5}{s} + \left(2.4 + 0.4s + \frac{10}{s}\right)I$$

or

$$I = \frac{2.5(s-5)}{s^2 + 6s + 25}$$

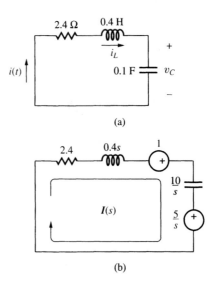

(a)

(b)

Figure 16.22 Circuits of Example 16.29.

Its inverse Laplace transform is

$$i_{\text{natural}} = i(t > 0) = \sqrt{31.25}e^{-3t}\cos{(4t + 63.43°)} \text{ A}$$

◄

▶**Example 16.30**

Assuming the op amp of Figure 16.23 is ideal, find $v_O(t)$ if
(a) $v_I = 10^{-3}\delta(t)$ V and (b) $v_I = 10u(t)$ V.

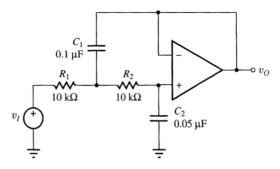

Figure 16.23 Circuit of Example 16.30.

Solution

(a) We readily find that the transfer function is

$$H(s) = \frac{2 \times 10^6}{s^2 + 2 \times 10^3 s + 2 \times 10^6}$$

This function has the pole pair $p_{1,2} = (-1 \pm j1)10^3$ complex Np/s. The response to an impulse with a strength of 10^{-3} V is thus

$$v_O(t) = \mathcal{L}^{-1}\{H(s)10^{-3}\} = 2e^{-10^3 t}(\sin 10^3 t)u(t) \text{ V}$$

(b) The response to a 10-V step is

$$v_O(t) = \mathcal{L}^{-1}\left\{H(s)\frac{10}{s}\right\} = 10[1+\sqrt{2}e^{-10^3 t}\sin(10^3 t - 135°)]u(t) \text{ V} \blacktriangleleft$$

Exercise 16.33   In the circuit of Figure 16.23 let $R_1 = R_2 = 10$ kΩ and $C_1 = C_2 = 10$ nF. The source v_I, after having been 5 V for a long time, is switched to 10 V at $t = 0$. Find $v_O(t)$.

ANSWER $v_O(t \le 0) = 5$ V, $v_O(t > 0) = 5[2 - (10^4 t + 1)e^{-10^4 t}]$ V.

16.7 CONVOLUTION

Laplace techniques transform a time-domain problem to a frequency-domain one. If the circuit under scrutiny is in its *zero state,* its *s*-domain response is found as

$$Y(s) = H(s)X(s)$$

where $H(s)$ is the network function of the circuit, and $X(s)$ and $Y(s)$ are, respectively, the Laplace transforms of the applied signal and the response. The *t*-domain response, which is the *forced response,* is then found as $y(t) = \mathcal{L}^{-1}\{Y(s)\}$, or

$$y(t) = \mathcal{L}^{-1}\{H(s)X(s)\}$$

There are situations in which a circuit must be analyzed entirely in the time domain. A common example is when the applied signal $x(t)$ or the impulse response $h(t)$ of the circuit are known only through experimental data and may thus lack explicit Laplace transforms. Exploiting the fact that $\mathcal{L}^{-1}\{H(s)X(s)\} = h(t)*x(t)$, these situations are more easily handled via the convolution operation, which is exclusively a time-domain operation,

$$y(t) = h(t) * x(t) = \int_{0^-}^{t} h(\xi)x(t - \xi)\, d\xi \qquad \text{(16.84a)}$$

Since the convolution operation is *commutative,* the response can also be found as

$$y(t) = x(t) * h(t) = \int_{0^-}^{t} x(\xi)h(t - \xi)\, d\xi \qquad \text{(16.84b)}$$

Convolution is pictured in Figure 16.24. We again stress that the response provided by this method is the *zero-state* or *forced response,* that is, the response for the case of zero initial stored energy in the reactive elements of the circuit.

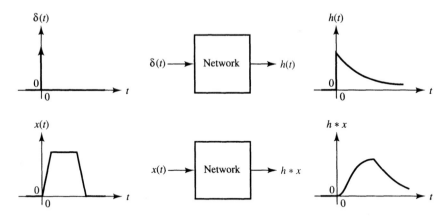

Figure 16.24 Once we know the response $h(t)$ to the impulse $\delta(t)$, the response $y(t)$ to an arbitrary signal $x(t)$ is found by convolving $h(t)$ and $x(t)$.

Graphical Convolution

We can better appreciate the convolution of two functions $f(t)$ and $g(t)$ via the graphical sequence of Figure 16.25. Its steps are:

(1) Plot g versus the dummy time-integration variable ξ, as shown in the example of Figure 16.25(a).

(2) *Fold* g about the vertical axis by changing its argument from ξ to $-\xi$, as in Figure 16.25(b). Whatever value $g(\xi)$ achieves at ξ, $g(-\xi)$ achieves at $-\xi$, since $g[-(-\xi)] = g(\xi)$.

(3) *Slide* g along the ξ axis by augmenting its argument by t, as in Figure 16.25(c). Whatever value $g(-\xi)$ achieves at $\xi = 0$, $g(t - \xi)$ achieves at $t - \xi = 0$, or $\xi = t$. Clearly, $t > 0$ implies a shift toward the *right,* and $t < 0$ a shift toward the *left.*

(4) Plot f versus the dummy time-integration variable ξ, as shown in the example of Figure 16.25(d).

(5) Take the *product* $f(\xi)g(t - \xi)$ and plot it versus ξ, point by point, as shown in Figure 16.25(e). The *area* under this curve represents the value of $f(t) * g(t)$ for the given t.

(6) To find the profile of $f(t) * g(t)$ versus t, calculate the areas of Step 5 for *different* values of t, starting with $t = 0$ and gradually increasing t, and then plot them point by point. The result is shown in Figure 16.25(f).

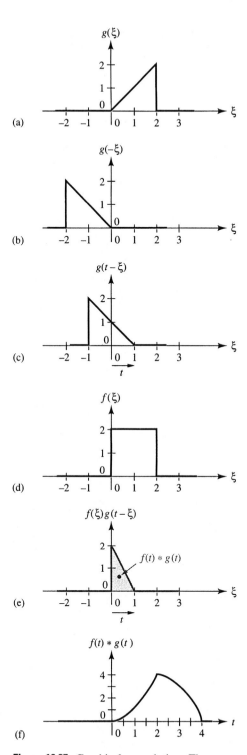

Figure 16.25 Graphical convolution. The curve of (f) represents the convolution of the functions of (a) and (d).

In this illustration we have kept $f(\xi)$ fixed while folding and sliding $g(\xi)$. By the commutativity property, we obtain the same result by folding and sliding $f(\xi)$ with $g(\xi)$ kept fixed.

▶ Example 16.31

Use Equation (16.84b) to find the response of the RC circuit of Figure 16.26 to a unity pulse of width W.

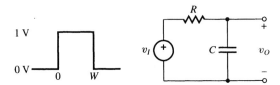

Figure 16.26 Finding the pulse response of an RC circuit.

Solution

The applied input and the impulse response are

$$v_I(t) = u(t) - u(t - W) \text{ V} \tag{16.85}$$

$$h(t) = \frac{1}{\tau} e^{-t/\tau} \text{ V} \tag{16.86}$$

where $\tau = RC$. Applying Equation (16.84b), we get

$$v_O(t) = v_I(t) * h(t) = \int_O^t v_I(\xi) h(t - \xi) \, d\xi \tag{16.87}$$

Folding and sliding h, and then plotting $v_I(\xi) = u(\xi) - u(\xi - W)$ and $h(t - \xi) = (1/\tau)e^{-(t-\xi)/\tau}$ versus ξ for different values of t yields the sequence of Figure 16.27. We distinguish three cases:

(1) $t \leq 0$. As shown in Figure 16.27(a), there is no overlapping between $h(t - \xi)$ and $v_I(\xi)$. The area under their product is thus zero, so

$$v_O(t \leq 0) = 0 \tag{16.88a}$$

This is not surprising, as $v_I(t)$ and $h(t)$ are causal signals.

(2) $0 \leq t \leq W$. The area of interest is the shaded area of Figure 16.27(b), whose value is

$$v_O(0 \leq t \leq W) = \int_0^t 1 \times \frac{1}{\tau} e^{-(t-\xi)/\tau} \, d\xi = \frac{1}{\tau} e^{-t/\tau} \left(\tau e^{\xi/\tau} \Big|_0^t \right)$$

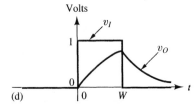

Figure 16.27 Graphical convolution for Example 16.31.

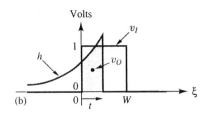

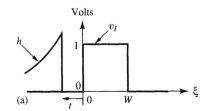

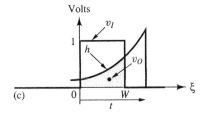

or

$$v_O(0 \leq t \leq W) = 1 - e^{-t/\tau} \text{ V} \qquad \textbf{(16.88b)}$$

(3) $t \geq W$. The shaded area of Figure 16.27(c) is now

$$v_O(t \geq W) = \int_0^W 1 \times \frac{1}{\tau} e^{-(t-\xi)/\tau} \, d\xi = \frac{1}{\tau} e^{-t/\tau} \left(\tau e^{\xi/\tau} \Big|_0^W \right)$$

or

$$v_O(t \geq W) = (e^{W/\tau} - 1)e^{-t/\tau} \text{ V} \qquad \textbf{(16.88c)}$$

As expected, these results agree with those obtained via a different procedure in Section 8.3, Equation (8.22).

◀

Exercise 16.34   Repeat Example 16.31, but with R and C interchanged with each other.

ANSWER $v_O(t \leq 0) = 0$; $v_O(0 \leq t \leq W) = e^{-t/\tau}$ V; $v_O(t \geq W) = (e^{-W/\tau} - 1)e^{-(t-W)/\tau}$ V.

▶ **Example 16.32**

Repeat Example 16.31, but using Equation (16.84a). Clearly, the result must be the same.

Solution

We now have

$$v_O(t) = h(t) * v_I(t) = \int_0^t h(\xi)v_I(t - \xi) \, d\xi \qquad \textbf{(16.89)}$$

Since $h(t)$ and $v_I(t)$ are causal signals, we must again have

$$v_O(t \leq 0) = 0 \qquad \textbf{(16.90a)}$$

The two cases to consider are thus those of Figure 16.28. Referring to Figure 16.28(a), which shows the shaded area for $0 \leq t \leq W$, we have

$$v_O(0 \leq t \leq W) = \int_0^t \frac{1}{\tau} e^{-\xi/\tau} \times 1 \, d\xi = \frac{1}{\tau} \left(-\tau e^{-\xi/\tau} \Big|_0^t \right)$$

or

$$v_O(0 \leq t \leq W) = 1 - e^{-t/\tau} \text{ V} \qquad \textbf{(16.90b)}$$

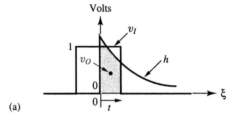

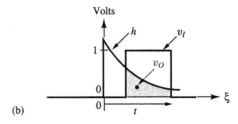

Figure 16.28 Alternate graphical convolution for Example 16.31.

Referring to Figure 16.28(b), which shows the shaded area for $t \geq W$, and noting that the pulse begins at $\xi = t - W$, we have

$$v_O(t \geq W) = \int_{t-W}^{t} \frac{1}{\tau} e^{-\xi/\tau} \times 1 \, d\xi = \frac{1}{\tau} \left(-\tau e^{-\xi/\tau} \Big|_{t-W}^{t} \right)$$

or

$$v_O(t \geq W) = (e^{W/\tau} - 1) e^{-t/\tau} \text{ V} \qquad \textbf{(16.90c)}$$

As expected, Equation (16.90) agrees with Equation (16.88).   ◀

Exercise 16.35   Solve Exercise 16.34 using Equation (16.84a).

Exercise 16.36

(a) Derive expressions for $f(t) * g(t)$ of Figure 16.25(f).

(b) Repeat, but using the form $g(t) * f(t)$.

ANSWER $y(t) = f(t) * g(t) = g(t) * f(t) = 0$ for $t \leq 0$ and $t \geq 4$, $y(t) = t^2$ for $0 \leq t \leq 2$, and $y(t) = 4t - t^2$ for $2 \leq t \leq 4$.

• Numerical Convolution

As indicated previously, the convolution method is especially useful when $h(t)$ and $x(t)$ are available in *tabulated form* as, for instance, in the case of laboratory measurements,

$$h(nT) = \{h(0T), h(1T), h(2T), h(3T), \ldots\} \qquad \textbf{(16.91a)}$$

$$x(nT) = \{x(0T), x(1T), x(2T), x(3T), \ldots\} \qquad \textbf{(16.91b)}$$

where T represents the time interval between consecutive data. Such a situation is depicted in Figure 16.29(a) and (b), where the data points have been numbered sequentially for easy identification.

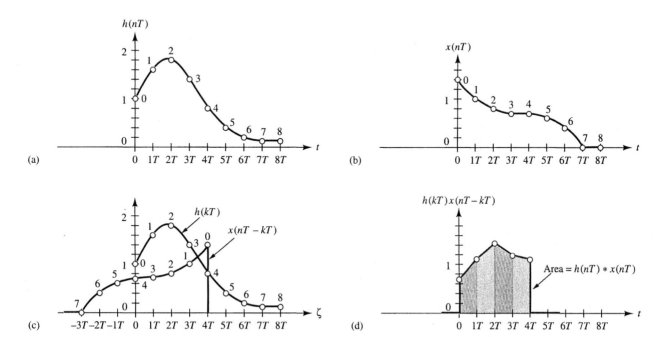

Figure 16.29 Numerical convolution.

Proceeding as usual, we fold and slide one of the two functions, say, x, to obtain the function $x(nT - kT)$. Figure 16.29(c) depicts the case $n = 4$. Next, we take the *product* $h(kT)x(nT - kT)$ point by point and plot it as in Figure 16.29(d). The area under this curve represents the convolution

$$y(nT) = h(nT) * x(nT) \qquad \textbf{(16.92)}$$

We wish to find and tabulate it for different values of nT,

$$y(nT) = \{y(0T), y(1T), y(2T), y(3T), \ldots\} \qquad \textbf{(16.93)}$$

The area under the curve $h(kT)x(nT - kT)$ can be approximated by summing the areas of the individual trapezoids. For the case $n = 4$ depicted in the figure, these areas are $0.5T[h(0T)x(4T) + h(1T)x(3T)]$, $0.5T[h(1T)x(3T) + h(2T)x(2T)]$, $0.5T[h(2T)x(2T) + h(3T)x(1T)]$, and $0.5T[h(3T)x(1T) + h(4T)x(0T)]$. Adding them up and combining equal terms yields

$$y(4T) = T\left[\frac{1}{2}h(0T)x(4T) + h(1T)x(3T) + h(2T)x(2T) + h(3T)x(1T)\right.$$

$$\left. + \frac{1}{2}h(4T)x(0T)\right]$$

This result is readily generalized to an arbitrary instant nT as

$$y(nT) = T \left\{ \frac{1}{2}h(0T)x(nT) + h(1T)x[(n-1)T] + h(2T)x[(n-2)T] \right.$$

$$\left. + \cdots + h[(n-1)T]x(1) + \frac{1}{2}h(nT)x(0T) \right\} \qquad \textbf{(16.94)}$$

Note that the h terms appear in *ascending* order and the x terms in *descending* order. The interested reader may find it instructive to write a computer program that accepts the data tables of Equation (16.91) and then uses Equation (16.94) to generate and plot the table of Equation (16.93).

 ## 16.8 LAPLACE TECHNIQUES USING PSPICE

The ability of PSpice to accommodate VCVSs and VCCSs of the *transfer function* type was exploited in Section 14.8 to plot frequency responses. Here we use it to plot impulse and step responses.

 ## ▶Example 16.33

Use PSpice to verify the impulse response of Example 16.20, where $H(s) = 50(s+5)/[(s+3)^2 + 16]$.

Solution

Use the circuit of Figure 14.41 with a pulse source having a width of 1 ms and an amplitude of 1 kV to ensure unity area.

```
IMPULSE RESPONSE OF H(S)
VS 1 0 PULSE(0 1K 0.1 1N 1N 1M 2)
R1 1 0 1
E1 2 0 LAPLACE {V(1,0)} = {50*(S+5)/((S+3)*(S+3)+16)}
R2 2 0 1
.TRAN 0.1 2
.PROBE
.END
```

After directing the Probe post-processor to plot V(1) and V(2) we obtain the traces of Figure 16.30. ◀

Exercise 16.37   Use PSpice to plot both the impulse and the step response of Example 16.21.

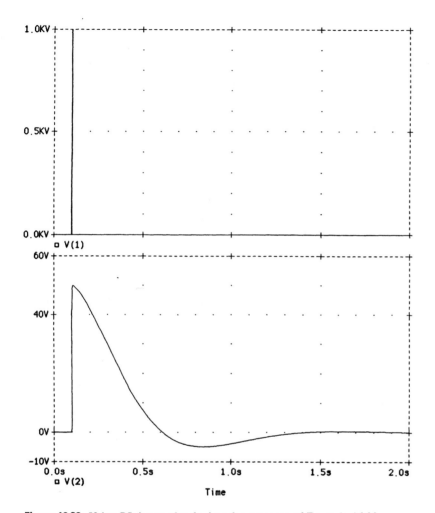

Figure 16.30 Using PSpice to plot the impulse response of Example 16.20.

▼ SUMMARY

- The *Laplace transform* effects a transformation from a *causal* function of time $f(t)$ to a function of the complex frequency $F(s)$. The *inverse Laplace transform* effects a transformation from $F(s)$ back to $f(t)$. The two functions are said to form a *transform pair,* and this is denoted as $f(t) \leftrightarrow F(s)$.

- The distinguishing features of the Laplace transform are that (a) it transforms an *integrodifferential* equation into an *algebraic* equation and (b) it automatically accounts for the *initial conditions,* thus providing the solution to the equation *explicitly.*

- The *step function* $u(t)$ and the *impulse function* $\delta(t)$ are related as $\delta(t) = du(t)/dt$. An impulse can be regarded as a pulse with *infinite* amplitude, *zero* width, and *unity* area.

- Subjecting a circuit to a *step function* causes it to display its response to a *unit dc forcing function.* Subjecting it to an *impulse* function injects energy into the circuit and causes it to display its *natural response.*

- Some important *transform pairs* are $\delta(t) \leftrightarrow 1$, $u(t) \leftrightarrow 1/s$, $tu(t) \leftrightarrow 1/s^2$, $e^{-at}u(t) \leftrightarrow 1/(s+a)$, $(\cos \omega t)u(t) \leftrightarrow s/(s^2+\omega^2)$, $(e^{-at}\cos \omega t)u(t) \leftrightarrow (s+a)/[(s+a)^2+\omega^2]$, and $te^{-at}u(t) \leftrightarrow 1/(s+a)^2$.

- Important *functional transforms* are $df/dt \leftrightarrow sF(s) - f(0^-)$, $\int_{0^-}^{t} f(\xi)\,d\xi \leftrightarrow F(s)/s$, $f(t-a)u(t-a) \leftrightarrow e^{-as}F(s)$, $f(at) \leftrightarrow (1/a)F(s/a)$, and $f(t) * g(t) \leftrightarrow F(s)G(s)$, where $*$ represents the operation of *convolution.*

- The *initial* and *final values* of a function $f(t)$ can be found from its transform $F(s)$ as $f(0^+) = \lim_{s \to \infty} sF(s)$, and $f(\infty) = \lim_{s \to 0} sF(s)$, provided these limits exist.

- The Laplace transforms arising in circuit analysis are *rational functions* of s. This feature allows us to find inverse transforms via *partial fraction expansions.* The poles p_1, p_2, $\ldots$, p_n and the residues A_1, A_2, $\ldots$, A_n of $F(s)$ determine the terms of $f(t)$.

- If $F(s)$ has a *real* nonrepeated pole at $s = p_k$, then $f(t)$ contains the term $f_k(t) = A_k e^{p_k t}$, where A_k is the residue at $s = p_k$.

- If $F(s)$ has a nonrepeated *complex-conjugate* pole pair at $s = \alpha_k \pm j\beta_k$, then $f(t)$ contains the term $f_k(t) = 2|A_k|e^{\alpha_k t}\cos(\beta_k t + \sphericalangle A_k)$, where A_k is the residue at $s = \alpha_k + j\beta_k$.

- If $F(s)$ has a *repeated real* pole with *multiplicity* r, then the corresponding term of $f(t)$ takes on the form $f_k(t) = (A_{k,r-1}t^{r-1}/(r-1)! + \cdots + A_{k,1})e^{p_k t}$. Similar considerations hold for the case of *repeated complex pole pairs.*

- The response of a circuit to a forcing signal $x(t)$ is $y(t) = y_{\text{forced}} + y_{\text{natural}}$. These components can be found as $y_{\text{forced}} = \mathcal{L}^{-1}\{H(s)X(s)\}$, and $y_{\text{natural}} = \mathcal{L}^{-1}\{Y_{\text{ic}}(s)\}$, where $H(s)$ is the *network function*, $X(s)$ is the Laplace transform of $x(t)$, and $Y_{\text{ic}}(s)$ is a suitable *initial-conditions* polynomial in s.

- The *impulse response* of a circuit is found from its network function as $h(t) = \mathcal{L}^{-1}\{H(s)\}$.

- The Laplace forms of the terminal characteristics of the basic circuit elements are $V(s) = RI(s)$, $V(s) = sLI(s) - Li_L(0^-)$, and $V(s) = (1/sC)I(s) + v_C(0^-)/s$. This allows us to work with a transformed circuit in which the initial conditions in its reactive elements are modeled via suitable dc sources.

- When $H(s)$ or $X(s)$ are not available, the forced response can be found via exclusively time-domain operations as $y_{\text{forced}} = x(t) * h(t)$.

- Given a transfer function $H(s)$, we can direct PSpice to plot the *frequency* as well as the *impulse* and *step* responses.

▼ PROBLEMS

16.1 The Step and Impulse Functions

16.1 Let the R-C circuit of Figure 16.3 be subjected to an impulse and a step function *simultaneously,* that is, let the applied source be $v_S = \delta(t) + u(t)$ V. Find $v(t)$ if $C = 10$ μF and (a) $R = 50$ kΩ, (b) $R = 200$ kΩ, and (c) $R = 100$ kΩ.

16.2 (a) Sketch and label the impulse response $v(t)$ for the C-R circuit of Figure of P16.2. (b) Outline a procedure for observing this response experimentally if $C = 1$ μF and $R = 1$ kΩ.

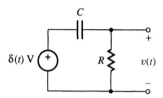

Figure P16.2

16.3 Repeat Problem 16.2 for the circuit of Figure P16.3. Assume $L = 25$ mH and $R = 100$ Ω.

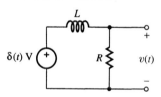

Figure P16.3

16.4 For the circuit of Figure P16.4, sketch and label $v(t)$ if (a) $i_S = 1u(t)$ A and (b) $i_S = 1\delta(t)$ A.

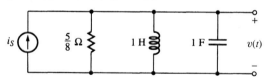

Figure P16.4

16.5 Sketch and label the following functions versus time: (a) $f_1 = 5u(t-1)u(t)u(3-t)$, (b) $f_2 = 2(\sin \pi t)\delta(4t-1)$, (c) $f_3 = 10e^{-(t-1)/2}u(t-2)$, and (d) $f_4 = (t-1)u(t-1) - (t-2)u(t-2) - (t-3)u(t-3) + (t-4)u(t-4)$.

16.2 The Laplace Transform

16.6 Using direct integration, find (a) $\mathcal{L}\{(\sinh at)u(t)\}$ and (b) $\mathcal{L}\{[\sin(\omega t + \phi)]u(t)\}$. What is the abscissa of convergence of each transform?

16.7 Using direct integration, find (a) $\mathcal{L}\{5\delta(t-1)\}$, (b) $\mathcal{L}\{5u(t-1)\}$, (c) $\mathcal{L}\{\delta(t^2-4t+3)\}$, and (d) $\mathcal{L}\{(t-2) \times u(t-3)\}$.

16.8 Using direct integration, find the Laplace transform as well as the abscissa of convergence of (a) $f(t) = e^{-3t}u(t-1)$ and (b) $f(t) = te^{-3t}u(t-1)$.

16.9 Find the Laplace transforms of the functions of Problem 16.5.

16.10 Find $f(t)$, given that $F(s)$ is (a) $-\pi$, (b) $-e^{-s}$, (c) $1/(0.1 + s)$, (d) $[2/(2 + s)]^2$, (e) $1/(0.5s^2 + s + 1)$, (f) $(2/s)^5$, and (g) $F(s)$ has a pole pair at $-3 \pm j4$ complex Np/s, and $F(1) = 1$.

16.3 Operational Transforms

16.11 Find the Laplace transforms of (a) $5(2t^2 - t + 1) \times e^{-3t}u(t)$, (b) $(\sin^2 10^3 t)u(t)$, and (c) $(\sin 10t \cos 10t)u(t)$.

16.12 Find $\mathcal{L}\{e^{-at} \sin(\omega t + \phi)u(t)\}$, $a > 0$.

16.13 (a) Find $\mathcal{L}\{f(10t)\}$ if $\mathcal{L}\{f(t)\} = 1/(s^4 + 1)$; (b) find $f(t)$ and $\mathcal{L}\{tf(t)\}$ if $\mathcal{L}\{f(t)\} = 1/(s + 1)^3$; (c) find $f(10^3 t)$ if $\mathcal{L}\{f(t)\} = 1/(s^2 + 2s + 2)$.

16.14 (a) Use two methods to find $\mathcal{L}\{f(t)u(t)\}$, where $f(t) = d(e^{-at} \cos \omega t)/dt$. (b) Repeat, but for $f(t) = \int_{0^-}^{t} e^{-a\xi} \sin \omega \xi \, d\xi$.

16.15 Find the Laplace transform of the function of Figure P16.15 by two methods: (a) using $df(t)/dt$; (b) using $d^2 f(t)/dt^2$.

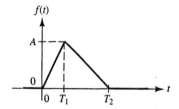

Figure P16.15

16.16 The Laplace transform of the function of Figure P16.16 can be put in the form $F(s) = F_1(s) + F_2(s)e^{-sT} + F_3(s)e^{-2sT}$, where F_1, F_2, and F_3 are rational functions of s. Find these functions.

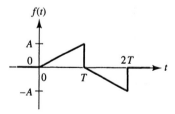

Figure P16.16

16.17 Use convolution to find $f(t)$ if $F(s) = 10/[(s + 1) \times (s + 3)]$.

16.18 Use convolution to find $f(t)$ if (a) $F(s) = 10/[(s+1)^2(s+3)]$ and (b) $F(s) = 2/s^3$.

16.19 Find the Laplace transform of the *triangular wave* of Figure P16.19.

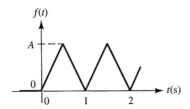

$f(t)$

Figure P16.19

16.20 Find the Laplace transform of the *half-wave rectified* sine wave of Figure P16.20.

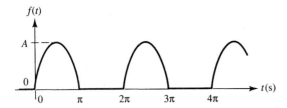

$f(t)$

Figure P16.20

16.21 Find the Laplace transform of the *sawtooth wave* of Figure P16.21, where $0 < d < 1$.

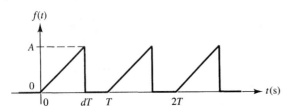

$f(t)$

Figure P16.21

16.22 (a) Show that if f is a function of time t as well as some other parameter a, then

$$\mathscr{L}\left\{\frac{\partial f(t,a)}{\partial a}\right\} = \frac{\partial F(s,a)}{\partial a}$$

(b) Apply this property to (1) $f(t,a) = e^{-at}u(t)$ and (2) $f(t,\omega) = (\cos\omega t)u(t)$. (c) Check your results by some other method.

16.23 Find $F(s)$ if $f(t) = [1 + \sqrt{2}e^{-t}\cos(t - 135°)]u(t)$. Check your results via the initial-value and final-value theorems.

16.24 Find $f(t)$ if $F(s) = (s-1)/(s^2+2s+5)+5/(s+2)+2/s$, and check via the initial-value and final-value theorems.

16.25 Given a causal function $f(t)$ with

$$\mathscr{L}\{f(t)\} = \frac{5(s+2)(s+3)}{s(s+1)(s+4)^2}$$

find the initial and final values of $f(t)$ and $df(t)/dt$.

16.4 The Inverse Laplace Transform

16.26 Find $f(t)$ if

(a) $F(s) = \dfrac{2s+10}{s^2+3s+2}$

(b) $F(s) = \dfrac{3}{s(s+1)(s+3)}$

(c) $F(s) = \dfrac{10s+1}{s^2+2s+1}$

16.27 Find $f(t)$ if

(a) $F(s) = \dfrac{s}{s^2+2s+5}$

(b) $F(s) = 6\dfrac{s^2+3}{(s^2+1)(s^2+4)}$

(c) $F(s) = \dfrac{5}{s(s+1)^4}$

16.28 Find $f(t)$ if

(a) $F(s) = \dfrac{12(s+1)^3}{s^4}$

(b) $F(s) = 10\dfrac{s^2+8s+5}{s^2+4s+5}$

(c) $F(s) = \dfrac{1}{s+1} + \dfrac{s-1}{s^2+2s+2}$

16.29 Find $f(t)$ if

(a) $F(s) = \dfrac{2s^2+s+1}{(s^2+2s+2)^2}$

(b) $F(s) = \dfrac{s^2+25s+150}{2(s+10)}$

(c) $F(s) = \dfrac{3s^2+2}{s+4}$

16.30 Find $f(t)$ if

(a) $F(s) = \dfrac{e^{-s}}{s+1}$

(b) $F(s) = \dfrac{e^{-2s}}{s^2+2s+1}$

(c) $F(s) = \dfrac{e^{-s} - e^{-2s}}{s^2}$

16.31   Find $f(t)$ if

(a) $F(s) = \dfrac{1 - e^{-s}}{s + 2}$

(b) $F(s) = \dfrac{se^{-s}}{s^2 + 3s + 2}$

(c) $F(s) = \dfrac{1 - e^{-\pi s}}{s^2 + 4}$

16.5 Application to Differential Equations

16.32   Use Laplace transforms to solve the differential equations

(a) $\dfrac{dy(t)}{dt} + 2y(t) = 3e^{-t}u(t)$, $\dfrac{dy(0^-)}{dt} = 1$

(b) $\dfrac{dy(t)}{dt} + y(t) = (e^{-t}\cos 2t)u(t)$, $y(0^-) = -1$

16.33   Use Laplace transforms to solve the differential equation

$$\frac{d^2 y(t)}{dt^2} + 7\frac{dy(t)}{dt} + 12y(t) = (t + 4)u(t),$$

$$y(0^-) = -\frac{dy(0^-)}{dt} = 1$$

Show both the natural and forced components of $y(t)$.

16.34   Use Laplace transforms to solve the differential equation

$$\frac{d^2 y(t)}{dt^2} + 4\frac{dy(t)}{dt} + 3y(t) = 5(\sin 2t)u(t),$$

$$y(0^-) = -\frac{dy(0^-)}{dt} = 1$$

Show both the natural and forced components of $y(t)$.

16.35   Assuming all initial conditions are zero, use Laplace transforms to solve the differential equation

$$\frac{d^3 y(t)}{dt^3} + 5\frac{d^2 y(t)}{dt^2} + 6\frac{dy(t)}{dt} = 6e^{-t}u(t)$$

16.36   Use Laplace transforms to solve the integrodifferential equations

(a) $y(t) + 3\displaystyle\int_{0^-}^{t} y(\xi)\,d\xi = 10e^{-2t}u(t)$, $y(0^-) = 0$

(b) $\dfrac{dy(t)}{dt} + 2y(t) + \displaystyle\int_{0^-}^{t} y(\xi)u(t - \xi)\,d\xi = 5u(t)$, $y(0^-) = 1$

16.37   Use Laplace transforms to solve the integrodifferential equations

(a) $\dfrac{dy(t)}{dt} + 9\displaystyle\int_{0^-}^{t} y(\xi)\,d\xi = 8(\sin t)u(t)$, $y(0^-) = 2$

(b) $\dfrac{dy(t)}{dt} + 2\displaystyle\int_{0^-}^{t} y(\xi)e^{-3(t-\xi)}\,d\xi = 2tu(t)$, $y(0^-) = 5$

16.38   A circuit has the network function $H(s) = 1/(s^2 + 2s + 2)$. Find its response $y(t)$ to $x(t) = u(t)$ if the initial conditions are (a) $y(0^-) = y^{(1)}(0^-) = 0$ and (b) $y(0^-) = -y^{(1)}(0^-) = 1$.

16.39   Use Laplace transforms to solve the simultaneous equations

$$\frac{dy_1(t)}{dt} + 2y_1(t) - 2y_2(t) = e^{-4t}$$

$$y_1(t) - \frac{dy_2(t)}{dt} - y_2(t) = 0$$

subject to the initial conditions $y_1(0^-) = -3$ and $y_2(0^-) = 0$.

16.40   Use Laplace transforms to solve the simultaneous equations

$$\frac{d^2 y_1(t)}{dt^2} - 2\frac{dy_2(t)}{dt} + 36y_1(t) = 3$$

$$3\frac{dy_1(t)}{dt} + 2y_2(t) = e^{-2t}$$

subject to the initial conditions $y_1(0^-) = 0$, $y_1^{(1)}(0^-) = -2$, and $y_2(0^-) = 1$.

16.41   Find the impulse and step responses of a circuit whose network function is $H(s) = 6(s + 2)/[(s + 1)(s + 3)]$.

16.42   Find the impulse and step responses of a circuit, given that its network function has a pole at the origin and a pole pair at $(-4 \pm j3)10^3$ complex Np/s; moreover, $H(1\text{ kNp/s}) = 1/34$.

16.43   Given that the impulse response of a circuit is $h(t) = (10e^{-10^3 t}\sin 10^3 t)u(t)$, find $H(s)$; hence, predict its response to the ramp $tu(t)$.

16.44   Given that the impulse response of a circuit is $h(t) = 10^6 te^{-10^3 t}$, find its step response subject to the initial conditions $y(0^-) = 1$, and $y^{(1)}(0^-) = 10^3$.

16.6 Application to Circuit Analysis

16.45   Use Laplace transforms to find $v(t > 0)$ in the circuit of Figure P16.45.

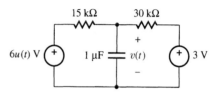

Figure P16.45

16.46   Repeat Problem 16.45, but with the capacitance replaced by a 10-mH inductance.

16.47   Use Laplace transforms to find $v(t)$ in the circuit of Figure P16.47.

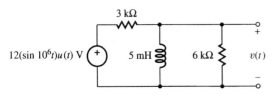

Figure P16.47

16.48   Find $v(t)$ in the circuit of Figure P16.48 if (a) $v_S = 5\delta(t)$ V, (b) $v_S = 2u(t)$ V, and (c) $v_S = 10tu(t)$ V.

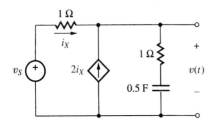

Figure P16.48

16.49   Use Laplace transforms to find $v(t > 0)$ in the circuit of Figure P16.49, subject to the initial conditions $v_C(0^-) = 10$ V, and $i_L(0^-) = 1$ A.

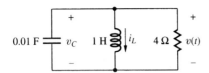

Figure P16.49

16.50   Use Laplace transforms to find $i(t > 0)$ in the circuit of Figure P16.50. Assume the circuit is in steady state at $t = 0^-$.

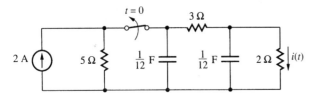

Figure P16.50

16.51   Repeat Problem 16.50 (a) with the leftmost capacitance replaced by a 0.3-H inductance, and (b) with both capacitances replaced by 1-H inductances.

16.52   In the circuit of Figure P16.52 find $v(t)$ via Laplace methods if (a) $v_S = 5\delta(t)$ V, (b) $v_S = 10u(t)$ V, and (c) $v_S = 2tu(t)$ V.

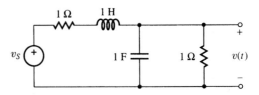

Figure P16.52

16.53   Use Laplace transforms to find $v(t > 0)$ in the circuit of Figure P16.52 if v_S, after having been 10 V for a long time, is switched to 5 V at $t = 0$.

16.54   Assuming the circuit of Figure P16.54 is in steady-state at $t = 0^-$, use Laplace transforms to find $i(t > 0)$.

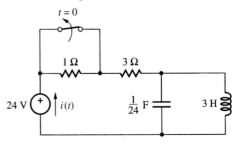

Figure P16.54

16.55   In the circuit of Figure P16.55 use Laplace transforms to find $v(t > 0)$ if $i_S = 2u(t)$ A.

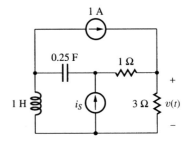

Figure P16.55

16.56   Suppose the 1-A source in the circuit of Figure P16.55 is replaced by a 6-V source, positive at the left. Use Laplace transforms to find $v(t > 0)$ if $i_S = (3\sin t)u(t)$ A.

16.57   (a) Find R and k_i so that the impulse response of the circuit of Figure P16.57 is a 1-rad/s sinusoid with a 1-Np/s damping coefficient. Hence, find the response to (b) $v_S = 5u(t)$ V, and (c) $v_S = 10te^{-t}u(t)$ V.

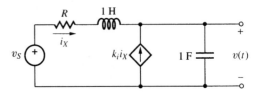

Figure P16.57

16.58 Use Laplace transforms to find $v(t)$ in the circuit of Figure P16.58.

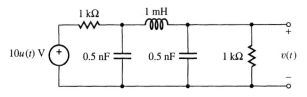

Figure P16.58

16.59 Assuming the circuit of Figure P16.59 is in steady state at $t = 0^-$, find $v(t > 0)$.

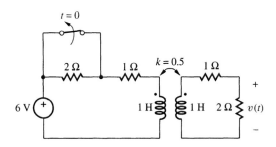

Figure P16.59

16.60 Use Laplace transforms to find $v(t)$ in the circuit of Figure P16.60.

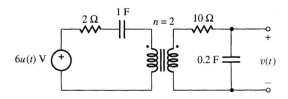

Figure P16.60

16.61 Find the energy trapped in the circuit of Figure P16.61 at $t = 0^+$. (*Hint:* Use the initial-value theorem.)

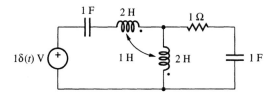

Figure P16.61

16.62 Use Laplace transforms to find $v(t)$ in the circuit of Figure P16.62, subject to the condition $v_C(0^-) = 4$ V.

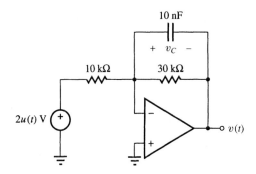

Figure P16.62

16.63 Use Laplace transforms to find $v_O(t)$ in the circuit of Figure P16.63 if $v_I = 2(\sin 5000t)u(t)$ V.

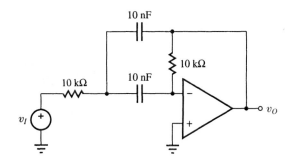

Figure P16.63

16.64 Find $v_O(t > 0)$ in the circuit of Figure P16.63 if v_I, after having been 1 V for a long time, is switched off at $t = 0$.

16.7 Convolution

16.65 Given $f(t) = 5tu(t)$ and $g(t) = 10u(t)$, find $F(s)$, $G(s)$, and $\mathscr{L}^{-1}\{F(s)G(s)\}$. Confirm the latter by performing the time-domain calculation of $f(t) * g(t)$.

16.66 Perform the graphical convolution of the signals shown in Figure P16.66 and plot the result.

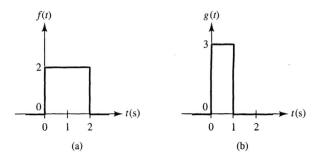

Figure P16.66

16.67   Repeat Problem 16.66 for the signals of Figure P16.67.

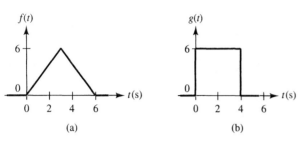

(a)   (b)

Figure P16.67

16.68   In the circuit of Figure P16.68(b), use convolution to find $v_O(1\,\text{s})$ and $v_O(4\,\text{s})$ if $v_I(t)$ is the signal of Figure P16.68(a).

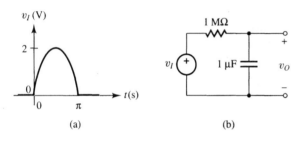

(a)   (b)

Figure P16.68

16.69   In the circuit of Figure P16.69(b), use convolution to find $v_O(t)$ if $v_I(t)$ is the signal of Figure P16.69(a).

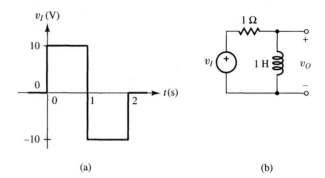

(a)   (b)

Figure P16.69

16.70   In the circuit of Figure P16.70(b), use convolution to find $v_O(t)$ if $v_I(t)$ is the signal of Figure P16.70(a).

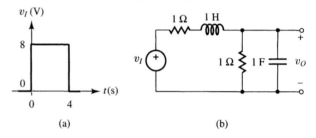

(a)   (b)

Figure P16.70

⑤ 16.8 Laplace Techniques Using PSpice

16.71   Use PSpice to check Problem 16.41.

16.72   Use PSpice to check Problem 16.49.

16.73   Use PSpice to check Problem 16.56.

16.72   Use PSpice to check Problem 16.70.

FOURIER ANALYSIS TECHNIQUES

A s we know, a periodic function is a function that satisfies the property

$$f(t) = f(t \pm nT)$$

for any t and $n = 1, 2, 3, \ldots$, where T is the period, in *seconds*. The most common periodic functions are $\sin \omega t$, $\cos \omega t$, and $e^{j\omega t}$, which are the building blocks of ac signals and complex exponential signals.

There are, however, many other periodic functions of sufficient practical importance to warrant our attention. The first examples to come to our mind are the *triangle, rectangle, sawtooth,* and *pulse-train* waves produced by function generators in the laboratory for testing and control. Also periodic is the waveform produced by the *sweep generator* controlling the electron beam of a cathode-ray tube of the type found in oscilloscopes and television sets. The *clock generator* controlling the sequence of operations in personal computers, pocket calculators, and wristwatches produces a periodic pulse train. Electrical power, generated in approximately sinusoidal form at the source, is subsequently processed by nonlinear devices such as *rectifiers* and practical *transformers* and *motors,* which distort the original wave, either intentionally or unintentionally, to yield nonsinusoidal waveforms that are nevertheless still periodic. Other examples of nearly periodic signals are the output of a microphone in response to a sustained *musical note* and the heart signal displayed by an *electrocardiograph.*

Periodic functions are important also in nonelectrical phenomena such as *mechanical vibrations, fluid flow,* and *heat flow.* In fact, it was precisely the investigation of heat flow in a metal rod that led the French mathematician Jean Baptiste J. Fourier to the extraordinary discovery that the periodic functions of physical interest can be expressed as *sums* of basic *sinusoidal components* known as *harmonic components,* each having a suitable *frequency, amplitude,* and *phase angle.*

This conclusion has two important implications. First, it provides an indication of how the power associated with a periodic signal distributes among its harmonic components; such a distribution is visualized via diagrams known as *power spectra* and is of fundamental importance in signal communication and processing. Second, it allows us to find the response to *any* periodic signal, regardless of its waveform. To this end, we first find the responses to its individual sinusoidal components; then, using the superposition principle, we add up these responses to find the overall response to the nonsinusoidal signal of interest.

The first half of the chapter focuses upon different but equivalent forms of the **Fourier series** and the effect of functional symmetries upon its parameters. We then use the concepts of frequency response and input and output spectra to study the effect of a circuit upon a periodic applied signal.

In the second half of the chapter we pursue a heuristic development of the **Fourier transform** in order to extend to *aperiodic* signals the frequency-domain representation provided by the Fourier series for *periodic* signals. Though the concept of transformation from the time domain to the frequency domain has already been encountered in connection with the Laplace transform, the Fourier transform offers a physical interpretation that is not immediately apparent from the Laplace transform. Specifically, the Fourier transform provides information about the *energy density* of a signal. This interpretation makes the Fourier transform a powerful tool in communication theory and signal processing, where the primary focus is on the frequency characteristics of signals and systems.

17.1 THE FOURIER SERIES

In his studies of heat flow, Fourier discovered that a periodic function $f(t)$ having **period** T and, hence, **angular frequency**

$$\omega_0 = \frac{2\pi}{T} \tag{17.1}$$

can be expressed as an infinite summation of basic sinusoidal components. Though there are various ways of expressing this summation, we shall focus upon the **amplitude-phase** Fourier series representation of $f(t)$, also called the **cosine** Fourier series,

$$f(t) = A_0 + \sum_{n=1}^{\infty} A_n \cos(n\omega_0 t + \theta_n) \tag{17.2}$$

where

> A_0 is the **full-cycle average** of $f(t)$
>
> A_n, $n\omega_0$, and θ_n are, respectively, the **amplitude, angular frequency,** and **phase angle** of the nth sinusoidal component.

The frequency ω_0 is called the **fundamental frequency,** and the corresponding component, $A_1 \cos(\omega_0 t + \theta_1)$, is called the **fundamental component.** The frequency $n\omega_0$, $n = 1, 2, 3, \ldots$, is referred to as the nth **harmonic frequency,** and the corresponding component, $A_n \cos(n\omega_0 + \theta_n)$, as the nth **harmonic component.** Its period is clearly $2\pi/(n\omega_0) = T/n$. In musical parlance the frequencies $n\omega_0$, $n > 1$, are referred to as *overtones*.

The parameters of Equation (17.2) depend on the particular function $f(t)$. They can be found *analytically* using Equations (17.5) and (17.6) to be discussed below, or they can be found *numerically* using the SPICE facility known as the .FOUR statement, discussed in Section 17.8. If $f(t)$ is a physical signal, they can also be found *experimentally* using an instrument known as *spectrum analyzer*.

Not all periodic functions admit a Fourier series. Mathematicians have shown that a *sufficient* set of conditions for a function $f(t)$ to admit a convergent Fourier series are the conditions known as *Dirichlet's conditions*:

(1) $f(t)$ must be *single valued.*

(2) $f(t)$ must have a *finite number of discontinuities* within any one period.

(3) $f(t)$ must have a *finite* number of *maxima and minima* within any one period.

(4) The *integral* $\int_{t_0}^{t_0+T} |f(t)|\, dt$ *must exist* for any choice of t_0.

Fortunately, the periodic functions of physical interest do satisfy Dirichlet's conditions and are thus Fourier-series expandable.

The Fourier Coefficients

The calculation of the amplitudes and phase angles of the individual components in Equation (17.2) is made easier if we work with the alternate representation, called the **cosine-sine** Fourier series,

$$f(t) = a_0 + \sum_{n=1}^{\infty} a_n \cos n\omega_0 t + b_n \sin n\omega_0 t \qquad \textbf{(17.3)}$$

The quantities a_n and b_n are known as the **Fourier coefficients.** Using the identity $\cos(\alpha + \beta) = \cos\alpha \cos\beta - \sin\alpha \sin\beta$, we have

$$A_n \cos(n\omega_0 t + \theta_n) = (A_n \cos\theta_n)\cos n\omega_0 t - (A_n \sin\theta_n)\sin n\omega_0 t$$

indicating that the amplitude-phase and the cosine-sine representations of

Equations (17.2) and (17.3) are equivalent, provided

$$a_0 = A_0 \qquad \textbf{(17.4a)}$$
$$a_n = A_n \cos \theta_n \qquad \textbf{(17.4b)}$$
$$b_n = -A_n \sin \theta_n \qquad \textbf{(17.4c)}$$

or, vice versa, provided

$$A_0 = a_0 \qquad \textbf{(17.5a)}$$
$$A_n = \sqrt{a_n^2 + b_n^2} \qquad \textbf{(17.5b)}$$
$$\theta_n = \tan^{-1}\left(-b_n/a_n\right) \qquad \textbf{(17.5c)}$$

Thus, once one of the two Fourier series representations is known, the other is readily found via Equations (17.4) or (17.5).

We now wish to show that once $f(t)$ has been defined over its fundamental period ranging from an arbitrarily chosen instant t_0 to $t_0 + T$, its Fourier coefficients can be found as

$$a_0 = \frac{1}{T} \int_{t_0}^{t_0+T} f(t)\, dt \qquad \textbf{(17.6a)}$$
$$a_n = \frac{2}{T} \int_{t_0}^{t_0+T} f(t) \cos n\omega_0 t\, dt \qquad \textbf{(17.6b)}$$
$$b_n = \frac{2}{T} \int_{t_0}^{t_0+T} f(t) \sin n\omega_0 t\, dt \qquad \textbf{(17.6c)}$$

To derive Equation (17.6a) we integrate both sides of Equation (17.3) from t_0 to $t_0 + T$. After simple manipulations, we obtain

$$\int_{t_0}^{t_0+T} f(t)\, dt = a_0 \int_{t_0}^{t_0+T} dt$$
$$+ \sum_{n=1}^{\infty} \left(a_n \int_{t_0}^{t_0+T} \cos n\omega_0 t\, dt + b_n \int_{t_0}^{t_0+T} \sin n\omega_n t\, dt \right)$$

The first integral at the right-hand side evaluates to $a_0 T$. In the remaining integrals, the integrands are sinusoidal functions with period T/n. Since the integration interval is T, which is an integral multiple of T/n, these integrals evaluate to zero, thus leaving $\int_{t_0}^{t_0+T} f(t)\, dt = a_0 T + 0$. Solving for a_0 yields Equation (17.6a). Clearly, a_0 represents the *full-wave average* or the *dc value* of $f(t)$.

To derive Equation (17.6b) we multiply both sides of Equation (17.3) by $\cos m\omega_0 t$ and then integrate throughout from t_0 to $t_0 + T$,

$$\int_{t_0}^{t_0+T} f(t) \cos m\omega_0 t \, dt = a_0 \int_{t_0}^{t_0+T} \cos m\omega_0 t \, dt$$

$$+ \sum_{n=1}^{\infty} \left(a_n \int_{t_0}^{t_0+T} \cos n\omega_0 t \cos m\omega_0 t \, dt + b_n \int_{t_0}^{t_0+T} \sin n\omega_0 t \cos m\omega_0 t \, dt \right)$$

The first integral on the right-hand side evaluates to zero. Using the trigonometric identities $\cos \alpha \cos \beta = \frac{1}{2}[\cos(\alpha + \beta) + \cos(\alpha - \beta)]$, and $\sin \alpha \cos \beta = \frac{1}{2}[\sin(\alpha + \beta) + \sin(\alpha - \beta)]$, we can express the nth term in the summation as

$$\frac{a_n}{2} \left[\int_{t_0}^{t_0+T} \cos(n+m)\omega_0 t \, dt + \int_{t_0}^{t_0+T} \cos(n-m)\omega_0 t \, dt \right]$$

$$+ \frac{b_n}{2} \left[\int_{t_0}^{t_0+T} \sin(n+m)\omega_0 t \, dt + \int_{t_0}^{t_0+T} \sin(n-m)\omega_0 t \, dt \right]$$

For $m \neq n$, all integrals evaluate to zero. For $m = n$, the second integral evaluates to $\int_{t_0}^{t_0+T} \cos 0 \, dt = T$ and the remaining integrals evaluate again to zero. Thus, for $m = n$, we obtain $\int_{t_0}^{t_0+T} f(t) \cos n\omega_0 t \, dt = (a_n/2)T$, or Equation (17.6b).

It is left as an exercise for the reader (see Problem 17.1) to derive Equation (17.6c). Let us now find the Fourier coefficients of some popular waveforms.

▶ Example 17.1

Find the cosine-sine as well as the amplitude-phase Fourier series for the *sawtooth* wave of Figure 17.1.

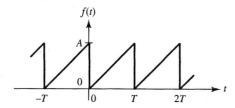

Figure 17.1 Sawtooth wave of Example 17.1.

Solution

Conveniently choosing $t_0 = 0$ we have, for $0 \leq t \leq T$,

$$f(t) = \frac{A}{T} t$$

Using Equation (17.6a), we get

$$a_0 = \frac{1}{T} \int_0^T \frac{A}{T} t \, dt = \frac{A}{2}$$

Using Equation (17.6b), along with the integral tables, we find

$$a_n = \frac{2}{T} \int_0^T \frac{A}{T} t \cos n\omega_0 t \, dt = \frac{2A}{T^2} \left(\frac{\cos n\omega_0 t}{(n\omega_0)^2} + \frac{t \sin n\omega_0 t}{n\omega_0} \Big|_0^T \right)$$

$$= \frac{2A}{T^2} \left(\frac{\cos 2n\pi - 1}{(n\omega_0)^2} + \frac{T \sin 2n\pi - 0}{n\omega_0} \right) = 0$$

indicating that the cosine terms are missing for this particular waveform. Using Equation (17.6c), along with the integral tables, we get

$$b_n = \frac{2}{T} \int_0^T \frac{A}{T} t \sin n\omega_0 t \, dt = \frac{2A}{T^2} \left(\frac{\sin n\omega_0 t}{(n\omega_0)^2} - \frac{t \cos n\omega_0 t}{n\omega_0} \Big|_0^T \right)$$

$$= \frac{2A}{T^2} \left(\frac{\sin 2n\pi - 0}{(n\omega_0)^2} - \frac{T \cos 2n\pi - 0}{n\omega_0} \right) = \frac{2A}{T^2} \left(-\frac{T}{n\omega_0} \right) = -\frac{A}{n\pi}$$

Thus, the cosine-sine and the amplitude-phase series are, respectively,

$$f(t) = A \left(\frac{1}{2} - \frac{1}{\pi} \sum_{n=1}^{\infty} \frac{1}{n} \sin n\omega_0 t \right) \qquad \textbf{(17.7a)}$$

$$f(t) = A \left(\frac{1}{2} + \frac{1}{\pi} \sum_{n=1}^{\infty} \frac{1}{n} \cos (n\omega_0 t + 90°) \right) \qquad \textbf{(17.7b)}$$

indicating that for the *sawtooth* wave the amplitude of each component is *inversely proportional* to its harmonic number n. Moreover, all components have the same phase angle.   ◀

▶ Example 17.2

Find the cosine-sine as well as the amplitude-phase Fourier representations for the *half-wave rectified sinewave* of Figure 17.2.

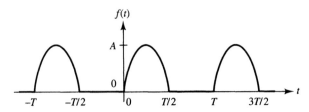

Figure 17.2 Half-wave rectified sinewave of Example 17.2.

Solution

We conveniently choose $t_0 = 0$. However, since $f(t) = 0$ for $T/2 \le t \le T$, we only integrate from 0 to $T/2$, where we have

$$f(t) = A \sin \frac{2\pi}{T} t = A \sin \omega_0 t$$

Using Equation (17.6a), we get

$$a_0 = \frac{1}{T} \int_0^{T/2} A \sin \frac{2\pi}{T} t \, dt = \frac{A}{T} \left(-\frac{T}{2\pi} \right) \cos \frac{2\pi}{T} t \Big|_0^{T/2} = \frac{A}{\pi}$$

Using Equation (17.6b), along with the integral tables, we get

$$a_n = \frac{2}{T} \int_0^{T/2} A \sin \omega_0 t \cos n\omega_0 t \, dt$$

$$= -\frac{A}{T} \left(\frac{\cos(1-n)\omega_0 t}{(1-n)\omega_0} + \frac{\cos(1+n)\omega_0 t}{(1+n)\omega_0} \right) \Big|_0^{T/2}$$

$$= -\frac{A}{2\pi} \left(\frac{\cos(1-n)\pi - 1}{1-n} + \frac{\cos(1+n)\pi - 1}{1+n} \right)$$

Since

$$\cos(1-n)\pi = \cos(1+n)\pi = \begin{cases} +1 & \text{for } n \text{ odd} \\ -1 & \text{for } n \text{ even} \end{cases}$$

it follows that

$$a_n = 0 \quad \text{for } n \text{ odd } (n > 1)$$

$$a_n = -\frac{A}{2\pi} \left(\frac{-2}{1+n} + \frac{-2}{1-n} \right) = \frac{2A}{\pi} \frac{1}{1-n^2} \quad \text{for } n \text{ even}$$

For $n = 1$ we obtain the indeterminate result $a_1 = 0/0$. However, applying l'Hôpital's rule we get $a_1 = 0$. We can represent even integers by replacing n with $2k$ so that

$$a_{2k} = \frac{2A}{\pi} \frac{1}{1 - 4k^2}$$

$k = 1, 2, 3, \ldots$

Using Equation (17.6c), along with the integral tables, we get

$$b_n = \frac{2}{T} \int_0^{T/2} A \sin \omega_0 t \sin n\omega_0 t \, dt$$

$$= \frac{A}{T} \left(\frac{\sin(1-n)\omega_0 t}{(1-n)\omega_0} - \frac{\sin(1+n)\omega_0 t}{(1+n)\omega_0} \right) \Big|_0^{T/2}$$

$$= \frac{A}{2\pi} \left(\frac{\sin(1-n)\pi}{1-n} - \frac{\sin(1+n)\pi}{1+n} \right) = 0$$

for all $n > 1$. For $n = 1$ we obtain the indeterminate result $b_1 = 0/0$. However, applying l'Hôpital's rule we get $b_1 = A/2$. In summary, the coefficients are

$$a_0 = A/\pi$$

$$b_1 = A/2$$

$$a_2 = -2A/3\pi$$

$$a_4 = -2A/15\pi$$

$$a_6 = -2A/35\pi$$

and so forth. Thus, the cosine-sine and the amplitude-phase series are, respectively,

$$f(t) = A \left(\frac{1}{\pi} + \frac{1}{2} \sin \omega_0 t - \frac{2}{\pi} \sum_{k=1}^{\infty} \frac{1}{4k^2 - 1} \cos 2k\omega_0 t \right) \quad \textbf{(17.8a)}$$

$$f(t) = A \left[\frac{1}{\pi} + \frac{1}{2} \cos(\omega_0 t - 90°) \right.$$

$$\left. + \frac{2}{\pi} \sum_{k=1}^{\infty} \frac{1}{4k^2 - 1} \cos(2k\omega_0 t - 180°) \right] \quad \textbf{(17.8b)} \blacktriangleleft$$

Exercise 17.1 Show that the cosine-sine and the amplitude-phase Fourier series for the *square* wave of Figure 17.3 are, respectively,

$$f(t) = A \left(\frac{1}{2} + \frac{2}{\pi} \sum_{k=1}^{\infty} \frac{1}{2k - 1} \sin(2k - 1)\omega_0 t \right) \quad \textbf{(17.9a)}$$

$$f(t) = A \left(\frac{1}{2} + \frac{2}{\pi} \sum_{k=1}^{\infty} \frac{1}{2k - 1} \cos[(2k - 1)\omega_0 t - 90°] \right) \quad \textbf{(17.9b)}$$

indicating that the *square* wave has only *odd* harmonic components. The amplitude of each component is *inversely proportional* to its harmonic number, and all components have the same phase angle.

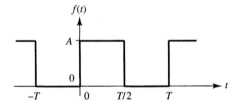

Figure 17.3 Square wave of Exercise 17.1.

Exercise 17.2 Show that the Fourier series for the *full-wave rectified sinewave* of Figure 17.4 is

$$f(t) = A \left(\frac{2}{\pi} + \frac{4}{\pi} \sum_{n=1}^{\infty} \frac{1}{1 - 4n^2} \cos n\omega_0 t \right) \quad \textbf{(17.10)}$$

indicating that for this particular function the expressions for the cosine-sine and the amplitude-phase series are coincident, aside from a phase reversal for $n \geq 1$.

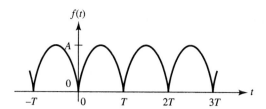

Figure 17.4 Full-wave rectified sinewave of Exercise 17.2.

It is certainly remarkable that nonsinusoidal waves such as the sawtooth and square waves can be synthesized by summing sinusoids. To better appreciate how each harmonic component contributes to the overall makeup of a function, let us try synthesizing the square wave. As a first approximation we might start with a sinusoid having the same frequency as the square wave. This component is the fundamental, and is shown as curve 1 in Figure 17.5(a). To

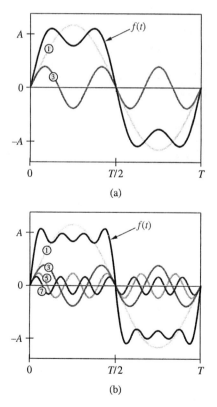

(a)

(b)

Figure 17.5 Illustrating the Fourier synthesis of a square wave using (a) harmonics 1 and 3, and (b) harmonics 1, 3, 5, and 7.

make it look more like a square wave, we need to flatten the peak and bolster the sides within each half-cycle. This is achieved by adding the third harmonic, shown as curve 3 in the same figure. Though the result is a closer approximation to the square wave, it is still a crude one, and to improve it we must add higher harmonics. How do we choose their individual amplitudes, phase angles, and frequencies? The answer is provided by Equation (17.9a), which, adapted to the present case of a square wave alternating between $+A$ and $-A$, becomes

$$f(t) = \frac{4A}{\pi} \left(\sin \omega_0 t + \frac{1}{3} \sin 3\omega_0 t + \frac{1}{5} \sin 5\omega_0 t + \frac{1}{7} \sin 7\omega_0 t + \cdots \right)$$

Figure 17.5(b) shows the improvement brought about by including just two additional harmonics, shown as curves 5 and 7. It is apparent that by adding a sufficient number of harmonics we can approximate a periodic wave to any degree we wish. To achieve a *perfect* square wave we would, of course, need an *infinite* number of terms.

We wish to point out that the Fourier series is not just a mathematical convenience. All periodic functions encountered in the physical world do indeed consist of sums of sinusoidal components, and their harmonic makeup can actually be visualized and measured with a spectrum analyzer.

The Power of a Periodic Signal

Applying a periodic signal $f(t)$ to a resistance R results in dissipation of power. The instantaneous power is either $p(t) = Rf^2(t)$ or $p(t) = f^2(t)/R$, depending on whether $f(t)$ is a current or a voltage. To avoid specifying the nature of the signal, it is convenient to choose a *unity resistance,* after which the instantaneous power simplifies as $p_{1\,\Omega}(t) = f^2(t)$. The *average power* delivered to a 1-Ω resistance is then

$$P_{1\Omega} = \frac{1}{T} \int_{t_0}^{t_0+T} f^2(t)\, dt \qquad \textbf{(17.11)}$$

Using Equation (17.2), we have

$$P_{1\Omega} = \frac{1}{T} \int_{t_0}^{t_0+T} \left(A_0 + \sum_{n=1}^{\infty} A_n \cos(n\omega_0 t + \theta_n) \right)^2 dt$$

As we known, products involving different harmonics integrate to zero. The only terms surviving the integration over one period are A_0^2, $A_1^2 \cos^2(\omega_0 t + \theta_1)$, $A_2^2 \cos^2(2\omega_0 t + \theta_2)$, $A_3^2 \cos^2(3\omega_0 t + \theta_3)$, and so forth. The result is

$$P_{1\Omega} = \frac{1}{T} \left(A_0^2 T + \sum_{n=1}^{\infty} A_n^2 \frac{T}{2} \right)$$

or

$$P_{1\Omega} = A_0^2 + \sum_{n=1}^{\infty} (A_n/\sqrt{2})^2 \qquad \textbf{(17.12)}$$

This equation, known as **Parseval's theorem** for period signals and named for the French mathematician Marc-Antoine Parseval-Deschenes (1755–1836), states that the average power dissipated by a signal into a 1-Ω resistance is *the sum of the squares of its dc value and the rms values of its harmonic components.*

In signal communication theory $P_{1\,\Omega}$ is referred to as the **power of a periodic signal.** Equation (17.12) indicates that this power depends only on the amplitudes of the harmonic components, regardless of their phase angles. Consequently, changing the phase angles has no effect on power, though it may significantly alter the shape of the wave. In signal communication the manner in which the power of a signal distributes among its harmonic components is far more important than its actual waveshape, and Equation (17.12) provides precisely this kind of information.

Once we know $P_{1\,\Omega}$, we can find F_{rms}, the *rms value* of $f(t)$, as

$$F_{\text{rms}} = \sqrt{P_{1\,\Omega}} \qquad\qquad \textbf{(17.13)}$$

This alternative to the closed-form definition of rms value of Section 4.4 emphasizes how the harmonic makeup of a signal contributes to its rms value.

▶ **Example 17.3**

Use Equation (17.12) with a sufficient number of terms to estimate the rms value of the half-wave rectified sinewave of Figure 17.2.

Solution

Using Equation (17.8), we can write

$$F_{\text{rms}} = \left[\left(\frac{A}{\pi} \right)^2 + \frac{1}{2}\left(\frac{A}{2} \right)^2 + \frac{1}{2}\left(\frac{2A}{3\pi} \right)^2 + \frac{1}{2}\left(\frac{2A}{15\pi} \right)^2 \right.$$
$$\left. + \frac{1}{2}\left(\frac{2A}{35\pi} \right)^2 + \cdots \right]^{1/2} = \frac{A}{2} \qquad\qquad ◀$$

Exercise 17.3   Find the terms needed in Equation (17.12) to predict the true rms value of the *sawtooth wave* of Figure 17.1 with 1% accuracy.

ANSWER A_0 through A_7.

Mean Square Error

Though the Fourier series consists of an *infinite* number of terms, when used in practice it must, of necessity, be truncated to a *finite* number N of terms.

Denoting the *truncated series* as

$$f_N(t) = A_0 + \sum_{n=1}^{N} A_n \cos(n\omega_0 t + \theta_n)$$

we define the *truncation error* as

$$\varepsilon_N(t) = f(t) - f_N(t)$$

where $f(t) = \lim_{N \to \infty} f_N(t)$. A convenient means for assessing how well $f_N(t)$ approximates $f(t)$ over an entire period is offered by the *mean square error,* defined as

$$E_N^2 = \frac{1}{T} \int_{t_0}^{t_0+T} \varepsilon_N^2(t)\, dt$$

The integration sums the time-dependent error $\varepsilon_N(t)$ over an entire period, and the squaring operation assigns more weight to the larger errors, while simultaneously preventing positive and negative errors from cancelling each other out.

It can be proved mathematically that E_N^2 is *minimized* when the parameters A_0, A_n, and θ_n in the expression for $f_N(t)$ are precisely the Fourier parameters predicted by Equations (17.5) and (17.6). This statement, whose proof is beyond our scope, reassures us that the Fourier series representation of a periodic signal is a valuable one even when used in truncated form.

17.2 THE EFFECT OF SHIFTING AND SYMMETRY

Shifting a periodic function $f(t)$ along either the vertical or the horizontal axis will generally affect its Fourier coefficients. The coefficients are also affected by the presence of certain types of functional symmetries in $f(t)$. Of particular interest are *even-function, odd-function,* and *half-period* symmetry because some coefficient sets vanish altogether, expediting our calculations.

Shifting

We wish to investigate how shifting a periodic function $f(t)$ along either the vertical or the horizontal axis affects its amplitude-phase Fourier series. It is apparent that shifting $f(t)$ up or down the vertical axis changes only the *average value* A_0, leaving the amplitude A_n and phase angle θ_n of each harmonic component unaffected.

To investigate the effect of shifting a periodic function $f(t)$ by a time t_0, we simply replace t by $t + t_0$ in Equation (17.2),

$$f(t + t_0) = A_0 + \sum_{n=1}^{\infty} A_n \cos[n\omega_0(t + t_0) + \theta_n]$$

Since the value achieved by $f(t)$ at $t = 0$ is the same as that achieved by $f(t + t_0)$ at $t + t_0 = 0$, or $t = -t_0$, it follows that the shift is toward the *left* if $t_0 > 0$, and toward the *right* if $t_0 < 0$. Regrouping the argument of the cosine function yields

$$f(t + t_0) = A_0 + \sum_{n=1}^{\infty} A_n \cos(n\omega_0 t + \theta_n + \phi_n)$$ **(17.14a)**

$$\phi_n = n\omega_0 t_0$$ **(17.14b)**

indicating that shifting a periodic function in time changes only the *phase angle* of each harmonic component, leaving the average value A_0, as well as the amplitude A_n and frequency $n\omega_0$ of each harmonic component unaffected. The phase shift ϕ_n of each component is directly proportional to its harmonic number n.

We are especially interested in the case in which the shift of $f(t)$ is expressed as a *fraction α* of its period T,

$$t_0 = \alpha T$$ **(17.15)**

Substituting in Equation (17.14b), the *phase shift* of the nth component becomes $\phi_n = n\omega_0 t_0 = n\omega_0 \alpha T$, or

$$\phi_n = n\alpha 2\pi$$ **(17.16)**

radians, or $\phi_n = n\alpha 360°$.

▶ Example 17.4

Find the Fourier series for the *square wave* of Figure 17.6.

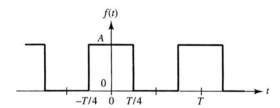

Figure 17.6 Square wave of Example 17.4.

Solution

Since this function can be obtained by shifting the function of Figure 17.3 toward the *left* by $\frac{1}{4}T$, we can reuse Equation (17.9b), but with the phase angle of each component *increased* by $\phi_{2k-1} = \frac{1}{4}(2k-1)360° = (2k-1)90°$. The new phase angle is thus $-90° + (2k-1)90° = (k-1)180°$. Since $\cos[\beta + (k-1)180°] = \cos\beta\cos(k-1)180° - \sin\beta\sin(k-1)180° = \cos\beta \times (-1)^{k-1}$, Equation (17.9b) becomes

$$f(t) = A\left(\frac{1}{2} + \frac{2}{\pi}\sum_{k=1}^{\infty} \frac{(-1)^{k-1}}{2k-1}\cos(2k-1)\omega_0 t\right)$$ **(17.17)**

Clearly, in this case the expressions for the cosine-sine and the amplitude-phase series are coincident, aside from a phase reversal for k even. ◀

Exercise 17.4 Show that the Fourier series for the *sawtooth wave* of Figure 17.7 is

$$f(t) = \frac{2A}{\pi} \sum_{n=1}^{\infty} \frac{(-1)^{n-1}}{n} \sin n\omega_0 t \qquad (17.18)$$

(*Hint:* Use Equation (17.7) as the basis for your derivations.)

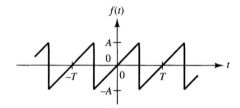

Figure 17.7 Sawtooth wave of Exercise 17.4.

Even and Odd Functions

A function $f(t)$ is said to be an **even function** if

$$f(t) = f(-t) \qquad (17.19)$$

Examples of even functions are $\cos \omega t$, $|t|$, t^2, and a constant A. Replacing t with $-t$ does not change the value of any of these functions. Graphically, $f(t)$ is even if the portion of $f(t)$ for $t < 0$ is the *mirror image* of the portion of $f(t)$ for $t > 0$. An example of an even waveform is that of Figure 17.6. Even functions are also said to be *symmetrical with respect to the y axis.*

A function $f(t)$ is said to be an **odd function** if

$$f(t) = -f(-t) \qquad (17.20)$$

Examples of odd functions are $\sin \omega t$, t, t^3, and the function $\text{sgn}\,(t)$, defined as $\text{sgn}\,(t) = 1$ for $t > 0$, and $\text{sgn}\,(t) = -1$ for $t < 0$. Replacing t with $-t$ yields the *negative* value of any of these functions. Graphically, $f(t)$ is odd if the function obtained by *rotating* $f(t)$ through by $180°$ using the origin as pivot coincides with $f(t)$ itself. An example of an odd waveform is that of Figure 17.7. Odd functions are also said to be *skew-symmetric about the origin.*

A distinguishing feature of an even function $f_{\text{even}}(t)$ is

$$\int_{-t_0}^{t_0} f_{\text{even}}(t)\,dt = 2 \int_{0}^{t_0} f_{\text{even}}(t)\,dt \qquad (17.21)$$

This property stems from the fact that the area under $f_{\text{even}}(t)$ from $-t_0$ to 0 is the *same* as that from 0 to t_0. By contrast, the area under an odd function $f_{\text{odd}}(t)$

from $-t_0$ to 0 is the *negative* of that from 0 to t_0. Hence, the sum of the two is zero,

$$\int_{-t_0}^{t_0} f_{\text{odd}}(t)\, dt = 0 \tag{17.22}$$

The *product* of two *even* functions or of two *odd* functions is *even*. This is readily seen, as

$$f_{\text{even}}(-t) \times g_{\text{even}}(-t) = f_{\text{even}}(t) \times g_{\text{even}}(t)$$

$$f_{\text{odd}}(-t) \times g_{\text{odd}}(-t) = [-f_{\text{odd}}(t)][-g_{\text{odd}}(t)] = f_{\text{odd}}(t) \times g_{\text{odd}}(t)$$

Hence, these products satisfy Equation (17.21). By contrast, the *product* of an *even* function and an *odd* function is *odd*,

$$f_{\text{even}}(-t) \times g_{\text{odd}}(-t) = [f_{\text{even}}(t)][-g_{\text{odd}}(t)] = -[f_{\text{even}}(t) \times g_{\text{odd}}(t)]$$

Hence, this product satisfies Equation (17.22).

When evaluating the Fourier coefficients of a periodic function $f(t)$, we integrate $f(t)$ as well as the products $f(t)\cos n\omega_0 t$ and $f(t)\sin n\omega_0 t$. Since $\cos n\omega_0 t$ is *even* and $\sin n\omega_0 t$ is *odd*, it follows that for an *even* function $f_{\text{even}}(t)$ we have

$$a_0 = \frac{2}{T} \int_0^{T/2} f_{\text{even}}(t)\, dt \tag{17.23a}$$

$$a_n = \frac{4}{T} \int_0^{T/2} f_{\text{even}}(t)\cos n\omega_0 t\, dt \tag{17.23b}$$

for $n = 1, 2, 3, \ldots$, and

$$b_n = 0 \tag{17.23c}$$

for $n = 1, 2, 3, \ldots$. Likewise, for an *odd* function $f_{\text{odd}}(t)$ we have

$$a_n = 0 \tag{17.24a}$$

for $n = 0, 1, 2, \ldots$, and

$$b_n = \frac{4}{T} \int_0^{T/2} f_{\text{odd}}(t)\sin n\omega_0 t\, dt \tag{17.24b}$$

for $n = 1, 2, 3, \ldots$. In either case, one coefficient set is *zero* and the other set is obtained by taking *twice* the integral over *half* the period. Moreover, the *average* or *dc value* of an *odd* function is always *zero*.

We observe that shifting a periodic function along the t axis generally alters its symmetry, so it is sometimes possible to establish even or odd symmetry by suitable choice of the origin of time and thus exploit the computational advantages accruing from such a symmetry. As an example, the square wave of

Figure 17.3 is neither even nor odd; however, shifting it leftward by $T/4$ yields the wave of Figure 17.6, which is even. Its Fourier series of Equation (17.17) confirms that this function has $b_n = 0$.

Likewise, the sawtooth wave of Figure 17.1 is neither even nor odd; however, shifting it leftward by $T/2$, and then downward by $A/2$ yields the odd function of Figure 17.7. Its Fourier series of Equation (17.18) confirms that this function has $a_n = 0$.

Half-Wave Symmetry

The symmetries just discussed are also said to be *full-wave* symmetries. Another important type of symmetry is the *half-wave symmetry,* also called *odd half-wave symmetry.* A periodic function possesses **half-wave symmetry** (hws) if

$$f(t) = -f\left(t - \frac{T}{2}\right) \tag{17.25}$$

Shifting such a function by *half a period* and then *inverting* it yields again the original function. Examples of half-symmetric functions are the waveforms shown in Figures 17.8 and 17.9. Unlike even and odd symmetry, half-wave symmetry is not affected by the choice of the origin of time.

We now wish to show that a function with half-wave symmetry contains only *odd* harmonics and that it has *zero* average or dc value. To this end, let $t_0 = -T/2$ in Equation (17.6). Then,

$$a_n = \frac{2}{T} \int_{-T/2}^{T/2} f_{\text{hws}}(t) \cos n\omega_0 t \, dt$$

$$= \frac{2}{T} \left(\int_{-T/2}^{0} f_{\text{hws}}(t) \cos n\omega_0 t \, dt + \int_{0}^{T/2} f_{\text{hws}}(t) \cos n\omega_0 t \, dt \right)$$

Let us now change the variable in the first integral from t to $t - T/2$, so that Equation (17.25) can be used. Exploiting the fact that the lower and upper limits of integration are now 0 and $T/2$, and that $d(t - T/2) = dt$, we can write

$$a_n = \frac{2}{T} \left(\int_{0}^{T/2} -f_{\text{hws}}(t) \cos n\omega_0(t - T/2) \, dt + \int_{0}^{T/2} f_{\text{hws}}(t) \cos n\omega_0 t \, dt \right)$$

Since $\cos n\omega_0(t - T/2) = \cos(n\omega_0 t - n\pi) = (-1)^n \cos n\omega_0 t$, we can write

$$a_n = \frac{2}{T} \int_{0}^{T/2} f_{\text{hws}}(t)[1 - (-1)^n] \cos n\omega_0 t \, dt$$

This expression indicates that for a function with half-wave symmetry we have

$$\boxed{a_n = 0} \tag{17.26a}$$

for *n* even, and

$$\boxed{a_n = \frac{4}{T} \int_{0}^{T/2} f_{\text{hws}}(t) \cos n\omega_0 t \, dt} \tag{17.26b}$$

for n odd. Equation (17.26a) holds also for $n = 0$, indicating a *zero* average or dc value. By a similar procedure, it is easy to show that for a function with half-wave symmetry we also have

$$b_n = 0 \qquad \textbf{(17.27a)}$$

for n even and

$$b_n = \frac{4}{T} \int_0^{T/2} f_{\text{hws}}(t) \sin n\omega_0 t \, dt \qquad \textbf{(17.27b)}$$

for n odd. These properties can be exploited to expedite our calculations, as we are now ready to demonstrate.

▶ **Example 17.5**

Find the Fourier series for the *triangle* wave of Figure 17.8.

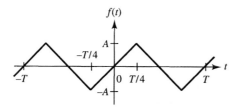

Figure 17.8 Triangle waveform of Example 17.5.

Solution

By inspection, this is an odd function, so $a_n = 0$ for all n. It is also half-wave symmetric, so $b_n = 0$ for n even. Thus, we only need to find b_n for n odd. Even though Equation (17.27b) specifies $T/2$ as the upper integration limit, we shall integrate only up to $T/4$ and double the result. It is readily seen that for $0 \leq t \leq T/4$ we have $f(t) = 4At/T$. Thus, for n odd,

$$b_n = \frac{8}{T} \int_0^{T/4} \frac{4A}{T} t \sin n\omega_0 t \, dt = \frac{32A}{T^2} \times \frac{\sin n\omega_0 t - n\omega_0 t \cos n\omega_0 t}{n^2 \omega_0^2} \Bigg|_0^{T/4}$$

$$= \frac{8A}{n^2 \pi^2} \left(\sin \frac{n\pi}{2} - \frac{n\pi}{2} \cos \frac{n\pi}{2} \right) = \frac{8A}{n^2 \pi^2} \sin \frac{n\pi}{2}$$

where we have exploited the fact that $\cos n\pi/2 = 0$ for n odd. To represent odd numbers, we replace n with $2k - 1$. Then, using the fact that $\sin[(2k-1)\pi/2] = -\cos k\pi = (-1)^{k-1}$, we obtain

$$f(t) = \frac{8A}{\pi^2} \sum_{k=1}^{\infty} \frac{(-1)^{k-1}}{(2k-1)^2} \sin(2k-1)\omega_0 t \qquad \textbf{(17.28)}$$

The first few terms are

$$f(t) = \frac{8A}{\pi^2} \left(\sin \omega_0 t - \frac{1}{9} \sin 3\omega_0 t + \frac{1}{25} \sin 5\omega_0 t - \frac{1}{49} \sin 7\omega_0 t + \cdots \right) \blacktriangleleft$$

Exercise 17.5   Show that the Fourier series for the waveform of Figure 17.9 is

$$f(t) = \frac{2A}{\pi} \sum_{k=1}^{\infty} \frac{-2/\pi}{(2k-1)^2} \cos(2k-1)\omega_0 t + \frac{1}{2k-1} \sin(2k-1)\omega_0 t$$

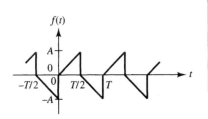

$f(t)$

Figure 17.9 Waveform of Exercise 17.5.

Effect of Discontinuities

We conclude with two useful observations whose proof, however, we shall omit for brevity.

(1) If $f(t)$ has a discontinuity at $t = a$, then the Fourier series converges to the *average value,*

$$f(a) = \frac{f(a^-) + f(a^+)}{2}$$

where a^- and a^+ are, respectively, the instants immediately *preceding* and *following a.*

(2) Denoting the function $f(t)$ as $f^{(0)}(t)$, and its kth derivative $d^k f(t)/dt^k$ as $f^{(k)}(t)$, the following holds: If $f^{(0)}(t)$, $f^{(1)}(t)$, $f^{(2)}(t)$, ..., $f^{(k-1)}(t)$ are *continuous* functions of time whereas $f^{(k)}(t)$ is *discontinuous,* then the coefficients a_n and b_n decrease with the harmonic number n as $1/n^{k+1}$.

For instance, since both the *sawtooth* and *square* waves are discontinuous, their coefficients decrease as $1/n^{0+1} = 1/n$. The *triangle* and *rectified sine* waves are continuous, but their first derivatives are discontinuous; hence, their coefficients decrease as $1/n^{1+1} = 1/n^2$. A sinelike wave made up of *parabolic* sections in such a way as to be continuous up to and including the first derivative would have coefficients that decrease as $1/n^3$.

17.3 Frequency Spectra and Filtering

The harmonic makeup of a signal can be visualized quite effectively by means of frequency diagrams known as *frequency spectra.* Applying a periodic signal to a circuit containing reactive elements yields a response whose spectrum will generally *differ* from that of the applied signal. Circuits specifically designed to alter frequency spectra are referred to as *filters*.

Frequency Spectra

Plotting the amplitudes A_n and phase angles θ_n of Equation (17.2) versus frequency yields a pair of diagrams known, respectively, as the **amplitude spectrum** and the **phase spectrum** of $f(t)$. Because the harmonic components have *discrete* frequency values, the spectra of periodic signals consist of *lines* and are thus called *line spectra*.

Figure 17.10 shows the spectra of the sawtooth of Figure 17.1. According to Equation (17.7), this wave has $A_0 = A/2$, $A_n = (A/\pi)/n$, and $\theta_n = 90°$, $n = 1$, 2, 3, ... It is readily seen that the envelope of the harmonic amplitudes is the curve $M(\omega) = A\omega_0/\pi\omega$, and the envelope of the harmonic phase angles is the curve $\theta(\omega) = 90°$.

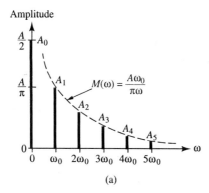

(a)

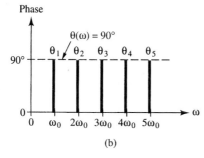

(b)

Figure 17.10 Amplitude and phase spectra of the sawtooth wave of Figure 17.1.

Plotting the squared rms values A_0^2 and $(A_n\sqrt{2})^2$ versus frequency yields, by Equation (17.12), the **power spectrum** of $f(t)$. This spectrum provides a visual indication of how the power of a periodic signal distributes among its harmonic components, a feature of paramount importance in signal communications. Figure 17.11 shows the power spectrum of the same sawtooth wave. The envelope is now the curve $P(\omega) = \frac{1}{2}M^2(\omega) = \frac{1}{2}(A\omega_0/\pi\omega)^2$. We observe that most of the power is concentrated in the dc component. If we shift the sawtooth wave downward until its dc component is reduced to zero, most of the power will

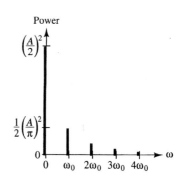

Figure 17.11 Power spectrum of the sawtooth wave of Figure 17.1.

come only from the first few harmonics, because it decreases quadratically with the harmonic number. Once again, we stress that the power spectrum depends only on the harmonic amplitudes, irrespective of the phase angles.

Exercise 17.6   Sketch the amplitude, phase, and power spectra of (a) the square wave of Figure 7.3, (b) the square wave of Figure 17.6, and (c) the triangle wave of Figure 17.8.

Steady-State Response to a Periodic Signal

We wish to find the response $y(t)$ of a circuit to a periodic signal $x(t)$. We are interested in the *steady-state response,* that is, the response after all transients have died out. We know that if the signal is *purely sinusoidal,* the response can be found via phasor techniques as $Y = H(j\omega_0)X$, where X and Y are, respectively, the *phasors* of the applied signal and the response, and $H(j\omega_0)$ is the *network function* calculated at the frequency ω_0 of the applied signal. This approach can be extended to a *nonsinusoidal* periodic signal as follows.

First, we express the signal as a Fourier sum of sinusoidal components,

$$x(t) = X_0 + \sum_{n=1}^{\infty} X_n \cos(n\omega_0 t + \theta_{xn}) \tag{17.29}$$

Next, using standard ac analysis techniques, we find the phasors of the responses to the individual components,

$$\boxed{\begin{aligned} Y_n &= |H(jn\omega_0)| \times X_n \\ \theta_{yn} &= \sphericalangle H(jn\omega_0) + \theta_{xn} \end{aligned}} \qquad \begin{aligned} &\textbf{(17.30a)} \\ &\textbf{(17.30b)} \end{aligned}$$

$n = 0, 1, 2, \ldots$. Finally, we apply the superposition principle and add up these responses to obtain the Fourier series of the overall response,

$$y(t) = Y_0 + \sum_{n=1}^{\infty} Y_n \cos(n\omega_0 t + \theta_{yn}) \tag{17.31}$$

In general both $|H(jn\omega_0)|$ and $\sphericalangle H(jn\omega_0)$ will vary with $n\omega_0$, indicating that the harmonic makeup of $y(t)$ will generally *differ* from that of $x(t)$. We describe this phenomenon by saying that as it progresses through the circuit, the applied signal will generally experience both **amplitude distortion** and **phase distortion.**

▶ Example 17.6

The *R-C* circuit of Figure 17.12(a) is subjected to the square wave v_I shown in Figure 17.12(b). Find the Fourier series of its steady-state response v_O.

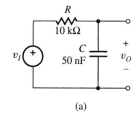

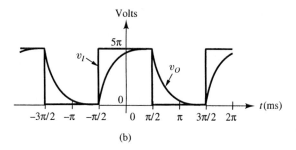

Figure 17.12 (a) *R-C* circuit of Example 17.6, and (b) its input and output waveforms.

Solution

Adapting Equation (17.17) to the present case, we have

$$v_I(t) = 7.854 + 10.000 \cos \omega_0 t + 3.333 \cos (3\omega_0 t + 180°)$$
$$+ 2.000 \cos 5\omega_0 t + 1.429 \cos (7\omega_0 t + 180°) + \cdots \text{ V}$$

where $\omega_0 = 2\pi/T = 2\pi/(2\pi \, 10^{-3}) = 10^3$ rad/s.

The transfer function of the *R-C* network is, by the generalized voltage divider formula, $H(j\omega) = (1/j\omega C)/(R + 1/j\omega C)$, or

$$H(j\omega) = \frac{1}{1 + j(\omega/\omega_c)}$$

$$\omega_c = \frac{1}{RC}$$

where ω_c is the *cutoff frequency*, also called the *half-power frequency*, the −3 dB *frequency*, the *corner frequency*, the *characteristic frequency*, or the *critical frequency* of the *R-C* network. We denote it as ω_c to avoid confusion with the *fundamental frequency* ω_0 of the input. Using

$$|H(jn\omega_n)| = \frac{1}{\sqrt{1 + (n\omega_0/\omega_c)^2}}$$

$$\sphericalangle H(jn\omega_0) = -\tan^{-1}(n\omega_0/\omega_c)$$

with $\omega_0 = 10^3$ rad/s and $\omega_c = 1/(10^4 \times 50 \times 10^{-9}) = 2 \times 10^3$ rad/s, we calculate

$$H(0) = 1\underline{/0°}$$

$$H(j1\omega_0) = 0.8944\underline{/-26.57°}$$

$$H(j3\omega_0) = 0.5547\underline{/-56.31°}$$

$$H(j5\omega_0) = 0.3714\underline{/-68.20°}$$

$$H(j7\omega_0) = 0.2747\underline{/-74.05°}$$

and so forth. Multiplying the amplitude of the nth component of $v_I(t)$ by $|H(jn\omega_0)|$ and augmenting its phase angle by $\sphericalangle H(jn\omega_0)$, we obtain the nth component of $v_O(t)$. Adding up all components,

$$v_O(t) = 7.854 + 8.944\cos(\omega_0 t - 26.57°)$$
$$+ 1.849\cos(3\omega_0 t + 123.69°) + 0.7428\cos(5\omega_0 t - 68.20°)$$
$$+ 0.3925\cos(7\omega_0 + 105.95°) + \cdots \text{ V}$$

Pause a moment to compare each component of $v_O(t)$ against the corresponding component of $v_I(t)$. ◀

Exercise 17.7 Repeat Example 17.6, but with R and C interchanged with each other so as to form a C-R circuit.

ANSWER $v_O(t) = 0 + 4.472\cos(\omega_0 t + 63.43°) + 2.774\cos(3\omega_0 t - 146.31°) + 1.857\cos(5\omega_0 t + 21.80°) + 1.374\cos(7\omega_0 - 164.05°) + \cdots$ V.

Even though Equation (17.31) provides an analytical expression for the response, it is not immediately apparent what kind of *waveform* v_O of Example 17.6 will exhibit. This can be found by calculating and plotting its Fourier series for different values of t within one cycle. A much quicker alternative is to find the response via the time-domain techniques of Section 8.3. The result is the waveform denoted as v_O in Figure 17.12b.

We wish to stress that in many applications the appearance of the waveform is less important than its harmonic makeup. It is precisely its ability to convey the harmonic content that distinguishes Fourier analysis from other techniques such as time-domain analysis and Laplace analysis.

Input-Output Spectra

The effect of a circuit upon a periodic signal, expressed *analytically* by Equation (17.30), can also be visualized *graphically* by means of frequency spectra. As illustrated in Figure 17.13 for the case of the *amplitude spectra,* we first sketch the line spectrum of the input signal. Next, we plot $|H(j\omega)|$ versus ω. Finally, we construct the line spectrum of the output by *multiplying* out each input line by the value of $|H(j\omega)|$ at the corresponding harmonic frequency.

It is apparent that the harmonic makeup of the output depends not only on the makeup of the input, but also on the frequency *profile* of $|H(j\omega)|$ as well as the location of the *cutoff frequency* ω_c relative to the *fundamental frequency* ω_0.

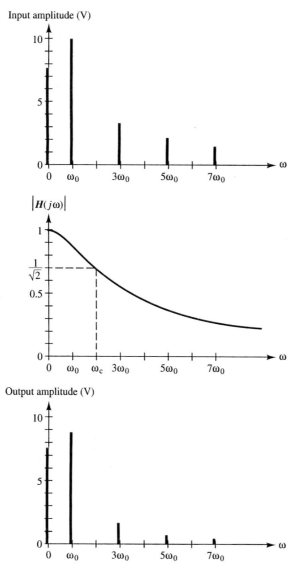

Figure 17.13 Input and output amplitude spectra for the *R-C* circuit of Figure 17.12.

As we know, the *R-C* circuit tends to pass low-frequency components while attenuating high-frequency components.

If we increase ω_c so as to make $\omega_c \gg \omega_0$, then a substantial number of harmonic components will be passed, yielding an output that is virtually an *undistorted* version of the input. A large ω_c can be achieved, for instance, by making C suitably small.

By contrast, increasing C so as to make $\omega_c \ll \omega_0$ will attenuate all harmonic components. Since the input amplitudes decrease with frequency as $1/n$ and $|H(jn\omega_0)|$ also decreases as $1/n$ for $n\omega_0 \gg \omega_c$, the output amplitudes will decrease as $1/n^2$. In light of Equation (17.28), this is the amplitude spectrum of a triangular wave, indicating a substantial amount of *distortion*.

We observe that the input dc component is fully passed to the output, regardless of ω_c. By making ω_c sufficiently small we can virtually block out all harmonic components and thus pass only the dc component. Needless to say, these observations are consistent with our findings of Sections 8.3 and 14.4. You are encouraged to verify also for the high-pass case, which is obtained by interchanging R with C.

A similar procedure can be used for the *phase spectra,* except that in this case the output lines are obtained by *adding* the input lines to the values of $\sphericalangle H(j\omega)$ at the corresponding frequencies. This is shown in Figure 17.14 for the R-C circuit of Example 17.6.

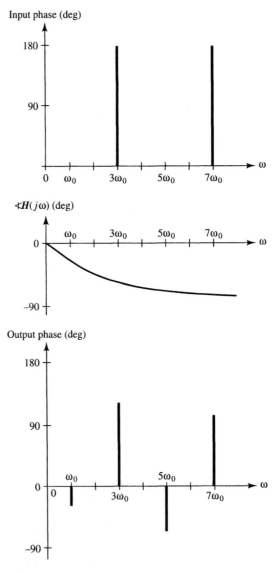

Figure 17.14 Input and output phase spectra for the R-C circuit of Figure 17.12.

Filters

The profiles of the $|H(j\omega)|$ and $\sphericalangle H(j\omega)$ curves, collectively referred to as the *frequency response,* provide a visual indication of the manner in which a circuit affects the spectrum of the applied signal. Circuits intended to serve this specific function are referred to as **filters.** Following are the *idealized* specifications of the most common filter functions.

(1) **Low-Pass Filter.** This filter is characterized by a frequency ω_c, called the *cutoff frequency,* such that

$$|H(j\omega)| = \begin{cases} 1 & \text{for } 0 \le \omega < \omega_c \\ 0 & \text{for } \omega > \omega_c \end{cases} \qquad \textbf{(17.32)}$$

As shown in Figure 17.15(a), this filter passes the dc component and all harmonic components whose frequencies lie within the *pass band* $0 \le \omega < \omega_c$, and it rejects all components with frequencies within the *stop band* $\omega_c < \omega \le \infty$.

(2) **High-Pass Filter.** This filter is the *opposite* of the low-pass filter, for

$$|H(j\omega)| = \begin{cases} 0 & \text{for } 0 \le \omega < \omega_c \\ 1 & \text{for } \omega > \omega_c \end{cases} \qquad \textbf{(17.33)}$$

As shown in Figure 17.15(b), it rejects all components within the band

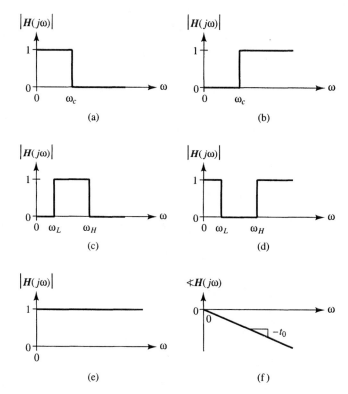

Figure 17.15 Idealized filter responses: (a) *low pass;* (b) *high pass;* (c) *band pass;* (d) *band reject;* (e), (f) *all pass.*

$0 \leq \omega < \omega_c$, now called the *stop band,* and it passes all components within the band $\omega_c < \omega \leq \infty$, now called the *pass band.*

(3) **Band-Pass Filter.** This filter is characterized by a frequency ω_L, called the *lower cutoff frequency,* and a frequency ω_H, called the *upper cutoff frequency,* such that

$$|H(j\omega)| = \begin{cases} 1 & \text{for } \omega_L < \omega < \omega_H \\ 0 & \text{for } 0 \leq \omega < \omega_L \quad \text{and} \quad \omega_H < \omega \leq \infty \end{cases} \quad \textbf{(17.34)}$$

As shown in Figure 17.15(c), the harmonic components within the *pass band* $\omega_L < \omega < \omega_H$ are passed; those within the *stop bands* $0 \leq \omega < \omega_L$ and $\omega_H < \omega \leq \infty$ are rejected.

(4) **Band-Reject Filter:** This filter, also referred to as a **notch filter,** is the *opposite* of the band-pass filter,

$$|H(j\omega)| = \begin{cases} 0 & \text{for } \omega_L < \omega < \omega_H \\ 1 & \text{for } 0 \leq \omega < \omega_L \quad \text{and} \quad \omega_H < \omega \leq \infty \end{cases} \quad \textbf{(17.35)}$$

As shown in Figure 17.15(d), it rejects the components within the *stop band* $\omega_L < \omega < \omega_H$ and it passes all those within the *pass bands* $0 \leq \omega < \omega_L$, and $\omega_H < \omega \leq \infty$.

(5) **All-Pass Filter:** This filter, also known as a **delay filter,** passes *all* components so that

$$|H(j\omega)| = 1 \quad \textbf{(17.36)}$$

regardless of ω, but it effects their phase angles as

$$\measuredangle H(j\omega) = -t_0\omega \quad \textbf{(17.37)}$$

where t_0, having the dimensions of time, is called the *time delay* of the filter. The frequency plots of such a filter are shown in Figures 17.15(e) and (f). Its response to a signal $x(t)$ is simply $x(t-t_0)$, a delayed version of $x(t)$ itself.

The effect of the various filters is depicted in Figure 17.16, using the signal

$$x(t) = 1 \sin \omega_0 t + 0.5 \sin 5\omega_0 t + 0.3 \sin 20\omega_0 t$$

as an example. Figure 17.16(a) shows the individual harmonic components of $x(t)$, while Figure 17.16(b) shows the waveform and amplitude spectrum of $x(t)$. Figure 17.16, parts (c) through (f), shows the waveforms and spectra after the signal has been sent, respectively, through a *low-pass, high-pass, band-pass,* and *band-reject* filter.

The idealized responses of Figure 17.15 can in practice only be *approximated.* A classical approach to filter synthesis is by means of resistances, capacitances, and inductances. Since these components are passive, these filters are called **passive filters.** However, since the inductor is the least ideal of the three elements, it is desirable to avoid it whenever possible. **Active filters** meet this goal by using resistances, capacitances, and amplifiers, but no inductances. Both active and passive filters are referred to as **analog filters** because they operate directly on physical signals, which are by nature analog or continuous.

A drawback of analog filters is the tolerance and drift of their physical components, which limits the degree of accuracy with which the idealized responses of Figure 17.16 can be approximated and maintained with temperature and time.

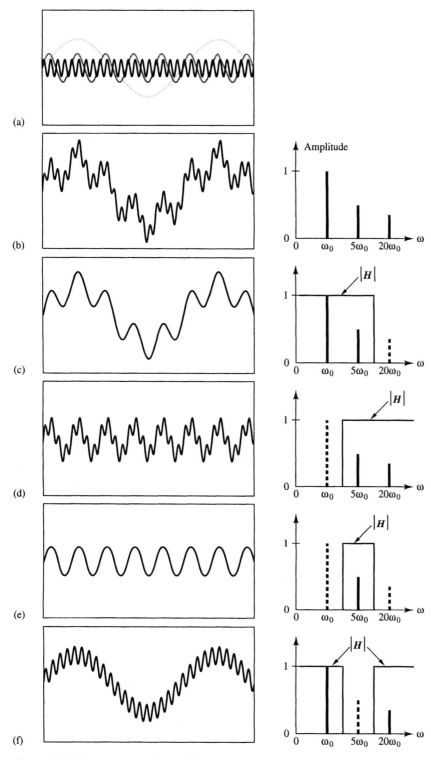

Figure 17.16 Illustrating the effect of filtering: (a) harmonic components and (b) waveform of the filter input; filter outputs: (c) low-pass, (d) high-pass, (e) band-pass, and (f) band-reject.

These limitations are overcome by an alternate class of filters known as **digital filters.** In this case, the analog signal is first converted to digital form, where it is represented as a sequence of numbers, just as a movie action is converted to a sequence of picture frames. Then, a digital processor operates upon this numerical representation according to algorithms designed to have the desired effect upon the input spectrum. Finally, the results are converted back to analog form to reconstruct the processed signal. Audio digital recording, mixing, editing, and mastering are common applications of this concept.

17.4 THE EXPONENTIAL FOURIER SERIES

The amplitude-phase and cosine-sine Fourier series of Equations (17.2) and (17.3) are jointly referred to as *trigonometric* Fourier series. An alternate Fourier representation utilizes exponential instead of trigonometric functions, and is thus referred to as the *exponential Fourier series.* Its form is

$$f(t) = \sum_{n=-\infty}^{\infty} c_n e^{jn\omega_0 t} \tag{17.38}$$

where

$$c_n = \frac{1}{T} \int_{t_0}^{t_0+T} f(t) e^{-jn\omega_0 t} \, dt \tag{17.39}$$

Because the coefficients c_n are in general complex, the exponential Fourier series is also referred to as the *complex Fourier series,* even when $f(t)$ happens to be a real signal, as is usually the case.

All the coefficients of the exponential series are found via the common formula of Equation (17.39), whereas the coefficients of the trigonometric series require the three separate formulas of Equation (17.6). Consequently, the exponential series is mathematically more concise and lends itself to quicker and more elegant manipulations, a feature that will facilitate our introduction to the *Fourier transform* in the next section.

We now wish to prove that Equations (17.38) and (17.39) follow directly from Equations (17.3) and (17.6). Letting $\cos n\omega_0 t = (e^{jn\omega_0 t} + e^{-jn\omega_0 t})/2$ and $\sin n\omega_0 t = (e^{jn\omega_0 t} - e^{-jn\omega_0 t})/2j$ in Equation (17.3), we obtain

$$f(t) = a_0 + \sum_{n=1}^{\infty} \frac{a_n}{2}(e^{jn\omega_0 t} + e^{-jn\omega_0 t}) + \frac{b_n}{2j}(e^{jn\omega_0 t} - e^{-jn\omega_0 t})$$

$$= a_0 + \sum_{n=1}^{\infty} \frac{a_n - jb_n}{2} e^{jn\omega_0 t} + \frac{a_n + jb_n}{2} e^{-jn\omega_0 t}$$

$$= c_0 + \sum_{n=1}^{\infty} c_n e^{jn\omega_0 t} + c_n^* e^{-jn\omega_0 t} \tag{17.40}$$

where we have set

$$c_0 = a_0 = A_0 \tag{17.41a}$$

$$c_n = \frac{a_n - jb_n}{2} = \frac{A_n}{2}\underline{/\theta_n} \tag{17.41b}$$

$n = 1, 2, 3, \ldots$; moreover, c_n^* denotes the complex conjugate of c_n. Substituting the expressions for a_n and b_n as given in Equation (17.6) and using Euler's identity, we obtain

$$c_n = \frac{1}{2}\left[\frac{2}{T}\int_{t_0}^{t_0+T} f(t)\cos n\omega_0 t\, dt - j\frac{2}{T}\int_{t_0}^{t_0+T} f(t)\sin n\omega_0 t\, dt\right]$$

$$= \frac{1}{T}\int_{t_0}^{t_0+T} f(t)(\cos n\omega_0 t - j\sin n\omega_0 t)\, dt = \frac{1}{T}\int_{t_0}^{t_0+T} f(t)e^{-jn\omega_0 t}\, dt$$

This proves Equation (17.39). To prove Equation (17.38), we replace n with $-n$ in Equation (17.39) to obtain

$$c_{-n} = \frac{1}{T}\int_{t_0}^{t_0+T} f(t)e^{jn\omega_0 t}\, dt$$

that is,

$$c_{-n} = c_n^* \tag{17.42}$$

This allows us to rewrite Equation (17.40) as

$$f(t) = \sum_{n=0}^{\infty} c_n e^{jn\omega_0 t} + \sum_{n=1}^{\infty} c_{-n} e^{-jn\omega_0 t} = \sum_{n=-\infty}^{\infty} c_n e^{jn\omega_0 t}$$

where we have exploited the fact that summing $c_{-n}e^{-jn\omega_0 t}$ from 1 to ∞ is equivalent to summing $c_n e^{jn\omega_n t}$ from -1 to $-\infty$. This completes the proof of Equation (17.38).

We observe that besides the *positive-valued* harmonic frequencies ω_0, $2\omega_0$, $3\omega_0$, ..., the exponential Fourier series contains the *negative-valued* frequencies $-\omega_0$, $-2\omega_0$, $-3\omega_0$, ... Physical frequencies are, of course, always positive. However, when using the exponential form we are in effect expressing *two* sums, one from 0 to ∞ and the other from 1 to ∞, as a *single* sum from $-\infty$ to $+\infty$; hence the need for negative frequencies.

We also observe that the coefficients c_n and c_{-n} are *conjugate* to each other and are related to the coefficients of the trigonometric series by Equation (17.41). Moreover, we note a duality between Equations (17.38) and (17.39): Interchanging $f(t)$ with c_n, the operation of summation with that of integration, n with $-n$, and division by T with multiplication by T leads from one equation to the other, and vice versa. This symmetry becomes more significant in the next section, when we introduce the concept of Fourier transform *pairs*.

▶ Example 17.7

Find the exponential Fourier series for the *pulse train* of Figure 17.17.

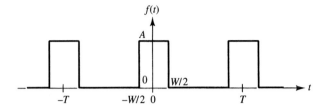

Figure 17.17 Pulse train waveform of Example 17.7.

Solution

We conveniently choose $-W/2 \le t \le W/2$ as the interval of integration, where $f(t) = A$. Then, Equation (17.39) yields

$$c_n = \frac{1}{T} \int_{-W/2}^{W/2} A e^{-jn\omega_0 t}\, dt = \frac{A}{T} \frac{e^{-jn\omega_0 t}}{-jn\omega_0}\Big|_{-W/2}^{W/2}$$

$$= \frac{A}{T} \frac{e^{jn\omega_0 W/2} - e^{-jn\omega_0 W/2}}{jn\omega_0} = \frac{AW}{T} \frac{\sin(n\omega_0 W/2)}{n\omega_0 W/2}$$

Since $n\omega_0 W/2 = n(2\pi/T)W/2 = n\pi W/T$, the coefficients are

$$c_n = A \frac{W}{T} \frac{\sin(n\pi W/T)}{n\pi W/T} \qquad \textbf{(17.43)}$$

The exponential Fourier series for the pulse train is thus

$$f(t) = A\frac{W}{T} \sum_{n=-\infty}^{\infty} \frac{\sin(n\pi W/T)}{n\pi W/T} e^{jn\omega_0 t} \qquad \textbf{(17.44)} \blacktriangleleft$$

Exercise 17.8   Find the coefficients of the exponential Fourier series for the *sawtooth* wave of Figure 17.1.

ANSWER $c_0 = A/2$; $c_n = jA/2\pi n$, $n = \pm 1, \pm 2, \pm 3, \ldots$

Power

The average power delivered to a 1-Ω resistance by a real signal $f(t)$ is, by Equations (17.11) and (17.38),

$$P_{1\,\Omega} = \frac{1}{T} \int_{t_0}^{t_0+T} f^2(t)\, dt = \frac{1}{T} \int_{t_0}^{t_0+T} f(t) \left[\sum_{n=-\infty}^{\infty} c_n e^{jn\omega_0 t}\right] dt$$

$$= \sum_{n=-\infty}^{\infty} c_n \left[\frac{1}{T} \int_{t_0}^{t_0+T} f(t) e^{jn\omega_0 t}\, dt\right]$$

Recognizing that the expression within brackets is c_n^*, we obtain

$$P_{1\Omega} = \sum_{n=-\infty}^{\infty} c_n c_n^* = \sum_{n=-\infty}^{\infty} |c_n|^2 \qquad \text{(17.45)}$$

This result, again referred to as *Parseval's theorem* for periodic signals, states that *the average power delivered by a periodic signal to a unity resistance is found by summing the moduli squared of all its complex Fourier series coefficients.* Considering that $|c_{-n}| = |c_n|$, Equation (17.45) can also be expressed as

$$P_{1\Omega} = c_0^2 + 2\sum_{n=1}^{\infty} |c_n|^2 \qquad \text{(17.46)}$$

Needless to say, this result could also have been obtained using Equations (17.12) and (17.41).

Exercise 17.9   Use Equation (17.46) with a sufficient number of terms to estimate the rms value of a *pulse train* alternating between 0 V and 5 V with a 25% duty cycle.

ANSWER   2.5 V.

Frequency Spectra

Plotting $|c_n|$, $\sphericalangle c_n$, and $|c_n|^2$ versus frequency yields, respectively, the **amplitude, phase,** and **power spectra** of $f(t)$. Because n ranges from $-\infty$ to ∞, the spectral lines now extend over *positive* as well as *negative* frequencies. Moreover, since $c_{-n} = c_n^*$, it follows that $|c_{-n}| = |c_n|$ and $\sphericalangle c_{-n} = -\sphericalangle c_n$, indicating that the amplitude and power spectra have *even* symmetry, whereas the phase spectrum has *odd* symmetry.

As an example, let us examine the amplitude spectrum of the pulse train of Figure 17.17 for different values of the duty cycle $d \triangleq W/T$. By Equation (17.43), the coefficients can be expressed as

$$c_n = A\frac{W}{T}\text{Sa}\left(n\pi\frac{W}{T}\right) \qquad \text{(17.47)}$$

where

$$\text{Sa}(x) \triangleq \frac{\sin x}{x} \qquad \text{(17.48)}$$

is a function of special importance in communication theory known as the

sampling function. The plot of Figure 17.18 indicates that $Sa(0) = 1$, and $Sa(\pm m\pi) = 0$, $m = 1, 2, 3, \ldots$.

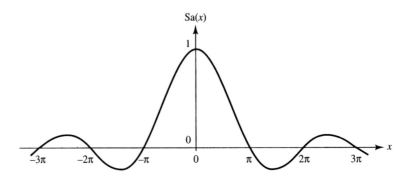

Figure 17.18   The sampling function $Sa(x) = (\sin x)/x$.

Letting $f = n/T$ in Equation (17.47) yields the envelope of the amplitude spectrum as a function of the *cyclical frequency f,*

$$|c(f)| = A\frac{W}{T}\,|Sa(\pi W f)| \tag{17.49}$$

This function is plotted in Figure 17.19 for two different values of the duty cycle $d = W/T$. We make the following observations:

(1) The maximum value of $|c(f)|$, reached for $f = 0$, represents the dc coefficients $c_0 = AW/T$. This maximum is proportional to the duty cycle $d = W/T$ of the pulse train.

(2) Since $Sa(\pi W f) = 0$ for $\pi W f = m\pi$, or $f = m/W$, the envelope goes to zero for $f = \pm 1/W$, $\pm 2/W$, $\pm 3/W$, $\ldots$.

(3) The spectral lines are located at $f = \pm 1/T$, $\pm 2/T$, $\pm 3/T$, $\ldots$. If T is an *integral multiple* of W, say, $T = mW$, then the harmonics $\pm m$, $\pm 2m$, $\pm 3m$, $\ldots$ will be missing. With a 50% duty cycle as in Figure 17.19(a), all *even* harmonics are missing. With a 20% duty cycle as in Figure 17.19(b), the *fifth, tenth, fifteenth, twentieth,* $\ldots$ harmonics are missing.

We note that if we increase the period T while keeping the pulse width W constant, the spacing $1/T$ between adjacent lines will decrease, thus squeezing more lines between consecutive zeros of the envelope. Moreover, the amplitudes of all components will decrease in proportion.

We can also use Equation (17.47) to sketch the *phase* and *power* spectra. For this particular waveform, the coefficients corresponding to the *positive* lobes of the function $Sa(\pi W f)$ have $\angle c_n = 0°$, and those corresponding to the *negative* lobes have $\angle c_n = \pm 180°$. Finally, the profile of the power spectrum is the curve $|c(f)|^2 = A^2(W/T)^2 \times Sa^2(\pi W f)$.

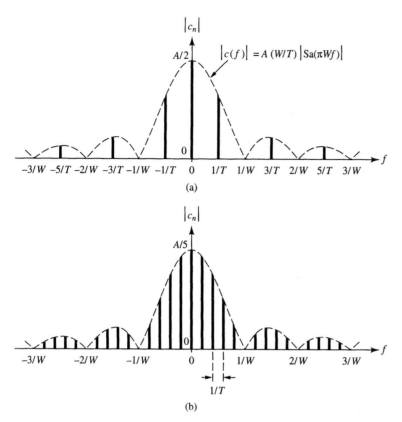

Figure 17.19 Amplitude spectra of a pulse train with (a) $W/T = 1/2$, and (b) $W/T = 1/5$.

Exercise 17.10 Consider the pulse train obtained from that of Figure 17.17 by shifting it toward the right by $W/2$ so that the leading edge of the central pulse occurs at $t = 0$. Obtain the coefficients c_n of the new waveform. Hence, sketch its amplitude and phase spectra for the case $W/T = 1/5$.

ANSWER $c_n = (AW/T)e^{-jn\pi W/T} \, \text{Sa}(n\pi W/T)$.

If a circuit with the transfer function $H(s)$ is subjected to the periodic signal

$$x(t) = \sum_{n=-\infty}^{\infty} c_n e^{jn\omega_0 t} \tag{17.50}$$

then its steady-state response $y(t)$ is readily obtained as

$$y(t) = \sum_{n=-\infty}^{\infty} H(jn\omega_0)c_n e^{jn\omega_0 t} \tag{17.51}$$

> **Exercise 17.11**   A periodic voltage signal having $\omega_0 = 1$ krad/s and
> $c_n = (j10/n)[\cos(n\pi/2) - 1]$ V is applied to a simple R-C low-pass
> filter having $R = 1$ kΩ and $C = \frac{1}{3}$ μF. Sketch the amplitude, phase, and
> power spectra of the output from the filter.

17.5 THE FOURIER TRANSFORM

In the last sections we have found that the periodic functions of physical interest
admit Fourier series representations. What if a function is not periodic? Mathematically, does it still admit some form of Fourier representation? Physically,
does it still exhibit a frequency spectrum? We shall answer these questions using the following strategy. With reference to Figure 17.20, where the aperiodic
function of interest is denoted as $f(t)$, we define its *periodic extension* as the
periodic function

$$f_p(t) = f_p(t + nT)$$

coinciding with $f(t)$ over some arbitrary time interval T, say,

$$f_p(t) = f(t) \qquad \text{for} -T/2 \leq t \leq T/2$$

We then consider the exponential Fourier series of $f_p(t)$ and examine it in the
limit $T \to \infty$. We clearly have

$$\lim_{T \to \infty} f_p(t) = f(t) \qquad \text{for} -\infty \leq t \leq \infty \tag{17.52}$$

By this process we extend the Fourier series concept to the aperiodic function
$f(t)$ by regarding it as periodic, but with an *infinite* period.

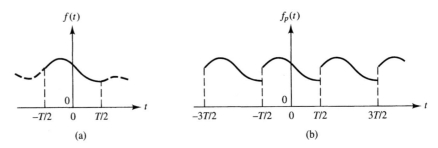

Figure 17.20 (a) An aperiodic function $f(t)$ and (b) its periodic extension $f_p(t)$.

By Equations (17.38) and (17.39) we can express $f_p(t)$ as

$$f_p(t) = \sum_{n=-\infty}^{\infty} c_n e^{jn\omega_0 t} = \sum_{n=-\infty}^{\infty} \left[\frac{1}{T} \int_{-T/2}^{T/2} f_p(t) e^{-jn\omega_0 t} \, dt \right] e^{jn\omega_0 t}$$

Exploiting the fact that $T = 2\pi/\omega_0$, this can be written as

$$f_p(t) = \sum_{n=-\infty}^{\infty} \left[\frac{1}{2\pi} \int_{-T/2}^{T/2} f_p(t) e^{-jn\omega_0 t} \, dt \right] e^{jn\omega_0 t} \omega_0 \tag{17.53}$$

To find a clue on how to handle this expression in the limit $T \to \infty$, we refer to the frequency spectrum of $f_p(t)$, consisting of lines spaced $1/T$ Hz apart from each other. As shown in the example of Figure 17.19, increasing T decreases the fundamental frequency as well as the line spacing. As T is increased *indefinitely,* the individual spectral components merge into a continuous spectrum and the fundamental frequency becomes vanishingly small. Consequently, the frequency $n\omega_0$ of each harmonic component becomes the *continuous* frequency variable ω, the line spacing ω_0 becomes the *infinitesimal $d\omega$*, and the operation of summation becomes the operation of *integration*. Thus, letting

$$T \to \infty$$

$$n\omega_0 \to \omega$$

$$\omega_0 \to d\omega$$

$$\sum \to \int$$

$$f_p \to f$$

in Equation (17.53) and rearranging yields

$$f(t) = \frac{1}{2\pi} \int_{-\infty}^{\infty} \left[\int_{-\infty}^{\infty} f(t)e^{-j\omega t}\, dt \right] e^{j\omega t}\, d\omega \qquad \textbf{(17.54)}$$

The integral within brackets is called the **Fourier transform** of $f(t)$, and is denoted as

$$\mathscr{F}\{f(t)\} = \int_{-\infty}^{\infty} f(t)e^{-j\omega t}\, dt \qquad \textbf{(17.55)}$$

It is apparent that once we evaluate this integral, we are left with a function of $j\omega$, which we denote as $F(j\omega)$,

$$F(j\omega) = \mathscr{F}\{f(t)\} \qquad \textbf{(17.56)}$$

Just as we define $F(j\omega)$ to be the Fourier transform of $f(t)$, we define $f(t)$ to be the **inverse Fourier transform** of $F(j\omega)$, and we denote this as

$$f(t) = \mathscr{F}^{-1}\{F(j\omega)\} \qquad \textbf{(17.57)}$$

From Equation (17.54) we have

$$f(t) = \frac{1}{2\pi} \int_{-\infty}^{\infty} F(j\omega)e^{j\omega t}\, d\omega \qquad \textbf{(17.58)}$$

The function $f(t)$ and its Fourier transform $F(j\omega)$ are said to constitute a **transform pair,** and we denote this as

$$\boxed{f(t) \leftrightarrow F(j\omega)} \tag{17.59}$$

One can readily verify that both the Fourier transform and its inverse are *linear* operations.

Mathematicians have shown that a *sufficient* set of conditions for a function $f(t)$ to admit a Fourier transform are again *Dirichlet's conditions,* with the integrability condition now rephrased as

$$\int_{-\infty}^{\infty} |f(t)| \, dt < \infty \tag{17.60}$$

Moreover, no two distinct functions have the same Fourier transform and vice versa, a property referred to as the *uniqueness* of the Fourier transform and its inverse.

Fourier Transform Pairs

Using Equation (17.55), we can readily find the Fourier transforms of commonly encountered signals. Conversely, given a function of $j\omega$, we can find its inverse transform using Equation (17.58).

The Fourier transform of the *impulse* function is

$$\boxed{\mathcal{F}\{\delta(t)\} = \int_{-\infty}^{\infty} \delta(t) e^{-j\omega t} \, dt = e^{-j\omega 0} = 1} \tag{17.61}$$

where we have exploited the sifting property of Equation (16.13). Exploiting the linearity property, the Fourier transform of an impulse of *strength A* is

$$\boxed{\mathcal{F}\{A\delta(t)\} = A} \tag{17.62}$$

The Fourier transform of a *delayed time impulse* is

$$\boxed{\mathcal{F}\{\delta(t - t_0)\} = \int_{-\infty}^{\infty} \delta(t - t_0) e^{-j\omega t} \, dt = e^{-j\omega t_0}} \tag{17.63}$$

The inverse Fourier transform of a *delayed frequency impulse* is

$$\boxed{\begin{aligned} \mathcal{F}^{-1}\{\delta(\omega - \omega_0)\} &= \frac{1}{2\pi} \int_{-\infty}^{\infty} \delta(\omega - \omega_0) e^{j\omega t} \, d\omega \\ &= \frac{1}{2\pi} e^{j\omega_0 t} \end{aligned}} \tag{17.64}$$

This allows us to state that the Fourier transform of a *complex exponential* is

$$\mathscr{F}\{e^{j\omega_0 t}\} = 2\pi\,\delta(\omega - \omega_0) \qquad\qquad \textbf{(17.65)}$$

Exploiting once again the linearity property of the Fourier transform, we can write $\mathscr{F}\{Ae^{j\omega_0 t}\} = 2\pi A\delta(\omega - \omega_0)$, where A is a constant. Letting $\omega_0 = 0$ so that $Ae^{j\omega_0 t} = Ae^{j0t} = A$, we find that the Fourier transform of a *constant* function is

$$\mathscr{F}\{A\} = 2\pi A\delta(\omega) \qquad\qquad \textbf{(17.66)}$$

The apparent duality between Equations (17.62) and (17.66) is evidenced in Figure 17.21.

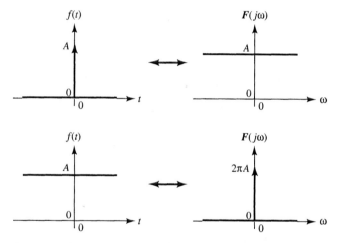

Figure 17.21 The Fourier transform of an impulse is a constant, and the Fourier transform of a constant is an impulse.

We can use Equation (17.65), along with the linearity property, to find the transform of the *cosine* function,

$$\mathscr{F}\{\cos\omega_0 t\} = \mathscr{F}\left\{\frac{e^{j\omega_0 t} + e^{-j\omega_0 t}}{2}\right\} = \frac{1}{2}\mathscr{F}\{e^{j\omega_0 t}\} + \frac{1}{2}\mathscr{F}\{e^{-j\omega_0 t}\}$$

that is,

$$\mathscr{F}\{\cos\omega_0 t\} = \pi[\delta(\omega + \omega_0) + \delta(\omega - \omega_0)] \qquad\qquad \textbf{(17.67)}$$

This Fourier pair is depicted in Figure 17.22. Similarly, the transform of the *sine* function is

$$\mathscr{F}\{\sin\omega_0 t\} = j\pi[\delta(\omega + \omega_0) - \delta(\omega - \omega_0)] \qquad\qquad \textbf{(17.68)}$$

It is interesting to observe that the Fourier transform, introduced in order to deal with aperiodic signals, can also be applied to periodic signals such as $\cos \omega_0 t$ and $\sin \omega_0 t$. Moreover, we find that these functions possess Fourier transforms even though they do not satisfy the condition of Equation (17.60), which is a sufficient but not a necessary condition for the transform to exist.

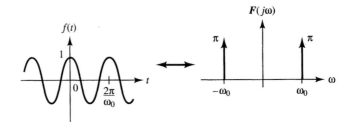

Figure 17.22 The Fourier transform of the cosine is an impulse pair.

The Fourier transform of the *positive-time exponential* is

$$\mathscr{F}\{e^{-at}u(t)\} = \int_{-\infty}^{\infty} e^{-at}u(t)e^{-j\omega t}\,dt = \int_{0}^{\infty} e^{-(a+j\omega)t}\,dt = \frac{-e^{-(a+j\omega)t}}{a+j\omega}\bigg|_{0}^{\infty}$$

or

$$\mathscr{F}\{e^{-at}u(t)\} = \frac{1}{a+j\omega} \qquad \text{(17.69)}$$

for $a > 0$. Likewise, the Fourier transform of the *negative-time exponential* is

$$\mathscr{F}\{e^{at}u(-t)\} = \frac{1}{a-j\omega} \qquad \text{(17.70)}$$

for $a > 0$.

Another useful function is the *signum* function, defined as

$$\operatorname{sgn}(t) = \begin{cases} +1 & \text{for } t > 0 \\ -1 & \text{for } t < 0 \end{cases} \qquad \text{(17.71)}$$

To find its Fourier transform, we write $\operatorname{sgn}(t) = \lim_{a \to 0}[e^{-at}u(t) - e^{at}u(-t)]$. Then,

$$\mathscr{F}\{\operatorname{sgn}(t)\} = \lim_{a \to 0}\left[\mathscr{F}\{e^{-at}u(t)\} - \mathscr{F}\{e^{at}u(-t)\}\right] = \frac{2}{j\omega} \qquad \text{(17.72)}$$

The signum function and its transform are depicted in Figure 17.23.

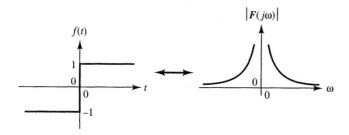

Figure 17.23 The signum function and its Fourier transform.

Equation (17.72), along with linearity and Equation (17.66), can be used to find the Fourier transform of the *step* function,

$$\mathscr{F}\{u(t)\} = \mathscr{F}\left\{\frac{1}{2} + \frac{1}{2}\operatorname{sgn}(t)\right\} = \pi\delta(\omega) + \frac{1}{j\omega} \qquad \textbf{(17.73)}$$

These transform pairs are summarized in Table 17.1.

TABLE 17.1 Some Fourier Transform Pairs

Name	$f(t)$	$F(j\omega)$
Impulse	$\delta(t)$	1
Constant	A	$2\pi A\delta(\omega)$
Signum	$\operatorname{sgn}(t)$	$\dfrac{2}{j\omega}$
Step	$u(t)$	$\pi\delta(\omega) + \dfrac{1}{j\omega}$
Pos-t exponential	$e^{-at}u(t), a > 0$	$\dfrac{1}{a + j\omega}$
Neg-t exponential	$e^{at}u(-t), a > 0$	$\dfrac{1}{a - j\omega}$
Abs-t exponential	$e^{-a\lvert t\rvert}, a > 0$	$\dfrac{2a}{a^2 + \omega^2}$
Complex exponential	$e^{j\omega_0 t}$	$2\pi\delta(\omega - \omega_0)$
Cosine	$\cos\omega_0 t$	$\pi[\delta(\omega + \omega_0) + \delta(\omega - \omega_0)]$
Sine	$\sin\omega_0 t$	$j\pi[\delta(\omega + \omega_0) - \delta(\omega - \omega_0)]$

▶ **Example 17.8**

Find the Fourier transform for the single *pulse* shown at the left of Figure 17.24.

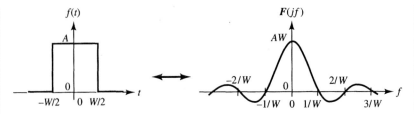

Figure 17.24 A pulse and its Fourier transform.

Solution

Since $f(t) = A$ for $-W/2 \leq t \leq W/2$, and $f(t) = 0$ elsewhere, we can change the limits of integration in Equation (17.55) from $-\infty$ to $-W/2$, and from ∞ to $W/2$. Then,

$$F(j\omega) = \int_{-W/2}^{W/2} Ae^{-j\omega t}\, dt = A\frac{e^{-j\omega t}}{-j\omega}\Big|_{-W/2}^{W/2} = \frac{2A}{\omega}\frac{e^{j\omega W/2} - e^{-j\omega W/2}}{j2}$$

$$= \frac{2A}{\omega}\sin\frac{\omega W}{2}$$

This can be put in either of the following forms

$$F(j\omega) = AW\,\mathrm{Sa}\left(\frac{\omega W}{2}\right) \qquad \textbf{(17.74a)}$$

$$F(jf) = AW\,\mathrm{Sa}(\pi W f) \qquad \textbf{(17.74b)}$$

where $\mathrm{Sa}(x) = (\sin x)/x$ is the sampling function introduced in Section 17.4, and $f = \omega/2\pi$ is the *cyclical frequency,* not to be confused with the function $f(t)$ itself! As shown at the right of Figure 17.24, the transform has the same functional form as the envelope of the amplitude spectrum of the pulse train. However, instead of discrete spectral lines, we now have a *continuous* spectrum. ◀

Exercise 17.12   Find the Fourier transform of the single *sawtooth pulse,* defined as $f(t) = (A/W)t$ for $0 \leq t \leq W$, and $f(t) = 0$ elsewhere.

ANSWER $F(j\omega) = (jA/\omega)[e^{-j\omega W} - e^{-j\omega W/2}\,\mathrm{Sa}(\omega W/2)]$.

Exercise 17.13   Find the Fourier transform of the *doublet,* defined as $f(t) = A$ for $-W \leq t < 0$, $f(t) = -A$ for $0 < t \leq W$, and $f(t) = 0$ elsewhere.

ANSWER $F(j\omega) = (j2A/\omega)(1 - \cos\omega W)$.

Energy Density Spectrum

We now wish to investigate the *physical significance* of the Fourier transform. To this end, suppose $f(t)$ represents the voltage across or the current through a 1-Ω resistor, so that the *instantaneous dissipated power* is $p_{1\Omega}(t) = f^2(t)$. The **total energy of the signal** $f(t)$ is then

$$W_{1\Omega} = \int_{-\infty}^{\infty} p_{1\Omega}(t)\, dt = \int_{-\infty}^{\infty} f^2(t)\, dt \qquad \text{(17.75)}$$

Using Equation (17.58), we can write

$$W_{1\Omega} = \int_{-\infty}^{\infty} f(t) \left[\frac{1}{2\pi} \int_{-\infty}^{\infty} F(j\omega)e^{j\omega t} d\omega \right] dt$$

Interchanging the order of integration and rearranging, we get

$$W_{1\Omega} = \frac{1}{2\pi} \int_{-\infty}^{\infty} F(j\omega) \left[\int_{-\infty}^{\infty} f(t)e^{j\omega t} dt \right] d\omega$$

It is readily seen that the term within brackets represents $F^*(j\omega)$, the complex conjugate of $F(j\omega)$. The integrand is thus $F(j\omega)F^*(j\omega)$, or $|F(j\omega)|^2$, so that

$$W_{1\Omega} = \frac{1}{2\pi} \int_{-\infty}^{\infty} |F(j\omega)|^2 d\omega \qquad \text{(17.76a)}$$

This result, known again as **Parseval's theorem,** holds both for periodic and aperiodic signals, and it can also be expressed in terms of the *cyclical frequency* $f = \omega/2\pi$ as

$$W_{1\Omega} = \int_{-\infty}^{\infty} |F(jf)|^2 df \qquad \text{(17.76b)}$$

In words, *the total energy delivered by a signal to a* 1-Ω *resistor equals the total area under the magnitude squared of its Fourier transform* $F(jf)$. Considering that $|F(j\omega)|^2$ is an even function of ω, the total energy of $f(t)$ can also be expressed as

$$W_{1\Omega} = \frac{1}{\pi} \int_{0}^{\infty} |F(j\omega)|^2 d\omega = 2 \int_{0}^{\infty} |F(jf)|^2 df \qquad \text{(17.77)}$$

Since $W_{1\Omega}$ is expressed in *joules* (J), $|F(jf)|^2$ is expressed in *joules/hertz* (J/Hz), which are the units of *energy density,* or energy per hertz. The curve obtained by plotting $|F^2|$ versus frequency is called the *energy density spectrum.* While the energy of a periodic signal is concentrated at the frequencies of its harmonic components, the energy of an aperiodic signal is generally *spread* over the entire frequency spectrum.

It is informative to discuss these findings in terms of simple transform pairs. Let us begin with the *impulse* function. Since $\mathscr{F}\{\delta(t)\} = 1$, the energy density of $\delta(t)$ is uniform with frequency, indicating the presence of *all* frequency components in the makeup of an impulse.

By contrast, because $\mathcal{F}\{A\} = 2\pi A\delta(\omega)$, the energy of a *dc signal A* is concentrated at just one frequency point, $\omega = 0$. Moreover, its energy density must be in the form of an *impulse* in order to yield a nonzero area when integrated as in Equation (17.76).

Since $\mathcal{F}\{\cos\omega_0 t\} = \pi[\delta(\omega + \omega_0) + \delta(\omega - \omega_0)]$, the energy of a purely sinusoidal signal is concentrated at *two* frequency points, $\omega = \pm\omega_0$. Likewise, the energy density must be impulsive in order to yield a nonzero total area.

According to Equation (17.74b), the energy density of a *pulse* of amplitude A and width W is $|F(f)|^2 = [AW \times \text{Sa}(\pi Wf)]^2$. The energy is concentrated primarily within the main lobe, $-1/W < f < 1/W$ (see Problem 17.37). The narrower W, the wider the lobe. If we let $A = 1/W$ and take the limit $W \to 0$, the pulse approaches the *impulse* function $\delta(t)$, and the lobe will become infinitely wide while retaining the amplitude $AW = (1/W) \times W = 1$. This confirms Equation (17.61).

▶ **Example 17.9**

Find the total energy dissipated in a 10-Ω resistance by the current $i(t) = 5e^{-10^3 t}u(t)$ A.

Solution

$W_{10\,\Omega} = 10W_{1\,\Omega}$. We can find $W_{1\,\Omega}$ either via Equation (17.75) or via Equation (17.76). Let us do both.

$$W_{1\,\Omega} = \int_{-\infty}^{\infty} f^2(t)\,dt = \int_0^{\infty} 25e^{-2\times 10^3 t}\,dt$$

$$= \frac{-25}{2\times 10^3}e^{-2\times 10^3 t}\Big|_0^{\infty} = 12.5 \text{ mJ}$$

$$W_{1\,\Omega} = \frac{1}{\pi}\int_0^{\infty} |F^2(j\omega)|\,d\omega = \frac{1}{\pi}\int_0^{\infty} \frac{25}{10^6 + \omega^2}\,d\omega$$

$$= \frac{25}{\pi}\frac{\tan^{-1}(\omega/10^3)}{10^3}\Big|_0^{\infty} = 12.5 \text{ mJ}$$

Clearly, $W_{10\,\Omega} = 0.125$ J.   ◀

Exercise 17.14   Use Parseval's theorem to find $W_{1\,\Omega}$ if $f(t) = Ae^{-a|t|}$.

ANSWER A^2/a.

17.6 PROPERTIES OF THE FOURIER TRANSFORM

The Fourier transform exhibits unique *functional* properties that we can exploit to expedite our calculations or to check our results. Moreover, it affects not only functions but also *operations* upon functions. Similarities with the Laplace transform are investigated at the closing of this section.

Fourier Transform Properties

Being a complex function, the Fourier transform admits both polar and rectangular representations,

$$F(j\omega) = A(\omega) + jB(\omega) = |F(j\omega)|e^{j\sphericalangle F(j\omega)} \qquad \text{(17.78)}$$

To find expressions for $A(\omega)$ and $B(\omega)$, we let $e^{-j\omega t} = \cos\omega t - j\sin\omega t$ in Equation (17.55). Then, it is apparent that

$$A(\omega) = \int_{-\infty}^{\infty} f(t)\cos\omega t \, dt \qquad \text{(17.79a)}$$

$$B(\omega) = -\int_{-\infty}^{\infty} f(t)\sin\omega t \, dt \qquad \text{(17.79b)}$$

Based on these expressions, one can readily prove the following:

   (1) $A(\omega)$ is an *even* function of ω; $B(\omega)$ is an *odd* function of ω.

   (2) $|F(j\omega)| = \sqrt{A^2(\omega) + B^2(\omega)}$ is an *even* function of ω; $\sphericalangle F(j\omega) = \tan^{-1} B(\omega)/A(\omega)$ is an *odd* function of ω.

   (3) *Negating* ω yields the *conjugate* of the transform, $F(-j\omega) = F^*(j\omega)$.

   (4) If $f(t)$ is an *even* function, $B(\omega) = 0$, so $F(j\omega)$ is *real*; if $f(t)$ is an *odd* function, $A(\omega) = 0$, so $F(j\omega)$ is *imaginary*.

   (5) If $f(t)$ is an *even* function, then $f(t)$ and $F(j\omega)$ have the *same functional form,* aside from the constant 2π.

The last property stems from the fact that if $f(t)$ is even, then $F(j\omega)e^{j\omega t} = [A(j\omega) + j0] \times [\cos\omega t + j\sin\omega t]$. Substituting into Equation (17.58) and exploiting the fact that $A(j\omega)$ is also even, we have

$$f(t) = \frac{1}{2\pi} \int_{-\infty}^{\infty} A(j\omega)\cos\omega t \, dt$$

Except for the term $1/2\pi$, this has the same form as Equation (17.79a), indicating that $f(t)$ and $F(j\omega)$ are *interchangeable.* For instance, because $\mathscr{F}\{\delta(t)\} = 1$, it follows that $\mathscr{F}^{-1}\{\delta(\omega)\} = 1/2\pi$. Likewise, because the transform of a *time pulse* is a function of the form $(\sin\omega)/\omega$, it follows that the *time function* whose spectrum is a *frequency pulse* is a function of the type $(\sin t)/t$.

Operational Transforms

Because operational Fourier transforms are in many respects similar to the operational Laplace transforms of Section 16.3, we list only the results, leaving their proofs as end-of-chapter problems.

(1) Linearity: The Fourier transform of a *linear combination* of two functions $f_1(t)$ and $f_2(t)$ is the *linear combination* of their Fourier transforms $F_1(j\omega)$ and $F_2(j\omega)$,

$$\mathscr{F}\{af_1(t) + bf_2(t)\} = aF_1(j\omega) + bF_2(j\omega) \qquad \textbf{(17.80)}$$

(2) Differentiation: *Time differentiation* of a function $f(t)$ corresponds to *multiplication* of its Fourier transform $F(j\omega)$ by $j\omega$,

$$\mathscr{F}\left\{\frac{df(t)}{dt}\right\} = j\omega F(j\omega) \qquad \textbf{(17.81)}$$

In order for this property to hold $f(t)$ must vanish as $t \to \pm\infty$. Unlike the Laplace transform, the Fourier transform does not take into account the initial value $f(0^-)$, owing to the fact that the lower limit in its defining integral is $-\infty$ instead of 0^-.

(3) Integration: *Time integration* of a function $f(t)$ corresponds to *division* of its Fourier transform $F(j\omega)$ by $j\omega$,

$$\mathscr{F}\left\{\int_{-\infty}^{t} f(\xi)\,d\xi\right\} = \frac{F(j\omega)}{j\omega} \qquad \textbf{(17.82)}$$

This property holds if $\int_{-\infty}^{+\infty} f(\xi)\,d\xi = 0$.

(4) Scaling: *Scaling* by $a > 0$ in the time domain corresponds to *scaling* by $1/a$ in the frequency domain, accompanied by division by a,

$$\mathscr{F}\{f(at)\} = \frac{1}{a}F\left(\frac{j\omega}{a}\right) \qquad \textbf{(17.83)}$$

(5) Time shifting: *Shifting* by t_0 in the time domain corresponds to *multiplication* by $e^{-j\omega t_0}$ in the frequency domain,

$$\mathscr{F}\{f(t - t_0)\} = e^{-j\omega t_0}F(j\omega) \qquad \textbf{(17.84)}$$

(6) Frequency shifting: *Shifting* by ω_0 in the frequency domain corresponds to *multiplication* by $e^{j\omega_0}$ in the time domain,

$$\mathscr{F}^{-1}\{F(j\omega - j\omega_0)\} = e^{j\omega_0 t} f(t) \tag{17.85}$$

(7) Modulation: The process of controlling the amplitude of a sinusoidal *carrier,* $\cos \omega_0 t$, by means of a *modulating signal,* $f(t)$, is called *amplitude modulation,* and the resulting signal is the product $f(t) \times \cos \omega_0 t$. Using Euler's identity, along with Equation (17.85), it is readily seen that *modulation* in the time domain corresponds to *shifting* by $\pm\omega_0$ in the frequency domain, where ω_0 is the carrier frequency,

$$\mathscr{F}\{f(t) \cos \omega_0 t\} = \frac{1}{2}[F(j\omega - j\omega_0) + F(j\omega + j\omega_0)] \tag{17.86}$$

The effect of modulation is depicted in Figure 17.25.

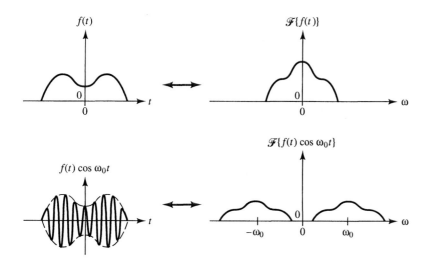

Figure 17.25 When a signal $f(t)$ with spectrum $F(j\omega)$ modulates a sinusoidal carrier having frequency ω_0, the resulting spectrum is half the spectrum of $f(t)$, but shifted by $\pm\omega_0$.

(8) Convolution: The *time convolution* of two functions $h(t)$ and $x(t)$ corresponds to *multiplication* of their Fourier transforms $H(j\omega)$ and $X(j\omega)$,

$$h(t) * x(t) \leftrightarrow H(j\omega)X(j\omega) \tag{17.87}$$

where convolution in the time domain is now defined as

$$
\begin{aligned}
h(t) * x(t) &= \int_{-\infty}^{\infty} h(\xi) x(t - \xi)\, d\xi \\
&= \int_{-\infty}^{\infty} h(t - \xi) x(\xi)\, d\xi
\end{aligned}
\tag{17.88}
$$

Likewise, $1/2\pi$ times the *frequency convolution* of two functions $F(j\omega)$ and $G(j\omega)$ corresponds to *multiplication* of their inverse Fourier transforms $f(t)$ and $g(t)$,

$$
f(t)g(t) \leftrightarrow \frac{1}{2\pi} F(j\omega) * G(j\omega)
\tag{17.89}
$$

where convolution in the frequency domain is defined as

$$
F(j\omega) * G(j\omega) = \int_{-\infty}^{\infty} F(j\xi) G(j\omega - j\xi)\, d\xi
\tag{17.90}
$$

These operational transforms are summarized in Table 17.2.

TABLE 17.2 Some Operational Transforms

Name	Time Domain	Frequency Domain
Linearity	$af_1(t) + bf_2(t)$	$aF_1(j\omega) + bF_2(j\omega)$
t differentiation	$\dfrac{df(t)}{dt}$	$j\omega F(j\omega)$
t integration	$\displaystyle\int_{-\infty}^{t} f(\xi)\, d\xi$	$\dfrac{F(j\omega)}{j\omega}$
t scaling	$f(at),\ a > 0$	$\dfrac{1}{a} F\left(\dfrac{j\omega}{a}\right)$
t shifting	$f(t - t_0)$	$e^{-j\omega t_0} F(j\omega)$
ω shifting	$e^{j\omega_0 t} f(t)$	$F(j\omega - j\omega_0)$
modulation	$f(t) \cos \omega_0 t$	$\dfrac{1}{2} F(j\omega - j\omega_0) + \dfrac{1}{2} F(j\omega + j\omega_0)$
t convolution	$h(t) * x(t)$	$H(j\omega) X(j\omega)$
ω convolution	$f(t)g(t)$	$\dfrac{1}{2\pi} F(j\omega) * G(j\omega)$

▶ **Example 17.10**

Find the Fourier transform of the function defined as $f(t) = A \cos t$ for $-\pi/2 \le t \le \pi/2$, and $f(t) = 0$ elsewhere.

Solution

Figure 17.26 shows the function, along with its first two derivatives. By inspection,

$$f^{(2)}(t) = A\delta(t + \pi/2) + A\delta(t - \pi/2) - f(t)$$

Because both $f(t)$ and $f^{(1)}$ vanish as $t \to \pm\infty$, we can apply Equation (17.81) twice. Using linearity, along with $\mathcal{F}\{\delta(t-t_0)\}=e^{-j\omega t_0}$, we get

$$(j\omega)^2 F(j\omega) = A(e^{j\omega\pi/2} + e^{-j\omega\pi/2}) - F(j\omega)$$

Using Euler's identity and simplifying, we have

$$F(j\omega) = \frac{2A}{1 - \omega^2} \cos \frac{\pi}{2}\omega$$

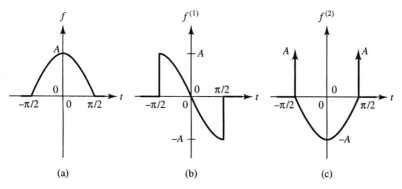

Figure 17.26 (a) Signal of Example 17.10, and its (b) first and (c) second derivatives.

Exercise 17.15   Confirm the Fourier transforms of the single sawtooth pulse of Exercise 17.12 and of the doublet of Exercise 17.13 by finding the Fourier transforms of their derivatives and then dividing by $j\omega$.

Relation to the Laplace Transform

It is apparent by now that there are similarities between the Fourier and Laplace transforms, repeated here for convenience,

$$\mathcal{F}\{f(t)\} = \int_{-\infty}^{\infty} f(t)e^{-j\omega t}\, dt \qquad \textbf{(17.91)}$$

$$\mathcal{L}\{f(t)\} = \int_{0-}^{\infty} f(t)e^{-st}\, dt \qquad \textbf{(17.92)}$$

To compare the two transforms we must first require that $f(t)$ be a *causal* or *positive-time* function. Then, letting $s = \sigma + j\omega$ in Equation (17.92), we can write

$$\mathscr{L}\{f(t)u(t)\} = \int_{0^-}^{\infty} f(t)u(t)e^{-(\sigma+j\omega)t}\, dt = \int_{-\infty}^{\infty} [f(t)u(t)e^{-\sigma t}]e^{-j\omega t}\, dt$$

or

$$\boxed{\mathscr{L}\{f(t)\} = \mathscr{F}\{f(t)e^{-\sigma t}\}} \qquad \text{(17.93)}$$

In words, the Laplace transform of the causal function $f(t)$ is the Fourier transform of the causal function $f(t)e^{-\sigma t}$. Alternately, the Fourier transform of the causal function $f(t)$ is the Laplace transform of $f(t)$ calculated for $\sigma = 0$, that is, on the $j\omega$ axis,

$$\boxed{\mathscr{F}\{f(t)\} = \mathscr{L}\{f(t)\}_{s=j\omega}} \qquad \text{(17.94)}$$

Thanks to the $e^{-\sigma t}$ term, the Laplace integral converges for a *wider range* of functions than the *Fourier* integral. Equation (17.94) thus holds only for functions for which the Fourier integral *converges*. This integral will converge when all the poles of $F(s) = \mathscr{L}\{f(t)\}$ lie in the *left half* of the s plane. If $F(s)$ has poles on the imaginary axis or in the right half of the s plane, then the integral $\int_{-\infty}^{+\infty} |f(t)|\, dt$ will not converge, and Equation (17.94) no longer holds. Yet, we have seen in the previous section that if $F(s)$ has simple poles on the imaginary axis, but not in the right half of the s plane, the Fourier transform nevertheless exists. To accommodate this special case we modify Equation (17.94) as

$$\boxed{\mathscr{F}\{f(t)\} = \mathscr{L}\{f(t)\}_{s=j\omega} + \sum_{k=1}^{N_i} \pi A_k \delta(\omega - \omega_k)} \qquad \text{(17.95)}$$

where ω_k is the ordinate of the kth imaginary pole, A_k is the residue of $F(s)$ at $s = j\omega_k$, and N_i is the number of imaginary poles. Let us illustrate these concepts with actual examples.

▶ **Example 17.11**

Find the Fourier transform of

$$f(t) = e^{-at}(\sin \omega_0 t)u(t)$$

Solution

The Laplace transform of this causal function is

$$F(s) = \frac{\omega_0}{(s+a)^2 + \omega_0^2}$$

This function has a pole pair at $s = -a \pm j\omega_0$. As long as $a > 0$, the pair lies in the left half of the s plane. We can thus evaluate $F(s)$ at $s = j\omega$ to find the Fourier transform of $f(t)$,

$$F(j\omega) = \frac{\omega_0}{(j\omega + a)^2 + \omega_0^2}$$

◄

Exercise 17.16   Find $\mathcal{F}\{3(t + 1)e^{-5t}u(t)\}$.

ANSWER $3(6 + j\omega)/(5 + j\omega)^2$.

► **Example 17.12**

Find the Fourier transform of

$$f(t) = (\sin \omega_0 t)u(t)$$

Solution

The Laplace transform of this causal function is

$$F(s) = \frac{\omega_0}{s^2 + \omega_0^2}$$

Since it has a complex pole pair at $s = \pm j\omega_0$, we expand it as

$$F(s) = \frac{j/2}{s + j\omega_0} - \frac{j/2}{s - j\omega_0}$$

Applying Equation (17.95), we find the Fourier transform as

$$F(j\omega) = \frac{\omega_0}{\omega_0^2 - \omega^2} + \frac{j\pi}{2}[\delta(\omega + \omega_0) - \delta(\omega - \omega_0)]$$

Comparing with the Fourier transform of the noncausal signal $\sin \omega_0 t$, we observe that making this signal causal by multiplying it by $u(t)$ halves the strength of its spectral impulses and also adds the term $\omega_0/(\omega_0^2 - \omega^2)$. ◄

Exercise 17.17   If $\mathcal{L}\{f(t)\} = 2(3s + 1)/[(s^2 + 1)(s + 2)]$, find (a) $f(t)$ and (b) $\mathcal{F}\{f(t)\}$.

ANSWER   (a) $2[\sqrt{2}\cos(t - 45°) - e^{-2t}]u(t)$; (b) $2(1 + 3j\omega)/[(1 - \omega^2)(2 + j\omega)] + \pi[(1 + j1)\delta(\omega + 1) + (1 - j1)\delta(\omega - 1)]$.

► **Example 17.13**

Show that for the causal function

$$f(t) = e^{at}u(t)$$

$a > 0$, $\mathscr{L}\{f(t)\}$ exists, but the Fourier transform does not.

Solution

It is apparent that for $a > 0$ the Fourier integral

$$\int_{-\infty}^{\infty} e^{at}u(t)e^{-j\omega t}\,dt$$

does not exist. We justify physically by noting that generating the diverging signal $e^{at}u(t)$ would require a diverging amount of energy, which is impossible. However, the Laplace integral

$$\int_{0^-}^{\infty} e^{at}u(t)e^{-st}\,dt = \int_{-0^-}^{\infty} e^{at}e^{-(\sigma+j\omega)t}\,dt = \int_{0^-}^{\infty} e^{(a-\sigma)t}e^{-j\omega t}\,dt$$

does exist provided $(a - \sigma) < 0$, or $\sigma > a$, and it converges to

$$F(s) = \frac{1}{s - a}$$

which has a pole at $s = a$, that is, in the *right half* of the s plane. As expected, what makes the Laplace integral converge is the term $e^{-\sigma t}$, which is not available in the Fourier case.   ◄

17.7 Fourier Transform Applications

Exploiting the linearity, differentiation, and integration properties of the Fourier transform, one can readily show that the constitutive laws of the basic circuit elements transform as

$$V(j\omega) = Z(j\omega)I(j\omega) \tag{17.96}$$

where $V(j\omega)$ and $I(j\omega)$ are, respectively, the *Fourier transforms* of voltage and current, in V-s and A-s, and

$$Z_R(j\omega) = R \tag{17.97a}$$
$$Z_L(j\omega) = j\omega L \tag{17.97b}$$
$$Z_C(j\omega) = 1/j\omega C \tag{17.97c}$$

Moreover, thanks again to the linearity property, KVL and KCL hold also in Fourier form. The formal similarity with phasors allows us to state that if we subject a *linear circuit* to a forcing signal $x(t)$, its response $y(t)$ will be such that

$$Y(j\omega) = H(j\omega)X(j\omega) \qquad\text{(17.98)}$$

where $Y(j\omega) = \mathscr{F}\{y(t)\}$, $X(j\omega) = \mathscr{F}\{x(t)\}$, and $H(j\omega)$ is the familiar **network function.** This function can be found via ω-domain circuit analysis, except that we are now working with *Fourier transforms* (in V-s or A-s), instead of phasors (in V or A). Alternatively, we can find it via s-domain techniques, and then let $s \rightarrow j\omega$.

It is apparent that the Fourier approach generalizes the phasor technique to *aperiodic signals,* subject, of course, to the constraint that they be Fourier transformable. Thanks to the convolution property, the Fourier transform approach also provides an alternate method for finding the response, which is not available with phasors, namely,

$$y(t) = h(t) * x(t) \qquad\text{(17.99)}$$

where $h(t)$ is the *impulse response* of the circuit. As we know, this method is especially useful when $h(t)$ and $x(t)$ are known, but their transforms are not available analytically and the computations must therefore be carried out entirely in the time domain. Let us demonstrate the use of Fourier techniques with actual examples.

▶ **Example 17.14**

Using the Fourier transform approach, find the response $v_o(t)$ of the circuit of Figure 17.27 to the applied signal

$$v_i(t) = 10\cos 2t \text{ V}$$

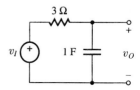

Figure 17.27 *R-C* Circuit of Examples 17.14 through 17.16.

Solution

Though this problem could be solved via the phasor method, we shall use the Fourier transform method to demonstrate how it works. The Fourier transform of the output is related to that of the input by the ac voltage divider formula,

$$V_o(j\omega) = \frac{1/j\omega 1}{3 + 1/j\omega 1} V_i(j\omega) = \frac{1}{1 + j3\omega} V_i(j\omega)$$

Applying Equation (17.67),

$$V_o(j\omega) = \frac{1}{1 + j3\omega} 10\pi[\delta(\omega + 2) + \delta(\omega - 2)]$$

Thus,

$$v_o(t) = \mathscr{F}^{-1}\{V_o(j\omega)\} = \frac{1}{2\pi} \int_{-\infty}^{\infty} 10\pi \frac{\delta(\omega - 2) + \delta(\omega + 2)}{1 + j3\omega} e^{j\omega t}\, dt$$

$$= 5\left(\frac{e^{j2t}}{1 + j6} + \frac{e^{-j2t}}{1 - j6}\right) = 5 \times 2\mathrm{Re}\left[\frac{e^{j2t}}{1 + j6}\right]$$

$$= \frac{10}{\sqrt{37}} \cos(2t - 80.54°)\ \text{V}$$

You may want to check this result using the phasor method. ◀

Exercise 17.18   Repeat Example 17.14 if the circuit includes also a 0.5-H inductance in parallel with the capacitance. Hence, check your result using the phasor method.

ANSWER $v_o(t) = \sqrt{10} \cos(2t - 71.57°)$ V.

▶ **Example 17.15**

Using the Fourier transform approach, find $v_O(t)$ in the circuit of Figure 17.27 if

$$v_I(t) = 10\,\mathrm{sgn}\,(t)\ \text{V}$$

Solution

Since the applied input is not of the ac type, we cannot use phasors. We could, of course, use the time-domain techniques of Chapter 8; however, we want to demonstrate the use of the Fourier approach. By Equation (17.72), we now have

$$V_O(j\omega) = \frac{1}{1 + j3\omega} V_I(j\omega) = \frac{1/3}{1/3 + j\omega} \frac{20}{j\omega} = \frac{20}{j\omega} - \frac{20}{1/3 + j\omega}$$

where we have performed a partial fraction expansion. Using Table 17.1 we find, by inspection,

$$v_O(t) = \mathscr{F}^{-1}\{V_O(j\omega)\} = 10\,\mathrm{sgn}\,(t) - 20e^{-t/3}u(t)\ \text{V}$$

or $v_O(t \le 0) = -10$ V, and $v_O(t \ge 0) = 10(1 - 2e^{-t/3})$ V. You may want to check this result using time-domain techniques. ◀

Exercise 17.19   Repeat Example 17.15, but with the resistance and capacitance interchanged with each other.

ANSWER $v_O(t) = 20e^{-t/3}u(t)$ V.

▶ Example 17.16

In the circuit of Figure 17.27 find the percentage η of the input signal energy transferred to the output if

$$v_I(t) = 10e^{-t/4}u(t) \text{ V}$$

Solution

Combining Equations (17.69) and (17.77), we find that the total energies of the input and output signal are

$$W_i = \frac{1}{\pi}\int_0^\infty |V_I(j\omega)|^2\,d\omega = \frac{1}{\pi}\int_0^\infty \left|\frac{10}{1/4 + j\omega}\right|^2 d\omega$$

$$= \frac{100}{\pi}\int_0^\infty \frac{d\omega}{(1/4)^2 + \omega^2} = \frac{100}{\pi}4\tan^{-1}4\omega\Big|_0^\infty = 200 \text{ J}$$

$$W_o = \frac{1}{\pi}\int_0^\infty |V_O(j\omega)|^2 d\omega = \frac{1}{\pi}\int_0^\infty \left|\frac{1/3}{1/3 + j\omega} \times \frac{10}{1/4 + j\omega}\right|^2 d\omega$$

$$= \frac{1}{\pi}\frac{100}{9}\int_0^\infty \frac{1}{(1/3)^2 + \omega^2} \times \frac{1}{(1/4)^2 + \omega^2}\,d\omega$$

$$= \frac{100}{9\pi}\int_0^\infty \left[\frac{144/7}{(1/4)^2 + \omega^2} - \frac{144/7}{(1/3)^2 + \omega^2}\right]d\omega$$

$$= \frac{14,400}{63\pi}\left[4\tan^{-1}4\omega - 3\tan^{-1}3\omega\right]_0^\infty = 114.29 \text{ J}$$

Finally, $\eta = 100(114.29/200) = 57.14\%$.   ◀

Exercise 17.20   Repeat Example 17.16, but with the resistance and capacitance interchanged with each other.

ANSWER $\eta = 42.86\%$

▶ Example 17.17

An ideal *band-pass* filter with $\omega_L = 5$ rad/s and $\omega_H = 20$ rad/s is subject to the input signal

$$v_I(t) = 30e^{-10t}u(t) \text{ V}$$

(a) Sketch the input and output energy density spectra.

(b) Find the percentage η of the input signal energy that is transferred to the output.

Solution

(a) The input energy density spectrum is, by Equation (17.69),

$$|V_I(j\omega)|^2 = \frac{900}{100 + \omega^2} \; \text{J/(rad/s)}$$

The plot of Figure 17.28, top, confirms the evenness of this function. Since the band-pass filter passes all frequency components within the pass band and rejects all components outside this band, the output energy density spectrum will appear as in Figure 17.28, bottom.

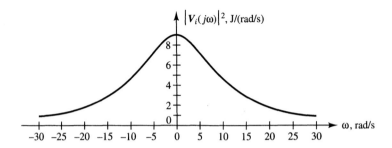

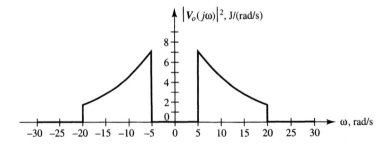

Figure 17.28 Input and output energy spectra for Example 17.17.

(b) The total energies of the input and output signals are

$$W_i = \frac{1}{\pi} \int_0^\infty \frac{900}{100 + \omega^2} \, d\omega = \frac{900}{\pi} \frac{1}{10} \tan^{-1} \frac{\omega}{10} \Big|_0^\infty = 45 \text{ J}$$

$$W_o = \frac{1}{\pi} \int_{\omega_L}^{\omega_H} |V_o(j\omega)|^2 \, d\omega = \frac{1}{\pi} \int_5^{20} \frac{900}{100 + \omega^2} \, d\omega$$

$$= \frac{900}{\pi} \frac{1}{10} \tan^{-1} \frac{\omega}{10} \Big|_5^{20} = 18.43 \text{ J}$$

Consequently, $\eta = 100(18.43/45) = 40.97\%$.

> **Exercise 17.21**   Repeat Example 17.17, but for the case of a *band-reject* filter with $\omega_L = 5$ rad/s and $\omega_H = 15$ rad/s.
>
> **ANSWER** $\eta = 66.95\%$.

Comparing the Fourier and Laplace Approaches

Having demonstrated that the Fourier transform approach extends the phasor approach to aperiodic signals, we want to clarify its relationship to the Laplace transform approach. To this end, we observe the following:

(1) The Laplace approach to *circuit analysis* is used more widely than the Fourier approach because (a) the Laplace integral converges for a *wider range* of applied signals and (b) it automatically accounts for the *initial conditions*.

(2) The Fourier transform accommodates signals that are nonzero over *all time,* that is, signals that are turned on at $t = -\infty$. Consequently, it is better suited to the study of *steady-state responses,* that is, responses after all transients have had sufficient time to die out.

(3) The unilateral Laplace transform accommodates *positive-time* signals, that is, signals that are turned on at $t = 0$. Consequently, it is better suited to the study of *complete responses,* that is, responses to particular forcing signals on the basis of the energy already present in the reactive elements when the source is turned on.

(4) The Fourier transform finds application especially in the areas of communication and signal processing, where the emphasis is on *frequency response* and *frequency spectra.*

17.8 Fourier Techniques Using SPICE

SPICE can be used to perform both the Fourier analysis and the Fourier synthesis of periodic signals.

Fourier Synthesis

Fourier synthesis is based on SPICE's ability to simulate sinusoidal sources. As seen in Chapter 14, these sources have the forms

```
VXXX N+ N- SIN(VO VA FREQ TD ALPHA THETA)
IXXX N+ N- SIN(IO IA FREQ TD ALPHA THETA)
```

By specifying a sufficient number of harmonically related sources we can synthesize and display any periodic waveform we wish. Voltage sources are connected in series, current sources in parallel.

▶ Example 17.18

Use PSpice to perform the Fourier synthesis of a *sawtooth voltage waveform* of the type of Figure 17.1, with $A = 10$ V, and $T = 1$ ms. Use Equation (17.7a) with the first six harmonics.

Solution

With reference to Figure 17.29, the input file is

```
FOURIER SYNTHESIS OF A SAWTOOTH
V0 1 0 DC 5
V1 1 2 SIN(0 3.1831 1.0E3 0 0 0)
V2 2 3 SIN(0 1.5915 2.0E3 0 0 0)
V3 3 4 SIN(0 1.0610 3.0E3 0 0 0)
V4 4 5 SIN(0 0.7958 4.0E3 0 0 0)
V5 5 6 SIN(0 0.6366 5.0E3 0 0 0)
V6 6 7 SIN(0 0.5305 6.0E3 0 0 0)
RL 7 0 100
.TRAN 0.05M 2M 0
.PROBE
.END
```

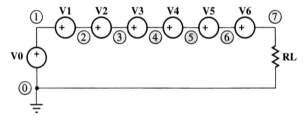

Figure 17.29 Waveform synthesis circuit of Example 17.18

After directing the Probe post-processor to display $V(7)$, we obtain the waveform of Figure 17.30. ◀

Exercise 17.22   Use PSpice to synthesize and display a *full-wave rectified sinusoidal current waveform* of the type of Figure 17.4, with $A = 5$ mA and $T = 1$ μs. Use Equation (17.10), and display the waveform first using the first two harmonics, then using the first six harmonics.

Fourier Analysis

Given a periodic signal, SPICE can be used to find the coefficients of a trigonometric series representation of the form

$$f(t) = A_0 + \sum_{n=1}^{\infty} A_n \sin(n\omega_0 t + \psi_n) \qquad \text{(17.100)}$$

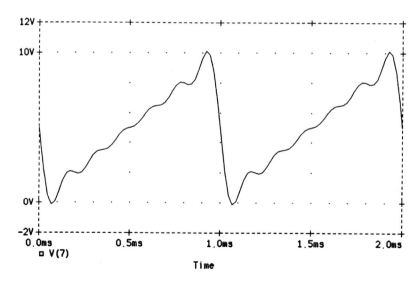

Figure 17.30 Fourier synthesis of the sawtooth wave.

This can readily be converted to the standard amplitude-phase representation of Equation (17.2) by letting

$$\theta_n = \psi_n - 90° \tag{17.101}$$

Besides the Fourier coefficients, SPICE calculates also the **total harmonic distortion** (THD) of $f(t)$, defined as

$$\text{THD} = 100 \frac{\sqrt{A_1^2 + A_2^2 + A_3^2 + \cdots}}{A_0} \tag{17.102}$$

This parameter provides an indication of the degree of departure of a periodic waveform from a pure sine wave.

SPICE Fourier analysis is initiated by the statement

```
.FOUR  FREQ  OUTVAR1  OUTVAR2 ... OUTVAR8
```
(17.103)

where FREQ is the *fundamental frequency,* in Hz, and OUTVAR1 through OUTVAR8 are the voltage or current variables of interest, for a maximum of eight. The .FOUR statement must be used in connection with the .TRAN statement, repeated here for convenience,

```
.TRAN  TSTEP  TSTOP  TSTART  TMAX  UIC
```
(17.104)

SPICE computes the dc component as well as the amplitudes and phase angles of the first *nine* harmonics over the interval from (TSTOP $-T$) to TSTOP, where T is the fundamental period. To ensure a reasonable degree of accuracy, choose TMAX $\simeq T/100$.

A CONVERSATION WITH
DENICE DENTON
UNIVERSITY OF WISCONSIN, MADISON

Denice Denton combines electrical engineering, chemistry, and materials science as she researches using polymers as insulators in memory chips.

Denice Denton is an associate professor of electrical and computer engineering at the University of Wisconsin, Madison. She did both her undergraduate work and her graduate work at Massachusetts Institute of Technology, receiving her PhD in 1987. Dr. Denton attended public schools in Houston, Texas. She was interested in science and math in high school and was a member of JETS (Junior Engineering Technology Society). She also participated in a summer program in engineering at Rice University in Houston. These activities led her to choose engineering as a career.

What is your field of research, Dr. Denton?

My major interest is microelectronics and microelectric-mechanical systems (MEMS). My work in graduate school was a combination of chemistry, materials science, and electrical engineering, and now I'm using all of these fields in my research. Electrical engineering has become a very interdisciplinary field. Microelectronics focuses on the fabrication of microchips. In integrated circuit (IC) fabrication, we work extensively with a wide variety of materials and processes. Knowledge of chemistry is essential to develop new processes for IC fabrication.

MEMS is an exciting field because of all the potential applications. One analogy I like for MEMS is the "machine shop on a chip." Imagine being able to build a motor on a microchip the diameter of three human hairs. In five to ten years, we might see such motors being used to clear clogged arteries the same way a Roto-Rooter®cleans tree roots out of pipes or to "weed whack" as a form of retinal surgery.

What's your advice to undergraduates studying EE?

Get as broad an education as you can. The field is changing so quickly that everyone has to be ready to move from one subdiscipline to another, and having an engineering degree is an entry to medicine, law, whatever.

Force yourself to slog through the early material in your courses. I promise things get more interesting down the road. If you stick it out you are going to have a lot more options. Go for breadth. Take lots of humanities courses, literature courses, writing courses. My minors in undergraduate school were art and music, and I firmly believe that being exposed

> *One analogy I like for MEMS is the "machine shop on a chip."*

to fields so far removed from the physical sciences taught me a different way of thinking.

One of the main tasks of any working engineer is to convince management to support his or her projects. If your managers are not from a technical background, you need to understand their way of thinking if you are going to plead your cause successfully.

Where do you see the fields of electrical and electronic engineering going in the next decade?

I think the field is going to get broader. What is now called the Institute of Electrical and Electronics Engineers (IEEE) used to be known as The Institute of Radio Engineers (IRE) fifty years ago. Back then it was a relatively narrow field. The transformation from IRE to IEEE still continues. Telecommunications and satellite research are hot. Some futurologists are predicting that what they call the "electronic room" will totally change the way business is

> *Telecommunications and satellite research are hot.*

done in this country. Such a room would be accessible via the computer. People from around the world could simultaneously "log on" to the same electronic room and it would be as though they were physically together. They would have access to the same scratch pads and projection screens in order to share ideas and to manipulate those ideas using the computer.

The MEMS devices I mentioned earlier may take off, and someone may design MEMS to act as antibodies that could be injected into people to fend off diseases we now consider incurable.

Another area I'm intrigued by is photonics, which is running computers with light rather than with electric current. The big challenge right now in photonics is size. The optical switches we're using today are much larger than transistors. Someone's got to shrink these switches a lot if light is ever going to replace electricity in computers. There are some novel MEMS devices for fiber-optic switching, but they require relatively high operating voltages (70 V). We need to find ways to shrink optical switches and lower power requirements.

▶ Example 17.19

Use SPICE to verify the input and output spectra of Example 17.6.

Solution

We have $f_0 = \omega_0/2\pi = 10^3/2\pi = 159.154$ Hz, and $T = 1/f = 6.2832$ ms. With reference to Figure 17.31, the input file is

```
FOURIER ANALYSIS OF R-C RESPONSE TO A SQUARE WAVE
V1 1 0 PULSE(15.708 0 1.5708M 1N 1N 3.1416M 6.2832M)
R 1 2 10K
C 2 0 50N
.TRAN 0.05M 12.5664M 0 0.05M
.FOUR 159.154 V(1) V(2)
.END
```

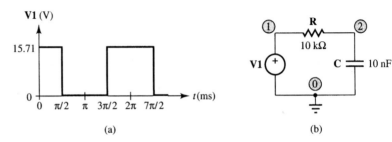

Figure 17.31 Input waveform and circuit of Example 17.19

After SPICE is run, the output file contains the following

```
FOURIER COMPONENTS OF TRANSIENT RESPONSE V(1)

DC COMPONENT = 7.931762E+00
```

HAR-MONIC NO	FRQNCY (HZ)	FOURIER COMPONENT	NORMALIZED COMPONENT	PHASE (DEG)	NORMALIZED PHASE (DEG)
1	1.592E+02	1.000E+01	1.000E+00	9.000E+01	0.000E+00
2	3.183E+02	1.257E-01	1.257E-02	-9.000E+01	-1.800E+02
3	4.775E+02	3.334E+00	3.334E-01	-9.000E+01	-1.800E+02
4	6.366E+02	1.258E-01	1.258E-02	9.000E+01	6.679E-13
5	7.958E+02	2.001E+00	2.001E-01	9.000E+01	5.684E-14
6	9.549E+02	1.260E-01	1.260E-02	-9.000E+01	-1.800E+02
7	1.114E+03	1.430E+00	1.430E-01	-9.000E+01	-1.800E+02
8	1.273E+03	1.263E-01	1.263E-02	9.000E+01	1.222E-12
9	1.432E+03	1.113E+00	1.113E-01	9.000E+01	2.984E-13

```
TOTAL HARMONIC DISTORTION = 4.297682E+01 PERCENT
```

```
FOURIER COMPONENTS OF TRANSIENT RESPONSE V(2)

DC COMPONENT = 7.853358E+00

HAR-    FRQNCY    FOURIER     NORMALIZED    PHASE      NORMALIZED
MONIC    (HZ)     COMPONENT   COMPONENT     (DEG)      PHASE  (DEG)
 NO
  1    1.592E+02  8.946E+00   1.000E+00    6.344E+01   -0.000E+00
  2    3.183E+02  1.218E-03   1.361E-04    4.566E+01   -1.778E+01
  3    4.775E+02  1.850E+00   2.068E-01   -1.463E+02   -2.098E+02
  4    6.366E+02  9.474E-04   1.059E-04   -1.658E+02   -2.293E+01
  5    7.958E+02  7.426E-01   8.301E-02    2.172E+01   -4.172E+01
  6    9.549E+02  8.257E-04   9.230E-05    2.782E+00   -6.066E+01
  7    1.114E+03  3.920E-01   4.382E-02   -1.642E+02   -2.276E+02
  8    1.273E+03  7.818E-04   8.739E-05    1.712E+02    1.077E+02
  9    1.432E+03  2.404E-01   2.687E-02    1.238E+01   -5.106E+01

TOTAL HARMONIC DISTORTION = 2.286496E+01 PERCENT
```

After subtracting 90° from the phase angles, we find a reasonable agreement with the data of Example 17.6. The slight discrepancies are due to roundoff errors in the calculations.

Exercise 17.23   Use SPICE to verify the input and output spectra of Exercise 17.7.

Piecewise-Linear Waveforms

SPICE allows for independent source waveforms to be specified in piecewise-linear fashion. The general forms of the statements for these sources are

```
VXXX  N+  N-  PWL(T1  V1  T2  V2  T3  V3 ...)
```  **(17.105a)**
```
IXXX  N+  N-  PWL(T1  I1  T2  I2  T3  I3 ...)
```  **(17.105b)**

where VK and IK (K = 1, 2, 3, ...) are the source values, in V or in A, at the instant TK, in seconds. SPICE calculates the source values between consecutive data points by linear interpolation.

▶ **Example 17.20**

Use SPICE to perform the Fourier analysis of the waveform of Figure 17.32.

Solution

The input file is

```
FOURIER ANALYSIS OF A PIECEWISE LINEAR WAVEFORM
V1 1 0 PWL(0 -1 1 -1 4 3)
RL 1 0 100
```

```
.FOUR 0.25 V(1)
.TRAN 0.04 4 0 0.04
.END
```

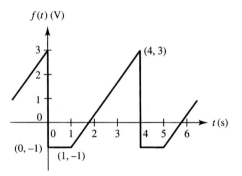

Figure 17.32 Waveform of Example 17.20.

After SPICE is run, the output file contains the following:

```
FOURIER COMPONENTS OF TRANSIENT RESPONSE V(1)

DC COMPONENT = 4.800000E-01
```

HAR-MONIC NO	FRQNCY (HZ)	FOURIER COMPONENT	NORMALIZED COMPONENT	PHASE (DEG)	NORMALIZED PHASE (DEG)
1	2.500E-01	1.560E+00	1.000E+00	1.715E+02	0.000E+00
2	5.000E-01	6.429E-01	4.121E-01	1.715E+02	-3.475E-02
3	7.500E-01	3.932E-01	2.520E-01	-1.786E+02	-3.501E+02
4	1.000E+00	3.191E-01	2.046E-01	-1.728E+02	-3.443E+02
5	1.250E+00	2.651E-01	1.699E-01	-1.737E+02	-3.452E+02
6	1.500E+00	2.112E-01	1.353E-01	-1.733E+02	-3.448E+02
7	1.750E+00	1.767E-01	1.133E-01	-1.688E+02	-3.403E+02
8	2.000E+00	1.608E-01	1.031E-01	-1.656E+02	-3.371E+02
9	2.250E+00	1.458E-01	9.343E-02	-1.655E+02	-3.370E+02

```
TOTAL HARMONIC DISTORTION = 5.954000E+01 PERCENT
```

Exercise 17.24 Use SPICE to perform the Fourier analysis of the waveform of Figure 17.9.

▼ Summary

- A periodic signal can be expressed as an infinite summation of sinusoidal components called *harmonic components,* plus a constant term representing the *average* or *dc component* of the signal.

- The Fourier series can take on three different but equivalent forms: (a) the *amplitude-phase* series and (b) the *cosine-sine* series, both known as *trigonometric* series, and (c) the *exponential* series. We use the *amplitude-phase* series when we wish to visualize the harmonic makeup of a signal via its *frequency spectra*, the *cosine-sine* series to calculate the *Fourier coefficients*, and the *exponential series* when we seek mathematical conciseness.

- The *power* of a periodic signal is the sum of the squares of the *dc* component and the *rms* values of its harmonic components. The *power spectrum* consists of lines conveying information on how the power of a signal distributes among its dc and harmonic components.

- Subjecting a periodic signal to a *time shift* changes the *phase angle* of each component in proportion to its harmonic number.

- The cosine-sine series of an *even* periodic function contains only *cosine* terms and a constant term, and the series of an *odd* periodic function contains only *sine* terms. A function with *half-wave symmetry* contains only *odd* harmonics.

- The effect of a circuit upon a periodic signal is visualized by means of *input* and *output spectra*. The *amplitude* (or *phase*) spectrum of the response is obtained by *multiplying* (or *adding*), line by line, the *amplitude* (or *phase*) *spectrum* of the applied signal, and the *magnitude* (or *phase angle*) of the *network function* $H(j\omega)$.

- Based on their frequency response, filters are classified as *low-pass, high-pass, band-pass, band-reject,* and *all-pass* filters.

- The *Fourier transform* extends the frequency-domain attributes of *amplitude* and *phase angle* to *aperiodic* signals. The *Fourier transform* effects a transformation from a function of time $f(t)$ to a function of frequency $F(j\omega)$. The *inverse Fourier transform* effects a transformation from $F(j\omega)$ back to $f(t)$.

- The physical significance of the Fourier transform is that its magnitude squared $|F(j\omega)|^2$ represents the *energy density* of the corresponding signal $f(t)$. Moreover, the area under this curve represents the *total energy* of $f(t)$.

- Some important Fourier transform *pairs* are $A \leftrightarrow 2\pi A\delta(\omega)$, $\delta(t) \leftrightarrow 1$, $u(t) \leftrightarrow [\pi\delta(\omega) + 1/j\omega]$, $e^{-at}u(t) \leftrightarrow 1/(a + j\omega)$, $e^{j\omega_0 t} \leftrightarrow 2\pi\delta(\omega - \omega_0)$, and $\cos\omega_0 t \leftrightarrow \pi[\delta(\omega + \omega_0) + \delta(\omega - \omega_0)]$.

- Some important *functional transforms* are as follows: $df/dt \leftrightarrow j\omega F(j\omega)$, $\int_{-\infty}^{t} f(\xi)\,d\xi \leftrightarrow F(j\omega)/j\omega$, $f(t)\cos(\omega_0 t) \leftrightarrow \frac{1}{2}[F(j\omega - j\omega_0) + F(j\omega + j\omega_0)]$, $f(at) \leftrightarrow (1/a)F(j\omega/a)$, $f(t - t_0) \leftrightarrow e^{-j\omega t_0}F(j\omega)$, and $h(t) * x(t) \leftrightarrow H(j\omega)X(j\omega)$.

- The Fourier transform of a *causal* signal can be obtained from its Laplace transform by letting $s \rightarrow j\omega$, provided all the poles of the Laplace transform lie in the *left half* of the s plane.

- Laplace techniques are better suited to *circuit analysis* because the Laplace integral converges for a wider range of functions and the Laplace transform also accounts for the initial conditions. Fourier techniques are better suited to *communication* and *signal processing* because they emphasize the steady-state *frequency-domain* characteristics of signals and systems.

- SPICE can be directed to perform the Fourier analysis of a periodic signal using the `.FOUR` statement. By using the `PWL` feature, we can perform the Fourier analysis of a wide variety of piecewise-linear waveforms.

▼ PROBLEMS

17.1 The Fourier Series

17.1   Derive Equation (17.6c) for the coefficient b_n.

17.2   Find the trigonometric Fourier series for the waveform of Figure P17.2.

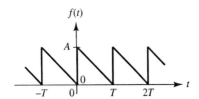

Figure P17.2

17.3   Show that for the periodic function of Figure P17.3,

$$a_0 = \frac{1}{4}, a_n = \frac{3}{n\pi}\sin\frac{3n\pi}{2},$$

$$b_n = \frac{3}{n\pi}\left(1 - \cos\frac{3n\pi}{2}\right), n = 1, 2, \ldots$$

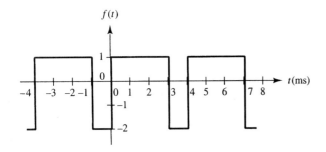

Figure P17.3

17.4   Show that for the periodic function of Figure P17.4,

$$a_0 = \frac{A}{4}, a_n = \frac{A}{\pi^2 n^2}[(-1)^n - 1],$$

$$b_n = \frac{A}{\pi n}(-1)^{n+1}, n = 1, 2, \ldots$$

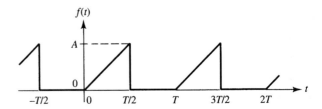

Figure P17.4

17.5   Sketch and label the periodic function defined as $f(t) = A|t|$ for $-1 < t < 1$, $f(t) = 0$ for $-2 \le t \le -1$ and $1 \le t \le 2$, and $f(t + 4) = f(t)$; hence, show that

$$a_0 = \frac{A}{4}, a_n = \frac{4A}{n^2\pi^2}\left(\cos\frac{n\pi}{2} - 1\right) + \frac{2A}{n\pi}\sin\frac{n\pi}{2},$$

$$b_n = 0, n = 1, 2, \ldots$$

17.6   Find the trigonometric Fourier series for the waveform of Figure P17.6.

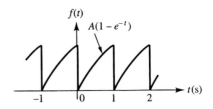

Figure P17.6

17.7   Sketch and label the periodic function defined as $f(t) = A(1 - t^2)$ for $-1 < t < 1$, and $f(t + 2) = f(t)$; hence, show that its trigonometric Fourier series is

$$f(t) = A\left\{\frac{2}{3} - \frac{4}{\pi^2}\sum_{n=1}^{\infty}\frac{(-1)^n}{n^2}\cos n\pi t\right\}$$

17.8 Find the trigonometric Fourier series for the partial sine wave of Figure P17.8.

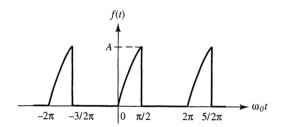

Figure P17.8

17.9 Find the rms value of the waveform of Figure P17.4 using both Parseval's theorem and the rms defining formula of Section 4.4.

17.10 Let the waveform of Figure P17.3 be a voltage waveform, so that the vertical scale is in V. (a) Write the first five nonzero terms of its amplitude-phase Fourier series. (b) Find the power $P_{1\,\Omega}$ associated with each term. (c) Find the total average power delivered by the first five nonzero terms to a 10-Ω resistor. (d) What percentage of the total average power is delivered by these terms?

17.11 Find the percentage of the total average power delivered to a 1-Ω resistance by the fundamental if the waveform is a voltage waveform of the type of (a) Figure P17.2, (b) Figure P17.5, and (c) Problem 17.7.

17.12 Find the number of Fourier components needed to predict the rms value of the waveform of Problem 17.7 with 0.1% accuracy.

17.2 The Effect of Shifting and Symmetry

17.13 Suitably modify Equation (17.9b) to obtain the amplitude-phase Fourier series for the voltage waveform of Figure P17.13.

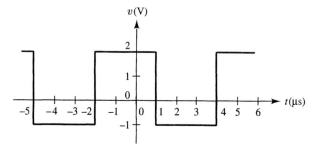

Figure P17.13

17.14 Figure P17.14 shows a portion of a periodic waveform $f(t)$. Sketch the complete waveform if (a) $f(t)$ is even and

has only even harmonics, (b) $f(t)$ is odd and has only odd harmonics, (c) $f(t)$ is even and has only odd harmonics, (d) $f(t)$ is odd and has only even harmonics, (e) $f(t)$ is even and has all harmonics, and (f) $f(t)$ is odd and has all harmonics.

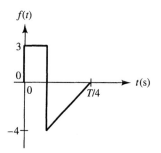

Figure P17.14

17.15 Show that the waveform of Figure P17.15 admits the trigonometric Fourier series

$$f(t) = \frac{12A}{\pi^2} \sum_{k=1}^{\infty} \frac{\sin\left[(2k-1)\pi/3\right]}{(2k-1)^2} \sin(2k-1)\frac{\pi}{3}10^3 t$$

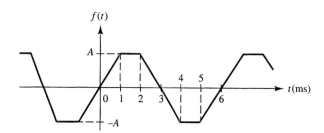

Figure P17.15

17.16 Find the trigonometric Fourier series for the waveform of Figure P17.16.

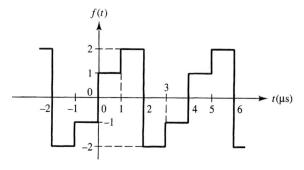

Figure P17.16

17.17   Find the percentage of the total average power delivered to a 1-Ω resistance by the first five nonzero terms of the Fourier series of the waveform of Figure P17.17.

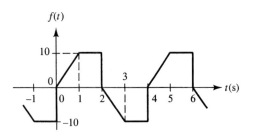

Figure P17.17

17.18   Estimate the rms value of the waveform of Figure P17.18 using both Parseval's theorem and the rms defining formula of Section 4.4.

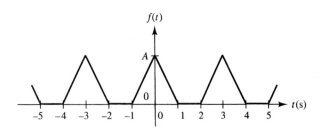

Figure P17.18

17.3 Frequency Spectra and Filtering

17.19   Sketch and label the amplitude spectrum of (a) $v(t) = 120\sqrt{2}\sin 2\pi 60t$ V, (b) the waveform obtained by half-wave rectifying $v(t)$, and (c) the waveform obtained by full-wave rectifying $v(t)$. Compare the spectra and comment on the unique features of each.

17.20   In the circuit of Figure P17.20(b) find the Fourier series of the steady-state response $v_O(t)$ if $v_I(t)$ is the sawtooth wave of Figure P17.20(a).

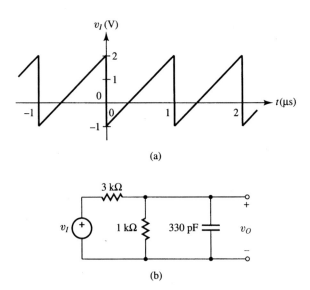

(a)

(b)

Figure P17.20

17.21   In the circuit of Figure P17.21(b) find the Fourier series of the steady-state response $v_O(t)$ if $i_I(t)$ is the waveform of Figure P17.21(a).

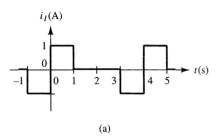

(a)

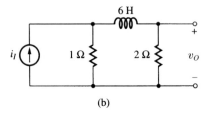

(b)

Figure P17.21

17.22   In the circuit of Figure P17.22(b) find the Fourier series of the steady-state response $v_O(t)$ if $v_I(t)$ is the square wave of Figure P17.22(a).

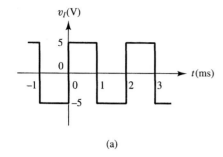

(a)

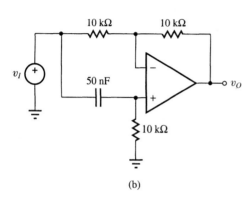

(b)

Figure P17.22

17.23 As illustrated in Figure P17.23, subjecting an integrator to a *square wave* input with zero dc value yields a *triangle wave* output. Justify this behavior both via time-domain and via Fourier analysis.

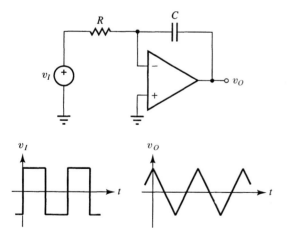

Figure P17.23

17.24 In the circuit of Figure P17.24(b) calculate the first four nonzero terms in the Fourier series of the steady-state response $v_O(t)$ if $v_I(t)$ is the square wave of Figure P17.24(a). Hence, justify the claim that this circuit converts its square wave input to an almost sinusoidal output.

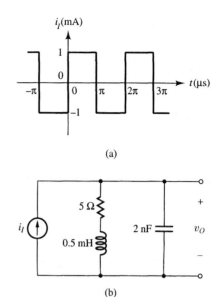

(a)

(b)

Figure P17.24

17.25 In the circuit of Figure P17.25(b) calculate the first four nonzero terms in the Fourier series of the load voltage v_L, given that the source voltage v_S is the full-wave rectified sinewave of Figure P17.25(a). Hence, justify the claim that this circuit converts the pulsating source voltage to an almost constant load voltage.

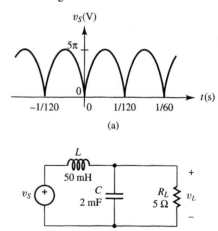

(a)

(b)

Figure P17.25

17.26 In the circuit of Figure P17.20 estimate the total average power supplied by the source.

17.4 The Exponential Fourier Series

17.27 Find the exponential Fourier series for the half-wave rectified sinewave of Figure 17.2.

17.28 Find the exponential Fourier series for the waveform of Figure 17.9.

17.29 Find the exponential Fourier series for the sawtooth waveform of Figure P17.2; sketch and label its amplitude and phase spectra.

17.30 Repeat Problem 17.29 for the waveform of Figure P17.4.

17.31 Find the exponential Fourier series for the output v_O in the circuit of Figure P17.21; sketch and label both the input and output amplitude spectra.

17.32 Using an ideal op amp, design a circuit that accepts the input signal

$$v_I(t) = \frac{4A}{\pi^2} \sum_{k=-\infty}^{\infty} \frac{1}{(2k-1)^2} e^{j(2k-1)10^3 t} \text{ V}$$

representing a *triangle* waveform, and yields the output signal

$$v_O(t) = \frac{2A}{j\pi} \sum_{k=-\infty}^{\infty} \frac{1}{(2k-1)} e^{j(2k-1)10^3 t} \text{ V}$$

representing a *square* wave. What type of circuit is this?

17.5 The Fourier Transform

17.33 Sketch and label the single *triangle* pulse, defined as

$$f(t) = \begin{cases} A(1 - |t|/W) & \text{for } |t| \le W \\ 0 & \text{for } |t| \ge W \end{cases}$$

Hence, find its Fourier transform.

17.34 (a) Find the Fourier transform of the single *trapezoidal* pulse of Figure P17.34. (b) Find the exponential Fourier series of the periodic waveform whose shape over half a period is that of Figure P17.34.

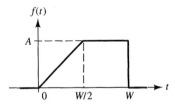

Figure P17.34

17.35 Find $F(j\omega)$ if (a) $f(t) = te^{-at}u(t)$, $a > 0$; (b) $f(t) = e^{-at^2}$, $a > 0$, and (c) $e^{-at}u(t - t_0)$, $a > 0$.

17.36 (a) Sketch and label the function

$$F(j\omega) = \begin{cases} A & \text{for } |\omega| < 1 \\ B & \text{for } 1 < |\omega| < 2 \\ 0 & \text{for } |\omega| > 2 \end{cases}$$

and find its inverse Fourier transform $f(t)$. (b) Discuss the cases $B = 0$, $B = A$, and $B = 2A$.

17.37 Given the pulse of Figure 17.24(a), find its total 1-Ω energy, as well as the percentage of this energy contained within the main lobe of its Fourier transform, that is, within the frequency range $-1/W \le f \le 1/W$.

17.38 Sketch and label the Fourier transform whose magnitude, in V-s, is

$$|V(j\omega)| = \begin{cases} 10(1 - |\omega|/10^3\pi) & \text{for } |\omega| \le 10^3\pi \text{ rad/s} \\ 0 & \text{for } |\omega| \ge 10^3\pi \end{cases}$$

What is the total energy delivered by $v(t)$ to a 2-kΩ resistance?

17.39 Given that $f(t) = 5(e^{-t} \sin t)u(t)$, find (a) $F(j\omega)$, and the 1-Ω energy of $f(t)$ contained within (b) the time interval $|t| \le 1$ s, and (c) the frequency range $|\omega| < 1$ rad/s.

17.40 Given that $F(j\omega) = e^{-|\omega|}$, find (a) $f(t)$, (b) the time interval $|t| \le t_1$, and (c) the frequency interval $|\omega| \le \omega_1$ in which 50% of $W_{1\,\Omega}$ lies.

17.6 Properties of the Fourier Transform

17.41 Prove the *differentiation* and *integration* properties of the Fourier transform as expressed in Equations (17.81) and (17.82). In each case state the restrictions on $f(t)$ for the corresponding property to hold.

17.42 Prove the *scaling* and *shifting* properties of the Fourier transform as expressed by Equations (17.83) through (17.85).

17.43 Find $F(j\omega)$ if (a) $f(t) = 12\cos^2(2\pi 10^3 t)$, (b) $f(t) = 10\cos(\omega_0 t + \theta)$, and (c) $f(t) = A[e^{-\alpha t}\cos(\omega_d t + \theta)]u(t)$ $\alpha > 0$.

17.44 Starting with the Fourier transform of the single pulse of Example 17.8 and using the linearity and shifting properties, find the Fourier transforms of $f_1(t)$ and $f_2(t)$ of Figure P17.44.

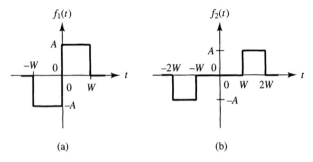

(a)   (b)

Figure P17.44

17.45 Find the Fourier transform of the function of Figure P17.45, defined as $f(t) = A \sin \pi 10^3 t$ for $|t| \le 1$ ms, and $f(t) = 0$ elsewhere. (*Hint:* Differentiate the function twice and apply the differentiation property of the Fourier transform.)

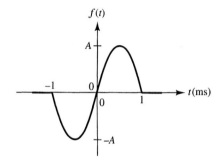

Figure P17.45

17.46 Confirm the results of Problems 17.33 and 17.34(a) by applying the differentiation property of the Fourier transform.

17.47 The function of Figure P17.45 can be regarded as the product $f(t) = f_1(t)f_2(t)$, where $f_1(t) = A \sin \pi 10^3 t$, and $f_2(t) = 1$ for $|t| < W/2$, and $f_2(t) = 0$ for $|t| > W/2$, $W = 2$ ms. (a) Use convolution in the frequency domain to find $F(j\omega)$. (b) Discuss what happens to $F(j\omega)$ as W is increased to make $f(t)$ include more and more cycles of $f_1(t)$. (c) Verify in the limit $W \to \infty$.

17.48 Find $f(t)$ and $\mathcal{F}\{f(t)\}$ if (a) $\mathcal{L}\{f(t)\} = 10/[s(s + 1)^2]$ and (b) $\mathcal{L}\{f(t)\} = (2s + 1)/[s(s^2 + 1)]$.

17.7 Fourier Transform Applications

17.49 The signal $v_I = V_m e^{-at} u(t)$ V, $a > 0$, is applied to an ideal low-pass filter having a cutoff frequency of 1 Hz. Find a so that the energy of the output v_O from the filter is exactly *half* that of the input v_I.

17.50 The signal $v_I = V_m e^{-t/(1\,\text{ms})} u(t)$ V is applied to an ideal high-pass filter having cutoff frequency f_c. Find f_c so that a specified percentage η of the input signal is transmitted to the output. Discuss the cases $\eta \to 0$, $\eta = 0.5$, and $\eta \to 1$.

17.51 Find the energy dissipated by the resistance in the circuit of Figure P17.51.

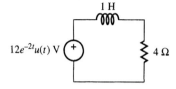

Figure P17.51

17.52 In the circuit of Figure P17.52 let $v_I(t) = 2\delta(t)$ V. Find the percentage of the total energy of $v_O(t)$ contained within the range $0 \le \omega \le 20$ krad/s.

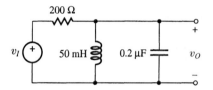

Figure P17.52

17.53 Repeat Problem 17.52 if $v_I = 2.5$ sgn (t) V and R is increased to 250 Ω.

17.54 In the circuit of Figure P17.54 let $R_1 = R_2 = 10$ kΩ and $C_1 = 2C_2 = 1$ μF. (a) Use Fourier analysis techniques to find the steady-state response v_O if $v_I = 10 \sin 100t$ V. (b) Check your result using phasors.

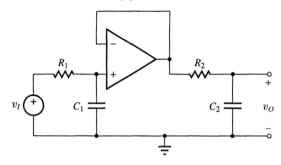

Figure P17.54

17.55 In the circuit of Figure P17.54 let $R_1 = R_2 = 10$ kΩ, $C_1 = C_2 = 1$ μF, and $v_I = 10u(t)$ V. (a) Assuming zero initial stored energy in C_1 and C_2, find $v_O(t \ge 0)$ using Fourier analysis techniques. (b) Check your result using time-domain techniques.

17.56 In the circuit of Figure P17.54 let $R_1 = R_2 = 10$ kΩ, $C_1 = 2C_2 = 1\mu$F, and $v_I = 5e^{-200t} u(t)$ V. Assuming both C_1 and C_2 are initially discharged, find the 1-Ω energy of v_O.

17.57 White noise refers to an unwanted, randomly varying signal $v_n(t)$ having a constant energy density spectrum, $|V_n(j\omega)|^2 = V_n^2$, where V_n is frequency independent. If a white-noise voltage $v_{ni}(t)$ with energy density spectrum V_{ni}^2 is applied to a low-pass R-C filter as in Figure P17.57, find the total 1-Ω energy of the noise voltage v_{no} emerging from the filter as a function of the filter cutoff frequency $\omega_0 = 1/RC$.

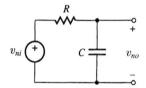

Figure P17.57

17.58 Integrated-circuit (IC) noise refers to the randomly varying signals found in integrated circuits such as IC op amps.

This type of noise exhibits an energy density spectrum of the type

$$|V_n(j\omega)|^2 = V_n^2 \left(\frac{\omega_c}{\omega} + 1\right)$$

Find the 1-Ω energy of $v_n(t)$ contained within a given frequency range $\omega_L \le |\omega| \le \omega_H$.

▼ 17.8 Fourier Techniques Using SPICE

17.59 Use SPICE to find the Fourier components of the waveform of Figure P17.4. Hence, direct PSpice to display the waveforms consisting, respectively, of the first nonzero term, the first two nonzero terms, the first three nonzero terms, the first four nonzero terms, and so forth, until all available terms are exhausted.

17.60 Use SPICE to find whether the half-wave or the full-wave rectified sinusoid exhibits more total harmonic distortion. Can you justify your finding intuitively?

17.61 Use SPICE to find the input and output spectra of the circuit of Figure P17.21.

17.62 Repeat Problem 17.61 for the circuit of Figure P17.22.

17.63 Use PSpice to display v_L in the circuit of Figure P17.25.

17.64 Use SPICE to compare the total harmonic distortion of the input and output for the square–to–sinewave converter of Figure P17.24. Comment on your findings.

ANSWERS TO ODD-NUMBERED PROBLEMS

CHAPTER 1

1.1

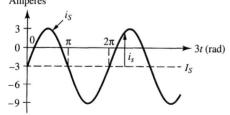

1.3 19.224×10^{-19} J, electron energy increases, battery releases energy.

1.5 (a) $5(1 - 10^3 t)e^{-10^3 t}$ A;
(b) 2 ms, -0.677 A.

1.7 (a) \$0.56;
(b) \$21.90.

1.9 10.834 J, absorbed.

1.11 (a) 20 W, X_1 to X_2;
(b) 10 W, X_2 to X_1;
(c) 120 W, X_2 to X_1;
(d) 24 W, X_1 to X_2;
(e), (f) 0 W.

1.13 (a)

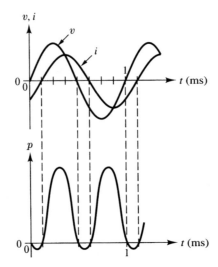

(b) $p > 0 \Rightarrow$ power from X_1 to X_2; $p < 0 \Rightarrow$ power from X_2 to X_1;

(c) 42.678 W.

1.15 (a) $i_S = -2.5 + 2.5 \sin 2\pi 10^3 t$ mA;

(b) $v_S = -12.5 + 2.5 \sin 2\pi 10^6 t$ V.

1.17 3.642 V.

1.19 27.87 mA.

1.21 $V_{av} = V_m / \pi$.

1.23 $T = 0.5$ ms, $I_{av} = 5$ A.

1.25 7.5 V.

1.27 (a)

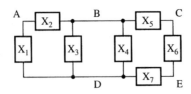

Nodes: A, B, C, D, and E; simple nodes: A, C, and E.

(b) Meshes: $X_1 X_2 X_3$, $X_3 X_4$, $X_4 X_5 X_6 X_7$. Loops: $X_1 X_2 X_4$, $X_1 X_2 X_5 X_6 X_7$, $X_3 X_5 X_6 X_7$.

(c) Series: $X_1 X_2$, $X_5 X_6 X_7$; parallel: $X_3 X_4$.

1.29

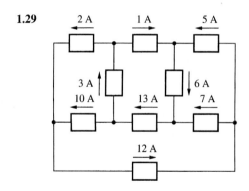

1.31

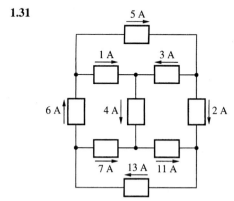

1.33

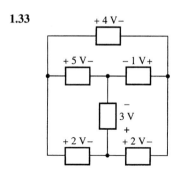

1.35

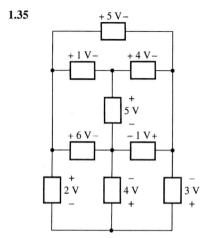

1.37 (a)

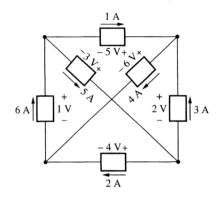

(b) $p_{rel} = p_{abs} = 32$ W.

1.39 (a)

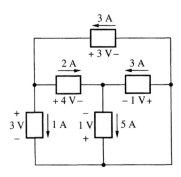

(b) $p_{rel} = p_{abs} = 14$ W.

1.41 26 mA, 26 μA, 26 nA.

1.43 (a) 2.5 mA;
(b) 6.6 V.

1.45 (a) 632.5 μA/V;
(b) 7.5 V.

1.47 0.25 kΩ, 2 V.

1.49 (a) 3 A (counterclockwise);
(b) $p_{6\,V} = 18$ W (absorbed), $p_{X_1} = 12$ W (delivered).

1.51 (a) 2 V (+@ top);
(b) $p_{3\,A} = 6$ W (absorbed), $p_{X_1} = 4$ W (absorbed).

1.53 (a) 20 W (absorbed by the 10-V source and delivered by the 2-A source);
(b) 20 W (absorbed by the 2-A source, delivered by the 10-V source).

CHAPTER 2

2.1 (a) 12 V, 6 A, 2 A;
(b) $p_{rel} = p_{abs} = 72$ W.

2.3   (a)   100 V, 0.1 A; 31.6 V, 31.6 mA; 11.2 V, 11.2 mA.
          (b)   10 V, 1 A; 3.16 V, 0.316 A; 1.12 V, 112 mA.

2.5   (a)   35.65 mΩ;
          (b)   35.65 mV, 35.65 mV/m.

2.7   0.955 μm.

2.9   (a)   25 kΩ ± 9%;
          (b)   4 kΩ ± 6%.

2.11 (a)   17.5 kΩ;
          (b)   40 kΩ;
          (c)   16.18 kΩ.

2.13 (a)   10//(10 + 10 + 10) kΩ;
          (b)   10 + (10//10) kΩ;
          (c)   10 + 10 + (10//10) kΩ.

2.15 (a)   25 Ω;
          (b)   9 W.

2.17 (a)   4 Ω;
          (b)   48 V.

2.19 2.5 A.

2.21 (a)   1 V, 2 V;
          (b)   10/3 Ω.

2.23 (a)   15 V;
          (b)   13.5 V.

2.25 0.731 V/V ≤ gain ≤ 0.768 V/V.

2.27 R_1 = 840 Ω, R_2 = 600 Ω.

2.29

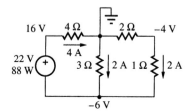

2.31 15 kΩ, 5 kΩ.

2.33

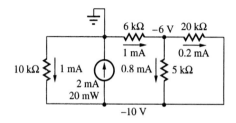

2.35 1 Ω, 4 Ω.

2.37 (b)   100 Ω, 800 Ω.

2.39   General circuit:

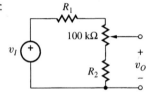

(a) $R_1 = 33.3$ kΩ, $R_2 = 0$;
(b) $R_1 = 0$, $R_2 = 25$ kΩ;
(c) $R_1 = R_2 = 12.5$ kΩ.

2.41   (a)   [kΩ]: $R_1 = 10(30)$, $R_2 = 30(10)$, $R_3 = 40(120)$, $R_4 = 120(40)$, $R_{eq} = 32$; $R_1 = 10(40)$, $R_2 = 40(10)$, $R_3 = 30(120)$, $R_4 = 120(30)$, $R_{eq} = 37.5$.
(b)   second bridge pair; first bridge pair.

2.43   3 V (+ @ C), 4.8 W.

2.45   89/11 V, 89/44 mW.

2.47   (a) $R = 1$ kΩ, $R_1 = 9$ kΩ, $R_2 = 10/9$ kΩ;
(b) $R = 2.5$ kΩ, $R_1 = 7.5$ kΩ, $R_2 = 10/3$ kΩ.

2.49   15 mV, 2 MΩ

2.51   12 mA, 10 kΩ.

2.53   400 kΩ, 100 kΩ.

2.55 $R_i \leq 20$ Ω.

2.57 $0.4\pi \times 10^6$ V/s

2.59   (a)

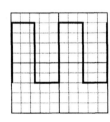   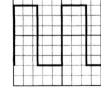

(b) $d = 1/3$.

CHAPTER 3

3.1   (a)   8 mA;
(b) $-8/3$ mA.

3.3   0.8 V, 16 V.

3.5 $X_2 = 4$ Ω, $X_3 = 2$ Ω, $X_4 = 3$ Ω, $X_5 = 1$ Ω.

3.7

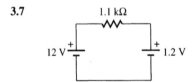

3.9

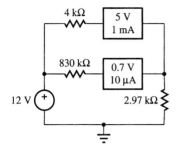

3.11   (a)

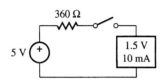

(b)

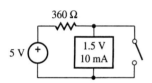

(c)   Circuit of (b) is more wasteful.

3.13   (a) $v_{4\,\Omega} = 4$ V (+ @ top), $v_{2\,\Omega} = 2$ V (+ @ bottom);
         (b) $p_{2\,A} = 12$ W (released), $p_{3\,V} = 0$.

3.15   (a) $v_{4\,\Omega} = 1.2$ V (+ @ top), $v_{2\,\Omega} = 3$ V (+ @ top);
         (b) $p_{3\,V} = 5.4$ W (released), $p_{1.8\,V} = 2.7$ W (released).

3.17 $v_{30\,k\Omega} = 12$ V (+ @ top), $v_{60\,k\Omega} = 17$ V (+ @ top).

3.19 $v_{16\,k\Omega} = 268/49$ V (+ @ top), $v_{40\,k\Omega} = 50/49$ V (+ @ left).

3.21   48 W.

3.23   (a)   1.8 W (absorbed);
         (b)   4.2 W (delivered).

3.25   40,401/1369 W.

3.27   36 V (+ @ top).

3.29   1.2 A ($\downarrow$).

3.31   16/9 W.

3.33   1.6 mW.

3.35   16/3 Ω.

3.37   [V], [A], [Ω]:

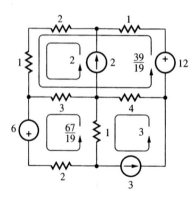

3.39 $p_{15\ V} = 105$ mW (released), $p_{5\ mA} = 35$ mW (absorbed).

3.41   2 W.

3.43   17.4 V (+ @ top).

3.45   675 mW.

3.47 $p_{5\ A} = 5$ W (delivered), $p_{6\ A} = 600/7$ W (delivered).

3.49 $p_{12\ V} = 60$ W (delivered), $p_{2\ A} = 4$ W (absorbed); (c) is easiest.

3.51   (a)   0.17 A (↓);
         (b)   0.8 V (+ @ left).

3.53 $p_{3\ mA} = 6$ mW (released), $p_{2\ V} = 4$ mW (released),
         $p_{10\ V} = 20$ mW (released).

3.55   41/45 A (↑).

3.57   16.25 mW (released).

CHAPTER 4

4.1   (a)   2 V (+ @ A), 1/7 A (A → B), 14 Ω;
         (b)   14 V (+ @ A), 1 A (A → B), 14 Ω.

4.3   (a)   4 V (+ @ A), 0.2 mA (A → B), 20 kΩ;
         (b)   28 V (+ @ B), 28/45 mA (B → A), 45 kΩ.

4.5   15 V (+ @ A), 8 A (A → B), 1.875 Ω.

4.7   450/19 V (+ @ A), 5 A (A → B), 90/19 Ω.

4.9   (a) $v_{OC} = 8$ V (+ @ A), $R_{eq} = 0$;
         (b) $v_{OC} = 20/3$ V (+ @ A), $i_{SC} = 10$ A (A → C), $R_{eq} = 2/3$ Ω;
         (c) $v_{OC} = i_{SC} = 0$, $R_{eq} = 12/7$ Ω;
         (d) $v_{OC} = 4/3$ V (+ @ C), $i_{SC} = 2$ A (C → B), $R_{eq} = 2/3$ Ω;
         (e)   8 V (+ @ D), 14/3 A (D → B), 12/7 Ω.

4.11   5 Ω.

4.13   (a)   2.5 V (+ @ bottom), 0.5 A (↑);
         (b)   1 A (↑);
         (c)   20 V (+ @ bottom);
         (d)   1.25 V (+ @ top), 0.75 A (↑).

4.15 12-V source (+ @ A) in series with 15 kΩ; 0.8-mA source (B → A) in parallel with 15 kΩ.

4.17 33 V (+ @ top).

4.19 0.8 V (+ @ CG), 11/60 A (G → C and D → H).

4.21 (a) 1/3 mA;
     (b) 0.3322 mA.

4.23 (a) 1/3 A;
     (b) 0.3252 A.

4.25 5.406 V, 0.9081 mA.

4.27 8.851 mA, 1.149 V.

4.29 3.50 mA, 2.47 V.

4.31 $10/\sqrt{6}$ V.

4.33 (2/3)10 V.

4.35 $0.5V_m$.

4.37 $V_{\text{rms}} = (V_m/\sqrt{2})\sqrt{1 - (\theta - 0.5\sin 2\theta)/\pi}$;
     $\theta = 0 \Rightarrow V_{\text{rms}} = V_m/\sqrt{2}$, $\theta = \pi/2 \Rightarrow V_{\text{rms}} = V_m/2$,
     $\theta = \pi \Rightarrow V_{\text{rms}} = 0$.

4.39 (a) 2.25 W;
     (b) 1/6 W.

4.41 60 kΩ, 5/3 mW.

4.43 (a) $R = \left[3(R_s - R_L) + \sqrt{9(R_s - R_L)^2 + 32R_s R_L}\right]/16$;
     (b) 7.215 kΩ;
     (c) 3.465 kΩ.

CHAPTER 5

5.1 1 Ω, −3.8 Ω.

5.3 $p_{5\text{ V}} = 25$ W (absorbed), $p_{2\ \Omega} = 50$ W (absorbed), $p_{3\ \Omega} = 75$ W (absorbed), $p_{\text{CCCS}} = 150$ W (released).

5.5 −10 Ω.

5.7 $R_B = 215.3$ kΩ, $R_C = 2.5$ kΩ.

5.9 $R_1 = 56.57$ kΩ, $R_E = 3.311$ kΩ.

5.11 $R_1 = 34.9$ kΩ, $R_C = 1.89$ kΩ.

5.13 2/3 mW.

5.15 $v_{\text{CCVS}} = 20$ V, $v_{8\ \Omega} = 8$ V, $v_{2\ \Omega} = 6$ V (all + @ top).

5.17 12 V (+ @ top).

5.19 2.88 mW (released).

5.21 $p_{12\text{ V}} = 12$ W (released), $p_{5\text{ A}} = 100$ W (released).

5.23 14 Ω.

5.25 4 Ω,

5.27 18 mW.

5.29   (a)   36 W;

(b) $R = 224/19 \ \Omega$, 47.58 W.

5.31   (a)   7.000 V;

(b)   6.542 V;

5.33   (a) $\sqrt{6.4}$ V;

(b)   62.5%.

5.35   115.2 W.

5.37   (a)   10, 180 W;

(b)   5, 115.2 W.

5.39   51,840/7,921 W.

5.41 $p_{500 \ \Omega} = 4.150$ W, $p_{20 \ \Omega} = 2.656$ W, $p_{10 \ \Omega} = 1.000$ W, $p_{250 \ \Omega} = 42.569$ W; $p_{100 \ v} = 50.375$ W.

5.43   1250/9 V/V.

5.45   500 Ω, 2.024 A/V, 400 Ω.

5.47   5,000/11 V/V.

5.49   2750/3 V/V, 5,500/3 A/A, 61.84 dB.

5.51 $v_L/v_S = [R_{i1}/(R_s + R_{i1})]A_{oc1}[R_{i2}/(R_{o1} + R_{i2})]A_{oc2}[R_L/(R_{o2} + R_L)]$; $v_L/v_S \rightarrow A_{oc1} \times A_{oc2}$ if $R_{i1} \gg R_s$, $R_{o1} \ll R_{i2}$, and $R_{o2} \ll R_L$.

CHAPTER 6

6.1   (a) -3 mV;

(b)   1.010 V;

(c) -4 V;

(d)   3.5035 V.

6.3   (a)   3 V/V;

(b)   2 V/V

(c)   5 V/V.

6.5   (a) $R_1 = 20$ kΩ, $R_2 = 80$ kΩ;

(b) $R_1 = 16.67$ kΩ, $R_2 = 83.33$ kΩ.

6.7   As v_I alternates between -1 V and $+1$ V, v_O alternates between $+9.891$ V and -9.891 V, and v_N between -9.891 mV and $+9.891$ mV.

6.9   (a) -0.151%;

(b) -0.150%;

(c) -0.0015%.

6.11 $v_N = v_P = 3$ V; $v_O = 5$ V.

6.13 $v_N = v_P = -1.5$ V; $v_O = -4.5$ V.

6.15   14 kΩ.

6.17 $A = 2k - 1$; 0, 1, -1 V/V.

6.19   Use circuit of Figure 16.16(a) with $R = 1$ kΩ, $R_2 = 11.11$ kΩ, and R_1 consisting of an 11.11-kΩ resistance in series with the potentiometer connected as a variable resistance from 0 to 100 kΩ.

6.21 −8 V/V.

6.23 $R_i = \infty$.

6.25 (a) −10 V;
(b) −10/3 V;
(c) −2.25 V.

6.27 $v_O = v_2 + v_4 + v_6 - v_1 - v_3 - v_5$.

6.29 (a) 14.4 V;
(b) 25/9 V.

6.31 Derivation.

6.33 Derivations.

6.35 $R_1 = R_2 = 402$ kΩ; R_G: 806 Ω in series with the potentiometer connected as a variable resistance from 0 to 100 kΩ.

6.37

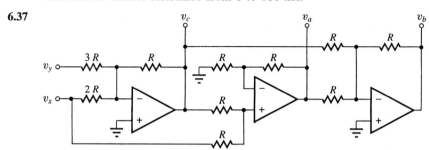

6.39 (b) $R_2 = 100$ kΩ, $R_3 = 1$ kΩ.

6.41 $R = R_5(1 + R_3/R_4)/(1 + R_2/R_1)$, $R_o = R_5(1 + R_4/R_3)(R_4/R_3 - R_2/R_1)$.

6.43 Derivation.

CHAPTER 7

7.1 57.04 pF.

7.3 (a) 4.84 in;
(b) 122.8 m, greater than a football field.

7.5 (a) $C = C_1(1 + \sqrt{5})/2$;
(b) $C = C_1(\sqrt{5} - 1)/2$.

7.7 (a) 3 pF;
(b) 50/13 pF.

7.9 (a) 56 μJ, 96.43%;
(b) 56 μJ, 42.86%.

7.11 (b) 2 nF.

7.13 (c) 100 MΩ, 100 kΩ.

7.15 (a) $5(t^2 + 1)$ V;
(b) $5(5 - 4e^{-10^5 t})$ V;
(c) $5(1 + \sin 10^5 t)$ V;

(d) $5(1 + t)$ V;

(e) 5 V.

7.17

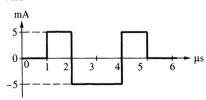

7.19 $i = 2.5 \cos 10^3 t$ mA, $p = 6.25 \sin(2 \times 10^3 t)$ mW,
$w = 3.125[1 - \cos(2 \times 10^3 t)]$ μJ.

7.21 (a) $L = L_1(1 + \sqrt{5})/2$;

(b) $L = L_1(\sqrt{5} - 1)/2$.

7.23 (a) -25 V;

(b) $12.5e^{-10^2 t}$ V;

(c) 0 V;

(d) $5 \cos 10^2 t$ V.

7.25

mA graph and mW graph

7.27 $i(0) = i(1 \text{ ms}) = 0$; $i(2 \text{ ms}) = 2.5$ A; $i(3 \text{ ms}) = 7.5$ A;
$i(4 \text{ ms}) = 11.25$ A; $i(t \geq 5 \text{ ms}) = 12.5$ A.

7.29 (a) $7.2 \sin(4 \times 10^4 t)$ mA;

(b) $-8.8 \sin(6 \times 10^4 t)$ mA;

(c) 0 mA.

7.31 -400 Np/s.

7.33 1.622 ms, 8.789 ms.

7.35 (a) $y(t) = 9e^{-1.5t} + 2(t - 2)e^{-t}$;

(b) $y(t) = [14e^{-1.5t} + 3(\cos 2t + 4 \sin 2t)e^{-t}]/17$.

7.37 -30.64 mA, 2.347 μJ.

7.39 (a) $-5e^{-t/\tau}$, $10(1 - e^{-t/\tau})$, $-15e^{-t/\tau}$, 10;

(b) $10e^{-t/\tau}$, $-5(1 - e^{-t/\tau})$, $15e^{-t/\tau}$, -5;

(c) 0, $10(1 - e^{-t/\tau})$, $-10e^{-t/\tau}$, 10;

(d) $5e^{-t/\tau}$, 0, $5e^{-t/\tau}$, 0. All in Amperes.

7.41 (a) $i(t) = 7.071 \cos(10^6 t - 45°) - 5e^{-t/(1 \ \mu s)}$ mA;

(b) $i_1(t) = 7.071 \cos(10^6 t - 45°)$ mA, and $i_2(t) = i_1(t) - 10e^{-t/(1 \ \mu s)}$ mA.

7.43 (a) 50 krad/s;

(b) 2 V;

(c) No.

7.45 $22.14\cos(3 \times 10^3 t - 71.57°)$ V.

7.47   (a) $5e^{-t/\tau}$, 0, $5e^{-t/\tau}$, 0;
      (b) $4.903e^{-5000t}$, $5\cos(10^3 t - 11.31°) - 4.903e^{-5000t}$, 0,
         $5\cos(10^3 t - 11.31°)$. All in volts.

CHAPTER 8

8.1   (a) 1 V;
     (b) 1 A, 1 kA.

8.3   5 V, 1.839 V.

8.5   200 kΩ, 5 nF, 1 ms, 250.0 nJ, 158.0 nJ.

8.7   1.125 μJ, 20.61 nJ.

8.9   (a) $3e^{-t/(500 \text{ ns})}$ V;
     (b) $-7.5e^{-t/(200 \text{ ns})}$ V.

8.11   (a) $v(t) = 10[3 - 2e^{-t/(7 \text{ ms})}]$ V; $10e^{-t/(7 \text{ ms})}$ V, $30[1 - e^{-t/(7 \text{ ms})}]$ V,
        $-20e^{-t/(7 \text{ ms})}$ V, 30 V.
     (b) $v(t) = 10[1 + 2e^{-t/(3 \text{ ms})}]$ V; $30e^{-t/(3 \text{ ms})}$ V, $10[1 - e^{-t/(3 \text{ ms})}]$ V,
        $20e^{-t/(3 \text{ ms})}$ V, 10 V.

8.13 $i(0^-) = 2$ mA, $i(t \geq 0^+) = -(2/3)e^{-t/(1.5 \text{ ms})}$ mA.

8.15 $v(0^-) = 18$ V, $v(t \geq 0^+) = [594 + 176e^{-t/(180 \text{ ns})}]/35$ V.

8.17 $4[2 - e^{-t/(62.5 \ \mu s)}]$ V.

8.19 $10[24e^{-t/(4 \ \mu s)} - 1]/23$ V.

8.21 $5(1 - e^{-t})$ V.

8.23 $12[1 - e^{-t/(80 \ \mu s)}]$ V.

8.25 $i(0^-) = 0.75$ mA, $i(t \geq 0^+) = [20 + 7e^{-t/(1 \text{ ms})}]/15$ mA.

8.27 $v(0^+ \leq t \leq 2 \text{ ms}) = 40e^{-t/(2.4 \text{ ms})}$ V, $v(t \geq 2 \text{ ms}) = 40e^{-0.5}e^{-t/(6 \text{ ms})}$ V.

8.29   (a) $v(0^-) = 0$, $v(t \geq 0^+) = -10e^{-t/(80 \ \mu s)}$ V;
     (b) $i_{40 \text{ mH}} = 20[4 + e^{-t/(80 \ \mu s)}]$ mA ($\downarrow$),
        $i_{10 \text{ mH}} = 80[e^{-t/(80 \ \mu s)} - 1]$ mA ($\rightarrow$).

8.31   (b) $T = 16.7$ s.

8.33   (a) $v_O(t > 0) = 6e^{-t/(2.5 \ \mu s)}$ V;
     (b) $v_O(0 < t < 3 \ \mu s) = 4e^{-t/(2.5 \ \mu s)}$ V;
        $v_O(t > 3 \ \mu s) = 4(1 - e^{3/2.5})e^{-t(2.5 \ \mu s)}$ V.

8.35

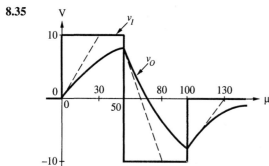

8.37   (a) $v(0^-) = 0$, $v(t \geq 0^+) = 6e^{-t/(20 \text{ ns})}$ V;

        (b) $v(0^-) = 0$, $v(t \geq 0^+) = -5e^{-t/(20 \text{ ns})}$ V.

8.39 $i_X(0^-) = 7/25$ mA, $i_X(t \geq 0^+) = (7/150)[3 - 2e^{-t/(18 \text{ ms})}]$ mA.

8.41 $v_O(0^-) = 2$ V, $v_O(t \geq 0^+) = 5 - 8e^{-t/(10 \text{ ms})}$ V;

        $-3e^{-t/(10 \text{ ms})}$ V, $5[1 - e^{-t/(10 \text{ ms})}]$ V, $-8e^{-t/(10 \text{ ms})}$ V, 5V.

8.43   (a) $v_O(0^-) = 0$, $v_O(t \geq 0^+) = 6[e^{-t/(3 \text{ ms})} - 1]$ V;

        (b) $v_O(0^-) = 0$, $v_O(0^+ \leq t \leq 5 \text{ ms}) = 6[e^{-t/(3 \text{ ms})} - 1]$ V,

           $v_O(t \geq 5 \text{ ms}) = 6(1 - e^{5/3})e^{-t/(3 \text{ ms})}$ V.

8.45   (a) $v_O(0^-) = 0$, $v_O(t \geq 0^+) = -4e^{-t/(3 \text{ ms})}$ V;

        (b) $v_O(0^-) = 0$, $v_O(t \geq 0^+) = 8e^{-t/(3 \text{ ms})}$ V.

8.47 $v_O(t \geq 0^-) = 2[3e^{-t/(60 \text{ }\mu s)} - 1]$ V; 65.92 μs.

8.49 $v_O(0^-) = 0$, $v_O(t \geq 0^+) = -10[e^{-t/(50 \text{ }\mu s)} - 1]$ V.

8.51   (a) $(10 \text{ ms})dv_O/dt + v_O = v_I - (10 \text{ ms})dv_I/dt$

        (b) $(10 \text{ ms})dv_O/dt + v_O = (10 \text{ ms})dv_I/dt - v_I$.

CHAPTER 9

9.1   Derivation.

9.3   (a) $R = 1$ kΩ, $L = 0.7$ H, $C = 1/7$ μF;

        (b) $R = 1$ kΩ, $L = 1/7$ H, $C = 0.7$ μF.

9.5 $w_L(0^+) = 8$ J, $w_C(0^+) = 6.4$ J.

9.7 $3(5e^{-6t} - 2e^{-t})$ V.

9.9 $w_L(t \geq 0^+) = 1.922e^{-5t}(1 - \cos 19.36t)$ J,

        $w_C(t \geq 0^+) = 1.922e^{-5t}[1 + \cos(19.36t - 28.96°)]$ J.

9.11   (a) $L = 50$ mH, $C = 2$ nF;

        (b) $L = 200$ mH, $C = 500$ pF.

9.13   (a)   0.1 s, 67.67%;

        (b)   10 μs, 67.67%.

9.15   4 Ω, $(4 + 8t)e^{-4t}$ V.

9.17   400 Ω → 625 Ω, $(116/3)e^{-16,000t} \cos(12,000t - 53.13°)$ mA.

9.19 $12.07 \cos(3t - 173.66°)$ V.

9.21   3, $90[e^{-15t} - e^{-(5/3)t}]$ V.

9.23   0, $(20/9)te^{-t}$ A.

9.25   Derivation.

9.27 $2,000te^{-10,000t} + 0.4$ A.

9.29   (a) $2 - 9.014e^{-t} \cos(2t + 56.31°)$ V;

        (b)   2 A, -6 V,

        (c)   1.25 A, $6.760e^{-t} \cos(2t - 123.69°)$ V.

9.31 $1.196e^{-7.854 \times 10^6 t} - 8.196e^{-1.146 \times 10^6 t} + 16$ V, 3.435 μs.

9.33 $(2/3)(e^{-5t} - 10e^{-2t}) + 12$ V.

9.35 $6.501e^{-(3625/3)t} \cos(199.8t - 22.64°) + 3$ V.

9.37 (a) 207.6 μH, 106.4 pF;
  (b) 4%, 747.9 kHz;
  (c) 2.794 kΩ.

9.39 (a) $18[1 - (2 + 3t)e^{-3t}]$ V;
  (b) 2.478 s.

9.41 $v_O(\text{max}) = 108$ V at $t = 0^+$,
  $v_O(\text{min}) = 45.45$ V at $t = 0.9226$ s.

9.43 $1.5(2 - t)e^{-0.5t}$ V.

9.45 (a) $6(1 - 2te^{-t})$ V;
  (b) $3 - 12.39e^{-0.9t} \sin 0.4359t$ V.

9.47 $\omega_0 = 10^4/\sqrt{2}$ rad/s, $\zeta = 1/\sqrt{2}$; 4.32%, 0.6283 ms

9.49 $\omega_0 = 10\sqrt{2}$ rad/s, $\zeta = 25\sqrt{2}$.

CHAPTER 10

10.1 169.7 V, 2 ms, $10^3\pi$ rad/s, 500 Hz, $\pi/6$ rad; 147.0 V, -84.85 V, 84.85 V.

10.3 $10\sqrt{2}$ A, $10^5/2\pi$ Hz, $2\pi/10^5$ s, $\pi/4$ rad; 7.854 μs, 23.56 μs.

10.5 v_1 lags v_2 by 75°; v_1 leads v_3 by 150°; v_2 lags v_3 by 135°.

10.7 (a) $10\underline{/-75°}$ V;
  (b) $10\underline{/-135°}$ A;
  (c) $15\underline{/-170°}$ mA;
  (d) $12\underline{/-45°}$ μV;
  (e) $100\underline{/-22.5°}$ μA.

10.9 (a) $2\pi 10^3$ rad/s, 120°;
  (b) 12 mW, 36 mW.

10.11 $60\cos(10^5t - 60°)$ V, $60\underline{/-60°}$ V; $40\cos(10^5t - 150°)$ V, $40\underline{/-150°}$ V.

10.13 (a) 7.5 W, absorbed;
  (b) 43.3 W, released;
  (c) 173.2 W, absorbed.

10.15 (a) 10.65 mW, released;
  (b) 10.32 mW, absorbed.

10.17 4 kΩ in series with 3 mH.

10.19 60.36 mW, 50 mW, 25 mW; 1.178 ms, 0.7854 ms, 1.571 ms.

10.21 $V_2/V_0 = (\omega L/R)\sqrt{1 + (\omega L/R)^2}$, $\phi_2 = 90° - \tan^{-1}(\omega L/R)$.

10.23 $(200/\sqrt{481})\cos(10^3t - 24.73°)$ V, $(200/\sqrt{481})\underline{/-24.73°}$ V.

10.25 $\sqrt{20}\cos(10^3t - 153.43°)$ V, $\sqrt{20}\underline{/-153.43°}$ V.

10.27 $30[1 - e^{-t/(625\ \mu s)}]$ mA.

10.29 (a) 25 kΩ;
  (b) 10 krad/s;
  (c) 951.49 krad/s, 1,050.98 krad/s.

10.31 100 krad/s, 40 V.

10.33 $V_2/V_0 = \omega^2 LC/\sqrt{(1 - \omega^2 LC)^2 + (\omega RC)^2}$, $\phi_2 = 90° + \tan^{-1}[(1 - \omega^2 LC)/(\omega RC)]$.

10.35 $11.09\cos(10^6 t + 56.31°)$ mA.

10.37 $2\cos 10^3 t$ mA.

CHAPTER 11

11.1 $50\underline{/105°}$ V/s, $2\underline{/-75°}$ V-s, $250\underline{/-165°}$ V/s^2, $0.4\underline{/-165°}$ V-s^2.

11.3   (a) $6\cos(5t - 90°)$;
       (b) $5\sqrt{29}\cos(5t + 111.80°)$, $25\sqrt{29}\cos(5t - 158.20°)$;
       (c) $(1/\sqrt{5})\cos(5t + 108.43°)$, $(1/5\sqrt{5})\cos(5t + 18.43°)$.

11.5   (a) $10\sqrt{19}\cos(\omega t + 36.59°)$;
       (b) $31.27\cos(10t - 112.84°)$;
       (c) $5.020\cos(2\pi 10^3 t - 75.79°)$.

11.7   (a) $1\underline{/120°}$, $-0.5 + j\sqrt{0.75}$;
       (b) $1\underline{/-120°}$, $-0.5 - j\sqrt{0.75}$;
       (c) $\sqrt{3}\underline{/30°}$, $1.5 + j\sqrt{0.75}$;
       (d) $\sqrt{3}\underline{/-30°}$, $1.5 - j\sqrt{0.75}$.

11.9 $\boldsymbol{A} + \boldsymbol{B} = 4.039\underline{/57.98°} = 2.141 + j3.424$;
       $\boldsymbol{AB} = 1.612\underline{/127.88°} = -0.9900 + j1.273$;
       $\boldsymbol{A} - \boldsymbol{B} = 3.176\underline{/54.19°} = 1.859 + j2.576$;
       $\boldsymbol{A}/\boldsymbol{B} = 8.062\underline{/-15.26°} = 7.778 - j2.122$.

11.11 $4\cos(\omega t - 171.87°)$.

11.13 $d^2 y/dt^2 + 5dy/dt + 2y = dx/dt - x$; $1.501\cos(5t - 61.30°)$.

11.15   (a) 15.92 μH, 0.1 Ω;
       (b) 3.183 μF, 50 mΩ;
       (c) 100 Ω.

11.17   (a) $12.5 + j12.5$ Ω, inductive;
       (b) $5 - j5\sqrt{3}$ Ω, capacitive.

11.19   (a) $R = R_1 + R_2/(1 + \omega^2 R_2^2 C^2)$, $X = \omega[L - (R_2^2 C)/(1 + \omega^2 R_2^2 C^2)]$;
       (b) $\omega = 0$, $R_1 + R_2$; $\omega = \sqrt{[1 - (L/R_2)/(R_2/C)]/(LC)}$, $R_1 + L/(R_2 C)$.

11.21   2 rad/s, 10/7 A.

11.23 $3\cos(2\pi 120t - 23.13°)$ A.

11.25   (a) $(1/2R)(1 - \omega^2 LC + j\omega 3RC)/(1 - \omega^2 LC + j\omega RC)$;
       (b) $1/2R$, $1/2R$;
       (c) $15\underline{/0°}$ mΩ^{-1}.

11.27   (a) $R_s = 2.5$ kΩ, $L_s = 4.330$ mH; $R_p = 10$ kΩ, $L_p = 5.774$ mH;
       (b) $R_s = 8.660$ kΩ, $C_s = 2$ nF; $R_p = 11.55$ kΩ, $C_p = 500$ pF.

11.29   2 Ω in series with 2.5 μF.

11.31 $0.3320\underline{/35.86°}$ Ω^{-1}, 0.2690 Ω^{-1}, 0.1945 Ω^{-1}.

11.33   (a) $1.961\cos(2\pi 10^5 t - 57.56°)$ mA;
       (b) $6.703\cos(2\pi 10^5 t + 12.95°)$ mA.

11.35 $47.27\cos(10^3 t - 99.93°)$ V.

11.37   Both $2.5\underline{/-90°}$ V (+ @ top).

11.39 $16.64\underline{/-56.31°}$ V.

11.41 $I_x = 7.428\underline{/68.20°}$ A.

11.43 $v(t) = 17.65\cos(10^6 t + 11.31°)$ V (+ @ top).

11.45 $-4 + j11$ A ($\downarrow$).

11.47 $1026\underline{/-24.88°}$ A (out of positive terminal), $8.836\underline{/27.74°}$ V (+ @ top).

11.49 $100 + j500$ Ω, inductive.

11.51 $1.213\underline{/75.96°}$ A.

11.53 $V_{oc} = (50/\sqrt{2})\underline{/-15°}$ V (+ @ A), $Z_{eq} = \sqrt{325/18}\underline{/-11.31°}$ Ω, $I_{sc} = (30/\sqrt{13})\underline{/-3.69°}$ A (out of A).

11.55 $V_{oc} = -j6$ V (+ @ A), $Z_{eq} = (4/\sqrt{10})\underline{/18.43°}$ Ω.

11.57   (a) $44.65\cos(10^5 t - 7.13°)$ mA;
      (b) $26.38\cos(10^5 t - 98.43°)$ mA;
      (c) $144\cos(10^5 t + 143.13°)$ mA.

11.59   (b)   1 kΩ in series with 0.1 μF; 2 kΩ in parallel with 50 nF.

11.61   (b)   34.54 Ω, 2.345 mH.

11.63   (a)   Unbalanced;
      (b)   balanced.

11.65   (b) $R = 100$ kΩ, $C = 1$ nF.

11.67   (b) $R_1 = R_2 = R = 100$ kΩ, $C = 50$ nF.

11.69   (a) $9.9995\cos(10^2 t - 8.13°)$ V;
      (b) $7.07\cos(10^3 t - 90°)$ V;
      (c) $99.995\cos(10^4 t - 171.87°)$ mV.

11.71 $V_o = [-j\omega RC/(1 - \omega^2 RC + j\omega RC)]V_i$; 1 krad/s, 1.

CHAPTER 12

12.1 $p_{v_s} = 6.923$ W (delivered), $p_R = 2.130$ W (absorbed), $p_L = 7.988$ W (absorbed), $p_C = 3.195$ W (delivered).

12.3   (a)   948.9 VA, 688.3 W, 0.725 (leading);
      (b)   812.9 VA, 688.3 W, 0.847 (lagging).

12.5 $P_{source} = 519.2$ mW, $P_{800\ \Omega} = 173.1$ W, $P_{100\ \Omega} = 346.1$ mW.

12.7   52.02 Ω in series with 12.83 mH; 72 Ω in parallel with 46.23 mH.

12.9   (a)   14.42 W;
      (b)   20.83 W.

12.11 $50 - j20$ Ω, 36 W.

12.13   90 W, $1 + j1$ Ω.

12.15   Derivation.

12.17   (b)   1395 VA, -344.4 VA.

12.19   150 W (delivering), 259.8 VAR (absorbing).

12.21   (a)   20 W, −15 VAR, $20 − j15$ VA;
 · (b)   25/3 H.

12.23   (a)   141.2 μF;
(b)   251.9 μF.

12.25 $237.4\underline{/6.32°}$ V.

12.27 $17.04\underline{/51.95°}$ kVA.

12.29 $278.7 + j40.05$ mΩ; 0.9898, lagging.

12.31   (a)   0.8688, lagging.
(b)   244.7 μF.

12.33   551.7 mW.

12.35   −3 mW (downscale).

12.37   (a) $V_{ab} = 208\underline{/120°}$ V (rms), $V_{bc} = 208\underline{/0°}$ V (rms),
$V_{ca} = 208\underline{/−120°}$ V (rms);
(b) $V_{an} = 120.1\underline{/30°}$ V (rms), $V_{bn} = 120.1\underline{/−90°}$ V (rms), $V_{cn} = $
$120.1\underline{/150°}$ V (rms).

12.39   48/7 Ω.

12.41 $1 + j2$ Ω.

12.43   (a) $Z_{AB} = j2$ Ω, $Z_{BC} = 0.4$ Ω, $Z_{CA} = −j2$ Ω;
(b) $Z_{AN} = 5 − j5$ Ω, $Z_{BN} = 5 + j5$ Ω, $Z_{CN} = −j5$ Ω.

12.45 $Z_{Y(composite)} = \frac{1}{3}Z_{\Delta(composite)} = 10(1 + j1)$ Ω.

12.47   (a)   6.405 A (rms);
(b)   410.8 V (rms).

12.49 $0.6101 + j0.5906$ Ω.

12.51 $I_{nN} = 11.43\underline{/−16.67°}$ A (rms).

12.53   (a) $I_{aA} = 6.690\underline{/−8.20°}$ A (rms), $I_{bB} = 6.690\underline{/−128.20°}$ A (rms),
$I_{cC} = 6.690\underline{/111.80°}$ A (rms);
(b) $V_{ab} = 209.5\underline{/3.53°}$ V (rms), $V_{bc} = 209.5\underline{/−116.47°}$ V (rms),
$V_{ca} = 209.5\underline{/123.53°}$ V (rms).

12.55   (a) $I_{aA} = 14.00\underline{/−26.57°}$ A (rms), $I_{bB} = 14.00\underline{/93.43°}$ A (rms), $I_{cC} = $
$14.00\underline{/−146.57°}$ A (rms);
(b) $V_{AB} = 180.7\underline{/30°}$ V (rms), $V_{BC} = 180.7\underline{/150°}$ V (rms), $V_{CA} = $
$180.7\underline{/−90°}$ V (rms).

12.57   (a) $121.24\underline{/20°}$ V (rms);
(b) $1.250 + j1.323$ Ω;
(c) $75.00 + j41.50$ Ω
(d) $197.95\underline{/48.96°}$ V (rms).

12.59   Derivation.

12.61   (a) $1.844 + j1.383$ kVA;
(b)   96.38%.

12.63   (a) $C_{\Delta} = 49.74$ μF;
(b) $C_Y = 149.2$ μF.

12.65   0.8253, lagging.

12.67   283.6 V (rms); 0.9283, lagging.

12.69   (a) 0.6130, leading;
         (b) 0.1833, leading.

12.71 $P_{AB} = 1{,}113.4$ W, $P_{CB} = 614.5$ W

12.73   (a) 64.9 Ω;
         (b) $-j37.46$ Ω;
         (c) $32.45 + j56.20$ Ω;
         (d) $32.45 - j56.20$ Ω;
         (e) $64.90 - j37.47$ Ω.

12.75   (a) 17.96 W;
         (b) $C_\Delta = 17.68$ μF.

CHAPTER 13

13.1   5 W, 1 Mrad/s, 1 kV.

13.3   (a) 50 mH, 0.8 μF;
       (b) 4950.25 rad/s, 5050.25 rad/s, 100 rad/s.
       (c) $5\sqrt{2}\underline{/-45^\circ}$ Ω, $5\sqrt{2}\underline{/45^\circ}$ Ω.

13.5   1.0513 Mrad/s, 9.990, 100.1 krad/s, 10.01 Ω, 100 μH.

13.7   (a) 400 Ω, 972.5 μH, 2.605 nF;
       (b) 58.26 kHz, 171.65 kHz.

13.9   160 Ω, 8 mH, 316.6 pF.

13.11   Derivation.

13.13   (b) $\omega = \omega_0 : |V_R| = 10$ V, $|V_L| = |V_C| = 12.5$ V, $pf = 1$;
          $\omega = \omega_m : |V_R| = 8.997$ V, $|V_L| = 9.274$ V, $|V_C| = 13.64$ V,
          $pf = 0.8997$, leading.

13.15   (a) 100 kΩ, 15.92 μH, 15.92 pF;
       (b) 9.976 Mrad/s $\le \omega \le$ 10.024 Mrad/s, $25.84^\circ \ge \theta_i \ge -25.84^\circ$.

13.17   (a) 200 μH, 50 μF;
       (b) 25 μJ, 3.142 μJ.

13.19   (a) 2.5 krad/s, 8, 2.349 krad/s, 2.661 krad/s;
       (b) $|I_{R_1}| = |I_{R_2}| = 1$ mA; $|I_L| = |I_C| = 10$ mA.

13.21   8.660 krad/s, 200 Ω.

13.23   100 kΩ, 1.9992 nF.

13.25   (b) 1.2 kΩ, $6\cos \omega t$ V.

13.27 $\omega_0 = 1/(\sqrt{R_1 R_2}C)$, $Q = \sqrt{R_2/R_1}/2$, $-\frac{1}{2}(R_2/R_1)$.

13.29   Derivation.

13.31   80 kΩ, 20 kΩ, 0.2 H, 50 nF.

13.33   10 kΩ $\to$ 30 kΩ; 100 kΩ $\to$ 300 kΩ; 1 μF $\to$ 167.8 μF;
       100 pF $\to$ 16.78 nF.

CHAPTER 14

14.1   (a) $5\underline{/0°}$ V, 0 complex Np/s;

       (b) $2\underline{/0°}$ V, -0.1 Np/s;

       (c) $5\underline{/30°}$ A, $-3 + j4$ complex Np/s;

       (d) $2\underline{/-120°}$ A, $j10^5$ rad/s.

14.3   (a) $-e^{-5t}/11$ A;

       (b) $134.5 \cos(5t - 160.35°)$ mA;

       (c) $82.59e^{-3t} \cos(4t - 67.33°)$ mA.

14.5 $\sqrt{50}e^{-t} \cos(t - 45°)$ V.

14.7   (a) $\mathbf{v}_o = \mathbf{v}_i/(2 + 3s + 2s^2 + s^3)$;

       (b) $(1/\sqrt{8})e^{-t} \cos(2t + 45°)$ V.

14.9 $v_{2\,\Omega} = (16/7)e^{-2t}$ V (+ @ top), $v_{1\,\Omega} = (-48/7)e^{-2t}$ V (+ @ top).

14.11 $\mathbf{V}_{oc} = 10.31\underline{/14.04°}$ V (+ @ A), $\mathbf{Z}_{eq} = 5.154\underline{/14.04°}$ Ω.

14.13 (a) $z_1 = -0.25$ Np/s; $p_1 = -1$ Np/s, $p_{2,3} = -1 \pm j2$ complex Np/s;

       (b) $d^3y/dt^3 + 3d^2y/dt^2 + 7dy/dt + 5y = 20dx/dt + 5x$.

14.15 (a) $50/(s + 5)$;

       (b) $10^{-3}s + 1$;

       (c) $(0.9s^2 + 2.7s)/(s^3 + 9s^2 + 15s + 7)$;

       (d) $-(4s^2 + 36)/(s^5 + 6s^4 + 26s^3 + 6s^2 + 25s)$.

14.17 $2 \times 10^5(s + 6 \times 10^3)/(s^2 + 6 \times 10^3 s + 10^7)$ V/A;

      $z_1 = -6$ kNp/s, $p_{1,2} = -3 \pm j1$ complex kNp/s.

14.19 $s/(s^2 + 2s + 5)$; $z_1 = 0$ complex Np/s, $p_{1,2} = -1 \pm j2$ complex Np/s.

14.21 $2 \times 10^6/(s^2 + 10^3 s + 10^6)$; $p_{1,2} = 500(-1 \pm j\sqrt{3})$ complex Np/s.

14.23 $-10^4 s/(s^2 + 1,250s + 6.25 \times 10^6)$; $z_1 = 0$ complex Np/s,

      $p_{1,2} = -625 \pm j2421$ complex Np/s.

14.25 $v(t) = 3.354e^{-3\times 10^5 t} \cos(4 \times 10^5 t + 26.57°)$ V.

14.27 $11.71e^{-382t} - 1.71e^{-2618t}$ V.

14.29 $3(4e^{-2t} - 3e^{-3t})$ A.

14.31 $17.01(e^{-0.7847t} - e^{-2.549t})$ V; $(60/\sqrt{35})e^{-t/6} \sin(\sqrt{35}t/6)$ A.

14.33 (a) $5.386e^{-1,340t} - 0.3868e^{-18,660t}$ V;

       (b) $5(1 + 10^4 t)e^{-10^4 t}$ V.

14.35 $-4.492e^{-6,667t} + 8.515 \cos(2\pi 10^3 t - 58.16°)$ V.

14.37 $1.224e^{-7t} \cos(\sqrt{15}t + 36.16°) + 2.5 + 3.993 \cos(5t - 150.88°)$ V.

14.39 $2.5 \left[2 - (1 + 10^4 t)e^{-10^4 t} - \cos 10^4 t\right]$ V.

14.41 (a) 52 dB;

       (b) 49 dB;

       (c) -27 dB;

       (d) $2\sqrt{20}$;

       (e) $1/50$;

       (f) $\sqrt{20}/10^5$.

14.43 (a) $1 + j(\omega/10^5 - 10^5/\omega)$ kΩ;

(b) $(1\ k\Omega)/\left[1 + j\left(\omega/10^5 - 10^5/\omega\right)\right]$.

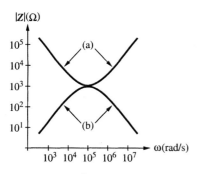

14.45 $0.1/(1 + s/10^3)$; $p_1 = -1$ kNp/s, $\omega_{p_1} = 1$ krad/s.

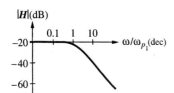

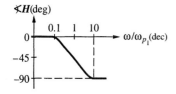

14.47   (a) $z_1 = 0$, $\omega_{z_1} = -10$ rad/s; $p_1 = -100$ Np/s, $\omega_{p_1} = 100$ rad/s;

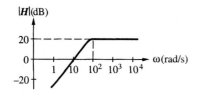

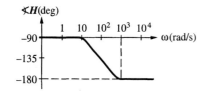

(b) $z_1 = 0$, $\omega_{z_1} = -10$ rad/s.

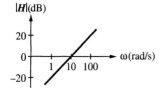   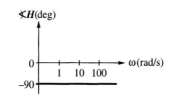

14.49 $p_1 = 0$, $\omega_{p_1} = 1/[RC(1 + R_2/R_1 + R_2/R)]$.

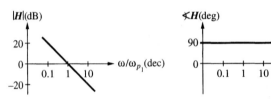

14.51 (a) $\omega = 0$, $\omega = 67.82$ rad/s;
(b) 48.99 rad/s;
(c) 8.136 dB.

14.53

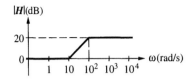

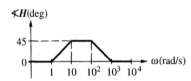

14.55 (a)

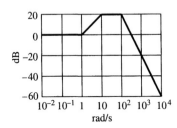 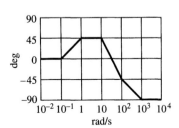

(b) 19.18 dB (20 dB), $-2.01°$ (0°).

14.57

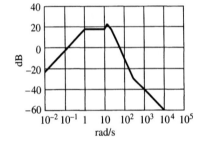 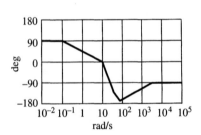

14.59 (a)

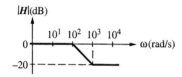

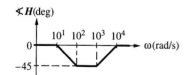

(b)

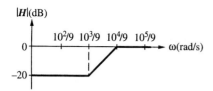

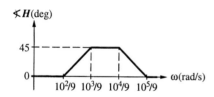

14.61 (a) $4,244\pi\ (s + 100\pi)/(s^2 + 4,344\pi s + 4.244\pi^2 \times 10^5)$;
(b) 17.03 dB, 2.732 dB, −2.778 dB.

14.63 (a)

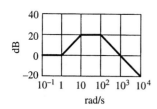

(b)

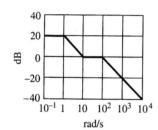

(c)

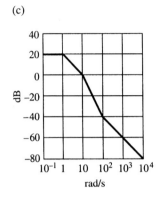

CHAPTER 15

15.1 $Z_{TA} = 1\ \Omega$ in series with 1 μF, $Z_{TB} = 1\ \Omega$ in series with 1 μH, $Z_{TC} = 1\ \Omega$.

15.3 $Z_{\pi A} = j5\ \Omega$, $Z_{\pi B} = -j5\ \Omega$, $Z_{\pi C} = 25\ \Omega$.

15.5   (a) 4 V;
(b) 4 V, reciprocal network.

15.7 $z_{11} = 20/9\ \Omega$, $z_{12} = z_{21} = 25/18\ \Omega$, $z_{22} = 157/72\ \Omega$; reciprocal.

15.9 $z_{21} = 2z_{11} = 2z_{22} = 60\ k\Omega$, $z_{12} = 0$; not reciprocal.

15.11 $z_{11} = 3\ k\Omega$, $z_{12} = 10\ \Omega$, $z_{21} = 2\ M\Omega$, $z_{22} = 20\ k\Omega$.

15.13 $y_{11} = s + n^2/(n^2 + 1)$, $y_{12} = -\left[s + n/(n^2 + 1)\right] = y_{21}$,
$y_{22} = s + 1/(n^2 + 1)$; reciprocal.

15.15 $a_{11} = 0.625$, $a_{12} = 1.5\ \Omega$, $a_{21} = 0.1875\ \Omega^{-1}$, $a_{22} = 1.25$.

15.17 $a_{11} = -2/9$, $a_{12} = 19/9\ \Omega$, $a_{21} = 1/9\ \Omega^{-1}$, $a_{22} = 13/9$.

15.19 $h_{11} = (111/11)10^4\ \Omega$, $h_{12} = 10/11$, $h_{21} = 9999990/11$,
$h_{22} = 10011001/(11 \times 10^5)\ \Omega^{-1}$.

15.21   (a) $h_{11} = R_3$, $h_{12} = 1$, $h_{21} = -1$, $h_{22} = -R_2/(R_1 R_4)$;
(b) $R_4/R_2 = R_3/R_1$.

15.23 $g_{11} = 1/3\ \Omega^{-1}$, $g_{12} = -2/3$, $g_{21} = 11/3$, $g_{22} = 5/3\ \Omega$.

15.25   (a) $v_{OC} = 0$, $R_{eq} = 124/7\ \Omega$;
(b) $v_{OC} = 2000/159$ V, $R_{eq} = 6.8\ \Omega$.

15.27   (b) 10 nF.

15.29   (a) $1000/19\ \Omega$;
(b) 13.16 W.

15.31 $y_{11} = y_{22} = (s^2 + 1)/\left[s(s^2 + 2)\right] + s(s^2 + 1)/(2s^2 + 1)$,
$y_{12} = y_{21} = -1/\left[s(s^2 + 2)\right] - s^3/(2s^2 + 1)$.

15.33 $a_{11} = s^2+3.5s+2$, $a_{12} = s+2\ \Omega$, $a_{21} = s^2+2.5s+0.5\ \Omega^{-1}$, $a_{22} = s+1$.

15.35   3538.6 V/V.

15.37 $1.5s/(s^2 + 7s + 6)$; $z_1 = 0$, $p_1 = -1$ Np/s, $p_2 = -6$ Np/s.

15.39 $i_2(t \geq 0^+) = -267.3 e^{-1.25 \times 10^5 t} \cos(66.14 \times 10^3 t - 62.11°) + 125$ mA.

15.41 $I_{3\,\Omega} = (0.3279\underline{/0.05°}\,V_s\,(\rightarrow)$,
$I_{1\,F} = (0.1646\underline{/8.53°})V_s\,(\rightarrow)$, $I_{5\,\Omega} = \left[(2/\sqrt{3730})\underline{/92.82°}\right]V_s\,(\downarrow)$.

15.43 $7.5 + j7\ \Omega$.

15.45 (a) 1 and 3;
(b) 1 and 4;
(c) 1 and 3.

15.47 (a) 250 nJ;
(b) 187.5 nJ;
(c) 190 nJ.

15.49 $49.87 - j269.8\ \Omega$, 3.787 W.

15.51 (a) $5.701\cos(100t + 82.87°)$ A, $3.162\cos(100t - 153.40°)$ V;
(b) 5.005 W, from right to left.

15.53 (a) 10/3 krad/s;
(b) 5 Ω;
(c) $|V_{L_2}| = 2|V_{L_1}| = 200/3$ V.

15.55 (a) $20 + j30\ \Omega$, $1.576\underline{/-23.20°}$ A, $2.228\underline{/-158.20°}$ A, $56.81\underline{/33.11°}$
V, $31.51\underline{/156.80°}$ V;
(b) Z_1, V_1, and I_1 remain the same; V_2 and I_2 change polarity.

15.57 $45 \times 10^4 k^2/(4706 + 738k^2 + 81k^4)$.

15.59 $v(t \geq 0^+) = 8.770\left(e^{-23.44 \times 10^3 t} - e^{-426.6 \times 10^3 t}\right)$ V.

15.61 0.866, 150 mA.

15.63 43.40 krad/s, 62.84 krad/s.

15.65 $204.3\underline{/6.49°}$ V (rms).

CHAPTER 16

16.1 (a) $(1 + e^{-t/0.5})u(t)$ V;
(b) $(1 - 0.5e^{-t/2})u(t)$ V;
(c) $1u(t)$ V.

16.3 (a) $v(t) = 4e^{-t/(250\ \mu s)}$ kV.

16.5 (a)

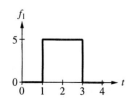

(b)

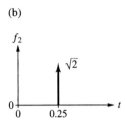

(c)

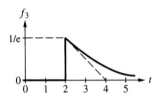

(d)

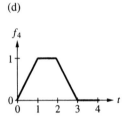

16.7   (a) $5e^{-s}$;

(b) $5e^{-s}/s$, $s > 0$;

(c) $e^{-s} - e^{-3s}$;

(d) $e^{-3s}(s + 1)/s^2$, $s > 0$.

16.9   (a) $5(e^{-s} - e^{-3s})/s$;

(b) $\sqrt{2}e^{-s/4}$;

(c) $20e^{-2s}/\left[\sqrt{e}(2s + 1)\right]$, $s > -0.5$;

(d) $(1 - e^{-s} - e^{-2s} + e^{-3s})/s^2$.

16.11   (a) $5(s^2 + 5s + 8)/(s + 3)^3$, $s > -3$;

(b) $2 \times 10^6/[s(s^2 + 4 \times 10^6)]$;

(c) $10/(s^2 + 400)$.

16.13   (a) $10^3/(s^4 + 10^4)$;

(b) $3/(s + 1)^4$;

(c) $(e^{-10^3 t} \sin 10^3 t)u(t)$.

16.15 $(A/s^2)\left\{1/T_1 + \left[e^{-sT_2} - (T_2/T_1)e^{-sT_1}\right]/(T_2 - T_1)\right\}$.

16.17 $5(e^{-t} - e^{-3t})u(t)$.

16.19 $(20/s^2)(1 - 2e^{-s/2} + e^{-s})/(1 - e^{-s})$.

16.21 $(A/s^2 dT)\left[1 - (sdT + 1)e^{-sdT}\right]/(1 - e^{-sT})$.

16.23 $2(s + 1)/\left[s(s^2 + 2s + 2)\right]$.

16.25 $f(0^+) = 0$, $f(\infty) = 1.875$, $f^{(1)}(0^+) = 5$, $f^{(1)}(\infty) = 0$.

16.27 (a) $[1.118e^{-t}\cos(2t + 26.57°)]u(t)$;

(b) $(4\sin t + \sin 2t)u(t)$;

(c) $5[1 - (t^3/6 + t^2/2 + t + 1)e^{-t}]u(t)$.

16.29 (a) $[e^{-t}(2 - 1.5t)\sin t]u(t)$;

(b) $7.5\delta(t) + 0.5d\delta(t)/dt$;

(c) $3d\delta(t)/dt - 12\delta(t) + 50e^{-4t}u(t)$.

16.31 (a) $e^{-2t}u(t) - e^{-2(t-1)}u(t - 1)$;

(b) $(2e^{-2(t-1)} - e^{-(t-1)})u(t - 1)$;

(c) $(0.5\sin 2t)[u(t) - u(t - \pi)]$.

16.33 $y_{\text{natural}} = (3e^{-3t} - 2e^{-4t})u(t)$;

$y_{\text{forced}} = (1/144) \times (12t + 41 - 176e^{-3t} + 135e^{-4t})u(t)$.

16.35 $(1 - 3e^{-t} + 3e^{-2t} - e^{-3t})u(t)$.

16.37 (a) $y(t > 0) = \cos 3t + \cos t$;

(b) $y(t > 0) = 0.5(6t - 7 + 28e^{-t} - 11e^{-2t})$.

16.39 $y_1(t > 0) = -(1/12)(11 + 16e^{-3t} + 9e^{-4t})$;

$y_2(t > 0) = -(1/12)(11 - 8e^{-3t} - 4e^{-4t})$.

16.41 $3(e^{-t} + e^{-3t})u(t)$; $(4 - 3e^{-t} - e^{-3t})u(t)$.

16.43 $10^4/(s^2 + 2 \times 10^3 s + 2 \times 10^6)$;

$[5 \times 10^{-3}t + 5 \times 10^{-6}(e^{-10^3 t}\cos 10^3 t - 1)]u(t)$.

16.45 $v(t > 0) = 5 - 4e^{-100t}$ V.

16.47 $\left[(40/\sqrt{29})\cos(10^6 t - 68.20°) - (80/29)e^{-4 \times 10^5 t}\right]u(t)$ V.

16.49 $v(t > 0) = 10(2e^{-20t} - e^{-5t})$ V.

16.51 (a) $-2e^{-8t}(\sin 6t)u(t)$ A;

(b) $0.4(e^{-t} - 6e^{-6t})u(t)$ A.

16.53 $v(t > 0) = 2.5\left[1 + \sqrt{2}e^{-t}\cos(t - 45°)\right]$ V.

16.55 $v(t > 0) = 3(3 - 8te^{-2t})$ V.

16.57 (a) $2\ \Omega, 1$;

(b) $5\left[1 + \sqrt{2}e^{-t}\cos(t - 135°)\right]u(t)$ V;

(c) $20e^{-t}(t - \sin t)u(t)$ V.

16.59 $4(e^{-6t} - e^{-2t})u(t)$ V.

16.61 $1/3$ J.

16.63 $\left[(8 \times 10^3 t + 0.48)e^{-10^4 t} + 0.8\cos(5 \times 10^3 t + 126.87°)\right]u(t)$ V.

16.65 $5/s^2$, $10/s$, $25t^2 u(t)$.

16.67 $f * g = 6t^2$ for $0 \le t \le 3$,

$6(12t - t^2 - 18)$ for $3 \le t \le 4$, $6(20t - 2t^2 - 34)$ for $4 \le t \le 6$,

$6(8t - t^2 + 2)$ for $6 \le t \le 7$,

$6(t^2 - 20t + 100)$ for $7 \le t \le 10$, 0 for $t \le 0$ and $t \ge 10$.

16.69 $v_O = 0$ for $t < 0$, $10e^{-t}$ V for $0 < t < 1$,
$10(1 - 2e)e^{-t}$ V for $1 < t < 2$, and
$10(1 + e^2 - 2e)e^{-t}$ V for $2 < t \leq \infty$.

CHAPTER 17

17.1   Derivation.

17.3   Derivation.

17.5   Derivation.

17.7   Derivation.

17.9 $A/\sqrt{6}$.

17.11   (a)   15.20%;
            (b)   16.05%;
            (c)   15.40%.

17.13 $v(t) = 0.5 + \frac{6}{\pi} \sum_{k=1}^{\infty} \frac{1}{2k-1} \cos \left[(2k-1)\frac{10^6\pi}{3}t + (5-4k)30° \right]$ V.

17.15   Derivation.

17.17   97.55%

17.19   (a)

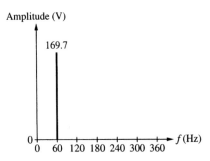

(b)

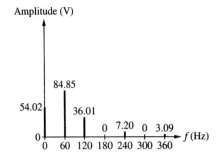

(c)

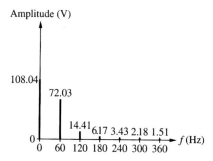

Amplitude (V)

108.04

72.03

14.41 6.17 3.43 2.18 1.51

0

0 60 120 180 240 300 360 f (Hz)

17.21 $\frac{4}{3\pi} \sum_{n=1}^{\infty} \frac{1-\cos(n\pi/2)}{n\sqrt{1+n^2\pi^2}} \sin\left(\frac{n\pi}{2}t\right)$ V.

17.23 $v_I(t) = \pm A_i;\ v_O(t) = (\mp A_i/RC)t + v_O(0);$
$v_I(t) = \frac{4A}{\pi}i \sum_{k=1}^{\infty} \frac{1}{2k-1} \cos\left[(2k-1)\,\omega_0 t - 90°\right]$ V;
$v_O(t) = \frac{4A_i}{\pi\omega_0 RC} \sum_{k=1}^{\infty} \frac{1}{(2k-1)^2} \cos(2k-1)\omega_0 t$ V.

17.25 $10 + 0.1183 \sin(240\pi t - 7.69°) + 5.876 \times 10^{-3} \sin(480\pi t - 3.81°) + 1.118 \times 10^{-3} \sin(720\pi t - 2.54°)$ V.

17.27 $A(\cdots - e^{-j4\omega_0 t}/15\pi - e^{-j2\omega_0 t}/3\pi + je^{-j\omega_0 t}/4 + 1/\pi - je^{j\omega_0 t}/4 - e^{j2\omega_0 t}/3\pi - e^{j4\omega_0 t}/15\pi - \cdots).$

17.29 $f(t) = A(\cdots + je^{-j2\omega_0 t}/4\pi + je^{-j\omega_0 t}/2\pi + 1/2 - je^{j\omega_0 t}/2\pi - je^{j2\omega_0 t}/4\pi - \cdots).$

17.31 $v_O(t) = \frac{2}{3\pi} \sum_{n=-\infty}^{\infty} \frac{j[\cos(n\pi/2)-1]}{n(1+jn\pi)} e^{jn\pi t/2}$ V, $n \neq 0$.

17.33 $2A(1 - \cos \omega W)/\omega^2 W$.

17.35 (a) $1/(a + j\omega)^2$;
(b) $\sqrt{\pi/a}\,e^{-\omega^2/4a}$;
(c) $e^{-(a+j\omega)t_0}/(a + j\omega)$.

17.37 $A^2 W$, 90.28%.

17.39 (a) $5/(2 - \omega^2 + 2j\omega)$;
(b) 3.561 W;
(c) 1.902 W.

17.41 $f(t \to \pm\infty) = 0;\ \int_{-\infty}^{\infty} f(\xi)\, d\xi = 0.$

17.43 (a) $6\pi \left[2\delta(\omega) + \delta(\omega + 4\pi\,10^3) + \delta(\omega - 4\pi\,10^3)\right]$;
(b) $(10\pi\underline{/-\theta}\,)\delta(\omega + \omega_0) + (10\pi\underline{/\theta}\,)\delta(\omega - \omega_0)$;
(c) $A\left[(\alpha + j\omega)\cos\theta - \omega_d \sin\theta\right] / \left[(\alpha + j\omega_d)^2 + \omega_d^2\right]$.

17.45 $(j2A/10^3\pi)\left[\sin(\omega/10^3)\right]/\left[(\omega/10^3\pi)^2 - 1\right]$.

17.47 (a) $j2A\pi\,10^3(\sin 10^{-3}\omega)/(\omega^2 - \pi^2 10^6)$;
(b) $\lim_{W\to\infty} \mathcal{F}\{f_1(t)f_2(t)\} = \mathcal{F}\{f_1(t)\}$.

17.49 $a = 2\pi$ Np/s.

17.51 6 J.

17.53 95.95%

17.55 $10[1 - (100t + 1)e^{-100t}]u(t)$ V.

17.57 $V_n^2 \omega_0/2$.

STANDARD RESISTANCE VALUES

As a rule, you should always specify *standard* resistance values for the circuits you design (see Table A.1). In many applications 5% resistors are adequate; however, when a higher precision is required, 1% resistors should be used. When even this tolerance is insufficient, the alternatives are either 0.1% (or better) resistors, or less precise resistors in conjunction with potentiometers to allow for exact adjustments.

The numbers in the table are multipliers. As an example, suppose your calculations indicate the need for a resistance value of 3.1415 kΩ. Then, Table A.1 indicates that the closest usable resistance in the 5% tolerance range is 3 kΩ, while the closest 1% resistance is 3.16 kΩ.

To facilitate the identification of their values and tolerance, resistors are often equipped with multicolored bands as shown in Figure A.1. The numerical values associated with these bands are shown in Table A.2. Bands x, y, and z give the nominal resistance value, while the percentage band gives the maximum percentage deviation from this value. The rule for calculating a resistance value from its color bands is

$$R = (10x + y)10^z \pm \% \text{ tolerance} \qquad \textbf{(A.1)}$$

Figure A.1 Color coding for resistors.

TABLE A.1. Standard Resistance Values

5 percent resistor values	1 percent resistor values			
10	100	178	316	562
11	102	182	324	576
12	105	187	332	590
13	107	191	340	604
15	110	196	348	619
16	113	200	357	634
18	115	205	365	649
20	118	210	374	665
22	121	215	383	681
24	124	221	392	698
27	127	226	402	715
30	130	232	412	732
33	133	237	422	750
36	137	243	432	768
39	140	249	442	787
43	143	255	453	806
47	147	261	464	825
51	150	267	475	845
56	154	274	487	866
62	158	280	499	887
68	162	287	511	909
75	165	294	523	931
82	169	301	536	953
91	174	309	549	976

TABLE A.2 Color Code for Resistances

Bands x, y, and z		% Tolerance band	
Color	Value	Color	Value
Black	0	Gold	± 5%
Brown	1	Silver	±10%
Red	2		
Orange	3		
Yellow	4		
Green	5		
Blue	6		
Violet	7		
Gray	8		
White	9		

▶ **Example A.1**

If the color bands of a resistor are yellow, violet, orange, and gold, what is its resistance?

Solution

$R = (10 \times 4 + 7)10^3 \pm 5\% = 47$ k$\Omega \pm 2.35$ kΩ, indicating that the actual value may lie anywhere between 44.65 kΩ and 49.35 kΩ. ◀

Exercise A.1 List the color bands of the following resistances: (a) 3.9 kΩ ± 10%; (b) 10 kΩ ± 5%; (c) 100 kΩ ± 10%; (d) 2.2 kΩ ± 5%.

ANSWER (a) orange, white, red, silver; (b) brown, black, orange, gold; (c) brown, black, yellow, silver; (d) red, red, red, gold.

SOLUTION of SIMULTANEOUS LINEAR ALGEBRAIC EQUATIONS

Circuit analysis via either the node or loop method involves the formulation and solution of a set of n linear algebraic equations in n unknowns, having the general form

$$a_{11}x_1 + a_{12}x_2 + a_{13}x_3 + \cdots + a_{1n}x_n = b_1$$
$$a_{21}x_1 + a_{22}x_2 + a_{23}x_3 + \cdots + a_{2n}x_n = b_2$$
$$a_{31}x_1 + a_{32}x_2 + a_{33}x_3 + \cdots + a_{3n}x_n = b_3$$
$$\vdots \qquad \vdots \qquad \vdots \qquad \qquad \vdots \qquad \vdots$$
$$a_{n1}x_1 + a_{n2}x_2 + a_{n3}x_3 + \cdots + a_{nn}x_n = b_n$$

where x_1 through x_n are *unknown node voltages* (or *mesh currents*), a_{ij} $(i, j = 1, 2, \ldots, n)$ are *known admittances* (or *resistances*), and b_1 through b_n are *known source voltages* (or *currents*). Note that the double-subscript notation uses the first subscript to identify a row and the second subscript a column. The two most common methods of solving these equations are the *Gaussian elimination method* and *Cramer's rule*.

GAUSSIAN ELIMINATION

The Gaussian elimination technique repeatedly combines different rows in such a way as to transform the original system of equations into a system of the type

$$\alpha_{11}x_1 + \alpha_{12}x_2 + \alpha_{13}x_3 + \cdots + \alpha_{1n}x_n = \beta_1$$
$$\alpha_{22}x_2 + \alpha_{23}x_3 + \cdots + \alpha_{2n}x_n = \beta_2$$
$$\alpha_{33}x_3 + \cdots + \alpha_{3n}x_n = \beta_3$$
$$\vdots \qquad \vdots$$
$$\alpha_{nn}x_n = \beta_n$$

that is, into a system in which all coefficients with $i > j$ are zero. This allows us to solve for x_n directly, and to solve for $x_{n-1}, x_{n-2}, \ldots$, all the way down to x_1 by successive back-substitutions. As an example, consider the following system of equations:

$$2x_1 - 1x_2 - 1x_3 = -5 \qquad \textbf{(A.1a)}$$

$$4x_1 - 3x_2 + 2x_3 = -8 \qquad \textbf{(A.1b)}$$

$$-3x_1 + 1x_2 + 1x_3 = 6 \qquad \textbf{(A.1c)}$$

Multiplying the first equation through by -2 and adding it pairwise to the second equation, and then multiplying the first equation through by 2/3 and adding it pairwise to the third equation, we obtain

$$-1x_2 + 4x_3 = 2 \qquad \textbf{(A.2a)}$$

$$-\frac{1}{2}x_2 + \frac{1}{2}x_3 = -\frac{3}{2} \qquad \textbf{(A.2b)}$$

Multiplying the new first equation through by $-1/2$ and adding it pairwise to the second we obtain

$$-\frac{5}{2}x_3 = -\frac{5}{2} \qquad \textbf{(A.3)}$$

This yields $x_3 = 1$. Back-substituting into Equation (A.2a) yields $x_2 = 4x_3 - 2 = 4 \times 1 - 2 = 2$. Back-substituting into Equation (A.1a) yields $x_1 = (1/2)(x_2 + x_3 - 5) = 1/2(2 + 1 - 5) = -1$.

CRAMER'S RULE

Cramer's rule allows us to find the j-th unknown ($j = 1, 2, 3, \ldots, n$) directly as

$$x_j = \frac{\Delta_j}{\Delta}$$

where Δ is the determinant of the matrix of the coefficients $a_{ij}(i, j = 1, 2, 3, \ldots, n)$, and Δ_j is the determinant of the matrix obtained from the matrix of the coefficients a_{ij} by replacing the coefficients of the j-th column with the coefficients $b_j(j = 1, 2, 3, \ldots, n)$. It is apparent that for Cramer's rule to work it is necessary that $\Delta \neq 0$.

For $n = 2$ and $n = 3$, the Δ determinants are found as

$$\Delta = \begin{vmatrix} a_{11} & a_{12} \\ a_{21} & a_{22} \end{vmatrix} = a_{11}a_{22} - a_{21}a_{22}$$

$$\Delta = \begin{vmatrix} a_{11} & a_{12} & a_{13} \\ a_{21} & a_{22} & a_{23} \\ a_{31} & a_{32} & a_{33} \end{vmatrix}$$

$$= a_{11}a_{22}a_{33} + a_{12}a_{23}a_{31} + a_{13}a_{21}a_{32} - (a_{31}a_{22}a_{13} + a_{32}a_{23}a_{11} + a_{33}a_{21}a_{12})$$

Similar groupings apply to the calculation of the Δ_j determinants.

A convenient method of visualizing the above patterns is by means of arrows, as follows:

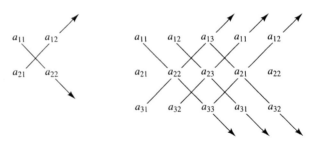

Figure A.2

To obtain Δ, we first sum the products of the coefficients on the diagonals identified by the upward arrows; next, we sum the products of the coefficients on the diagonals identified by the downward arrows; finally, we subtract the second sum from the first. Note that for this visual method to work for $n = 3$, the first two columns must be repeated to the right of the third column, as shown.

As an example, let us solve Equation (A.1) via Cramer's rule.

$$\Delta = \begin{vmatrix} 2 & -1 & -1 \\ 4 & -3 & 2 \\ -3 & 1 & 1 \end{vmatrix}$$

$$= 2(-3)1 + (-1)2(-3) + (-1)4(1)$$
$$- [(-3)(-3)(-1) + 1(2)2 + 1(4)(-1)] = 5$$

$$\Delta_1 = \begin{vmatrix} -5 & -1 & -1 \\ -8 & -3 & 2 \\ 6 & 1 & 1 \end{vmatrix}$$

$$= (-5)(-3)1 + (-1)2(6) + (-1)(-8)1$$
$$- [6(-3)(-1) + 1(2)(-5) + 1(-8)(-1)] = -5$$

$$\Delta_2 = \begin{vmatrix} 2 & -5 & -1 \\ 4 & -8 & 2 \\ -3 & 6 & 1 \end{vmatrix}$$

$$= 2(-8)1 + (-5)2(-3) + (-1)4(6)$$
$$- [(-3)(-8)(-1) + 6(2)2 + 1(4)(-5)] = 10$$

$$\Delta_3 = \begin{vmatrix} 2 & -1 & -5 \\ 4 & -3 & -8 \\ -3 & 1 & 6 \end{vmatrix}$$

$$= 2(-3)6 + (-1)(-8)(-3) + (-5)4(1)$$
$$- [(-3)(-3)(-5) + 1(-8)2 + 6(4)(-1)] = 5$$

Consequently, $x_1 = \Delta_1/\Delta = -5/5 = -1$, $x_2 = \Delta_2/\Delta = 10/5 = 2$, and $x_3 = \Delta_3/\Delta = 5/5 = 1$, thus confirming the results obtained via Gaussian elimination.

EULER'S IDENTITY AND THE UNDERDAMPED RESPONSE

EULER'S IDENTITY

This important identity, named for the Swiss mathematician Leonhard Euler (pronounced *Oiler*) (1707–1783), can be proved using the *power series expansion* of $\cos\theta$, $\sin\theta$, and $e^{j\theta}$,

$$\cos\theta = 1 - \frac{\theta^2}{2!} + \frac{\theta^4}{4!} - \cdots$$

$$\sin\theta = \theta - \frac{\theta^3}{3!} + \frac{\theta^5}{5!} - \cdots$$

$$e^{j\theta} = 1 + j\theta + \frac{(j\theta)^2}{2!}\frac{(j\theta)^3}{3!} + \frac{(j\theta)^4}{4!} + \frac{(j\theta)^5}{5!} + \cdots$$

Since $j^2 = -1$, the last identity can be written as

$$e^{j\theta} = 1 + j\theta - \frac{\theta^2}{2!} - j\frac{\theta^3}{3!} + \frac{\theta^4}{4!} + j\frac{\theta^5}{5!} - \cdots$$

Finally,

$$\cos\theta + j\sin\theta = 1 + j\theta - \frac{\theta^2}{2!} - j\frac{\theta^3}{3!} + \frac{\theta^4}{4!} + j\frac{\theta^5}{5!} - \cdots = e^{j\theta}$$

that is, Euler's identity,

$$e^{j\theta} = \cos\theta + j\sin\theta \qquad \textbf{(A.4)}$$

The above identity, in turn, allows us to express the sine and cosine functions in terms of exponentials, as follows. Consider the function obtained by replacing j with $-j$ in Equation (A.4),

$$e^{-j\theta} = \cos\theta - j\sin\theta \qquad \textbf{(A.5)}$$

Adding Equations (A.4) and (A.5) pairwise and dividing through by 2,

$$\cos\theta = \frac{1}{2}\left(e^{j\theta} + e^{-j\theta}\right) \qquad \textbf{(A.6)}$$

Likewise, subtracting Equation (A.5) from (A.4) pairwise and dividing through by $j2$,

$$\sin\theta = \frac{1}{j2}\left(e^{j\theta} - e^{-j\theta}\right) \tag{A.7}$$

We use these expressions in various parts of the book.

THE UNDERDAMPED RESPONSE

The expression for the underdamped response is obtained by substituting Equations (9.20) into Equation (9.10),

$$y(t) = A_1 e^{-(\alpha - j\omega_d)t} + A_2 e^{-(\alpha + j\omega_d)t} \tag{A.8}$$

Factoring out the common term $e^{-\alpha t}$,

$$y(t) = e^{-\alpha t}\left(A_1 e^{j\omega_d t} + A_2 e^{-j\omega_d t}\right) \tag{A.9}$$

We wish to prove that this expression can be put in the form

$$y(t) = Ae^{-\alpha t}\cos(\omega_d t + \phi) \tag{A.10}$$

The proof is based on Euler's identity of Equation (A.4), which in the present case becomes

$$e^{j\omega_d t} = \cos\omega_d t + j\sin\omega_d t \tag{A.11a}$$

$$e^{-j\omega_d t} = \cos\omega_d t - j\sin\omega_d t \tag{A.11b}$$

Substituting into Equation (A.9) and collecting yields, on the one hand,

$$y(t) = e^{-\alpha t}\left[(A_1 + A_2)\cos\omega_d t + j(A_1 - A_2)\sin\omega_d t\right]$$

Expanding Equation (A.10) yields, on the other hand,

$$y(t) = e^{-\alpha t} A\left[\cos\phi\cos\omega_d t - \sin\phi\sin\omega_d t\right]$$

For the above expressions to be equivalent, the coefficients of the terms $\cos\omega_d t$ and $\sin\omega_d t$ must be equal,

$$A\cos\phi = A_1 + A_2 \tag{A.12a}$$

$$A\sin\phi = j(A_2 - A_1) \tag{A.12b}$$

Squaring each side of Equations (A.12a, b) and adding terms pairwise,

$$A^2\left(\cos^2\phi + \sin^2\phi\right) = (A_1 + A_2)^2 + j^2(A_2 - A_1)^2$$

Using the identities $\cos^2\phi + \sin^2\phi = 1$ and $j^2 = -1$, and simplifying,

$$A = 2\sqrt{A_1 A_2} \tag{A.13a}$$

Likewise, dividing Equation (A.12b) by Equation (A.12a) pairwise, we obtain $\tan\phi = j(A_2 - A_1)/(A_1 + A_2)$, or

$$\phi = \tan^{-1} j\frac{A_2 - A_1}{A_1 + A_2} \tag{A.13b}$$

The initial-condition constants A and ϕ can be determined either by finding A_1 and A_2 via Equation (9.14) and then substituting into Equations (A.13), or by direct calculation as in Example 9.4.

SUMMARY OF COMPLEX ALGEBRA

Phasors, impedances, and ac transfer functions belong to a class of variables known as **complex variables**, so called because they involve real numbers as well as the *imaginary unit* $j = \sqrt{-1}$.

- A complex variable Z can be represented in either one of three forms:

1. The **shorthand** form, also called the **polar** form:

$$Z = Z_m \underline{/\phi}$$

2. The **rectangular** form, also called the **cartesian** form:

$$Z = Z_r + j Z_i$$

3. The **exponential** form:

$$Z = Z_m e^{j\phi}$$

- In the above expressions, Z_m is the **length** or **magnitude** of Z, and ϕ is the **angle** or **argument** of Z. They are denoted as

$$Z_m = |Z|$$
$$\phi = \sphericalangle Z$$

and are called the **polar coordinates** of Z. Moreover, Z_r is the **real part** of Z, and Z_i is the **imaginary part** of Z, expressed as

$$Z_r = \text{Re} \, [Z]$$
$$Z_i = \text{Im} \, [Z]$$

They are called the **rectangular** or **cartesian coordinates** of Z.

- Given Z_m and ϕ, Z_r and Z_i are found as

$$Z_r = Z_m \cos \phi$$
$$Z_i = Z_m \sin \phi$$

Conversely, given Z_r and Z_i, Z_m and ϕ are found as

$$Z_m = \sqrt{Z_r^2 + Z_i^2}$$

$$\phi = \tan^{-1}\frac{Z_i}{Z_r} \quad \text{if } Z_r > 0$$

$$\phi = \tan^{-1}\frac{Z_i}{Z_r} \pm 180° \quad \text{if } Z_r < 0$$

The $\pm$ sign is chosen in such a way as to keep $-180° \le \phi \le 180°$.

- Given two complex variables, $X = X_r + jX_i$ and $Y = Y_r + jY_i$, their **sum** and **difference** are

$$X + Y = (X_r + Y_r) + j(X_i + Y_i)$$

$$X - Y = (X_r - Y_r) + j(X_i - Y_i)$$

- Given two complex variables, $X = X_m\underline{/\theta_x}$ and $Y = Y_m\underline{/\theta_y}$, their **product** and their **ratio** are

$$XY = X_m Y_m\underline{/\theta_x + \theta_y}$$

$$\frac{X}{Y} = \frac{X_m}{Y_m}\underline{/\theta_x - \theta_y}$$

- Given a complex variable $Z = Z_r + jZ_i = Z_m\underline{/\phi}$, its **complex conjugate** is

$$Z^* = Z_r - jZ_i = Z_m\underline{/-\phi} = Z_m e^{-j\phi}$$

A complex variable and its conjugate satisfy the properties:

$$ZZ^* = Z_m^2$$

$$Z/Z^* = 1\underline{/2\phi}$$

$$Z + Z^* = 2Z_r$$

$$Z - Z^* = 2Z_i$$

- The reciprocal of a complex variable $Z = Z_r + jZ_i = Z_m\underline{/\phi}$ is

$$\frac{1}{Z} = \frac{1}{Z_m}\underline{/-\phi} = \frac{Z^*}{|Z|^2} = \frac{Z_r - jZ_i}{Z_r^2 + Z_i^2}$$

- If a complex variable $Z = Z_r + jZ_i$ is such that $|Z_r| \gg |Z_i|$, then it can be approximated as

$$Z \simeq Z_r\underline{/0°} \quad \text{if } Z_r > 0$$

$$Z \simeq |Z_r|\underline{/180°} \quad \text{if } Z_r < 0$$

Conversely, if $|Z_i| \gg |Z_r|$, then

$$Z \simeq Z_i\underline{/90°} \quad \text{if } Z_i > 0$$

$$Z \simeq |Z_i|\underline{/-90°} \quad \text{if } Z_i < 0$$

- Given an ac signal $x(t)$ having angular frequency ω and phasor X,
  1. The phasor of the **derivative** dx/dt is $j\omega X$
  2. The phasor of the **integral** $\int x(t)dt$ is $(1/j\omega)X$

INDEX

Useful Integrals

$$\int u\, dv = uv - \int v\, du$$

$$\int \frac{dx}{x^2 + a^2} = \frac{1}{a}\tan^{-1}\frac{x}{a}$$

$$\int \frac{dx}{(x^2 + a^2)^2} = \frac{1}{2a^2}\left(\frac{x}{x^2 + a^2} + \frac{1}{a}\tan^{-1}\frac{x}{a}\right)$$

$$\int \sin^2 ax\, dx = \frac{x}{2} - \frac{\sin 2ax}{4a}$$

$$\int \cos^2 ax\, dx = \frac{x}{2} + \frac{\sin 2ax}{4a}$$

$$\int x \sin ax\, dx = \frac{1}{a^2}(\sin ax - ax\cos ax)$$

$$\int x \cos ax\, dx = \frac{1}{a^2}(\cos ax + ax\sin ax)$$

$$\int x^2 \sin ax\, dx = \frac{2x}{a^2}\sin ax + \frac{2 - a^2 x^2}{a^3}\cos ax$$

$$\int x^2 \cos ax\, dx = \frac{2x}{a^2}\cos ax - \frac{2 - a^2 x^2}{a^3}\sin ax$$

$$\int e^{ax} \sin bx\, dx = \frac{e^{ax}}{a^2 + b^2}(a\sin bx - b\cos bx)$$

$$\int e^{ax} \cos bx\, dx = \frac{e^{ax}}{a^2 + b^2}(a\cos bx + b\sin bx)$$

$$\int e^{ax} \sin^2 bx\, dx = \frac{e^{ax}}{a^2 + 4b^2}\left((a\sin bx - 2b\cos bx)\sin bx + \frac{2b^2}{a}\right)$$

$$\int e^{ax} \cos^2 bx\, dx = \frac{e^{ax}}{a^2 + 4b^2}\left((a\cos bx - 2b\sin bx)\cos bx + \frac{2b^2}{a}\right)$$

$$\int \sin ax \sin bx\, dx = \frac{\sin(a - b)x}{2(a - b)} - \frac{\sin(a + b)x}{2(a + b)}, \quad a^2 \neq b^2$$

$$\int \cos ax \cos bx\, dx = \frac{\sin(a - b)x}{2(a - b)} + \frac{\sin(a + b)x}{2(a + b)}, \quad a^2 \neq b^2$$

$$\int \sin ax \cos bx\, dx = -\frac{\cos(a - b)x}{2(a - b)} - \frac{\cos(a + b)x}{2(a + b)}, \quad a^2 \neq b^2$$

$$\int xe^{ax}dx = \frac{e^{ax}}{a^2}(ax - 1)$$

$$\int x^2 e^{ax}dx = \frac{e^{ax}}{a^3}(a^2 x^2 - 2ax + 2)$$

$$\int_0^\pi \sin mx \sin nx \, dx = \int_0^\pi \cos mx \cos nx \, dx = 0, \quad m \neq n, \quad m \text{ and } n \text{ integers}$$

$$\int_0^\pi \sin mx \cos nx \, dx = \begin{cases} 0, & m+n \text{ even} \\ 2m/(m^2 - n^2), & m+n \text{ odd} \end{cases}$$

$$\int_0^\infty \frac{\sin ax}{x} \, dx = \begin{cases} \pi/2, & a > 0 \\ 0, & a = 0 \\ -\pi/2, & a < 0 \end{cases}$$

USEFUL TRIGONOMETRIC IDENTITIES

$$\sin(\alpha \pm 90°) = \pm \cos \alpha$$

$$\cos(\alpha \pm 90°) = \mp \sin \alpha$$

$$\sin(\alpha \pm \beta) = \sin \alpha \cos \beta \pm \cos \alpha \sin \beta$$

$$\cos(\alpha \pm \beta) = \cos \alpha \cos \beta \mp \sin \alpha \sin \beta$$

$$\sin \alpha \sin \beta = \tfrac{1}{2}[\cos(\alpha - \beta) - \cos(\alpha + \beta)]$$

$$\cos \alpha \cos \beta = \tfrac{1}{2}[\cos(\alpha + \beta) + \cos(\alpha - \beta)]$$

$$\sin \alpha \cos \beta = \tfrac{1}{2}[\sin(\alpha + \beta) + \sin(\alpha - \beta)]$$

$$\sin \alpha + \sin \beta = 2 \sin \frac{\alpha + \beta}{2} \cos \frac{\alpha - \beta}{2}$$

$$\sin \alpha - \sin \beta = 2 \cos \frac{\alpha + \beta}{2} \sin \frac{\alpha - \beta}{2}$$

$$\cos \alpha + \cos \beta = 2 \cos \frac{\alpha + \beta}{2} \cos \frac{\alpha - \beta}{2}$$

$$\cos \alpha - \cos \beta = -2 \sin \frac{\alpha + \beta}{2} \sin \frac{\alpha - \beta}{2}$$

$$A \cos \alpha + B \sin \alpha = \sqrt{A^2 + B^2} \cos[\alpha + \tan^{-1}(-B/A)]$$

$$\sin 2\alpha = 2 \sin \alpha \cos \alpha$$

$$\cos 2\alpha = 2 \cos^2 \alpha - 1 = 1 - 2 \sin^2 \alpha$$

$$\sin^2 \alpha = \tfrac{1}{2}(1 - \cos 2\alpha)$$

$$\cos^2 \alpha = \tfrac{1}{2}(1 + \cos 2\alpha)$$

$$\cos \alpha = 1/\sqrt{1 + \tan^2 \alpha}$$

$$\sin \alpha = (e^{j\alpha} - e^{-j\alpha})/j2$$

$$\cos \alpha = (e^{j\alpha} + e^{-j\alpha})/2$$

$$e^{\pm j\alpha} = \cos \alpha \pm j \sin \alpha$$